ELECTRICAL
MOTOR CONTROLS
FOR INTEGRATED SYSTEMS

Fifth Edition

AMERICAN TECHNICAL PUBLISHERS
Orland Park, Illinois 60467-5756

Gary J. Rockis
Glen A. Mazur

American Technical Publishers, Editorial Staff

Editor in Chief:
    Jonathan F. Gosse
Vice President—Production:
    Peter A. Zurlis
Art Manager:
    Jennifer M. Hines
Multimedia Manager:
    Carl R. Hansen
Technical Editor:
    Scott C. Bloom
Copy Editor:
    Dane K. Hamann
Cover Design:
    Melanie G. Doornbos
Illustration/Layout:
    Nicholas W. Basham
    Joshua P. Hugo
    Mark S. Maxwell
    Robert M. McCarthy
    Thomas E. Zabinski
DVD Development:
    Cory S. Butler
    Kathryn C. Deisinger
    Daniel Kundrat
    Kathleen A. Moster
    Nicole S. Polak
    Robert E. Stickley

5 6 7 8 9 – 14 – 9 8 7 6 5

Printed in the United States of America

ISBN 978-0-8269-1226-8
eISBN 978-0-8269-9522-3

 This book is printed on recycled paper.

Acknowledgments

The authors and publisher are grateful to the following companies and organizations for providing information, photographs, and technical assistance.

- ABB Inc., Drives and Power Electronics
- ABB Power T&D Company Inc.
- AEMC® Instruments
- Alemite Corp.
- Amprobe | Advanced Test Products
- ASI Robicon
- Bacharach, Inc.
- Baldor Electric Company
- Banner Engineering
- Canadian National Railway Company
- Carlo Gavazzi Inc.
- Carlo Gavazzi Inc. Electromatic Business Unit
- Cutler-Hammer
- Datastream Systems, Inc.
- Eagle Signal Industrial Controls
- East Penn Manufacturing Co., Inc.
- Electrical Apparatus Service Association, Inc.
- FANUC Robotics North America
- Fluke Corporation
- Furnas Electric Co.
- GE Motors & Industrial Systems
- General Electric Company
- Grayhill Inc.
- Greenlee Textron Inc.
- Guardian Electric Mfg. Co.
- Hand Held Products
- Harrington Hoists, Inc.
- Heidelberg Harris, Inc.
- Honeywell's MICRO SWITCH Division
- Honeywell Sensing and Control
- IBM
- Ideal Industries, Inc.
- International Rectifier
- Justrite Manufacturing Company
- The Lincoln Electric Company
- March Manufacturing, Inc.
- Milwaukee Tool Corporation
- Omron Electronics LLC
- Pacific Bearing Company
- Panduit Corp.
- Predict/DLI
- Products Unlimited
- Rockwell Automation, Allen-Bradley Company, Inc.
- Rockwell Automation/Reliance Electric
- Rofin Sinar
- Saftronics Inc.
- Siemens
- SMA America, Inc.
- Sprecher & Scuh
- Square D Company
- The Stanley Works
- Teco
- UE Systems, Inc.

Contents

Learner Resources

- Quick Quizzes®
- Illustrated Glossary
- Flash Cards
- Checkpoints

- Applying Your Knowledge
- Interactive Motor Controls
- Media Library
- Internet Resources

Book Features

Electrical Motor Controls for Integrated Systems is the industry-leading textbook covering electrical, motor, and mechanical devices and their use in industrial control circuits. This textbook provides the structure and content for acquiring the knowledge and skills required in an advanced manufacturing environment. In these fast-changing environments, technicians must be competent in various aspects of mechanical, electrical, and fluid power systems for successful productivity. The textbook also serves as a practical resource for maintenance technicians responsible for production and HVAC equipment.

Featuring a new open design, the textbook offers proven and thoroughly updated content in a new format organized into sections based on specific topics. The textbook begins with basic electrical and motor theory, builds on circuit fundamentals, and reinforces comprehension through examples of industrial applications. Special emphasis is placed on the development of troubleshooting skills. Expanded content in this edition includes the following:

- Electrical safety information on NFPA® 70E, PPE, arc flashes, and arc blasts
- Smart grid systems
- Energy efficiency applications
- Electrical test instruments

Each chapter includes checkpoint questions and Energy Efficiency Practices. Applying Your Knowledge questions in the learner resources reinforce key motor control concepts. Answers to the checkpoint questions are located in the Answer Keys. The Appendix contains many useful tables, charts, and formulas.

Sections segment each chapter into discrete topics.

Chapter introductions preview content to be covered.

Section objectives identify the main concepts addressed within that section of the chapter.

Checkpoint questions reinforce comprehension of section content.

QR Codes at the end of each chapter enable quick access to digital resources.

Energy Efficiency Practices provide information on topics related to energy efficiency.

Industrial test instruments are shown in common troubleshooting applications.

Tech Facts provide supplemental information related to topics presented.

Industrial application photos supplement text and illustrations.

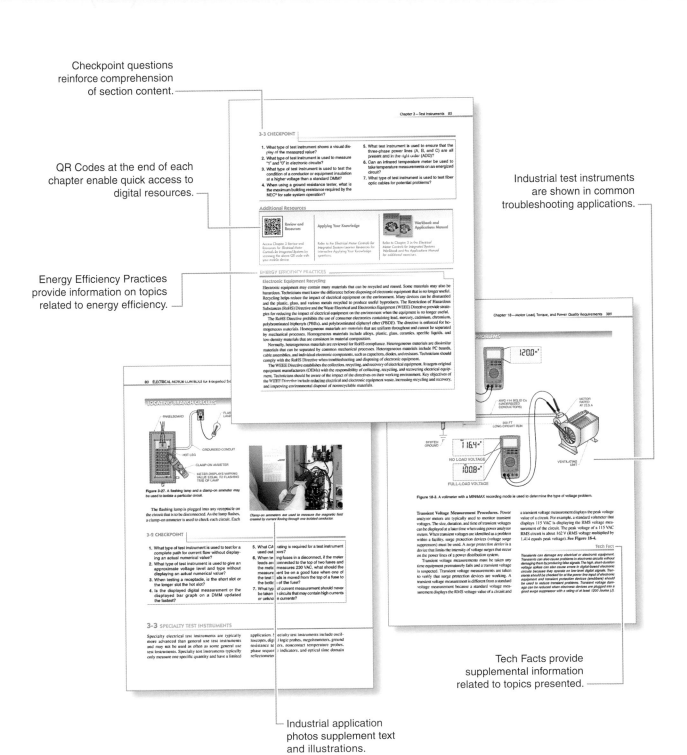

Online Learner Resources

The learner resources include the following:

- **Quick Quizzes®** that provide interactive questions for each chapter, with embedded links to highlighted content within the textbook and to the Illustrated Glossary
- **Illustrated Glossary** that serves as a helpful reference to commonly used terms, with selected terms linked to textbook illustrations
- **Flash Cards** that provide a self-study/review of common terms and their definitions
- **Checkpoints** that provide learners the opportunity to demonstrate comprehension of chapter concepts
- **Applying Your Knowledge** that provides interactive motor control activities
- **Interactive Motor Control Enclosure** that provide troubleshooting experiences with electronic components and circuits
- **Media Library** that consists of videos and animations that reinforce textbook content
- **Internet Resources** that provide access to additional resources to support continued learning

Electrical Motor Controls for Integrated Systems includes access to online learner resources that reinforce content and enhance learning. These online resources can be accessed using either of the following methods:

- Key atplearningresources.com into a web browser and enter **Access Code 362245**
- Use a Quick Response (QR) reader app to scan the QR Code with a mobile device.

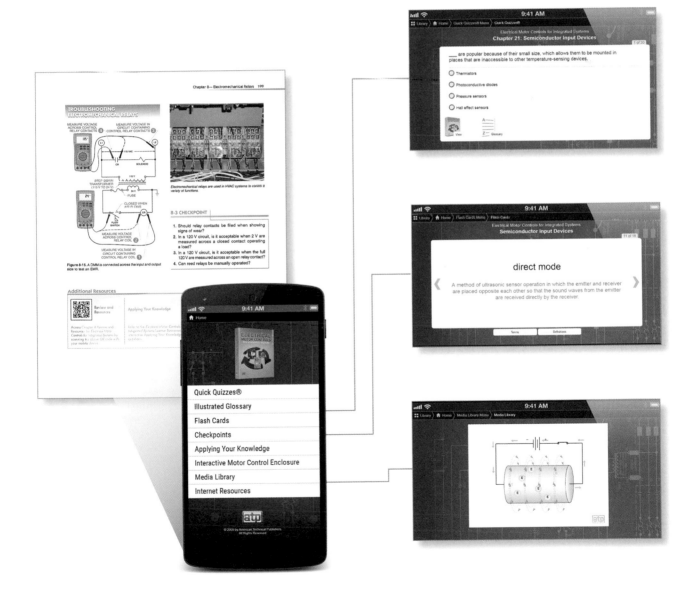

Resources

Workbook

The *Electrical Motor Controls for Integrated Systems Workbook* reinforces concepts and provides design activities for systems included in the fifth edition of *Electrical Motor Controls for Integrated Systems*. The *Workbook* contains both Tech-Cheks and Worksheets for each chapter.

- **Tech-Cheks** provide multiple choice, completion, and/or matching questions that reinforce comprehension of key concepts.
- **Worksheets** provide opportunities to apply electrical motor control concepts to practical design problems.

The *Electrical Motor Controls for Integrated Systems Workbook* also includes Certificates of Completion that can be used to document the knowledge and skills learned in four areas: motor control principles, motor control devices, motor control circuits, and motor control methods.

Applications Manual

The *Electrical Motor Controls for Integrated Systems Applications Manual* expands upon the concepts and provides both applications and activities that build on the knowledge and skills acquired using *Electrical Motor Controls for Integrated Systems*. The *Applications Manual* contains both applications and activities.

- **Applications** present technical, manufacturing, and troubleshooting data as it appears in service manuals used by industrial electricians. Selected motor control topics highlight the proper use, sizing, connection, and troubleshooting of electrical motor control devices.
- **Activities** offer opportunities for learners to order, install, maintain, and troubleshoot electrical motor control devices and circuits in real-world scenarios. Questions within the activities include identification, completion, short answer, calculations, diagrams, and schematics.

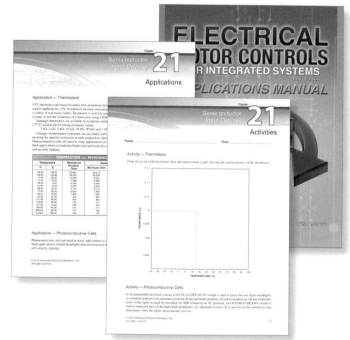

Online Instructor Resources

The *Electrical Motor Controls for Integrated Systems* online instructor resources serves as the primary organizational tool for the instructor. The instructor resources and guide provide a comprehensive teaching resource including PowerPoint® Presentations, an Instructional Guide, an Image Library, Assessments, and Answer Keys.

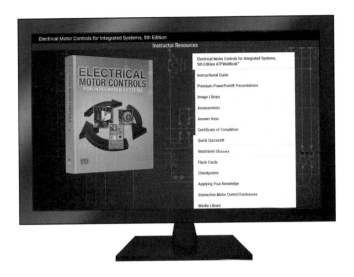

The online instructor resources include the following:

- **Instructional Guide** explains how to best use all the learning tools and resources provided, including detailed Instructional Plans for each chapter.
- **PowerPoint® Presentations** review key concepts in each section and chapter of *Electrical Motor Controls for Integrated Systems*. PowerPoint® Presentation notes are also provided.
- **Image Library** provides all of the numbered figures in *Electrical Motor Controls for Integrated Systems* in a classroom-friendly format.
- **Assessments** are sets of questions based on objectives and key concepts from each chapter and consist of a pretest, posttest, and test banks. The test banks can be used with most test development software packages.
- **Answer Keys** list answers to questions in the pretest, posttest, textbook, workbook, and applications manual.
- **Certificate of Completion** that can be used to document the knowledge and skills learned in four areas: motor control principles, motor control devices, motor control circuits, and motor control methods.

To obtain information on related products visit the American Technical Publishers website at www.atplearning.com.

The Publisher

Sections

Objectives

1-1

- State the three fundamental parts of an atom and identify their states of charge.
- Define and describe conductors, insulators, and semiconductors.
- State the operating function of a diode in a circuit.
- State the two forms of energy and give examples of each.
- Define voltage and state its unit of measure and common abbreviation.
- Define current and state its unit of measure and common abbreviation.
- Define resistance and state its unit of measure and common abbreviation.
- Determine an unknown voltage, current, and resistance with Ohm's law.

1-2

- Calculate resistance at any point in a series or parallel circuit.
- Calculate voltages at any point in a series or parallel circuit.
- Calculate current at any point in a series or parallel circuit.

1-3

- Define the molecular theory of magnetism and electromagnetism.
- Define inductance and state how it affects an AC circuit.
- Define capacitance and state how it affects an AC circuit.

1-4

- Define true power and state its unit of measure and common abbreviation.
- Determine an unknown power, voltage, and current with the power formula.
- Calculate power at any point in a series or parallel circuit.
- Define reactive power and state its unit of measure and common abbreviation.
- Define apparent power and state its unit of measure and common abbreviation.
- Define power factor and explain its relationship to efficiency.

Review and Resources
atplearningresources.com/Quicklinks
Access Code: 362245

2

Electrical Quantities and Basic Circuits

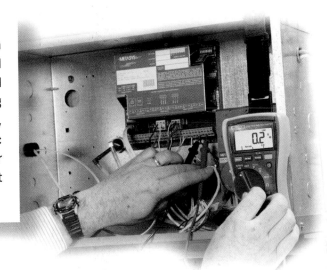

Electrical components and circuits are designed to operate in a predetermined manner to safely produce light, heat, rotary and linear motion, and sound; transfer and store information; and provide many other uses. Electricity always operates by following basic scientific principles that can be tested, explained, utilized, and measured. Understanding these basic laws and principles is the first step for anyone who is going to work safely in any electrical or electronic field and wants to design, build, install, and troubleshoot circuits and components.

1-1 ELECTRICAL THEORY

It can be difficult to describe electricity. The effects of electricity can be seen, but electricity itself is invisible. Electricity must be expressed in terms of positive and negative charges, voltage, current, and resistance. To understand these effects, the basic structure of matter must be examined.

Atomic Theory

All matter consists of an organized collection of atoms. An *atom* is the smallest particle that an element can be reduced to and still keep the properties of that element. The three fundamental particles contained in atoms are protons, neutrons, and electrons. Protons and neutrons make up the nucleus, and electrons whirl about the nucleus in orbits of shells. **See Figure 1-1.** The number of electrons in a shell determines many of the characteristics of the atom.

The *nucleus* is the heavy, dense center of an atom and has a positive electrical charge. A *proton* is a particle contained in the nucleus of an atom that has a positive electrical charge. A *neutron* is a particle contained in the nucleus of an atom that has no electrical charge. The nucleus is surrounded by one or more electrons. An *electron* is a negatively charged particle that whirls around the nucleus at great speed in shells. Each shell can hold a specific number of electrons. The innermost shell can hold two electrons. The second shell can hold eight electrons. The third shell can hold 18 electrons, etc. The shells are filled starting with the inner shell and working outward, so that when the inner shells are filled with as many electrons as they can hold, the next shell is started. Electrons and protons have equal amounts of opposite charges. There are as many electrons as there are protons in an atom, which leaves the atom electrically neutral.

ATOMIC THEORY

ATOM

ELECTRON SHELLS

Shell	Maximum Electrons
1	2
2	8
3	18

Figure 1-1. In an atom, electrons orbit the nucleus in shells that can hold a specific number of electrons.

Valence Electrons

Electrons in the outer shell of the atom are not held as tightly to the atom as those in the inner shells. These outer electrons, called valence electrons, are responsible for movement in electricity. Because valance electrons can be moved from atom to atom, they are also known as free electrons.

An attraction called a charge exists between the positive protons and the negative electrons. It is this attraction that prevents the electrons from flying off into space as they orbit the nucleus at high speed. This attraction introduces one of the basic rules of electricity: unlike charges attract each other. The second basic rule of electricity is that like charges repel each other. Therefore, electrons repel each other.

Movement in electricity involves separating some electrons from their atoms and moving them to other atoms. The movement of electrons takes place by applying an electromotive force (EMF) to a given material. EMF is also called voltage and is expressed in volts (V).

Conductors, Insulators, and Semiconductors

Conductors, insulators, and semiconductors are materials that either allow or resist the flow of electricity. The atomic structure of each material is the deciding factor in whether the material allows or resists the flow of electricity. Conductors have very low resistance and allow the flow of electricity, and insulators have very high resistance and resist the flow of electricity. Semiconductors may allow the flow of electricity in some circumstances and resist the flow of electricity in other circumstances.

Conductors. A *conductor* is a material that has very little resistance and permits electrons to move through it easily. Conductors offer very little resistance to electron flow and conduct electricity very well. By applying a negative charge to one side of the conductor and a positive charge to the other side of the conductor, electrons are forced to move through the conductor. This movement is called current and is expressed in amperes (A). **See Figure 1-2.**

Copper is the most widely used type of conductor because of its flexibility and low resistance.

CONDUCTORS

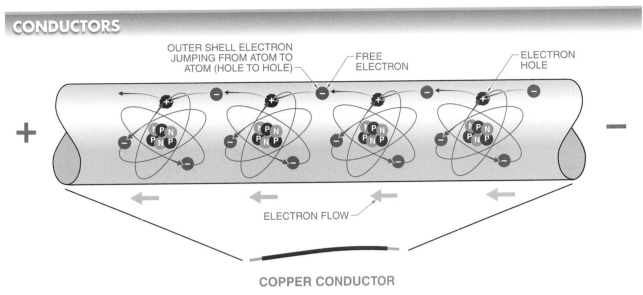

Figure 1-2. A conductor allows free electrons to pass readily through it.

If the outer shell of electrons in an atom is less than half complete, that material is a conductor. For example, if the outer shell contains one, two, or three electrons, those electrons are held to the atom with minimal force. **See Figure 1-3.** These electrons can be moved easily from atom to atom. Some examples of conductors are metals such as silver, copper, gold, and aluminum.

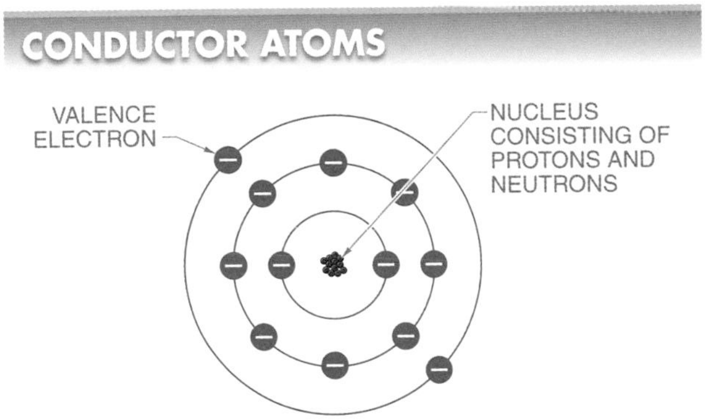

CONDUCTOR ATOMS

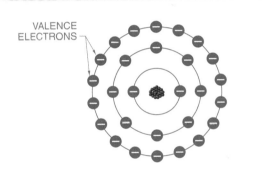

Figure 1-3. Electrons in a conductor atom are held to the atom with minimal force.

Insulators. An *insulator* is a material with an atomic structure that allows few free electrons to pass through it. Insulators offer high resistance to electron flow and do not conduct electricity very well. If there are more than four electrons in the outer shell of an atom, those electrons are held to the atom with a relatively strong force and cannot be moved very easily. **See Figure 1-4.** Some examples of insulators are rubber, plastic, glass, and paper.

INSULATOR ATOMS

Figure 1-4. Electrons in an insulator atom are held to the atom with a relatively strong force and cannot be moved very easily.

Tech Fact

Electrical devices include a label that indicates the conductor material that the device is compatible with. Electrical devices are labeled with AL for use with aluminum, CU for use with copper, or CU-CLAD for use with copper-clad conductors. Incompatible material connections can loosen, causing a circuit problem or fire hazard.

The electrical term used to describe the opposition to electron flow is resistance. Resistance is expressed in ohms (Ω). The letter R may also be used to represent resistance. Nonmetallic-sheathed cable contains conductors and insulators that are used in common electrical circuits. **See Figure 1-5.** Printed circuit boards, on the other hand, have conductors laminated on an insulated material that makes up the board.

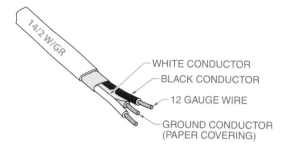

NONMETALLIC-SHEATHED CABLE

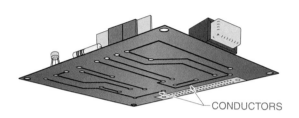

PRINTED CIRCUIT BOARD

Figure 1-5. Nonmetallic-sheathed cable is manufactured in various wire sizes and with a specified number of conductors. Printed circuit boards have conductors laminated on an insulated material that makes up the board.

Semiconductors. Semiconductor materials fall between the low resistance offered by a conductor and the high resistance offered by an insulator. Semiconductors are made from atoms that have only four valence electrons. **See Figure 1-6.**

SEMICONDUCTOR ATOMS

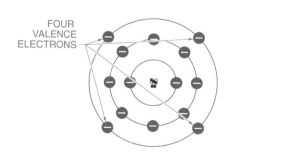

FOUR VALENCE ELECTRONS

Figure 1-6. Semiconductors are made from materials that have four valence electrons.

The basic materials used in most semiconductor devices are germanium and silicon. In their natural state, germanium and silicon are pure crystals. These pure crystals do not have enough free electrons to support a significant current flow. To prepare these crystals for use as a semiconductor device, their structure must be altered to permit significant current flow through a process called doping.

Doping is the addition of impurities to the crystal structure of a semiconductor. In doping, some of the atoms in the crystal are replaced with atoms of other materials. The addition of new atoms in the crystal structure creates N-type material and P-type material.

N-type material is material created by doping a region of a crystal with atoms of a material that have more electrons in their outer shells than the crystal. Adding these atoms to the crystal results in more free electrons. Free electrons (carriers) support current flow. Current flows from negative to positive through the crystal when voltage is applied to N-type material. The material is N-type material because electrons have a negative charge. **See Figure 1-7.**

Materials commonly used for creating N-type material are arsenic, bismuth, and antimony. The quantity of doping material used ranges from a few parts per billion to a few parts per million. By controlling these small quantities of impurities in a crystal, a manufacturer can control the operating characteristics of the semiconductor.

N-TYPE MATERIAL

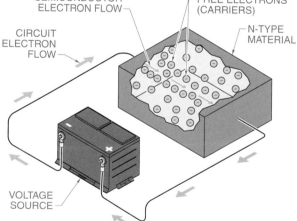

Figure 1-7. Current flows from negative potential to positive potential and is assisted by free electrons when voltage is applied to N-type material.

P-type material is material with empty spaces (holes) in its crystal structure. To create P-type material, a crystal is doped with atoms of a material that have fewer electrons in their outer shells than the crystal. *Holes* are the missing electrons in the structure of the crystal. These holes are represented as positive charges.

In P-type material, holes act as carriers. The holes are filled with free electrons when voltage is applied, and the free electrons move from negative potential to positive potential through the crystal. **See Figure 1-8.** Movement of the electrons from one hole to the next makes the holes appear to move in the opposite direction. Hole flow is equal to and opposite of electron flow. Typical materials used for doping a crystal to create P-type material are gallium, boron, and indium.

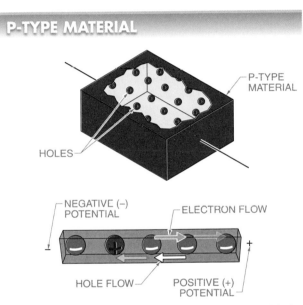

Figure 1-8. When voltage is applied to P-type material, the holes are filled with free electrons that move from the negative potential to the positive potential through the crystal.

A *diode* is an electronic component that allows current to pass through it in only one direction. This is made possible by the doping process, which creates N-type material and P-type material in the same component. The P-type and N-type materials exchange carriers at the junction of the two materials, creating a thin depletion region. **See Figure 1-9.**

When voltage is applied to a diode, the action occurring in the depletion region either blocks current flow or passes current. *Forward-bias voltage* is the application of the proper polarity to a diode. Forward bias results in forward current. *Reverse-bias voltage* is the application

of the opposite polarity to a diode. Reverse bias results in a reverse current, which should be very small (normally 1 mA).

DIODES

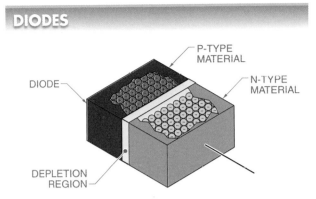

Figure 1-9. In a diode, P-type and N-type materials exchange carriers at the junction of the two materials, creating a thin depletion region.

A diode has a relatively low resistance in the forward-bias direction and a high resistance in the reverse-bias direction. The anode (depicted by a triangle) represents the P-type material, and the cathode (depicted by a straight line) represents the N-type material. Electrons flow from the cathode to the anode, or against the triangle, when the diode is in forward bias. When a negative polarity is applied to the anode and a positive polarity is applied to the cathode, the diode is in reverse bias and there is no electron flow. **See Figure 1-10.**

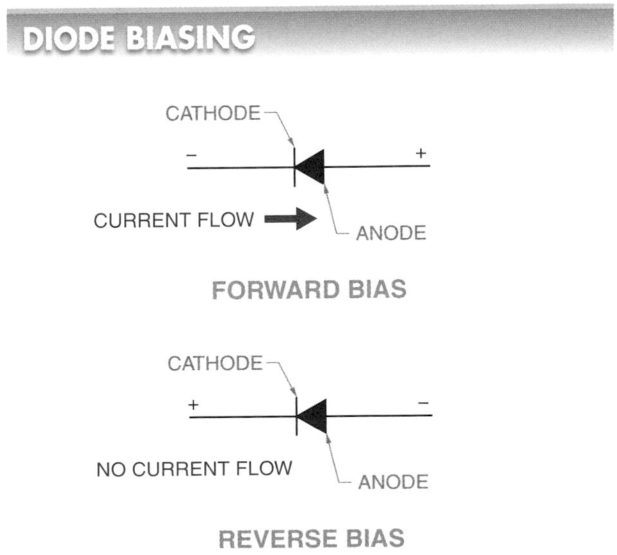

DIODE BIASING

Figure 1-10. In forward bias, electrons flow from cathode to anode, but in reverse bias, electrons do not flow.

Manufacturers mark diodes in different ways to indicate the cathode and the anode. **See Figure 1-11.** Diodes may be marked with the schematic symbol, or there may be a band at one end to indicate the cathode. Some manufacturers use the shape of the diode package to indicate the cathode end. Typically the cathode end is marked or enlarged to ensure proper installation and connection into a circuit.

DIODE CATHODES

CATHODE MARKED WITH SCHEMATIC SYMBOL

CATHODE MARKED WITH A BAND

CATHODE PHYSICALLY LARGER

STUD-MOUNTED DIODE—BOLT END USED FOR MOUNTING

CATHODE MARKED WITH SCHEMATIC SYMBOL

STUD-MOUNTED DIODE—BOLT END USED FOR MOUNTING

CATHODE BEVELED

Figure 1-11. Manufacturers use a variety of methods to indicate the cathode end of a diode.

Energy

Energy is used for producing electricity. *Energy* is the capacity to do work. The two forms of energy are potential energy and kinetic energy. *Potential energy* is the stored energy a body has due to its position, chemical state, or physical condition. For example, water behind a dam has potential energy because of its position. A battery has potential energy based on its chemical state. A compressed spring has potential energy because of its physical condition.

Kinetic energy is the energy of motion. Examples of kinetic energy include falling water, a rotating motor, or a released spring. Kinetic energy is released potential energy. Energy released when water falls through a dam is used to generate electricity. Energy released when a battery is connected to a motor is used to produce a rotating mechanical force. Energy released by a compressed spring is used to apply a braking force on a motor shaft.

The sources of energy used to produce electricity are coal, nuclear power, natural gas, and oil. Wind, solar power, and water also provide energy. These energy sources are used for producing work when converted to electricity. Some energy sources, such as coal, oil, and natural gas, are consumed during use. Energy sources such as wind, solar power, and water are not consumed during use. **See Figure 1-12.**

Coal is used to produce approximately 46% of the electricity produced. Nuclear power is used to produce approximately 20%, natural gas is used to produce approximately 20%, and oil is used to produce approximately 1% of the electricity produced. Wind, solar power, and water account for approximately 13% of the electricity produced. Wind and solar power are growing as energy sources for producing electricity.

The recent growth of wind power is expected to continue as a clean energy source for producing electricity.

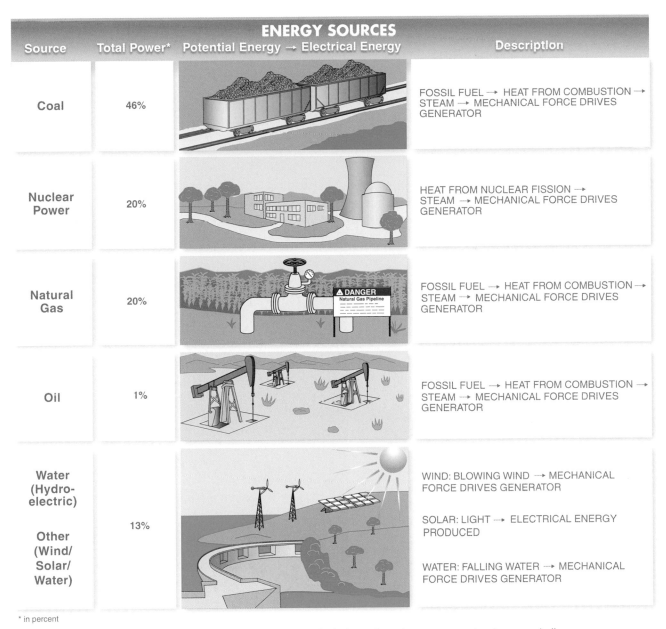

Source	Total Power*	Potential Energy → Electrical Energy	Description
ENERGY SOURCES			
Coal	46%		FOSSIL FUEL → HEAT FROM COMBUSTION → STEAM → MECHANICAL FORCE DRIVES GENERATOR
Nuclear Power	20%		HEAT FROM NUCLEAR FISSION → STEAM → MECHANICAL FORCE DRIVES GENERATOR
Natural Gas	20%		FOSSIL FUEL → HEAT FROM COMBUSTION → STEAM → MECHANICAL FORCE DRIVES GENERATOR
Oil	1%		FOSSIL FUEL → HEAT FROM COMBUSTION → STEAM → MECHANICAL FORCE DRIVES GENERATOR
Water (Hydro-electric) Other (Wind/ Solar/ Water)	13%		WIND: BLOWING WIND → MECHANICAL FORCE DRIVES GENERATOR SOLAR: LIGHT → ELECTRICAL ENERGY PRODUCED WATER: FALLING WATER → MECHANICAL FORCE DRIVES GENERATOR

* in percent

Figure 1-12. The forms of energy used to produce electricity include coal, nuclear power, natural gas, and oil.

Electricity is converted into motion, light, heat, sound, and visual outputs. **See Figure 1-13.** Approximately 62% of all electricity is converted into rotary motion by electric motors. Three-phase motors use the largest amount of electricity in commercial and industrial applications. Three-phase motors are used because they are the most energy-efficient motors.

Approximately 20% of all electricity is converted into light by lamps. The most common lamp used in residential lighting was the incandescent lamp. However, incandescent lamps are rapidly being replaced with compact fluorescent lamps (CFLs) and light-emitting diodes (LEDs). The most common lamps used in commercial and industrial lighting are fluorescent lamps for office installations and high-intensity discharge (HID) lamps for warehouse and factory installations. HID lamps include low-pressure sodium, mercury-vapor, metal-halide, and high-pressure sodium lamps. HID lamps are also the most common lamps used for exterior lighting applications. However, new technologies such as LEDs are also being used in many of these areas.

Approximately 18% of all electricity is used to produce heat, linear motion, audible signals, and visual outputs. When the total number of individual electrical loads is considered, this group is the largest number of different electricity-using components. It also includes the largest number of loads that consume very little power compared to motors.

Voltage

All electrical circuits must have a source of power to produce work. The source of power used depends on the application and the amount of power required. All sources of power produce a set voltage level or voltage range.

Voltage (E) is the amount of electrical pressure in a circuit. Voltage is measured in volts (V). Voltage is also known as electromotive force (EMF) or potential difference. Voltage is produced when electrons are freed from atoms.

Voltage may be produced when electrons are freed from atoms by electromagnetism (generators), heat (thermocouples), light (photocells), chemical reaction (batteries/fuel cells), pressure (piezoelectricity in strain gauges), and friction (static electricity). **See Figure 1-14.**

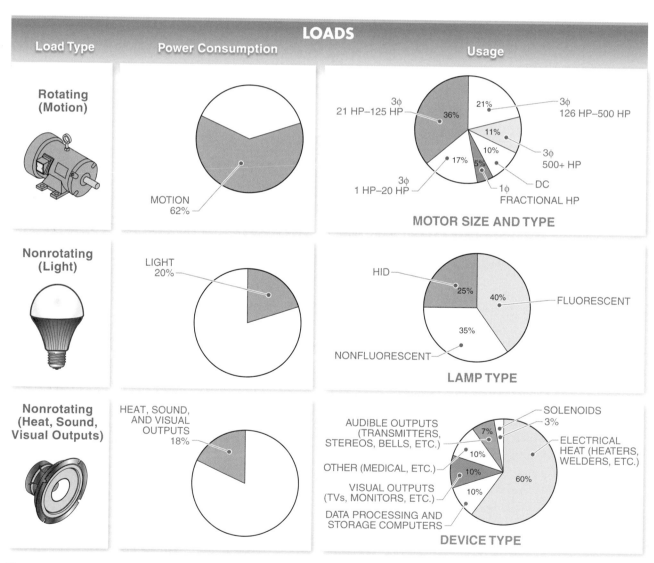

Figure 1-13. Electrical energy is used to produce motion, light, heat, sound, and visual outputs.

VOLTAGE PRODUCTION

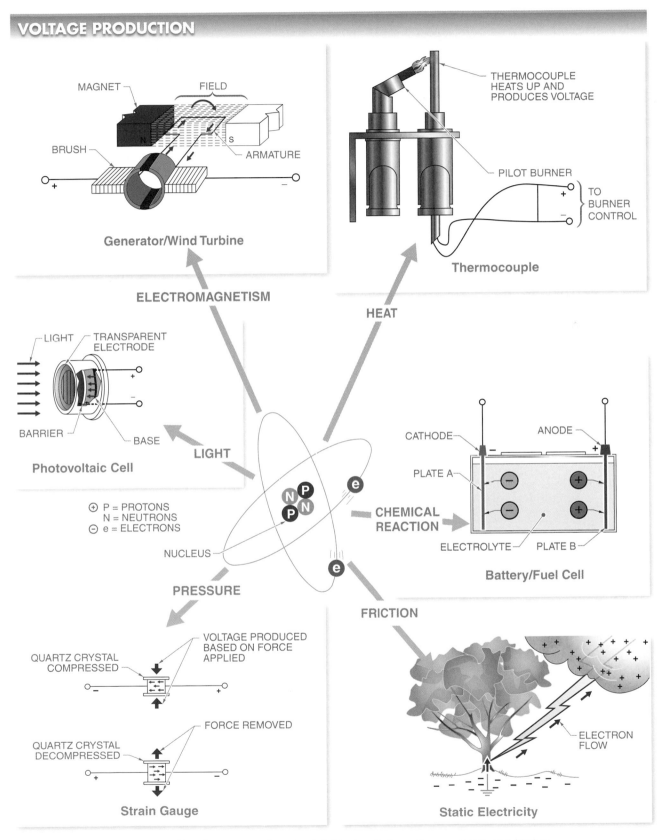

Figure 1-14. Voltage is produced by electromagnetism, heat, light, chemical reaction, pressure, and friction.

Voltage is either direct current (DC) or alternating current (AC). *DC voltage* is voltage that flows in one direction only. *AC voltage* is voltage that reverses its direction of flow at regular intervals. DC voltage is used in almost all portable equipment (automobiles, golf carts, flashlights, cameras, etc.). AC voltage is used in residential, commercial, and industrial lighting and power distribution systems.

DC Voltage. All DC voltage sources have a positive terminal and a negative terminal. The positive and negative terminals establish polarity in a circuit. Polarity is the positive (+) or negative (−) state of an object. All points in a DC circuit have polarity.

The most common power sources that directly produce DC voltage are batteries, fuel cells, and photovoltaic cells. In addition to obtaining DC voltage directly from batteries and photovoltaic cells, DC voltage is also obtained from a rectified AC voltage supply. **See Figure 1-15.** DC voltage is obtained any time an AC voltage is passed through a diode. Diodes convert AC voltage to DC voltage by allowing the voltage and current to flow in only one direction. DC voltage obtained from a rectified AC voltage supply can vary from almost pure DC voltage to half-wave DC voltage. Common DC voltage levels include 1.5 V, 6 V, 9 V, 12 V, 24 V, 36 V, and 125 V.

A photovoltaic cell is a semiconductor device that converts light energy directly to electrical energy.

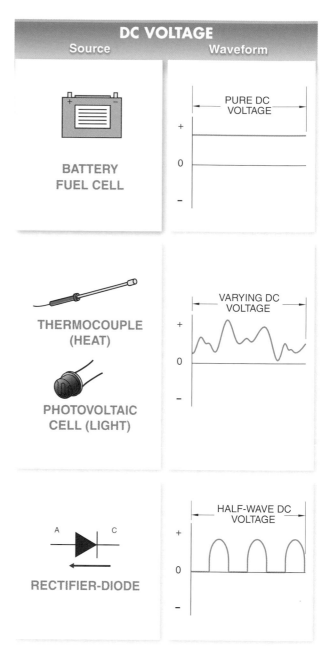

Standard DC Voltages	
Device	Level*
Flashlights, watches, etc.	1.5, 3
Toys, automobiles, trucks	6, 9, 12, 24, 36
Printing presses, small electric railway systems	125, 250, 600
Large electric railway systems	1200, 1500, 3000

* in V

Figure 1-15. DC voltage is produced from batteries, photovoltaic cells, and rectified AC voltage supplies and can vary from almost pure DC voltage to half-wave DC voltage.

AC Voltage. AC voltage is the most common voltage used to produce work. AC voltage is produced by generators, which create AC sine waves as they rotate. An *AC sine wave* is a symmetrical waveform that contains 360 electrical degrees. The wave reaches its peak positive value at 90°, returns to 0 V at 180°, increases to its peak negative value at 270°, and returns to 0 V at 360°.

A *cycle* is one complete positive and negative alternation of a wave form. An *alternation* is half of a cycle. A sine wave has one positive alternation and one negative alternation per cycle. **See Figure 1-16.**

AC voltage is either single-phase (1ϕ) or three-phase (3ϕ). Single-phase AC voltage contains only one alternating voltage waveform. Three-phase AC voltage is a combination of three alternating voltage waveforms, each displaced 120 electrical degrees (one-third of a cycle) apart. Three-phase voltage is produced when three coils are simultaneously rotated in a generator.

Low AC voltages (6 V to 24 V) are used for doorbells and security systems. Medium AC voltages (110 V to 120 V) are used in residential applications for lighting, heating, cooling, cooking, running motors, etc. High AC voltages (208 V to 480 V) are used in commercial/residential applications for cooking, heating, cooling, etc. High AC voltages are also used in industrial applications to convert raw materials into usable products, in addition to providing lighting, heating, and cooling for plant personnel.

Current

Current flows through a circuit when a source of power is connected to a device that uses electricity. *Current (I)* is the amount of electrons flowing through an electrical circuit. Current is measured in amperes (A). An *ampere* is the number of electrons passing a given point in one second. The more power a load requires, the larger the amount of current flow. For example, a 10 horsepower (HP) motor draws approximately 28 A when wired for 230 V. A 20 HP motor draws approximately 56 A when wired for 230 V.

Current Levels. Different voltage sources produce different amounts of current. For example, standard AAA, AA, A, C, and D size batteries all produce 1.5 V, but each size is capable of delivering a different amount of current. Size AAA batteries are capable of delivering the smallest amount of current, and size D batteries are capable of delivering the highest amount of current. For this reason, a load connected to a size D battery operates longer than the same load connected to a size AAA battery. **See Figure 1-17.**

Current may be direct current or alternating current. *Direct current (DC)* is current that flows in only one direction. Direct current flows in any circuit connected to a power supply producing a DC voltage. *Alternating current (AC)* is current that reverses its direction of flow at regular intervals. Alternating current flows in any circuit connected to a power supply producing an AC voltage.

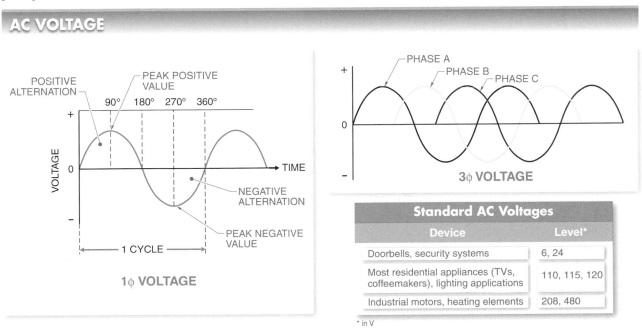

Figure 1-16. AC voltage has one positive alternation and one negative alternation per cycle and is either single-phase (1ϕ) or three-phase (3ϕ).

CELL AND BATTERY CURRENT

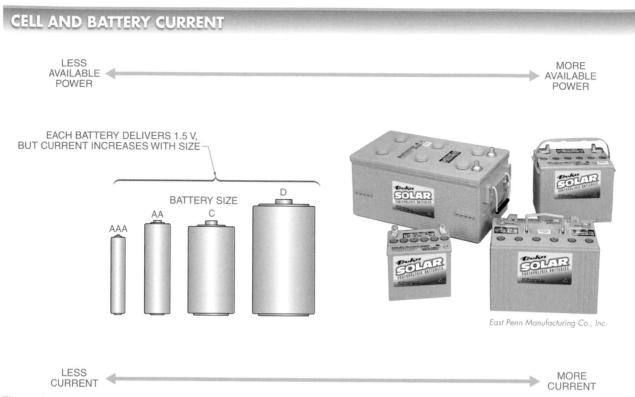

Figure 1-17. The amount of current a cell or battery can supply depends on the cell or battery size.

Current Flow. Early scientists believed that electrons flowed from positive to negative. Later, when atomic structure was studied, electron flow from negative to positive was introduced. *Conventional current flow* is the movement of electrons from positive to negative. *Electron current flow* is the movement of electrons from negative to positive. Both current flow theories are still used. The current flow theory used depends on the industry. For example, the automobile industry usually uses the conventional current flow theory when explaining electricity. The electrical and electronics industry usually uses the electron current flow theory when explaining electricity.

Resistance

Resistance (R) is the opposition to the flow of electrons. Resistance is measured in ohms. The Greek letter omega (Ω) is used to represent ohms. Higher resistance measurements are expressed using prefixes, as in kilohms (kΩ) and megohms (MΩ).

Resistance limits the flow of current in an electrical circuit. The higher the resistance, the lower the current

flow. Likewise, the lower the resistance, the higher the current flow. Components designed to insulate, such as rubber or plastic, should have a very high resistance. Components designed to conduct, such as conductors (wires) or switch contacts, should have a very low resistance. The resistance of insulators decreases when they are damaged by moisture and/or overheating. The resistance of conductors increases when they are damaged by burning and/or corrosion. Factors that affect the resistance of conductors are the size of the wire, length of the wire, conductor material, and temperature.

A conductor with a large cross-sectional area has less resistance than a conductor with a small cross-sectional area. **See Figure 1-18.** A large conductor may also carry more current. The longer the conductor, the greater the resistance is as well. Short conductors have less resistance than long conductors of the same size. Copper (Cu) is a better conductor (less resistance) than aluminum (Al) and may carry more current for a given size. Temperature also affects resistance. For metals, the higher the temperature, the greater the resistance.

AWG SIZES

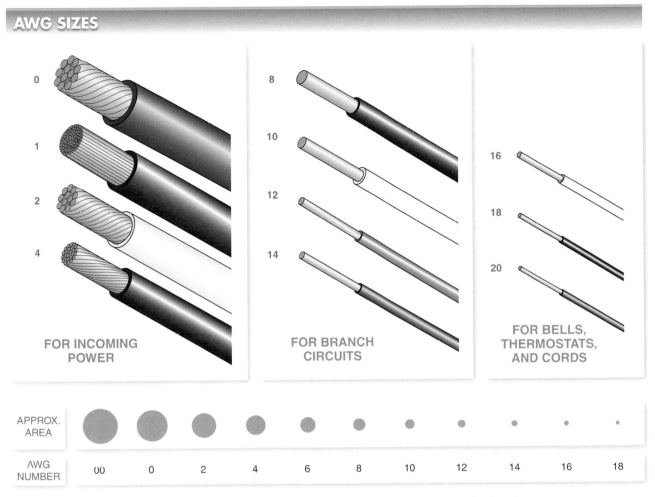

Figure 1-18. The smaller the AWG number, the greater the cross-sectional area and the heavier the wire.

Electrical Abbreviations/Prefixes

Electrical abbreviations are used to simplify the expression of common electrical terms and quantities. An *abbreviation* is a letter or combination of letters that represents a word. The exact abbreviation used normally depends on the use of the electrical unit. For example, voltage may be abbreviated using a capital letter E or V. A capital letter V is used to indicate voltage quantity because voltage is measured in volts (V). These abbreviations are often interchanged and both can be used to represent voltage. **See Figure 1-19.**

Tech Fact

The most common prefixes encountered when using electrical test instruments to take circuit voltage (V), current (A), and resistance (Ω) measurements are micro (μ), milli (m), kilo (k), and mega (M).

COMMON ELECTRICAL QUANTITIES

Equation Variable	Name	Unit of Measure and Abbreviation
E	voltage	volt—V
I	current	ampere—A
R	resistance	ohm—Ω
P	power	watt—W
P_A	power (apparent)	volt-amp—VA
C	capacitance	farad—F
L	inductance	henry—H
Z	impedance	ohm—Ω
G	conductance	siemens—S
f	frequency	hertz—Hz
T	period	second—s

Figure 1-19. Abbreviations are used to simplify the expression of common electrical terms and quantities.

Prefixes are used to avoid long expressions of units that are smaller or larger than the base unit. A base unit is a number that does not include a metric prefix. To convert between different units, the decimal point is moved to the left or right, depending on the unit. The decimal point is moved to the left and a prefix is added to convert a large base value to a simpler term. For example, 1000 V can be written as 1 kV. The decimal point is moved to the right and a prefix is added to convert a small base value to a simpler term. For example, 0.001 V can be written as 1 mV.

Ohm's Law

Ohm's law is the relationship between voltage, current, and resistance in a circuit. Ohm's law states that current in a circuit is proportional to the voltage and inversely proportional to the resistance. Any value in this relationship can be found when the other two values are known. The relationship between voltage, current, and resistance may be visualized by presenting Ohm's law in pie chart form. **See Figure 1-20.**

OHM'S LAW

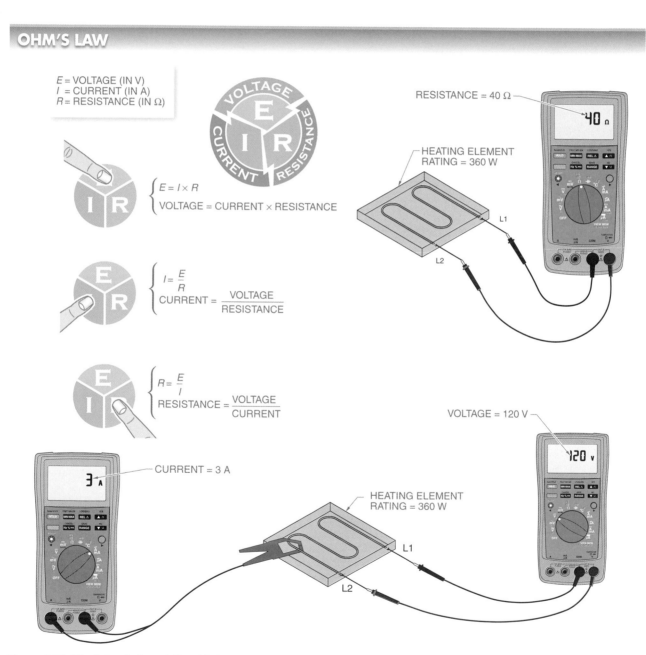

E = VOLTAGE (IN V)
I = CURRENT (IN A)
R = RESISTANCE (IN Ω)

$E = I \times R$
VOLTAGE = CURRENT × RESISTANCE

$I = \dfrac{E}{R}$
CURRENT = $\dfrac{\text{VOLTAGE}}{\text{RESISTANCE}}$

$R = \dfrac{E}{I}$
RESISTANCE = $\dfrac{\text{VOLTAGE}}{\text{CURRENT}}$

RESISTANCE = 40 Ω

HEATING ELEMENT RATING = 360 W

VOLTAGE = 120 V

CURRENT = 3 A

HEATING ELEMENT RATING = 360 W

Figure 1-20. Ohm's law is the relationship between voltage (E), current (I), and resistance (R) in a circuit.

Calculating Voltage Using Ohm's Law. Ohm's law states that voltage (E) in a circuit is equal to current (I) times resistance (R). To calculate voltage using Ohm's law, apply the following formula:

$$E = I \times R$$

where

E = voltage (in V)

I = current (in A)

R = resistance (in Ω)

For example, what is the voltage in a circuit that includes a 40 Ω heating element that draws 3 A?

$$E = I \times R$$
$$E = 3 \times 40$$
$$E = \textbf{120 V}$$

Calculating Current Using Ohm's Law. Ohm's law states that current (I) in a circuit is equal to voltage (E) divided by resistance (R). To calculate current using Ohm's law, apply the following formula:

$$I = \frac{E}{R}$$

where

I = current (in A)

E = voltage (in V)

R = resistance (in Ω)

For example, what is the current in a circuit with a 40 Ω heating element connected to a 120 V supply?

$$I = \frac{E}{R}$$
$$I = \frac{120}{40}$$
$$I = \textbf{3 A}$$

Calculating Resistance Using Ohm's Law. Ohm's law states that resistance (R) in a circuit is equal to voltage (E) divided by current (I). To calculate resistance using Ohm's law, apply the following formula:

$$R = \frac{E}{I}$$

where

R = resistance (in Ω)

E = voltage (in V)

I = current (in A)

For example, what is the resistance of a circuit in which a load that draws 3 A is connected to a 120 V supply?

$$R = \frac{E}{I}$$
$$R = \frac{120}{3}$$
$$I = \textbf{40 Ω}$$

1-1 CHECKPOINT

1. What are the three fundamental particles contained in atoms?

2. Which particle has a negative charge?

3. What unit is used to measure resistance?

4. What device allows current to flow in only one direction?

5. What type of fuel is used to produce the most amount of electricity?

6. What electrical devices consume the largest share of produced electricity?

7. Voltage is measured in volts (V), but what letter is used to represent voltage?

8. What device converts AC voltage to DC voltage?

9. What are the two types of AC voltage?

10. Current is measured in amperes (A), but what letter is used to represent current?

11. What are the two types of current?

12. Resistance is measured in ohms (Ω), but what letter is used to represent resistance?

13. If resistance is increased in a circuit, does current increase or decrease?

14. If 12 V is applied to a circuit that has a resistance of 500 Ω, how many milliamperes (mA) will flow through the circuit?

15. If 230 V and 6.25 A are measured in a heating element, how much resistance (in Ω) does the heating element have?

1-2 CIRCUITS

A circuit is a continuous path that allows current flow. Circuits may consist of conductors and electrical components. Types of circuit connections include series, parallel, and series/parallel.

Series Circuits

Fuses, switches, loads, and other electrical components can be connected in series. A *series connection* is a connection that has two or more components connected so there is only one path for current flow. Opening the circuit at any point stops the flow of current. Current stops flowing any time a fuse blows, a circuit breaker trips, or a switch or load opens. An example of a series connection is a DC series motor. **See Figure 1-21.** A *DC series motor* is a DC motor that has the series field coils connected in series with the armature. The armature wires are marked A1 and A2. The series coil wires are marked S1 and S2.

Electronic circuit boards may contain series, parallel, and series/parallel circuits.

Resistance in Series Circuits. The total resistance in a circuit containing series-connected loads equals the sum of the resistances of all loads. The resistance in the circuit increases if loads are added in series and decreases if loads are removed. To calculate total resistance of a series circuit, apply the following formula:

$$R_T = R_1 + R_2 + R_3 + ...$$

where

R_T = total resistance (in Ω)

R_1 = resistance 1 (in Ω)

R_2 = resistance 2 (in Ω)

R_3 = resistance 3 (in Ω)

For example, what is the total resistance of a circuit that has 2 Ω, 4 Ω, and 6 Ω resistors connected in series?

$$R_T = R_1 + R_2 + R_3$$
$$R_T = 2 + 4 + 6$$
$$R_T = 12\ \Omega$$

Voltage in Series Circuits. The total voltage applied across loads connected in series is divided across the individual loads. Each load drops a set percentage of the applied voltage. The exact voltage drop across each load depends on the resistance of that load. The voltage drops across any two loads are the same if the resistance values are the same. To calculate total voltage of a series circuit when the voltage across each load is known or measured, apply the following formula:

$$E_T = E_1 + E_2 + E_3 + ...$$

where

E_T = total applied voltage (in V)

E_1 = voltage drop across load 1 (in V)

E_2 = voltage drop across load 2 (in V)

E_3 = voltage drop across load 3 (in V)

For example, what is the total applied voltage of a circuit containing 4 V, 8 V, and 12 V drops across three loads?

$$E_T = E_1 + E_2 + E_3$$
$$E_T = 4 + 8 + 12$$
$$E_T = 24\ \mathbf{V}$$

Current in Series Circuits. The current in a circuit containing series-connected loads is the same throughout the circuit. The current in the circuit will decrease if the circuit resistance increases, and the current will increase if the circuit resistance decreases. To calculate total current of a series circuit, apply the following formula:

$$I_T = I_1 = I_2 = I_3 = ...$$

where

I_T = total circuit current (in A)

I_1 = current through load 1 (in A)

I_2 = current through load 2 (in A)

I_3 = current through load 3 (in A)

For example, what is the total current through a series circuit if the current measured at each load is 2 A?

$$I_T = I_1 = I_2 = I_3$$
$$I_T = 2 = 2 = 2$$
$$I_T = \mathbf{2\ A}$$

SERIES CONNECTIONS

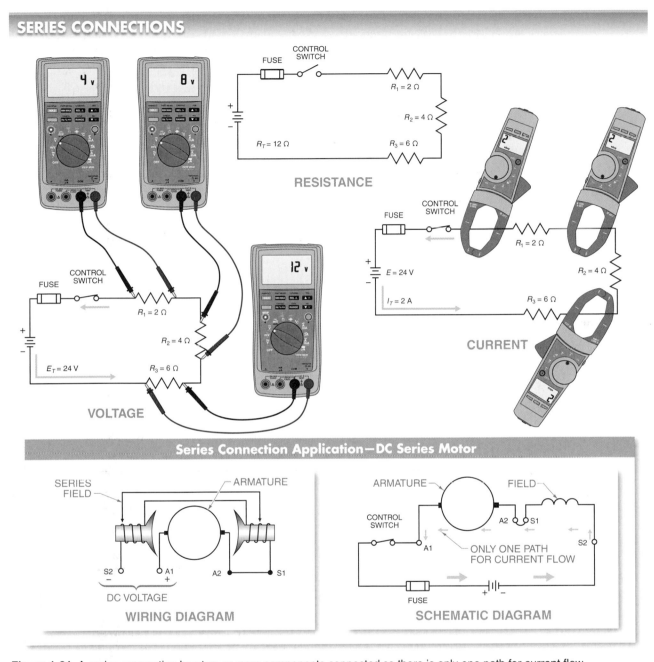

RESISTANCE

VOLTAGE

CURRENT

Series Connection Application—DC Series Motor

SERIES FIELD — ARMATURE

S2 A1 A2 S1
−   +

DC VOLTAGE

WIRING DIAGRAM

ARMATURE — FIELD

CONTROL SWITCH

A2 S1

A1 S2

ONLY ONE PATH FOR CURRENT FLOW

FUSE

SCHEMATIC DIAGRAM

Figure 1-21. A series connection has two or more components connected so there is only one path for current flow.

Parallel Circuits

Fuses, switches, loads, and other components can be connected in parallel. A *parallel connection* is a connection that has two or more components connected so there is more than one path for current flow. An example of a parallel connection is a DC shunt motor. **See Figure 1-22.** A *DC shunt motor* is a DC motor that has the field connected in parallel (shunt) with the armature. The armature wires are marked A1 and A2. The parallel (shunt) coil wires are marked F1 and F2.

Care must be taken when working with parallel circuits because current can be flowing in one part of the circuit even though another part of the circuit is OFF. Understanding and recognizing parallel-connected components and circuits enables a technician or troubleshooter to take proper measurements, make circuit modifications, and troubleshoot the circuit.

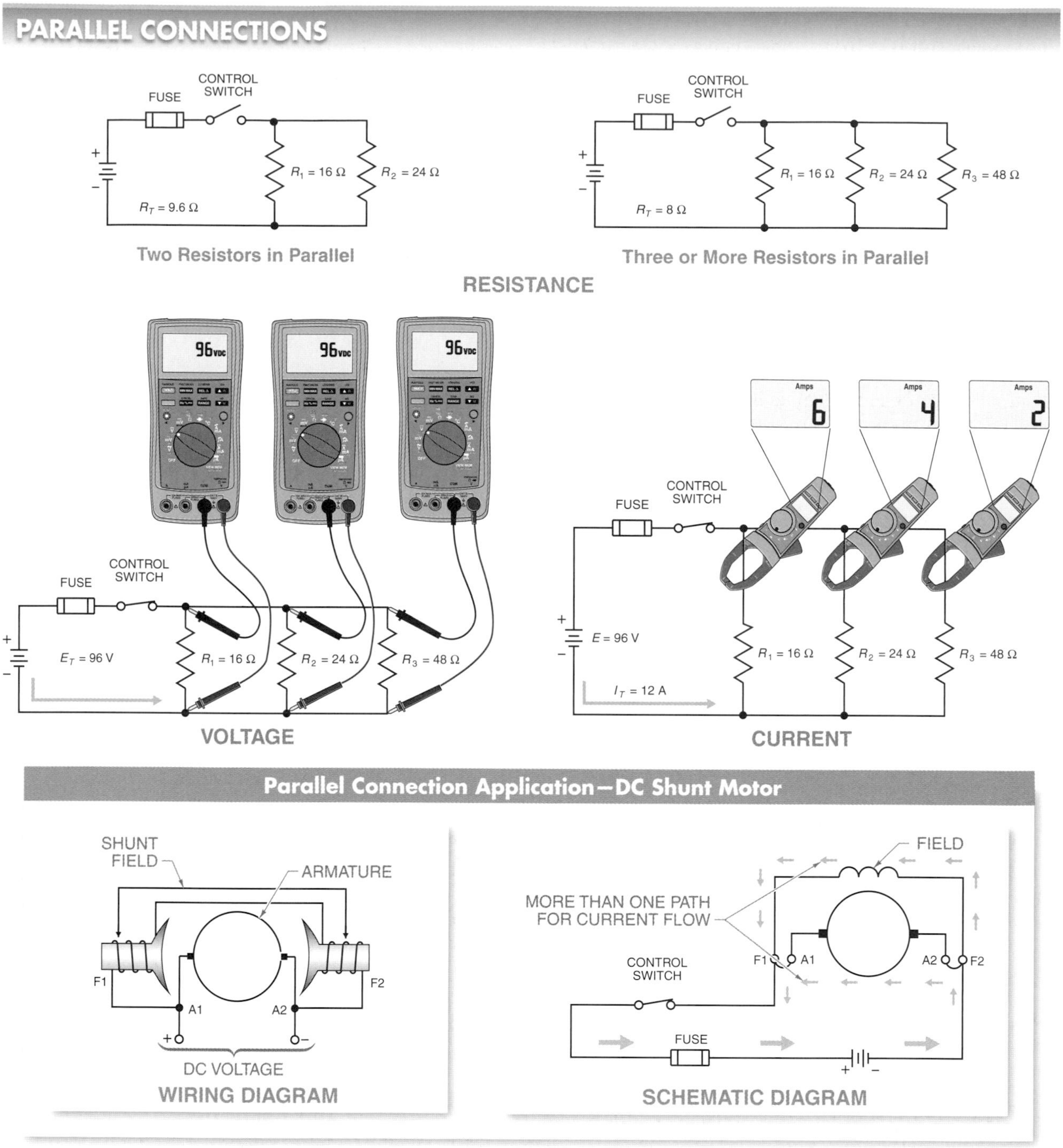

PARALLEL CONNECTIONS

Two Resistors in Parallel

$R_1 = 16\ \Omega$ $R_2 = 24\ \Omega$
$R_T = 9.6\ \Omega$

Three or More Resistors in Parallel

$R_1 = 16\ \Omega$ $R_2 = 24\ \Omega$ $R_3 = 48\ \Omega$
$R_T = 8\ \Omega$

RESISTANCE

$E_T = 96\ V$ $R_1 = 16\ \Omega$ $R_2 = 24\ \Omega$ $R_3 = 48\ \Omega$

VOLTAGE

$E = 96\ V$ $R_1 = 16\ \Omega$ $R_2 = 24\ \Omega$ $R_3 = 48\ \Omega$
$I_T = 12\ A$

CURRENT

Parallel Connection Application—DC Shunt Motor

SHUNT FIELD   ARMATURE
F1   A1   A2   F2
DC VOLTAGE
WIRING DIAGRAM

FIELD
MORE THAN ONE PATH FOR CURRENT FLOW
CONTROL SWITCH
F1 A1   A2 F2
FUSE
SCHEMATIC DIAGRAM

Figure 1-22. A parallel connection has two or more components connected so there is more than one path for current flow.

Resistance in Parallel Circuits. The total resistance in a circuit containing parallel-connected loads is less than the smallest resistance value. The total resistance decreases if loads are added in parallel and increases if loads are removed. To calculate total resistance in a parallel circuit containing two resistors, apply the following formula:

$$R_T = \frac{R_1 \times R_2}{R_1 + R_2}$$

where
R_T = total resistance (in Ω)
R_1 = resistance 1 (in Ω)
R_2 = resistance 2 (in Ω)

For example, what is the total resistance in a circuit containing resistors of 16 Ω and 24 Ω connected in parallel?

$$R_T = \frac{R_1 \times R_2}{R_1 + R_2}$$

$$R_T = \frac{16 \times 24}{16 + 24}$$

$$R_T = \frac{384}{40}$$

$$R_T = \mathbf{9.6\ \Omega}$$

To calculate total resistance in a parallel circuit with three or more resistors, the formula for two resistors can be used by solving the problem for two resistors at a time. In addition, to calculate the total resistance in a parallel circuit with three or more resistors, apply the following formula:

$$R_T = \frac{1}{\frac{1}{R_1}} + \frac{1}{\frac{1}{R_2}} + \frac{1}{\frac{1}{R_3}} + ...$$

where

R_T = total resistance (in Ω)
R_1 = resistance 1 (in Ω)
R_2 = resistance 2 (in Ω)
R_3 = resistance 3 (in Ω)

For example, what is the total resistance in a circuit containing resistors of 16 Ω, 24 Ω, and 48 Ω connected in parallel?

$$R_T = \frac{1}{\frac{1}{R_1}} + \frac{1}{\frac{1}{R_2}} + \frac{1}{\frac{1}{R_3}} + ...$$

$$R_T = \frac{1}{\frac{1}{16}} + \frac{1}{\frac{1}{24}} + \frac{1}{\frac{1}{48}} + ...$$

$$R_T = \frac{1}{0.06250} + \frac{1}{0.04166} + \frac{1}{0.02083} + ...$$

$$R_T = \mathbf{8\ \Omega}$$

Voltage in Parallel Circuits. The voltage across each load is the same when loads are connected in parallel. The voltage across each load remains the same if parallel loads are added or removed. To calculate total voltage in a parallel circuit when the voltage across a load is known or measured, apply the following formula:

$$E_T = E_1 = E_2 = E_3 = ...$$

where

E_T = total applied voltage (in V)
E_1 = voltage across load 1 (in V)
E_2 = voltage across load 2 (in V)
E_3 = voltage across load 3 (in V)

For example, what is the total applied voltage if the voltage across three parallel connected loads is 96 VDC?

$$E_T = E_1 = E_2 = E_3$$
$$E_T = 96 = 96 = 96$$
$$E_T = \mathbf{96\ VDC}$$

Current in Parallel Circuits. Total current in a circuit containing parallel-connected loads equals the sum of the current through all the loads. Total current increases if loads are added in parallel and decreases if loads are removed. To calculate total current in a parallel circuit, apply the following formula:

$$I_T = I_1 + I_2 + I_3 + ...$$

where

I_T = total circuit current (in A)
I_1 = current through load 1 (in A)
I_2 = current through load 2 (in A)
I_3 = current through load 3 (in A)

For example, what is the total current in a circuit containing three loads connected in parallel if the current through the three loads is 6 A, 4 A, and 2 A?

$$I_T = I_1 + I_2 + I_3$$
$$I_T = 6 + 4 + 2$$
$$I_T = \mathbf{12\ A}$$

Tech Fact

A building grounding electrode conductor (GEC) must have a resistance to the earth of less than 25 Ω. When one ground rod, pipe, or plate electrode has more than 25 Ω resistance to the earth, a second or third GEC must be added in parallel with the first to reduce the total resistance. The additional GEC(s) must be added in parallel because the total resistance in a circuit decreases when devices are connected in parallel.

Series/Parallel Circuits

Fuses, switches, loads, and other components can be connected in a series/parallel connection. A *series/parallel connection* is a combination of series- and parallel-connected components. An example of a series/parallel connection is a DC compound motor. **See Figure 1-23.** A *DC compound motor* is a DC motor with the field connected in both series and parallel (shunt) with the armature. The armature wires are marked A1 and A2. The parallel (shunt) coil wires are marked F1 and F2. The series coil is marked S1 and S2.

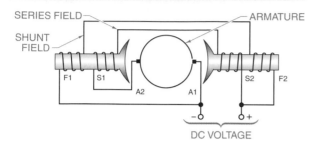

SERIES/PARALLEL CONNECTIONS

DC COMPOUND MOTOR WIRING DIAGRAM

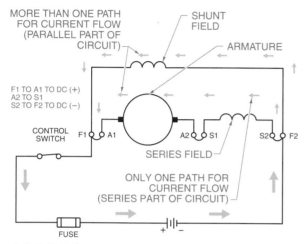

DC COMPOUND MOTOR SCHEMATIC DIAGRAM

Figure 1-23. A series/parallel connection is a combination of series- and parallel-connected components.

Resistance in Series/Parallel Circuits. A series/parallel circuit may contain any number of individual resistors (loads) connected in any number of different series/parallel circuit combinations. The series/parallel combination is always equal to one combined total resistance value. The total resistance in a circuit containing series/parallel connected resistors equals the sum of the series loads and the equivalent resistance of the parallel combinations. To calculate total resistance in a series/parallel circuit that contains two resistors in series connected to two resistors in parallel, apply the following formula:

$$R_T = \left(\frac{R_{P1} \times R_{P2}}{R_{P1} + R_{P2}} \right) + R_{S1} + R_{S2}$$

where

R_T = total resistance (in Ω)
R_{P1} = parallel resistance 1 (in Ω)
R_{P2} = parallel resistance 2 (in Ω)
R_{S1} = series resistance 1 (in Ω)
R_{S2} = series resistance 2 (in Ω)

For example, what is the total resistance of a 150 Ω and 50 Ω resistor connected in parallel with a 25 Ω and 100 Ω resistor connected in series?

$$R_T = \left(\frac{R_{P1} \times R_{P2}}{R_{P1} + R_{P2}} \right) + R_{S1} + R_{S2}$$

$$R_T = \left(\frac{150 \times 50}{150 + 50} \right) + 25 + 100$$

$$R_T = \left(\frac{7500}{200} \right) + 125$$

$$R_T = 37.5 + 125$$

$$R_T = \mathbf{162.5 \ \Omega}$$

Current in Series/Parallel Circuits. The total current and current in individual parts of a series/parallel circuit follow the same laws of current as in a basic series and a basic parallel circuit. Current is the same in each series part of the series/parallel circuit. Current is equal to the sum of each parallel combination in each parallel part of the series/parallel circuit.

Voltage in Series/Parallel Circuits. The total voltage applied across resistors (loads) connected in a series/parallel circuit is divided across the individual resistors (loads). The higher the resistance of any one resistor or equivalent parallel resistance, the higher the voltage drop. Likewise, the lower the resistance of any one resistor or equivalent parallel resistance, the lower the voltage drop.

Applications of Series and Parallel Circuits

The principles of series and parallel circuits can be used to produce several different heat outputs in heating element circuits. For example, one principle is that heat is produced any time electricity passes through a wire that has resistance. This principle is used to produce heat in devices such as toasters, portable space heaters, hair dryers, electric ovens, coffee makers, irons, and electric hot water heaters. This principle also applies to industrial heating applications when heating solids, gases, liquids, surfaces, and pipes.

Heating elements are made from special resistance wire that is capable of withstanding the temperature produced by the electricity flowing through the element. The amount of heat a wire produces is proportional to the resistance of the wire. The lower the resistance, the greater the current flow in a wire (Ohm's law). The greater the current flow in a wire, the higher the temperature of the wire (power formula). Likewise, the higher the resistance, the smaller the current flow (Ohm's law). The smaller the current flow, the lower the temperature of the wire.

By varying the total resistance of a heating element, the heat output of the heating element can be varied. The total resistance can be varied by using several individual heating elements that can be connected in series, in parallel, or in a series/parallel combination. **See Figure 1-24.**

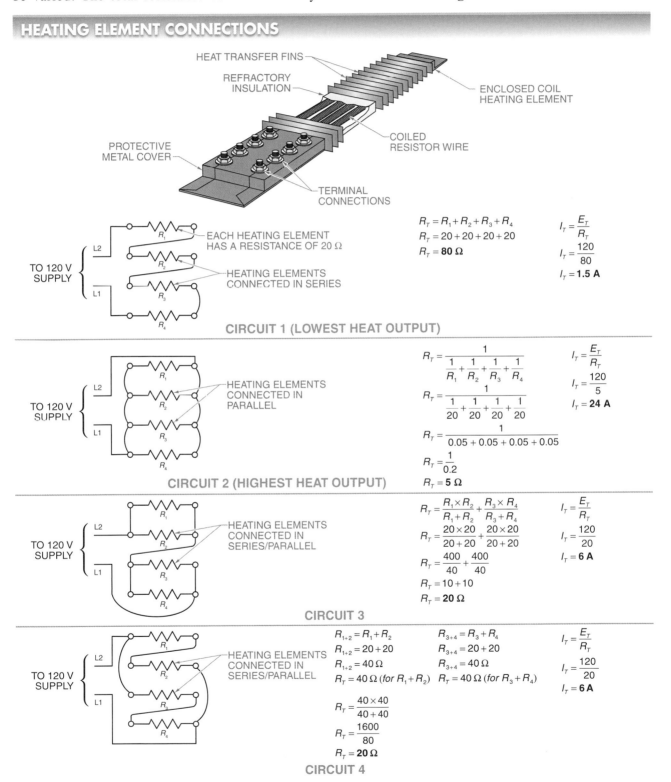

HEATING ELEMENT CONNECTIONS

HEAT TRANSFER FINS

REFRACTORY INSULATION

ENCLOSED COIL HEATING ELEMENT

PROTECTIVE METAL COVER

COILED RESISTOR WIRE

TERMINAL CONNECTIONS

CIRCUIT 1 (LOWEST HEAT OUTPUT)

TO 120 V SUPPLY

R_1 — EACH HEATING ELEMENT HAS A RESISTANCE OF 20 Ω

R_2, R_3, R_4 — HEATING ELEMENTS CONNECTED IN SERIES

$R_T = R_1 + R_2 + R_3 + R_4$
$R_T = 20 + 20 + 20 + 20$
$R_T = \mathbf{80\ \Omega}$

$I_T = \dfrac{E_T}{R_T}$
$I_T = \dfrac{120}{80}$
$I_T = \mathbf{1.5\ A}$

CIRCUIT 2 (HIGHEST HEAT OUTPUT)

TO 120 V SUPPLY

HEATING ELEMENTS CONNECTED IN PARALLEL

$R_T = \dfrac{1}{\dfrac{1}{R_1} + \dfrac{1}{R_2} + \dfrac{1}{R_3} + \dfrac{1}{R_4}}$

$R_T = \dfrac{1}{\dfrac{1}{20} + \dfrac{1}{20} + \dfrac{1}{20} + \dfrac{1}{20}}$

$R_T = \dfrac{1}{0.05 + 0.05 + 0.05 + 0.05}$

$R_T = \dfrac{1}{0.2}$

$R_T = \mathbf{5\ \Omega}$

$I_T = \dfrac{E_T}{R_T}$
$I_T = \dfrac{120}{5}$
$I_T = \mathbf{24\ A}$

CIRCUIT 3

TO 120 V SUPPLY

HEATING ELEMENTS CONNECTED IN SERIES/PARALLEL

$R_T = \dfrac{R_1 \times R_2}{R_1 + R_2} + \dfrac{R_3 \times R_4}{R_3 + R_4}$

$R_T = \dfrac{20 \times 20}{20 + 20} + \dfrac{20 \times 20}{20 + 20}$

$R_T = \dfrac{400}{40} + \dfrac{400}{40}$

$R_T = 10 + 10$

$R_T = \mathbf{20\ \Omega}$

$I_T = \dfrac{E_T}{R_T}$
$I_T = \dfrac{120}{20}$
$I_T = \mathbf{6\ A}$

CIRCUIT 4

TO 120 V SUPPLY

HEATING ELEMENTS CONNECTED IN SERIES/PARALLEL

$R_{1+2} = R_1 + R_2$
$R_{1+2} = 20 + 20$
$R_{1+2} = 40\ \Omega$
$R_T = 40\ \Omega\ (for\ R_1 + R_2)$

$R_{3+4} = R_3 + R_4$
$R_{3+4} = 20 + 20$
$R_{3+4} = 40\ \Omega$
$R_T = 40\ \Omega\ (for\ R_3 + R_4)$

$R_T = \dfrac{40 \times 40}{40 + 40}$

$R_T = \dfrac{1600}{80}$

$R_T = \mathbf{20\ \Omega}$

$I_T = \dfrac{E_T}{R_T}$
$I_T = \dfrac{120}{20}$
$I_T = \mathbf{6\ A}$

Figure 1-24. Principles of series and parallel circuits can be used to produce several different heat outputs in heating element circuits.

Series and Parallel Photovoltaic Cell Output. Photovoltaic cells are rated by the amount of energy they convert. Most manufacturers rate photovoltaic cell output in terms of voltage (V) and current (mA). Photovoltaic cells produce a limited amount of voltage and current. For example, each photovoltaic cell may produce up to 0.6 V. To increase the voltage output, cells are connected in series. **See Figure 1-25.**

SERIES CONNECTIONS—PHOTOVOLTAIC CELLS

+
−
INDIVIDUAL CELLS
+
−
SYMBOLS
CELLS

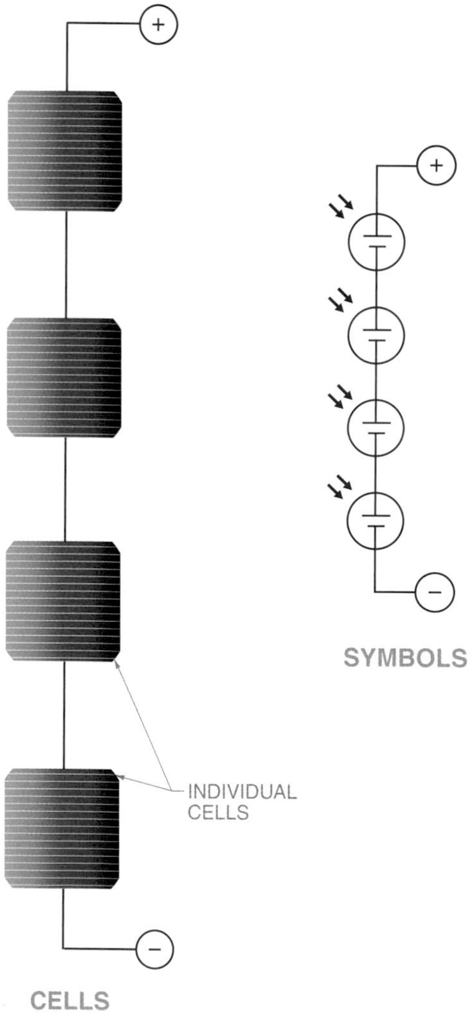

Figure 1-25. Photovoltaic cells are placed in series to increase the voltage output from a set of photovoltaic cells.

Tech Fact

In 1974, photovoltaic cells were used to power Skylab 1 as it orbited Earth. Today, photovoltaic cells are widely used to power everything from portable calculators to small cities.

In addition to the maximum voltage, each photovoltaic cell can produce up to 40 mA of current. To increase the current output, cells are connected in parallel. **See Figure 1-26.** To increase both voltage and current, the individual cells are connected in both series and parallel.

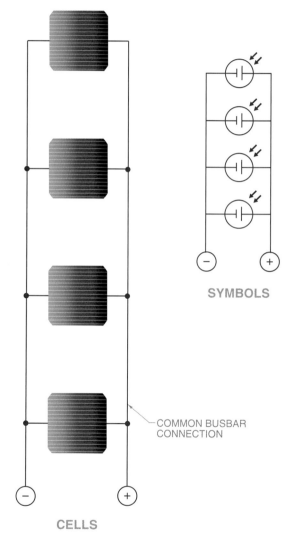

PARALLEL CONNECTIONS—PHOTOVOLTAIC CELLS

−
+
SYMBOLS
COMMON BUSBAR CONNECTION
−
+
CELLS

Figure 1-26. Photovoltaic cells are placed in parallel to increase the current output from a set of photovoltaic cells.

The photovoltaic effect is measured using a high-impedance voltage-measuring device such as a digital multimeter. In the dark, there is no open-circuit voltage present. When sunlight strikes the cell, the light is absorbed and, if the photon energy is large enough, it frees electron-hole pairs.

1-2 CHECKPOINT

1. What is the total voltage of six 1.5 V batteries connected in series?

2. If 200 mA total is measured at the 60 V power supply that includes three 100 Ω resistors connected in series, how much current is flowing through each resistor?

3. What is the total voltage of six 1.5 V/10 mA-rated batteries connected in parallel?

4. In a circuit that contains a 100 Ω, a 200 Ω, and a 300 Ω resistor connected in parallel, which resistor would have the largest amount of current flowing through it?

5. If six 3 V/200 mA-rated photovoltaic cells that are connected in series and six more 3 V/200 mA-rated photovoltaic cells also connected in series and then placed in parallel with the first set of six, what is the total voltage of the combination of photovoltaic cells?

6. If six 3 V/200 mA-rated photovoltaic cells that are connected in series and six more 3 V/200 mA-rated photovoltaic cells also connected in series and then placed in parallel with the first set of six, what is the total available current of the combination of photovoltaic cells?

1-3 MAGNETISM

Magnetism was first discovered by the ancient Greeks when they noticed that a certain type of stone attracted bits of iron. This stone was first found in Asia Minor in the province of Magnesia. The stone was named magnetite after this province.

A *magnet* is a substance that produces a magnetic field and attracts iron. Magnets are either permanent or temporary. A *permanent magnet* is a magnet that can retain its magnetism after the magnetizing force has been removed. Permanent magnets include natural magnets, such as magnetite, and manufactured magnets.

A *temporary magnet* is a magnet that retains only trace amounts of magnetism after the magnetizing force has been removed. Temporary magnets have a low retentivity. A magnet with low retentivity has very little residual magnetism (leftover magnetism) remaining once the magnetizing force has been removed.

Molecular Theory of Magnetism

The *molecular theory of magnetism* is the theory that states that all substances are made up of a number of molecular magnets that can be arranged in either an organized or disorganized manner. **See Figure 1-27.** A material is demagnetized if it has disorganized molecular magnets. A material is magnetized if it has organized molecular magnets.

The molecular theory of magnetism explains how certain materials used in control devices react to magnetic fields. For example, it explains why hard steel is used for permanent magnets, while soft iron is used for the temporary magnets found in control devices.

The dense molecular structure of hard steel does not easily disorganize once the magnetizing force has been removed. Hard steel is difficult to magnetize and demagnetize, making it a good permanent magnet. Hard steel is considered to have high retentivity. However, permanent magnets may be demagnetized by a sharp blow or by heat.

The loose molecular structure of soft iron can be magnetized and demagnetized easily. Soft iron is ideal for use as a temporary magnet in control devices because it does not retain residual magnetism very easily. Soft iron is considered to have low retentivity.

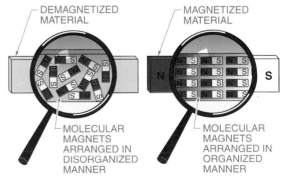

MOLECULAR THEORY OF MAGNETISM

Figure 1-27. The molecular theory of magnetism states that all substances are made up of a number of molecular magnets that can be arranged in either an organized or disorganized manner.

Electromagnetism

In 1819, the Danish physicist Hans C. Oersted discovered that a magnetic field is created around an electrical conductor when electric current flows through the conductor. *Electromagnetism* is the magnetism produced when electric current passes through a conductor. **See Figure 1-28.**

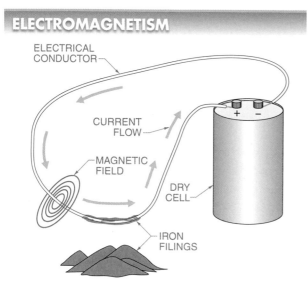

Figure 1-28. In 1819, the Danish physicist Hans C. Oersted discovered that a magnetic field is created around an electrical conductor when electric current flows through the conductor.

The direction in which current flows through a conductor determines the direction of the magnetic field around it. Lines of force (lines of induction) are present all along the full length of the conductor. **See Figure 1-29.**

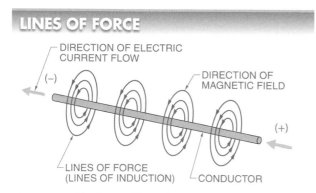

Figure 1-29. The lines of force (lines of induction) are present along the full length of a conductor.

One line of force is called a maxwell, and the total number of lines is called flux. The total number of lines of force in a space of 1 square centimeter (sq cm) equals the flux density (in gauss) of the field. For example, 16 lines of force in 1 sq cm equal 16 gauss. **See Figure 1-30.**

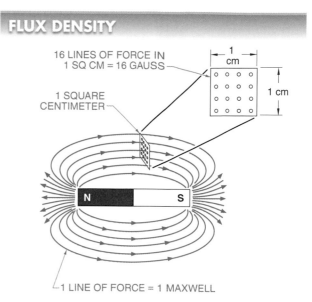

Figure 1-30. The total number of lines of force (maxwells) in a one sq cm section of a magnetic field equals the flux density of the field (gauss).

If a conductor is bent to form a loop, all of the lines of force circling the conductor enter one side of the loop and leave from the other side of the loop. Thus, a north pole is created on one side of the loop and a south pole is created on the other side of the loop. The side of the loop into which the lines of force enter is the south pole, and the side from which the lines of force leave is the north pole. A loop of wire has poles just like a bar magnet.

If a conductor is wound into multiple loops (a coil), the magnetic lines of force combine. Thus, the magnetic force of a coil with multiple turns is stronger than the magnetic force of a coil with a single loop. **See Figure 1-31.**

Oersted attempted several other experiments to increase the strength of the magnetic field. He found three ways to increase the strength of the magnetic field in a coil: increase the amount of current by increasing the voltage, increase the number of turns in the coil, and insert an iron core through the coil. **See Figure 1-32.** These early experiments led to the development of a huge control industry, which depends on magnetic coils to convert electrical energy into usable magnetic energy.

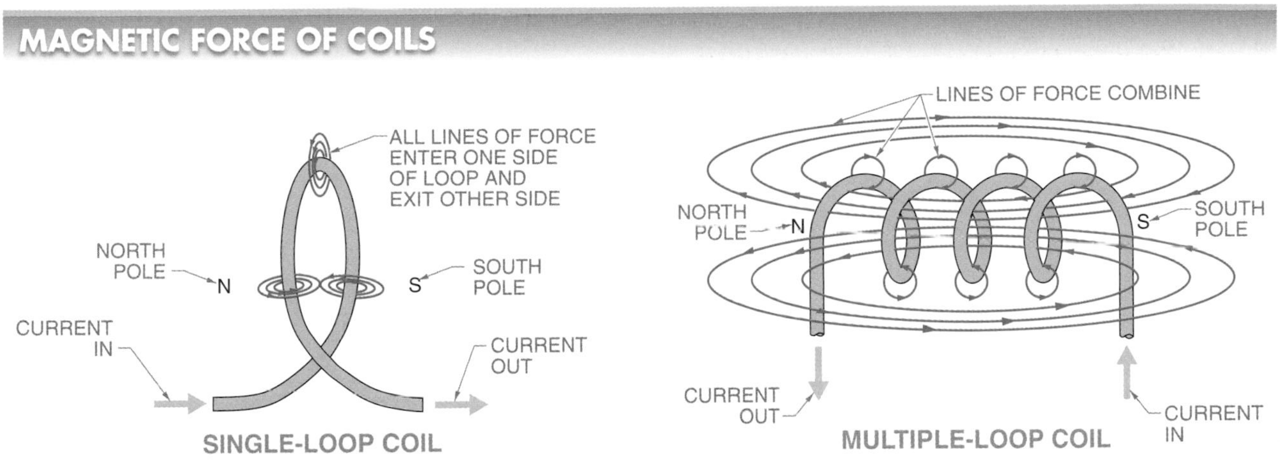

MAGNETIC FORCE OF COILS

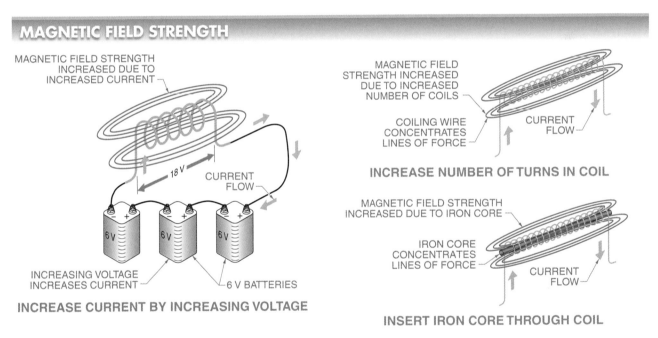

Figure 1-31. If a conductor is formed into a coil, the lines of force combine, forming a stronger field than the lines of force from a single loop.

MAGNETIC FIELD STRENGTH

Figure 1-32. The strength of a magnetic field produced by a conductor may be increased by increasing the voltage, increasing the number of coils, or inserting an iron core through the coil.

Inductive Circuits. An *inductive circuit* is a circuit in which current lags voltage. In a DC circuit, a magnetic field is created and remains at maximum potential until the circuit (switch) is opened. At this point the inductor is storing energy in the circuit. Once the circuit is opened, the magnetic field collapses. At this point the inductor is releasing energy back into the circuit. In an AC circuit, the magnetic field is continuously building and collapsing until the circuit is opened. The magnetic field also changes direction with each change in sine wave alternation. **See Figure 1-33.**

In coils commonly found in motor windings, transformers, and solenoids, the expansion and contraction of the magnetic field creates an effect in the coil called inductance. *Inductance (L)* is the property of a circuit that causes it to oppose a change in current due to energy stored in a magnetic field. Inductance is measured in henrys (H). Inductance is normally stated in henrys (H), millihenrys (mH), or microhenrys (µH).

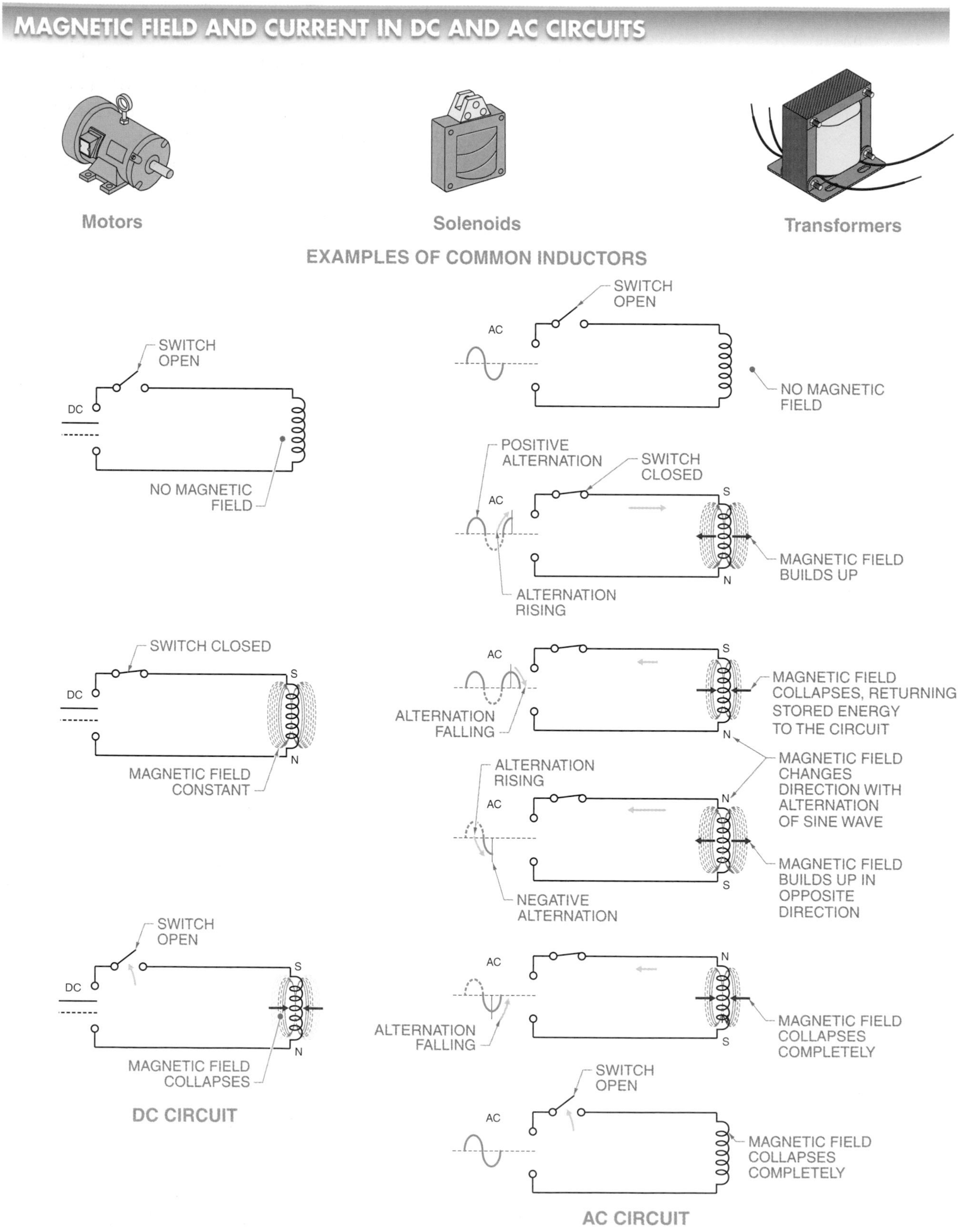

MAGNETIC FIELD AND CURRENT IN DC AND AC CIRCUITS

Motors

Solenoids

Transformers

EXAMPLES OF COMMON INDUCTORS

SWITCH OPEN

DC

NO MAGNETIC FIELD

SWITCH CLOSED

DC

S

N

MAGNETIC FIELD CONSTANT

SWITCH OPEN

DC

S

N

MAGNETIC FIELD COLLAPSES

DC CIRCUIT

AC

SWITCH OPEN

NO MAGNETIC FIELD

POSITIVE ALTERNATION

SWITCH CLOSED

AC

ALTERNATION RISING

S

N

MAGNETIC FIELD BUILDS UP

AC

ALTERNATION FALLING

S

N

MAGNETIC FIELD COLLAPSES, RETURNING STORED ENERGY TO THE CIRCUIT

ALTERNATION RISING

AC

NEGATIVE ALTERNATION

N

S

MAGNETIC FIELD CHANGES DIRECTION WITH ALTERNATION OF SINE WAVE

MAGNETIC FIELD BUILDS UP IN OPPOSITE DIRECTION

AC

ALTERNATION FALLING

N

S

MAGNETIC FIELD COLLAPSES COMPLETELY

SWITCH OPEN

AC

MAGNETIC FIELD COLLAPSES COMPLETELY

AC CIRCUIT

Figure 1-33. A conductor formed into a coil produces a strong magnetic field around the coil when current flows through the coil.

In an inductive circuit, the current lags the voltage. When current and voltage are not synchronized, they are said to be out of phase with each other. The greater the inductance in a circuit, the larger the phase shift. **See Figure 1-34.** In-phase AC sine waves occur in resistive circuits. Phase shifts in AC sine waves occur in inductive circuits.

PHASE SHIFT

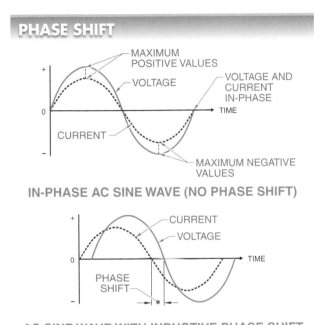

IN-PHASE AC SINE WAVE (NO PHASE SHIFT)

AC SINE WAVE WITH INDUCTIVE PHASE SHIFT

Figure 1-34. Phase shift occurs when voltage and current in an AC circuit do not reach their maximum amplitude and zero level simultaneously.

Capacitive Circuits. A *capacitive circuit* is a circuit in which current leads voltage (voltage lags current). *Capacitance (C)* is the ability of a component or circuit to store energy in the form of an electrical charge. A *capacitor* is an electric device that stores electrical energy by means of an electrostatic field. Small capacitors may be manufactured in several shapes and sizes for use in electronic control boards. **See Figure 1-35.** Larger capacitors are manufactured for use in bigger devices like electrical motors.

The unit of capacitance is the farad (F). However, the farad is too large a unit to express capacitance for most electrical applications. Capacitance and capacitor values are normally stated in microfarads (μF) or picofarads (pF).

A capacitor consists of two conducting surfaces separated by an insulating material called a dielectric. When a DC voltage is applied across two plates they will charge to a level corresponding to the difference of potential between the two terminals of the source. An electrostatic

force is produced in the dielectric between the two plates. At this point the capacitor is storing energy in the circuit. There is no movement of electrons from one plate to the other. However, there is a displacement of charges, and an electrostatic field exists in the dielectric material. This electrostatic field represents stored electrical energy. How much energy is stored depends upon the applied voltage, the area of the plates, the separation between the plates, and the type of dielectric material.

CAPACITORS

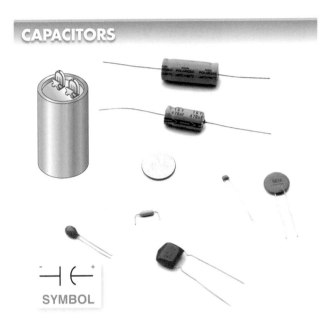

SYMBOL

Figure 1-35. Capacitors are available in different shapes and sizes.

When the charging voltage is removed and the shorting switch S2 is closed, the excess electrons on the left plate will move through the switch to the right plate. Now the capacitor acts as a voltage source with the left plate as the negative terminal and the right plate as the positive terminal. At this point the capacitor is releasing energy into the circuit. The supply of electricity is limited to the electrical energy stored in the dielectric. The movement of electrons off the left plate reduces the negative charge, and their arrival at the right plate reduces the positive charge. **See Figure 1-36.**

The motion of electrons will continue until there is no charge on either plate and the difference of potential is zero. At this point, all of the energy originally stored in the dielectric material will have been used to move the electrons from the left plate to the right plate. No electrostatic field exists between the plates at that time. When capacitance is created in an electrical circuit, a phase shift occurs between the voltage and the current in the circuit. **See Figure 1-37.**

CHARGING CAPACITORS

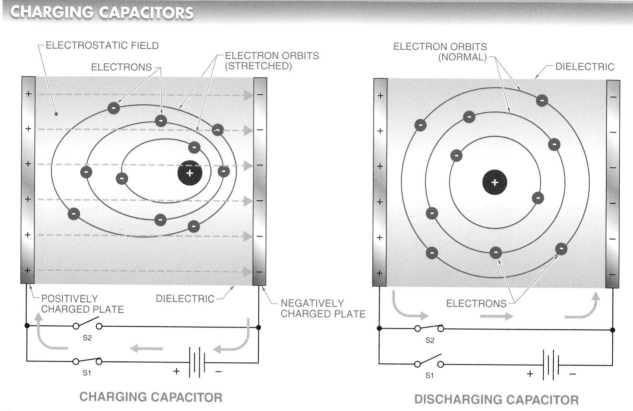

Figure 1-36. With a charged capacitor, the electron orbits become stretched toward the positively charged plate.

CAPACITIVE CIRCUITS

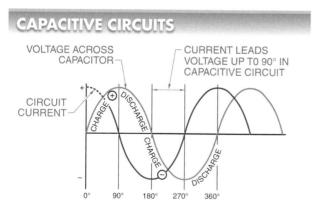

Figure 1-37. Current leads voltage in AC capacitive circuits.

Capacitor Water Pump Analogy. An elastic diaphragm represents the internal action of a capacitor. The diaphragm insulates one side of the water supply from the other. Since it is elastic, it can move and stretch to allow the water to push it back and forth. The elastic diaphragm opposes the flow of the water in one direction but flexes back and helps water flow in the reverse direction. This reverse action is faster than the pressure developed when using a pump. The water (current) "leads" the pressure (voltage). This results in a phase shift where current leads voltage. **See Figure 1-38.**

WATER PUMP ANALOGY

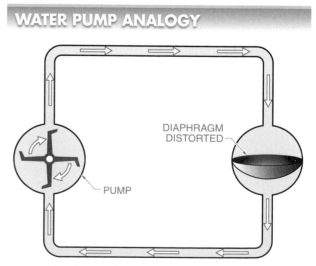

Figure 1-38. An elastic diaphragm and water can be used to represent the internal action of a capacitor.

Effects of Induction and Capacitance in a Circuit. Electrical devices that are designed primarily to produce heat offer only resistance to the flow of the current and are known as resistors. Electric heating elements are of this type.

Devices that are designed to convert electrical energy to magnetic energy are referred to as inductive. Components of this type include solenoids, relays, transformers, and motors. Coil-type devices offer inductive reactance to the flow of AC in addition to resistance.

Devices that function primarily to store electrical charges are referred to as capacitive. Among these devices are the starting and running capacitors used with some motors. Capacitors offer capacitive reactance to the flow of AC.

Inductive Reactance. *Inductive reactance* (X_L) is the opposition to current flow of an inductor in an AC circuit. Like resistance, inductive reactance is measured in ohms. The amount of inductive reactance in a circuit depends on the amount of inductance (in henrys) of the coil and the frequency of the current. Inductance is normally a fixed amount. Frequency may be a fixed or a variable amount. Inductive reactance is expressed by the following formula:

$$X_L = 2\pi f l$$

where

X_L = inductive reactance (in Ω)

2π = constant

f = frequency (in Hz)

l = inductance (in H)

Ohm's law applies equally to inductive AC circuits as it does to a resistive circuit. The current (I) flowing in an inductive AC circuit is directly proportional to the applied voltage (E) and inversely proportional to the inductive reactance (X_L). **See Figure 1-39.** This relationship is represented mathematically by the following expression:

$$I = \frac{E}{X_L}$$

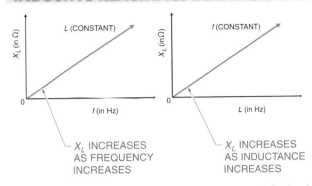

INDUCTIVE REACTANCE BEHAVIOR

Figure 1-39. The current flowing in an inductive AC circuit is directly proportional to the applied voltage and inversely proportional to the inductive reactance.

In this expression, the current is in amperes (A), the applied voltage is in volts (V), and the reactance is in ohms (Ω). Increasing the voltage or decreasing the reactance will cause an increase in the current. Decreasing the applied voltage or increasing the reactance will cause a decrease in the current.

Capacitive Reactance. *Capacitive reactance* (X_C) is the opposition to current flow by a capacitor. Capacitive reactance is measured in ohms (Ω). In an AC circuit, the capacitor is constantly charging and discharging. The voltage across the capacitor is in constant opposition to the applied voltage. This constant opposition to changes in the applied voltage creates an opposition to current flow in the circuit. The amount of opposition offered to current flow in an AC circuit by a capacitor is a function of the capacitance and the frequency of the voltage. The capacitive reactance is inversely proportional to the capacitance and the frequency. This means that increasing the capacitance or frequency causes the reactance to decrease. **See Figure 1-40.**

Capacitive reactance is calculated by using the following formula:

$$X_C = \frac{1}{2\pi f c}$$

where

X_C = capacitive reactance (Ω)

2π = constant

f = frequency (Hz)

c = capacitance (F)

Ohm's law can be used to find voltage, current, and reactance if two of the three values are known.

Tech Fact

Capacitors were originally referred to as condensers. The word "condenser" was used extensively in the automotive industry where a tune-up meant changing the points (switch), plugs, and condenser. As automotive technology changed and because the word "capacitor" was used in the electronic field, capacitor eventually replaced the word "condenser," except in some older devices such as the condenser microphone.

CAPACITIVE REACTANCE BEHAVIOR

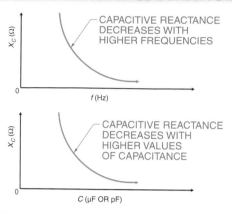

Figure 1-40. Capacitive reactance is inversely proportional to the capacitance and the frequency.

1-3 CHECKPOINT

1. Does increasing the number of loops in a coil increase or decrease the electromagnetic field when current passes through the coil?
2. Inductance is measured in henrys (H), but what letter is used to represent inductance?
3. Capacitance is measured in farads (F), but what letter is used to represent capacitance?
4. Inductive reactance and capacitive reactance oppose a flow of current in a circuit and are stated or measured in what electrical unit?
5. In an AC circuit that includes a coil (inductance), is the voltage and current in phase or out of phase?

1-4 POWER

Power is the rate of doing work or using energy. Types of power include true, reactive, and apparent power.

True Power

True power (P_T) is the actual power used in an electrical circuit. True power is the power that is converted into work for use by devices, such as heating elements. True power is measured in watts (W), kilowatts (kW), or megawatts (MW). In DC circuits or AC circuits in which voltage and current are in phase, such as resistive loads, true power is equal to the voltage *(E)* times the current *(I)*. Heating elements are resistive loads and are rated in true power (watts).

True Power Calculations. The *power formula* is the relationship between power *(P)*, voltage *(E)*, and current *(I)* in an electrical circuit. Any value in this relationship may be found using the power formula when the other two values are known. The relationship between power, voltage, and current may be visualized by presenting the power formula in pie chart form. **See Figure 1-41.**

Calculating Power Using the Power Formula. The power formula states that power *(P)* in a circuit is equal to voltage *(E)* times current *(I)*. To calculate power using the power formula, apply the following formula:

$$P = E \times I$$

where

P = power (in W)
E = voltage (in V)
I = current (in A)

For example, what is the power of a load that draws 5 A when connected to a 120 V supply?

$$P = E \times I$$
$$P = 120 \times 5$$
$$P = \textbf{600 W}$$

Calculating Voltage Using the Power Formula. The power formula states that voltage *(E)* in a circuit is equal to power *(P)* divided by current *(I)*. To calculate voltage using the power formula, apply the following formula:

$$E = \frac{P}{I}$$

where

E = voltage (in V)
P = power (in W)
I = current (in A)

For example, what is the voltage in a circuit in which a 600 W load draws 5 A?

$$E = \frac{P}{I}$$
$$E = \frac{600}{5}$$
$$E = \textbf{120 V}$$

Tech Fact

Wind turbines are rated for their power output in watts. The average power consumption at any one time for a common residence is approximately 3.33 kW. Thus, a 1 megawatt (1 MW) wind turbine can generate enough power for approximately 300 homes.

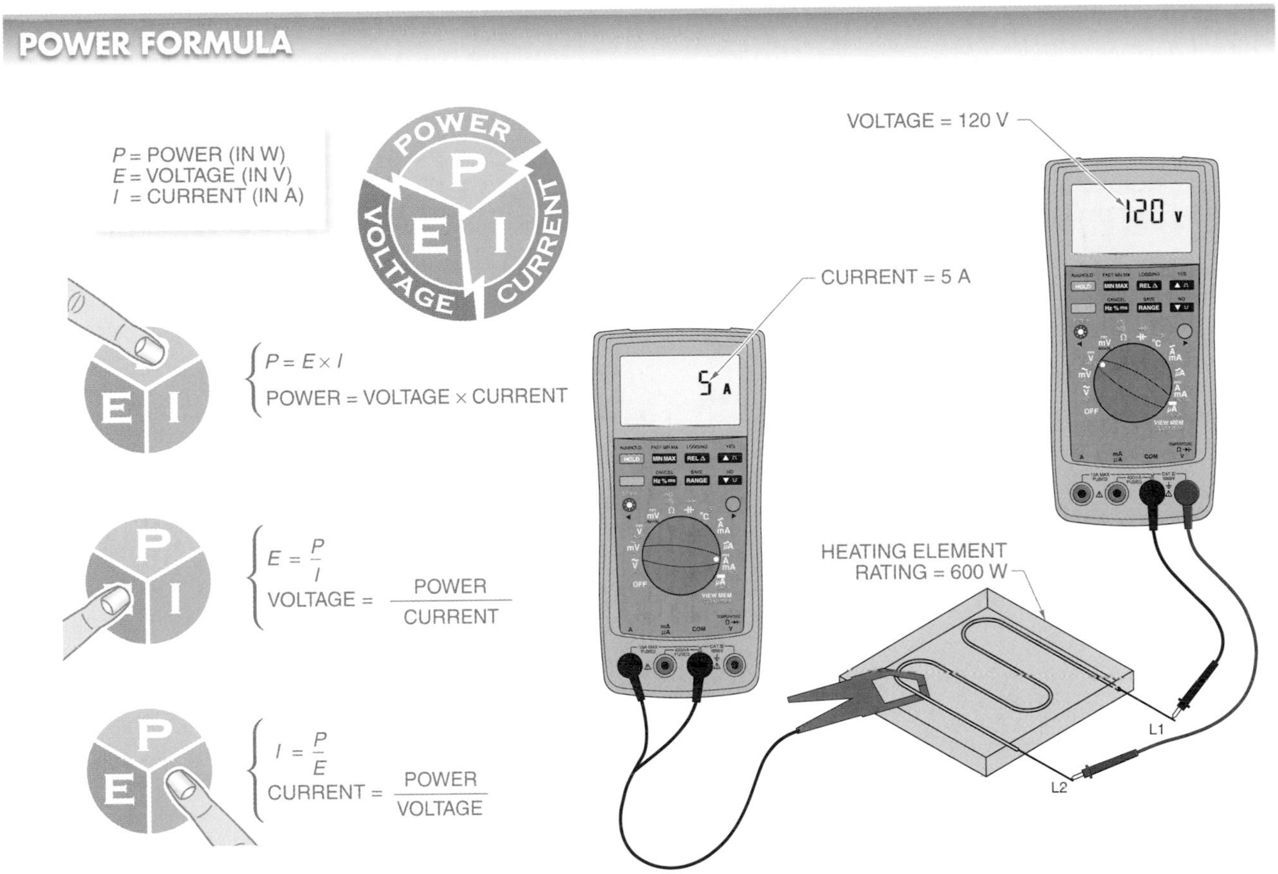

POWER FORMULA

P = POWER (IN W)
E = VOLTAGE (IN V)
I = CURRENT (IN A)

$$P = E \times I$$
POWER = VOLTAGE × CURRENT

$$E = \frac{P}{I}$$
$$VOLTAGE = \frac{POWER}{CURRENT}$$

$$I = \frac{P}{E}$$
$$CURRENT = \frac{POWER}{VOLTAGE}$$

VOLTAGE = 120 V

CURRENT = 5 A

HEATING ELEMENT
RATING = 600 W

L1

L2

Figure 1-41. The power formula is the relationship between power *(P)*, voltage *(E)*, and current *(I)* in an electrical circuit.

Calculating Current Using the Power Formula. The power formula states that current (*I*) in a circuit is equal to power (*P*) divided by voltage (*E*). To calculate current using the power formula, apply the following formula:

$$I = \frac{P}{E}$$

where

I = current (in A)

P = power (in W)

E = voltage (in V)

For example, what is the current in a circuit in which a 600 W load is connected to a 120 V supply?

$$I = \frac{P}{E}$$

$$I = \frac{600}{120}$$

$$I = \mathbf{5\,A}$$

Power in Series Circuits. Power is produced when voltage is applied to a load and current flows through the load. The lower the resistance of the load or the higher the applied voltage, the more power is produced. The higher the resistance of the load or the lower the applied voltage, the less power is produced. Total power produced in a series circuit is equal to the sum of the power produced in each load. To calculate total power in a series circuit when the power across each load is known or measured, apply the following formula:

$$P_T = P_1 + P_2 + P_3 + ...$$

where

P_T = total applied power (in W)

P_1 = power drop across load 1 (in W)

P_2 = power drop across load 2 (in W)

P_3 = power drop across load 3 (in W)

For example, what is the total power in a series circuit if three loads are connected in series and load 1 equals 8 W, load 2 equals 16 W, and load 3 equals 24 W?

$$P_T = P_1 + P_2 + P_3$$

$$P_T = 8 + 16 + 24$$

$$P_T = \mathbf{48\,W}$$

Power in Parallel Circuits. Power is produced when voltage is applied to a load and current flows through the load. Total power produced in a parallel circuit is equal to the sum of the power produced by each load. To calculate total power in a parallel circuit when the power across each load is known or measured, apply the following formula:

$$P_T = P_1 + P_2 + P_3 + \ldots$$

where

P_T = total circuit power (in W)

P_1 = power of load 1 (in W)

P_2 = power of load 2 (in W)

P_3 = power of load 3 (in W)

For example, what is the total circuit power if three loads are connected in parallel and the loads produce 576 W, 384 W, and 192 W?

$$P_T = P_1 + P_2 + P_3$$
$$P_T = 576 + 384 + 192$$
$$P_T = \mathbf{1152\ W}$$

Power in Series/Parallel Circuits. Power is produced when current flows through any load or component that has resistance. The lower the resistance or higher the amount of current, the more power is produced. The higher the resistance or lower the amount of current, the less power is produced. As with any series or parallel circuit, the total power in a series/parallel circuit is equal to the sum of the powers produced by each load or component.

Reactive and Apparent Power

Up to this point in this chapter, only true power used in resistive loads has been discussed. There are, however, two other types of power. The two other types of power are reactive power and apparent power.

Reactive Power. *Reactive power* is power absorbed and returned to a load due to its inductive and/or capacitive properties. Reactive power is indicated by the letter Q and is measured in volt-amperes reactive (VAR). Pure reactive power uses no true power. This is because pure reactive power performs no actual work, such as the production of heat. The reason reactive power does no work is because most of the reactive power drawn from a source is returned to that source. Pure inductive and capacitive components merely store energy temporarily in the device and then return it to the source. Inductors store energy in a magnetic field and capacitors store energy in an electrostatic field.

Apparent Power. Apparent power is a combination of true and reactive power. Since all circuits have some resistance, the total power in an AC circuit is a combination of resistive and reactive properties. Apparent power is the product of the voltage and current in a circuit calculated without considering the phase shift that may be present between the voltage and the current in a circuit.

Apparent power represents a load or circuit that includes both true power and reactive power. Apparent power is expressed in volt-amperes (VA). Apparent power is a measure of the system capacity. This is true because calculating apparent power considers all circuit current regardless of how it exists in the circuit.

Power Factor. Power factor (PF) is a ratio between true power and apparent power. True power is measured in watts (W) and apparent power is measured in volt-amperes (VA). The ratio of the true power to the apparent power in an AC circuit is expressed by the following formula:

$$power\ factor = \frac{true\ power}{apparent\ power}$$

Effect of Power Factor. The nameplate information on a ¼ HP inductive motor shows the difference between true power and apparent power. **See Figure 1-42.** The ¼ HP AC motor (resistive/reactive load) is required to lift a 60 lb load 30′ in 15 sec. To lift the load, the motor must deliver 186.5 W (true power). The motor nameplate lists motor current at 5 A and voltage at 115 V. The rated current (5 A) multiplied by the rated voltage (115 V) equals 575 VA. The difference between true power and apparent power exists because the coil in the motor must produce a rotating magnetic field for the motor to perform the work. Reactive power and true power are required from the power source because the motor coil is a reactive load. In small 1ϕ AC motor circuits, apparent power is much higher than true power.

To determine the power factor of a motor, apply the following formula:

$$PF = \left(\frac{P_T}{P_A} \right) \times 100$$

where

PF = power factor (in %)

P_T = true power (in W)

P_A = apparent power (in VA)

100 = constant (decimal to percentage conversion)

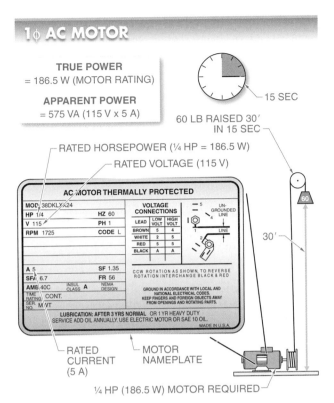

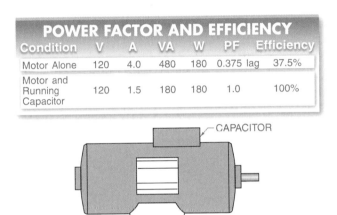

Figure 1-42. True power is always less than apparent power in a circuit with a phase shift between voltage and current.

For example, what is the power factor of the motor described above?

$$PF = \left(\frac{P_T}{P_A}\right) \times 100$$

$$PF = \left(\frac{186.5}{575}\right) \times 100$$

$$PF = 0.324 \times 100$$

$$PF = \textbf{32.4\%}$$

Power factor is an indication of the relative amounts of resistance and reactance in a given circuit. In a purely resistive circuit, the true power and the apparent power will be equal. Therefore the power factor will be equal to one and is referred to as "unity." In a purely reactive circuit, such as one containing only a capacitor or inductor, the true power will be zero. The power factor will also be zero. For circuits containing both resistance and reactance, the power factor will be some value between zero and one. The greater the power factor, the more resistive the circuit. The lower the power factor, the greater the circuit reactance. The reason power factor is so important is that when the power factor is less than 100%, or unity, the circuit is less efficient and has higher operating cost because not all current provide by the source is performing work.

Power Factor and Efficiency. An example of power factor and efficiency can be shown with a small 60 Hz 1ϕ AC induction motor. The motor may be operated alone or a running capacitor can be added. When the motor is operated on its own, it has a lagging power factor of 37.5% efficiency. This is due to the effect of inductive reaction within the motor. When capacitors and capacitive reaction are introduced into the circuit, the current draw by the motor drops 2.5 A. The drop in current draw results from the corrected power factor. This results in less line voltage drop from the power source and a higher efficiency of the power source. The motor still uses 180 W of power to do its work but the overall efficiency of the system is significantly improved. **See Figure 1-43.**

POWER FACTOR AND EFFICIENCY					
Condition	V	A	VA	W	PF	Efficiency
Motor Alone	120	4.0	480	180	0.375 lag	37.5%
Motor and Running Capacitor	120	1.5	180	180	1.0	100%

Figure 1-43. A running capacitor can be added to a motor to achieve a power factor of 1.0 and 100% efficiency.

The power factor was moved to 1.0 because of the balancing value of the running capacitor with the induction of the motor. Their opposing effects canceled each other out and left a 1.0 power factor.

This means that the current neither leads nor lags the voltage and is "in-phase." The true power is now numerically equal to volt-amperes. The fact that the apparent power and the true power are equal means that the reactive power has been reduced by the corrected power factor.

Ohm's Law and Impedance. *Impedance (Z)* is the combined opposition to the flow of current in circuits that contain resistance and reactance. *Reactance (X)* is the opposition to the flow of current in circuits that contain inductance (X_L) or capacitance (X_C). The lower the impedance, the easier it is for current to flow. Impedance is expressed in ohms (Ω).

Ohm's law is used in circuits that contain impedance. However, the letter Z is substituted for the letter R in the formula. The letter Z represents the total resistive force (resistance and reactance) opposing current flow. The relationship between voltage (E), current (I), and impedance (Z) may be visualized by presenting the relationship in pie chart form where the known variables can be used to calculate the unknown variable. **See Figure 1-44.**

AC CIRCUIT FORMULAS

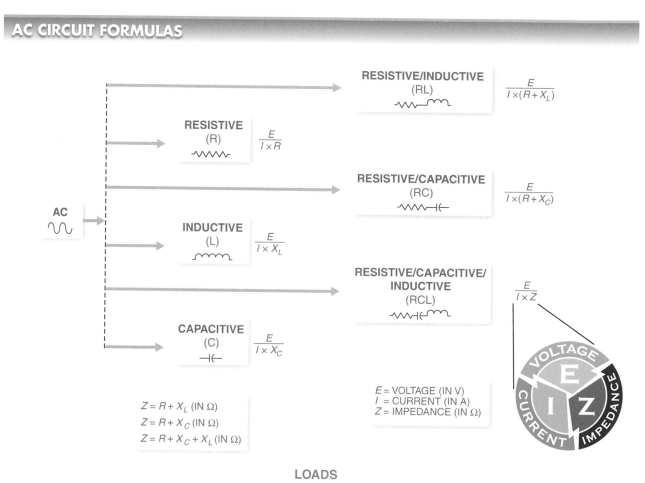

LOADS

Figure 1-44. Ohm's law can be used on circuits with impedance by substituting Z (impedance) for R (resistance) in the formula.

1-4 CHECKPOINT

1. If 8 A are measured in a 120 V circuit, how much power (in W) is the circuit using?
2. If two 25 W and four 60 W lamps are connected into a parallel circuit, what is the total power (in W) used by the circuit?
3. If a string of sixty 2 W holiday lights are connected in series, what is the total power (in W) of all the lights?

4. What electrical unit is reactive power measured in?
5. What electrical unit is apparent power measured in?
6. What is the ratio between true power and apparent power called?

Additional Resources

Review and Resources

Access Chapter 1 Review and Resources for *Electrical Motor Controls for Integrated Systems* by scanning the above QR code with your mobile device.

Applying Your Knowledge

Refer to the *Electrical Motor Controls for Integrated Systems* Learner Resources for interactive Applying Your Knowledge questions.

Workbook and Applications Manual

Refer to Chapter 1 in the *Electrical Motor Controls for Integrated Systems Workbook* and the *Applications Manual* for additional exercises.

ENERGY EFFICIENCY PRACTICES

Improving Energy Efficiency

Improving energy efficiency by using advanced industrial processes and high-efficiency motors can significantly reduce electricity use. Energy efficiency is the ratio of output energy (energy released by a process) to input energy (energy used as input to run a process) and may be expressed as a percentage. Energy efficiency cannot exceed 100% in a closed system. A device that is 100% efficient converts all energy supplied (input power) to output power. An electrical device with 100% efficiency is not realistic because all electrical equipment, especially motors, transformers, and generators, consumes some energy in order to produce work. This energy is normally given off as heat.

Motors are rated by output power, which is normally a percentage of their full-load rating. If a 9 HP motor is 90% efficient, it would require 10 HP of input power at full load for the motor to output 9 HP. Motor efficiency varies with the load. Large motors are normally more efficient than small motors.

Motor efficiency can vary widely between manufacturers. For this reason, motor efficiency should be checked before purchasing a motor. The initial purchase price of a motor is a fraction of its total lifetime cost. The increased cost associated with the purchase of a high-efficiency motor is paid back during the lifetime of the motor because high-efficiency motors consume less energy over their lifetime than standard motors.

Electrical output is normally calculated in horsepower (HP) or watts (W). One horsepower is equal to 746 watts. Horsepower is normally used when referring to motors. Watts is normally used when referring to other types of electrical devices such as lights, solenoids, and heating elements.

Objectives

2-1

- Identify the differences between pictorial drawings, wiring diagrams, schematic diagrams, line diagrams, block diagrams, and flow charts.
- Identify the difference between the switch symbols normally open; normally closed; normally open, held closed; and normally closed, held open.

2-2

- State the five basic components of a basic electrical circuit.
- Understand how to draw a circuit using symbols that illustrate the five basic components of an electrical circuit.
- Explain the difference between a manual control circuit and an automatic control circuit.
- Explain what a solenoid is used for and draw the symbol for a solenoid.
- Define contactor, give an example of its usage, and state the two types of electrical contactors.
- Define magnetic motor starter and give an example of its usage.
- Explain how prints are read and how they are used during troubleshooting.

Review and Resources
atplearningresources.com/Quicklinks
Access Code: 362245

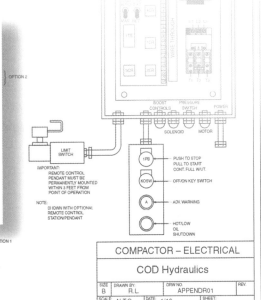

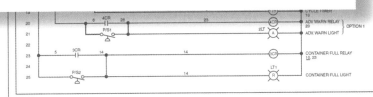

Symbols and Diagrams

Understanding electrical principles and circuits as covered in Chapter 1 is an important starting point for understanding what electricity is, where it is used, and the basic laws that apply to all electrical circuits. Understanding the different types of diagrams and symbols used in electrical circuits is the next step in being able to see how electrical circuits are drawn and how their components are interconnected in circuits and systems. Understanding circuit types and symbols is a basic requirement when designing, installing, troubleshooting, and repairing any electrical circuit.

2-1 LANGUAGE OF CONTROL

All trades have a certain language that must be understood in order to transfer information efficiently. This language may include symbols, drawings or diagrams, words, phrases, or abbreviations. Work in the electrical industry requires an understanding of this language, an understanding of the function of electrical components, and an understanding of the relationship between each component in a circuit. With this understanding, an electrician is able to read drawings and diagrams, understand circuit operation, and troubleshoot problems. Drawings and diagrams used to convey electrical information include pictorial drawings, wiring diagrams, schematic diagrams, and line diagrams.

Pictorial Drawings

A *pictorial drawing* is a drawing that shows the length, height, and depth of an object in one view. Pictorial drawings show physical details of an object as seen by the eye. **See Figure 2-1.**

PICTORIAL DRAWINGS

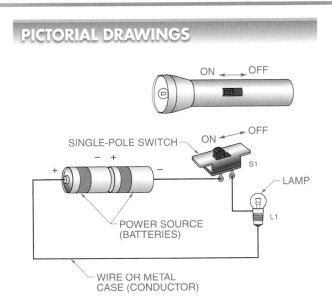

Figure 2-1. A pictorial drawing shows the physical details of components as seen by the eye.

Electrical Symbols and Abbreviations

A *symbol* is a graphic element that represents a quantity or unit. Symbols are used to represent electrical components on electrical and electronic diagrams. An *abbreviation* is a letter or combination of letters that represents a word. **See Figure 2-2. See Appendix.**

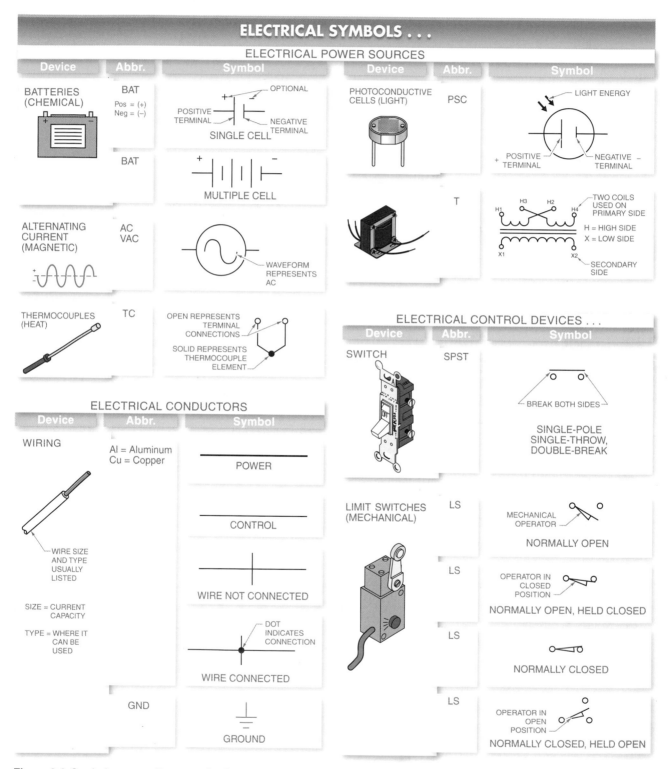

Figure 2-2. Symbols are used to conveniently represent electrical components in diagrams of most electrical and electronic circuits.

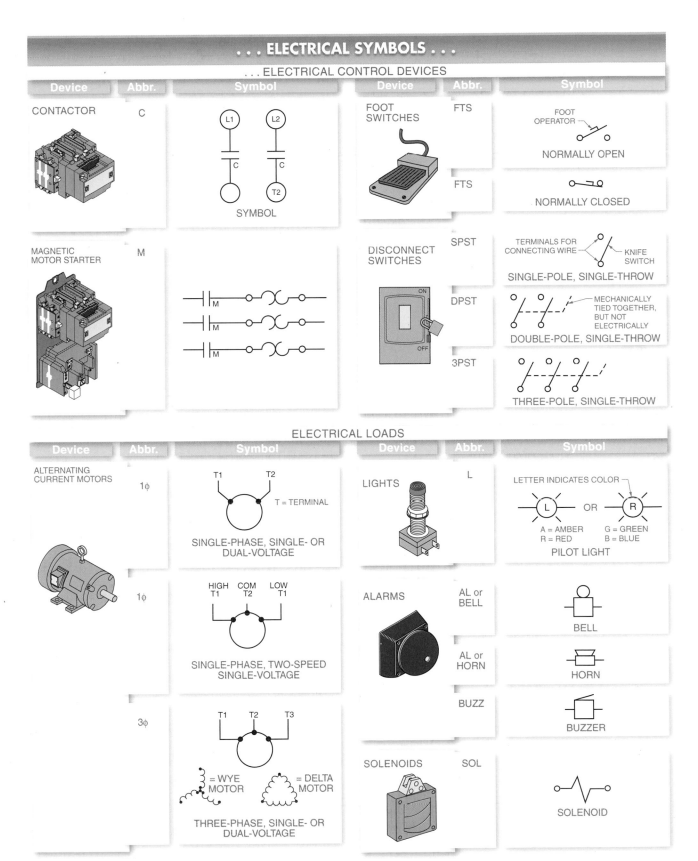

Figure 2-2. Continued...

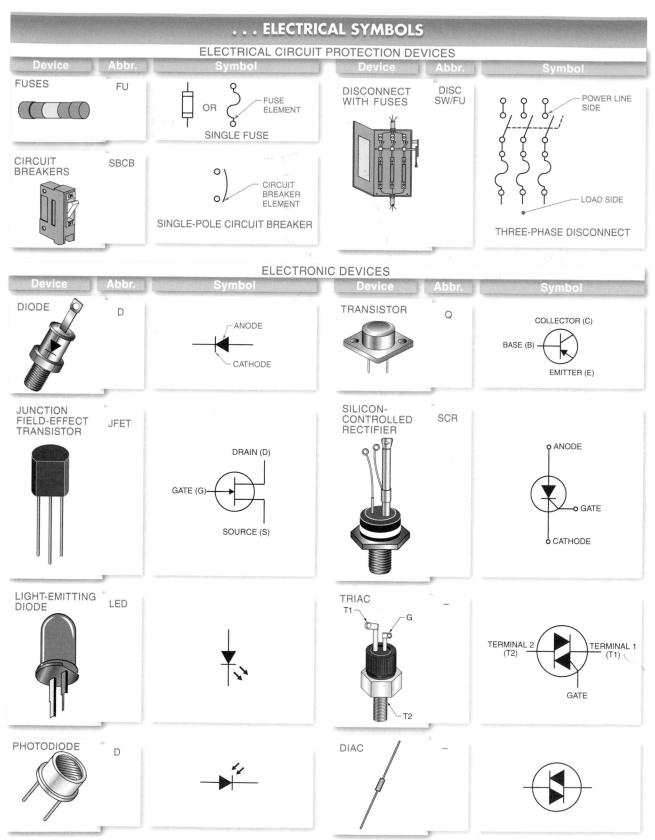

. . . ELECTRICAL SYMBOLS

ELECTRICAL CIRCUIT PROTECTION DEVICES

Device	Abbr.	Symbol	Device	Abbr.	Symbol
FUSES	FU	OR — FUSE ELEMENT — SINGLE FUSE	DISCONNECT WITH FUSES	DISC SW/FU	POWER LINE SIDE — LOAD SIDE — THREE-PHASE DISCONNECT
CIRCUIT BREAKERS	SBCB	CIRCUIT BREAKER ELEMENT — SINGLE-POLE CIRCUIT BREAKER			

ELECTRONIC DEVICES

Device	Abbr.	Symbol	Device	Abbr.	Symbol
DIODE	D	ANODE — CATHODE	TRANSISTOR	Q	COLLECTOR (C) — BASE (B) — EMITTER (E)
JUNCTION FIELD-EFFECT TRANSISTOR	JFET	DRAIN (D) — GATE (G) — SOURCE (S)	SILICON-CONTROLLED RECTIFIER	SCR	ANODE — GATE — CATHODE
LIGHT-EMITTING DIODE	LED		TRIAC	—	TERMINAL 2 (T2) — TERMINAL 1 (T1) — GATE
PHOTODIODE	D		DIAC	—	

Figure 2-2. ...Continued

Wiring Diagrams

A *wiring diagram* is a diagram that shows the connection of all components in a piece of equipment. Wiring diagrams show, as closely as possible, the actual location of each component in a circuit. Wiring diagrams often include details of the type of wire and the kind of hardware by which wires are fastened to terminals. **See Figure 2-3.**

A wiring diagram is similar to a pictorial drawing except that the components are shown as rectangles or circles. The location or layout of the parts is accurate for the particular equipment. All connecting wires are shown connected from one component to another. Wiring diagrams are used widely by electricians when installing electronic equipment, and by technicians when maintaining such equipment.

Schematic Diagrams

A *schematic diagram* is a diagram that shows the electrical connections and functions of a specific circuit arrangement with graphic symbols. Schematic diagrams do not show the physical relationship of the components in a circuit. The term schematic diagram is normally associated with electronic circuits.

Schematic diagrams are intended to show the circuitry that is necessary for the basic operation of a device. Schematic diagrams are not intended to show the physical size or appearance of the device. In troubleshooting, schematic diagrams are essential because they enable an individual to trace a circuit and its functions without regard to the actual size, shape, or location of the component, device, or part. **See Figure 2-4.**

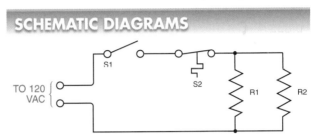

Figure 2-4. Schematic diagrams are essential in troubleshooting because they enable an individual to trace a circuit and its functions without regard to the actual size, shape, or location of the component, device, or part.

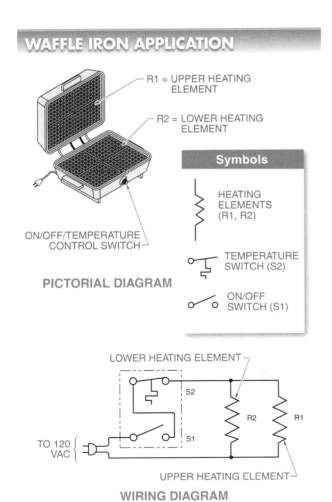

Figure 2-3. In a wiring diagram, the location of components is generally shown as close to the actual circuit configuration as possible.

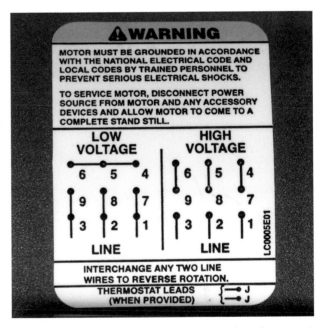

A motor wiring diagram located on the nameplate of a motor is used to determine the wiring arrangement when connecting power to the motor.

Line Diagrams

A *line (ladder) diagram* is a diagram that shows the logic of an electrical circuit or system using standard symbols. A line diagram is used to show the relationship between circuits and their components but not the actual location of the components. Line diagrams provide a fast, easy understanding of the connections and use of components. **See Figure 2-5.**

The arrangement of a line diagram should promote clarity. Graphic symbols, abbreviations, and device designations are drawn per standards. The circuit should be shown in the most direct path and logical sequence. Lines between symbols can be horizontal or vertical but should be drawn to minimize line crossing.

Line diagrams are often incorrectly referred to as one-line diagrams. A *one-line diagram* is a diagram that uses single lines and graphic symbols to indicate the path and components of an electrical circuit. One-line diagrams have only one line between individual components. A line diagram, on the other hand, often shows multiple lines leading to or from a component (parallel connections). **See Figure 2-6.**

Care must be taken when using electrical symbols to design or communicate electrical circuit operation. Electrical circuit operation may be changed and hazardous situations may be created by using incorrect electrical symbols. One problem that occurs is that limit switch operation is commonly misinterpreted when using electrical circuit diagrams.

ONE-LINE DIAGRAMS

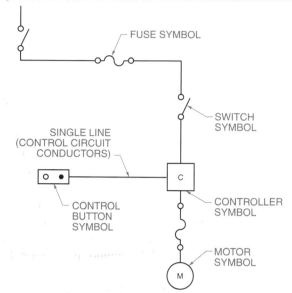

Figure 2-6. A one-line diagram is a diagram that uses single lines and graphic symbols to indicate the path and components of an electrical circuit.

Tech Fact

The line diagram was the first diagram designed to clearly show electrical circuit logic. The line diagram was designed to be read the same as one reads a printed page: from left to right and from top to bottom. In the past, line diagrams enabled circuit information to be easily communicated via paper and pencil. Today, line diagrams are produced by computer software.

LINE (LADDER) DIAGRAMS

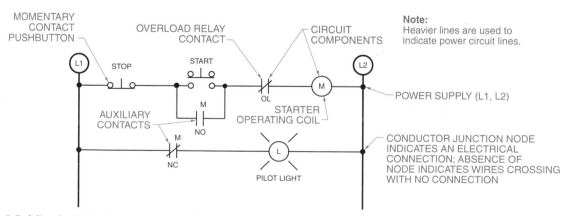

Figure 2-5. A line (ladder) diagram consists of a series of symbols interconnected by lines that are laid out like rungs on a ladder to indicate the flow of current through the various components of a circuit.

For example, a circuit contains four limit switches that are used to control four lamps. **See Figure 2-7.** Lamp 1 (red) is controlled by limit switch 1 (LS1). Limit switch 1 includes a normally open (NO) contact. Thus, lamp 1 is not energized (not turned on) until an object presses on the limit switch operator and closes the NO limit switch contacts. Lamp 2 (green) is controlled by limit switch 2 (LS2). Limit switch 2 also includes a NO contact. However, the NO contacts are shown in their held closed position. This is often done when a switch would normally be found in the held position, such as a limit switch that detects when a door is closed. Anytime the door is closed, the limit switch NO contacts are held closed and lamp 2 is energized. Thus, for any limit switch symbol, the limit switch contacts are always NO when the moving part of the symbol is drawn below the terminal connections.

Likewise, lamp 3 (blue) is controlled by limit switch 3 (LS3). Limit switch 3 includes a normally closed (NC) contact. Thus, lamp 3 is energized (turned on) before an object presses on the limit switch operator. Lamp 4 (yellow) is controlled by limit switch 4 (LS4). Limit switch 4 also includes a NC contact. However, the NC contacts are shown in their held open position. This is often done when a switch would normally be found in the held position. Thus, for any limit switch symbol, the limit switch contacts are always NC when the moving part of the symbol is drawn above the terminal connections. Line diagrams are designed to show circuit operation and include switches in their "normal" position and their "held" (actuated) position.

Block Diagrams

A *block diagram* is a diagram that shows the relationship between individual sections, or blocks, of a circuit or system. The primary function of a block diagram is to show how the distinct parts of a system relate to each other. Block diagrams are used with schematic diagrams to help troubleshoot a system, such as a regulated power supply. It should be noted that each block is labeled for the function it performs. The diagram does not explain the actual construction of the circuits in the system. Instead, the block diagram visually shows the systematic processing of an AC source into a DC output. **See Figure 2-8.** This approach is intended to be simple to easily explain the system.

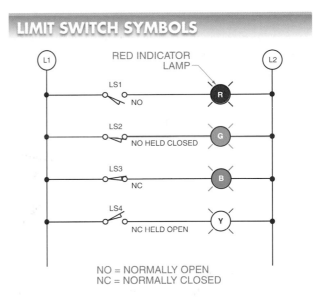

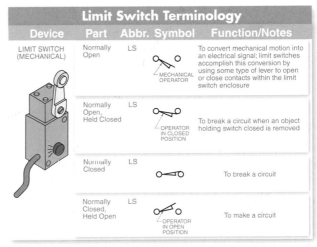

Figure 2-7. Care should be taken when using electrical symbols to design or communicate electrical circuit operations because electrical circuit operations may be changed.

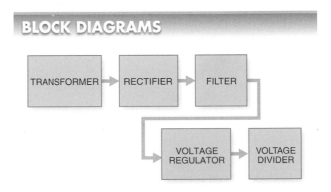

Figure 2-8. A block diagram can show the basic components of a DC power supply.

Block diagrams are usually read from left to right. This is because the signal input comes in on the left side of the diagram and exits on the right side of the diagram. This block diagram uses arrows to show this progression but many do not. The regulated power supply block diagram would be read as follows: an AC voltage (signal) leaves the transformer and goes to the rectifier where it is converted to DC. The DC voltage is then filtered and enters the voltage regulator to be stabilized. The output is then distributed as a DC power source. The block diagram is a useful tool in understanding and troubleshooting systems because it uses a simplistic representation of the sections found within a system.

Flow Charts

A *flow chart* is a diagram that shows a logical sequence of steps for a given set of conditions. Flow charts help a troubleshooter to follow a logical path when trying to solve a problem. Flow charts use symbols and interconnecting lines to provide direction. **See Figure 2-9.**

Symbols used with flow charts include an ellipse, rectangle, diamond, and arrow. An ellipse indicates the beginning and end of a flow chart or section of a flow chart. A rectangle contains a set of instructions. A diamond contains a question stated so that a yes or no answer is achieved. The yes or no answer determines the direction to follow through the flow chart. An arrow indicates the direction to follow based on the answers.

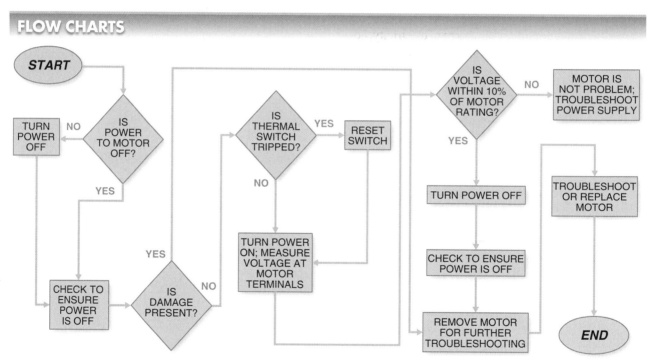

FLOW CHARTS

Figure 2-9. Flow charts use symbols and interconnecting lines to show a logical sequence of steps for a given set of conditions when troubleshooting motors.

2-1 CHECKPOINT

1. Which type of diagram helps when troubleshooting because it shows a logical path to take that helps solve a problem?

2. Which type of diagram shows physical details, such as the length and height of an object, as seen by the eye?

3. Which type of diagram uses symbols to show the interconnections of components in electronic circuits?

4. Which type of diagram shows the logic of an electrical circuit drawn with vertical power lines and circuit components drawn on horizontal lines between the power lines?

2-2 ELECTRICAL CIRCUITS

An *electrical circuit* is an assembly of conductors and electrical devices through which current flows. When an electrical circuit is complete (closed circuit), current makes a complete trip through the circuit. If the circuit is not complete (open circuit), current does not flow. A broken wire, loose connection, or switch in the OFF position stops current from flowing in an electrical circuit.

Electrical Circuit Components

All electrical circuits include five basic components. Electrical circuits must include a load that converts electrical energy into some other usable form of energy such as light, heat, or motion; a source of electricity; conductors to connect the individual components; a method of controlling the flow of electricity (switch); and a protection device (fuse or circuit breaker) to ensure that the circuit operates safely and within electrical limits.

Electrical circuit components may be shown using line diagrams, pictorial drawings, and/or wiring diagrams. For example, an automobile interior lighting circuit includes the five components of a typical electrical circuit. **See Figure 2-10.** The source of electricity is the battery, the conductors may be the chassis wires or the car frame, the control device is the plunger-type door switch, the load is the interior light, and the fuse is the protection device.

A *power source* is a device that converts various forms of energy into electricity. The power source in an electrical circuit is normally the point at which to start when reading or troubleshooting a diagram. The components in electrical circuits are connected using conductors. A *conductor* is a material that has very little resistance and permits electrons to move through it easily. Copper is the most commonly used conductor material.

A *control switch* is a switch that controls the flow of current in a circuit. Switches can be activated manually, mechanically, or automatically. A *load* is any device that converts electrical energy to motion, heat, light, or sound. Common loads include lights, heating elements, speakers, and motors.

An *overcurrent protection device (OCPD)* is a disconnect switch with circuit breakers (CBs) or fuses added to provide overcurrent protection for the switched circuit. A *fuse* is an overcurrent protection device with a fusible link that melts and opens the circuit when an overload condition or short circuit occurs. A *circuit breaker* is an overcurrent protection device with a mechanism that may manually or automatically open the circuit when an overload condition or short circuit occurs.

ELECTRICAL CIRCUIT COMPONENTS

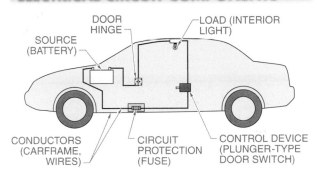

PICTORIAL DRAWING

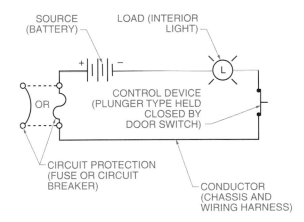

WIRING DIAGRAM

Figure 2-10. All electrical circuits include the source, load, control device, and conductors. Most circuits also include fuses or circuit breakers to provide protection for the circuit.

Tech Fact

Electrical circuits are always being improved. For example, in 1850, Isaac Singer produced a sewing machine that used a foot-operated treadle, allowing both hands to be free, making the machine an instant success. Later, in 1930, the Singer Sewing Machine Company again revolutionized sewing machines by adding an electric motor, enabling fast and easy sewing. Later, an electric material-feed mechanism and lights were added.

Manual Control Circuits

A *manual control circuit* is any circuit that requires a person to initiate an action for the circuit to operate. A line diagram may be used to illustrate a manual control circuit of a pushbutton controlling a pilot light. In a line diagram, the lines labeled L1 and L2 represent the power circuit. **See Figure 2-11.** The voltage of the power circuit is normally indicated on the circuit near these lines. In this circuit, the voltage is 115 VAC, but may be 12 VAC, 18 VAC, 24 VAC, or some other voltage. A common control voltage for many motor control circuits is 24 VAC. The control voltage may also be DC when DC components are used.

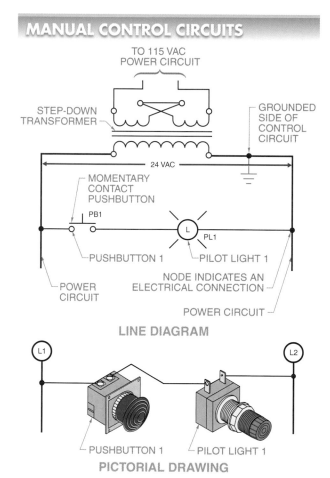

Figure 2-11. A line diagram may be used to illustrate a manual control circuit of a pushbutton controlling a pilot light.

The dark black nodes on a circuit indicate an electrical connection. If a node is not present, the wires only cross each other and are not electrically connected.

Line diagrams are read from left (L1) to right (L2). In this circuit, pressing pushbutton 1 (PB1) allows cur-

rent to pass through the closed contacts of PB1, through pilot light 1 (PL1) and on to L2, forming a complete circuit that activates PL1. Releasing PB1 opens the PB1 contacts, stopping the current flow to the pilot light, and turning the pilot light (PL1) off.

A line diagram may be used to illustrate the control and protection of a 1ϕ motor using a manual starter with overload protection. **See Figure 2-12.** The manual starter is represented in the line diagram by the set of normally open (NO) contacts S1 and by the overload contacts OL1. The line diagram is drawn for ease of reading and does not indicate where the devices are physically located. For this reason, the overloads are shown between the motor and L2 in the line diagram but are physically located in the manual starter.

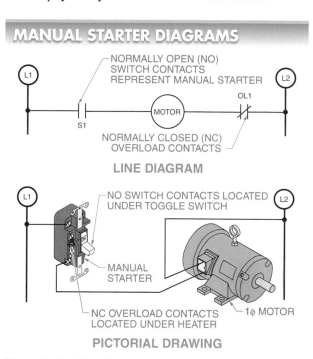

Figure 2-12. A line diagram may be used to illustrate the control and protection of a 1ϕ motor using a manual starter with overload protection.

In this circuit, current passes through contacts S1, the motor, the overloads, and on to L2 when the manual starter (S1 NO contacts) is closed. This starts the motor. The motor runs until contacts S1 are opened, a power failure occurs, or the motor experiences an overload. In the case of an overload, the OL1 contacts open and the motor stops. The motor cannot be restarted until the overload is removed and the overload contacts are reset to their normally closed (NC) position. *Note:* For consistency, the overload symbol is often drawn in a line diagram after the motor. In the actual circuit, the overload is located before the motor. It does not matter that the overload is shown after the motor in the line diagram because the overload is a series device and opens the motor control circuit in either position.

Automatic Control Circuits

Automatically controlled devices have replaced many functions that were once performed manually. As a part of automation, control circuits are intended to replace manual devices. Any manual control circuit may be converted to automatic operation. For example, an electric motor on a sump pump can be turned on and off automatically by adding an automatic control device such as a float switch. **See Figure 2-13.** This control circuit is used in basements to control a sump pump to prevent flooding. When water reaches a predetermined level, the float switch senses the change in water level and automatically starts the pump, which removes the water.

In this circuit, float switch contacts FS1 determine if current passes through the circuit when switch contacts S1 are closed. Current passes through contacts S1 and float switch contacts FS1, the motor, and the motor overload contacts to L2 when the float switch contacts FS1 are closed. This starts the pump motor. The pump motor pumps water until the water level drops enough to open contacts FS1 and shut off the pump motor. A motor overload, a power failure, or the manual opening of contacts S1 would prevent the pump motor from automatically pumping water even after the water reaches the predetermined level.

Control devices such as float switches are normally designed with NO and NC contacts. Variations in the application of the float switch are possible because the NO and NC contacts can close or open when changes in liquid level occur. For example, the NC contacts of the float switch may be used for a pump operation to maintain a certain water level in a livestock water tank. **See Figure 2-14.** The NC contacts close and start the pump motor when the water level drops due to evaporation or the livestock drinking the water. The float opens the NC contacts, shutting the pump motor off when the water level rises to a predetermined level.

In this circuit, when contacts S1 and float switch contacts FS1 are closed, current passes from L1 through the S1 and FS1 contacts, the motor, and the overloads to L2. This starts the pump motor. The pump motor pumps water until the water level rises high enough to open contacts FS1 and shut off the pump. A motor overload, a power failure, or the manual opening of contacts S1 would prevent the pump from automatically filling the tank to a predetermined level.

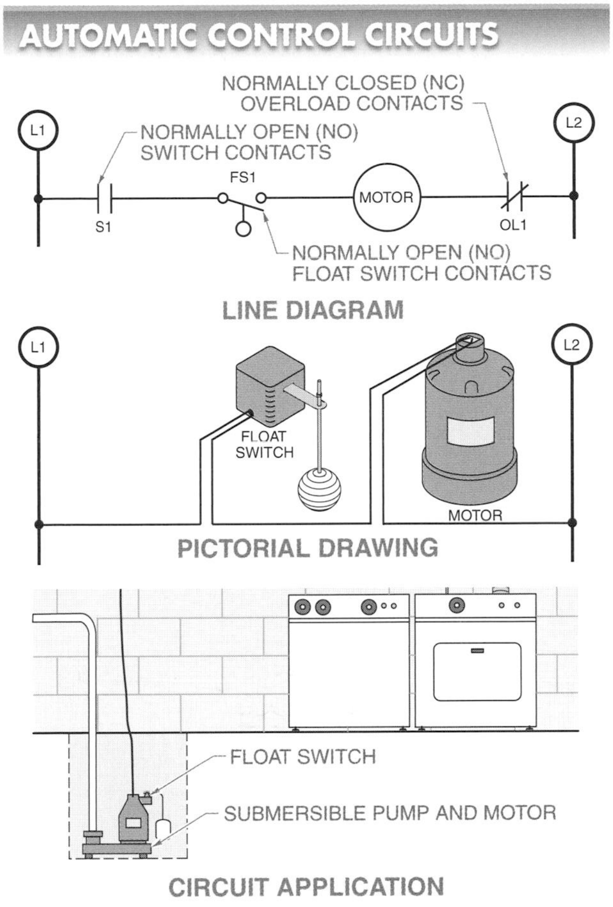

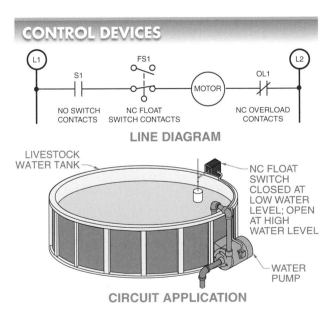

Figure 2-13. An electric motor on a sump pump can be turned on and off by using an automatic control device such as a float switch.

Figure 2-14. The NC contacts of a float switch may be used for a pump operation to maintain a certain level of water in a livestock water tank.

Magnetic Control Circuits

Although manual controls are compact and sometimes less expensive than magnetic controls, industrial and commercial installations often require that electrical control equipment be located in one area while the load device is located in another. Solenoids, contactors, and magnetic motor starters are used for remote control of devices.

Solenoids. A *solenoid* is an electric output device that converts electrical energy into a linear mechanical force. A solenoid consists of a frame, plunger, and coil. **See Figure 2-15.** A magnetic field is set up in the frame when the coil is energized by an electric current passing through it. This magnetic field causes the plunger to move into the frame. The result is a straight-line force, normally a push or pull action.

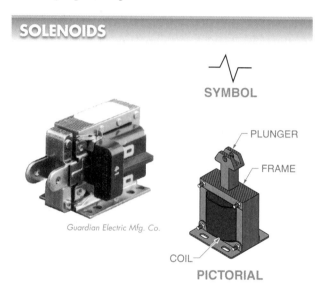

SOLENOIDS

SYMBOL

PLUNGER

FRAME

Guardian Electric Mfg. Co.

COIL

PICTORIAL

Figure 2-15. A solenoid is an electric output device that converts electrical energy into a linear mechanical force.

A solenoid may be used to control a door lock that opens only when a pushbutton is pressed. **See Figure 2-16.** In this circuit, pressing pushbutton 1 allows an electric current to flow through the solenoid, creating a magnetic field. The magnetic field, depending on the solenoid construction, causes the plunger to push or pull. In this circuit, the door may open as long as the pushbutton is pressed. The door is locked when the pushbutton is released. This circuit provides security access to a building or room.

Contactors. A *contactor* is a control device that uses a small control current to energize or de-energize the load connected to it. A contactor does not include overload protection. A contactor is constructed and operates

similarly to a solenoid. **See Figure 2-17.** Like a solenoid, a contactor has a frame, plunger, and coil. The action of the plunger, however, is directed to close (or open) sets of contacts. The closing of the contacts allows electrical devices to be controlled from remote locations.

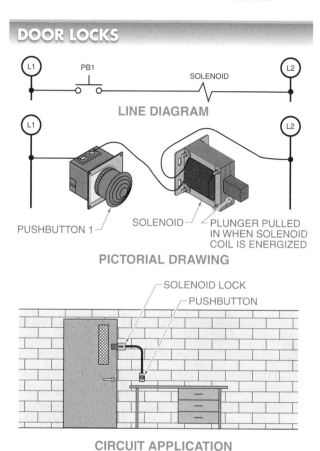

DOOR LOCKS

L1   PB1   SOLENOID   L2

LINE DIAGRAM

L1   L2

PUSHBUTTON 1   SOLENOID   PLUNGER PULLED IN WHEN SOLENOID COIL IS ENERGIZED

PICTORIAL DRAWING

SOLENOID LOCK
PUSHBUTTON

CIRCUIT APPLICATION

Figure 2-16. A solenoid may be used to control a door lock that is opened only when a pushbutton is pressed.

The electrical operation of a contactor can be shown using a line diagram, a pictorial drawing, and/or a wiring diagram. **See Figure 2-18.** In this circuit, pressing pushbutton 1 (PB1) allows current to pass through the switch contacts and the contactor coil (C1) to L2. This energizes the contactor coil (C1). The activation of C1 closes the power contacts of the contactor. The contactor contains power contacts that close each time the control circuit is activated. These contacts are not normally shown in the line diagram but are shown in the pictorial drawing and wiring diagram.

Releasing PB1 stops the flow of current to the contactor coil and de-energizes the coil. The power contacts return to their NO condition when the coil de-energizes. This shuts off the lights or other loads connected to the power contacts.

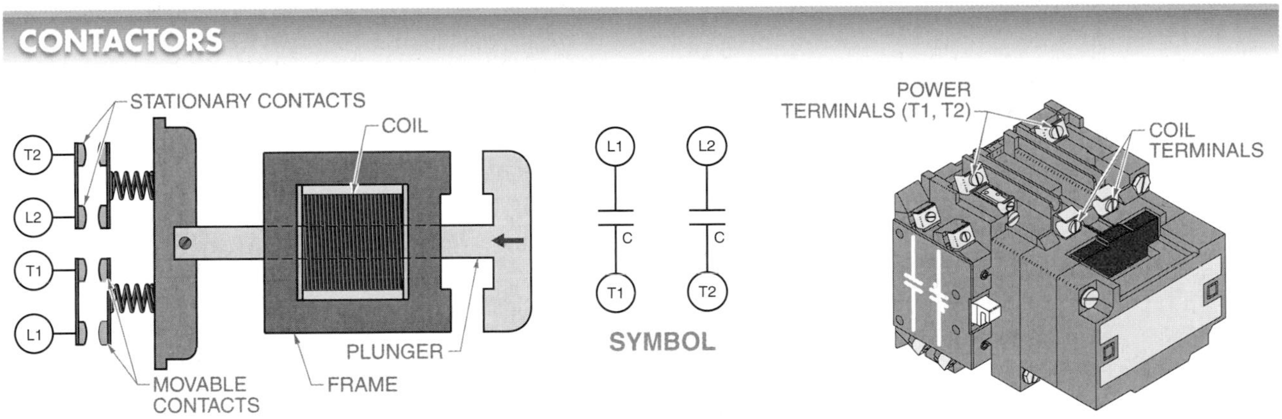

CONTACTORS

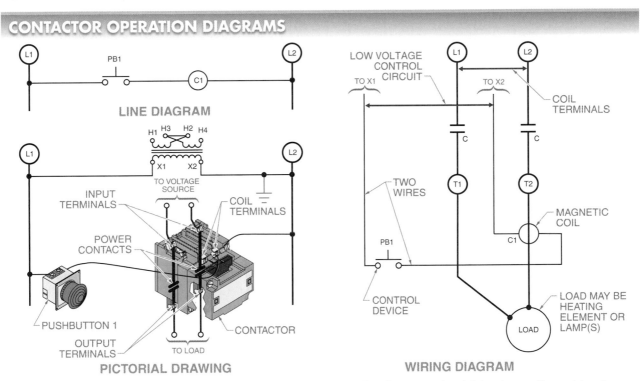

Figure 2-17. A contactor is a control device that uses a small control current to energize or de-energize the load connected to it.

CONTACTOR OPERATION DIAGRAMS

Figure 2-18. The electrical operation of a contactor can be shown using a line diagram, a pictorial drawing, and/or a wiring diagram.

This circuit works well in turning on and off various loads remotely. It does, however, require someone to hold the contacts closed (pressing PB1) if the coil must be continuously energized. Auxiliary contacts may be added to a contactor to form an electrical holding circuit and to eliminate the necessity of someone holding the pushbutton continuously. **See Figure 2-19.** The auxiliary contacts are attached to the side of the contactor and are opened and closed with the power contacts as the coil is energized or de-energized. These contacts are shown on the line diagram because they are part of the control circuit, not the power circuit.

A line diagram may be used to show the logic of the electrical holding circuit. In this circuit, pressing the start pushbutton (PB2) allows current to pass through the closed contacts of the stop pushbutton (PB1), through the closed contacts of the start pushbutton, and through coil C1 to L2. This energizes coil C1. With coil C1 energized, auxiliary contacts C1 close and remain closed as long as the coil is energized. This forms a continuous electrical path around the start pushbutton (PB2) so that, even if the start pushbutton is released, the circuit remains energized because the coil remains energized.

AUXILIARY CONTACTS

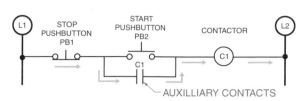

LINE DIAGRAM

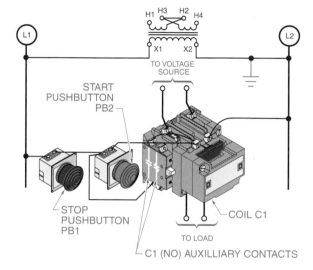

PICTORIAL DRAWING

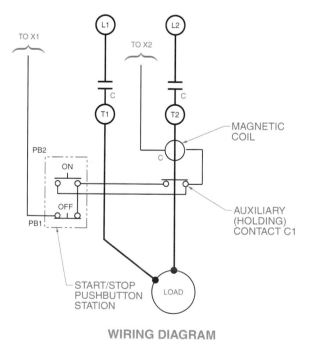

WIRING DIAGRAM

Figure 2-19. Auxiliary contacts may be added to a contactor to form an electrical holding circuit.

The circuit is de-energized by a power failure or by pressing the NC stop pushbutton (PB1). In either case, the current flow to coil C1 stops and the coil is de-energized, causing auxiliary contacts C1 to return to their NO position. The start pushbutton (PB2) must be pressed to re-energize the circuit.

The two main types of contactors are heating contactors and lighting contactors. Heating contactors are used to control high-power electrical heating elements. Lighting contactors are used to control banks of lamps, such as those used in malls and sport stadiums.

Magnetic Motor Starters. A *magnetic motor starter* is an electrically operated switch (contactor) that includes motor overload protection. Magnetic motor starters are used to start and stop motors. Magnetic motor starters are identical to contactors except that they have overloads attached to them. **See Figure 2-20.**

MAGNETIC MOTOR STARTERS

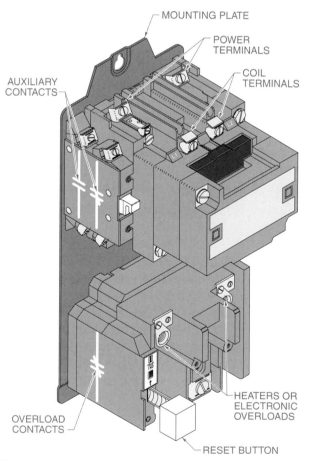

Figure 2-20. A magnetic motor starter is an electrically operated switch (contactor) that includes motor overload protection.

The magnetic motor starter overloads have heaters or electronic overloads (located in the power circuit) that sense excessive current flow to the motor. The heaters open the NC overload contacts (located in the control circuit) when the overload becomes dangerous to the motor.

The electrical operation of a magnetic motor starter may be shown using a line diagram, a pictorial drawing, and/or a wiring diagram. **See Figure 2-21.** The only difference between the drawings and diagrams of a contactor circuit and the drawings and diagrams of a magnetic motor starter circuit is the addition of the heaters or electronic overloads. In this circuit, pressing

the start pushbutton allows current to pass through coil M1 and the overload contacts. This energizes coil M1. With coil M1 energized, auxiliary contacts M1 close and the circuit remains energized even if the start pushbutton is released.

This circuit is de-energized if the stop pushbutton is pressed, a power failure occurs, or any one of the overloads senses a problem in the power circuit. Coil M1 de-energizes, causing auxiliary contacts M1 to return to their NO condition if one of these situations occurs. When a motor stops because of an overload, the overload must be removed, the overload device reset, and the start pushbutton pressed to restart the motor.

MAGNETIC MOTOR STARTER DIAGRAMS

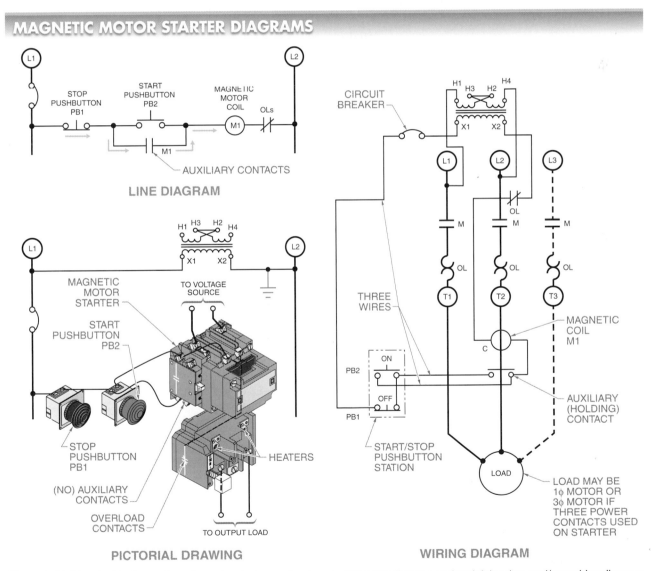

Figure 2-21. The electrical operation of a motor starter can be shown using a line diagram, a pictorial drawing, and/or a wiring diagram.

Printreading

Electrical components are represented on prints as symbols and pictorial drawings. The symbols are used to simplify the understanding of the circuit operation and individual component function. However, because symbols represent a physical component and the physical component is installed and checked when troubleshooting, a clear understanding must exist between the component symbol and the physical shape and location of the component. **See Figure 2-22.**

A typical electrical print uses standard symbols and a pictorial drawing to show the location of the circuit components. Some components, such as transformers, motor starters, relays, timers, or fuses, are mounted inside the control panel enclosure. Other components, such as motors, solenoids, limit switches, pressure switches, or pushbuttons, are located outside the control panel enclosure.

Printreading involves identifying an electrical symbol, understanding where the component is located within a system, and understanding how it is used. For example, a control transformer is shown on a print as a symbol and in a control panel as a pictorial drawing. The symbol shows that the control transformer is a step-down transformer with a 208/240/480 V primary and a 120 V secondary. H1, H2, H3, and H4 are used to show the high-voltage side and circuit connections. X1 and X2 are used to show the low-voltage side and circuit connections. The pictorial drawing of the control transformer shows that it is located in the upper right side of the control panel, that the high-voltage side is on the left side of the transformer, and that the low-voltage side is on the right side of the transformer.

ELECTRICAL PRINTS

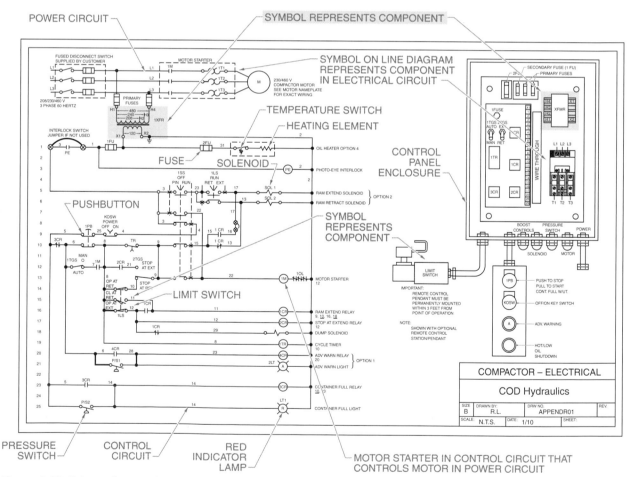

Figure 2-22. Printreading requires understanding the relationship between a component symbol and the physical shape and location of the component.

Troubleshooting Using Symbols and Diagrams

Troubleshooting requires the testing of individual components within a system to determine a circuit or component problem. Troubleshooting requires reading a print to determine the location and use of individual components and an understanding of how to use test equipment. For example, a digital multimeter (DMM) can be used to test a circuit to verify that individual components are working properly. In this example, a DMM is used to troubleshoot the oil heater circuit. **See Figure 2-23.**

The DMM is set to measure AC voltage and the leads are placed in position 1 across the output of the control transformer (X1 and X2). In position 1, the DMM measures the voltage out of the transformer. If the voltage out of the transformer is not 120 VAC ±10%, the input

side of the transformer (H1 and H4) must be tested to determine if the transformer or voltage source coming into the transformer is the problem. Once the transformer is tested, one DMM lead is moved to position 2 (from the input side of fuse 2FU to the output side of the fuse). If there is no voltage out of the fuse, the fuse is bad (open). In the same manner, the DMM test lead is moved along the circuit components, such as the temperature switch or heating element, testing each one.

In an electrical system, the print normally shows all of the circuit components and connections. Understanding all symbols and components of a print is required when installing, modifying, testing, and troubleshooting a component or part of a system. Locating the actual places to connect the DMM test leads requires transferring the circuit symbol information to the actual location of the component in the control panel.

TROUBLESHOOTING USING PRINTS

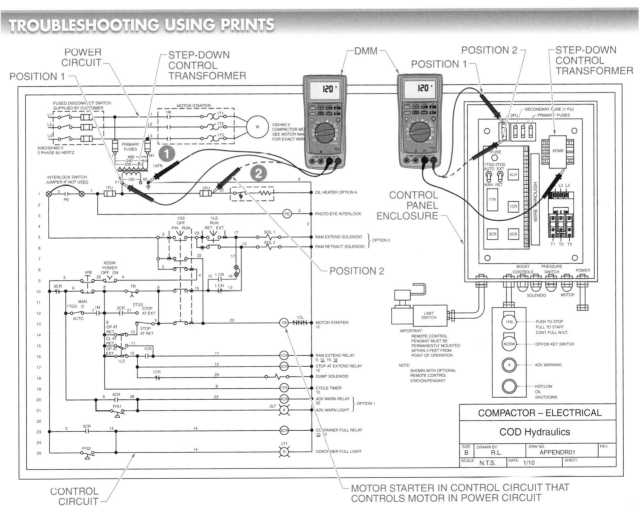

Figure 2-23. Troubleshooting requires reading a print to determine the location and use of individual components and understanding how to use test equipment.

2-2 CHECKPOINT

1. What are the five parts of a basic electrical circuit?
2. What are the two basic types of circuit protection devices?
3. In electrical switches, what do the abbreviations NO and NC stand for?
4. What electrical device is used to turn a motor on and off and includes overload protection?
5. What electrical device converts electrical energy into a linear mechanical force?
6. What are the two basic types of contactors?

Additional Resources

Review and Resources

Access Chapter 2 Review and Resources for *Electrical Motor Controls for Integrated Systems* by scanning the above QR code with your mobile device.

Applying Your Knowledge

Refer to the *Electrical Motor Controls for Integrated Systems* Learner Resources for interactive Applying Your Knowledge questions.

Workbook and Applications Manual

Refer to Chapter 2 in the *Electrical Motor Controls for Integrated Systems Workbook* and the *Applications Manual* for additional exercises.

ENERGY EFFICIENCY PRACTICES

Energy Efficient Motors

Energy-efficient motors use less energy than conventional motors because they are manufactured with higher-quality materials and techniques. Energy-efficient motors have high service factors and bearing lives and less waste heat output and vibration than conventional motors. This reliability is often reflected by longer manufacturer warranties.

The Energy Policy Act (EP Act) of 1992 requires most general-purpose motors between 1 HP and 200 HP for sale in the United States to meet NEMA standards. To be considered energy-efficient, motor performance must equal or exceed the nominal full-load efficiency values provided by the National Electrical Manufacturers Association (NEMA) standard MG-1. The EP Act also provided grandfather protection to existing motors.

In June 2001, NEMA created the NEMA Premium™ designation to distinguish motors that are more efficient than required by EP Act standards. The NEMA Premium™ designation applies to single-speed polyphase, 1 HP to 500 HP, 2-, 4-, and 6-pole (3600 rpm, 1800 rpm, and 1200 rpm) squirrel-cage induction, NEMA Designs A or B, 600 V or less (5 kV or less for medium voltage), and continuous rated motors.

The Consortium for Energy Efficiency (CEE), a nonprofit organization that includes many electric utilities among its members, recognizes NEMA Premium™ motors up to 200 HP as meeting their criteria for possible energy-efficiency rebates.

Sections

Objectives

3-1

- Understand the stated rules for proper and safe usage of hand and power tools.
- Understand the stated rules for proper and safe usage of test instruments.

3-2

- Explain how to troubleshoot fuses using a continuity tester.
- Explain how to properly use a receptacle tester to test a receptacle.
- Explain how to properly use a voltage tester to test for voltage in a circuit.
- Explain how a measurement is displayed on an analog multimeter using linear and nonlinear scales.
- Explain how measurements are displayed on a digital multimeter (DMM).
- Describe bar graph and wraparound bar graph as displayed on a digital multimeter.
- Describe what each CAT rating means on a test instrument.
- Describe the procedure required to set a multimeter to take a resistance measurement.
- Explain how to set a multimeter to test diodes.
- Describe the procedure required to set a multimeter to take a DC voltage measurement.
- Describe the procedure required to set a multimeter to take an AC voltage measurement.
- Understand how to properly and safely test fuses in an energized disconnect switch.
- Understand how to properly and safely test to ensure that a disconnect switch is grounded.
- Describe the procedure required to set a multimeter to take an in-line current measurement.
- Describe the procedure required to use a clamp-on ammeter to take a current measurement.

3-3

- Explain how an oscilloscope is used.
- Explain how a digital logic probe is used.
- Explain how a megohmmeter is used.
- Define ground resistance tester.
- Define infrared temperature meter and noncontact temperature probe.
- Explain how a phase sequence indicator is used.
- Explain how an optical time domain reflectometer is used.

Review and Resources
atplearningresources.com/Quicklinks
Access Code: 362245

Test Instruments

3

All electrical and electronic systems must be tested for proper and safe operation when installed, during preventive maintenance checks, and whenever calibrations or repairs are required. Understanding which test instrument is the best to use for each component and system test ensures the correct measurements and data are provided to determine proper operation or a fault. Understanding what the measured values actually mean requires an understanding of the system and function of each component within the system. Often, several different measurements and types of measurements (voltage, current, temperature, etc.) must be taken to determine exactly how a system is operating or to find a fault. Therefore, a knowledge of different types of test instruments and their capabilities is important for anyone working on or around electrical and electronic equipment.

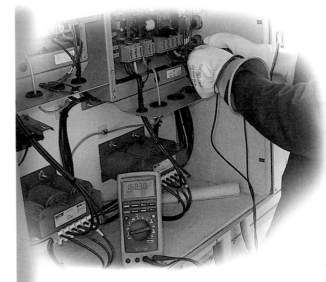

3-1 SAFETY

Various hand and power tools are used by electricians for the maintenance, troubleshooting, and installation of electrical equipment. *Troubleshooting* is the systematic elimination of the various parts of a system or process to locate a malfunctioning part. Each tool is designed for the efficient and safe completion of a specific job. Proper use of tools is required for safe and efficient electrical work.

Hand Tool Safety

Tools must be used properly to prevent injury. Tools should not be forced or used beyond their rated capacity. The proper tool does the job it was designed for quickly and safely. The time spent searching for the correct tool or the expense of purchasing it is less costly than a serious accident.

Tools should be kept in good working condition. Periodic inspection of tools helps to keep them in good condition. A tool should always be inspected for worn, chipped, or damaged surfaces prior to its use. A tool that is in poor or faulty condition must not be used. Tool handles should be free of cracks and splinters and should be fastened securely to the working part. Damaged tools are dangerous and less productive than those in good working condition and should not be used. A tool should be repaired or replaced immediately when inspection shows a dangerous condition.

Cutting tools should be sharp and clean. Dull tools arc dangerous. The extra force exerted while using dull tools often results in losing control of the tool. Dirt, oil, or debris on a tool may cause slippage while it is used and cause injury.

Tools should be kept in a safe place. Any tool can be dangerous when left in the wrong place. Many accidents are caused by tools falling off ladders, shelves, and scaffolds.

Each tool should have a designated place in a toolbox, chest, or cabinet. Tools should not be carried in clothing pockets unless the pocket is designed to carry a specific tool. Pencils should be kept in a pocket designed for them.

All tools should be kept away from the edge of a bench or work area. Brushing against the tool may cause it to fall and injure a leg or foot. Sharp-edged and pointed tools should be carried with the cutting edge or point down and away from the body. Work tools should not be placed in environments with solvents, prolonged moisture, or excessive heat, which can cause permanent damage.

Power Tool Safety

Power tools should only be used after gaining knowledge of their principles of operation, methods of use, and safety precautions. Authorization should always be obtained from supervisors before using any power tool. Full safety and health standards for the safe operation of power and hand tools are published by OSHA in Title 29, Code of Federal Regulations (CFR), Part 1910, Subpart P.

All power tools should be grounded unless they are an approved double-insulated design. Power tools must have a grounded three-wire cord. A three-prong plug connects into a grounded electrical outlet (receptacle). Approved receptacles may be locking or nonlocking. OSHA, the National Electrical Code® (NEC)®, and local codes may be consulted for proper grounding requirements. **See Figure 3-1.**

It is dangerous to use an adapter to plug a three-prong plug into a nongrounded receptacle unless a separate ground wire or strap is connected to an approved ground. The ground ensures that any short circuit trips the circuit breaker or blows the fuse. **WARNING:** An ungrounded power tool can cause fatal accidents.

Double-insulated tools have two prongs and a notation on the specification plate that they are double insulated. Electrical parts in the motor of a double-insulated tool are surrounded by extra insulation to help prevent electrical shock. For this reason, the tool is not required to be grounded. Both the interior and exterior should be kept clean of grease and dirt that may conduct electricity. Safety rules must be followed when using power tools. Electrical power tool safety rules include the following:

- Review and understand all manufacturer safety recommendations.
- Read the owner's manual before using any power tool.
- Ensure that all safety guards are properly in place and in working order.
- Wear safety goggles at all times and a dust mask when required.
- Ensure that the workpiece is free of obstructions and securely clamped.
- Ensure that the tool switch is in the OFF position before connecting a tool to the power source.
- Keep attention focused on the work.
- A change in sound during tool operation normally indicates trouble. Investigate immediately.
- Power tools should be inspected and serviced by a qualified repair person at regular intervals as specified by the manufacturer or by OSHA.
- Inspect electrical cords regularly to ensure that they are in good condition.
- Shut off the power when work is completed. Wait until all movement of the tool stops before leaving a stationary tool or laying down a portable tool.
- Clean and lubricate all tools as needed.
- Remove all defective power tools from service. Alert others to the situation.
- Take extra precautions when working on damp or wet surfaces. Use additional insulation to prevent any body part from coming into contact with a wet or damp surface.
- Always work with at least one coworker in hazardous or dangerous locations.

APPROVED RECEPTACLES

NEMA L5-15R
15 A 125 V
UL\CSA
0.5 HP

NEMA L6-30R
30 A 250 V
UL\CSA
2 HP

LOCKING RECEPTACLES

NEMA 5-15R
15 A 125 V
UL\CSA
0.5 HP

NEMA 6-30R
30 A 250 V
UL\CSA
2 HP

NONLOCKING RECEPTACLES

LEGEND: G = GROUND N = NEUTRAL H, X, and Y = HOT

Figure 3-1. Approved receptacles may be locking or nonlocking.

Tech Fact

Underwriters Laboratories® (UL®) requires ground continuity testing on all corded electrical devices sold in the United States. Therefore, only UL®-labeled power tools should be used.

Test Instrument Safety

Proper equipment is required when taking measurements with test instruments. To take and interpret a measurement, the test instrument must be properly set to the correct measuring position and properly connected into the circuit to be tested. There is always the possibility that the test instrument or meter will not be properly set to the correct function or to the correct range and/or will be misread. Test instruments not properly connected to a circuit increase the likelihood of an improper measurement, an improper meter reading, or the creation of an unsafe condition.

Electrical Measurement Safety. Electrical measurements are taken with test instruments designed for specific tasks. Test instruments such as voltage testers, analog and digital multimeters, and clamp-on ammeters measure electrical quantities, such as voltage, current, power, and frequency. Each meter has specific features and limits. User manuals detail specifications and features, proper operating procedures, safety precautions, warnings, and allowed applications.

Conditions can change quickly as voltage and current levels vary in individual circuits. General safety precautions for using test instruments include the following:

• Never assume a test instrument is operating correctly. Check the test instrument that will be measuring voltage on a known (energized) voltage source before taking a measurement on an unknown voltage source. After taking a measurement on an unknown voltage source, retest the test instrument on a known source to verify the meter still operates properly.

• Always assume that equipment and circuits are energized until positively identified as de-energized by taking proper measurements.

• Never work alone when working on or near exposed energized circuits that may cause an electrical shock.

• Always wear personal protective equipment (PPE) appropriate for the test area.

• Never assume that a circuit is de-energized or equipment is fully discharged. Capacitors can hold a charge for a long time—several minutes or more. Always check for the presence of voltage before taking any other measurements.

• Check test leads for frayed or broken insulation. Electrical shock can occur from accidental contact with live components.

• Use meters that conform to the IEC 1010 category in which they will be used.

• Use one hand when working on a live circuit to reduce the chance of an electrical shock passing through the heart and lungs.

5S Organization System. The 5S organization system is a workplace organizational system that stands for sort, straighten, sweep or shine, standardize, and sustain. The system is used to reduce waste, optimize productivity, and promote safety through maintaining an orderly workplace. Visual cues are used to achieve more consistent operational results. Routines and systems, such as the 5S organization system, that maintain organization and orderliness are essential for maintaining a smooth and efficient flow of activities when performing work. **See Figure 3-2.**

The main goals of the 5S organization system are improved workplace morale, safety, and efficiency. The system is based on the concept that by assigning a location for each item needed, time is not wasted looking for needed items. In addition, it is easy to identify when an important or expensive tool is missing from its designated location by using the system.

5S ORGANIZATION SYSTEM

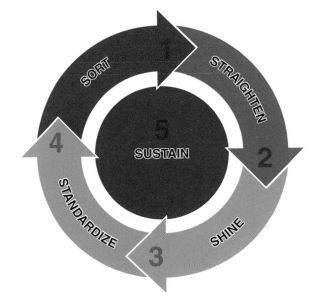

Figure 3-2. The 5S organization system is used to optimize productivity and promote safety in the workplace.

3-1 CHECKPOINT

1. Are double-insulated power tools required to have a plug with a grounding connection?
2. Before and after using any test instrument, what important step should be taken?
3. What is the only way to ensure that power is OFF in a system, circuit, or equipment?
4. What does the 5S organization system stand for?

3-2 GENERAL USE TEST INSTRUMENTS

Test instruments are used by technicians to aid in the taking of various electrical measurements. Care must be taken when using electronic test instruments because damage to the instrument or personal injury may result from improper or unsafe usage. The owner's manual must be consulted and all functions of a test instrument fully understood before using it.

Many types of test instruments are used by technicians. Most test instruments are available in either analog or digital versions and are designed to perform specific functions. General use electrical/electronic test instruments include continuity testers, receptacle testers, voltage testers, branch circuit identifiers, multimeters, clamp-on ammeters, and oscilloscopes.

Processes require the use of instrumentation to measure variables such as temperature, humidity, pressure, vacuum, flow, conductivity, voltage, current, resistance, speed, time, and other process operating conditions. Without the ability to accurately measure process variables, standards, quality control, and documentation would be limited.

Instruments installed as part of making a product are permanent instruments used to monitor that piece of equipment. Test instruments used as part of a troubleshooting system or when performing preventative maintenance tasks are of the portable type.

Continuity Testers

A *continuity tester* is a test instrument that tests for a complete path for current to flow. A continuity tester is an economical tester that is used to test switches, fuses, and grounds and is also used for identifying individual conductors in a multiwire cable. Continuity testers give an audible indication when there is a complete path. Most multimeters have a built-in continuity test mode. **See Figure 3-3.**

WARNING: A continuity tester must only be used on de-energized circuits or components. Voltage ap-

plied to a continuity tester can cause damage to the test instrument and/or harm to the electrician. Always check a circuit for voltage before taking a continuity test.

Troubleshooting Fuses with a Continuity Tester. A *fuse* is an overcurrent protection device (OCPD) with a fusible link that melts and opens the circuit on an overcurrent condition. Fuses are connected in series with a circuit to protect the circuit from overcurrents or short circuits. Fuses may be one-time or renewable. One-time fuses are fuses that cannot be reused after they have opened. One-time fuses are the most common. Renewable fuses are OCPDs designed so that the fusible link can be replaced. Fuses can be checked using a continuity tester or a DMM set to measure voltage.

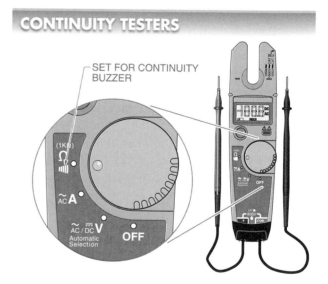

CONTINUITY TESTERS

Figure 3-3. A continuity tester tests for a complete path for current to flow.

Fuses can be checked using a continuity tester placed across a fuse that has been removed from a circuit. **See Figure 3-4.** When testing a good fuse, the continuity tester beeps because the circuit inside the fuse is closed, has a low resistance, and allows current flow through the fuse and continuity test circuit. When testing a bad fuse, the continuity tester does not beep because the circuit inside the fuse is open, has an infinite resistance (OL), and does not allow current to flow through the fuse and continuity test circuit.

TROUBLESHOOTING FUSES— CONTINUITY TESTERS

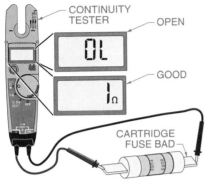

Figure 3-4. Fuses can be checked using a continuity tester set to measure continuity placed across a fuse that has been removed from a circuit.

Receptacle Testers

A *receptacle tester* is a device that is plugged into a standard receptacle to determine if the receptacle is properly wired and energized. **See Figure 3-5.** Some receptacle testers include a ground fault circuit interrupter (GFCI) test button that allows the receptacle tester to be used on GFCI receptacles.

When testing a receptacle, the indicator light code indicates whether the receptacle is wired correctly. The situation of having the hot and neutral wires reversed is a safety hazard and must be corrected. The hot slot on a receptacle is the short slot, while the neutral slot is the wide slot. The round hole is the ground. Improper grounds are also a safety hazard and must be corrected.

Voltage Testers

A *voltage tester* is a device that indicates approximate voltage level and type (AC or DC) by the movement and vibration of a pointer on a scale. Voltage testers contain a scale marked 120 VAC, 240 VAC, 480 VAC, 600 VAC, 120 VDC, 240 VDC, and 600 VDC. Some voltage testers include a colored plunger to indicate the polarity of test leads. If the red indicator is up, the red test lead is connected to the positive DC voltage. If the black indicator is up, the black test lead is connected to the positive DC voltage. **See Figure 3-6.**

RECEPTACLE TESTERS

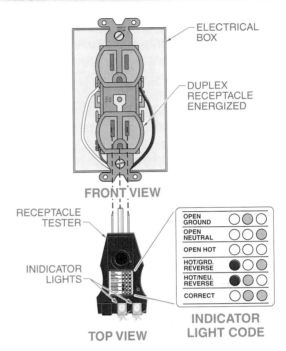

Figure 3-5. Receptacle testers are plugged into standard receptacles to determine if the receptacle is properly wired and energized.

VOLTAGE TESTERS

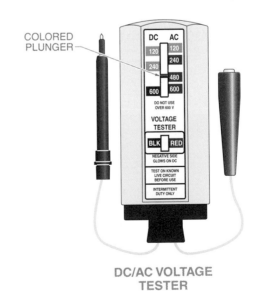

Figure 3-6. A voltage tester indicates approximate voltage level and type (AC or DC) by the movement and vibration of a pointer on a scale.

Voltage testers are used to take voltage measurements anytime voltage of a circuit to be tested is within the rating of the tester and an exact voltage measurement is not required. Exact voltage measurements are not required when testing to determine whether a receptacle is energized (hot), a system is grounded, fuses or circuit breakers are good or bad, or a circuit is a 115 VAC, 230 VAC, or 460 VAC circuit.

Before using a voltage tester or any voltage measuring instrument, it is important to always check the voltage tester on a known energized circuit that is within the voltage rating of the voltage tester to verify proper operation.

WARNING: If a voltage tester does not indicate a voltage, a voltage that can cause an electrical shock may still be present. The voltage tester may also have been damaged during the test (from too high a voltage). Always retest a voltage tester on a known energized circuit after testing a suspect circuit to ensure the voltage tester is still operating correctly.

Troubleshooting GFCIs with a Voltage Tester. Voltage testers are used to take voltage measurements any time the voltage of a circuit being tested is within the rating of the tester and an exact voltage measurement is not required. Exact voltage measurements are not required to determine when a receptacle is energized (hot), when a system is grounded, when fuses and/or circuit breakers are functioning properly, and when a system is a 115 VAC, 230 VAC, or 460 VAC circuit.

Voltage testers are considered the best test instrument for testing GFCI receptacles because test lights, voltage indicators, and DMMs do not draw enough current to trip a GFCI receptacle. Properly wired GFCI receptacles trip when the test button on the receptacle is pressed. **See Figure 3-7.**

A ground fault circuit interrupter (GFCI) is a device that protects against electrical shock by detecting an imbalance of current in the normal conductor pathways and opening the circuit.

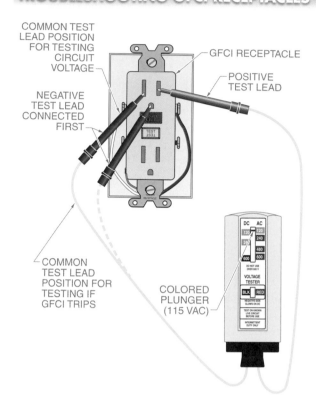

TROUBLESHOOTING GFCI RECEPTACLES

COMMON TEST LEAD POSITION FOR TESTING CIRCUIT VOLTAGE

GFCI RECEPTACLE

POSITIVE TEST LEAD

NEGATIVE TEST LEAD CONNECTED FIRST

COMMON TEST LEAD POSITION FOR TESTING IF GFCI TRIPS

COLORED PLUNGER (115 VAC)

VOLTAGE TESTER

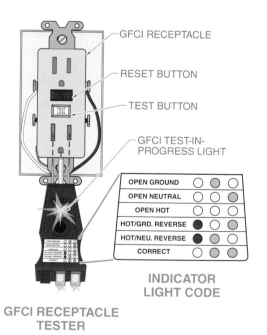

GFCI RECEPTACLE

RESET BUTTON

TEST BUTTON

GFCI TEST-IN-PROGRESS LIGHT

OPEN GROUND	○	○	○
OPEN NEUTRAL	○	○	○
OPEN HOT	○	○	○
HOT/GRD. REVERSE	●	○	○
HOT/NEU. REVERSE	●	○	○
CORRECT	○	○	○

INDICATOR LIGHT CODE

GFCI RECEPTACLE TESTER

Figure 3-7. Voltage testers and GFCI receptacle testers are the best test instruments to use when troubleshooting GFCI receptacles.

Branch Circuit Identifiers

Before working on any electrical circuit that does not require power, the circuit must be de-energized. Normally, branch circuits are de-energized by turning off the circuit breakers of the circuit and applying a lockout/tagout. Often, the circuit breaker in the circuit to be de-energized is not clearly marked or identifiable. Turning off an incorrect circuit breaker may unnecessarily require loads to be reset. To identify a particular circuit breaker, branch circuit identifiers are used. A *branch circuit identifier* is a two-piece test instrument that includes a transmitter that is plugged into a receptacle and a receiver that provides an audible indication when located near the circuit to which the transmitter is connected. **See Figure 3-8.**

AEMC® Instruments

Figure 3-8. A branch circuit identifier provides an audible indication to identify the circuit breaker or fuse for a particular receptacle without disconnecting the power.

Analog Multimeters

An *analog multimeter* is a meter that can measure two or more electrical properties and displays the measured properties along calibrated scales using a pointer. Analog multimeters use electromechanical components to display measured values. Most analog multimeters have several calibrated scales, which correspond to the different selector switch settings (AC, DC, and R) and placement of the test leads (mA jack and 10 A jack). **See Figure 3-9.**

When reading a measurement on an analog multimeter, the correct scale must be used. The most common measurements made with analog multimeters are voltage,

resistance, and current. Analog multimeters may also include scales for measuring decibels (dB) and checking batteries. Analog test instruments and meters are less susceptible to noise (magnetic field coupling) and do not display ghost readings as easily as digital meters.

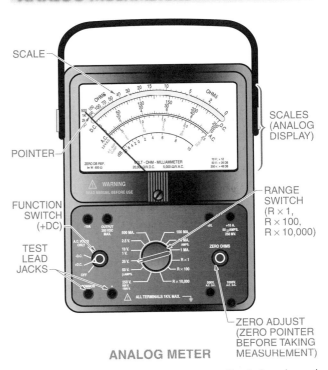

Figure 3-9. Analog multimeters use a calibrated scale and pointer to display electrical measurements.

Reading Analog Displays. An *analog display* is an electromechanical device that indicates a value by the position of a pointer on a scale. Analog scales may be linear or nonlinear. A *linear scale* is a scale that is divided into equally spaced segments. A *nonlinear scale* is a scale that is divided into unequally spaced segments. **See Figure 3-10.**

Analog scales are divided into segments using primary divisions, secondary divisions, and subdivisions. A *primary division* is a division with a listed value. A *secondary division* is a division that divides primary divisions in halves, thirds, fourths, fifths, etc. A *subdivision* is a division that divides secondary divisions in halves, thirds, fourths, fifths, etc.

Secondary divisions and subdivisions do not have listed numerical values. When reading an analog scale, the primary, secondary, and subdivision readings are added.

ANALOG DISPLAYS

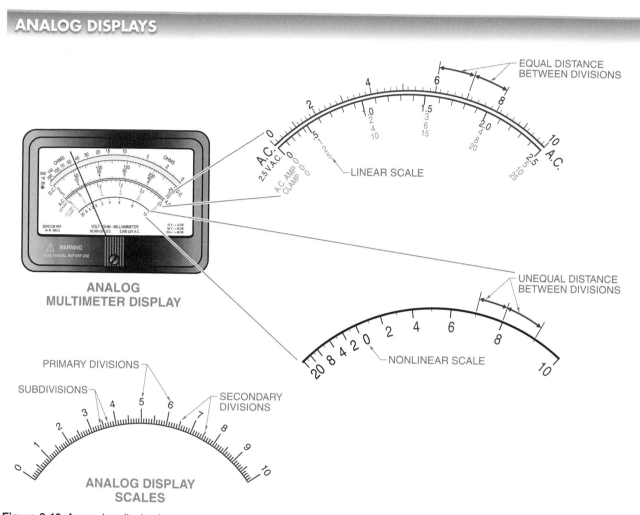

Figure 3-10. An analog display is an electromechanical device that indicates a value by the position of the pointer on a scale.

Digital Multimeters

A *digital multimeter (DMM)* is a meter that can measure two or more electrical properties and displays the measured properties as numerical values. Basic digital multimeters can measure voltage, current, and resistance. Digital multimeters can be equipped with attachments (adapters) that allow the multimeters to measure many types of variables. Advanced digital multimeters may include functions that measure capacitance and/or temperature. The main advantages of a digital multimeter over an analog multimeter are the ability to record measurements and ease in reading the displayed values. **See Figure 3-11.**

Digital multimeters display measurements as numerical numbers, not a scale position. They are not likely to be misread unless the prefixes and symbols accompanying numerical values are misapplied. Most digital multimeters are autoranging. Once a measuring function is selected, such as VAC, the meter will automatically select the best meter range for taking the measurement (400 mV, 4 V, 40 V, 400 V, or 4000 V). Digital multimeters are more accurate than analog multimeters (typical voltage accuracy specifications are between 0.01% and 1.5%). Digital multimeters have high input impedance (resistance), which will not load down sensitive circuits when taking a measurement.

Tech Fact

When a numerical value is read on a DMM, it is important to understand the available ranges that determine the resolution of a measured value. DMM ranges are preset by the meter when it is on the AUTO setting and can be manually set by pressing the RANGE button. The DMM changes range each time the RANGE button is pressed. A DMM with ranges of 400.0 mV, 4.000 V, 40.00 V, 400.0 V, and 1000 V will have a resolution of 0.0001 V on the 400 mV range, 0.001 V on the 4 V range, 0.01 on the 40 V range, 0.1 on the 400 V range, and 1 V on the 1000 V range.

DIGITAL MULTIMETERS

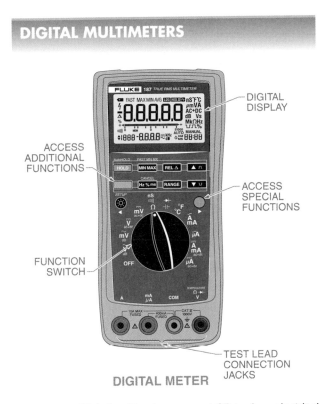

DIGITAL DISPLAY

ACCESS ADDITIONAL FUNCTIONS

ACCESS SPECIAL FUNCTIONS

FUNCTION SWITCH

TEST LEAD CONNECTION JACKS

DIGITAL METER

Figure 3-11. Digital multimeters use an LCD to show electrical measurements.

Reading Digital Displays. A *digital display* is an electronic device that displays readings on a meter as numerical values. Digital displays help eliminate human error when taking readings by displaying exact values measured. Errors occur when reading a digital display if the displayed prefixes, symbols, and/or decimal points are not properly applied.

Digital displays display values using either a light-emitting diode (LED) display or a liquid crystal display (LCD). LED displays are easier to read than LCDs, but use more power. Most portable digital meters use LCDs. The exact value on a digital display is determined from the numbers displayed and the position of the decimal point. A selector switch (range switch) determines the placement of the decimal point.

Typical voltage ranges on a digital display are 3 V, 30 V, and 300 V. The highest possible reading with the range switch on 3 V is 2.999 V. The highest possible reading with the range switch on 30 V is 29.99 V. The highest possible reading with the range switch on 300 V is 299.9 V. Accurate readings are obtained by using the range that gives the best resolution without overloading the meter. **See Figure 3-12.**

DIGITAL DISPLAYS

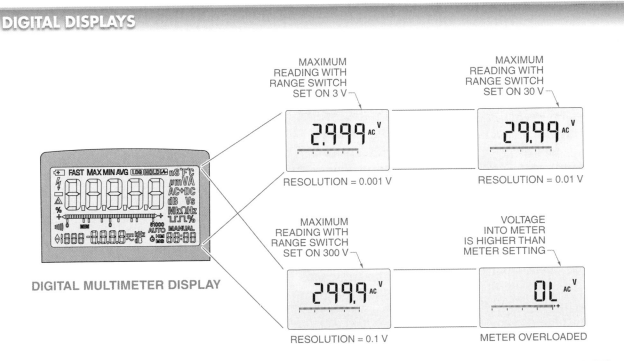

DIGITAL MULTIMETER DISPLAY

MAXIMUM READING WITH RANGE SWITCH SET ON 3 V
2.999 AC V
RESOLUTION = 0.001 V

MAXIMUM READING WITH RANGE SWITCH SET ON 30 V
29.99 AC V
RESOLUTION = 0.01 V

MAXIMUM READING WITH RANGE SWITCH SET ON 300 V
299.9 AC V
RESOLUTION = 0.1 V

VOLTAGE INTO METER IS HIGHER THAN METER SETTING
OL AC V
METER OVERLOADED

Figure 3-12. Digital displays display values using either a light-emitting diode (LED) display or a liquid crystal display (LCD).

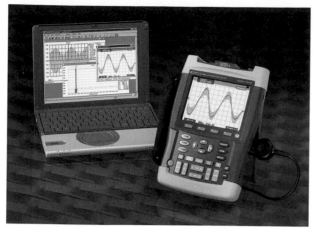

Fluke Corporation

A portable computer is often interfaced with an electrical test instrument to store and analyze data.

Bar Graphs. A *graph* is a diagram that shows a variable in comparison to other variables. Most digital displays include a bar graph to show changes and trends in a circuit. A *bar graph* is a graph composed of segments that function as an analog pointer. The displayed bar graph segments increase as the measured value increases and decrease as the measured value decreases. The polarity of test leads should be reversed if a negative sign is displayed at the beginning of a bar graph. A *wraparound*

bar graph is a bar graph that displays a fraction of the full range on the graph at one time. The pointer wraps around and starts over when the limit of the bar graph is reached. **See Figure 3-13.**

A bar graph reading is updated 30 times per second. A digital reading is updated four times per second. The bar graph is used when quickly changing signals cause the digital display to flash or when there is a change in the circuit that is too rapid for the digital display to detect.

Digital Multimeter Abbreviations and Symbols. An *abbreviation* is a letter or combination of letters that represent a word. Standard abbreviations are used on DMMs to represent a quantity or term for quick recognition. For example, quantities such as voltage and amperes are identified with abbreviations. **See Figure 3-14.** A symbol is a graphic element that represents a quantity, unit, or device. Symbols can be quickly recognized and interpreted regardless of the language.

Tech Fact

On most DMMs, the bar graph provides additional safety when using the MIN/MAX voltage operating mode because the bar graph displays the voltage at the test leads regardless of what the meter is displaying. For example, a meter might be displaying a 24 VAC minimum recorded value, with the bar graph displaying a 120 VAC value at the test lead connection.

WRAPAROUND BAR GRAPH DISPLAYS

05.00 AC V
0 1 2 3 4 5 6 7 8 9 0 — MAXIMUM READING OF 10 V BEFORE POINTER WRAPS AROUND AND STARTS OVER
0 1 2 3 4 5 6 7 8 9 0
— EACH PRIMARY DIVISION = 1 V
— EACH SECONDARY DIVISION = 0.2 V

15.00 AC V
0 1 2 3 4 5 6 7 8 9 0 — ADD 10 V TO READING AFTER FIRST WRAPAROUND 10 V + 5 V = 15 V
— ANALOG POINTER WRAPPED AROUND ONCE
0 1 2 3 4 5 6 7 8 9 0
0 1 2 3 4 5 6 7 8 9 0

25.00 AC V
0 1 2 3 4 5 6 7 8 9 0 — ADD 20 V TO READING AFTER SECOND WRAPAROUND 20 V + 5 V = 25 V
— ANALOG POINTER WRAPPED AROUND TWICE
0 1 2 3 4 5 6 7 8 9 0
0 1 2 3 4 5 6 7 8 9 0
0 1 2 3 4 5 6 7 8 9 0

WRAPAROUND BAR GRAPH

Figure 3-13. A bar graph is composed of segments that function as an analog pointer.

MULTIMETER ABBREVIATIONS

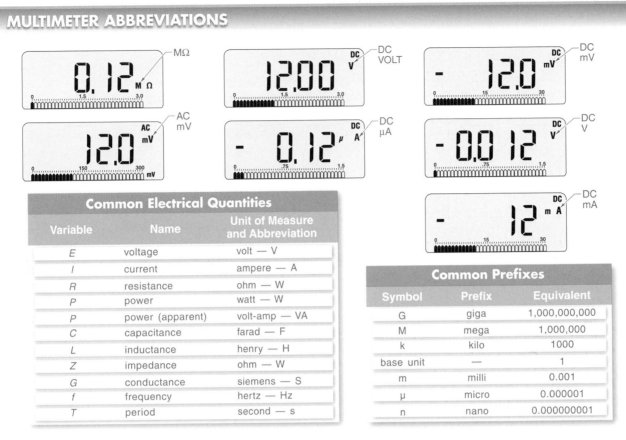

Common Electrical Quantities		
Variable	Name	Unit of Measure and Abbreviation
E	voltage	volt — V
I	current	ampere — A
R	resistance	ohm — W
P	power	watt — W
P	power (apparent)	volt-amp — VA
C	capacitance	farad — F
L	inductance	henry — H
Z	impedance	ohm — W
G	conductance	siemens — S
f	frequency	hertz — Hz
T	period	second — s

Common Prefixes		
Symbol	Prefix	Equivalent
G	giga	1,000,000,000
M	mega	1,000,000
k	kilo	1000
base unit	—	1
m	milli	0.001
μ	micro	0.000001
n	nano	0.000000001

Figure 3-14. Test instruments use multiple abbreviations and symbols when displaying a measurement.

Digital Multimeter Voltage Protection. To protect against transient voltages, protection must be built into the test equipment used. In the past, the industry standard followed was IEC 348. This standard has been replaced by IEC 1010. A DMM designed to the IEC standard offers a higher level of protection. A higher CAT rating indicates an electrical environment with higher power available, larger short-circuit current available, and higher energy transients. For example, a DMM designed to the CAT III standard is resistant to higher energy transients than a DMM designed to the CAT II standard. **See Figure 3-15.**

Power distribution systems are divided into categories because a dangerous high-energy transient voltage, such as a lightning strike, is attenuated or dampened as it travels through the impedance (AC resistance) of the system and the system grounds. Within an IEC 1010 standard category, a higher voltage rating denotes a higher rating for withstanding transient voltage. For example, a CAT III–1000 V (steady-state) DMM provides better protection than a CAT III–600 V (steady-state) DMM.

Between categories, a DMM with a higher voltage rating might not have higher transient voltage protection. For example, a CAT III– 600 V DMM has better transient protection compared to a CAT II–1000 V DMM. A DMM should be chosen based on the IEC overvoltage installation category first and the voltage second.

IEC 1010 Safety Standard for Digital Multimeters. The *International Electrotechnical Commission (IEC)* is an organization that develops international safety standards for electrical equipment. IEC standards reduce safety hazards that can occur from unpredictable circumstances when using electrical test equipment such as DMMs. For example, voltage surges on a power distribution system can cause a safety hazard. A *voltage surge* is a higher-than-normal voltage that temporarily exists on one or more power lines. Voltage surges vary in voltage amount and time present on power lines. One type of voltage surge is a transient voltage.

A *transient voltage (voltage spike)* is a temporary, unwanted voltage in an electrical circuit. Transient voltages typically exist for a very short time but are often larger in magnitude than voltage surges and very erratic. Transient voltages occur due to lightning strikes, unfiltered electrical equipment, and power being switched on and off. High transient voltages may reach several thousand volts. A transient voltage on a 120 V power line can reach 1000 V (1 kV) or more.

Multimeter Attachments. A standard multimeter can be equipped with attachments (adapters) that allow the multimeter to measure many types of variables. **See Figure 3-16.** Most attachments output 1 mV DC per unit of measurement. For example, a temperature adapter outputs 1 mV DC per °F/°C, a tachometer outputs 1 mV DC per rpm, and a humidity adapter outputs 1 mV DC per % relative humidity. Most multimeters include a mV DC setting in addition to the standard DC setting to allow taking measurements with attachments. The mV DC setting allows for a more precise measurement because of the high resolution. *Resolution* is the degree of measurement precision a test instrument is capable of making as it is used.

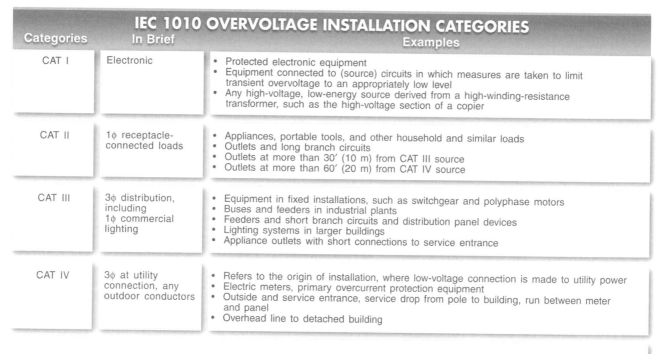

IEC 1010 OVERVOLTAGE INSTALLATION CATEGORIES		
Categories	**In Brief**	**Examples**
CAT I	Electronic	• Protected electronic equipment • Equipment connected to (source) circuits in which measures are taken to limit transient overvoltage to an appropriately low level • Any high-voltage, low-energy source derived from a high-winding-resistance transformer, such as the high-voltage section of a copier
CAT II	1φ receptacle-connected loads	• Appliances, portable tools, and other household and similar loads • Outlets and long branch circuits • Outlets at more than 30′ (10 m) from CAT III source • Outlets at more than 60′ (20 m) from CAT IV source
CAT III	3φ distribution, including 1φ commercial lighting	• Equipment in fixed installations, such as switchgear and polyphase motors • Buses and feeders in industrial plants • Feeders and short branch circuits and distribution panel devices • Lighting systems in larger buildings • Appliance outlets with short connections to service entrance
CAT IV	3φ at utility connection, any outdoor conductors	• Refers to the origin of installation, where low-voltage connection is made to utility power • Electric meters, primary overcurrent protection equipment • Outside and service entrance, service drop from pole to building, run between meter and panel • Overhead line to detached building

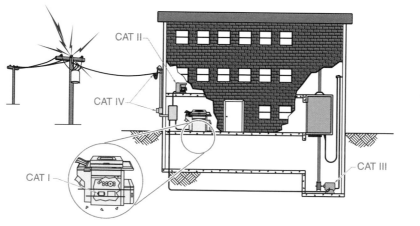

Figure 3-15. The applications in which a DMM may be used are classified by the IEC 1010 standard into four overvoltage installation categories.

MULTIMETER ATTACHMENTS

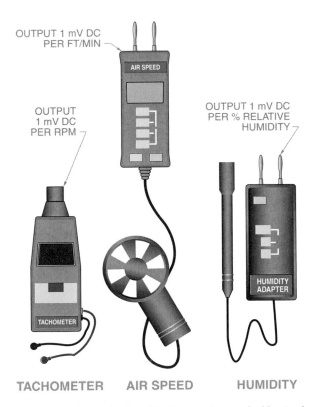

Figure 3-16. A standard multimeter can be used with attachments (adapters) that allow the multimeter to measure almost any type of variable.

Troubleshooting with Digital Multimeters

Digital multimeters are used to identify and correct problems in an electrical system. Resistance, DC voltage, AC voltage, and current measurements may need to be taken in order to troubleshoot an electrical system. Technicians must know how to properly operate a DMM and take accurate measurements.

Resistance Measurements. A DMM measures the resistance of a circuit or the resistance of a component removed from a circuit. **See Figure 3-17. WARNING:** Voltage applied to a DMM that is set to measure resistance will damage the DMM, even if the DMM has internal protection.

TROUBLESHOOTING CONTACTOR COILS

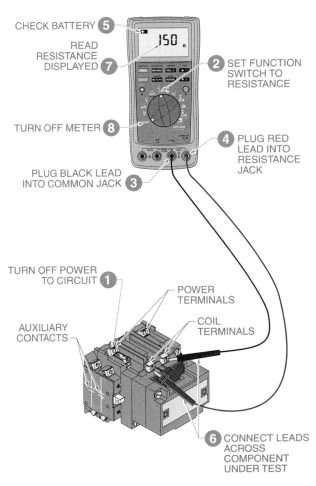

Figure 3-17. A DMM can be used to measure the resistance of a circuit or component removed from a circuit.

To measure resistance with a DMM, the following procedure is applied:

1. Check that all power is OFF to the circuit under test. Always remove any device under test from the circuit.

2. Set the function switch to resistance mode on the DMM. The DMM should display "OL" and the "Ω" symbol when the DMM is in the resistance mode.

3. Plug the black test lead into the common jack.

4. Plug the red test lead into the resistance jack.

5. Ensure that the DMM batteries are in good condition. The battery symbol is displayed when batteries are low.

6. Connect the leads across the device under test. Ensure that there is good contact between the test leads and the component leads.

7. Read the resistance displayed. Check the circuit schematic for parallel paths. Parallel paths with the resistance under test cause reading errors. Do not touch exposed metal parts of the test leads during the test. Resistance of a person's body can cause reading errors.

8. After completing all resistance measurements, turn off the DMM to prevent battery drain.

Tech Fact

When taking very small resistance measurements (less than 10 Ω), the resistance of the test leads will increase the displayed value. To eliminate the resistance of the test leads from the measurement, touch the test leads together (to display the resistance of the test leads) and press the REL (relative) button on the DMM to cancel out the test lead resistance before taking a measurement.

Troubleshooting Printed Circuit Boards. If a printed circuit (PC) board is suspected of having an open trace, an ohmmeter can be used to check each section of the trace for continuity. **See Figure 3-18.** If the resistance increases dramatically on a trace, an open circuit is likely. As the PC board is inspected, a certain amount of flexing may indicate a hairline crack that would not have been noticed at first.

Troubleshooting Diodes. In most cases, diodes are tested with a digital multimeter (DMM). The polarity of the diode can be determined with a DMM, and the diode can also be checked for opens and shorts. Diode troubleshooting includes determining polarity and checking for opens and shorts.

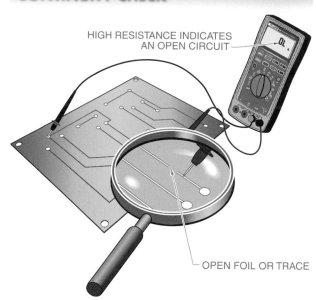

PRINTED CIRCUIT BOARD CONTINUITY CHECK

HIGH RESISTANCE INDICATES AN OPEN CIRCUIT

OPEN FOIL OR TRACE

Figure 3-18. An ohmmeter can be used to check for open circuits on a PC board.

CAUTION: Because DMMs vary somewhat from manufacturer to manufacturer, consult the operator's manual before conducting any diode troubleshooting with them.

A DMM set to measure resistance can be used to determine which end of a diode is the cathode and which end is the anode. This is possible because the DMM is a voltage source with a definite polarity.

Externally, the polarity of the DMM may be marked positive (+) and negative (–). It may also be identified by a color-coded system—usually red for positive and black for negative. Internally, however, the voltage source, or battery, actually determines the external polarity.

The forward and reverse bias of an unknown diode can be determined if the diode is placed between known polarities one way, and then placed in the opposite direction. The diode indicates a low resistance in forward bias and a high resistance in reverse bias. **See Figure 3-19.** Since the polarity of the source is known, the end connected to the negative lead during forward bias must be the cathode and the end connected to the positive lead must be the anode.

DC Voltage Measurements. It is important to exercise caution when measuring DC voltages over 60 V. **WARNING:** Ensure that no body parts contact any part of a live circuit, including the metal contact points at the tips of the test leads.

TROUBLESHOOTING DIODES

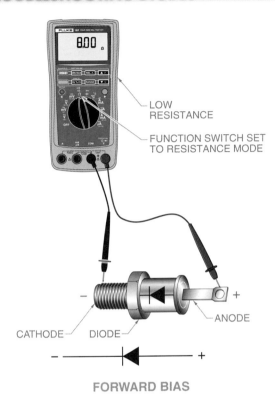

FORWARD BIAS

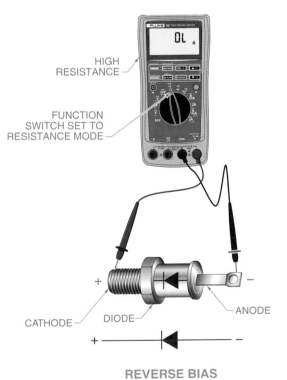

REVERSE BIAS

Figure 3-19. A diode indicates a low resistance in forward bias and high resistance in reverse bias.

A standard procedure is followed when taking DC voltage measurements. **See Figure 3-20.** To measure DC voltages with a digital multimeter (DMM), the following procedure is applied:

1. Set the function switch to DC voltage. If the DMM has more than one voltage position or if the circuit voltage is unknown, select a setting high enough to measure the highest possible circuit voltage.
2. Plug the black test lead into the common jack.
3. Plug the red test lead into the voltage jack.
4. Discharge any capacitors.
5. Connect the DMM test leads to the circuit. Connect the black test lead to circuit ground and the red test lead to the point at which the voltage is under test. Reverse the black and red test leads if a negative sign appears in front of the reading on the DMM.
6. Read the voltage displayed.

DC VOLTAGE MEASUREMENTS

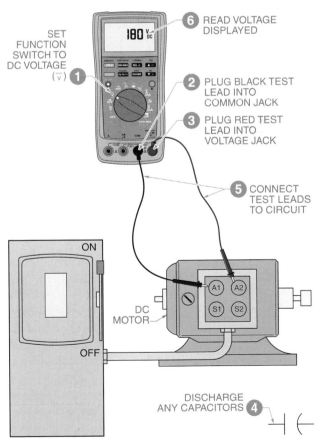

Figure 3-20. Caution must be exercised when measuring DC voltages over 60 V.

AC Voltage Measurements. It is important to exercise caution when measuring AC voltages over 24 V. Proper PPE should always be worn when working on any electrical system. **WARNING:** Ensure that no body parts contact any part of a live circuit, including the metal contact points at the tips of the test leads.

A standard procedure is followed when testing AC voltage measurements. **See Figure 3-21.** To measure AC voltages, the following procedure is applied:

1. Set the function switch to AC voltage. If the DMM has more than one voltage position or if the circuit voltage is unknown, select a setting high enough to measure the highest possible circuit voltage.
2. Plug the black test lead into the common jack.
3. Plug the red test lead into the voltage jack.
4. Connect the DMM test leads to the circuit. The position of the test leads is arbitrary. Common industrial practice is to connect the black test lead to the grounded (neutral) side of the AC voltage.
5. Read the voltage displayed.

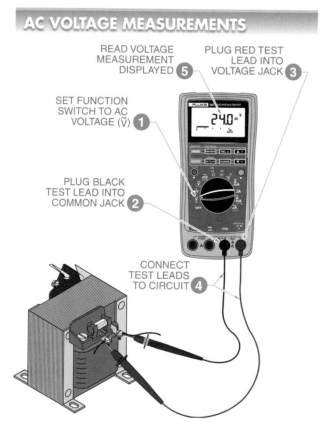

AC VOLTAGE MEASUREMENTS

READ VOLTAGE MEASUREMENT DISPLAYED **5**

PLUG RED TEST LEAD INTO VOLTAGE JACK **3**

SET FUNCTION SWITCH TO AC VOLTAGE (V̄) **1**

PLUG BLACK TEST LEAD INTO COMMON JACK **2**

CONNECT TEST LEADS TO CIRCUIT **4**

Figure 3-21. AC voltage measurements may vary slightly with different DMMs. Manufacturer procedural manuals should always be consulted for proper operating procedures.

Troubleshooting Fuses. Fuses can also be checked using a DMM set to measure voltage. **See Figure 3-22.** To troubleshoot fuses using a DMM set to measure voltage, the following procedure is applied:

1. Turn the handle of the safety switch or disconnect to the OFF position.
2. Open the door of the safety switch or disconnect. The operating handle must be capable of opening the switch. If the operating handle is not working properly, replace the switch.
3. Check the enclosure and interior parts for deformation, displacement of parts, and burns. Such damage may indicate a short circuit, fire, or lightning strike. Deformation requires replacement of the part or complete device. Any indication of arcing or overheating damage, such as discoloration or melting of insulation, requires replacement of the damaged part(s).
4. Check the incoming voltage between each pair of power leads. Incoming voltage should be within 10% of the voltage rating of the motor. A problem exists if voltage is not within 10%. This voltage variation may be the reason the fuses have blown.
5. Test the enclosure for grounding. To test for grounding, connect one test lead of a DMM to an unpainted metal part of the enclosure and touch the other test lead to each of the incoming power leads. A voltage difference is indicated if the enclosure is properly grounded. The line-to-ground voltage may not equal the line-to-line voltage reading taken in Step 4.
6. Check fuses. Turn the handle of the safety switch, disconnect, or combination starter to the ON position to test the fuses. **WARNING:** When turning on any electrical circuit, always wear proper PPE and follow all safety rules. One test lead of a DMM set to measure voltage is connected to one of the incoming power lines at the top of any fuse. The other test lead of the DMM is connected to the top of any other fuse. If no voltage is read, there is no power coming into the fuses. If voltage is read, move one of the leads from the top of a fuse to the bottom of the fuse. If the same voltage is measured, that fuse is good. If no voltage is read, that fuse is bad. Repeat this step for each fuse. If any fuse is bad, turn power off before replacing the fuse. After replacing a fuse, repeat the entire step.
7. Replace bad fuses. Turn the handle of the safety switch or combination starter to the OFF position to replace the fuses. Use a fuse puller to remove bad fuses. Replace all bad fuses with the correct type

and size replacement. Close the door on the safety switch or combination starter and turn the circuit to the ON position.

TROUBLESHOOTING FUSES — DIGITAL MULTIMETERS

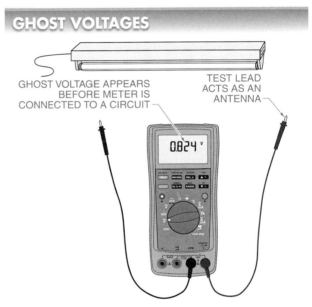

CHECK INCOMING VOLTAGE **4**

5 CHECK FOR GROUNDING

1 TURN HANDLE OFF

6 CHECK FUSES (WITH POWER ON)

7 REPLACE ANY BAD FUSES (AFTER POWER TURNED OFF)

A   B   C — GROUND

2 OPEN DOOR

3 CHECK ENCLOSURE FOR DAMAGE

WARNING:
The top power lugs in a disconnect are always hot, even when the disconnect is in the off position.

Note:
When checking voltage, check A-B, A-C, and B-C. also check A to ground, B to ground, and C to ground.

CHECK FUSES (WITH POWER ON) **6**

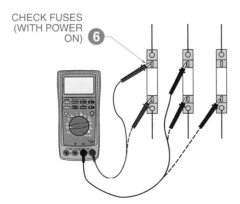

Figure 3-22. Fuses can be checked using a DMM set to measure voltage.

Ghost Voltages. A meter set to measure voltage may display a reading before the meter is connected to a powered circuit. The displayed voltage is a ghost voltage that appears as changing numbers on a digital display or as a vibrating analog display. A *ghost voltage* is a voltage that appears on a meter not connected to a circuit.

Ghost voltages are produced by the magnetic fields generated by current-carrying conductors, fluorescent lighting, and operating electrical equipment. Ghost voltages enter a meter through the test leads because test leads not connected to a circuit act as antennae for stray voltages. **See Figure 3-23.**

Ghost voltages do not damage a meter. Ghost voltages may be misread as circuit voltages when a meter is connected to a circuit that is believed to be powered. A circuit that is not powered can also act as an antenna for stray voltages. To ensure true circuit voltage readings, a meter is connected to a circuit for a long enough time that the meter displays a constant reading.

GHOST VOLTAGES

GHOST VOLTAGE APPEARS BEFORE METER IS CONNECTED TO A CIRCUIT

TEST LEAD ACTS AS AN ANTENNA

0824 ᵛ

Figure 3-23. Ghost voltages are produced by the magnetic fields generated by current-carrying conductors, fluorescent lighting, and operating electrical equipment.

Tech Fact

When a DMM is set to measure voltage, the meter automatically sets itself to the lowest measuring range (usually mV) and adjusts the measuring range higher (6 V, 60 V, 600 V, etc.) when the test leads are connected to the voltage under measurement, which eliminates the ghost voltage displayed. However, if the test leads are connected to a circuit that is not powered, a higher ghost voltage is usually displayed and must not be confused with an actual circuit voltage. For example, a 120 mV ghost voltage is not 120 V.

In-Line Current Measurements. Care is required to protect the DMM, the circuit, and the user when taking in-line AC or DC current measurements. To observe standard safety precautions when taking in-line current measurements, the following steps can be taken:

• Follow manufacturer recommended procedures when testing internal DMM fuses.

• Ensure that the expected load current measurement is less than the current setting (limit) of the DMM. Start with the highest current-measuring range if the load current is unknown. If the current measurement may exceed the limit of the DMM setting, use a clamp-on ammeter.

• Ensure that the DMM function switch is set to the proper setting for measuring current (AC or DC). Most DMMs include more than one current level, such as μA, mA, and A.

• Ensure that the test leads are connected to the proper jacks for measuring current. Most DMMs include more than one current jack.

   WARNING: Always ensure that the function switch position matches the connection of the test leads. The DMM can be damaged if the test leads are connected to measure current and the function switch is set for a different measurement such as voltage or resistance. Some DMMs have an alert feature, which provides a constant audible warning (beep) if the test leads are connected in the current jacks and a noncurrent mode is selected.

• Ensure that power to the test circuit is OFF before connecting and disconnecting test leads. If necessary, take a voltage measurement to ensure that the voltage is OFF.

• Do not change the function switch position on the DMM while the circuit under test is energized.

• Turn power to the DMM and the circuit off before changing any settings.

• Connect the DMM in series with the load(s) to be measured. Never connect a DMM in parallel with the load(s) to be measured.

   Many DMMs include a fuse in the current-measuring circuit to prevent damage caused by excessive current. A DMM should be checked before it is used to see if all the fuses on the current ranges being used are good. The DMM is marked as fused or not fused at the test lead current terminals. In-line current measurements are not recommended if the DMM is not fused. In-line AC or DC measurements are taken using a standard procedure.

See Figure 3-24. To take in-line current measurements, the following procedure is applied:

   WARNING: Ensure that no body parts contact any part of the live circuit, including the metal contact points at the tips of the test leads.

1. Set the function switch to the proper position for measuring the AC or DC and current level (A and mA, or μA). If the DMM has more than one position, select a setting high enough to measure the highest possible circuit current.

2. Plug the black test lead into the common jack.

3. Plug the red test lead into the current jack. The current jack may be marked A and mA, or μA.

4. Turn off the power to the circuit or device under test and discharge all capacitors if possible.

5. Open the circuit at the test point and connect the test leads to each side of the opening. For DC current, the black (negative) test lead is connected to the negative side of the opening, and the red (positive) test lead is connected to the positive side of the opening. Reverse the black and red test leads if a negative sign appears to the left of the measurement displayed.

6. Turn on the power to the circuit under test.

7. Read the current measurement displayed.

8. Turn off the power and remove the DMM from the circuit.

Ideal Industries, Inc.
A fuse puller is used to safely remove blade fuses and cartridge fuses from electrical boxes and cabinets.

DC IN-LINE CURRENT MEASUREMENTS

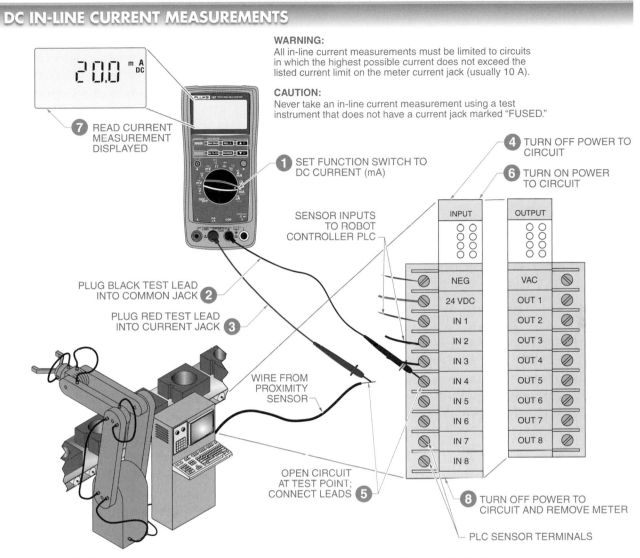

WARNING:
All in-line current measurements must be limited to circuits in which the highest possible current does not exceed the listed current limit on the meter current jack (usually 10 A).

CAUTION:
Never take an in-line current measurement using a test instrument that does not have a current jack marked "FUSED."

7 READ CURRENT MEASUREMENT DISPLAYED

1 SET FUNCTION SWITCH TO DC CURRENT (mA)

4 TURN OFF POWER TO CIRCUIT

6 TURN ON POWER TO CIRCUIT

SENSOR INPUTS TO ROBOT CONTROLLER PLC

PLUG BLACK TEST LEAD INTO COMMON JACK 2

PLUG RED TEST LEAD INTO CURRENT JACK 3

WIRE FROM PROXIMITY SENSOR

OPEN CIRCUIT AT TEST POINT; CONNECT LEADS 5

8 TURN OFF POWER TO CIRCUIT AND REMOVE METER

PLC SENSOR TERMINALS

INPUT | OUTPUT
NEG | VAC
24 VDC | OUT 1
IN 1 | OUT 2
IN 2 | OUT 3
IN 3 | OUT 4
IN 4 | OUT 5
IN 5 | OUT 6
IN 6 | OUT 7
IN 7 | OUT 8
IN 8

Figure 3-24. A DMM used to take in-line current measurements becomes part of the circuit being tested.

Clamp-On Ammeters

A *clamp-on ammeter* is a meter that measures the current in a circuit by measuring the strength of the magnetic field around a single conductor. Clamp-on ammeters measure currents from 0.01 A or less to 1000 A or more. A clamp-on ammeter is normally used to measure current in a circuit with over 1 A of current and in applications in which current can be measured by easily placing the jaws of the ammeter around one of the conductors. Most clamp-on ammeters can also measure voltage and resistance. To measure voltage and resistance, the clamp-on ammeter must include test leads and voltage and resistance modes. **See Figure 3-25.**

Current is a common troubleshooting measurement because only a current measurement can be used by an electrician to determine how much a circuit is loaded or if a circuit is operating correctly. Current measurements vary because current can vary at different points in parallel or series/parallel circuits. Current in series circuits is constant throughout the circuit. The largest amount of current in a parallel circuit is at the point of lowest resistance. Current decreases with any load that has a higher resistance. Any variation that is excessively high must be investigated because the current measurement may indicate that a partial short exists on one of the lines and a small amount of current is flowing to ground.

CLAMP-ON AMMETERS

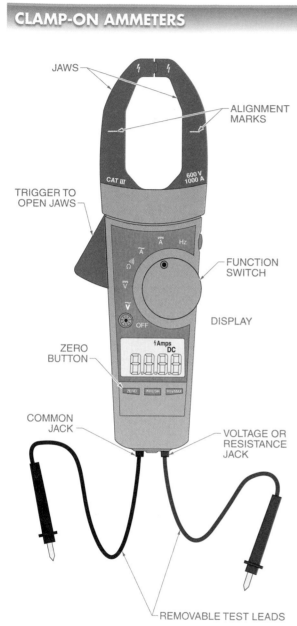

Figure 3-25. A clamp-on ammeter includes test leads and voltage and resistance modes.

Electricians must ensure that clamp-on ammeters do not pick up stray magnetic fields by separating conductors being tested as much as possible from other conductors during testing. If stray magnetic fields are possibly affecting a measurement, several measurements at different locations along the same conductor must be taken.

Clamp-On Ammeter Current Measurement. AC or DC measurements using a clamp-on ammeter or a

DMM with a clamp-on current probe accessory follow standard procedures. **See Figure 3-26.** To measure current using a clamp-on ammeter, the following procedure is applied:

1. Determine if AC or DC current is to be measured.

2. Select the ammeter required to measure the circuit current (AC or DC). If both AC and DC measurements are required, select an ammeter that can measure both AC and DC.

3. Determine if the ammeter range is high enough to measure the maximum current that may exist in the test circuit. If the ammeter range is not high enough, select an accessory that has a high enough current rating, or select an ammeter with a higher range. If the ammeter includes fused current terminals, check to ensure that the ammeter fuses are good.

4. Set the function switch to the proper current setting (600 A, 200 A, 10 A, 400 mA, etc.). If there is more than one current position or if the circuit current is unknown, select a setting greater than the highest possible circuit current.

5. Open the jaws by pressing against the trigger.

6. Enclose one conductor in the jaws. Ensure that the jaws are completely closed before taking readings. Care should be taken to ensure that the meter does not pick up stray magnetic fields. Whenever possible, conductors under test should be separated from other surrounding conductors by a few inches. If this is not possible, several readings should be taken at different locations along the same conductor.

7. Read the current measurement displayed.

8. If required, plug the clamp-on current probe accessory into the DMM. The black test lead of the clamp-on current probe accessory is plugged into the common jack. The red test lead is plugged into the mA jack for current measurement accessories that produce a current output. The red test lead is plugged into the voltage (V) jack for current measurement accessories that produce a voltage output. The current measurement accessories that produce a current output are designed to measure AC only and deliver 1 mA to the DMM for every 1 A of measured current (1 mA/A). Current accessories that produce a voltage output are designed to measure AC or DC and deliver 1 mV to the DMM for every 1 A of measured current (1 mV/A).

CLAMP-ON AMMETER CURRENT MEASUREMENT

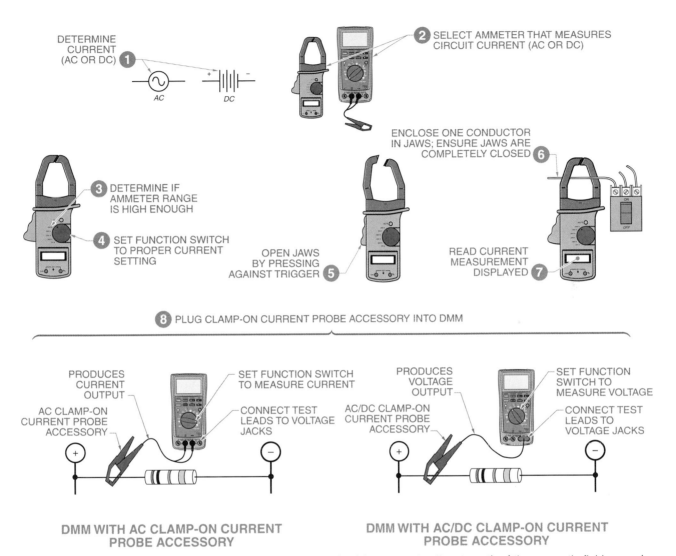

DETERMINE CURRENT (AC OR DC) **1**

AC   DC

2 SELECT AMMETER THAT MEASURES CIRCUIT CURRENT (AC OR DC)

3 DETERMINE IF AMMETER RANGE IS HIGH ENOUGH

4 SET FUNCTION SWITCH TO PROPER CURRENT SETTING

OPEN JAWS BY PRESSING AGAINST TRIGGER **5**

ENCLOSE ONE CONDUCTOR IN JAWS; ENSURE JAWS ARE COMPLETELY CLOSED **6**

READ CURRENT MEASUREMENT DISPLAYED **7**

8 PLUG CLAMP-ON CURRENT PROBE ACCESSORY INTO DMM

PRODUCES CURRENT OUTPUT

AC CLAMP-ON CURRENT PROBE ACCESSORY

SET FUNCTION SWITCH TO MEASURE CURRENT

CONNECT TEST LEADS TO VOLTAGE JACKS

PRODUCES VOLTAGE OUTPUT

AC/DC CLAMP-ON CURRENT PROBE ACCESSORY

SET FUNCTION SWITCH TO MEASURE VOLTAGE

CONNECT TEST LEADS TO VOLTAGE JACKS

DMM WITH AC CLAMP-ON CURRENT PROBE ACCESSORY

DMM WITH AC/DC CLAMP-ON CURRENT PROBE ACCESSORY

Figure 3-26. A clamp-on ammeter measures the current on a circuit by measuring the strength of the magnetic field around a single conductor.

Locating Branch Circuits Using Clamp-On Ammeters. A technician must often locate one circuit in a switchboard, panelboard, or load center to turn off the power before troubleshooting or working on a circuit. Switchboards, panelboards, and load centers are often crowded with wires that are not marked or that are mismarked. A technician cannot start turning off each circuit until the correct circuit is found because this disconnects all loads connected to that circuit. Timers, counters, clocks, starters, and other control devices must be reset, otherwise critical equipment such as alarms and safety circuits may be stopped. A flashing lamp and a clamp-on ammeter may be used to isolate a particular circuit. **See Figure 3-27.**

Tech Fact

Today, clamp-on ammeters are available that are capable of measuring currents as low as 4 mA and as high as thousands of amps. In addition, clamp-on ammeters are available with removable display heads for remote monitoring, flexible cables that open to enable connection around large conductors and bus bars, and small jaws with an attached lead that can be detached from the meter body and connected easily in tight areas.

LOCATING BRANCH CIRCUITS

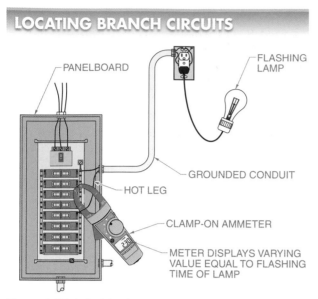

Figure 3-27. A flashing lamp and a clamp-on ammeter may be used to isolate a particular circuit.

circuit displays a constant current reading except the one with the flashing lamp. The circuit with the flashing lamp displays a varying value on the ammeter equal to the flashing time of the lamp. This circuit may then be turned off for troubleshooting.

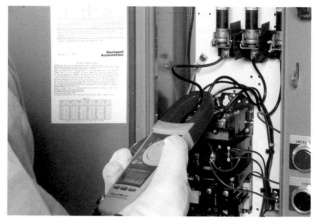

Clamp-on ammeters are used to measure the magnetic field created by current flowing through one isolated conductor.

The flashing lamp is plugged into any receptacle on the circuit that is to be disconnected. As the lamp flashes, a clamp-on ammeter is used to check each circuit. Each

3-2 CHECKPOINT

1. What type of test instrument is used to test for a complete path for current flow without displaying an actual numerical value?

2. What type of test instrument is used to give an approximate voltage level and type without displaying an actual numerical value?

3. When testing a receptacle, is the short slot or the longer slot the hot slot?

4. Is the displayed digital measurement or the displayed bar graph on a DMM updated the fastest?

5. What CAT rating is required for a test instrument used outdoors?

6. When testing fuses in a disconnect, if the meter leads are connected to the top of two fuses and the meter measures 230 VAC, what should the measurement be on a good fuse when one of the test leads is moved from the top of a fuse to the bottom of the fuse?

7. What type of current measurement should never be taken on circuits that may contain high currents or unknown currents?

3-3 SPECIALTY TEST INSTRUMENTS

Specialty electrical test instruments are typically more advanced than general use test instruments and may not be used as often as some general use test instruments. Specialty test instruments typically only measure one specific quantity and have a limited application. Specialty test instruments include oscilloscopes, digital logic probes, megohmmeters, ground resistance testers, noncontact temperature probes, phase sequence indicators, and optical time domain reflectometers.

Oscilloscopes

Most electronic circuits are designed to take a signal (power, voice, data, etc.) and change it by amplifying, filtering, storing, displaying, or converting the signal from analog to digital or digital to analog. The testing of electronic circuits and equipment often requires measuring or observing electrical signals. Electrical signals are measured by using meters (voltmeters and ammeters) or are observed by using oscilloscopes. An *oscilloscope* is a test instrument that provides a visual display of voltages. An oscilloscope displays the waveform for the voltage in a circuit and allows the voltage level, frequency, and phase to be measured. The two types of oscilloscopes are bench and handheld. Both types of oscilloscopes include the same basic features. **See Figure 3-28.**

A *bench oscilloscope* is a test instrument that displays the shape of a voltage waveform and is used mostly for bench testing electrical and electronic circuits. Bench testing is testing performed when equipment being tested is brought to a designated service area. Bench oscilloscopes are used to troubleshoot digital circuit boards, communication circuits, TVs, DVD players, computers, and other types of electronic circuits and equipment.

A *handheld oscilloscope* is a test instrument that displays the shape of a voltage waveform and is typically used for field testing. Field testing is testing performed when the test instrument is taken to the location of the equipment to be tested. Most handheld oscilloscopes are a combination oscilloscope and digital multimeter. Handheld oscilloscopes are also referred to as portable oscilloscopes, Scopemeters®, PowerPads®, power quality meters, power analyzers, or power meters.

Oscilloscope Operation. An oscilloscope is an instrument that graphically displays an instantaneous voltage. The oscilloscope is one of the most useful measuring instruments used in electronic circuit diagnostics and bench testing. **See Figure 3-29.** Oscilloscopes are used to troubleshoot digital circuits, communication circuits, factory process instrumentation, machine control circuitry, and computers. Besides showing a voltage waveform in a circuit, oscilloscopes also allow the voltage level, frequency, and phase to be measured. Oscilloscopes are available in basic and specialized types that can display different waveforms simultaneously.

Digital Logic Probes

A *digital logic probe* is a special DC voltmeter that detects the presence or absence of a signal. Displays on a digital logic probe include logic high, logic low, pulse light, memory, and TTL/CMOS. **See Figure 3-30.** The high LED lights when the logic probe detects a high logic level (1). The low LED lights when the logic probe detects a low logic level (0). The pulse LED flashes relatively slowly when the probe detects logic activity present in a circuit. Logic activity indicates that the circuit is changing between logic levels. The pulse light displays the changes between logic levels because the changes are usually too fast for the high and low LEDs to display.

OSCILLOSCOPES

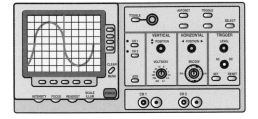

BENCH OSCILLOSCOPE

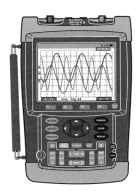

HANDHELD OSCILLOSCOPE

Figure 3-28. Bench oscilloscopes and handheld oscilloscopes provide a visual display of voltages.

Tech Fact

When selecting or using a handheld oscilloscope, it is important to ensure that the measured and recorded data can be downloaded to a PC for viewing and that the software allows adding documentation and display options such as overlapping or departing voltage, current, and power measurements.

OSCILLOSCOPE OPERATION

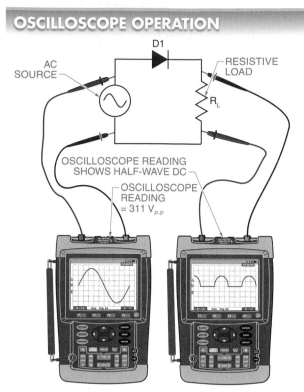

Figure 3-29. An oscilloscope is a test instrument that displays an instantaneous voltage image on a graph.

DIGITAL LOGIC PROBES

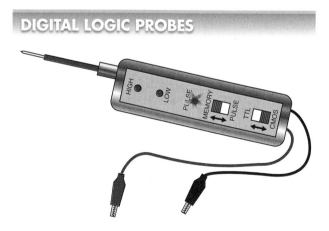

Figure 3-30. A digital logic probe is a special DC voltmeter that detects the presence or absence of a signal.

The memory switch sets the logic probe to detect short pulses, usually lasting a few nanoseconds. Any change from the original logic level causes the memory LED to light and remain on. The memory switch is manually moved to the pulse position and back to the memory position to reset the logic probe.

The TTL/CMOS switch selects the logic family of integrated circuits (ICs) to be tested. *Transistor-transistor logic (TTL) ICs* are a broad family of ICs that employ a two-transistor arrangement. The supply voltage for TTL ICs is 5.0 VDC, ±0.25 V.

Complementary metal-oxide semiconductor (CMOS) ICs are a group of ICs that employ MOS transistors. CMOS ICs are designed to operate on a supply voltage ranging from 3 VDC to 18 VDC.

Digital circuits fail because the signal is lost somewhere between the circuit input and output stages. Finding the point where the signal is missing and repairing that area usually solves the problem. Repairs normally involve replacing a component, module, or an entire PC board.

Megohmmeters

A *megohmmeter* is a device that detects insulation deterioration by measuring high resistance values under high test voltage conditions. Megohmmeter test voltages range from 50 V to 5000 V. A megohmmeter detects insulation failure or potential failure of insulation caused by excessive moisture, dirt, heat, cold, corrosive vapors or solids, vibration, and aging. **See Figure 3-31.**

Some insulation, such as that found on conductors used to wire branch circuits, has a thick insulation and is harder to damage or break down. Other insulation, such as the insulation used on motor windings, is very thin (to save weight and space) and breaks down much more easily. Megohmmeters are used to test for insulation breakdown in long wire runs or motor windings.

Insulation allows conductors to stay separated from each other and from earth ground. Insulation must have a high resistance to prevent current from leaking through the insulation. All insulation has a resistance value that is less than infinity, which allows some leakage current to occur. *Leakage current* is current that flows through insulation. Under normal operating conditions, the amount of leakage current is so small (only a few microamperes) that the leaking current has no effect on the operation or safety of a circuit.

Ground Resistance Testers

A *ground resistance tester* is a device used to measure ground connection resistance of electrical installations such as power plants, industrial plants, high-tension towers, and lightning arrestors. Ground resistance testers make routine ground tests as specified by NEC® Article 230.95(C). **See Figure 3-32.**

MEGOHMMETERS

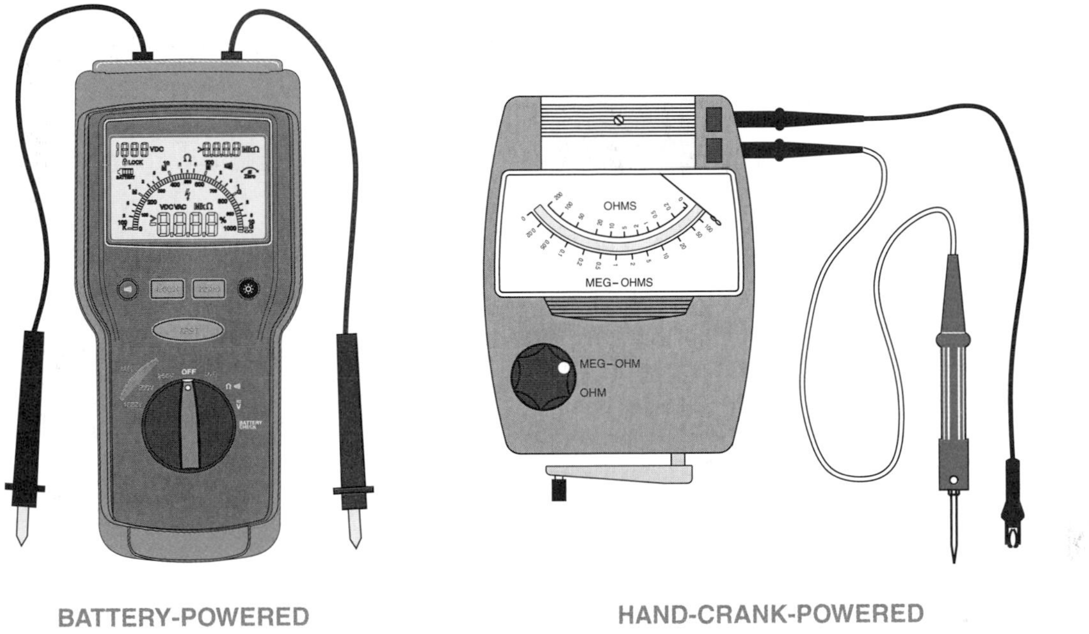

BATTERY-POWERED

HAND-CRANK-POWERED

Figure 3-31. A megohmmeter detects insulation deterioration by measuring high resistance values under high test voltage conditions.

GROUND RESISTANCE TESTERS

AEMC® Instruments

Figure 3-32. A ground resistance tester is a device used to measure ground connection resistance of electrical installations.

Infrared Temperature Meter

An *infrared temperature meter* is a noncontact temperature probe that senses the infrared energy emitted by a material. All materials emit infrared energy in proportion to the temperature at the surface of the material. Infrared temperature probes are commonly used to take temperature measurements of electrical distribution systems, motors, bearings, switching circuits, and other equipment where electrical heat buildup is critical. **See Figure 3-33.** A *noncontact temperature probe* is a device used for taking temperature measurements on energized circuits or on moving parts. A noncontact temperature probe may be an accessory for a DMM.

INFRARED TEMPERATURE METER

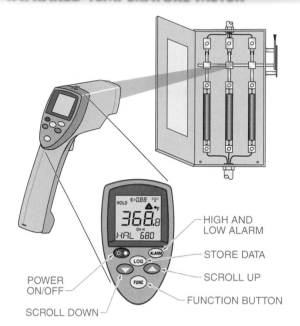

HIGH AND LOW ALARM

STORE DATA

SCROLL UP

FUNCTION BUTTON

POWER ON/OFF

SCROLL DOWN

CAUTION:
Make sure proper PPE is worn when taking noncontact temperature measurements around energized electrical circuits.

Figure 3-33. An infrared temperature meter can be used to assess the condition of electrical connections and operating equipment.

Phase Sequence Indicators

A *phase sequence indicator* is a device used to determine phase sequence and open phases. Phase sequence indicators help protect motors, generators, and other equipment from damage due to incorrect motor rotation. They also ensure that the three-phase power lines are present and in the correct order. Phase sequence indicators are available with either a rotating disk with a colored dot or with two or three LED lamps to indicate phase. **See Figure 3-34.**

Optical Time Domain Reflectometers

An *optical time domain reflectometer (OTDR)* is a test instrument that is used to measure fiber optic cable attenuation. An OTDR uses a laser light source that sends out short pulses into a fiber and analyzes the light scattered back. The light source decays with fiber attenuation. Attenuation is produced by reflections from splices, connectors, and any areas in the cable that cause problems. Based on the amount of signal reflected back, the type and location of a fault is displayed. **See Figure 3-35.**

PHASE SEQUENCE INDICATORS

Greenlee Textron Inc.

Figure 3-34. A phase sequence indicator is a device used to determine phase sequence and open phases.

OPTICAL TIME DOMAIN REFLECTOMETER (OTDR)

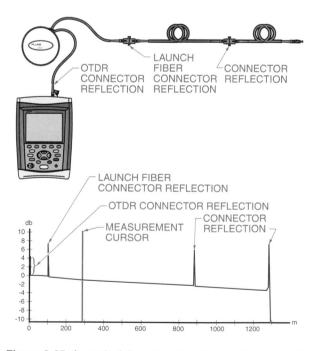

LAUNCH FIBER CONNECTOR REFLECTION

OTDR CONNECTOR REFLECTION

CONNECTOR REFLECTION

LAUNCH FIBER CONNECTOR REFLECTION

OTDR CONNECTOR REFLECTION

CONNECTOR REFLECTION

MEASUREMENT CURSOR

Figure 3-35. An optical time domain reflectometer is used to measure fiber optic signal loss (attenuation), length, and fiber endface integrity.

3-3 CHECKPOINT

1. What type of test instrument shows a visual display of the measured value?

2. What type of test instrument is used to measure "1" and "0" in electronic circuits?

3. What type of test instrument is used to test the condition of a conductor or equipment insulation at a higher voltage than a standard DMM?

4. When using a ground resistance tester, what is the maximum building resistance required by the NEC® for safe system operation?

5. What test instrument is used to ensure that the three-phase power lines (A, B, and C) are all present and in the right order (ABC)?

6. Can an infrared temperature meter be used to take temperature measurements on an energized circuit?

7. What type of test instrument is used to test fiber optic cables for potential problems?

Additional Resources

Review and Resources

Access Chapter 3 Review and Resources for *Electrical Motor Controls for Integrated Systems* by scanning the above QR code with your mobile device.

Applying Your Knowledge

Refer to the *Electrical Motor Controls for Integrated Systems* Learner Resources for interactive Applying Your Knowledge questions.

Workbook and Applications Manual

Refer to Chapter 3 in the *Electrical Motor Controls for Integrated Systems Workbook* and the *Applications Manual* for additional exercises.

ENERGY EFFICIENCY PRACTICES

Electronic Equipment Recycling

Electronic equipment may contain many materials that can be recycled and reused. Some materials may also be hazardous. Technicians must know the difference before disposing of electronic equipment that is no longer useful. Recycling helps reduce the impact of electrical equipment on the environment. Many devices can be dismantled and the plastic, glass, and various metals recycled to produce useful byproducts. The Restriction of Hazardous Substances (RoHS) Directive and the Waste Electrical and Electronics Equipment (WEEE) Directive provide strategies for reducing the impact of electrical equipment on the environment when the equipment is no longer useful.

The RoHS Directive prohibits the use of consumer electronics containing lead, mercury, cadmium, chromium, polybrominated biphenyls (PBBs), and polybrominated diphenyl ether (PBDE). The directive is enforced for homogeneous materials. Homogeneous materials are materials that are uniform throughout and cannot be separated by mechanical processes. Homogeneous materials include alloys, plastic, glass, ceramics, specific liquids, and low-density materials that are consistent in material composition.

Normally, heterogeneous materials are reviewed for RoHS compliance. Heterogeneous materials are dissimilar materials that can be separated by common mechanical processes. Heterogeneous materials include PC boards, cable assemblies, and individual electronic components, such as capacitors, diodes, and resistors. Technicians should comply with the RoHS Directive when troubleshooting and disposing of electronic equipment.

The WEEE Directive establishes the collection, recycling, and recovery of electrical equipment. It targets original equipment manufacturers (OEMs) with the responsibility of collecting, recycling, and recovering electrical equipment. Technicians should be aware of the impact of the directives on their working environment. Key objectives of the WEEE Directive include reducing electrical and electronic equipment waste, increasing recycling and recovery, and improving environmental disposal of nonrecyclable materials.

Objectives

4-1

- State four factors that determine the severity of an electrical shock.
- Explain the three approach boundaries as identified by NFPA 70E®.
- State three ways to help prevent an unwanted electrostatic discharge (ESD) from damaging equipment.
- Identify the meanings of the different colors used with safety labels.
- List the basic electric motor safety rules.

4-2

- State where and when a lockout/tagout device should be used.
- Describe the different types of lockout devices.
- State the purpose of applying NFPA 70E® standards.

4-3

- Describe the types of personal protective equipment (PPE).
- Define arc flash and arc blast and how to minimize their effects.
- List the different types of head, eye, and ear protection.
- State the different parts of hand protection used to prevent an electrical shock and the purpose of each part.
- List the different types of foot, back and knee protection.
- State the purpose of the National Electrical Code® (NEC®) and the Article that covers the requirements for motors.
- State the purpose of grounding.
- Explain how a ground fault circuit interrupter (GFCI) protects individuals.

4-4

- State the different classes of fires, specifically the electrical fire classification.
- State possible materials that create a hazardous location and how hazardous locations are identified.

4-5

- Define confined space and state several ways to help prevent an accident in a confined space.

4-6

- Explain the importance of safety when working on overhead power lines.

Review and Resources
atplearningresources.com/Quicklinks
Access Code: 362245

Potential hazards often can be seen before there is a problem. This allows time to take corrective action. For example, when scaffolding looks uneven or when a road is icy, there is time to take corrective action. However, electrical hazards usually cannot be seen until it is too late since electricity cannot be seen. There is no way of knowing, without testing or seeing if a load is ON, whether a device, circuit, or conductor is powered. Electrical power can and does kill individuals every day. Learning ways to help prevent electrical shock or minimize its effects is the single most important thing a person working in the electrical field or around electrical devices/circuits can learn. All the theoretical knowledge about how electrical circuits and systems work is meaningless if a person cannot prevent or minimize receiving an electrical shock or being electrocuted.

4-1 ELECTRICAL SAFETY

Technicians must work safely at all times. Electrical safety rules must be followed when working with electrical equipment to help prevent injuries from electrical energy sources. Electrical safety has been advanced by the efforts of the National Fire Protection Association (NFPA), Occupational Safety and Health Administration (OSHA), and state safety laws. The *National Fire Protection Association (NFPA)* is a national organization that provides guidance in assessing the hazards of the products of combustion. The NFPA sponsors the development of the National Electrical Code® (NEC®) and the NFPA 70E®.

Electrical Shock

An *electrical shock* is a shock that results any time a body becomes part of an electrical circuit. Electrical shock effects vary from a mild sensation, to paralysis, to death. Also, severe burns may occur where current enters and exits the body. The severity of an electrical shock depends on the amount of electric current in milliamps (mA) that flows through the body, the length of time the body is exposed to the current flow, the path the current takes through the body, and the physical size and condition of the body through which the current passes. **See Figure 4-1.**

Prevention is the best protection from electrical shock. Anyone working on electrical equipment should have respect for all voltages, have a knowledge of the principles of electricity, and follow safe work procedures. All technicians should be encouraged to take a basic course in cardiopulmonary resuscitation (CPR) so they can aid a coworker in emergency situations.

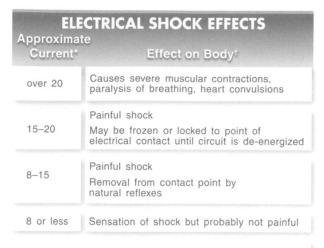

ELECTRICAL SHOCK EFFECTS

Approximate Current*	Effect on Body†
over 20	Causes severe muscular contractions, paralysis of breathing, heart convulsions
15–20	Painful shock May be frozen or locked to point of electrical contact until circuit is de-energized
8–15	Painful shock Removal from contact point by natural reflexes
8 or less	Sensation of shock but probably not painful

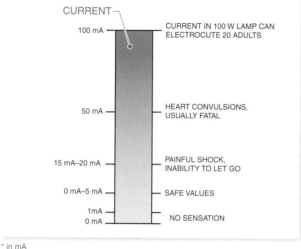

CURRENT

100 mA — CURRENT IN 100 W LAMP CAN ELECTROCUTE 20 ADULTS

50 mA — HEART CONVULSIONS, USUALLY FATAL

15 mA–20 mA — PAINFUL SHOCK, INABILITY TO LET GO

0 mA–5 mA — SAFE VALUES

1mA
0 mA — NO SENSATION

* in mA

† effects vary depending on time, path, amount of exposure, and condition of body

Figure 4-1. Electrical shock is a condition that results any time a body becomes part of an electrical circuit.

To reduce the chance of electrical shock, it is important to always ensure that portable electric tools are in safe operating condition and contain a third wire on the plug for grounding. If electric power tools are grounded and an insulation breakdown occurs, the fault current should flow through the third wire to ground instead of through the body of the operator to ground.

NFPA 70E® identifies approach boundaries to protect against electrical shock. These boundaries are the limited approach boundary and restricted approach boundary. Each boundary is viewed as a sphere, extending 360° around an exposed energized conductor or circuit part. The size of each boundary is based on the phase-to-phase nominal voltage of the energized conductor or circuit part. **See Figure 4-2.**

The *limited approach boundary* is the distance from an exposed energized conductor or circuit part at which a person can get an electric shock and is the closest distance an unqualified person can approach. The *restricted approach boundary* is the distance from an exposed energized conductor or circuit part where an increased risk of electric shock exists due to the close proximity of the person to the energized conductor or circuit part.

During an electrical shock, the body of a person becomes part of an electrical circuit. The resistance the body of a person offers to the flow of current varies. Sweaty hands have less resistance than dry hands. A wet floor has less resistance than a dry floor. The lower the resistance, the greater the current flow. As the current flow increases, the severity of the electrical shock increases.

If a person is receiving an electrical shock, power should be removed as quickly as possible. If power cannot be removed quickly, the victim must be removed from contact with live parts. Action must be taken quickly and cautiously. Delay may be fatal. An individual must keep themselves from also becoming a casualty while attempting to rescue another person. If the equipment circuit disconnect switch is nearby and can be operated safely, the power is shut off. Excessive time should not be spent searching for the circuit disconnect. In order to remove the energized part, insulated protective equipment such as a hot stick, rubber gloves, blankets, wood poles, plastic pipes, etc., can be used if such items are accessible.

After the victim is free from the electrical hazard, help is called and first aid (CPR, etc.) begun as needed. The injured individual should not be transported unless there is no other option and the injuries require immediate professional attention.

Electrostatic Discharge

Electrostatic discharge (ESD) is the movement of electrons from a source to an object across a gap. *Static electricity* is an electrical charge at rest. The most common way to build up static electricity is by friction. Friction causes electrons from one source to flow to another material, causing an electron buildup (negative static charge). When a person (negatively charged) contacts a positively charged or a grounded object, all the excess electrons flow (jump) to that object. **See Figure 4-3.**

APPROACH BOUNDARIES TO ENERGIZED PARTS FOR SHOCK PREVENTION

Nominal System (Voltage, Range, Phase to Phase*)	Limited Approach Boundary		Restricted Approach Boundary (Allowing for Accidential Movement)
	Exposed Movable Conductor	Exposed Fixed-Circuit Part	
Less than 50	N/A	N/A	N/A
50 to 300	10'-0"	3'-6"	Avoid contact
301 to 750	10'-0"	3'-6"	1'-0"
751 to 15,000	10'-0"	5'-0"	2'-2"

* in V

Figure 4-2. The limited and restricted approach boundaries protect personnel from electric shock.

ELECTROSTATIC DISCHARGE

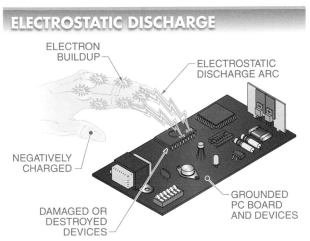

Figure 4-3. An excess of electrons produces an arc during electrostatic discharge (ESD).

Electrostatic discharge can be 35,000 V or more. People typically do not feel electrostatic discharges until the discharge reaches 3000 V. Solid-state devices and circuits may be damaged or destroyed by a 10 V electrostatic discharge. Electricians and maintenance personnel should wear a wrist grounding strap or other type of grounding device to avoid damage to solid-state devices and circuits from ESD.

Safety rules that are effective in preventing static damage include the following:
- Always keep work areas clean and clear of unnecessary materials, especially common plastics.
- Never handle electronic devices by their leads.
- Test grounding devices daily to make sure that they have not become loose or intermittent.
- Never work on ESD-sensitive objects without the proper grounding device.
- Always handle printed circuit (PC) boards by their outside corners.

- Always transport PC boards in antistatic trays or bags.
- Always keep single electronic devices sealed in conductive static shielding when transporting them.

Prevention of Electrostatic Discharge Damage. Certain devices are more susceptible to ESD damage than others. A PC board should be checked to see if it has an ESD label. A warning symbol can be used to identify ESD-sensitive devices. **See Figure 4-4.**

ELECTROSTATIC DISCHARGE LABELS

Figure 4-4. All ESD-sensitive components should be labeled so that they are handled properly.

When removing or replacing an ESD device or assembly in equipment, the device or assembly should be held with an electrostatic-free wrap, if possible. Otherwise, the device or assembly should be picked up by its body only. Component leads, connector pins, paths, PC boards, and any other electrical connections should not be touched, even though they are covered by a coating.

Safety Labels

A *safety label* is a label that indicates areas or tasks that can pose a hazard to personnel and/or equipment. Safety labels appear in several ways on equipment and in equipment manuals. Safety labels use signal words to communicate the severity of a potential problem. The three most common signal words are danger, warning, and caution. **See Figure 4-5.**

Danger Signal Word. A *danger signal word* is a word used to indicate an imminently hazardous situation which, if not avoided, results in death or serious injury. The information indicated by a danger signal word indicates the most extreme type of potential situation, and must be followed. The danger symbol is an exclamation mark enclosed in a triangle followed by the word "danger" written boldly in a red box.

Warning Signal Word. A *warning signal word* is a word used to indicate a potentially hazardous situation which, if not avoided, could result in death or serious injury. The information indicated by a warning signal word indicates a potentially hazardous situation and must be followed. The warning symbol is an exclamation mark enclosed in a triangle followed by the word "warning" written boldly in an orange box.

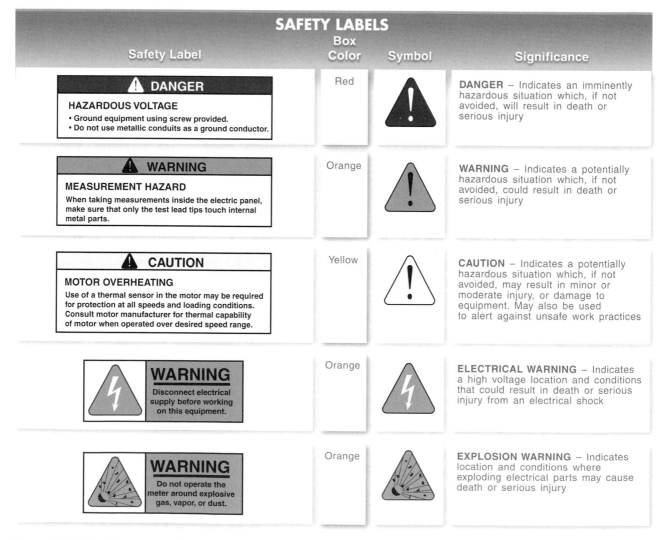

Figure 4-5. Safety labels are used to indicate a situation with different degrees of likelihood of death or injury to personnel.

Caution Signal Word. A *caution signal word* is a word used to indicate a potentially hazardous situation which, if not avoided, may result in minor or moderate injury. The information indicated by a caution signal word indicates a potential situation that may cause a problem to people and/or equipment. A caution signal word also warns of problems due to unsafe work practices. The caution symbol is an exclamation mark enclosed in a triangle followed by the word "caution" written boldly in a yellow box.

Other signal words may also appear with danger, warning, and caution signal words used by manufacturers. ANSI Z535.4, *Product Safety Signs and Labels,* provides additional information concerning safety labels. Additional signal words may be used alone or in combination on safety labels.

Electrical Warning Signal Word. *Electrical warning signal word* is a word used to indicate a high-voltage location and conditions that could result in death or serious personal injury from an electrical shock if proper precautions are not taken. An electrical warning safety label is usually placed where there is a potential for coming in contact with live electrical wires, terminals, or parts. The electrical warning symbol is a lightning bolt enclosed in a triangle. The safety label may be shown with no words or may be preceded by the word "warning" written boldly.

Explosion Warning Signal Word. *Explosion warning signal word* is a word used to indicate locations and conditions where exploding parts may cause death or serious personal injury if proper precautions and procedures are not followed. The explosion warning symbol is an explosion enclosed in a triangle. The safety label may be shown with no words or may be preceded by the word "warning" written boldly.

Electric Motor Safety

Two areas requiring attention when working with electric motors are the electrical circuit and rotating shaft. Basic electric motor safety rules include the following:
- Connect a motor to the correct grounding system.
- Ensure that guards or housings cover the rotating parts of a motor or anything connected to the motor. **See Figure 4-6.**
- Use the correct motor type for the location. For example, a DC or universal motor must never be used in a hazardous location that contains flammable materials because the sparking at the brushes can ignite the material.

- Connect a motor to the correct voltage and power source.
- Provide the motor with the correct overload and overcurrent protection to protect the motor when starting or shorted (overcurrent protection) and when running (overload protection).

MOTOR GUARDS

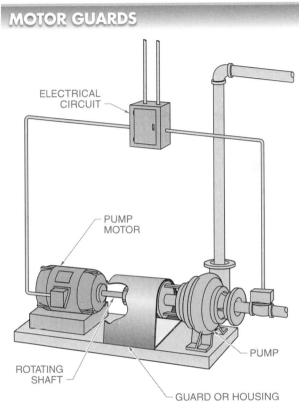

Figure 4-6. Motor guards or housings are used as protection from rotating parts of a motor or anything connected to the motor, such as the drive shaft.

4-1 CHECKPOINT

1. What are four factors that affect the severity of an electrical shock on a person?
2. When a safety sign or label indicates "ESD PROTECTED AREA," what does ESD stand for?
3. Which safety label indicates an imminently hazardous situation that, if not avoided, results in death or serious injury?
4. What two types of motor protection protect the motor when starting and running?

4-2 LOCKOUT/TAGOUT

Electrical power must be removed when electrical equipment is inspected, serviced, or repaired. To ensure the safety of personnel working with the equipment, power is removed and the equipment must be locked out and tagged out. *Lockout* is the process of removing the source of electrical power and installing a lock that prevents the power from being turned on. To ensure the safety of personnel working with equipment, all electrical power is removed and the equipment must be locked out and tagged out. Any stored energy in the pneumatic or hydraulic system must be released during lockout/tagout. *Tagout* is the process of placing a danger tag on the source of electrical power, which indicates that the equipment may not be operated until the danger tag is removed.

Per OSHA standards, equipment is locked out and tagged out before any installation or preventive maintenance is performed. **See Figure 4-7.** Lockout/tagout is used in the following situations:

- when power is not required to be ON for a piece of equipment
- when machine guards or other safety devices are removed or bypassed
- when the possibility exists of being injured or caught in moving machinery
- when jammed equipment is being cleared
- when danger exists of being injured if equipment power is turned on

A danger tag has the same importance and purpose as a lock and is used alone only when a lock does not fit the disconnect device. A danger tag shall be attached at the disconnect device with a tag tie or equivalent and shall have space for the technician's name, craft, and other company-required information. A danger tag must withstand the elements and expected atmosphere for the maximum period of time that exposure is expected.

Lockout and tagouts do not by themselves remove power from a machine or its circuitry. OSHA provides a standard procedure for equipment lockout/tagout. Lockout is performed and tagouts are attached only after the equipment is turned off and tested. OSHA provides the following procedure:

1. Prepare for machinery shutdown.
2. Shut down machinery or equipment.
3. Isolate machinery or equipment.
4. Apply lockout or tagout.
5. Release stored energy.
6. Verify isolation.

Panduit Corp.

Figure 4-7. Equipment must be locked out and/or tagged out before installation, preventive maintenance, or servicing is performed.

WARNING: Personnel should consult OSHA 29 CFR 1910.147—*The control of hazardous energy (lockout/tagout)* for industry standards on lockout/tagout.

A lockout/tagout must not be removed by any person other than the authorized person who installed the lockout/tagout, except in an emergency. In an emergency, only supervisory personnel may remove a lockout/tagout, and only upon notification of the authorized person. A list of company rules and procedures is given to authorized personnel and any person who may be affected by a lockout/tagout. The procedures to take when using a lockout/tagout often include the following general steps:

- Use a lockout and tagout when possible.
- Use a tagout when a lockout is impractical. A tagout is used alone only when a lock does not fit the disconnect device.
- Use a multiple lockout when individual employee lockout of equipment is impractical.
- Notify all employees affected before using a lockout/tagout.
- Remove all power sources including primary and secondary.
- Use a DMM set to measure voltage to ensure that the power is OFF.

When more than one technician is required to perform a task on a piece of equipment, each technician shall place a lockout and/or tagout on the energy-isolating device(s). A multiple lockout/tagout device (hasp) must be used because energy-isolating devices typically cannot accept more than one lockout/tagout. A *hasp* is a multiple lockout/tagout device.

Lockout Devices

Lockout devices are lightweight enclosures that allow the lockout of standard control devices. Lockout devices are available in various shapes and sizes that allow for the lockout of ball valves, gate valves, and electrical equipment such as plugs, disconnects, etc.

Lockout devices resist chemicals, cracking, abrasion, and temperature changes. They are available in colors to match ANSI pipe colors. Lockout devices are sized to fit standard size industry control devices. **See Figure 4-8.**

Locks used to lock out a device may be color-coded and individually keyed. The locks are rust-resistant and are available with various size shackles.

Danger tags provide additional lockout and warning information. Various danger tags are available. Danger tags may include warnings such as "Do Not Start" or "Do Not Operate," or they may provide space to enter worker name, date, and reason for lockout. Tag ties must be strong enough to prevent accidental removal and must be self-locking and nonreusable.

Lockout/tagout kits are also available. A lockout/tagout kit contains items required to comply with OSHA lockout/tagout standards. Lockout/tagout kits contain reusable danger tags, multiple lockouts, locks, magnetic signs, and information on lockout/tagout procedures. **See Figure 4-9.** A lockout/tagout should be checked to ensure power is removed when returning to work after leaving a job for any reason or when a job cannot be completed in the same day.

LOCKOUT DEVICES

MULTIPLE LOCKOUT/TAGOUT DEVICE

Panduit Corp.

Figure 4-8. Lockout devices are available in various shapes and sizes that allow for the lockout of standard control devices.

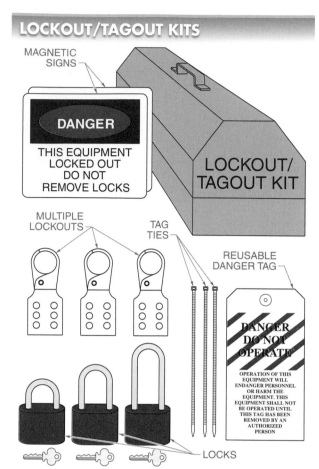

Figure 4-9. Lockout/tagout kits comply with OSHA lockout/tagout standards.

Restoring Equipment to Service. After servicing and/or maintenance work is completed on locked out or tagged out equipment and the equipment is ready to resume normal operation, the following steps must be taken before the lockout/tagout devices are removed:

- Ensure that all tools and nonessential items have been removed from the equipment and that all machine guards, components, etc., have been properly reinstalled.

- Perform a thorough visual check of the area around the equipment to ensure that all individuals are safely positioned or removed from the area and from equipment, circuits, etc., that are about to be reenergized.

- Notify all affected individuals in the area that lockout and/or tagout devices will be removed and the time frame for removal.

- Ensure that only the authorized individual who applied the lockout/tagout removes locks and/or tags from each energy-isolating device.

Note: If the authorized individual who applied the lock and/or tag is unavailable, only the supervisor of the individual may remove the lockout and/or tagout devices after necessity for removal has been positively established and all of the following conditions have been met:

- A removal of lockout device form has been completed.

- Verification has been made that the individual who applied the lockout and/or tagout device(s) is not at the facility or location.

- All reasonable efforts have been made to contact the individual who applied the lockout/tagout to inform the individual that their lockout and/or tagout device(s) will be removed.

- The individual's direct supervisor is certain that removal of the lock and/or tag will not endanger anyone.

- Prior to resuming work within the facility or location, the individual that placed the lockout/tagout device shall be notified that his/her lock and/or tag has been removed in the individual's absence.

NFPA 70E®

The National Fire Protection Association (NFPA) standard NFPA 70E®, Standard for Electrical Safety in the Workplace, addresses "work practices that are necessary to provide a practical safe workplace relative to the hazards associated with electrical energy." Per NFPA 70E®, "Only qualified persons shall perform tasks such as testing, troubleshooting, and voltage measuring within the limited approach boundary of energized electrical conductors or circuit parts operating at 50 volts or more or where an electrical hazard exists."

NFPA 70E® was written at the request of OSHA and has become the standard for electrical safety in the electrical industry. Its methods for protection are more detailed than OSHA requirements. NFPA 70E® is the basis for the OSHA 29 CFR 1910 Subpart S—*Electrical*. For technicians, NFPA 70E® addresses requirements such as personal protective equipment (PPE) and safe approach distance requirements that could be encountered in jobs such as installing temporary power. Safety rules that apply to commercial and industrial installations include the following:

- Always comply with the NEC®, state, and local codes.

- Use UL®-listed equipment, components, and test equipment.

- Before removing any fuse from a circuit, be sure the switch for the circuit is open or disconnected and properly verify that the circuit is de-energized. Never remove fuses from an energized circuit.
- Always use safety equipment.
- Perform the appropriate task required during an emergency situation.
- Be trained in cardiopulmonary resuscitation (CPR) and emergency rescue procedures.
- Always work with another individual when working in a dangerous area, on dangerous equipment, or with high voltages.
- Do not work when tired or taking medication that causes drowsiness unless specifically authorized by a physician.
- Do not reach blindly into areas that may contain exposed energized electrical conductors or circuit parts.
- Do not work in poorly lighted areas.
- Ensure there are no atmospheric hazards such as flammable dust or vapor in the area.
- Use one hand when working on a live circuit to reduce the chance of an electrical shock passing through the heart and lungs.
- Never bypass fuses, circuit breakers, or any other safety device.

Qualified Persons

To prevent an accident, electrical shock, or damage to equipment, all electrical work must be performed by qualified persons. A *qualified person* is a person who is trained and has special knowledge of the construction and operation of electrical equipment or a specific task and is trained to recognize and avoid electrical hazards that might be present with respect to the equipment or specific task. NFPA 70E® Section 110.2(D)(1), *Qualified Person,* provides additional information regarding the definition of a qualified person. A qualified person does the following:

- determines the voltage of energized electrical parts
- determines the degree and extent of hazards and uses the proper personal protective equipment and job planning to perform work safely on electrical equipment by following all NFPA, OSHA, equipment manufacturer, state, and company safety procedures and practices
- performs the appropriate task required during an accident or emergency situation
- understands electrical principles and follows all manufacturer procedures and approach distances specified by the NFPA
- understands the operation of test equipment and follows all manufacturer procedures
- informs other technicians and operators of tasks being performed and maintains all required records

4-2 CHECKPOINT

1. Except in an emergency, who is allowed to remove a lockout and/or tagout device?
2. At what voltage level do NFPA 70E® requirements apply to energized electrical conductors or circuit parts?

4-3 PERSONAL PROTECTIVE EQUIPMENT

Personal protective equipment (PPE) is clothing and/or equipment worn by a technician to reduce the possibility of injury in the work area. The use of PPE is required whenever work may occur on or near energized exposed electrical circuits. For maximum safety, PPE must be used as specified in NFPA 70E®, OSHA 29 CFR 1910 Subpart I—*Personal Protective Equipment (1910.132 through 1910.138),* and other applicable safety mandates.

All PPE and tools are selected for at least the operating voltage of the equipment or circuits to be worked on or near. Equipment, devices, tools, or test equipment must be suited for the work to be performed. PPE includes protective clothing, head protection, eye protection, ear protection, hand protection, foot protection, back protection, knee protection, and rubber insulated matting. **See Figure 4-10.**

PERSONAL PROTECTIVE EQUIPMENT

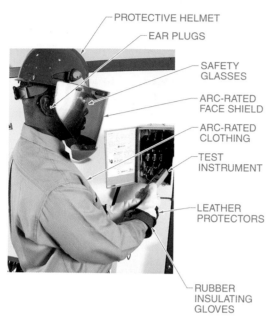

PROTECTIVE HELMET
EAR PLUGS
SAFETY GLASSES
ARC-RATED FACE SHIELD
ARC-RATED CLOTHING
TEST INSTRUMENT
LEATHER PROTECTORS
RUBBER INSULATING GLOVES

Figure 4-10. Personal protective equipment is used to reduce the possibility of an injury.

Protective Clothing

Protective clothing is clothing that provides protection from contact with sharp objects, hot equipment, and harmful materials. Protective clothing made of durable material such as denim should be snug, yet allow ample movement. Clothing should fit snugly to avoid danger of becoming entangled in moving machinery. Pockets should allow convenient access but should not snag on tools or equipment. Soiled protective clothing should be washed to reduce the flammability hazard.

Arc flash and arc blast are electrical hazards that may occur while working on electrical equipment. Approved arc-rated clothing must be worn for protection from arc flash and arc blast when performing certain operations on or near energized equipment or circuits. Arc-rated clothing must be kept as clean and sanitary as practical and must be inspected prior to each use. Defective clothing must be removed from service immediately and replaced. Defective arc-rated clothing must be tagged "unsafe" and returned to a supervisor.

Arc-rated clothing must be used when working with live high-voltage electrical circuits. Arc-rated clothing is made of materials such as Nomex®, Basofil®, and/or Kevlar® fibers. The arc-rated fibers can be coated with PVC to offer weather resistance and to increase arc rating. Arc-rated clothing must meet the three following requirements:

- Clothing must not ignite and continue to burn.
- Clothing must provide an insulating value to dissipate heat throughout the clothing and away from the skin.
- Clothing must provide resistance to the break-open forces generated by the shock wave of an arc.

The NFPA specifies boundary distances where arc protection is required. All personnel working within specified boundary distances require arc-rated clothing and equipment. Boundary distances vary depending on the voltage involved.

Arc Flash and Arc Blast

An *electric arc* is a discharge of electric current across an air gap. Arcs are caused by excessive voltage ionizing an air gap between two conductors or by accidental contact between two conductors and followed by reseparation. When an electric arc occurs, there is the possibility of an arc flash or an arc blast. An *arc flash* is an extremely high-temperature discharge produced by an electrical fault in the air. Arc flash temperatures reach 35,000°F. An *arc blast* is an explosion that occurs when the air surrounding electrical equipment becomes ionized and conductive. The threat of arc blast is greatest from electrical systems of 480 V and higher. Arc blasts are possible in systems of lesser voltage, but are not likely to be as destructive as in a high-voltage system.

Arc flash and arc blast are always a possibility when working with electrical equipment. Arc flash can occur when using a voltmeter or DMM to measure voltage in a 480 V or higher electrical system when a power line transient occurs, such as a lightning strike or power surge. A potential cause for arc flash and arc blast is improper test instrument and meter use. For example, an arc blast can occur by connecting an in-line ammeter across two points of a circuit that is energized with a voltage higher than the rating of the meter. To prevent an arc blast or arc flash, an electrical system needs to be de-energized, locked out, and tagged out prior to performing work. Only qualified electricians are allowed to work on energized circuits of 50 V or higher.

Arc Flash Boundary. The *arc flash boundary* is the distance from exposed energized conductors or circuit parts where bare skin would receive the onset of a second-degree burn. The arc flash boundary is dependent on the available short-circuit current, maximum total clearing time of the OCPD, the voltage of the circuit, and a standard factor that varies if the actual short-circuit current is known or not known. The arc flash boundary can be calculated using the equations in NFPA 70E® Informative Annex D.

Head Protection

Head protection requires using a protective helmet. A *protective helmet* is a hard hat that is used in the workplace to prevent injury from the impact of falling and flying objects and from electrical shock. Protective helmets resist penetration and absorb impact force. Protective helmet shells are made of durable, lightweight materials. A shock-absorbing lining keeps the shell away from the head to provide ventilation. Protective helmets are identified by class of protection against specific hazardous conditions.

Class G, E, and C helmets are used for construction and industrial applications. Class G (general) protective helmets protect against impact and voltage up to 2200 V and are commonly used in construction and manufacturing facilities. Class E (electrical) protective helmets protect against impact and voltage up to 20,000 V. Class C (conductive) protective helmets are manufactured with lighter materials and provide adequate impact protection, but provide no electrical protection.

Eye Protection

Eye protection must be worn to prevent eye or face injuries caused by flying particles, contact arcing, and radiant energy. Eye protection must comply with OSHA 29 CFR 1910.133—*Eye and face protection*. Eye protection standards are specified in ANSI Z87.1, *Occupational and Educational Eye and Face Protection Devices*. Eye protection includes safety glasses, face shields, goggles, and arc-rated hoods. **See Figure 4-11.**

Safety glasses are an eye protection device with special impact-resistant glass or plastic lenses, reinforced frames, and side shields. Plastic frames are designed to keep the lenses secured in the frame if an impact occurs and minimize the shock hazard when working with electrical equipment. Side shields provide additional protection from flying objects. Tinted-lens safety glasses protect against low-voltage arc hazards.

A *face shield* is an eye and face protection device that covers the entire face with a plastic shield and is used for protection from flying objects. An arc-rated face shield is tinted and protects against low-voltage arc hazards. *Goggles* are an eye protection device with a flexible frame that is secured on the face with an elastic headband. Goggles fit snugly against the face to seal the areas around the eyes and may be used over prescription glasses. Goggles with clear lenses protect against small flying particles or splashing liquids. Tinted goggles are used to protect against low-voltage arc hazards. An *arc-rated hood* is an eye and face protection device that covers the entire head and is used for protection from arc blast and arc flash.

Figure 4-11. Eye protection must be worn to prevent eye or face injuries caused by flying particles, contact arcing, or radiant energy.

Safety glasses, face shields, goggle lenses, and arc-rated hoods must be properly maintained to provide protection and clear visibility. Lens cleaners are available that clean without risk of lens damage. Pitted or scratched lenses reduce vision and may cause lenses to fail on impact.

Ear Protection

Ear protection is any device worn to limit the noise entering the ear and includes earplugs and earmuffs. An *earplug* is an ear protection device made of moldable rubber, foam, or plastic and inserted into the ear canal. An *earmuff* is an ear protection device worn over the ears. A tight seal around an earmuff is required for proper protection.

Power tools and equipment can produce excessive noise levels. Technicians subjected to excessive noise levels may develop hearing loss over a period of time. The severity of hearing loss depends on the intensity and duration of exposure. Noise intensity is expressed in decibels. A *decibel (dB)* is a unit of measure used to express the relative intensity of sound.

Ear protection devices are assigned a noise reduction rating (NRR) number based on the noise level reduced. For example, an NRR of 27 means that the noise level is reduced by 27 dB when tested at the factory. To determine approximate noise reduction in the field, 7 dB is subtracted from the NRR. For example, an NRR of 27 provides a noise reduction of approximately 20 dB in the field.

Hand Protection

Hand protection includes gloves worn to prevent injuries to hands caused by cuts or electrical shock. The appropriate hand protection required is determined by the duration, frequency, and degree of the hazard to the hands. *Rubber insulating gloves* are gloves made of latex rubber and are used to provide maximum insulation from electrical shock. Rubber insulating gloves are stamped with a working voltage range such as 500 V – 26,500 V. *Leather protectors* are gloves worn over rubber insulating gloves to prevent penetration of the rubber insulating gloves and provide added protection against electrical shock. Safety procedures for the use of rubber insulating gloves and leather protectors must be followed at all times. **See Figure 4-12.**

The primary purpose of rubber insulating gloves and leather protectors is to insulate the hands and lower arms from possible contact with live conductors. Rubber insulating gloves offer a high resistance to current flow to help prevent an electrical shock. Leather protectors help protect rubber insulating gloves and add additional insulation.

HAND PROTECTION

Rubber Insulating Glove Classes		
Class	Maximum Use Voltage*	Label Color
00	500	Beige
0	1000	Red
1	7500	White
2	17,000	Yellow
3	26,500	Green
4	36,500	Orange

* in V

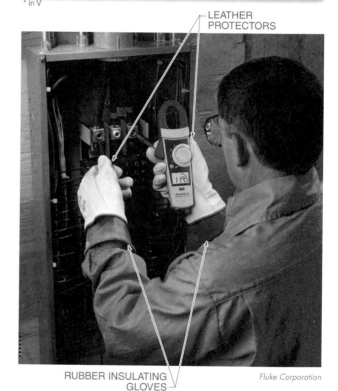

LEATHER PROTECTORS

RUBBER INSULATING GLOVES

Fluke Corporation

Figure 4-12. Hand protection includes gloves worn to prevent injuries to hands caused by cuts or electrical shock.

WARNING: Rubber insulating gloves are designed for specific applications. Leather protectors are required for protecting rubber insulating gloves and should not be used alone. Rubber insulating gloves offer the highest resistance and greatest insulation. Serious injury or death can result from improper use or using outdated and/or the wrong type of gloves for the application.

The proper care of leather protectors is essential to user safety. Leather protectors should be inspected when inspecting rubber insulating gloves. Metal particles or any substance that could physically damage rubber insulating gloves must be removed from a leather protector before it is used.

The entire surface of rubber insulating gloves must be field tested (visual inspection and air test) before each use. In addition, rubber insulating gloves should also be laboratory tested by an approved laboratory every six months. Visual inspection of rubber insulating gloves is performed by stretching a small area (particularly fingertips) and checking for defects such as punctures or pin holes, embedded or foreign material, deep scratches or cracks, cuts or snags, or deterioration caused by oil, heat, grease, insulating compounds, or any other substance that may harm rubber.

Rubber insulating gloves must also be air tested when there is cause to suspect damage. The entire surface of the glove must be inspected by rolling the cuff tightly toward the palm in such a manner that air is trapped inside the glove or by using a mechanical inflation device. When using a mechanical inflation device, care must be taken to avoid overinflation. The glove is examined for punctures and other defects. Puncture detection may be enhanced by listening for escaping air by holding the glove to the face or ear to detect escaping air. Gloves failing the air test should be tagged "unsafe" and returned to a supervisor.

Proper care of leather protectors is essential for user safety. Leather protectors are checked for cuts, tears, holes, abrasions, defective or worn stitching, oil contamination, and any other condition that might prevent them from adequately protecting rubber insulating gloves. Any substance that could physically damage rubber insulating gloves must be removed before use. Rubber insulating gloves or leather protectors found to be defective shall not be discarded or destroyed in the field, but shall be tagged "unsafe" and returned to a supervisor.

WARNING: It is the rubber insulating part of the glove that offers electrical shock protection and not the leather protector part. The leather protector can conduct electricity when wet/moist or dirty. The leather protective part of the glove must never be allowed higher than the rubber part on the arm because electricity can travel up the leather and into the arm.

Foot Protection

Foot protection is shoes worn to prevent foot injuries that are typically caused by objects falling less than 4′ and having an average weight of less than 65 lb. Safety shoes with reinforced steel toes protect against injuries caused by compression and impact. Insulated rubber-soled shoes are commonly worn during electrical work to prevent electrical shock. Protective footwear must comply with ASTM F2413, *Standard Specification for Performance Requirements for Protective (Safety) Toe Cap Footwear.*

Thick-soled work shoes may be worn for protection against sharp objects such as nails. Rubber boots may be used when working in damp locations.

Back Protection

A back injury is one of the most common types of injuries resulting in lost time in the workplace. Back injuries are the result of improper lifting procedures. Back injuries are prevented through proper planning and work procedures. Assistance should be sought when moving heavy objects. When lifting objects from the ground, ensure the path is clear of obstacles and free of hazards. When lifting objects, the knees are bent and the object is grasped firmly. The object is lifted by straightening the legs and keeping the back as straight as possible. Keep the load close to the body and keep the load steady.

Long objects such as conduit may not be heavy, but the weight might not be balanced. Long objects should be carried by two or more people whenever possible. When carried on the shoulder by one person, long objects should be transported with the front end pointing downward to minimize the possibility of injury to others when walking around corners or through doorways.

Knee Protection

A *knee pad* is a rubber, leather, or plastic pad strapped onto the knees for protection. Knee pads are worn by technicians who spend considerable time working on their knees or who work in close areas and must kneel for proper access to equipment. Knee pads are secured by buckle straps or Velcro® closures. **See Figure 4-13.**

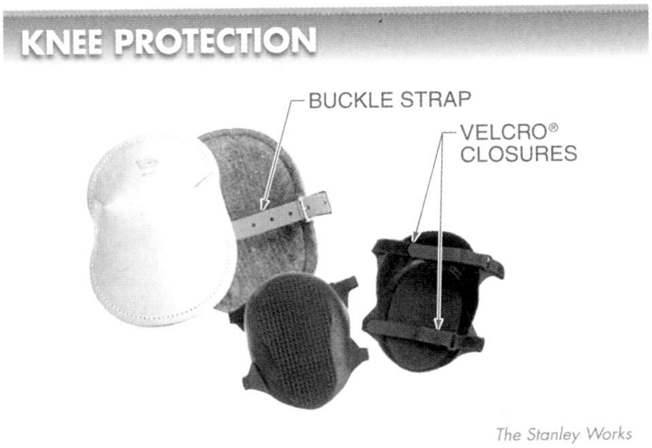

KNEE PROTECTION

BUCKLE STRAP

VELCRO® CLOSURES

The Stanley Works

Figure 4-13. Knee pads are used to provide protection and comfort to technicians who spend considerable time on their knees.

Rubber Insulating Matting

Rubber insulating matting is a floor covering that provides technicians protection from electrical shock when working on live electrical circuits. Dielectric black fluted rubber matting is specifically designed for use in front of open cabinets or high-voltage equipment. Matting is used to protect technicians when voltages are over 50 V. Two types of matting that differ in chemical and physical characteristics are designated as Type I natural rubber and Type II elastomeric compound matting. **See Figure 4-14.**

National Electrical Code®

The National Electrical Code® (NEC®) is one of the most widely used and recognized consensus standards in the world. The purpose of the NEC® is to protect people and property from hazards that arise from the use of electricity. Improper procedures when working with electricity can cause permanent injury or death. Many city, county, state, and federal agencies use the NEC® to set requirements for electrical installations. Article 430 of the NEC® covers requirements for motors, motor circuits, and controllers. **See Figure 4-15.** The NEC® is updated every three years. Electrical safety rules include the following:

- Always comply with the NEC®, state, and local codes.
- Use UL®-approved equipment, components, and test equipment.
- Before removing any fuse from a circuit, be sure the switch for the circuit is open or disconnected.

When removing fuses, use an approved fuse puller and break contact on the line side of the circuit first. When installing fuses, install the fuse first into the load side of the fuse clip, and then into the line side.

- Inspect and test grounding systems for proper operation. Ground any conductive component or element that is not energized.
- Turn off, lock out, and tag out any circuit that is not required to be energized when maintenance is being performed.
- Always use personal protective equipment and safety equipment.
- Perform the appropriate task required during an emergency situation.
- Use only a Class C-rated fire extinguisher on electrical equipment. A Class C fire extinguisher is identified by the color blue inside a circle.
- Always work with another individual when working in a dangerous area, on dangerous equipment, or with high voltages.
- Do not work when tired or taking medication that causes drowsiness unless specifically authorized by a physician.
- Do not work in poorly lighted areas.
- Ensure there are no atmospheric hazards such as flammable dust or vapor in the area.
- Use one hand when working on a live circuit to reduce the chance of an electrical shock passing through the heart and lungs.
- Never bypass fuses, circuit breakers, or any other safety device.

RUBBER INSULATING MATTING RATINGS

Safety Standards	Material Thickness		Material Width (in.)	Test Voltage	Maximum Working Voltage
	Inches	Millimeters			
BS921*	0.236	6	36	11,000	450
BS921*	0.236	6	48	11,000	450
BS921*	0.354	9	36	15,000	650
BS921*	0.354	9	48	15,000	650
VDE0680†	0.118	3	39	15,000	1000
ASTM D178‡	0.236	6	24	25,000	17,000
ASTM D178‡	0.236	6	30	25,000	17,000
ASTM D178‡	0.236	6	36	25,000	17,000
ASTM D178‡	0.236	6	48	25,000	17,000

* BSI— British Standards Institute
† VDE— Verband Deutscher Elektrotechniker Testing and Certification Institute
‡ ASTM International

Figure 4-14. Rubber insulating matting provides protection from electrical shock when work is performed on live electrical circuits.

MOTORS—NEC® ARTICLE 430

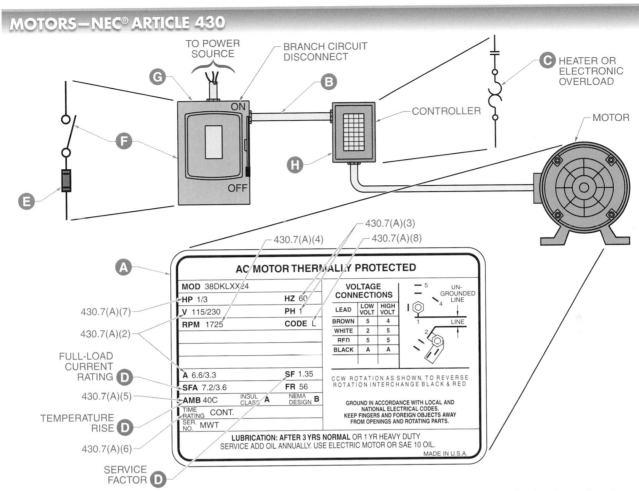

A. **430.7** Motors shall be marked with specific information.

B. **430.22** The ampacity of branch-circuit conductors shall be not less than 125% of the motor current rating.

C. **430.32(A)(1)** Continuous-duty motors over 1 HP shall be protected against overload by a separate overload device rated at not more than 125% of the full-load current rating for motors with a service factor not less than 1.15 or a temperature rise not over 40°C, and at not over 115% for all other motors.

430.32(D)(2)(a) Continuous-duty motors of 1 HP or less which are not permanently installed and are manually started and within sight from the controller are permitted to be protected by the branch-circuit, short-circuit, and ground-fault protective device.

430.32(D)(2)(a), Ex. Any motor in 430.32(B) is permitted on a 20 A, 120 V branch circuit.

430.32(B)(1) Continuous-duty motors of 1 HP or less which are automatically started shall be protected against overload by a separate overload device rated at not more than 125% of the full-load current rating for motors with a service factor not less than 1.15 or a temperature rise not over 40°C, and at not over 115% for all other motors.

D. **430.33** Any motor applications shall be considered as continuous duty unless the driven apparatus is such that the motor cannot operate continuously.

430.32 (C) Where the overcurrent relay selected does not allow the motor to start, the next higher size overload relay is selected provided the trip current does not exceed 140% of the motor's full-load current rating for motors with a service factor of not less than 1.15 or a temperature rise not over 40°C, and 130% for all other motors.

430.35(A) A running overcurrent device may be shunted at starting of a manual-start motor if no hazard is introduced and the branch-circuit device of not over 400% is operative in the circuit during the starting period.

430.35(b) Shunting is not permitted if the motor is automatically started. See Exception.

430.40 Thermal cutouts and overload relays for motor-running protection not capable of opening short circuits shall be protected per 430.52 unless approved for group installation and marked with the maximum size required protection.

E. **430.52** The motor branch-circuit overcurrent device shall be able to carry the starting current (150% – 300% per Table 430.52; absolute maximum 400% with NTDFs and Class CC fuses, and 225% with TDF's).

430.102(A) A disconnecting means shall be in sight from the controller location and shall disconnect the controller. See ex. 1 and 2.

F. **430.102(B)** A disconnecting means shall be in sight from the motor location and the driven machinery location. See Exception.

430.107 One of the disconnecting means shall be readily accessible.

430.108 All disconnecting means shall comply with 430.109 and 430.110.

G. **430.109** The disconnecting means shall be a type specified in 430.109(A) unless otherwise permitted in 430.109(B) through 430.109(G), under the conditions specified.

430.110(A) All disconnecting means shall have an ampere rating of at least 115% of the motor's FLC, taken from FLC tables per 430.6(A).

H. **430.111** A suitable switch or CB may serve as both the disconnecting means and the controller.

Figure 4-15. Article 430 of the NEC® covers requirements for motors, motor circuits, and controllers.

Grounding. *Grounding* is the connection of all exposed non-current-carrying metal parts to the earth. Grounding provides a direct low-resistance path to the earth. Electrical circuits are grounded to safeguard equipment and personnel against the hazards of electrical shock. Proper grounding of electrical tools, motors, equipment, enclosures, and other control circuitry helps prevent hazardous conditions. Conversely, improper electrical wiring or misuse of electricity causes destruction of equipment and fire damage to property as well as personal injury.

Grounding is accomplished by connecting the circuit to a metal underground water pipe, the metal frame of a building, a concrete-encased electrode, or a ground ring in accordance with the NEC®. To prevent problems, a grounding path must be as short as possible and of sufficient size as recommended by the manufacturer (minimum 14 AWG copper), never be fused or switched, be a permanent part of the electrical circuit, and be continuous and uninterrupted from the electrical circuit to the ground.

A ground is provided at the main service equipment or at the source of a separately derived system (SDS). A *separately derived system (SDS)* is a system that supplies electrical power derived (taken) from transformers, storage batteries, solar photovoltaic systems, or generators. The majority of separately derived systems are produced by the secondary of the distribution transformer.

The neutral ground connection must be made at the transformer or at the main service panel only. The neutral ground connection is made by connecting the neutral bus to the ground bus with a main bonding jumper. A *main bonding jumper (MBJ)* is a connection at the service equipment that connects the equipment grounding conductor, the grounding electrode conductor, and the grounded conductor (neutral conductor). **See Figure 4-16.**

WARNING: Neutral-to-ground connections must not be made in any subpanels, receptacles, or equipment. If a neutral-to-ground connection is made, a parallel path for the normal return current from loads is created and current will flow through the ground system, creating a dangerous shock potential situation.

An *equipment grounding conductor (EGC)* is an electrical conductor that provides a low-impedance ground path between electrical equipment and enclosures within the distribution system. A *grounding electrode conductor (GEC)* is a conductor that connects grounded parts of a power distribution system (equipment grounding conductors, grounded conductors, and all metal parts) to the NEC®-approved earth grounding system. A *grounded conductor* is a conductor that has been intentionally grounded.

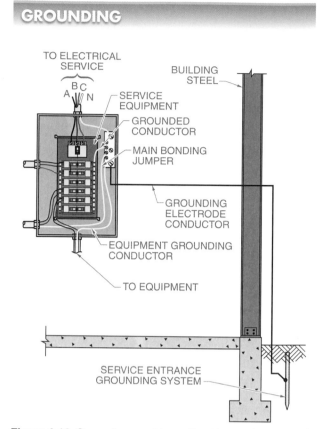

GROUNDING

Figure 4-16. Grounding provides a direct low-resistance path to earth in order to limit the voltage to ground.

Ground Fault Circuit Interrupters. A *ground fault circuit interrupter (GFCI)* is a device that protects against electrical shock by detecting an imbalance of current in the normal conductor pathways and opening the circuit. When current in the two conductors of an electrical circuit varies by more than 5 mA, a GFCI opens the circuit. A GFCI is rated to trip quickly enough (1/40 of a second) to prevent electrocution.

A potentially dangerous ground fault is any amount of current above the level that may deliver a dangerous shock. Any current over 8 mA is considered potentially dangerous depending on the path the current takes, the physical condition of the person receiving the shock, and the amount of time the person is exposed to the shock. Therefore, GFCIs are required in such places as dwellings, hotels, motels, construction sites, marinas, receptacles near swimming pools and hot tubs, underwater lighting, fountains, and other areas where a person may experience a ground fault.

A GFCI compares the amount of current in the ungrounded (hot) conductor with the amount of current in the neutral conductor. **See Figure 4-17.** If the current in the neutral conductor becomes less than the current in the hot conductor, a ground fault condition exists. The amount of current that is missing is returned to the source by some path other than the intended path (fault current).

GFCI protection may be installed at different locations within a circuit. Direct-wired GFCI receptacles provide ground fault protection at the point of installation. GFCI receptacles may also be connected to provide protection at all other receptacles installed downstream on the same circuit. GFCI circuit breakers, when installed in a load center or panelboard, provide GFCI protection and conventional circuit overcurrent protection for all branch-circuit components connected to the circuit breaker.

Plugin GFCIs provide ground fault protection for devices plugged into them. These plugin devices are often used by personnel working with power tools in an area that does not include GFCI receptacles.

Portable GFCIs are designed to be easily moved from one location to another. Portable GFCIs commonly contain more than one receptacle outlet protected by an electronic circuit module. Portable GFCIs should be inspected and tested before each use. GFCIs have a built-in test circuit to ensure that the ground fault protection is operational.

A GFCI protects against the most common form of electrical shock hazard, the ground fault. A GFCI does not protect against line-to-line contact hazards, such as a technician holding two hot wires or a hot and a neutral wire in each hand. GFCI protection is required in addition to NFPA grounding requirements.

WARNING: GFCIs have terminals marked LINE and LOAD. The incoming power lines to the GFCI should always be connected to the LINE terminals. The LOAD terminals are used for connecting to the next receptacle down the line.

4-3 CHECKPOINT

1. What three requirements must arc-rated clothing meet?
2. Building and equipment grounding systems must have a resistance of less than how many ohms?
3. How much current imbalance (fault current) does it take to trip a GFCI?

GROUND FAULT CIRCUIT INTERRUPTERS (GFCIs)

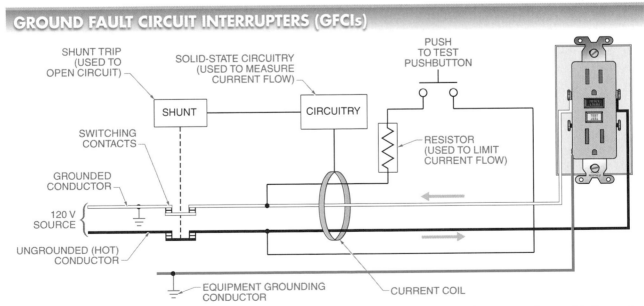

Figure 4-17. A GFCI compares the amount of current in the ungrounded (hot) conductor with the amount of current in the neutral conductor.

4-4 FIRE SAFETY

Fire safety requires established procedures to reduce or eliminate conditions that could cause a fire. Guidelines in assessing hazards of the products of combustion are provided by the NFPA. Prevention is the best strategy to protect against potential fire hazards. Technicians must take responsibility in preventing conditions that could result in a fire. This includes proper use and storage of lubricants, oily rags, and solvents, and immediate cleanup of combustible spills.

The chance of fire is greatly reduced by good housekeeping. Rags containing oil, gasoline, alcohol, shellac, paint, varnish, lacquer, or other solvents may spontaneously combust and should be kept in a covered metal container. A self-closing steel container specially designed for the disposal of rags containing oil, grease, and flammable liquids is recommended. For example, an oily waste can seals out oxygen to prevent spontaneous combustion. **See Figure 4-18.** To reduce the possibility of a fire, debris must be kept in a designated area away from the building.

OILY WASTE CAN

LID REMAINS
CLOSED
BY GRAVITY

EMPTY EVERY NIGHT

FOOT PEDAL
OPENS LID

Justrite Manufacturing Company

VENTILATION HOLES
AT BOTTOM ALLOW
AIR CIRCULATION

Figure 4-18. An oily waste can seals out oxygen to prevent spontaneous combustion.

In the event of a fire, a technician must act quickly to minimize injury and damage. An alarm is sounded if a fire occurs, all workers are alerted, and the fire department is called. Before starting any work, all individuals should be advised of the location of the nearest telephone and fire alarm reporting station for summoning emergency medical assistance. The telephone shall be reasonably close to the workplace, readily accessible, and functional throughout the work period. When the nearest telephone does not satisfy these requirements, two-way radios or some other positive means of rapid communication must be employed. Cellular telephones can be used only if they are checked to make sure they are operational in the area and approved by the supervisor. Cellular telephones are prohibited by law in an explosionproof environment. A procedure to evacuate the premises and account for all personnel after the fire department is called should be in place and practiced on a regular basis.

All facilities must have a fire safety plan. A fire safety plan establishes procedures that must be followed if a fire occurs. The fire safety plan lists the locations of the main electrical breaker, fire main, exits, fire alarms, and fire extinguishers for each area of a facility.

Classes of Fire

The five classes of fires are Class A, Class B, Class C, Class D, and Class K. Class A fires include burning wood, paper, textiles, and other ordinary combustible materials containing carbon. Class B fires include burning oil, gas, grease, paint, and other liquids that convert to a gas when heated. Class C fires include burning electrical devices, motors, and transformers. Class D is a specialized class of fires that includes burning metals such as zirconium, titanium, magnesium, sodium, and potassium. Class K fires include grease in commercial cooking equipment. Fire extinguishers are selected for the class of fire based on the combustibility of the material. **See Figure 4-19.**

Fuel, heat, and oxygen are required to start and sustain a fire. A fire goes out when any one of the three is taken away. Fire extinguishing equipment does not take the place of plant fire protection personnel or the local fire department. Proper authorities must be notified whenever there is a fire in the plant. Technicians must know the locations of all fire extinguishing equipment in a facility and be ready to direct firefighters to the location of the fire. In addition, technicians should be able to inform firefighters of any special problems or conditions that exist, such as downed electrical wires, leaks in gas lines, locations of gasoline or propane tanks, and locations of flammable materials.

FIRE EXTINGUISHER CLASSES

TRASH • WOOD • PAPER

BOXES

Ⓐ ORDINARY COMBUSTIBLES

LIQUIDS • GREASE

SOLVENT CEMENT

Ⓑ FLAMMABLE LIQUIDS

MOTORS • TRANSFORMERS

ELECTRICAL MOTOR

Ⓒ ELECTRICAL EQUIPMENT

ZIRCONIUM • TITANIUM

METAL

Ⓓ COMBUSTIBLE METALS

GREASE

DEEP FAT FRYER

K—COMMERCIAL COOKING GREASE

Figure 4-19. Fire extinguisher classes are based on the combustibility of the material.

Fire extinguishing equipment, such as fire extinguishers, water hoses, and sand buckets, must be routinely checked according to plant procedures. The instructions for use should be read before using a fire extinguisher, and the correct fire extinguisher must be used for the class of fire. Fire extinguishers are normally painted red but could also be painted yellow or be made of stainless steel. Fire extinguishers may be located on a red background with a bright red arrow directly above the location so that they can be located easily.

In-Plant Training

All personnel should be acquainted with all fire extinguisher types and sizes available in a plant or a specific work area. Training should include a tour of the facility indicating special fire hazard operations and should be practiced on a routine basis.

In addition, it is helpful to periodically discharge each type of extinguisher. Such practice is essential in learning how to activate each type, knowing the discharge ranges, realizing which types are affected by winds and drafts, familiarizing oneself with discharge duration, learning where to aim the discharge, and learning of any precautions to take as noted on the nameplate.

Tech Tip

Fire extinguishers labeled BC or ABC use dry chemicals, such as sodium or potassium bicarbonate or monoammonium phosphate, that produce a mildly corrosive residue. This residue can damage electrical components and should be cleaned off equipment immediately after a fire is extinguished.

Fire extinguishers should be inspected at least once a month. It is common to find units that are missing, damaged, or empty. Consider contracting for such a service. Contract for annual maintenance with a qualified service agency. Never attempt to make repairs to fire extinguishers.

Hazardous Locations

The use of electrical equipment in areas where explosion hazards are present can lead to an explosion and fire. This danger exists in the form of escaped flammable gases such as naphtha, benzene, propane, and others. Coal, grain, and other dust suspended in the air can also cause an explosion. Article 500 of the NEC® and Articles 497 and 499 of the NFPA cover hazardous locations. **See Figure 4-20.** Any hazardous location requires the maximum in safety and adherence to local, state, and federal guidelines and laws, as well as inplant safety rules. Hazardous locations are indicated by Class, Division, and Group.

HAZARDOUS LOCATIONS—NEC® ARTICLE 500

Hazardous Location – A location where there is an increased risk of fire or explosion due to the presence of flammable gases, vapors, liquids, combustible dusts, or easily-ignitable fibers or flyings.

Location – A position or site.

Flammable – Capable of being easily ignited and of burning quickly.

Gas – A fluid (such as air) that has no independent shape or volume but tends to expand indefinitely.

Vapor – A substance in the gaseous state as distinguished from the solid or liquid state.

Liquid – A fluid (such as water) that has no independent shape but has a definite volume. A liquid does not expand indefinitely and is only slightly compressible.

Combustible – Capable of burning.

Ignitable – Capable of being set on fire.

Fiber – A thread or piece of material.

Flyings – Small particles of material.

Dust – Fine particles of matter.

Classes	Likelihood that a flammable or combustible concentration is present
I	Sufficient quantities of flammable gases and vapors present in air to cause an explosion or ignite hazardous materials
II	Sufficient quantities of combustible dust are present in air to cause an explosion or ignite hazardous materials
III	Easily ignitable fibers or flyings are present in air, but not in a sufficient quantity to cause an explosion or ignite hazardous materials

Divisions	Location containing hazardous substances
1	Hazardous location in which hazardous substance is normally present in air in sufficient quantities to cause an explosion or ignite hazardous materials
2	Hazardous location in which hazardous substance is not normally present in air in sufficient quantities to cause an explosion or ignite hazardous materials

Class I Division I:

Spray booth interiors

Areas adjacent to spraying or painting operations using volatile flammable solvents

Open tanks or vats of volatile flammable liquids

Drying or evaporation rooms for flammable vents

Areas where fats and oils extraction equipment using flammable solvents is operated

Cleaning and dyeing plant rooms that use flammable liquids that do not contain adequate ventilation

Refrigeration or freezer interiors that store flammable materials

All other locations where sufficient ignitable quantities of flammable gases or vapors are likely to occur during routine operations

Class II Division I:

Grain and grain products

Pulverized sugar and cocoa

Dried egg and milk powders

Pulverized spices

Starch and pastes

Potato and wood flour

Oil meal from beans and seeds

Dried hay

Any other organic materials that may produce combustible dusts during their use or handling

Class III Division I:

Portions of rayon, cotton, or other textile mills

Manufacturing and processing plants for combustible fibers, cotton gins, and cotton seed mills

Flax processing plants

Clothing manufacturing plants

Woodworking plants

Other establishments involving similar hazardous processes or conditions

		Hazardous Locations	
Class	Group	Material
I	A	Acetylene
	B	Hydrogen, butadiene, ethylene oxide, propylene oxide
	C	Carbon monoxide, ether, ethylene, hydrogen sulfide, morpholine, cyclopropane
II	D	Gasoline, benzene, butane, propane, alcohol, acetone, ammonia, vinyl chloride
	E	Metal dusts
	F	Carbon black, coke dust, coal
	G	Grain dust, flour, starch, sugar, plastics
III	No groups	Wood chips, cotton, flax, and nylon

Figure 4-20. Article 500 of the NEC® covers hazardous locations.

When working with energized electrical equipment, it is recommended that an energized electrical work permit procedure be followed. This practice documents all electrical work performed on the premises and requires the signatures of supervisors and electricians involved in the work being performed. **See Figure 4-21.** Although an energized electrical work permit procedure is not an NEC® or NFPA requirement, both agencies recommend the procedure. Most commercial insurance companies require such policies be in place as a prerequisite for coverage.

ELECTRICAL WORK PERMITS

SEC I: WORK REQUESTED (TO BE COMPLETED BY REQUESTER)   Use additional sheets if necessary

1. Description of equipment & location: _____

_____

2. Description of work to be performed: _____

_____

3. Reason for request: _____

   Requested by: _____   Date: _____

SEC II: JOB PROCEDURE (TO BE COMPLETED BY ELECTRICIAN)

1. Detailed description of procedure to be used in performing above work: _____

_____

2. Safe work practice description: _____

3. Shock hazard analysis results: _____

4. Electrical shock/flash hazard protection boundary: _____

5. Flash-hazard analysis results: _____

6. PPE requirements: _____

7. Access restriction requirements: _____

8. Pre-work meeting documentation: _____

9. Can above job be performed safely?   yes _____   no _____
                                                                     (If no, return to requester)

   Electrician: _____   Date: _____
   Electrician: _____   Date: _____

SEC III: MANAGEMENT APPROVALS

_____   _____
Manufacturing Manager   Maintenance Manager

_____   _____
Safety Manager   Electrician Supervisor

_____   _____
Plant Manager   General Manager

                                 _____
                                 Date

White Copy: Office
Yellow Copy: Safety Manager
Pink Copy: Maintenance Manager

Figure 4-21. An energized electrical work permit documents all electrical work performed on the premises.

4-4 CHECKPOINT

1. What class of fire are electrical fires rated?

2. In a hazardous location, what class is combustible dust or grain rated?

4-5 CONFINED SPACES

A *confined space* is a space large enough and so configured that an employee can physically enter and perform assigned work, that has limited or restricted means for entry and exit, and is not designed for continuous employee occupancy. Confined spaces have a limited means of egress and are subject to the accumulation of toxic or flammable contaminants or an oxygen-deficient atmosphere. Confined spaces include, but are not limited to, storage tanks, process vessels, bins, boilers, ventilation or exhaust ducts, sewers, underground utility vaults, tunnels, pipelines, and open top spaces such as pits, tubes, ditches, and vaults more than 4' in depth.

Confined spaces cause entrapment hazards and life-threatening atmospheres through oxygen deficiency, combustible gases, and/or toxic gases. Oxygen deficiency is caused by the displacement of oxygen by leaking gases or vapors, the combustion or oxidation process, oxygen absorbed by the vessel or product stored, and/or oxygen consumed by bacterial action. Oxygen-deficient air can result in injury or death. **See Figure 4-22.**

Combustible gases in a confined space are commonly caused by leaking gases or gases produced in the space such as methane, carbon monoxide, carbon dioxide, and hydrogen sulfide. Air normally contains 21% oxygen. An increase in the oxygen level increases the explosive potential of combustible gases. Finely ground materials including carbon, grain, fibers, metals, and plastics can also cause explosive atmospheres.

WARNING: Confined space procedures vary in each facility. For maximum safety, it is important to always refer to specific facility procedures and applicable federal, state, and local regulations.

Confined Space Permits

Confined space permits are required for work in confined spaces based on safety considerations for workers. A permit-required confined space is a confined space that has specific health and safety hazards associated with it. OSHA 29 CFR 1910.146—*Permit-Required Confined Spaces* contains the requirements for practices and procedures to protect workers from the hazards of entry into permit-required confined spaces. These spaces are grouped into the categories of containing or having a potential to contain a hazardous atmosphere, containing a material that has the potential for engulfing an entrant, having an internal configuration such that an entrant could be trapped or asphyxiated by inwardly converging walls or a floor that slopes downward and tapers into a smaller cross-section, or containing any other recognized safety or health hazard.

POTENTIAL EFFECTS OF OXYGEN-DEFICIENT ATMOSPHERES*	
Oxygen Content[†]	Effects and Symptoms[‡]
19.5	Minimum permissible oxygen level
15–19.5	Decreased ability to work strenuously. May impair condition and induce early symptoms in persons with coronary, pulmonary, or circulatory problems
12–14	Respiration exert and pulse increases. Impaired coordination, perception, and judgement
10–11	Respiration further increases in rate and depth, poor judgement, lips turn blue
8–9	Mental failure, fainting, unconsciousness, ashen face, blue lips, nausia, and vomiting
6–7	Eight minutes, 100% fatal; 6 minutes, 50% fatal; 4–5 minutes, recovery with treatment
4–5	Coma in 40 seconds, convulsions, respiration ceases, death

Bacharach, Inc.

* values are approximate and vary with state of health and physical activities
† % by volume
‡ at atmospheric pressure

Figure 4-22. Oxygen-deficient atmospheres in confined spaces can cause life-threatening conditions.

Permit-required confined spaces require assessment of procedures in compliance with OSHA standards prior to entry. **See Figure 4-23.** A *non-permit confined space* is a confined space that does not contain or, with respect to atmospheric hazards, have the potential to contain any hazards capable of causing death or serious physical harm. These conditions can change with tasks such as welding, painting, or solvent use in the confined space.

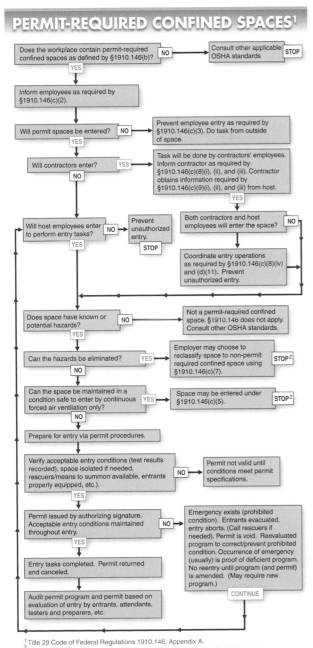

PERMIT-REQUIRED CONFINED SPACES[1]

[1] Title 29 Code of Federal Regulations 1910.146, Appendix A.
[2] Spaces may have to be evacuated and reevaluated if hazards arise during entry.

Figure 4-23. For maximum safety, procedures for entering a confined space must follow established OSHA standards.

Employers must evaluate the workplace to determine if spaces are permit-required confined spaces. If confined spaces exist in the workplace, the employer must inform exposed technicians of the existence, location, and danger posed by the spaces. This is accomplished by posting danger signs or by other equally effective means. In addition, the employer must develop a written permit-required confined space program. A written permit-required confined space program specifies procedures, identification of hazards in each permit-required confined space, restriction of access to authorized personnel, control of hazards, and monitoring of permit-required confined spaces during entry.

Entry Permit Procedures

An entry permit must be posted at confined space entrances or otherwise made available to entrants before entering a permit-required confined space. The permit is signed by the entry supervisor and verifies that pre-entry preparations have been completed and that the space is safe to enter. **See Figure 4-24.** A permit-required confined space must be isolated before entry. This prevents hazardous energy or materials from entering the space. Plant procedures for lockout/tagout of permit-required confined spaces must be followed.

Tech Tip

Assistance in understanding entry permit requirements and procedures for a particular location can be found by contacting the state Department of Labor commissioner, Division of Occupational Safety and Health. This department can provide documents and forms and may also provide sample permits to be used as a guide.

The duration of entry permits must not exceed the time required to complete an assignment. The entry supervisor must terminate entry and cancel permits when an assignment has been completed or when new conditions exist. New conditions are noted on the canceled permit and used in revising the permit-required confined space program. All canceled entry permits must be filed for at least one year.

Training is required for all technicians who are required to work in or around permit-required confined spaces. A certificate of training includes the technician's name, the signature or initials of trainer(s), and the dates of training. The certificate must be available for inspection by authorized officials.

CONFINED SPACE ENTRY PERMITS

ENTRY PERMIT

✔ CONFINED SPACE ✔ HAZARDOUS AREA

PERMIT VALID FOR 8 HOURS ONLY. ALL COPIES OF PERMIT WILL REMAIN AT JOB SITE UNTIL JOB IS COMPLETED

SITE LOCATION and DESCRIPTION *Electrical Vault*

PURPOSE OF ENTRY *Routine Maintenance/Inspection*

SUPERVISOR(S) in charge of crews.   Type of Crew   Phone #

Michael Green   *Maintenance   Shift II - X5924*

* BOLD DENOTES MINIMUM REQUIREMENTS TO BE COMPLETED AND REVIEWED PRIOR TO ENTRY*

REQUIREMENTS COMPLETED	DATE	TIME	REQUIREMENTS COMPLETED	DATE	TIME
Lock Out/De-energize/Try-out	10/2	09:00	Full Body Harness w/"D" ring	10/4	08:00
Line(s) Broken-Capped-Blanked	N/A	N/A	Emergency Escape Retrieval Equip	10/4	08:00
Purge-Flush and Vent	N/A	N/A	Lifelines	10/4	08:00
Ventilation	10/3	10:00	Fire Extinguishers	10/4	08:00
Secure Area (Post and Flag)	10/2	08:00	Lighting (Explosive Proof)	N/A	N/A
Breathing Apparatus	N/A	N/A	Protective Clothing	10/4	08:00
Resuscitator - Inhalator	N/A	N/A	Respirator(s) (Air Purifying)	N/A	N/A
Standby Safety Personnel	10/4	08:00	Burning and Welding Permit	N/A	N/A

Note: Items that do not apply enter N/A in the blank.

** RECORD CONTINUOUS MONITORING RESULTS EVERY 2 HOURS

CONTINUOUS MONITORING** TEST(S) TO BE TAKEN	Permissible Entry Level		10/4				
PERCENT OF OXYGEN	19.5% to 23.5%		20.5	20.6	20.7	20.5	20.5
LOWER FLAMMABLE LIMIT	Under 10%		0	0	0	0	0
CARBON MONOXIDE	+35 PPM		0	0	0	0	0
Aromatic Hydrocarbon	+ 1 PPM	* 5PPM	N/A				
Hydrogen Cyanide	(Skin)	* 4PPM	N/A				
Hydrogen Sulfide	+10 PPM	*15PPM	N/A				
Sulfur Dioxide	+ 2 PPM	* 5PPM	N/A				
Ammonia		* 35PPM	N/A				

* Short-term exposure limit:Employee can work in the area up to 15 minutes.

+ 8 hr. Time Weighted Avg.:Employee can work in area 8 hrs (longer with appropriate respiratory protection).

REMARKS:

GAS TESTER NAME & CHECK #	INSTRUMENT(S) USED	MODEL &/OR TYPE	SERIAL &/OR UNIT #
Marty James	*Combination Gas Meter*	*Industrial Scientific*	*15A*

SAFETY STANDBY PERSON IS REQUIRED FOR ALL CONFINED SPACE WORK

SAFETY STANDBY PERSON(S)	CHECK #	NAME OF SAFETY STANDBY PERSON(S)	CHECK #
Kate Washington	*3312*		
Tony Linder	*3318*		

SUPERVISOR AUTHORIZING ENTRY
ALL ABOVE CONDITIONS SATISFIED *Michael Green*   AMBULANCE 2800   FIRE 2900
Safety 4901   Gas Coordinator 4529/5387

Figure 4-24. Confined space entry permit forms document preparations, procedures, and required equipment.

4-5 CHECKPOINT

1. What is the typical percentage of oxygen in air?   **2.** What are the dangers of confined spaces?

4-6 OVERHEAD POWER LINE SAFETY

People are killed every day from accidental contact with overhead power lines. *Overhead power lines* are electrical conductors designed to deliver electrical power and are located in an aboveground aerial position. Overhead power lines are suspended from ceramic insulators that are attached to wood utility poles or metal structures. Overhead power lines are generally owned and operated by an electric utility company. Overhead power line conductors 600 V or higher are usually bare (uninsulated), while low-voltage systems such as service drops to buildings consist of insulated conductors.

Electrical power lines should be located far enough overhead or out of reach as to not pose an electrical hazard. Electrical equipment such as transformers and power panels are also isolated by fences, locked in buildings, or buried underground. Entrances to electrical rooms and other guarded locations containing exposed energized electrical parts must be marked "High Voltage – Do Not Enter," forbidding unqualified persons to enter.

Utility company electrical workers and linemen are skilled workers who have received extensive and specific training to safely work on and near energized overhead power lines and are equipped with the proper personal protection equipment and tools. Workers in other occupations, including residential/commercial/industrial electricians, technicians, engineers, and supervisors, are unqualified (unless trained) to approach overhead power lines closer than an established safe distance. Per NFPA 70E®, if the line voltage exceeds 50 kV, the minimum overhead line clearance for all nonqualified individuals is 10′ plus 4″ for every 10 kV over 50 kV.

Scaffolds

A *scaffold* is a temporary or movable platform and structure for workers to stand on when working at a height above the floor. Any person or item on a scaffold must also maintain a safe distance from power lines at all times including during the erection, use, and dismantling of scaffolds. All scaffolds, persons, and items on scaffolds must maintain the minimum distance from power lines.

4-6 CHECKPOINT

1. Do all overhead power lines have an insulated conductor coating?
2. If line voltage exceeds 50 kV, what is the minimum overhead line clearance?

Additional Resources

Review and Resources

Access Chapter 4 Review and Resources for *Electrical Motor Controls for Integrated Systems* by scanning the above QR code with your mobile device.

Applying Your Knowledge

Refer to the *Electrical Motor Controls for Integrated Systems* Learner Resources for interactive Applying Your Knowledge questions.

Workbook and Applications Manual

Refer to Chapter 4 in the *Electrical Motor Controls for Integrated Systems Workbook* and the *Applications Manual* for additional exercises.

ENERGY EFFICIENCY PRACTICES

Indoor Air Quality

The Occupational Safety and Health Administration (OSHA) requires all employers to provide a safe environment for their employees. To perform their jobs safely and effectively, individuals require an environment that is free from harmful levels of chemicals and particles. In addition, the temperature and humidity of the air must be comfortable to ensure optimum productivity.

Indoor air quality (IAQ) is the quality of indoor air based on a combination of temperature, humidity, airflow, and contaminant levels. Good IAQ occurs when the air in a building is free of harmful levels of chemicals and particles, and the temperature and humidity of the air is comfortable. Poor IAQ occurs when the temperature, humidity, chemical level, or contaminant level of the air in a building rise to harmful or uncomfortable levels. Poor IAQ can lead to serious health problems.

Early commercial building design and construction allowed a large percentage of outdoor air to infiltrate buildings, diluting contaminant levels. During the 1970s, the cost of energy increased dramatically, affecting the design, construction, operation, and control of buildings. Heating, ventilating, and air conditioning (HVAC) systems are often the source of and pathway for IAQ contaminants. But in many buildings, the HVAC system also became the sole source of fresh outdoor air.

The correct percentage of outdoor air must be brought into an HVAC system to provide fresh air and dilute contaminants in occupied spaces. Variable-air-volume (VAV) systems can be used to efficiently bring the correct percentage of outdoor air into building spaces. VAV systems normally use electric motor drives to control airflow. Electric motor drives conserve energy by operating fans at a certain speed to minimize contaminants and maintain comfort. VAV systems should not be set to a level low enough to allow contaminants to build up in building spaces.

Objectives

5-1
- State the basic rules that determine how and where two or more loads are connected into a control circuit.
- State the basic rules that determine how and where two or more switches are connected into a control circuit.
- Add line number references to any given control circuit drawn in line (ladder) diagram format.
- Add a numerical cross-reference system to any given control circuit drawn in line (ladder) diagram format.
- Add wire reference numbers to any given control circuit drawn in line (ladder) diagram format.
- Explain how to add manufacturer's terminal numbers to any given control circuit drawn in line (ladder) diagram format.

5-2
- Identify the components in an electrical circuit as being a part of the signal, decision, or action section of a control circuit.

5-3
- Define and give an example of switches connected for AND circuit control logic.
- Define and give an example of switches connected for OR circuit control logic.
- Define and give an example of how a switch is connected for NOT circuit control logic.
- Define and give an example of switches connected for NOR circuit control logic.
- Define and give an example of switches connected for NAND circuit control logic.
- Define and give an example of switches connected to develop memory control logic.
- List the four most common gates used in digital electronics.
- Identify the input and output pin numbers of digital logic gates within a digital integrated circuit (IC) chip.

5-4
- Draw a control circuit showing how additional stop switches can be connected into a control circuit.
- Draw a control circuit showing how additional start switches can be connected into a control circuit.
- Draw a control circuit showing how two motors can be started almost simultaneously.
- Draw a control circuit showing how a pilot light is used with a pressure switch to indicate device activation and how a pilot is used with a start/stop station to indicate device activation.
- Draw a control circuit showing how a selector switch is used to provide a common industrial jog/run circuit.

5-5
- Troubleshoot a control circuit using a digital multimeter (DMM) to determine problems with the switches and/or loads of the circuit.

Control Logic

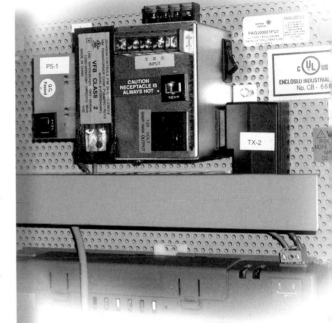

Before a lamp, motor, heating element, or any other electrical load turns ON, there must be a control circuit that determines just how, when, and what environmental and operating conditions must occur in order for that to happen. Manual, mechanical, and automatic control switches are used to turn loads on and off. Control switches can be normally open (NO) or normally closed (NC) and connected in series, parallel, or series/parallel combinations. The type of contacts used and the way the switches are connected determines the control logic of the circuit. Control switches can also be connected to digital electronic gates to produce the required operating logic of the circuit. Regardless of whether the switches are interconnected to develop the circuit logic, connected to logic gates, or even programmed using a programmable logic controller (PLC), the basic control logic functions of AND, OR, NOT, NAND, NOR and memory remain operationally the same in controlling the loads.

5-1 BASIC RULES OF LINE DIAGRAMS

The electrical industry has established a universal set of symbols and rules on how line diagrams (circuits) are laid out. By applying these standards, an electrician establishes a working practice with a language in common with all electricians.

One Load per Line

No more than one load should be placed in any circuit line between the power lines L1 and L2. For example, a pilot light can be connected into a circuit with a single-pole switch. **See Figure 5-1.** In this circuit, the power lines are drawn vertically on the sides of the drawing and marked L1 and L2. The space between L1 and L2 represents the voltage of the control circuit. This voltage appears across pilot light PL1 when switch S1 is closed. The pilot light glows when current flows through S1 and PL1 because the voltage between L1 and L2 is the proper voltage for the pilot light.

Two loads must not be connected in series in one line of a line diagram. If the two loads are connected in series, then the voltage between L1 and L2 must divide across both loads when S1 is closed. The result is that neither device receives the entire 120 V necessary for proper operation.

The load that has the highest resistance drops the highest voltage. The load that has the lowest resistance drops the lowest voltage.

ONE LOAD PER LINE

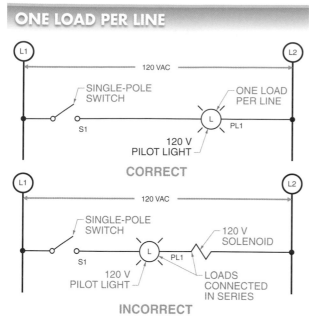

Figure 5-1. No more than one load should be placed in any circuit line between L1 and L2.

Loads must be connected in parallel when more than one load must be connected in the line diagram. **See Figure 5-2.** In this circuit, there is only one load for each line between L1 and L2, even though there are two loads in the circuit. The voltage from L1 and L2 appears across each load for proper operation of the pilot light and solenoid. This circuit has two lines, one for the pilot light and one for the solenoid.

LOADS CONNECTED IN PARALLEL

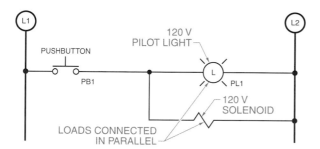

Figure 5-2. Loads must be connected in parallel when more than one load must be connected in the line diagram.

Load Connections

A *load* is any device that converts electrical energy to motion, heat, light, or sound. A load is the electrical device in a line diagram that uses the electrical power from L1 to L2. Control relay coils, solenoids, and pilot lights are loads that are connected directly or indirectly to L2. **See Figure 5-3.**

DIRECTLY CONNECTED LOADS

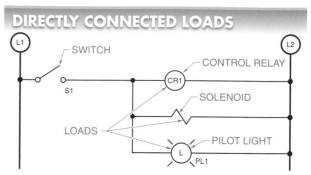

Figure 5-3. Control relay coils, solenoids, and pilot lights are loads that are connected directly or indirectly to L2.

Magnetic motor starter coils are connected to L2 indirectly through normally closed (NC) overload contacts. **See Figure 5-4.** An overload contact is normally closed and opens only if an overload condition exists in the motor. The number of NC overload contacts between the starter coil and L2 depends on the type of starter and power used in the circuit.

INDIRECTLY CONNECTED LOADS

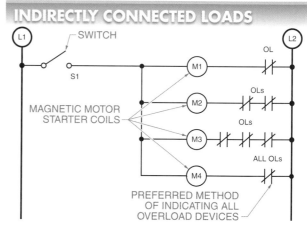

Figure 5-4. Magnetic motor starter coils are connected to L2 indirectly through NC overload contacts.

One to three NC overload contacts may be shown between the starter and L2 in all line diagrams. One to three NC overload contacts are shown because starters may include one, two, or three overload contacts, depending on the manufacturer and motor used. Early starters often included three overload contacts, one for each heater in the starter. Modern starters include only one overload contact. To avoid confusion, it is common practice to draw one set of NC overload contacts and mark these contacts all overloads (OLs). An overload marked this way indicates that the circuit is correct for any motor or starter used. The electrician knows to connect all the NC overload contacts that the starter is designed for in series if there is more than one on the starter.

Control Device Connections

Control devices are connected between L1 and the operating coil (or load). Operating coils of contactors and starters are activated by control devices such as pushbuttons, limit switches, and pressure switches. **See Figure 5-5.**

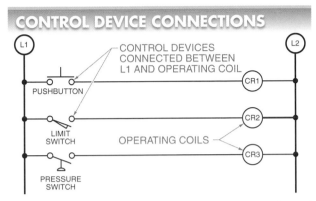

Figure 5-5. Control devices are connected between L1 and the operating coil.

Each line includes at least one control device. The operating coil is ON all the time if no control device is included in a line. A circuit may contain as many control devices as are required to make the operating coil function as specified. These control devices may be connected in series or parallel when controlling an operating coil. Although a circuit may include any number of loads, the total number of loads determines the required wire size and rating of the incoming power supply (typically a transformer). The total current increases as loads are added to the circuit.

Two control devices (a flow switch and a temperature switch) can be connected in series to control a coil in a magnetic motor starter. The flow switch and temperature switch must close to allow current to pass from L1, through the control device, the magnetic starter coil, and the overloads, to L2. Two control devices (a pressure switch and a foot switch) can be connected in parallel to control a coil in a magnetic motor starter. **See Figure 5-6.** Either the pressure switch or the foot switch can be closed to allow current to pass from L1 through the control device, the magnetic starter coil, and the overloads to L2. Regardless of how the control devices are arranged in a circuit, they must be connected between L1 and the operating coil (or load). The contacts of the control device may be either normally open (NO) or normally closed (NC). The contacts used and the way the control devices are connected into a circuit (series or parallel) determines the function of the circuit.

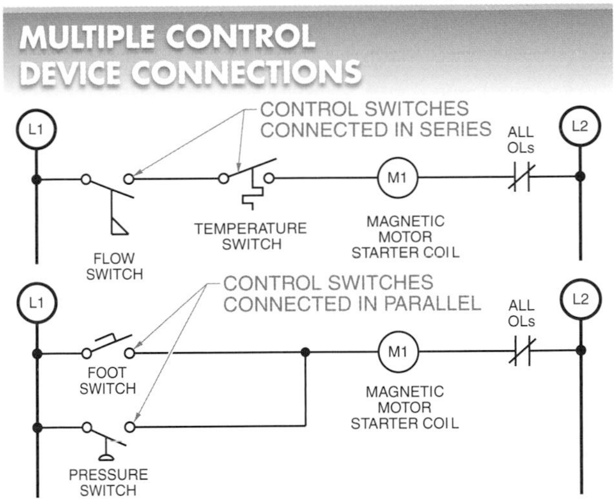

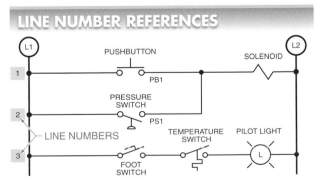

Figure 5-6. Two control devices can be connected in series or parallel to control a coil in a magnetic motor starter.

Line Number References

Each line in a line diagram should be numbered starting with the top line and reading down. **See Figure 5-7.**

Line 1 connects PB1 to the solenoid to complete the path from L1 to L2. Line 2 connects PS1 to the solenoid to complete the path from L1 to L2. PB1 and PS1 are marked as two separate lines even though they control the same load, because either the pushbutton or the pressure switch completes the path from L1 to L2. Line 3 connects a foot switch and a temperature switch to complete the path from L1 to L2. The foot switch and temperature switch both appear in the same line because it takes both the foot switch and the temperature switch to complete the path to the pilot light.

Numbering each line simplifies the understanding of the function of a circuit. The importance of this numbering system becomes clear as circuits become more complex and lines are added.

Figure 5-7. Each line in a line diagram should be numbered starting with the top line and reading down.

Numerical Cross-Reference Systems

Numerical cross-reference systems are required to trace the action of a circuit in complex line diagrams. Common rules help to simplify the operation of complex circuits.

Numerical Cross-Reference System (NO Contacts). Relays, contactors, and magnetic motor starters normally have more than one set of auxiliary contacts. These contacts may appear at several different locations in the line diagram. Numerical cross-reference systems quickly identify the location and type of contacts controlled by a given device. A numerical cross-reference system consists of numbers in parentheses to the right of the line diagram. NO contacts are represented by line numbers. The line numbers refer to the line on which the NO contacts are located. **See Figure 5-8.**

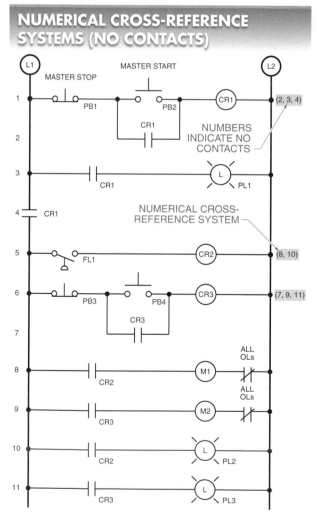

NUMERICAL CROSS-REFERENCE SYSTEMS (NO CONTACTS)

Figure 5-8. The locations of NO contacts controlled by a device are determined by the numbers on the right side of the line diagram.

In the circuit depicted, pressing master start pushbutton PB2 energizes control relay coil CR1. Control relay coil CR1 controls three sets of NO contacts. This is shown by the numerical codes (2, 3, 4) on the right side of the line diagram. Each number indicates the line in which the NO contacts are located.

In line 2, the NO contacts form the holding circuit (memory) for maintaining the coil CR1 after master start pushbutton PB2 is released. In line 3, the NO contacts energize pilot light PL1, indicating that the circuit has been energized. In line 4, the NO contacts allow the remainder of the circuit to be activated by connecting L1 to the remainder of the circuit. The numerical cross-reference system shows the location of all contacts controlled by coil CR1 as well as the effect each has on the operation of the circuit.

In line 5, control relay CR2 energizes if float switch FL1 closes. Control relay CR2 closes the NO contacts located in lines 8 and 10 as indicated by the numerical codes (8, 10). The magnetic motor starter controlled by coil M1 is energized when the NO contacts of line 8 close. Pilot light PL2 turns on, indicating the motor has started, when the NO contacts in line 10 close.

In line 6, several NO contacts located in lines 7, 9, and 11 are used to control other parts of the circuit through control relay CR3. In line 7, the NO contacts form the memory circuit for maintaining the circuit to control relay coil CR3 after pushbutton PB4 is released. The NO contacts in line 9 close, energizing the magnetic motor starter controlled by coil M2 when coil CR3 is energized. Simultaneously, the NO contacts in line 11 close, causing pilot light PL3 to light as an indicator that the motor has started.

The numerical cross-reference system allows the simplification of complex line diagrams. Each NO contact must be clearly marked because each set of NO contacts is numbered according to the line in which they appear.

Numerical Cross-Reference System (NC Contacts). In addition to NO contacts, there are also NC contacts in a circuit. To differentiate between NO and NC, NC contacts are indicated as a number that is underlined. The underlined number refers to the line on which the NC contacts are located. **See Figure 5-9.** For example, lines 9 and 11 contain devices that control NC contacts in lines 12 and 13 as indicated by the underlined numbers (12, 13) to the right of the line diagram.

In the circuit depicted, pressing master start pushbutton PB2 energizes control relay coil CR1. Control relay coil CR1 controls three sets of NO contacts. This is shown by the numerical codes (2, 3, 4) on the right side of the line diagram. In line 2, the NO contacts form the holding circuit (memory) for maintaining the coil CR1 after master start pushbutton PB2 is released. In line 3, the NO contacts energize pilot light PL1, indicating that the circuit has been energized. In line 4, the NO contacts allow the remainder of the circuit to be activated by connecting L1 to the remainder of the circuit.

NUMERICAL CROSS-REFERENCE SYSTEMS (NC CONTACTS)

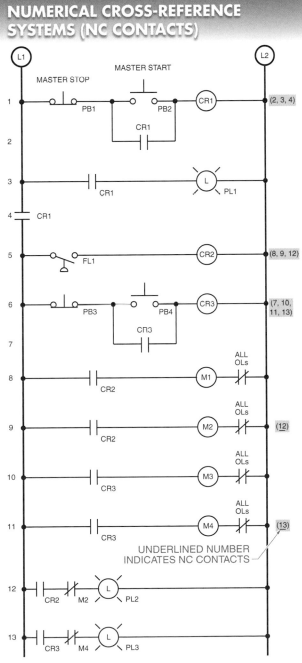

Figure 5-9. NC contacts are indicated by numbers that are underlined to distinguish them from NO contacts.

In line 5, when control relay coil CR2 is energized, the NO contacts in lines 8, 9, and 12 close. Closing these contacts energizes coils M1 and M2 and completes the circuit going to the pilot light in line 12. The pilot light in line 12 does not glow because the NC contacts controlled by coil M2 in line 12 are opened at the same time that the NO contacts are closed, leaving the circuit open. With the NO contacts of CR2 closed and the NC contacts of M2 open, the light stays OFF unless something happens to shut down line 9, which contains coil M2. For example, if coil M2 represents a safety cooling fan protecting the motor controlled by M1, the light would indicate the loss of cooling.

A similar sequence of events took place when line 6 was energized. In this case, pressing pushbutton PB4 energizes control relay coil CR3, which closes NO contacts in line 7, 10, 11, and 13. A memory circuit is formed in line 7, coils M3 and M4 are energized in lines 10 and 11, and part of the circuit to pilot light PL3 in line 13 is completed. Because coil M4 is energized, the NC contacts it controls open in line 13, forming a similar alarm circuit to the one in line 12.

If coil M4 in line 11 drops out for any reason, the NC contacts in line 13 return to their NC position and the pilot light alarm signal in line 13 is turned on. This circuit could be used where it is extremely important for the operator to know when something is not functioning.

Wire Reference Numbers

Each wire in a control circuit is assigned a reference point (number) on a line diagram to keep track of the different wires that connect the components in the circuit. Each reference point is assigned a reference number. Reference numbers are normally assigned from the top left to the bottom right. This numbering system can apply to any control circuit such as single-station, multistation, or reversing circuits. **See Figure 5-10.**

Any wire that is always connected to a common point is the same electrically and assigned the same number. The wires that are assigned a number vary from 2 to the number required by the circuit. Any wire that is prewired when the component is purchased is normally not assigned a reference number. The exact numbering system used varies for each manufacturer or design engineer. One common method used is to circle the wire reference numbers. Circling the wire reference numbers helps separate them from other numbering systems.

WIRE REFERENCE NUMBERS

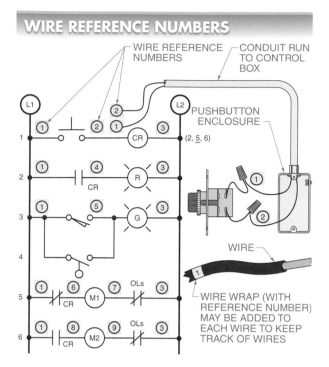

Figure 5-10. Each wire in a control circuit is assigned a reference point on a line diagram to keep track of the different wires that connect the components in the circuit.

Manufacturer's Terminal Numbers

Manufacturers of electrical relays, timers, counters, etc., include numbers on the terminal connection points. These terminal numbers are used to identify and separate the different component parts (coil, NC contacts, etc.) included on the individual pieces of equipment. Manufacturer's terminal numbers are often added to a line diagram after the specific equipment to be used in the control circuit is identified. **See Figure 5-11.**

Cross-Referencing Mechanically Connected Contacts

Control devices such as limit switches, flow switches, temperature switches, liquid level switches, and pressure switches normally have more than one set of contacts operating when the device is activated. These devices normally have at least one set of NO contacts and one set of NC contacts that operate simultaneously. For all practical purposes, the multiple contacts of these devices normally do not control other devices in the same lines of a control circuit.

The two methods used to illustrate how contacts found in different control lines belong to the same control switch are the dashed line method and the numerical cross-reference method. **See Figure 5-12.**

TERMINAL NUMBERS

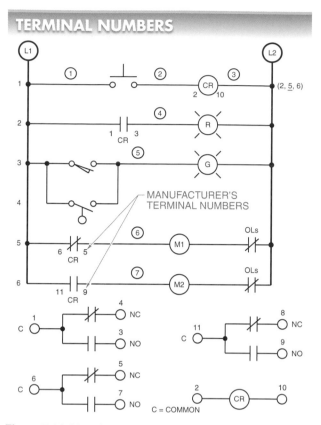

Figure 5-11. Manufacturers include terminal numbers to identify and separate the different component parts included on individual pieces of equipment.

In the dashed line method, the dashed line between the NO and NC contacts indicates that both contacts move from the normal position when the arm of the limit switch is moved. In this circuit, pilot light PL1 is ON and motor starter coil M1 is OFF. After the limit switch is actuated, pilot light PL1 turns off and the motor starter coil M1 turns on.

The dashed line method works well when the control contacts are close together and the circuit is relatively simple. If a dashed line must cut across many lines, the circuit becomes hard to follow.

Tech Tip

Wire reference numbers and manufacturer's terminal numbers can be easily marked on conductors and terminals by using self-sticking, preprinted labels. Preprinted labels are available on cards and in books for any combination of numbers and letters. It is a good idea to keep a book that contains a selection of numbers and letters handy when designing, testing, or troubleshooting a system. A wire numbering label maker/printer can be used to develop customized labels for large applications.

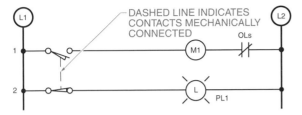

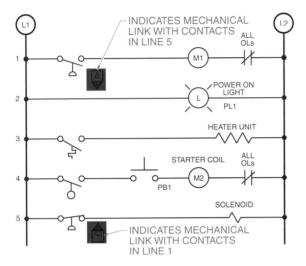

DASHED LINE METHOD

NUMERICAL CROSS-REFERENCE METHOD

Figure 5-12. Contacts found in different control lines that belong to the same control switch are illustrated using the dashed line or numerical cross-reference method.

The numerical cross-reference method is used on complex line diagrams where a dashed line cuts across several lines. In this circuit, a pressure switch with an NO contact in line 1 and an NC contact in line 5 is used to control a motor starter and a solenoid. The NO and NC contacts of the pressure switch are simultaneously actuated when a predetermined pressure is reached. A solid arrow pointing down is drawn by the NO contact in line 1 and is marked with a 5 to show the mechanical link with the contact in line 5. A solid arrow pointing up is drawn by the NC contact in line 5 and is marked with a numeral 1 to show the mechanical linkage with the contact in line 1. This cross-reference method eliminates the need for a dashed line cutting across lines 2, 3, and 4. This makes the circuit easier to follow and understand. This system may be used with any type of control switch found in a circuit.

Simplifying Printreading

An electrical print is much easier to use for installing, modifying, and troubleshooting an electrical circuit when line, cross-reference, wire reference, and manufacturer's terminal numbers are included with standard electrical symbols. **See Figure 5-13.**

For example, a typical electrical circuit used to control two pump motors can be found in a water tower application. The power circuit includes the main disconnect, disconnects for pumps 1 and 2, power contacts for motor starters 1 and 2, and the two pump motors. The control circuit includes components on the 120 VAC side of the control transformer. The line numbers indicate that there are 14 lines in the control circuit. The numerical cross-reference system allows easy identification of the location of timer, relay, and motor starter contacts in the control circuit and whether each is normally open (NO) or normally closed (NC). The wire reference numbers allow easy identification of control device connection points and where the leads of a DMM can be placed when troubleshooting. The manufacturer's terminal numbers identify and separate the different components included on each piece of equipment.

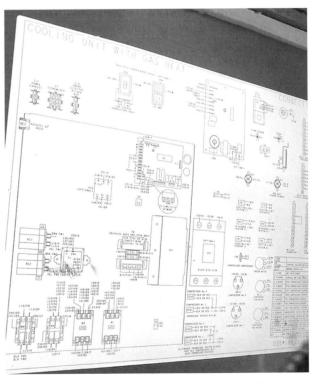

Electrical prints can be used for installing, modifying, and troubleshooting electrical circuits.

PRINT NUMBERING SYSTEMS

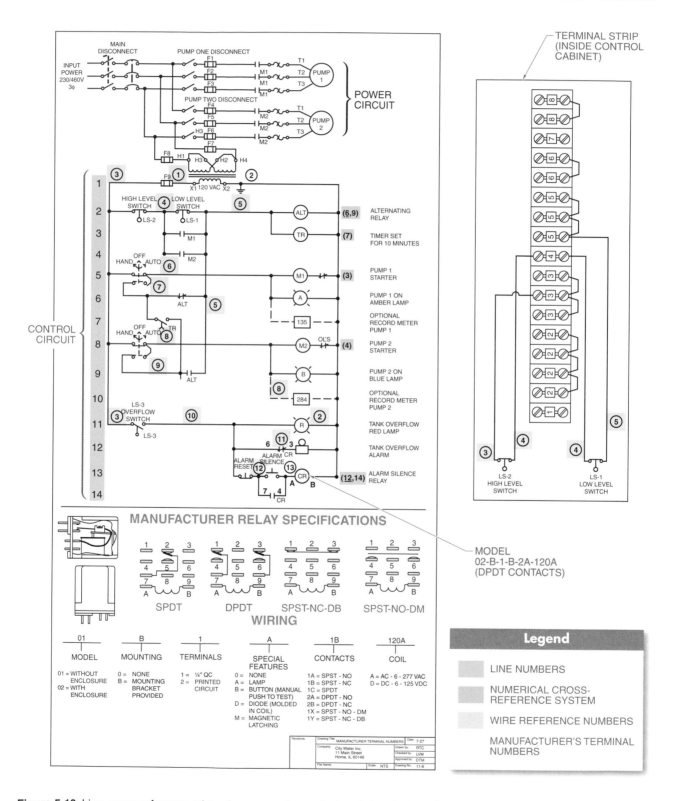

Figure 5-13. Line, cross-reference, wire reference, and manufacturer's terminal numbers are used to simplify electrical prints.

5-1 CHECKPOINT

1. How many loads can be connected in series per line in a control circuit in which a switch is used to control the load(s)?
2. How many loads can be connected in parallel per line in a control circuit in which a switch is used to control the load(s)?
3. Why are line numbers added to a control circuit?
4. Why are numerical cross-reference numbers added to a control circuit?
5. Why are wire reference numbers added to a control circuit?
6. Why are manufacturer's terminal numbers added to a control circuit?

5-2 LINE DIAGRAMS—SIGNALS, DECISIONS, AND ACTION

The concept of control is to accomplish specific work in a predetermined manner. A circuit must respond as designed, without any changes. To accomplish this consistency, all control circuits are composed of three basic sections: the signals, the decisions, and the action sections. **See Figure 5-14.** Complete understanding of these sections enables easy understanding of any existing industrial control circuit, as well as those that are created as systems become more mechanized and automated.

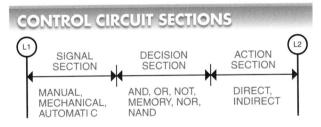

Figure 5-14. All control circuits are composed of signal, decision, and action sections.

Signals

A signal starts or stops the flow of current by closing or opening the contacts of the control device. Current is allowed to flow through the control device if the contacts are closed. Current is not allowed to flow through the control device if the contacts are opened. Pushbuttons, limit switches, flow switches, foot switches, temperature switches, and pressure switches may be used as the signal section of a control circuit.

All signals depend on some condition that must take place. This condition can be manual, mechanical, or automatic. A manual condition is any input into the circuit by a person. Foot switches and pushbuttons are control devices that respond to a manual condition. A mechanical condition is any input into the circuit by a mechanically moving part. A limit switch is a control device that responds to a mechanical condition. When a moving object, such as a box, hits a limit switch, the limit switch normally has a lever, roller, ball, or plunger actuator that causes a set of contacts to open or close. An automatic condition is any input that responds automatically to changes in a system. Flow switches, temperature switches, and pressure switches respond to automatic conditions. These devices automatically open and close sets of contacts when a change in the flow of a liquid is created, when a change in temperature is sensed, or when pressure varies. The signal accomplishes no work by itself; it merely starts or stops the flow of current in that part of the circuit.

Decisions

The decision section of a circuit determines what work is to be done and in what order the work is to occur. The decision section of a circuit adds, subtracts, sorts, selects, and redirects the signals from the control devices to the load. For the decision part of the circuit to perform a definite sequence, it must perform in a logical manner. The way the control devices are connected into the circuit gives the circuit logic. The decision section of the circuit accepts informational inputs (signals), makes logical decisions based on the way the control devices are connected into the circuit, and provides the output signal that controls the load.

Action

Once a signal is generated and the decision has been made within a circuit, some action (work) should result. In most cases it is the operating coil in the circuit that is responsible for initiating the action. This action is direct when devices such as motors, lights, and heating elements are turned on as a direct result of the signal and the decision. This action is indirect when the coils in solenoids, magnetic starters, and relays are energized. The action is indirect because the coil energized by the signal and the decision may energize a magnetic motor starter, which actually starts the motor. Regardless of how this action takes place, the load causes some action (direct or indirect) in the circuit and, for this reason, is the action section of the circuit.

5-2 CHECKPOINT

1. Name the circuit signals used in the signal section of the circuit in Figure 5-7.
2. Name the circuit loads used in the action section of the circuit in Figure 5-7.
3. Name the circuit control logic used in the decision section of the circuit in Figure 5-7.

5-3 LOGIC FUNCTIONS

Control devices such as pushbuttons, limit switches, and pressure switches are connected into a circuit so that the circuit can function in a predetermined manner. All control circuits are basic logic functions or combinations of logic functions. Logic functions are common to all areas of industry. This includes electricity, electronics, hydraulics, pneumatics, math, and other routine activities. Logic functions include AND, OR, AND/OR, NOT, NOR, and NAND.

Common logic functions have been used to develop circuit logic since the first electrical circuits were used. Line (ladder) diagrams are one of the oldest and most common methods of illustrating and understanding basic logic functions.

Programmable logic controller (PLC) programming diagrams decrease the design time needed for electrical circuits and add flexibility by reprogramming electrical circuits using software instead of rewiring. The basic logic functions of a circuit remain the same; however, the PLC programming diagram is drawn in a generic manner to allow for greater flexibility when the circuits are reprogrammed. Programmable logic relay (PLR) function block diagrams are another method of designing and drawing circuit logic using simple logic blocks. A PLR function block diagram is a simplified way of showing common circuit logic functions by connecting inputs and outputs to a logic block labeled with the desired logic function.

PLCs and PLRs can be programmed using standard line (ladder) programming or function block diagrams. Some PLCs and PLRs allow the user to select either method. Once the basic circuit design methods are understood, any method can be used to program a circuit. On each type of diagram, each of the basic logic functions is shown using each format. The function of the electrical circuit is exactly the same for each type of diagram.

AND Logic

AND logic is used in industry when two NO pushbuttons are connected in series to control a solenoid. **See Figure 5-15.** PB1 and PB2 must be pressed before the solenoid is energized. The logic function that makes up the decision section of this circuit is AND logic. The reason for using the AND function could be to build in safety for the operator of this circuit.

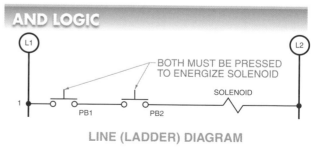

Figure 5-15. In AND logic, the load is ON if both of the control signal contacts are closed.

If the solenoid were operating a punch press or shear, the pushbuttons could be spaced far enough apart so that the operator would have to use both hands to make the machine operate. This ensures that the operator's hands are not near the machine when it is activated. With AND logic, the load is ON only if all the control signal contacts are closed. As with any logic function, the signals may be manually, mechanically, or automatically controlled. Any control device such as limit switches, pressure switches, etc., with NO contacts can be used in developing AND logic. The NO contacts of each control device must be connected in series for AND logic.

A simple example of AND logic takes place whenever an automobile that has an automatic transmission is started. The ignition switch must be turned to the start position and the transmission selector must be in the park position before the starter is energized. Before the action (load ON) in the automobile circuit can take place, the control signals (manual) must be performed in a logical manner (decision).

OR Logic

OR logic is used in industry when an NO pushbutton and an NO temperature switch are connected in parallel. **See Figure 5-16.** In this circuit, the load is a heating element that is controlled by two control devices.

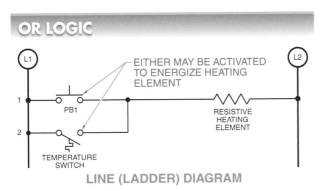

OR LOGIC

EITHER MAY BE ACTIVATED TO ENERGIZE HEATING ELEMENT

PB1

TEMPERATURE SWITCH

RESISTIVE HEATING ELEMENT

LINE (LADDER) DIAGRAM

Figure 5-16. In OR logic, the load is ON if any one of the control signal contacts is closed.

The logic of this circuit is OR logic because either the pushbutton or the temperature switch energizes the load. The temperature switch is an example of an automatic control device that turns the heating element on and off to maintain the temperature setting for which the temperature switch is set. The manually controlled pushbutton could be used to test or turn on the heating element when the temperature switch contacts are open.

In OR logic, the load is ON if any one of the contacts of the control signal is closed. The control devices are connected in parallel. Series and parallel refer to the physical relationship of each control device to other control devices or components in the circuit. This series and parallel relationship is only part of what determines the logic function of any circuit.

An example of OR logic is in a dwelling that has two pushbuttons controlling one bell. The bell (load) may be energized by pressing (signal ON) either the front or the back pushbutton (control device). In this example, as in the automobile circuit example, the control devices are connected to respond in a logical manner.

AND/OR Logic Combination

The decision section of any circuit may contain one or more logic functions. **See Figure 5-17.** In this circuit, both pressure and flow must be present in addition to the pushbutton or the foot switch being engaged to energize the starter coil (load). This provides the circuit with the advantage of both AND logic and OR logic. The machine is protected because both pressure and flow must exist before it is started, and there is a choice between using a pushbutton or a foot switch for final operation. The action taking place in this circuit is energizing a coil in a magnetic motor starter. The signal inputs for this circuit have to be two automatic and at least one manual.

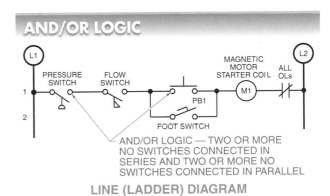

AND/OR LOGIC

PRESSURE SWITCH   FLOW SWITCH   MAGNETIC MOTOR STARTER COIL   ALL OLs

PB1

M1

FOOT SWITCH

AND/OR LOGIC — TWO OR MORE NO SWITCHES CONNECTED IN SERIES AND TWO OR MORE NO SWITCHES CONNECTED IN PARALLEL

LINE (LADDER) DIAGRAM

Figure 5-17. The decision section of any circuit may contain one or more logic functions.

Each control device responds to its own input signal and has its own decision-making capability. When multiple control devices are used in combination with other control devices making their own decisions, a more complex decision can be made through the combination of all control devices used in the circuit. All industrial control circuits consist of control devices capable of making decisions in accordance with the input signals received.

NOT Logic

NOT logic has an output if the control signal is OFF. For example, replacing NO contacts on a pushbutton with NC contacts energizes the solenoid and pilot light without pressing the pushbutton. **See Figure 5-18.** In this circuit, the loads are de-energized when the pushbutton is pressed. There must not be a signal if the loads are to remain energized. With NOT logic, the output remains on only if the control signal contacts remain closed.

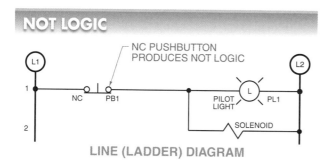

NOT LOGIC

LINE (LADDER) DIAGRAM

Figure 5-18. In NOT logic, the load is ON only if the control signal contacts are closed.

An example of NOT logic is the courtesy light in a refrigerator. The light is ON if the control signal is OFF. The control signal is the door of the refrigerator. Any time the door is open (signal OFF), the load (courtesy light) is ON. The condition that controls the signal can be manual, mechanical, or automatic. With the refrigerator door, the condition is mechanical.

NOR Logic

NOR logic is an extension of NOT logic in which two or more NC contacts in series are used to control a load. **See Figure 5-19.** In this circuit, additional operator safety is provided by adding several emergency stop pushbuttons (NOT logic) to the control circuit. The load (coil M1) is de-energized by pressing any emergency stop pushbutton. By incorporating NOR logic, each machine may be controlled by one operator, but any operator or supervisor can have the capability of turning off all the machines on the assembly line to protect individual operators or the entire system. With the knowledge of NOR logic, the electrician can readily add stop pushbuttons by wiring them in series to perform their necessary function.

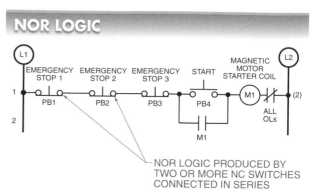

NOR LOGIC

NOR LOGIC PRODUCED BY TWO OR MORE NC SWITCHES CONNECTED IN SERIES

Figure 5-19. NOR logic is an extension of NOT logic in which two or more NC contacts in series are used to control a load.

NAND Logic

NAND logic is an extension of NOT logic in which two or more NC contacts are connected in parallel to control a load. **See Figure 5-20.** In the circuit depicted, two interconnected tanks are filled with a liquid. When pushbutton PB3 is pressed, coil M1 is energized and auxiliary contacts M1 close until both tanks are filled. Both tanks fill to a predetermined level because the float switches in tank 1 and tank 2 do not open until both tanks are full. Every NOT logic must be open (signal OFF) to stop the filling process based on the input of the float switches. NOR logic is also present in this circuit because the emergency stops (NOT) at tank 1 or tank 2 may be used to stop the process if an operator at either of the tanks sees a problem.

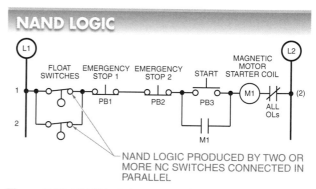

NAND LOGIC

NAND LOGIC PRODUCED BY TWO OR MORE NC SWITCHES CONNECTED IN PARALLEL

Figure 5-20. NAND logic is an extension of NOT logic in which two or more NC contacts are connected in parallel to control a load.

An example of a NAND circuit is the courtesy light in an automobile. In an automobile, the courtesy lights are ON if the control signal (door switches) is OFF (normally closed). This circuit is different from a refrigerator door in that an automobile may have two or more door switches, any of which will turn on the courtesy lights.

Memory

Today, many of the industrial circuits require their control circuits to not only make logic decisions such as AND, OR, and NOT but also be capable of storing, memorizing, or retaining the signal inputs to keep the load energized even after the signals are removed. A switch that controls house lights from only one location is an example of a memory circuit. Memory circuits are also known as holding or sealing circuits. When the memory circuit is ON, it remains on until it is turned off and remains off until it is turned on. It performs a memory function because the output corresponds to the last input information until new input information is received to change it. In the case of the house light switch, the memory circuit is accomplished by a switch that mechanically stays in one position or another.

In industrial control circuits, it is more common to find pushbuttons with return spring contacts (momentary contacts) than those that mechanically stay held in one position (maintained contacts). Auxiliary contacts are added to provide memory to circuits with pushbuttons. **See Figure 5-21.** Once coil M1 of the magnetic motor starter is energized, it causes coil contacts M1 to close and remain closed (memory) until the coil is de-energized.

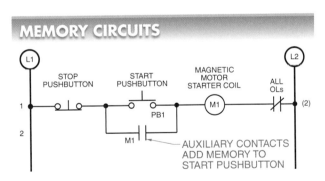

Figure 5-21. Auxiliary contacts are added to provide memory to circuits with pushbuttons.

Digital signals are a series of pulses that change levels between the OFF or ON state. The analog and digital processes can be seen by comparing a light dimmer and light switch. A light dimmer varies the intensity of light from fully OFF to fully ON. This is an example of an analog process. A standard light switch has only two positions: fully OFF or fully ON. This is an example of a digital process. Electronic circuits that process these quickly changing pulses are digital or logic circuits. The four most common gates used in digital electronics are the AND, OR, NAND, and NOR gates.

AND Gates. An *AND gate* is a device with an output that is high only when both of its inputs are high. The quad AND gate is one type of integrated circuit chip. **See Figure 5-22.** In this chip, the manufacturer has placed four AND gates in one package. By using the numbering system on the chip, any one or all four of the AND gates may be used. In this case, voltage is applied to the circuit at pins 14 and 7.

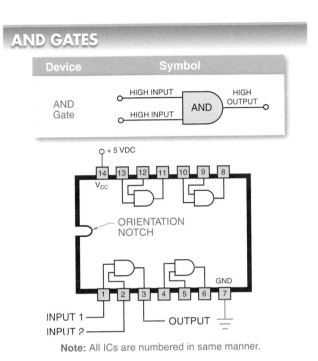

Figure 5-22. An AND gate is a device with an output that is high only when both of its inputs are high.

To connect an external circuit to an AND gate, pins 1, 2, and 3 of the quad AND gate chip could be used. Pins 1 and 2 are the input and pin 3 is the output. An application of an AND gate is in an elevator control circuit. **See Figure 5-23.** The elevator cannot move unless the inner and outer doors are closed. Once both doors are closed, the output of the AND gate could be fed to a circuit that would control the elevator motor.

OR Gates. An *OR gate* is a device with an output that is high when either or both inputs are high. **See Figure 5-24.** An application of an OR gate is in a burglar alarm circuit. A signal is sent to the burglar alarm circuit if the front door or the back door is opened. The electrical equivalent of an OR gate is two pushbuttons connected in parallel.

ELEVATOR CONTROL CIRCUITS

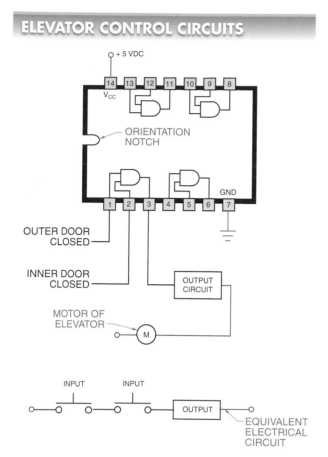

Figure 5-23. An AND gate may be used in an elevator control circuit.

OR GATES

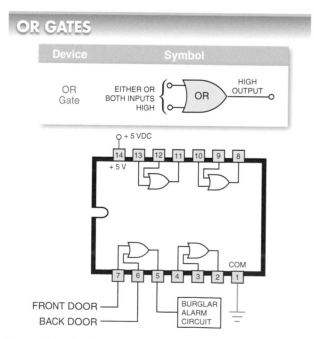

Figure 5-24. An OR gate is a device with an output that is high when either or both inputs are high.

NAND Gates. A *NAND gate* is a device that provides a low output when both inputs are high. A NAND (NOT-AND) gate is an inverted AND function. The NAND gate is represented by the AND symbol followed by a small circle indicating an inversion of the output. **See Figure 5-25.**

NAND GATES

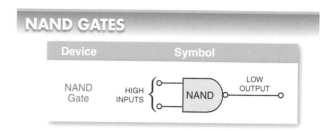

Figure 5-25. A NAND gate is a device that provides a low output when both inputs are high.

A NAND gate is a universal building block of digital logic. NAND gates are normally used in conjunction with other elements to implement more complex logic functions. NAND gates are also available in quad integrated circuit packaging.

NOR Gates. A *NOR gate* is a device that provides a low output when either or both inputs are high. A NOR (NOT-OR) gate is the same as an inverted OR function. A NOR gate is represented by the OR gate symbol followed by a small circle indicating an inversion of the output. **See Figure 5-26.** The NOR gate is a universal building block of digital logic. NOR gates are normally used in conjunction with other elements to implement more complex logic functions. NOR gates are also available in quad integrated circuit packaging.

NOR GATES

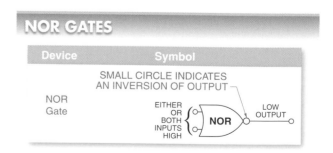

Figure 5-26. A NOR gate is a device that provides a low output when either or both inputs are high.

NOT logic may be added to memory logic to create a common start/stop control circuit. **See Figure 5-27.** When stop pushbutton PB1 is activated, current to coil M1 stops and contacts M1 open, returning the circuit to its original condition.

START/STOP CONTROL CIRCUITS

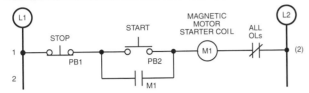

Figure 5-27. A common start/stop control circuit is created by adding the NOT logic of a stop pushbutton to the memory logic of magnetic coil contacts.

Digital Logic Integrated Circuits

An *integrated circuit (IC)* is a circuit composed of thousands of semiconductor devices, providing a complete circuit function in one small semiconductor package. Integrated circuits are popular because they provide a complete circuit function in one package. Integrated circuits are often referred to as chips, although chips are actually a part of the integrated circuit. **See Figure 5-28.**

INTEGRATED CIRCUITS

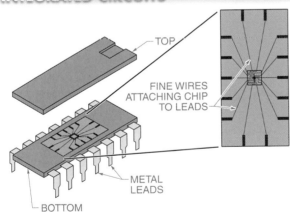

Figure 5-28. Integrated circuits are thousands of semiconductors, providing a complete circuit function in one small semiconductor package.

Because of the nature of integrated circuits, a technician must approach them in an entirely different manner from individual solid-state components. Integrated circuits consist of systems within a system. The entire system must be understood. Troubleshooting integrated circuits requires knowledge of the system functions and input and output characteristics. Defective integrated circuits must be replaced because they cannot be repaired.

Integrated Circuit Packages

There are different shapes and sizes of integrated circuits. **See Figure 5-29.** These typically include mini-DIP, dual in-line, and flat-pack integrated circuits for digital logic functions. The dual in-line package (DIP) with 14, 16, or 24 pins is the most widely used configuration. Mini-DIP is a smaller dual in-line package with 8 pins.

INTEGRATED CIRCUIT PACKAGES

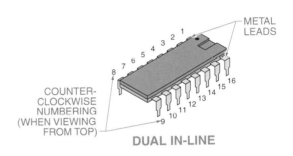

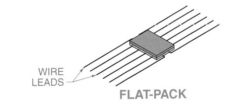

Figure 5-29. Integrated circuits range in shape and size and include mini-DIP, dual in-line, and flat-pack.

Pin Numbering System. All manufacturers use a standardized pin numbering system for their devices. Manufacturer data sheets should be consulted when unsure about pin numbering patterns.

Dual in-line packages and flat packs have index marks and notches at the top for reference. Before removing an integrated circuit, it should be noted where the index mark is in relation to the board or socket to aid in installation of the new unit. The numbering of the pins is always the same. The notch is at the top of the chip. To the left of the notch is a dot that is in line with pin 1. The pins are numbered counterclockwise around the chip when viewed from the top.

5-3 CHECKPOINT

1. What type of circuit logic is developed when two NO switches are connected in parallel?
2. What type of circuit logic is developed when two NC switches are connected in parallel?
3. What type of circuit logic is developed when two NO switches are connected in series?
4. What type of circuit logic is developed when two NC switches are connected in series?

5. Draw the symbol for an AND logic chip with two inputs and one output.
6. Draw the symbol for a NAND logic chip with two inputs and one output.
7. Draw the symbol for an OR logic chip with two inputs and one output.
8. Draw the symbol for a NOR logic chip with two inputs and one output.

5-4 COMMON CONTROL CIRCUITS

Various control circuits are commonly used in commercial and industrial electrical circuits. An electrician must understand the entire circuit operation to begin wiring or troubleshooting the circuit.

Start/Stop Stations Controlling Magnetic Starters

A load is often required to be started and stopped from more than one location. See Figure 5-30. In this circuit, the magnetic motor starter may be started or stopped from two locations. Additional stop pushbuttons are connected in series (NOR logic) with the existing stop pushbuttons.

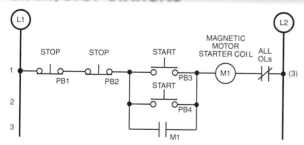

Figure 5-30. Two stop pushbuttons connected in series and two start pushbuttons connected in parallel are used to control a motor from two locations.

Additional start pushbuttons are connected in parallel (OR logic) with the existing start pushbuttons. Pressing any one of the start pushbuttons (PB3 or PB4) causes coil M1 to energize. This causes auxiliary contacts M1 to close, adding memory to the circuit until coil M1 is de-energized. Coil M1 may be de-energized by pressing stop pushbuttons PB1 or PB2, by an overload that would activate the OLs,

or by a loss of voltage to the circuit. In the case of an overload, the overload has to be removed and the circuit overload devices reset before the circuit would return to normal starting condition.

Two Magnetic Starters Operated by Two Start/Stop Stations with Common Emergency Stop

In almost all electrical systems, several devices can be found running off a common supply voltage. Two start/stop stations may be used to control two separate magnetic motor starter coils with a common emergency stop protecting the entire system. See Figure 5-31. Pressing start pushbutton PB3 causes coil M1 to energize and seal in auxiliary contacts M1. Pressing start pushbutton PB5 causes coil M2 to energize and seal in auxiliary contacts M2. Once the entire circuit is operational, emergency stop pushbutton PB1 can shut down the entire circuit or the individual stop pushbuttons PB2 or PB4 can de-energize the coils in their respective circuits. Each circuit is overload protected and does not affect the other when one magnetic motor starter experiences a problem.

Start/Stop Station Controlling Two or More Magnetic Starters

Steel mills, paper mills, bottling plants, and canning plants are industries that require simultaneous operation of two or more motors. In each industry, products or materials are spread out over great lengths but must be started together to prevent product separation or stretching. To accomplish this, two motors can be started almost simultaneously from one location. See Figure 5-32.

START/STOP STATIONS— COMMON EMERGENCY STOP

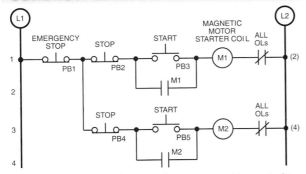

Figure 5-31. Two start/stop stations are used to control two separate magnetic motor starter coils with a common emergency stop protecting the entire system.

TWO MOTOR SIMULTANEOUS START—ONE LOCATION

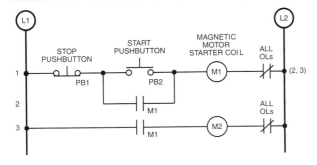

Figure 5-32. Two motors can be started almost simultaneously from one location to prevent product separation or stretching.

In this circuit, pressing start pushbutton PB2 energizes coil M1 and seals in both sets of auxiliary contacts M1. *Note:* It is acceptable to have more than one set of auxiliary contacts controlled by one coil. When both sets of contacts close, the first set of M1 contacts (line 2) provides memory for the start pushbutton and completes the circuit to energize coil M1. The second set of M1 contacts (line 3) completes the circuit to coil M2, energizing coil M2. The motors associated with these magnetic motor starters start almost simultaneously because both coils energize almost simultaneously. Pushing the stop pushbutton breaks the circuit (line 1), de-energizing coil M1. When coil M1 drops out, both sets of auxiliary contacts are deactivated. The motors associated with these magnetic motor starters stop almost simultaneously because both coils de-energize almost simultaneously. An overload in magnetic motor starter M2 affects only the operation of coil M2. The entire circuit is shut down if an overload exists in motor starter

M1. The entire circuit stops because de-energizing coil M1 also affects both sets of auxiliary contacts M1. This protection might be used where a machine such as an industrial drill would be damaged if the cooling liquid pump shuts off while the drill was still operating.

Pressure Switch with Pilot Light Indicating Device Activation

Pilot lights are manufactured in a variety of colors, shapes, and sizes to meet the needs of industry. The illumination of these lights signals an operator that any one of a sequence of events may be taking place. A pilot light may be used with a pressure switch to indicate when a device is activated. **See Figure 5-33.**

PRESSURE SWITCH WITH PILOT LIGHT

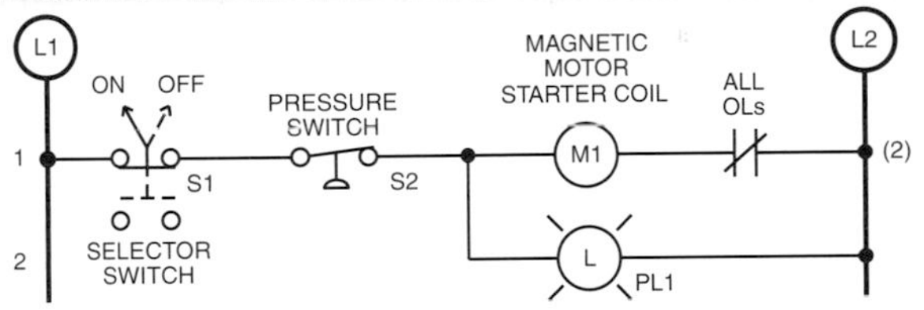

Figure 5-33. A pilot light is used with a pressure switch to indicate when a device is activated.

In this circuit, pressure switch S2 has automatic control over the circuit when switch S1 is closed. When the pressure to switch S2 drops, the switch closes and activates coil M1, which controls the magnetic starter of the compressor motor and starts the compressor. At the same time, contacts M1 close and pilot light PL1 turns on. The compressor continues to run and the pilot light stays ON as long as the motor runs. When pressure builds sufficiently to open pressure switch S2, coil M1 de-energizes and the magnetic motor starter drops out, stopping the compressor motor. The pilot light goes out because contact M1 controlled by coil M1 opens. The pilot light is ON only when the compressor motor is running. This circuit might be used in a garage to let the owner know when the air compressor is ON or OFF.

Tech Fact

When designing or troubleshooting a control circuit, it is important to remember that most control circuits are powered by a control transformer with a fixed power output (kVA rating). Adding additional control loads such as lamps and alarms can overload the control transformer. The control transformer should have a power rating 30% or more above the control circuit loads.

A pilot light may be used with a start/stop station to indicate when a device is activated. **See Figure 5-34.** In this circuit, pressing start pushbutton PB2 energizes coil M1, causing auxiliary contacts M1 to close. Closing contacts M1 provides memory for start pushbutton PB2 and maintains an electrical path for the pilot light. As long as coil M1 is energized, the pilot light stays on. Pressing stop pushbutton PB1 de-energizes coil M1, opening contacts M1 and turning off the pilot light. An overload in this circuit also de-energizes coil M1, opening contacts M1 and turning off the pilot light. A circuit like this can be used as a positive indicator that some process is taking place. The process may be in a remote place such as in a pump well or in another building.

START/STOP STATION WITH PILOT LIGHT

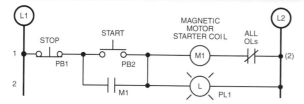

Figure 5-34. A pilot light is used with a start/stop station to indicate when a device is activated.

Start/Stop Station with Pilot Light Indicating Device Not Activated

Pilot lights may be used to show when an operation is stopped as well as when it is started. NOT logic is used in a circuit when a pilot light is used to show that an operation has stopped. NOT logic is established by placing one set of NC contacts in series with a device. **See Figure 5-35.**

NOT LOGIC INDICATING DEVICE NOT ACTIVATED

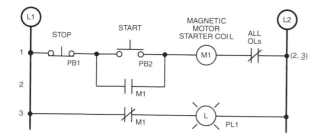

Figure 5-35. NOT logic is used to indicate when a device is not operating.

In this circuit, pressing start pushbutton PB2 energizes coil M1, causing both sets of auxiliary contacts M1 to energize. NO contacts M1 (line 2) close, providing memory for PB2, and NC contacts M1 (line 3) open, disconnecting pilot light PL1 from the line voltage, causing the light to turn off. Pressing stop pushbutton PB1 de-energizes coil M1, causing both sets of contacts to return to their normal positions. NO contacts M1 (line 2) return to their on position, and NC contacts M1 (line 3) return to their NC position, causing the pilot light to be reconnected to the line voltage and causing it to turn on. The pilot light is ON only when the coil to the magnetic motor starter is OFF. A bell or siren could be substituted for the pilot light to serve as a warning device. A circuit like this is used to monitor critical operating procedures such as a cooling pump for a nuclear reactor. When the cooling pump stops, the pilot light, bell, or siren immediately calls attention to the fact that the process has been stopped.

Pushbutton Sequence Control

Conveyor systems often require one conveyor system to feed boxes or other materials onto another conveyor system. If one conveyor is feeding a second conveyor, a circuit is needed to prevent the pileup of material on the second conveyor if the second conveyor is stopped. A sequence control circuit does not let the first conveyor operate unless the second conveyor has started and is running. **See Figure 5-36.**

SEQUENCE CONTROL CIRCUITS

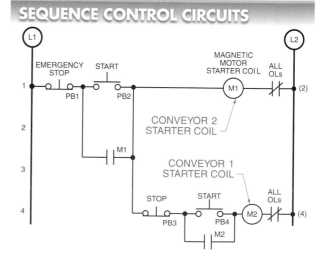

Figure 5-36. A sequence control circuit does not let the first conveyor operate unless the second conveyor has started and is running.

In this circuit, pressing start pushbutton PB2 energizes coil M1 and causes auxiliary contacts M1 to close. With auxiliary contacts M1 closed, PB2 has memory and provides an electrical path to allow coil M2 to be energized when start pushbutton PB4 is pressed.

With start pushbutton PB4 pressed, coil M2 energizes and closes contacts M2, providing memory for start pushbutton PB4 so that both conveyors run. Conveyor 1 (coil M2) cannot start unless conveyor 2 (coil M1) is energized. Both conveyors shut down if an overload occurs in the circuit with coil M1 or if stop pushbutton PB1 is pressed. Only conveyor 1 shuts down if conveyor 1 (coil M2) experiences an overload. A problem in conveyor 1 does not affect conveyor 2. This type of control is also known as cascade control or cascade protection.

Jogging with a Selector Switch

Jogging is the frequent starting and stopping of a motor for short periods of time. Jogging is used to position materials by moving the materials small distances each time the motor starts. A selector switch is used to provide a common industrial jog/run circuit. **See Figure 5-37.** The selector switch (two-position switch) is used to manually open or close a portion of the electrical circuit. In this circuit, the selector switch determines if the circuit is a jog circuit or run circuit. With selector switch S1 in the open (jog) position, pressing start pushbutton PB2 energizes coil M1, causing the magnetic motor starter to operate. Releasing start pushbutton PB2 de-energizes coil M1, causing the magnetic motor starter to stop. With selector switch S1 in the closed (run) position, pressing start pushbutton PB2 energizes coil M1, closing auxiliary contacts M1 and providing memory so that the magnetic starter operates and continues to operate until stop pushbutton PB1 is pressed.

When stop pushbutton PB1 is pressed, coil M1 de-energizes and all circuit components return to their original condition. The overloads may also open the circuit and must be reset after the overload is removed to return the circuit to normal operation. This circuit may be found where an operator may run a machine continuously for production, but it may stop it at any time for small adjustments or repositioning. Jogging may also be accomplished by other types of circuits. The common feature of all jog circuits is that they prevent the holding circuit from operating.

5-4 CHECKPOINT

1. When additional switches are to be wired into a control circuit to produce additional STOP functions, are NO or NC contacts used and are they connected in series or parallel?
2. When additional switches are to be wired into a control circuit to produce additional START functions, are NO or NC contacts used and are they connected in series or parallel?
3. What function does a JOG button perform in a control circuit?
4. How is a lamp added into a control circuit so it is ON anytime the motor is ON?

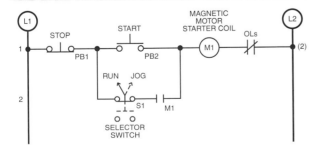

JOGGING WITH SELECTOR SWITCHES

Figure 5-37. A selector switch is used to provide a common industrial jog/run circuit.

Selector switches and pushbuttons are commonly used to operate manufacturing machines.

5-5 CONTROL CIRCUIT TROUBLESHOOTING

Troubleshooting is the systematic elimination of the various parts of a system, circuit, or process to locate a malfunctioning part. Troubleshooting electrical control circuits requires an organized, sequenced approach. Troubleshooting requires the use of electrical test equipment, drawings and diagrams, and manufacturer specifications.

Before troubleshooting an electrical circuit, an individual must understand the operation of the circuit, the sequence of events, timing or counting functions, and devices used to energize and de-energize the circuit. A line diagram shows the logic of an electrical circuit using single lines and symbols. Along with a line diagram, a DMM can be used to troubleshoot components in electrical circuits. Common electrical problems include open circuits and short circuits. The most common troubleshooting method is the tie-down troubleshooting method.

Tie-Down Troubleshooting Method

The *tie-down troubleshooting method* is a testing method in which one DMM probe is connected to either the L2 (neutral) or L1 (hot) side of a circuit and the other DMM probe is moved along a section of the circuit to be tested. The tie-down troubleshooting method allows a troubleshooter to work quickly on a familiar circuit that is small enough for the test probes to reach across the test points. When connecting a DMM to L2, the black test lead should always be used.

When using the tie-down troubleshooting method, one DMM test lead should be placed (tied down) on L2 (neutral conductor) and the other lead should be moved through the circuit starting with L1 (hot conductor). **See Figure 5-38.** If the correct voltage is not measured at L1 and L2, there is a power problem and the main power must be checked (for a possible fuse, circuit breaker, or main switch problem). If the proper voltage is present between L1 and L2, the DMM lead connected to L1 is moved along the circuit until the meter lead is directly at the load. If voltage is measured at the load but the load is not operating, the problem is the load on any circuit in which the load is connected directly to L2.

All loads are connected directly to L2 except when a magnetic motor starter overload contact is connected between the starter coil and L2. When a magnetic motor starter overload contact is used in a circuit, the DMM lead connected to L2 can be moved to the other side of the overload (side connected directly to the starter coil) to check if the overloads are open. However, caution must be exercised when doing this because one DMM lead is still connected to L1 (hot conductor). This means the tip of the other DMM lead (the one being moved) can cause an electrical shock if touched and there is a complete path to ground through the troubleshooter's body. **See Figure 5-39.**

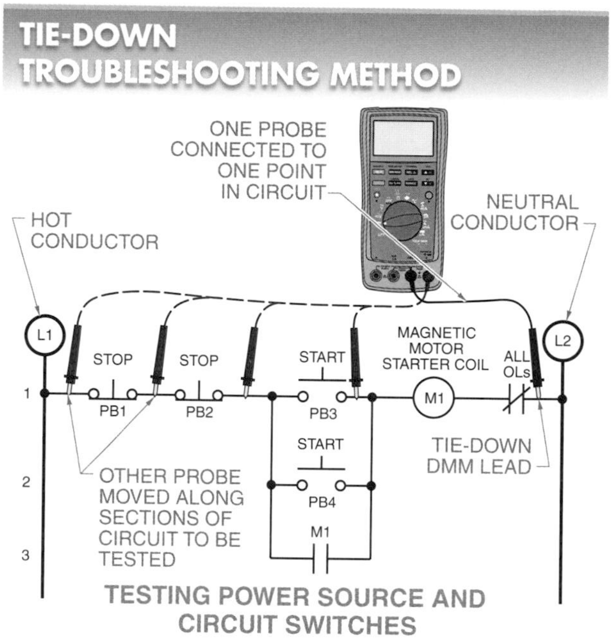

TIE-DOWN TROUBLESHOOTING METHOD

TESTING POWER SOURCE AND CIRCUIT SWITCHES

Figure 5-38. When using the tie-down troubleshooting method, one DMM test lead should be placed (tied down) on L2 (neutral conductor) and the other lead should be moved through the circuit starting with L1 (hot conductor).

TESTING OVERLOAD CONTACTS

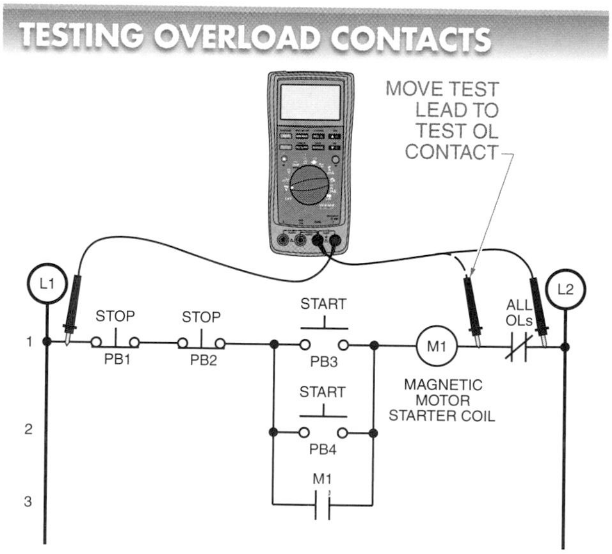

Figure 5-39. When a magnetic motor starter overload contact is used in a circuit, the DMM lead connected to L2 can be moved to the other side of the overload (side connected directly to the starter coil) to check if the overload is open.

Troubleshooting Open Circuits

An *open circuit* is an electrical circuit that has an incomplete path that prevents current flow. An open circuit represents a very high resistance path for current and is usually regarded as having infinite resistance. An open circuit in a series circuit de-energizes the entire circuit. Open circuits may be caused intentionally or unintentionally. An open circuit is caused intentionally when a switch is used to open a circuit. An open circuit may be caused unintentionally when the wiring between parts in a circuit is broken, when a component or device in a circuit malfunctions, or when a fuse blows.

A switch in the OFF position is an open circuit. Switches are tested by toggling the switch to check if the contacts open and close. A DMM set to measure voltage can be used to test a mechanical switch. A good switch indicates source voltage when open and 0 V when closed. **See Figure 5-40.** A faulty switch indicates source voltage both when open and when closed.

The proper operation of a switch must be known to determine when it is not operating properly because not all switches operate in the same manner. For example, a good solid-state switch indicates source voltage when open and a slight voltage drop when closed. This is normal due to the construction of the solid-state switch. **See Figure 5-41.**

A switch may also be checked with a jumper wire. A jumper wire is placed in parallel around the switch and the circuit is energized. The jumper wire closes the circuit, energizing the load.

WARNING: Jumper wires can cause equipment to start unexpectedly and must be removed from the circuit when no longer needed for testing.

An open circuit may occur unintentionally by a break in the wire of a circuit, a malfunctioning component, or a blown fuse. When a wire breaks, the path for current is interrupted and current flow stops. A broken wire in an individual line of a circuit de-energizes that line only. That branch of the circuit can be tested using a DMM set to measure voltage. For example, a DMM may be placed across a section of wire to determine if there is a break in that part of the wire. This test is often taken across wire connection points. The DMM indicates 0 V if the wire has no break. The DMM indicates source voltage if the wire is broken and there is no other open in the branch.

A faulty electrical component may also cause an unintentional open circuit. For example, a break in the conducting path of an electrical component, such as a burnt-out filament of a light bulb, also breaks the path for current and opens the circuit. In addition, when a fuse blows, the current flow in the circuit increases to a level that opens the conducting path inside the fuse.

TROUBLESHOOTING MECHANICAL SWITCHES

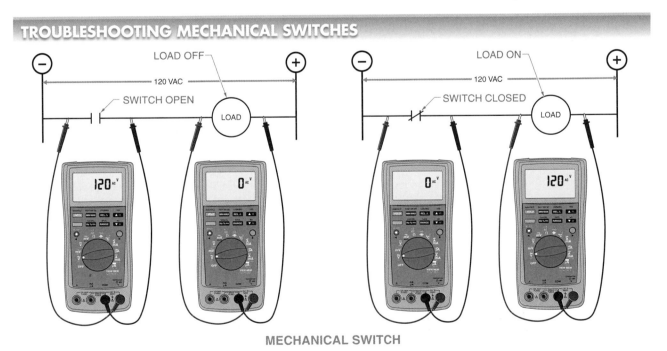

MECHANICAL SWITCH

Figure 5-40. A good mechanical switch indicates source voltage when open and 0 V when closed.

TROUBLESHOOTING SOLID-STATE SWITCHES

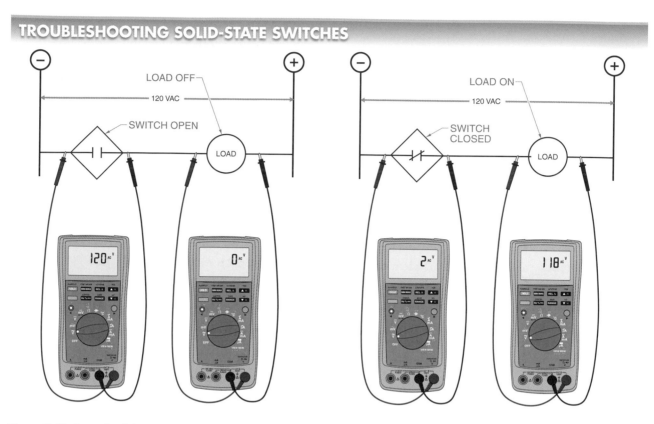

Figure 5-41. A good solid-state switch indicates source voltage when open and a slight voltage drop when closed.

Troubleshooting Short Circuits

A *short circuit* is a circuit in which current takes a shortcut around the normal path of current flow. In a short circuit, current leaves the normal current-carrying path and goes around the load and back to the power source or to ground. The low-resistance path can be due to failure of circuit components or failure in the wiring of the circuit. For example, if two pieces of wire accidentally contact each other, the wires produce a dead short across the circuit. A *dead short* is a short circuit that opens the circuit as soon as the circuit is energized or when the section of the circuit containing the short is energized. **See Figure 5-42.**

A dead short reduces the resistance of the short-circuited part of a circuit to nearly 0 Ω. A dead short produces a surge of current in the circuit, resulting in an overload device, such as a fuse, being blown or circuit breaker being tripped. In a circuit with a dead short, the fuse must be replaced or the circuit breaker reset. The circuit is inspected for the location of the short if the fuse blows or circuit breaker trips again when the

circuit is energized. The location of the short is usually indicated by signs of overheating, such as burn marks or discolored insulation. The location of a short can be determined using a continuity tester or a DMM set to measure resistance. **WARNING:** Ensure that the circuit is de-energized when measuring resistance.

A continuity tester can be used to test for short circuits. **See Figure 5-43.** A continuity tester uses its own power (usually a battery) to power the circuit to determine if a short circuit exists between a wire and its housing. Once the short circuit is located, the shorted wire must be replaced. It is important to ensure that the circuit is disconnected from its power source before testing using a continuity tester.

Tech Tip

Short circuits often cause sparks at the location of the fault. A circuit in which a circuit breaker is reset and immediately trips again still contains a short and may have damage caused by sparks. A circuit in which a short circuit occurs requires fixing the short and making sure that no fire hazards have developed that may only be evident after the fault has been fixed.

DEAD SHORTS

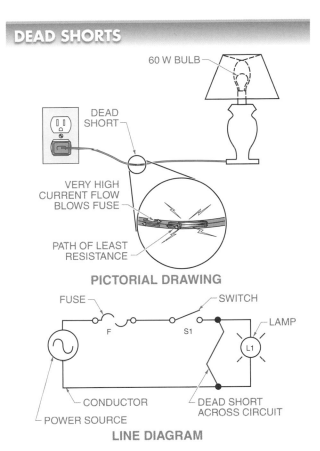

60 W BULB

DEAD SHORT

VERY HIGH CURRENT FLOW BLOWS FUSE

PATH OF LEAST RESISTANCE

PICTORIAL DRAWING

FUSE   SWITCH

F   S1

LAMP

L1

CONDUCTOR   DEAD SHORT ACROSS CIRCUIT

POWER SOURCE

LINE DIAGRAM

Figure 5-42. When two pieces of wire touch because of damaged insulation, the wires produce a dead short across the circuit.

TROUBLESHOOTING SHORT CIRCUITS— CONTINUITY TESTERS

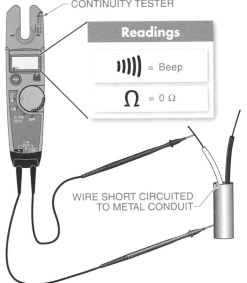

CONTINUITY TESTER

Readings

))))) = Beep

Ω = 0 Ω

WIRE SHORT CIRCUITED TO METAL CONDUIT

Figure 5-43. A continuity tester can be used to test for short circuits.

A DMM set to measure resistance can also be used to test for short circuits. A circuit is tested for a short circuit with all open contacts closed. In a good circuit, a DMM reads total circuit resistance when all open contacts are closed. In a circuit with a dead short, a DMM reads near 0 Ω. **See Figure 5-44.** To test each branch of a circuit, each branch is isolated by disconnecting a wire from the branch. The branch does not contain a short circuit if this produces no change in the DMM resistance reading. The branch is reconnected after the resistance reading is taken.

This process is continued by isolating each branch in succession. A branch contains a short if, when the branch is disconnected, the DMM resistance reading jumps from 0 Ω to a high resistance. This branch is inspected for signs of overheating and crossed, frayed, or loose wires. Further inspection is required to find the exact cause of the short. Large, complex circuits are tested one section at a time to determine which section contains the short. The individual branches of the section are then tested to find the exact location of the short.

WARNING: Power must always be removed from the circuit before a resistance check can be made.

When troubleshooting circuits using a DMM, PPE should always be selected based on the designed voltage for the equipment or circuit to be worked on.

TROUBLESHOOTING SHORT CIRCUITS—DIGITAL MULTIMETERS

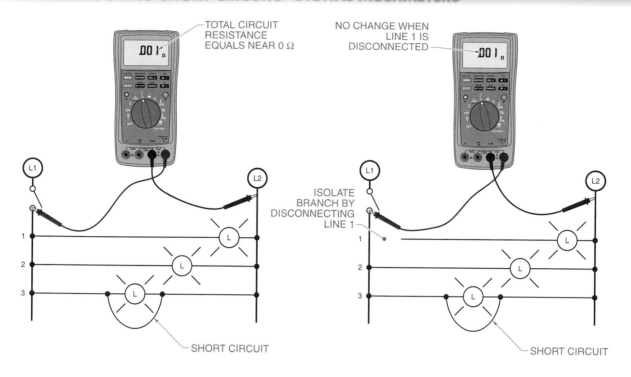

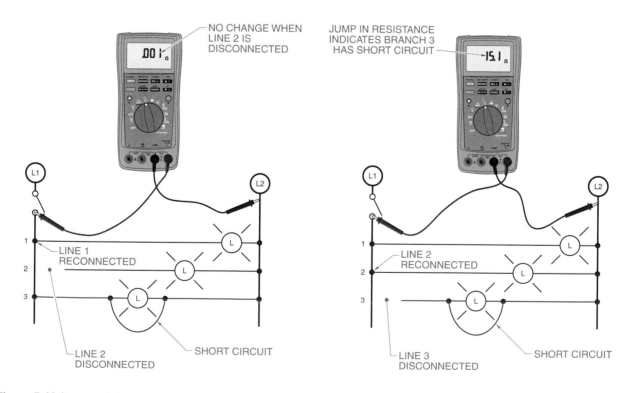

Figure 5-44. In a good circuit, a DMM reads total circuit resistance when all open contacts are closed. In a circuit with a short, a DMM reads near 0 Ω.

5-5 CHECKPOINT

1. When connecting a DMM to L2 of a control circuit, which color DMM test lead is always used?

2. When a switch is used to control a lamp or other load in a 24 VAC control circuit, what is the voltage that would be measured across the switch when the switch is open and the load is OFF?

3. When a switch is used to control a lamp or other load in a 24 VAC control circuit, what is the voltage that would be measured across the switch when the switch is closed and the load is ON?

4. When there is a short circuit in a control circuit, what function of the meter is used to test the circuit components before the faulty component is repaired and power is restored?

Additional Resources

Review and Resources

Access Chapter 5 Review and Resources for *Electrical Motor Controls for Integrated Systems* by scanning the above QR code with your mobile device.

Applying Your Knowledge

Refer to the *Electrical Motor Controls for Integrated Systems* Learner Resources for interactive Applying Your Knowledge questions.

Workbook and Applications Manual

Refer to Chapter 5 in the *Electrical Motor Controls for Integrated Systems Workbook* and the *Applications Manual* for additional exercises.

ENERGY EFFICIENCY PRACTICES

Energy-Efficient Lamps

According to the U.S. Department of Energy, lighting consumes approximately 22% of the total electricity generated in the United States. More than half the lighting energy is consumed in the commercial sector. Consumers and businesses spend approximately $58 billion per year on lighting.

Incandescent lamps have long been the standard in homes and as pilot lamps in industrial operator panels and pushbutton stations. However, incandescent lamps are inefficient and, over their lifetime, result in energy costs of five to ten times more than they cost to purchase. To increase energy efficiency, incandescent lamps can be replaced with compact fluorescent lamps (CFLs).

CFLs are, on average, four times more efficient than incandescent lamps, and last an average of ten times longer. CFLs are much less expensive to use over their lifetime than incandescent lamps because they require less electricity to produce the same amount of light. The use of CFLs reduces the demand for electricity generation. Generating less electricity reduces the amount of CO_2 gas, sulfur oxide, and nuclear waste produced.

Light-emitting diode (LED) lamps have become an even more efficient means of producing light for residential, commercial, and industrial applications. Although LED lamps do not effectively produce general lighting, they produce lighting for applications such as traffic signals, emergency signs, and industrial operator panels and pushbutton stations. LED lamps are normally grouped together to produce a high light output.

LED lamps commonly last approximately 125 times longer than incandescent lamps and approximately ten times longer than equivalent CFLs. In addition, LED lamps use approximately 85% less energy than incandescent lamps. Although an LED lamp has a higher initial cost than other lamp types, an LED lamp saves energy over its lifetime.

Objectives

6-1
- Describe the different parts of pushbuttons and their functions.
- Identify the NEMA and IEC enclosure location rating for each service location environmental condition.

6-2
- Identify two-position and three-position selector switches.
- Explain switch operation given a switch's truth table.

6-3
- Define joysticks and describe their most common positions.

6-4
- Explain the purpose of a limit switch.
- Define actuator and describe its typical applications.
- Explain the importance of properly installing limit switches.

6-5
- Explain the purpose of a foot switch.

6-6
- Identify and draw the symbols for normally open (NO) and normally closed (NC) pressure switches.
- Identify the different types of pressure switch sensing devices and how they work.
- Define deadband (differential) as applied to pressure and temperature switches.
- Explain the advantage and disadvantage of different deadband range settings.

6-7
- Explain the purpose of a temperature switch.

6-8
- Explain the purpose of a flow switch.

6-9
- Explain the purpose of a level switch.
- Explain the difference between charging and discharging level control circuits.
- Explain the difference between one- and two-level control circuits.

6-10
- Explain how mechanical contacts can be protected for longer operating life when switching higher currents.
- Explain how mechanical contacts can be protected for longer operating life when releasing higher pressure.
- State the procedure for testing mechanical switch contacts using a DMM in an operational circuit.

Sections

Review and Resources
atplearningresources.com/Quicklinks
Access Code: 362245

Mechanical Input Control Devices

Electrical power is used to operate loads that provide light, heat, linear motion, rotary motion, and sound and to power security, communication, and entertainment systems. All electrical loads must be controlled as to when and how they are to be turned on and off. Electrical loads can be controlled manually using switches such as pushbuttons, sector switches, and joysticks. Electrical loads can be controlled mechanically using limit switches that include different operators for different applications. Electrical loads can be controlled automatically by using switches that react to different operating and environmental conditions, such as pressure switches, temperature switches, flow switches, and level switches. Understanding each switch symbol, usage, and limit is important when designing, installing, and troubleshooting electrical systems that include manual, mechanical, and/or automatic switches.

6-1 INDUSTRIAL PUSHBUTTONS

Pushbuttons are the most common control switches used on industrial equipment. Almost all industrial machines and processes have a manually controlled position, even if the machine or process is designed to operate automatically. An industrial pushbutton consists of a legend plate, an operator, and one or more contact blocks (electrical contacts). **See Figure 6-1.**

Legend Plates

A *legend plate* is the part of a switch that includes the written description of the switch's operation. A legend plate indicates the pushbutton's function in the circuit. Legend plates are available indicating common circuit operations such as start, stop, jog, up, down,

ON, OFF, reset, and run, or they are available blank. The lettering on legend plates is normally uppercase for clarity and visibility.

Legend plates are also available in different colors. The color red is normally used for such circuit functions as stop, OFF, and emergency stop. The color black with white lettering is used for most other circuit functions. However, different colored legend plates can be used along with colored operators to highlight different circuit functions. When color is used, red is normally used to indicate a stop or OFF function, green is used to indicate an ON or open function, and amber is used to indicate a manual override or reset function.

INDUSTRIAL PUSHBUTTONS

CONTACT BLOCK

—||— NORMALLY OPEN (NO) CONTACT

—|/|— NORMALLY CLOSED (NC) CONTACT

LEGEND PLATE

DOWN

OPERATOR

Figure 6-1. An industrial pushbutton consists of a legend plate, an operator, and one or more contact blocks (electrical contacts).

Operators

An *operator* is the device that is pressed, pulled, or rotated by the individual operating the circuit. An operator activates the pushbutton's contacts. Operators are available in many different colors, shapes, and sizes. Standard pushbutton operators include the flush, half-shrouded, extended, and jumbo mushroom buttons. The operator used depends on the application. **See Figure 6-2.**

Half-Shrouded Button Operators. A *half-shrouded button operator* is a pushbutton with a guard ring that extends over the top half of the button. The guard ring helps prevent accidental operation but allows for easier operation with the thumb. The half-shrouded button operator is used where preventing accidental operation is preferred but where the technician may be wearing gloves. Wearing gloves makes depressing a flush button operator difficult.

OPERATORS

Figure 6-2. An operator is the device that is pressed, pulled, or rotated by the individual operating the circuit.

Flush Button Operators. A *flush button operator* is a pushbutton with a guard ring surrounding the button that prevents accidental operation. The flush button operator is the most common operator used in applications in which accidental turn-on may create a dangerous situation.

A three-position selector switch is a type of operator device that allows the user to select one of three circuit conditions.

Extended Button Operators. An *extended button operator* is a pushbutton that has the button extended beyond the guard. An extended button operator is easily accessible and the color of the operator may be seen from all angles. The extended button operator is the most common operator used in applications in which an accidental start is not dangerous, such as when turning on lights.

Jumbo Mushroom Button Operators. A *jumbo mushroom button operator* is a pushbutton that has a large curved operator extending beyond the guard. A jumbo mushroom button operator is easily seen because of its large size. It can be operated from any angle and is used in applications that require fast operation such as emergency stops, motor stops, and valve shutoffs.

Contact Blocks

A *contact block* is the part of the pushbutton that is activated when the operator is pressed. A contact block includes the switching contacts of the pushbutton. Contact blocks include normally open (NO), normally closed (NC), or both NO and NC contacts. The most common contact block includes one NO and one NC contact. NO contacts make the circuit when the pushbutton operator is pressed and are used mainly for start or ON functions. NC contacts break the circuit when the pushbutton operator is pressed and are used mainly for stop or OFF functions. More than one contact block may be added to an operator. **See Figure 6-3.**

Tech Tip

Enclosures with prestamped holes for mounting pushbuttons, selector switches, and joysticks are available in sizes of 30.5 mm (standard size) and 22 mm (miniature size). Since control device contact blocks are stackable to provide additional normally open (NO) and normally closed (NC) contacts on each device in the minimum amount of space, the required number of contacts, and thus, the correct enclosure depth, must be determined for each application.

Pushbuttons are housed in pushbutton stations. A *pushbutton station* is an enclosure that protects the pushbutton, contact block, and wiring from dust, dirt, water, and corrosive fluids. Enclosures are available in various sizes and with a number of punched holes

for mounting the operators. Every basic NEMA and IEC enclosure size is available because pushbutton stations need to be mounted where they can be conveniently operated.

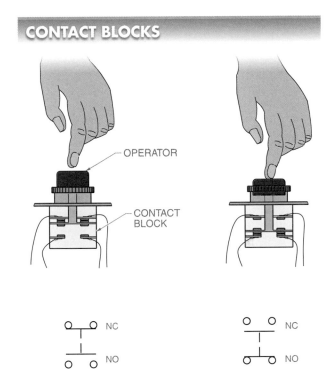

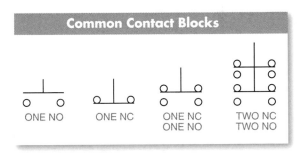

Figure 6-3. Contact blocks include normally open (NO), normally closed (NC), or both NO and NC contacts.

A pushbutton must be placed in the proper enclosure for continuous and safe operation. Pushbuttons are often required to operate in environments where dust, dirt, oil, vibration, corrosive material, extreme variations of temperature and humidity, as well as other damaging factors are present. The correct components and enclosure should always match the environment in which they will operate. **See Figure 6-4. See Appendix.**

NEMA ENCLOSURE CLASSIFICATION

Type	Use	Service Conditions	Test	Comments	Type
1	Indoor	No unusual	Rod entry, rust resistance		
3	Outdoor	Windblown dust, rain, sleet, and ice on enclosure	Rain, external icing, dust, and rust resistance	Does not provide protection against dust, internal condensation, or internal icing	1
3R	Outdoor	Falling rain and ice on enclosure	Rod entry, rain, external icing, and rust resistance	Does not provide protection against dust, internal condensation, or internal icing	
4	Indoor/outdoor	Windblown dust and rain, splashing water, hose-directed water, and ice on enclosure	Hosedown, external icing, and rust resistance	Does not provide protection against internal condensation or internal icing	4
4X	Indoor/outdoor	Corrosion, windblown dust and rain, splashing water, hose-directed water, and ice on enclosure	Hosedown, external icing, and corrosion resistance	Does not provide protection against internal condensation or internal icing	
6	Indoor/outdoor	Occasional temporary submersion at a limited depth			4X
6P	Indoor/outdoor	Prolonged submersion at a limited depth			
7	Indoor locations classified as Class I, Groups A, B, C, or D, as defined in the NEC®	Withstand and contain an internal explosion of specified gases, contain an explosion of specified gases, contain an explosion sufficiently so an explosive gas-air mixture in the atmosphere is not ignited	Explosion, hydrostatic, and temperature	Enclosed heat-generating devices shall not cause external surfaces to reach temperatures capable of igniting explosive gas-air mixtures in the atmosphere	7
9	Indoor locations classified as Class II, Groups E or G, as defined in the NEC®	Dust	Dust penetration, temperature, and gasket aging	Enclosed heat-generating devices shall not cause external surfaces to reach temperatures capable of igniting explosive gas-air mixtures in the atmosphere	9
12	Indoor	Dust, falling dirt, and dripping noncorrosive liquids	Drip, dust, and rust resistance	Does not provide protection against internal condensation	12
13	Indoor	Dust, spraying water, oil, and noncorrosive coolant	Oil explosion and rust resistance	Does not provide protection against internal condensation	

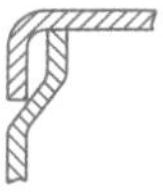

IEC ENCLOSURE CLASSIFICATION

IEC Publication 529 describes standard degrees of protection that enclosures of a product must provide when properly installed. The degree of protection is indicated by two letters, IP, and two numerals. International Standard IEC 529 contains descriptions and associated test requirements to define the degree of protection that each numeral specifies. The following table indicates the general degrees of protection. For complete test requirements refer to IEC 529.

FIRST NUMERAL*†	SECOND NUMERAL*†
Protection of persons against access to hazardous parts and protection against penetration of solid foreign objects	Protection against liquids‡ under test conditions specified in IEC 529
0 Not protected	0 Not protected
1 Protection against objects greater than 50 mm in diameter (hands)	1 Protection against vertically falling drops of water (condensation)
2 Protection against objects greater than 12.5 mm in diameter (fingers)	2 Protection against falling water with enclosure tilted 15°
3 Protection against objects greater than 2.5 mm in diameter (tools, wires)	3 Protection against spraying of falling water with enclosure tilted 60°
4 Protection against objects greater than 1.0 mm in diameter (tools, small wires)	4 Protection against splashing water
5 Protection against dust (dust may enter during test but must not interfere with equipment operation or impair safety)	5 Protection against low-pressure water jets
6 Dusttight (no dust observable inside enclosure at end of test)	6 Protection against powerful water jets
	7 Protection against temporary submersion
	8 Protection against continuous submersion

Example: IP41 describes an enclosure that is designed to protect against the entry of tools or objects greater than 1 mm in diameter, and to protect against vertically dripping water under specified test conditions.

* All first and second numerals up to and including numeral 6 imply compliance with the requirements of all preceding numerals in their respective series. Second numerals 7 and 8 do not imply suitability for exposure to water jets unless dual coded; e.g., IP_5/IP_7.
† The IEC permits use of certain supplementary letters with the characteristic numerals. If such letters are used, refer to IEC 529 for an explanation.
‡ The IEC test requirements for degrees of protection against liquid ingress refer only to water.

Figure 6-4. The correct components and enclosure should always match the environment in which they will operate.

6-1 CHECKPOINT

1. What is the best type of pushbutton operator to use in an emergency stop application?
2. Can one pushbutton operate both NO and NC contacts?
3. What type (number) NEMA enclosure can be used to protect internal electrical components against prolonged submersion at a limited depth, such as occasional flooding of an area?
4. What is the IEC number used to describe the enclosure type that can be used to protect internal electrical components against prolonged submersion at a limited depth, such as occasional flooding of an area?

6-2 SELECTOR SWITCHES

A *selector switch* is a switch with an operator that is rotated (instead of pushed) to activate the electrical contacts. Selector switches select one of several different circuit conditions. They are normally used to select either two or three different circuit conditions. However, selector switches are available that have more than three positions.

Two-Position Selector Switches

A two-position selector switch allows the operator to select one of two circuit conditions. For example, a two-position selector switch may be used to place a heating circuit in the manual (HAND) or automatic (AUTO) condition. **See Figure 6-5.** Only the manual control switch can turn the heating contactor on or off when the selector switch is placed in the HAND position. The heating contactor controls the high-power heating elements. Only the temperature switch can turn the heating contactor on or off when the selector switch is placed in the AUTO position. Circuit conditions controlled by two-position selector switches include ON/OFF, left/right, manual/automatic, up/down, slow/fast, run/stop, forward/reverse, jog/run, and open/close conditions.

Three-Position Selector Switches

A three-position selector switch allows the operator to select one of three circuit conditions. For example, a three-position selector switch may be used to place a heating circuit in the manual, automatic, or OFF position. The OFF position is added for safety. In the OFF position, the heating contactor (or other machine being controlled) cannot be energized by the manual or automatic switch. Circuit conditions controlled by

three-position selector switches include manual/OFF/automatic, heat/OFF/cool, forward/OFF/reverse, jog/OFF/run, slow/stop/fast, and up/stop/down conditions. **See Figure 6-6.**

TWO-POSITION SELECTOR SWITCHES

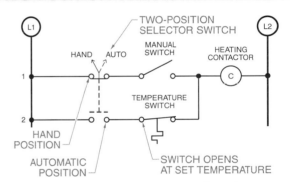

Figure 6-5. A two-position selector switch allows the operator to select one of two circuit conditions.

THREE-POSITION SELECTOR SWITCHES

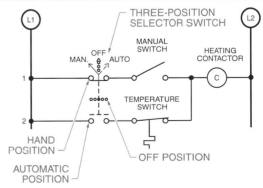

Figure 6-6. A three-position selector switch allows the operator to select one of three circuit conditions.

Truth Tables

Contact position on a selector switch may be illustrated using truth tables (target tables) or solid lines, dashed lines, and a series of small circles. **See Figure 6-7.** In truth tables, each contact on the line diagram is marked A, B, etc., and each position of the selector switch is marked 1, 2, etc. The truth table is made and positioned near the switch to illustrate each position and each contact.

An X is placed in the table if a contact is closed in any position. The table is easily read as to what contacts are closed in what positions. An O is placed in the table if a contact is open. Truth tables illustrate the selector switch contacts more clearly than the method of using solid and dashed lines and small circles when a selector switch has more than two contacts or more than three positions.

SELECTOR SWITCH CONTACT POSITIONS

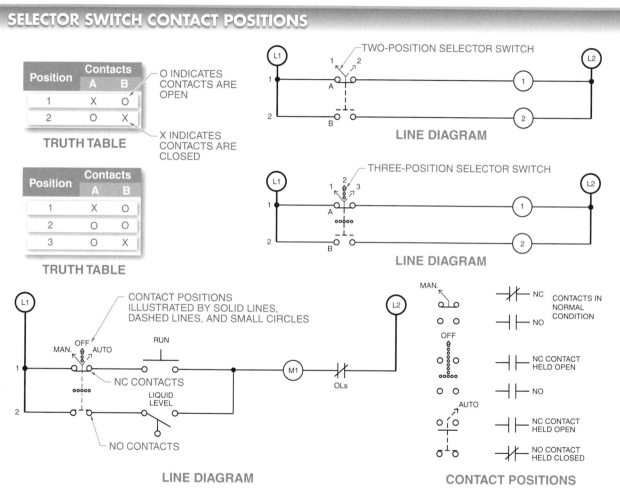

Figure 6-7. Contact position on a selector switch may be illustrated using truth tables (target tables) or solid lines, dashed lines, and a series of small circles.

6-2 CHECKPOINT

1. A selector switch with a legend plate labeled SLOW/FAST is an example of what type of selector switch?

2. A selector switch with a legend plate labeled SLOW/OFF/FAST is an example of what type of selector switch?

3. In a truth table, an X represents what condition of the contacts?

4. In a truth table, an O represents what condition of the contacts?

6-3 JOYSTICKS

A *joystick* is an operator that selects one to eight different circuit conditions when the joystick is shifted from the center position into one of the other positions. The most common joysticks can move from the center position into one of four different positions (up, down, left, or right). The advantage of a joystick is that a technician may control many operations without removing their hand from the joystick and without taking their eyes off the operation performed by the circuit.

The most common circuit condition controlled by a joystick is in controlling a hoist (or crane) in the raise, lower, left, right, or OFF position. **See Figure 6-8.** In the hoist application, two reversing motors move the hoist and pulleys. One forward and reversing motor starter controls the hoist drive motor, and another forward and reversing motor starter controls the pulley motor. The joystick can turn only one motor starter on at a time.

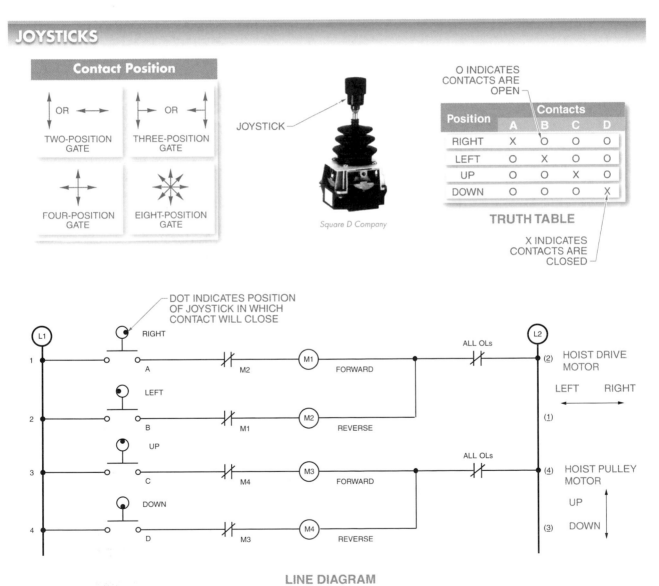

Figure 6-8. A joystick is used to control many different circuit operations from one location.

Two methods are used to indicate in which position the joystick must be placed to operate the contacts. In the first method, a dot is placed in the symbol of the joystick to indicate the position the joystick must be in to switch the contacts. The NO contacts close and the NC contacts open when the contacts are switched.

In the second method, a truth table is used to indicate which contacts are switched in each position. In the truth table, an X indicates when the contact is closed. Truth tables are normally given in manufacturer's catalogs showing joystick operation, and a dot in the symbol is normally used on the line diagram.

6-3 CHECKPOINT

1. What is the main advantage of using a joystick in an application such as moving the passenger loading ramp to and from an airplane door?

2. In a joystick symbol, what does the dot represent?

6-4 LIMIT SWITCHES

A *limit switch* is a mechanical input that requires physical contact of the object with the switch actuator. The physical contact is obtained from a moving object that comes in contact with the limit switch. The mechanical motion physically opens or closes a set of contacts within the limit switch enclosure. The contacts start or stop the flow of current in the electrical circuit. The contacts start, stop, operate in forward, operate in reverse, recycle, slow, or speed an operation. **See Figure 6-9.**

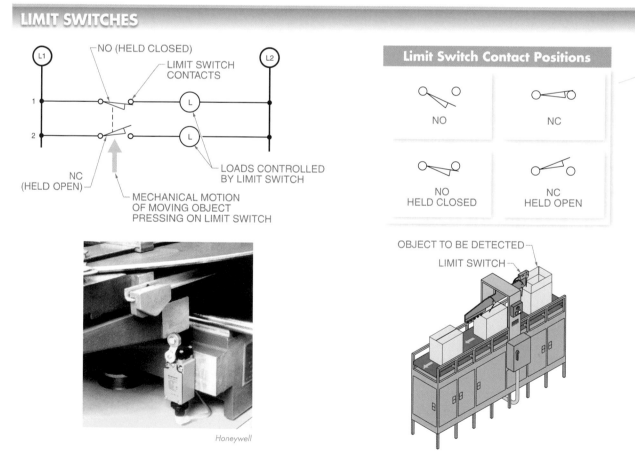

Figure 6-9. Limit switches are used to convert a mechanical motion into an electrical signal.

For example, a limit switch is used to automatically turn on a light in a refrigerator or prevent a microwave oven from operating with the door open. In a washing machine, a limit switch is used to automatically turn off the washer if the load is not balanced. In an automobile, limit switches are used to automatically turn on lights when a door is opened and prevent overtravel of automatically operated windows. In industry, limit switches are used to limit the travel of machine parts, sequence operations, detect moving objects, monitor an object's position, and provide safety by, for example, detecting guards in place.

Limit switch contacts are normally snap-acting switches, which quickly change position to minimize arcing at the contacts. Limit switch contacts may be NO, NC, or any combination of NO and NC contacts. Most limit switches include one NO contact and one NC contact. Contacts are rated for the maximum current and voltage they can safely control.

Limit switch contacts must be connected to the proper polarity. There is no arcing between the contacts when the contacts energize and de-energize the load as long as the contacts are at the same polarity. Arcing or welding of the contacts may occur from a possible short circuit if the contacts are connected to opposite polarity. **See Figure 6-10.** Contacts must be selected according to proper voltage and current size according to the load and manufacturer specifications.

A relay, contactor, or motor starter must be used to interface the limit switch with the load if the load current exceeds the contact rating. **See Figure 6-11.**

Limit Switch Actuators

An *actuator* is the part of a limit switch that transfers the mechanical force of the moving part to the electrical contacts. The basic actuators used on limit switches include lever, fork lever, push-roller, and wobble-stick. Most manufacturers offer several variations in addition to the basic actuators. **See Figure 6-12.**

Small limit switches are available with one fixed actuator. The fixed actuator may be any one of several different types, but it is neither removable nor interchangeable. Large limit switches are available with a knurled shaft that allows different actuators to be attached. The actuator used depends on the application.

Levers. A *lever actuator* is an actuator operated by means of a lever that is attached to the shaft of the limit switch. The lever actuator includes a roller on the end that helps to prevent wear. The length of the lever may be fixed or adjustable. The adjustable lever is used in applications in which the length of the arm, or actuator travel, may require adjustment. A lever actuator may be operated from either direction but is normally used in applications in which the actuating object is moving in only one direction. A typical application is on an assembly line conveyor system.

Fork Levers. A *fork lever actuator* is an actuator operated by either one of two roller arms. Fork lever actuators are used where the actuating object travels in two directions. A typical application is a grinder that automatically alternates back and forth.

CONTACT POLARITY CONNECTIONS

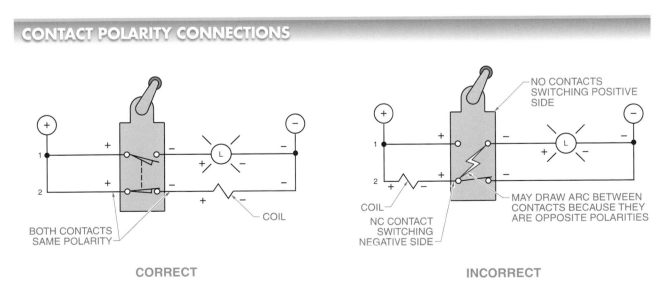

CORRECT   **INCORRECT**

Figure 6-10. Arcing or welding of the contacts may occur from a possible short circuit if the contacts are connected to opposite polarity.

CONTACT LOAD CONNECTIONS

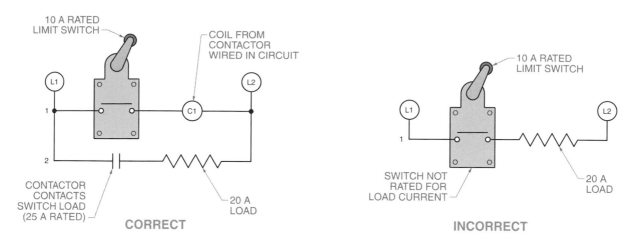

Figure 6-11. A relay, contactor, or motor starter must be used to interface the limit switch with the load if the load current exceeds the contact rating.

LIMIT SWITCH ACTUATORS

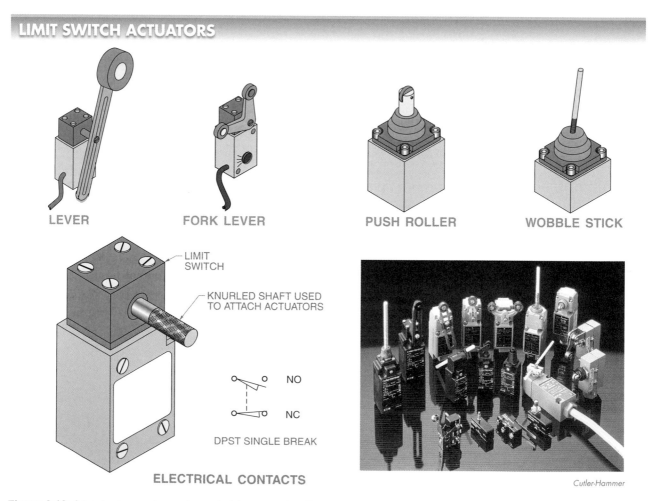

Cutler-Hammer

Figure 6-12. An actuator, such as a lever, fork lever, push roller, or wobble stick, is the part of a limit switch that transfers the mechanical force of the moving part to the electrical contacts.

Push Rollers. A *push-roller actuator* is an actuator operated by direct forward movement into the limit switch. A direct thrust with very limited travel is accomplished. Push-roller actuators are commonly used to prevent overtravel of a machine part or object. The switch contacts stop the forward movement of the object when the machine part comes in contact with the limit switch. A typical application is on a milling machine or an automatic turret lathe, because the travel of the work surface needs to be monitored.

Wobble Sticks. A *wobble-stick actuator* is an actuator operated by means of any movement into the switch, except a direct pull. The wobble-stick actuator normally has a long arm that may be cut to the required length. Wobble-stick actuators are used in applications that require detection of a moving object from any direction such as in the robotics section of an automated manufacturing facility.

Limit Switch Installation

Limit switches are actuated by a moving part. Limit switches must be placed in the correct position in relationship to the moving part. Limit switches should not be operated beyond the manufacturer's recommended travel specifications. **See Figure 6-13.**

Limit switch contacts do not operate if the actuating object does not force the limit switch actuator to move far enough. Overtravel may also damage the limit switch or force it out of position.

A rotary cam-operated limit switch must be installed according to manufacturer recommendations. A push-roller actuator should not be allowed to snap back freely. The cam should be tapered to allow a slow release of the lever. This helps to eliminate roller bounce and switch wear and allows for better repeat accuracy. **See Figure 6-14.**

Limit switches installed where relatively fast motions are involved must be installed so that the limit switch's lever does not receive a severe impact. The cam should be tapered to extend by the time it takes to engage the electrical contacts. This prevents wear on the switch and allows the contacts a longer closing time, ensuring that the circuit is complete.

Limit switches using push-roller actuators must not be operated beyond their travel limit in emergency conditions. A lever actuator is used instead of a push-roller actuator in applications where an override may occur. **See Figure 6-15.**

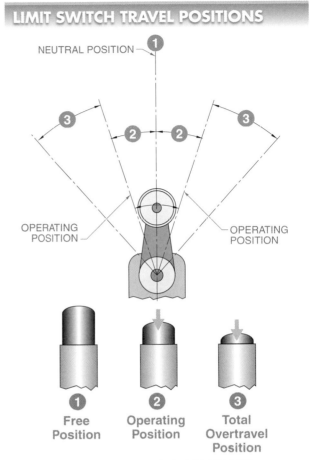

LIMIT SWITCH TRAVEL POSITIONS

NEUTRAL POSITION

OPERATING POSITION

OPERATING POSITION

1 Free Position

2 Operating Position

3 Total Overtravel Position

FORCES ACTING ON PLUNGER

Figure 6-13. Limit switches should not be operated beyond the manufacturer's recommended travel specifications.

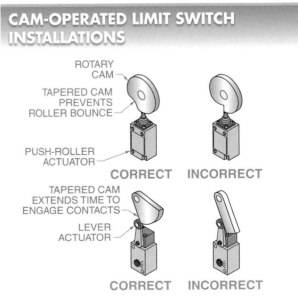

CAM-OPERATED LIMIT SWITCH INSTALLATIONS

ROTARY CAM

TAPERED CAM PREVENTS ROLLER BOUNCE

PUSH-ROLLER ACTUATOR

CORRECT   **INCORRECT**

TAPERED CAM EXTENDS TIME TO ENGAGE CONTACTS

LEVER ACTUATOR

CORRECT   **INCORRECT**

Figure 6-14. A cam-operated limit switch must be installed to prevent severe impact and allow a slow release of the lever.

LIMIT SWITCH OVERTRAVEL CONDITIONS

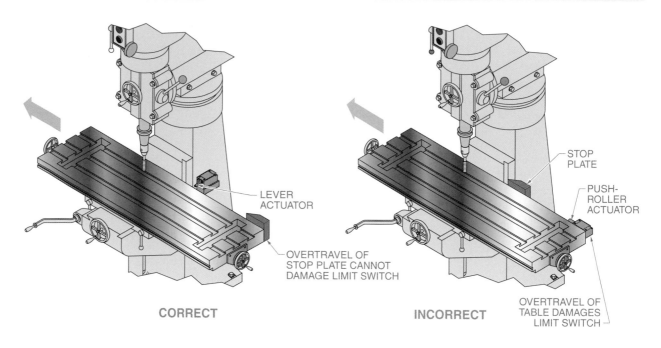

CORRECT

STOP PLATE

PUSH-ROLLER ACTUATOR

LEVER ACTUATOR

OVERTRAVEL OF STOP PLATE CANNOT DAMAGE LIMIT SWITCH

INCORRECT

OVERTRAVEL OF TABLE DAMAGES LIMIT SWITCH

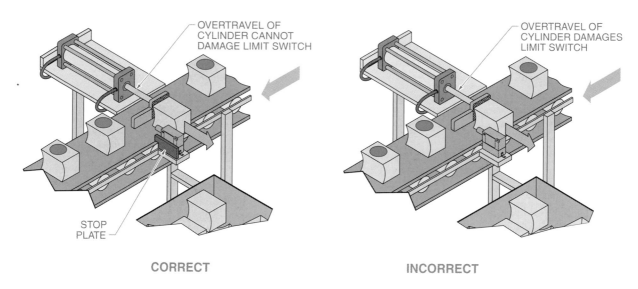

OVERTRAVEL OF CYLINDER CANNOT DAMAGE LIMIT SWITCH

OVERTRAVEL OF CYLINDER DAMAGES LIMIT SWITCH

STOP PLATE

CORRECT

INCORRECT

Figure 6-15. Limit switches using push-roller actuators must not be operated beyond their travel limit.

A lever actuator has an extended range, which prevents damage to the switch and mounting in case of an overtravel condition. A limit switch should never be used as a stop. A stop plate should always be added to protect the limit switch and its mountings from any damage due to overtravel.

Limit switches are designed to be used as automatic controllers that are mechanically activated. Care should be taken to prevent any human error. Limit switches should be mounted so that a technician cannot accidentally activate the limit switch. **See Figure 6-16.**

LIMIT SWITCH MOUNTING CONSIDERATIONS

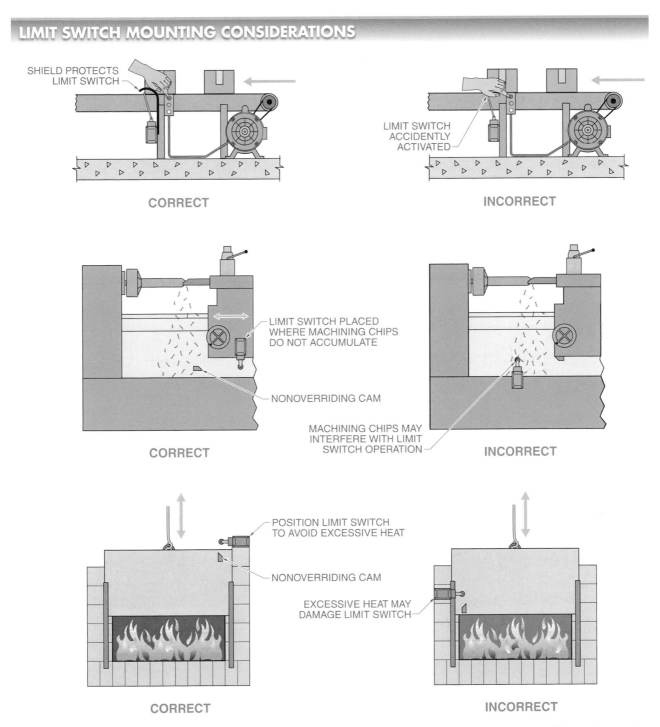

Figure 6-16. Limit switches should be mounted in a way that prevents accidental activation, accumulated materials, and excessive heat.

The atmosphere and surroundings of the limit switch must be considered when mounting a limit switch. A limit switch must be mounted in a location where machining chips or other materials do not accumulate. These could interfere with the operation of the limit switch and cause circuit failure. Submerging the limit switch or splashing it with oils, coolants, or other liquids must be avoided. Heat levels above the specified limits of the switch must also be avoided. A limit switch should always be positioned to avoid any excessive heat.

6-4 CHECKPOINT

1. If the moving part of a limit switch is drawn below the contacts in the symbol, is the switch NO/NO held closed or NC/NC held open?
2. If the moving part of the limit switch is drawn above the contacts in the symbol, is the switch NO/NO held closed or NC/NC held open?
3. What type of limit switch actuator allows detection of a moving object from any direction except head on?
4. What type of limit switch actuator allows the arm length to be adjusted to meet the application?

6-5 FOOT SWITCHES

A *foot switch* is a control switch that is operated by a person's foot. A foot switch is used in applications that require a person's hands to be free or that require an additional control point. Foot switch applications include sewing machines, drill presses, lathes, and other similar machines. Most foot switches have two positions: a toe-operated position and an OFF position.

The OFF position is normally spring-loaded so that the switch automatically returns to the OFF position when released. Foot switches with three positions include a pivot on a fulcrum to allow toe or heel control. Like the two-position foot switch, the three-position foot switch is normally spring-loaded so that the switch automatically returns to the OFF position when released. **See Figure 6-17.**

FOOT SWITCHES

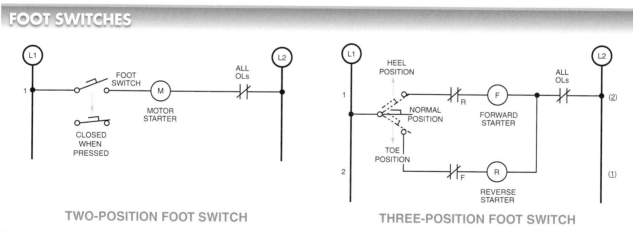

TWO-POSITION FOOT SWITCH   THREE-POSITION FOOT SWITCH

Figure 6-17. A foot switch is used to allow hands-free control or an additional control point.

6-5 CHECKPOINT

1. When a foot switch has three positions, does the toe usually operate all three positions?
2. Is the OFF position of a foot switch usually a maintained position or a spring-loaded position?

6-6 PRESSURE SWITCHES

Pressure is force exerted over a surface divided by its area. The exerted force always produces a deflection or change in the volume or dimension of the area to which it is applied. Pressure is expressed in pounds per square inch (psi). Low pressures are expressed in inches of water column (in. WC). One psi equals 27.68 in. WC.

A *pressure switch* is a switch that detects a set amount of force and activates electrical contacts when the set amount of force is reached. The contacts may be activated by positive, negative (vacuum), or differential pressures. Differential pressure switches are connected to two different system pressures.

Depending on the application, NC or NO contacts are used for a pressure switch. **See Figure 6-18.** NC contacts are used to maintain system pressure. The closed contacts energize a pump motor until system pressure is reached.

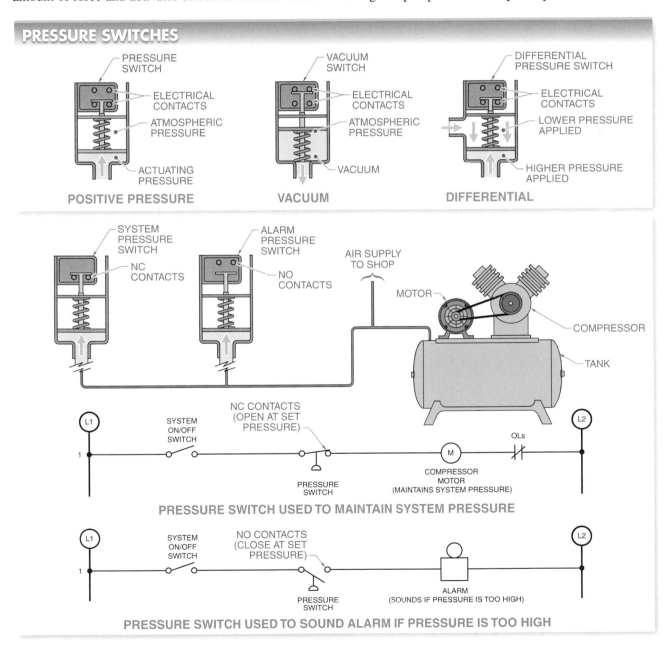

Figure 6-18. A pressure switch is a control switch that detects a set amount of force and activates electrical contacts when the set amount of force is reached.

When system pressure is reached, the contacts open and the pump motor is turned off. NO contacts are used to signal an overpressure condition. An alarm is sounded when the open contacts close. The alarm remains on until the pressure is reduced.

Pressure switches use different sensing devices to detect the amount of pressure. The pressure switch used depends on the application and system pressure. Most pressure switches use a diaphragm, bellows, or piston sensing device. Newer pressure switches incorporate solid-state technology.

Diaphragm and bellows sensing devices are used for low-pressure applications. Piston sensing devices are used for high-pressure applications. **See Figure 6-19.**

SENSING DEVICES

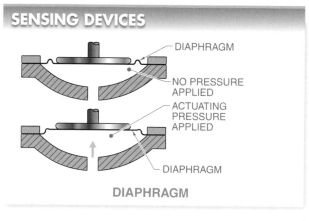

DIAPHRAGM

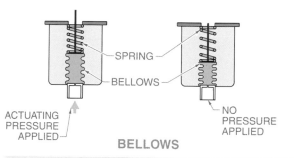

BELLOWS

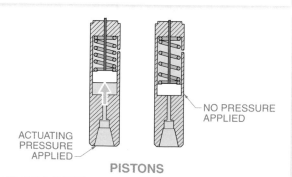

PISTONS

Figure 6-19. Pressure switches use different sensing devices, including diaphragms, bellows, and pistons, to detect the amount of pressure.

Diaphragms

A *diaphragm* is a deflecting mechanism that moves when a force (pressure) is applied. One side of the diaphragm is connected to the pressure to be detected (source pressure) and the other side is vented to the atmosphere.

The diaphragm moves against a spring switch mechanism that operates electrical contacts when the source pressure increases. The spring tension is adjustable to allow for different pressure settings. A diaphragm pressure switch is used with pressures of less than 200 psi, but some are designed to detect several thousand pounds of pressure.

Bellows

A *bellows* is a cylindrical device with several deep folds that expand or contract when pressure is applied. One end of the bellows is closed and the other end is connected to the source pressure. The expanding bellows moves against a spring switch mechanism that operates electrical contacts when the source pressure increases.

The spring tension is adjustable to allow for different pressure settings. A bellows pressure switch is used with pressures up to 500 psi, but some are designed for higher pressures.

Pistons

A *piston* is a cylinder that is moved back and forth in a tight-fitting chamber by the pressure applied in the chamber. A piston sensing device (pressure switch) uses a stainless steel piston moving against a spring tension to operate electrical contacts. The piston moves a switch mechanism that operates electrical contacts when the source pressure increases. The spring tension is adjustable to allow for different pressure settings. Piston sensing devices (pressure switches) are designed for high-pressure applications of 10,000 psi or more.

Deadband

When a change in pressure occurs causing the diaphragm to move far enough to actuate the switch contacts, some of the pressure must be removed before the switch resets for another cycle. *Deadband (differential)* is the amount of pressure that must be removed before the switch contacts reset for another cycle after the setpoint has been reached and the switch has been actuated. **See Figure 6-20.**

DEADBAND

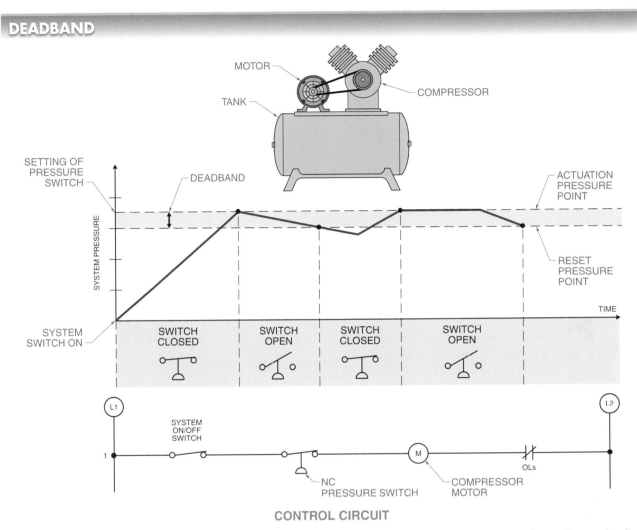

Figure 6-20. Deadband is the amount of pressure that must be removed before the switch contacts reset for another cycle after the setpoint has been reached and the switch has been actuated.

Deadband is inherent in all pressure, temperature, level, and flow switches and most automatically actuated switches. Deadband is not a fixed amount and is different at each setpoint. Deadband is minimum when the setpoint is at the low end of the switch range. Deadband is maximum when the setpoint is at the high end of the switch range.

Deadband may be beneficial or detrimental. Without a deadband range, or too small of one, electrical contacts chatter on and off as a pressure switch approaches the setpoint. However, a large deadband is detrimental in applications that require the pressure to be maintained within a very close range. Different switches have different deadband ratings. The amount of listed deadband should always be checked when using pressure switches in different applications.

Pressure Switch Applications

Most pressure switches are used to maintain a predetermined pressure in a tank or reservoir. Pressure switches may also be used to sequence the return of pneumatic or hydraulic cylinders. **See Figure 6-21.**

Tech Fact

Pressure switches are available with a variety of different enclosures. For example, weathertight enclosures are available that enable the enclosure and switch to withstand harsh outside weather conditions. Explosionproof enclosures are available that can contain an internal explosion. Hermetically sealed enclosures are available that completely seal the switch from the outside damaging environment. Pressure switch enclosures are also available with analog or digital pressure displays.

PRESSURE SWITCHES

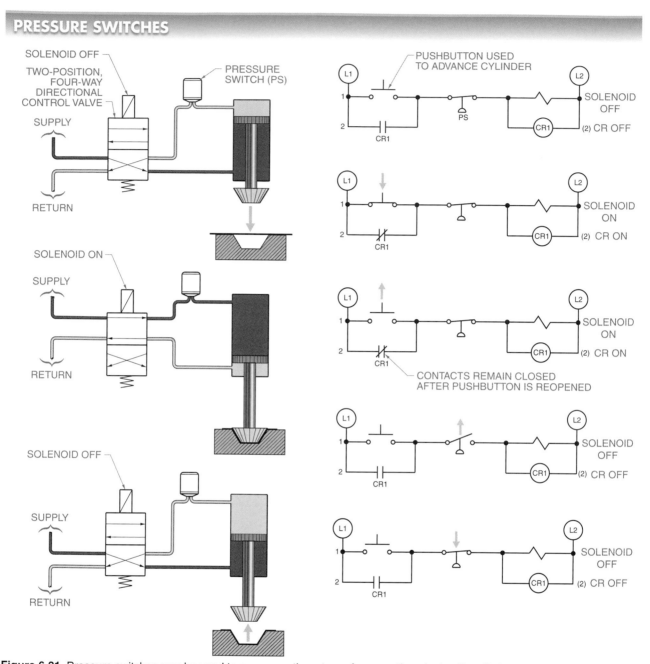

Figure 6-21. Pressure switches may be used to sequence the return of pneumatic or hydraulic cylinders.

The two-position, four-way, directional control valve solenoid is energized when the technician presses the start pushbutton. This changes the directional control valve from the spring position to the solenoid-actuated position. The cylinder advances because the flow of pressure is changed in the cylinder. The control relay energizes and its NO contacts close because it is in parallel with the solenoid. This adds memory to the circuit. The technician releases the pushbutton and the cylinder continues to advance. The cylinder advances until the

preset pressure is reached on the pressure switch, which signals the return of the cylinder by de-energizing the solenoid and relay. The de-energizing of the solenoid returns the directional control valve to the spring position. The return of the directional control valve to the spring position reverses the flow in the cylinder that returns it. The return of the cylinder occurs until the pushbutton is pressed. The setting of the pressure switch depends on the application of the cylinder. This setting may be low for packing fragile materials or high for forming metals.

The advantage of using a pressure switch over a pushbutton for returning the cylinder is that the load always receives the same amount of pressure before the cylinder returns. An emergency stop could be added to the control circuit for manual return of the cylinder. A low-range pressure switch may be used with a metal tubing arrangement in a fluidic sensor. **See Figure 6-22.** In this application, a constant low-pressure stream of air is directed at a sheet of material through the metal tubing. As long as the material in process is present, the air stream is deflected. The stream of air is sensed in the receiver tube if the material breaks. The pressure switch would signal corrective action through the control relay.

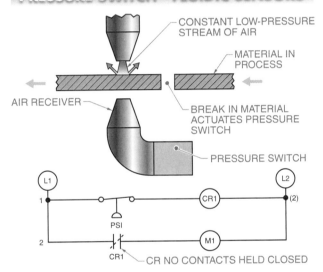

Figure 6-22. A low-range pressure switch may be used with a metal tubing arrangement in a fluidic sensor to determine the presence of a material in process.

A fluidic sensor has certain advantages over a photoelectric sensor. For example, the air stream flowing over the material in normal operation may perform a second function such as cooling, cleaning, or drying the material.

An air compressor uses a pressure switch to turn the compressor motor on or off based on the amount of pressure.

6-6 CHECKPOINT

1. When a pressure switch is rated in "in. WC," what does "in. WC" stand for?

2. If a pressure switch is used to maintain a set pressure, are NO or NC contacts used to control the compressor motor starter?

6-7 TEMPERATURE SWITCHES

A *temperature switch* is a control device that reacts to heat intensity. Temperature switches are used in heating systems, cooling systems, fire alarm systems, process control systems, and equipment/circuit protection systems. In most applications, temperature switches react to rising or falling temperatures. Cooling systems, alarm systems, and protection systems use temperature switches that react to rising temperatures. Heating systems use temperature switches that react to falling temperatures.

A heating system maintains a set temperature when the ambient temperature drops. In a heating system, as ambient temperature drops, the switch contacts close and turn on the heat-producing device. The heat-producing device may be an electric coil, gas furnace, heat pump, or any device that produces heat. **See Figure 6-23.**

The temperature switch energizes a heating contactor. The heating contactor energizes the heat-producing coils. By having the temperature switch control a contactor, the high current required by the heating coils does not pass through the contacts of the temperature switch. This reduces the size of the required temperature switch and increases the life of the contacts. A temperature switch may also sound an alarm if the temperature rises too high.

HEATING SYSTEMS

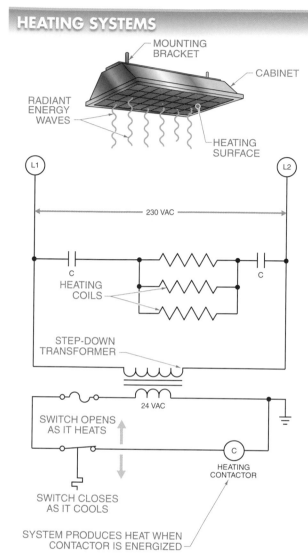

Figure 6-23. In a heating system, heat is produced when the temperature switch contacts cool and close.

COOLING SYSTEMS

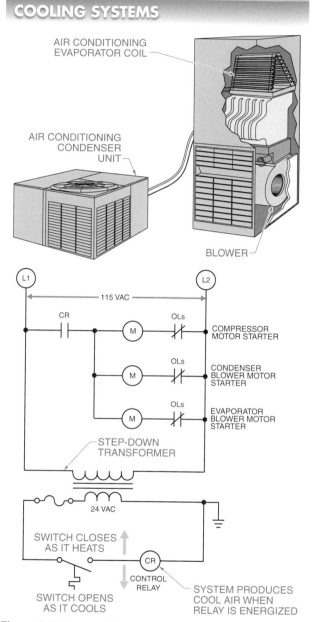

Figure 6-24. In a cooling system, cool air is produced when the temperature switch contacts heat and close.

A cooling system maintains a set temperature when the ambient temperature rises. In a cooling system, as the ambient temperature rises, the switch contacts close and turn on a cooling device. The cooling device may be a standard air conditioning unit, cooling tower, radiator, or any device that is used to cool. **See Figure 6-24.**

The temperature switch energizes a control relay. The control relay energizes the motor starters. The motor starters control the motors that produce cooling and provide overload protection for the motors. The control relay may also open a valve that allows water (or coolant) to circulate over the heated area.

6-7 CHECKPOINT

1. If a temperature switch is used in a heating system to maintain a set temperature, are NO or NC contacts used to control the heating contactor?

2. If a temperature switch is used in a cooling system to maintain a set temperature, are NO or NC contacts used to control the air condition system?

6-8 FLOW SWITCHES

Flow is the travel of fluid in response to a force caused by pressure or gravity. The fluid may be air, water, oil, or some other gas or liquid. Most industrial processes depend on fluids flowing from one location to another. Problems may occur if the flow is stopped or slowed. Flow may be stopped by a frozen pipe, a clogged pipe, or an improperly closed valve (manual or automatic).

A *flow switch* is a control switch that detects the movement of a fluid. **See Figure 6-25.** Applications that use flow switches to detect the presence or absence of flow include the following:

- boilers
- cooling lines
- air compressors
- fluid pumps
- food processing systems
- machine tools
- sprinkler systems
- water treatment systems

- heating processes
- refrigeration systems
- chemical processing and refining

Flow switches use different methods to detect whether the fluid is flowing. The methods used to detect whether the fluid is flowing include the paddle and transmitter/receiver methods. In the paddle method, a paddle extends into the pipe or duct. The paddle moves and actuates electrical contacts when the fluid flow is sufficient to overcome the spring tension on the paddle. The spring tension is adjustable on many flow switches, allowing for different flow rate adjustments.

In the transmitter/receiver method, a transmitter sends a signal through the pipe. The receiver picks up the transmitted signal. The strength of the signal changes when the product is flowing. One common transmitter/receiver method uses a sound transmitter to produce sound pulses through the fluid. Moving solids or bubbles in the fluid reflect a distorted sound back to the receiver. The transmitter/receiver unit is adjustable for detecting different flow rates.

FLOW SWITCHES

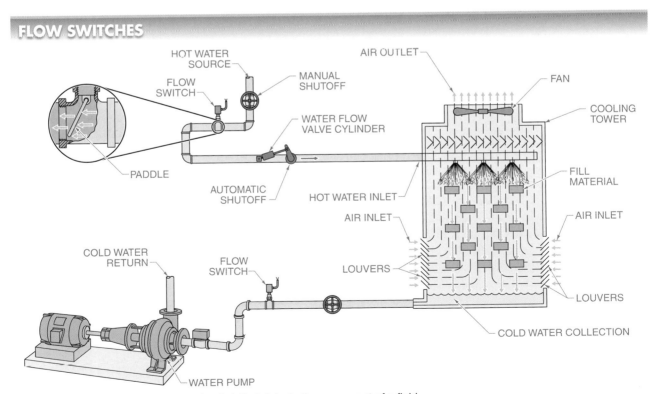

Figure 6-25. A flow switch is a control switch that detects the movement of a fluid.

Both NO and NC electrical contacts can be used with flow switches. In some applications, a flow of fluid indicates a problem. For example, in an automatic sprinkler system used as fire protection, the flow of water indicates a problem. In this application, an NO contact on the flow switch could be used to sound an alarm. **See Figure 6-26.** When a fire (or high heat) opens the sprinkler head, the water starts flowing through the pipe. The flow of water closes the NO contacts and the alarm sounds. The alarm sounds as long as the water is flowing.

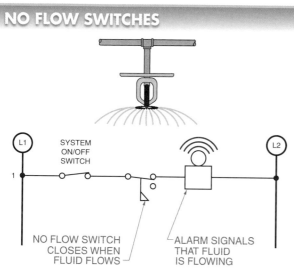

Figure 6-26. A NO electrical contact can be used with flow switches.

Tech Tip

Switch contacts that use low voltage or low current, or are used in a corrosive atmosphere, are subject to sticking and high-resistance switching. To reduce contact problems in such cases, flow switches that contain gold-plated contacts can be used.

An NC contact is used to signal when a fluid is not flowing. The NC contact may be used to sound an alarm if fluid stops flowing. When fluid is flowing, the NC contacts are held open by the fluid flow.

A flow switch may also be used to detect airflow across the heating elements of an electric heater. **See Figure 6-27.** The heating elements burn out if sufficient airflow is not present. The flow switch is used as an economical way to turn the heater off anytime there is not enough airflow. This circuit can also be applied to an air conditioning or refrigeration system. In this circuit,

the flow switch is used to detect insufficient airflow over the refrigeration coils. The restricted airflow is normally caused by the icing of the coils, which blocks the airflow. In this case, the flow switch would automatically start the defrost cycle of the refrigeration unit.

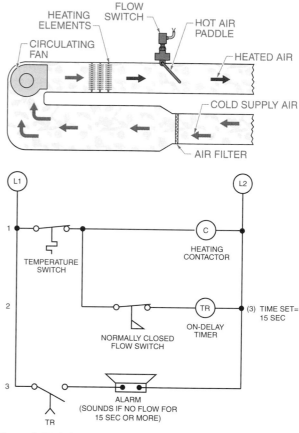

Figure 6-27. A flow switch may be used to determine whether there is sufficient airflow across the heating elements of an electric heater.

Flow switches may also be used to detect the proper airflow in a ventilation system. **See Figure 6-28.** The ventilation system may be directing dangerous gases away from the technician. Poisonous gases could overcome the technician or damage could occur to the process involved if there is insufficient airflow.

Flow switches are often used to protect large motion picture projectors used in theaters, where poor airflow would cause heat buildup and reduce the life expectancy of the expensive bulbs used in the projectors. Airflow may be restricted from a large draft caused by high winds outside the building or by clogged air filters in the intake system.

Figure 6-28. A flow switch may be used to maintain a critical ventilation process.

A flow switch may be used to advance a clogged filter based on restricted airflow. **See Figure 6-29.** The flow switch is used to start a gear-reduced motor that slowly advances the roll of filter material until sufficient airflow is present.

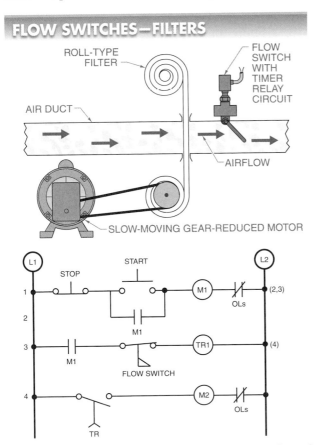

Figure 6-29. A flow switch may be used to advance a clogged filter when restricted airflow is sensed.

Flow switches are commonly used with boilers to detect if a fluid is moving.

6-8 CHECKPOINT

1. If a flow switch is used to sound an alarm when there is no flow, are NO or NC contacts used to control the alarm?
2. If a flow switch is used to detect flow in an automatic sprinkler system that has detected a fire and opened the sprinklers to flow, are NO or NC contacts used to control the alarm?

6-9 LEVEL SWITCHES

In most industrial plants there are tanks, vessels, reservoirs, and other containers in which process water, wastewater, raw materials, or product must be stored or mixed. The level of the product must be controlled. A *level switch* is a switch that detects the height of a liquid or solid (gases cannot be detected by level switches) inside a tank.

Systems that use level switches include processing systems for products such as milk, water, oil, beer, wine, solvents, plastic granules, coal, grains, sugar, chemicals, and many other products. Different level switches are used to detect each product. Factors that determine the correct level switch to use for an application include the following:

• Motion—Turbulence causes some level switches to chatter on and off or actuate falsely.

• Corrosiveness—Level switches are made of different materials such as stainless steel, copper, plastic, etc. It is important to always use a level switch made of a material that is compatible with the product to be detected.

• Density—All solid materials have a certain density. Capacitive level switches are designed to detect different amounts of density.

• Physical state—The physical state of a liquid depends on its type, temperature, and condition. Any liquid or solid may be detected if the correct level switch is used. Different level switches are designed to operate at different temperatures and to detect different types and thicknesses of liquids.

• Movement—A moving product may require a special level switch. A product that is stationary for a long period may cause certain mechanical level switches to stick.

• Conductivity—Some level switches with metal probes placed in a liquid depend on the liquid to be a conductor for proper operation.

• Abrasiveness—Noncontact level switches should be used with abrasive products.

• Sensing distance—Some level switches are designed to detect short distances and others are designed to detect long distances.

All level switches are designed to detect a certain range of materials. Some level switches can only detect liquids; others can detect both liquids and solids. Some level switches must come in direct contact with the product

to be detected; others do not need to make contact. The level switch used depends on the application, cost, life expectancy, and product to be detected. The different level switches include mechanical, magnetic, conductive probe, capacitive, optical, and ultrasonic level switches.

Mechanical

Mechanical level switches were the first level switches used and are still one of the most common. A *mechanical level switch* is a level switch that uses a float that moves up and down with the level of the liquid and activates electrical contacts at a set height. **See Figure 6-30.** Mechanical level switches may be used with many different liquids because the float is the only part of the switch that is in contact with the liquid. Mechanical level switches work well with water (even dirty water) and any other liquid that dries without leaving a crust. Mechanical limit switches are not used with paint because the paint builds up on the float as the paint dries. The dried paint weighs the float down and affects the operation of the switch.

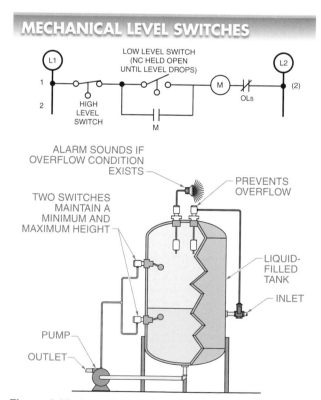

Figure 6-30. A mechanical level switch uses a float, which moves up and down with the level of the liquid and activates electrical contacts at a set height.

One of the most common applications of a mechanical level switch is in sump pumps found in most houses with basements. In a sump application, the level switch turns on a pump when the level reaches a set height.

Conductive Probe

A *conductive probe level switch* is a level switch that uses liquid to complete the electrical path between two conductive probes. The voltage that is applied to the probes is 24 V or less. The liquid must be a fair to good electrical conductor. All water-based solutions conduct electricity to some degree. The conductance of water increases as salts and acids are added to the water, which decreases resistance.

A fluid with an electrical resistance of less than 25 kΩ allows a sufficient amount of current to pass through it to actuate the relay inside the conductive probe level control. The electrical resistance is high and no current passes through the probes when the conductive liquid is no longer between the two probes. **See Figure 6-31.**

The number of probes used and their length depends on the application. Two probes of the same length may be used to detect a liquid at a given height. The relay inside the level control is activated when the liquid reaches the probes. The relay is no longer activated when the liquid is no longer in contact with the probes.

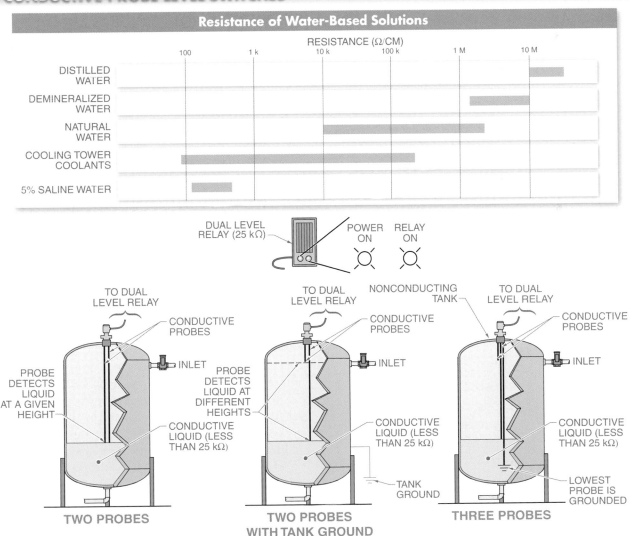

Figure 6-31. A conductive probe level switch uses liquid to complete the electrical path between two conductive probes.

Two probes of different lengths and a ground may be used to detect a liquid at different heights. The ground is connected to the conductive tank. The relay is activated when the liquid reaches the highest probe. The relay is not deactivated until the liquid no longer is in contact with the lowest probe. The probes may be any distance apart. Three probes of different lengths are used if the tank that holds the liquid is made of a nonconductive material. The lowest probe is connected to the ground wire of the level control.

Capacitive

A *capacitive level switch* is a level switch that detects the dielectric variation when the product is in contact (proximity) with the probe and when the product is not in contact with the probe. *Dielectric variation* is the range at which a material can sustain an electric field with a minimum dissipation of power. Capacitive level switches are used to detect solids or granules, such as sand, sugar, grain, and chemicals, in addition to some liquids. The capacitance of the sensor is changed when the product comes in proximity with the sensor. Some capacitive sensors are adjustable to allow only certain products to be sensed. Capacitive level switches are available that work well with hard-to-detect products such as plastic granules, shredded paper, copying machine toner, and fine powders. **See Figure 6-32.**

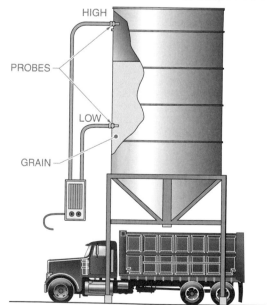

Figure 6-32. A capacitive level switch detects the dielectric variation when the product is in contact with the probe and when the product is not in contact with the probe.

Charging and Discharging

Level switches detect and respond to the level of a material in a tank. The response is normally to charge or discharge the tank. Charging a tank is also known as pump control and discharging a tank is also known as sump control.

In a charging application, the level in a tank is maintained. As liquid is removed from the tank, the level switch signals the circuit to add liquid. Liquid may be added by opening a valve or starting a pump motor. The liquid is added until the level switch detects the correct height. Flow is stopped when the level switch detects the correct height. Liquid is added when the level switch is no longer in contact with the liquid. **See Figure 6-33.**

In a discharging application, the liquid in a tank is removed once it reaches a predetermined level. Liquid may be removed by a pump or through gravity when a valve is opened. The liquid is removed until the level switch detects the tank is empty.

One- or Two-Level Control

In level control applications, the distance between the high and low level must be considered. This distance may be small or large. In applications using a one-level switch, the distance is small. In applications using a two-level switch, the distance may be any length. **See Figure 6-34.** Although it may appear that a small distance maintained in a system is the best, this may not be true because the smaller the distance to be maintained, the greater the number of times the pump motor must cycle on and off.

Since motors draw much more current when starting than when running, excess heat is produced in a motor that must turn on and off frequently. The faster the level in the tank drops, the faster the motor must cycle.

While it may appear that two-level registration is the best because the pump does not have to cycle as often, what is best for the motor may not be best for the total system. For example, if common house paint is to be maintained in a fill tank, problems develop if the length of time between the high level and low level is excessive. As the paint dries on the inside of the tank, it accumulates layer by layer. This causes skin to form, which may clog or impede the pump or fill action if the skin falls into the product.

Therefore, product type must be taken into consideration when determining the distance and time between the high and low level. In general, one-level control is best when the liquid is emptied very slowly from the tank. Two-level control is best when the liquid is emptied at a fast rate.

CHARGING AND DISCHARGING

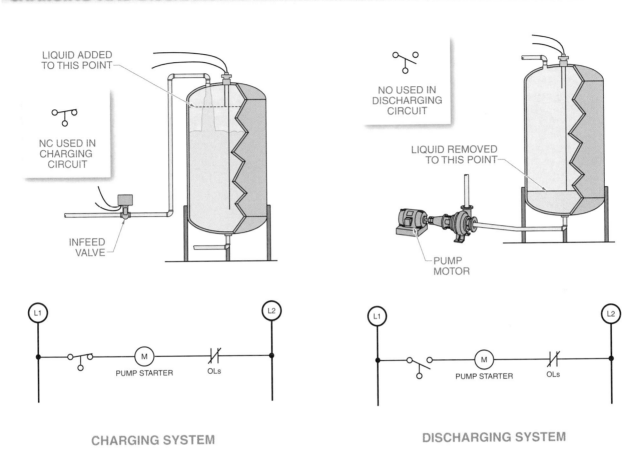

Figure 6-33. Level switches detect and respond to the level of a material in a tank.

LEVEL CONTROL

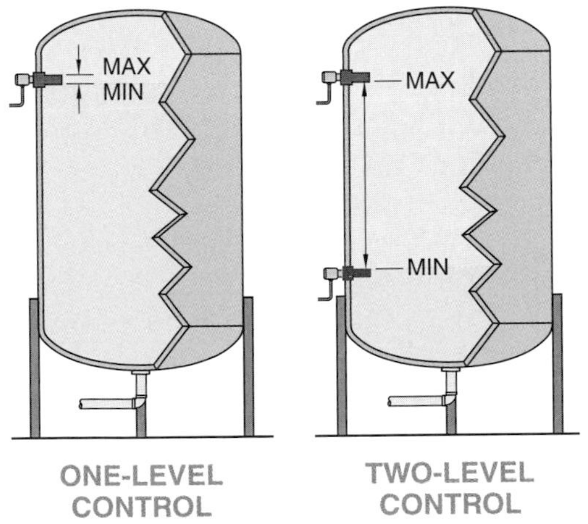

ONE-LEVEL CONTROL

TWO-LEVEL CONTROL

Figure 6-34. The distance controlled in one-level control is small, but any distance may be controlled in two-level control.

6-9 CHECKPOINT

1. If a level switch is used in a charging system, are NO or NC contacts used to control the pump motor starter?
2. If a level switch is used in a discharging system, are NO or NC contacts used to control the pump motor starter?

6-10 PREVENTING PROBLEMS WHEN INSTALLING CONTROL DEVICES

All electrical circuits must be controlled. For this reason, control devices are used in every type of control application. A control device must be properly protected and installed to ensure that the control device operates properly for a long time. Proper protection means that the switching contacts are operated within their electrical rating and are not subjected to destructive levels of current or voltage. Proper installation means that the control device is installed in such a manner as to ensure it operates as designed.

Protecting Switch Contacts

Control devices are used to switch loads on or off or redirect the flow of current in an electrical circuit. The control devices switch contacts that are rated for the amount of current they can safely switch. The switch rating is normally specified for switching a resistive load, such as small heating elements. Resistive loads are the least destructive loads to switch. However, most loads that are switched are inductive loads, such as solenoids and motor starter coils. Inductive loads are the most destructive loads to switch because of the collapsing magnetic field present due to CEMF when the contacts are opened.

A large induced voltage appears across the switch contacts when inductive loads are turned off. The induced voltage is the opposite polarity of the applied voltage. The induced voltage causes arcing at the switch contacts. Arcing may cause the contacts to burn, stick, or weld together. Contact protection should be added when frequently switching inductive loads to prevent or reduce arcing. **See Figure 6-35.**

A diode is added in parallel with the load to protect contacts that switch DC. The diode conducts only when the switch is open, providing a path for the induced voltage in the load to dissipate.

A resistor and capacitor (RC network or snubber) are connected across the switch contacts to protect contacts that switch AC. The capacitor acts as a high-impedance (resistor) load at 60 Hz, but becomes a short circuit at the high frequencies produced by the induced voltage of the load. This allows the induced voltage to dissipate across the resistor when the load is switched off.

Protecting Pressure Switches

A pressure switch is a switch that detects a set amount of force and activates electrical contacts when the set amount of force is reached. Pressure switches are designed to activate their contacts at a preset pressure. A pressure switch is rated according to its operating pressure range. A pressure switch may be damaged if its maximum pressure limit is exceeded.

Protection for a pressure switch should be added in any system in which a higher pressure than the maximum limit is possible. A pressure relief valve is installed to protect the pressure switch. **See Figure 6-36.** A pressure relief valve should be set just below the pressure switch's maximum limit. The valve opens when the system pressure increases to the setting of the relief valve.

CONTACT PROTECTION

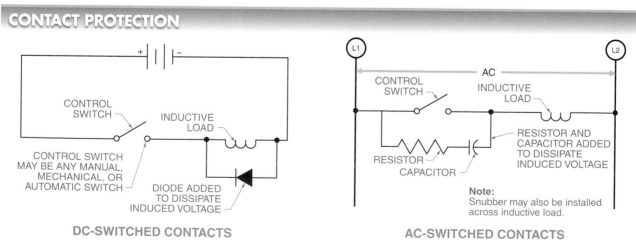

DC-SWITCHED CONTACTS

AC-SWITCHED CONTACTS

Figure 6-35. Contact protection should be added when switching large DC and AC inductive loads to prevent or reduce arcing at the switch contacts.

PRESSURE SWITCH RELIEF VALVES

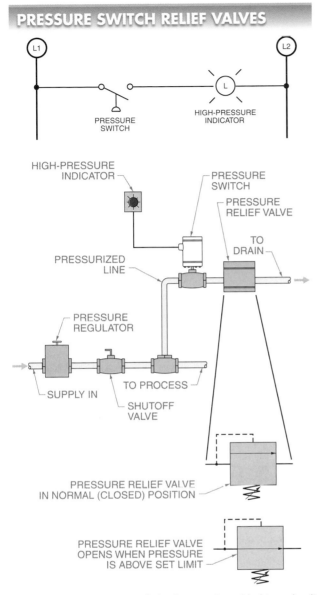

Figure 6-36. A pressure relief valve may be added to a circuit to protect a pressure switch from excessive pressure.

CAUTION: The output of the relief valve must be connected to a proper drain (or return line) if the product under pressure is a gas or fluid.

Installing Flow Switches

A *flow switch* is a switch that detects the movement of a fluid. Most flow switches use a paddle to detect the movement of the product. The paddle is designed to detect the product movement with the least possible pressure drop across the switch. A flow switch must be installed correctly to ensure it does not interfere with the movement of the product. Most flow switches are designed to operate in the horizontal position. There is a great deal of turbulence in flowing product at a distance of at least three pipe inside diameters (ID) on each side of the flow switch. **See Figure 6-37.** For example, the minimum horizontal distance of straight pipe required on each side of a flow switch is 4½″ (1½″ × 3 = 4½″) when used in an application that moves a product through a 1½″ diameter pipe.

FLOW SWITCH MOUNTING

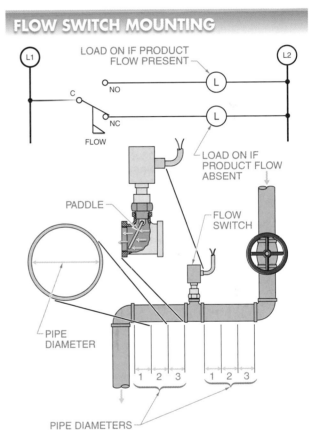

Figure 6-37. A distance of at least three pipe inside diameters (ID) should be allowed on each side of the flow switch when mounting a flow switch.

Tech Tip

Water hammer damages flow switches. Water hammer is the banging that is caused when flow is suddenly stopped, such as when a valve suddenly closes. To protect flow switches, a water hammer arrestor should be installed before the switch. A water hammer arrestor absorbs the shock, stops pipe banging, and prevents pipe and switch damage.

Testing Electromechanical Switches

A suspected fault with an electromechanical switch is tested using a DMM set to measure voltage. The voltage setting on the DMM is used to test the voltage flowing into and out of the switch. **See Figure 6-38.** To test a mechanical switch, the following procedure is applied:

1. Measure the voltage into the switch. Connect the DMM between the neutral and hot conductor feeding the switch. Set the DMM to the voltage setting. When working on a grounded system, the DMM lead may be connected to ground instead of neutral if the neutral conductor is not available in the same box in which the switch is located. The problem is located upstream from the switch when there is no voltage present or the voltage is not at the correct level. The problem may be a blown fuse or open circuit. Voltage must be reestablished to the switch before the switch may be tested.

2. Measure the voltage out of the switch. There should be a voltage reading when the switch contacts are closed. There should not be a voltage reading when the switch contacts are open. The switch has an open and must be replaced if there is no voltage reading in either switch position. The switch has a short and must be replaced if there is a voltage reading in both switch positions.

WARNING: Always ensure power is OFF before changing a control switch. Use a DMM set to measure voltage to ensure the power is OFF.

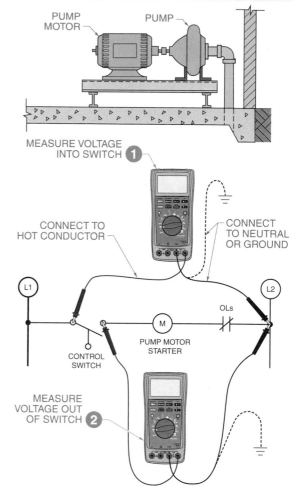

TESTING ELECTROMECHANICAL SWITCHES

Figure 6-38. A DMM set to measure voltage is used to test the operation of a switch.

6-10 CHECKPOINT

1. Is a diode used to help protect electrical contacts in an AC or DC load switching circuit?
2. What position are most flow switches designed to operate in?
3. If a voltage is measured going into a control switch and no voltage is measured out of the switch, what operating conditions might the switch be in?
4. If a voltage is measured going into a control switch and the same voltage is measured going out of the switch, what operating conditions might the switch be in?

Additional Resources

Review and Resources

Access Chapter 6 Review and Resources for *Electrical Motor Controls for Integrated Systems* by scanning the above QR code with your mobile device.

Applying Your Knowledge

Refer to the *Electrical Motor Controls for Integrated Systems* Learner Resources for interactive Applying Your Knowledge questions.

Workbook and Applications Manual

Refer to Chapter 6 in the *Electrical Motor Controls for Integrated Systems Workbook* and the *Applications Manual* for additional exercises.

ENERGY EFFICIENCY PRACTICES

Wind Energy

The majority of motors and motor control circuits are powered by electricity generated via power plants. Many remote applications exist where standard electrical power is not easily available or is too costly to provide. Such applications include agricultural pumping (watering and drainage), logging, and mining operations.

Alternative energy sources such as wind turbines can be used to supply environmentally friendly power to applications in remote locations. In such applications, alternating current (AC) generated from wind turbines can be used to power AC motors, motor starters, and control circuits. AC can be changed to direct current (DC) using a converter, which can then be used to power DC motors, motor starters, and control circuits.

Wind turbines convert kinetic energy from wind into electricity. In a wind turbine, the wind rotates turbine blades that are connected to a rotor. The rotor converts the kinetic energy from wind into rotary motion, which drives a generator. Most wind turbine systems have automatic overspeed governing systems that limit rotor speed in high wind conditions.

Wind turbine systems can be grid-connected (connected to an electrical distribution grid) or off-grid (stand-alone) systems. A grid-connected wind turbine system is used to reduce consumption of utility-supplied electricity for lighting loads, motor loads, and electric heat loads. If a grid-connected system cannot produce the required amount of electricity, additional electricity is available from the utility company. If a grid-connected system produces more electricity than is required, the excess electricity may be sold to the utility company. An off-grid wind turbine system is not connected to an electrical distribution grid. Off-grid systems are designed to provide adequate power for single applications.

Objectives

Sections

7-1
- Define electromagnet and explain the left-hand rule.
- State the main characteristics of electromagnets.

7-2
- Define solenoid and state the function of each part used to produce linear motion in a solenoid.
- Understand why solenoids are available in different operating configurations to meet different application requirements.
- Describe the construction of a solenoid.

7-3
- State common problems, such as inrush current, that solenoids produce in the electrical system.
- Describe the solenoid selection methods.

7-4
- State the number of positions, number of ways, and type of actuator a directional control valve has.
- List the different types of systems and applications solenoids may be used for.

7-5
- State how transients are caused and how they can be reduced within an electrical system.
- Describe the steps required when using a digital multimeter (DMM) to troubleshoot a solenoid within a system.

Review and Resources
atplearningresources.com/Quicklinks
Access Code: 362245

Electrical energy is used to produce light, heat, sound, rotary motion, and linear motion. Solenoids are the devices used to convert electrical energy into a linear mechanical force. Solenoids are an important part of most systems because they electrically operate fluid power valves, brakes, clutches, motor starters, contactors, relays, locks, and many other devices. Solenoids are a typical part of residential, commercial, industrial, medical, and military electrical systems as well as transportation systems such as automobiles, trucks, airplanes, boats, high-speed trains, and maglev trains. Learning the different types of solenoids, how they operate, and how to troubleshoot them is an important part of understanding most electrically controlled systems.

7-1 ELECTROMAGNETS

An *electromagnet* is a magnet whose magnetic energy is produced by the flow of electric current. Some electromagnets are so large and powerful that they can lift tons of scrap metal at one time. Other electromagnets used in some electrical and electronic circuits are very small, such as those found in solenoids and relays.

An electromagnet consists of an iron core inserted into a coil. The iron core concentrates the lines of force produced by the coil. With the core in place and the coil energized, the polarity of the magnet can be determined using the left-hand rule. The left-hand rule states that if the fingers of the left hand are wrapped in the direction of the current flow in a coil, the left thumb points to the magnetic north pole of the coil. **See Figure 7-1.**

LEFT-HAND RULE

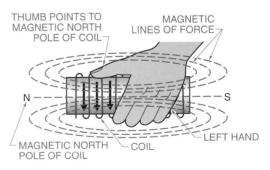

THUMB POINTS TO MAGNETIC NORTH POLE OF COIL

MAGNETIC LINES OF FORCE

N

S

MAGNETIC NORTH POLE OF COIL

COIL

LEFT HAND

Figure 7-1. The left-hand rule states that if the fingers of the left hand are wrapped in the direction of the current flow in a coil, the left thumb points to the magnetic north pole of the coil.

The advantages of electromagnets are that they can be made stronger than permanent magnets, and the magnetic strength can be easily controlled by regulating the electric current. The main characteristics of an electromagnet include the following:

- When electricity flows through a conductor, a magnetic field is created around that conductor.
- The field is stronger close to the wire and weaker further away.
- The strength of the magnetic field and the current are directly related. The more current, the stronger the magnetic field, the less current, the weaker the magnetic field.
- The direction of the magnetic field is determined by the direction of the current flowing through the conductor.
- The more permeable the core, the greater the concentration of magnetic lines of force.

7-1 CHECKPOINT

1. Does the added iron core of an electromagnet increase or decrease the magnetic lines of force?
2. In an electromagnet, what determines the strength of the magnetic field?

3. What is the left-hand rule used to determine when the direction of current flow is known?

7-2 SOLENOIDS

A *solenoid* is an electric output device that converts electrical energy into a linear mechanical force. The magnetic attraction of a solenoid may be used to transmit force. Solenoids may be combined with an armature, which transmits the force created by the solenoid into useful work. An *armature* is the movable part of a solenoid.

Solenoid Configurations

Solenoids are configured in various ways for different applications and operating characteristics. The five solenoid configurations are clapper, bell-crank, horizontal-action, vertical-action, and plunger. **See Figure 7-2.**

A clapper solenoid has the armature hinged on a pivot point. As voltage is applied to the coil, the magnetic effect produced pulls the armature to a closed position so that it is picked up (sealed in).

A bell-crank solenoid uses a lever attached to the armature to transform the vertical action of the armature into a horizontal motion. The use of the lever allows the shock of the armature to be absorbed by the lever and not transmitted to the end of the lever. This is beneficial when a soft but firm motion is required in the controls.

A horizontal-action solenoid is a direct-action device. The movement of the armature moves the resultant force in a straight line. Horizontal-action solenoids are one of the most common solenoid configurations.

A vertical-action solenoid also uses a mechanical assembly but transmits the vertical action of the armature in a straight-line motion as the armature is picked up.

Solenoids are commonly used in industrial control circuit applications such as hydraulic systems.

SOLENOID CONFIGURATIONS

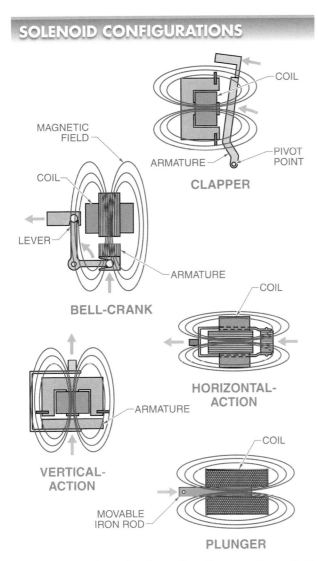

Figure 7-2. The five solenoid configurations are clapper, bell-crank, horizontal-action, vertical-action, and plunger solenoids.

A plunger solenoid contains only a movable iron cylinder, or rod. A movable iron rod placed within the electrical coil tends to equalize or align itself within the coil when current passes through the coil. The current causes the rod to center itself so that the rod ends line up with the ends of the solenoid if the rod and solenoid are of equal length. In a plunger solenoid, a spring is used to move the rod a short distance from its center in the coil. The rod moves against the spring tension to recenter itself in the coil when the current is turned on. The spring returns the rod to its off-center position when the current is turned off. The motion of the rod is used to operate any number of mechanical devices. **See Figure 7-3.**

PLUNGER SOLENOIDS

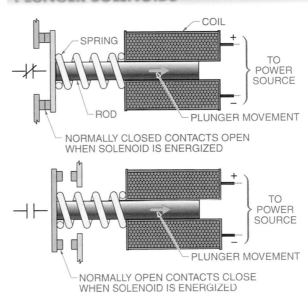

Figure 7-3. In a plunger solenoid, a spring is used to move the rod a short distance from its center in the coil. The rod moves against the spring tension to recenter itself when the current is turned on.

Solenoid Construction

Solenoids are constructed of many turns of wire wrapped around a magnetic laminate assembly. Passing electric current through the coil causes the armature to be pulled toward the coil. Devices may be attached to the solenoid to accomplish tasks like opening and closing contacts.

Eddy Current. *Eddy current* is unwanted current induced in the metal structure of a device due to the rate of change in the induced magnetic field. Strong eddy currents are generated in solid metal when used with alternating current. In AC solenoids, the magnetic assembly and armature consist of a number of thin pieces of metal laminated together. The thin pieces of metal reduce the eddy current produced in the metal. **See Figure 7-4.** Eddy current is confined to each lamination, thus reducing the intensity of the magnetic effect and subsequent heat buildup. For DC solenoids, a solid core is acceptable because the current is in one direction and continuous.

Armature Air Gap. To prevent chattering, solenoids are designed so that the armature is attracted to its sealed-in position so that it completes the magnetic circuit as completely as possible. To ensure this, both the faces on the magnetic laminate assembly and those on the armature are machined to a very close tolerance.

EDDY CURRENTS

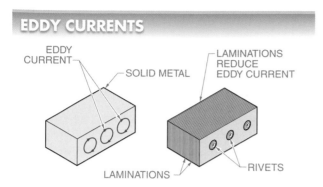

Figure 7-4. In AC solenoids, the magnetic assembly and armature consist of a number of thin pieces of metal laminated together.

As the coil is de-energized, some magnetic lines of force (residual magnetism) are always retained and could be enough to hold the armature in the sealed position. To eliminate this possibility, a small air gap is always left between the armature and the magnetic laminate assembly to break the magnetic field and allow the armature to drop away freely when de-energized. **See Figure 7-5.**

ARMATURE AIR GAPS

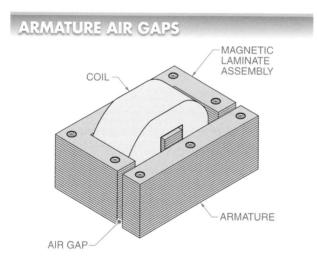

Figure 7-5. A small air gap is left in the magnetic laminate assembly to break the magnetic field and allow the armature to drop away freely after being de-energized.

Shading Coil. A *shading coil* is a single turn of conducting material (normally copper or aluminum) mounted on the face of the magnetic laminate assembly or armature. **See Figure 7-6.** A shading coil sets up an auxiliary magnetic field that helps hold in the armature as the main coil magnetic field drops to zero in an AC circuit.

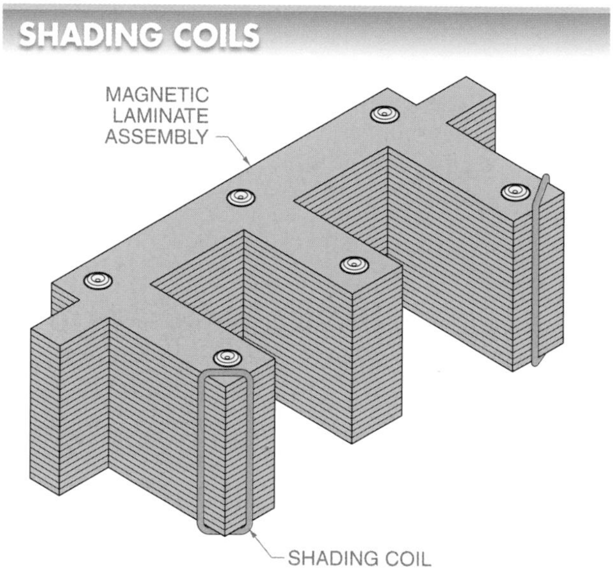

SHADING COILS

Figure 7-6. A shading coil sets up an auxiliary magnetic field that helps hold in the armature as the main coil magnetic field drops to zero in an AC circuit.

The magnetic field generated by alternating current periodically drops to zero. This makes the armature drop out or chatter. The attraction of the shading coil adds enough pull to the unit to keep the armature firmly seated. Without the shading coil, excessive noise, wear, and heat builds up on the armature faces, reducing the armature life expectancy.

7-2 CHECKPOINT

1. What is the movable part of a solenoid called?
2. Why are solenoid armatures laminated?
3. Why is a shading coil added to the armature of a solenoid?

7-3 SOLENOID CHARACTERISTICS

The two primary characteristics of a solenoid are the amount of voltage applied to the coil and the amount of current allowed to pass through the coil. Solenoid voltage characteristics include pick-up voltage, seal-in voltage, and drop-out voltage. Solenoid current characteristics include coil inrush current and sealed current.

Coils

Magnetic coils are normally constructed of many turns of insulated copper wire wound on a spool. The mechanical life of most coils is extended by encapsulating the coil in an epoxy resin or glass-reinforced alkyd material. **See Figure 7-7.** In addition to increasing mechanical strength, these materials greatly increase the moisture resistance of the magnetic coil. Because magnetic coils are encapsulated and cannot be repaired, they must be replaced when they fail.

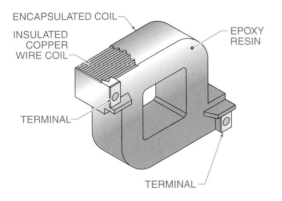

MECHANICAL COIL LIFE

Figure 7-7. The mechanical life of most coils is extended by encapsulating the coil in an epoxy resin or glass-reinforced alkyd material.

Coil Inrush and Sealed Currents. Solenoid coils draw more current when first energized than the amount that is required to keep them running. In a solenoid coil, the inrush current is approximately 6 to 10 times the sealed current. **See Figure 7-8.** After the solenoid has been energized for some time, the coil becomes hot, causing the coil current to fall and stabilize at approximately 80% of its value when cold. The reason for such a high inrush current is that the basic opposition to current flow when a solenoid is energized is only the resistance of the copper coil. Upon energizing, however, the armature begins to

move iron into the core of the coil. The large amount of iron in the magnetic circuit increases the magnetic opposition of the coil and decreases the current through the coil. This magnetic opposition is referred to as inductive reactance or total impedance. The heat produced by the coil further reduces current flow because the resistance of copper wire increases when hot, which limits some current flow.

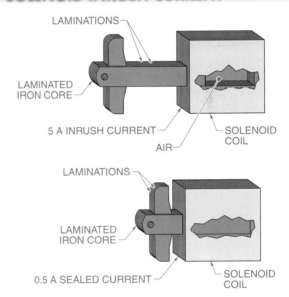

SOLENOID INRUSH CURRENT

Figure 7-8. Solenoid inrush current is approximately 6 to 10 times the sealed current.

Coil Inrush and Sealed Current Ratings. Magnetic coil data is normally given in volt amperes (VA). For example, a solenoid with a 120 V coil rated at 600 VA inrush and 60 VA sealed has an inrush current of 5 A ($^{600}/_{120} = 5$ A) and a sealed current of 0.5 A ($^{60}/_{120} = 0.5$ A). The same solenoid with a 480 V coil draws only 1.25 A ($^{600}/_{480} = 1.25$ A) inrush current and 0.125 A ($^{60}/_{480} = 0.125$ A) sealed current. The VA rating helps determine the starting and energized current load drawn from the supply line.

Tech Fact

Solenoids are rated for intermittent or continuous duty. An intermittent-duty solenoid is designed to produce a strong force in a small package but will overheat if current is continuously applied to the coil. A continuous-duty solenoid is designed to handle a continuous current but is larger to help dissipate the heat produced.

Coil Voltage Characteristics. All solenoids develop a magnetic field in their coil when voltage is applied. This magnetic field produces a force on the armature and tries to move it. The applied voltage determines the amount of force produced on the armature. The voltage applied to a solenoid should be ±10% of the rated solenoid value. A solenoid overheats when the voltage is excessive. The heat destroys the insulation on the coil wire and burns out the solenoid. The solenoid armature may have difficulty moving the load connected to it when the voltage is too low.

Pick-up voltage is the minimum voltage that causes the armature to start to move. *Seal-in voltage* is the minimum control voltage required to cause the armature to seal against the pole faces of the magnet. *Drop-out voltage* is the voltage that exists when voltage is reduced sufficiently to allow the solenoid to open. Seal-in voltage can be higher than pick-up voltage because a higher force may be required to seal in the armature than to just move the armature. Drop-out voltage is lower than pick-up voltage or seal-in voltage because it takes more force to hold the armature in place than to release the armature.

For most solenoids, the minimum pick-up voltage is about 80% to 85% of the solenoid rated voltage. The seal-in voltage is somewhat higher than the pick-up voltage and should be no less than 90% of the solenoid rated voltage. Drop-out voltage can be as low as 70% of the solenoid rated voltage. The exact pick-up, seal-in, and drop-out voltages depend on the load connected to the solenoid armature and the mounting position of the solenoid. The greater the applied armature load, the higher the required voltage values.

Voltage Variation Effects

Voltage variations are one of the most common causes of solenoid failure. Precautions must be taken to select the proper coil for a solenoid. Excessive or low voltage must not be applied to a solenoid coil.

Excessive Voltage. A coil draws more than its rated current if the voltage applied to the coil is too high. Excessive heat is produced, which causes early failure of the coil insulation. The magnetic pull is also too high and causes the armature to slam in with excessive force. This causes the magnetic faces to wear rapidly, reducing the expected life of the solenoid.

Low Voltage. Low voltage on the coil produces low coil current and reduced magnetic pull. The solenoid may pick up but does not seal in when the applied voltage is greater than the pick-up voltage but less than the seal-in voltage. The greater pick-up current (6 to 10 times sealed current) quickly heats up and burns out the coil because it is not designed to carry a high continuous current. The armature also chatters, which creates noise and increases the wear of the magnetic faces.

Solenoid Selection Methods

Solenoids are selected based on the outcome required. It is important to select the correct solenoid to achieve the desired outcome. Solenoid selection methods include push or pull, length of stroke, required force, duty cycle, mounting, and voltage rating.

Push or Pull. A solenoid may push or pull, depending on the application. In the case of a door latch, the unit must pull. In a clamping jig, the unit must push.

Length of Stroke. The length of the stroke is calculated after determining whether the solenoid must push or pull. For example, a door latch requires a ½″ maximum stroke length.

Required Force. Manufacturer specification sheets are used to determine the correct solenoid based on the required force. A solenoid is selected from the manufacturer specification sheets based on required solenoid function.

Duty Cycle. Solenoid characteristic tables are also used to check the duty cycle requirements of the application against the duty cycle information given for the solenoid. For example, an A 101 solenoid is required for an application requiring 190 operations per minute.

Mounting. Manufacturers provide letter or number codes to indicate the solenoid mount. **See Figure 7-9.** For example, an A solenoid is selected for a door latch application because the door latch application requires an end-mounting solenoid.

SOLENOID MOUNTING CODES

Code	Mounting
A	End
B	Right side
C	Throat
D	None (for thru-bolts)
E	Left side
F	Both sides

Figure 7-9. Manufacturers provide letter or number codes to indicate the solenoid mount.

Voltage Rating. Manufacturers provide letter or number codes to indicate the voltages that are available for a given solenoid. **See Figure 7-10.** For example, a 2 A solenoid may be used for an application that requires a 115 V coil.

Solenoids include class ratings. Solenoid class ratings refer to the insulation material temperature rating and include Class Y (90°C), Class A (105°C), Class E (120°C), Class B (130°C), Class F (155°C), Class H (180°C), and Class C (over 180°C).

SOLENOID VOLTAGE RATINGS	
Number	Voltage (in V)
2 X	115
3 X	230
4 X	460
5 X	575

Figure 7-10. Manufacturers provide letter or number codes to indicate the voltages that are available for a given solenoid.

7-3 CHECKPOINT

1. What is the initial current draw of a solenoid called?
2. What is the operating current of a solenoid called?
3. If the voltage applied to a solenoid is higher than the rating of the solenoid, will the current draw increase or decrease?
4. What is the number of times a solenoid can operate in a given time period (usually operations per minute) called?

7-4 SOLENOID APPLICATIONS

Solenoids can be found in a wide range of equipment. In residential equipment, solenoids can be found in doorbells, washing machines, and kitchen appliances. Solenoids are commonly used in commercial and industrial control circuit applications such as hydraulics/pneumatics, refrigeration, combustion, general-purpose controls, and pneumatic robotics.

Hydraulics/Pneumatics

Solenoid-operated valves typically control hydraulic and pneumatic equipment. A solenoid is used to move the valve spool that controls the flow of fluid (air or oil) in a directional control valve. A *directional control valve* is a valve that is used to direct the flow of fluid throughout a fluid power system. Directional control valves are identified by the number of positions, number of ways, and type of actuator.

Positions. A manual directional control valve is placed in different positions to start, stop, or change the direction of fluid flow. **See Figure 7-11.** A *position* is the number of locations within the valve in which the spool can be placed to direct fluid through the valve. A directional control valve normally has two or three positions.

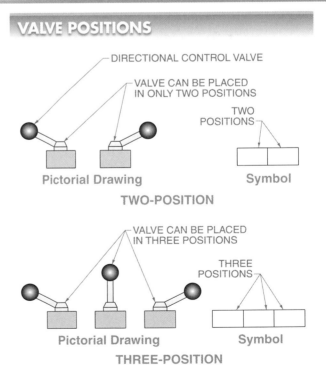

Figure 7-11. Positions are the number of locations within the valve in which the spool can be placed to direct fluid through the valve.

Ways. A way is a flow path through a valve. Most directional control valves are either two-way or three-way valves. The number of ways required depends on the application. Two-way directional control valves have two main ports that allow or stop the flow of fluid. Two-way valves are used as shutoff, check, and quick-exhaust valves. **See Figure 7-12.**

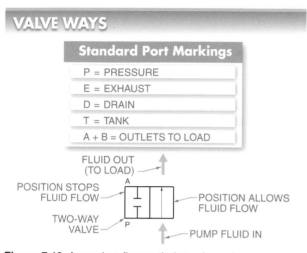

VALVE WAYS

Standard Port Markings

P = PRESSURE
E = EXHAUST
D = DRAIN
T = TANK
A + B = OUTLETS TO LOAD

FLUID OUT (TO LOAD)
POSITION STOPS FLUID FLOW
POSITION ALLOWS FLUID FLOW
TWO-WAY VALVE
PUMP FLUID IN

Figure 7-12. A way is a flow path through a valve.

Valve Actuators. A manual directional control valve uses a handle to change the valve spool position. An electrical control valve uses an actuator to change the position of a valve spool. In an electrical control valve, the solenoid acts as the actuator. **See Figure 7-13.**

Refrigeration

Direct-acting, two-way valves are commonly used in a refrigeration system. Two-way (shutoff) valves have one inlet and one outlet pipe connection. These units may be constructed as normally open (NO), where the valve is open when de-energized and closed when energized; or they may be constructed as normally closed (NC), where the valve is closed when de-energized and open when energized. **See Figure 7-14.**

A number of different solenoids may be used in a typical refrigeration system. The liquid line solenoid valves could be operated by two-wire or three-wire thermostats. The hot gas solenoid valve remains closed until the defrost cycle and then feeds the evaporator with hot gas for the defrost operation. **See Figure 7-15.**

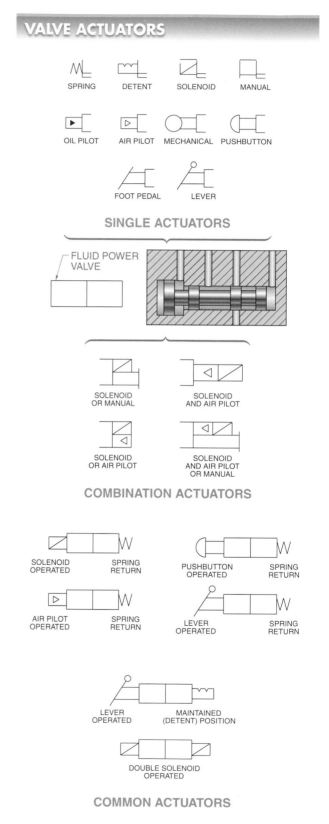

VALVE ACTUATORS

SPRING   DETENT   SOLENOID   MANUAL

OIL PILOT   AIR PILOT   MECHANICAL   PUSHBUTTON

FOOT PEDAL   LEVER

SINGLE ACTUATORS

FLUID POWER VALVE

SOLENOID OR MANUAL   SOLENOID AND AIR PILOT

SOLENOID OR AIR PILOT   SOLENOID AND AIR PILOT OR MANUAL

COMBINATION ACTUATORS

SOLENOID OPERATED   SPRING RETURN   PUSHBUTTON OPERATED   SPRING RETURN

AIR PILOT OPERATED   SPRING RETURN   LEVER OPERATED   SPRING RETURN

LEVER OPERATED   MAINTAINED (DETENT) POSITION

DOUBLE SOLENOID OPERATED

COMMON ACTUATORS

Figure 7-13. In an electrical control valve, the solenoid acts as the actuator.

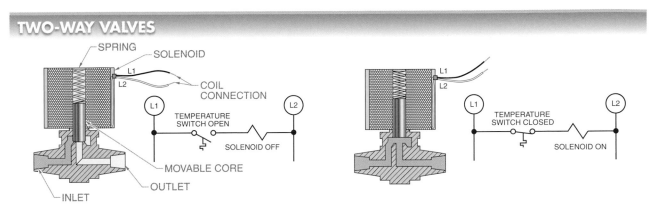

TWO-WAY VALVES

Figure 7-14. In a refrigeration system, direct-acting two-way valves may be constructed as normally open (NO), where the valve is open when de-energized and closed when energized; or they may be constructed as normally closed (NC), where the valve is closed when de-energized and open when energized.

HOT GAS AND LIQUID LINE SOLENOIDS

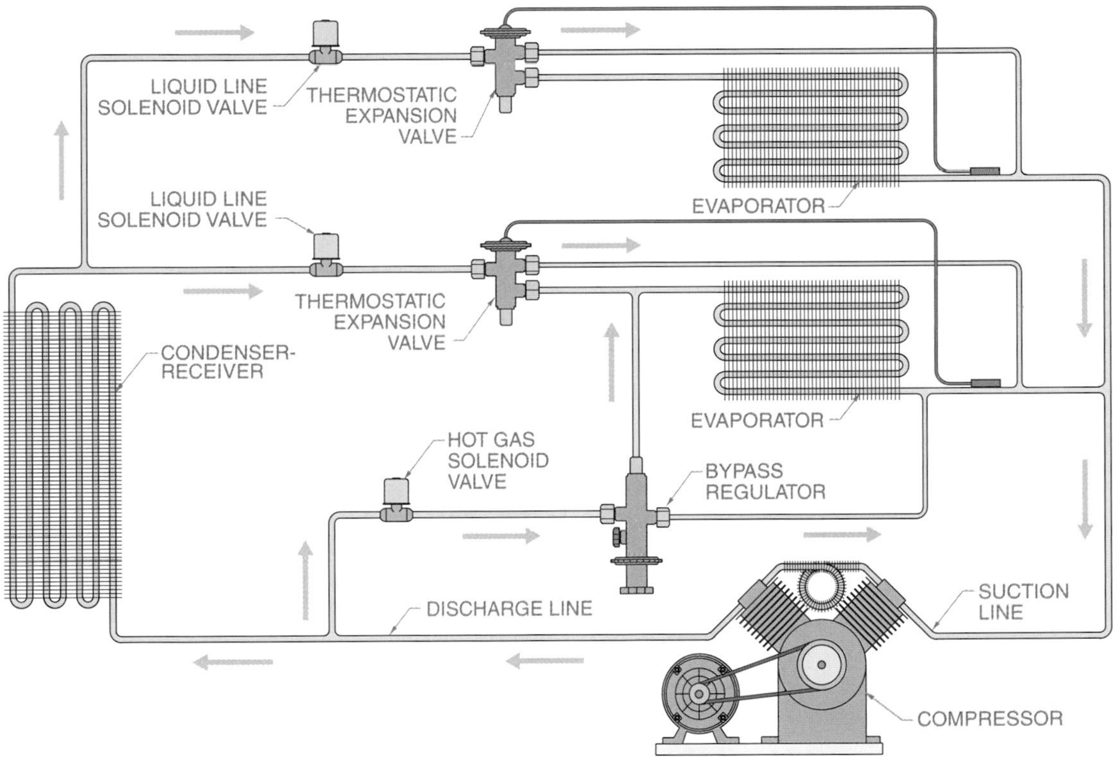

Figure 7-15. Refrigeration systems may use different solenoids, such as liquid line solenoids and hot gas solenoids.

Combustion

Solenoids may also be used in an oil-fired single-burner system. **See Figure 7-16.** The solenoids are crucial in the startup and normal operating functions of the system.

General Purpose

In addition to commercial and industrial use, solenoids are used for general-purpose applications. Typical general-purpose applications include products such as printing calculators, cameras, and airplanes. **See Figure 7-17.**

COMBUSTION SOLENOIDS

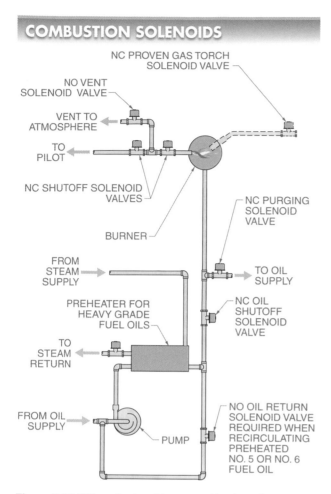

Figure 7-16. Different solenoids are used for the safe operation of an oil-fired single-burner system.

GENERAL-PURPOSE SOLENOIDS

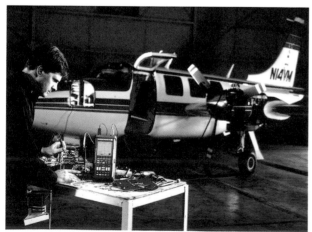

Fluke Corporation

Figure 7-17. Solenoids are used for general-purpose applications, such as those in airplanes.

Rofin Sinar

Solenoids are used to control the flow of gas during tube welding.

Pneumatic Robotics

Industrial robots are used in all kinds of applications from welding, painting, sorting, and assembling extremely small to extremely large parts. They can replicate human movement with the added advantage of being able to lift objects of almost any size or weight repeatedly in almost any type of environment. Industrial robots use fluid power (hydraulic and/or pneumatic) cylinders (linear motion), actuators (rotary motion), and grippers to provide the required power and movement. The cylinders, actuators, and grippers are controlled by solenoid-operated valves. **See Figure 7-18.**

In the robotic assembly example, cylinder 1 advances to move the arm out when the system starts. Rotary actuator 1 closes to grasp the part in the part feeder. Cylinder 2 advances to move the part up and out of the part feeder. Cylinder 1 retracts to move the part away from the part feeder. Rotary actuator 2 rotates counterclockwise to turn the part over. Cylinder 2 retracts to move the part over the subassembly. Rotary actuator 1 opens to release the part that drops into the subassembly. Rotary actuator 2 rotates clockwise to return the arm to the start position.

PNEUMATIC ROBOTICS

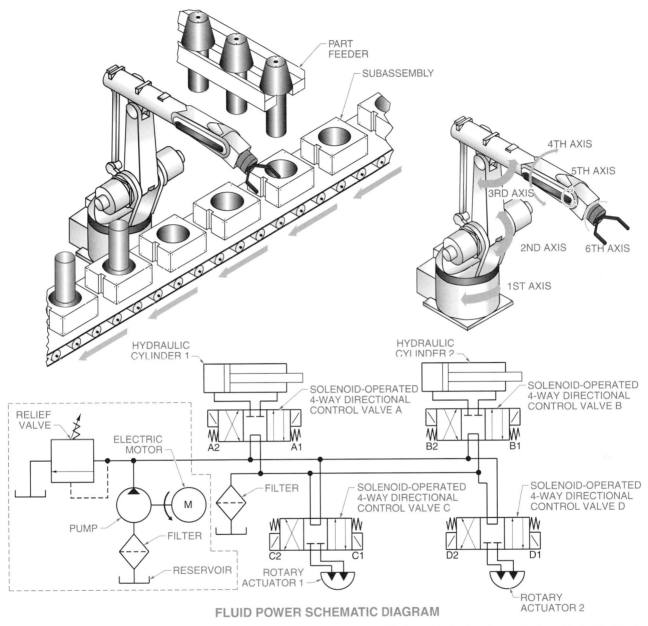

FLUID POWER SCHEMATIC DIAGRAM

Figure 7-18. Pneumatic robots can be used to replicate human movement with the added advantage of being able to lift objects of almost any size and weight repeatedly.

7-4 CHECKPOINT

1. What type of valve is used to direct the flow of fluid throughout a fluid power system?

2. What is the number of locations within a valve in which a spool is placed to direct fluid through the valve?

3. What is a flow path through a valve?

4. What does an electrical control valve use to change the position of a valve spool?

7-5 TROUBLESHOOTING SOLENOIDS

Solenoids usually fail due to electrical problems or mechanical damage. Electrical problems include coil burnout, erratic operation, or an open fuse or circuit breaker. Manufacturer charts are used to help determine the cause of solenoid failure. **See Figure 7-19.**

SOLENOID FAILURE CHARACTERISTICS

Problem	Possible Causes	Comments
Failure to operate when energized	Complete loss of power to solenoid	Normally caused by blown fuse or control circuit problem
	Low voltage applied to solenoid	Voltage should be at least 85% of solenoid rated value
	Burned out solenoid coil	Normally evident by pungent odor caused by burnt insulation
	Shorted coil	Normally a fuse is blown and continues to blow when changed
	Obstruction of plunger movement	Normally caused by a broken part, misalignment, or the presence of a foreign object
	Excessive pressure on solenoid plunger	Normally caused by excessive system pressure in solenoid-operated valves
Failure of spring-return solenoids to operate when de-energized	Faulty control circuit	Normally a problem of the control circuit not disengaging the solenoid's hold or memory circuit
	Obstruction of plunger movement	Normally caused by a broken part, misalignment, or the presence of a foreign object
	Excessive pressure on solenoid plunger	Normally caused by excessive system pressure in solenoid-operated valves
Failure of electrically operated return solenoids to operate when de-energized	Complete loss of power to solenoid	Normally caused by a blown fuse or control circuit problem
	Low voltage applied to solenoid	Voltage should be at least 85% of solenoid rated value
	Burned out solenoid coil	Normally evident by pungent odor caused by burnt insulation
	Obstruction of plunger movement	Normally caused by broken part, misalignment, or presence of a foreign object
	Excessive pressure on solenoid plunger	Normally caused by excessive system pressure in solenoid-operated valves
Noisy operation	Solenoid housing vibrates	Normally caused by loose mounting screws
	Plunger pole pieces do not make flush contact	An air gap may be present, causing the plunger to vibrate; these symptoms are normally caused by foreign matter
Erratic operation	Low voltage applied to solenoid	Voltage should be at least 85% of the solenoid rated voltage
	System pressure may be low or excessive	Solenoid size is inadequate for the application
	Control circuit is not operating properly	Conditions on the solenoid have increased to the point where the solenoid cannot deliver the required force

Figure 7-19. Manufacturer charts are used to help determine the cause of solenoid failure.

Incorrect Voltage

The voltage applied to a solenoid should be ±10% of the solenoid rated value. The voltage is measured directly at the valve when the solenoid is energized. A DMM set to measure AC voltage is used for AC solenoids. A DMM set to measure DC voltage is used for DC solenoids. The range setting must be greater than the applied voltage. **See Figure 7-20.**

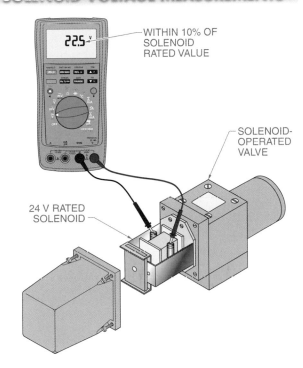

SOLENOID VOLTAGE MEASUREMENTS

WITHIN 10% OF SOLENOID RATED VALUE

SOLENOID-OPERATED VALVE

24 V RATED SOLENOID

Figure 7-20. The voltage applied to a solenoid should be ±10% of the solenoid rated value.

A solenoid overheats when the voltage is excessive. The heat destroys the insulation on the coil wire and burns out the solenoid. However, the solenoid has difficulty moving the spool inside the valve when the voltage is too low. The slow operation causes the solenoid to draw its high inrush current longer. Longer high inrush current also causes excessive heat.

Transients

In most industrial applications, the power supplying a solenoid comes from the same power lines that supply electric motors and other solenoids. High transient voltages are placed on the power lines as these inductive loads are turned on and off. Transient voltages may damage the insulation on the solenoid coil, nearby contacts, and other loads. The transient voltages may be suppressed by using snubber circuits. A *snubber circuit* is a circuit that suppresses noise and high voltage on the power lines.

Rapid Cycling

A solenoid draws several times its rated current when first connected to power. This high inrush current produces heat. In normal applications, the heat is low and dissipates over time. Rapid cycling does not allow the heat to dissipate quickly. The heat buildup burns the coil insulation and causes solenoid failure. To eliminate failure, a high-temperature solenoid should be used in applications requiring a solenoid to be cycled more than 10 times per minute.

Environmental Conditions

A solenoid must operate within its rating and not be mechanically damaged or damaged by the surrounding atmosphere. A solenoid coil is subject to heat during normal operation. This heat comes from the combination of fluid flowing through the valve, the temperature rise from the coil when energized, and the ambient temperature of the solenoid.

Solenoid Troubleshooting Procedure

A DMM set to measure voltage and resistance is required when troubleshooting a solenoid. **See Figure 7-21.** To troubleshoot a solenoid, the following procedure is applied:

1. Turn off electrical power to solenoid or circuit.

2. Measure the voltage at the solenoid to ensure the power is OFF.

3. Remove the solenoid cover and visually inspect the solenoid. Look for a burnt coil, broken parts, or other problems. Replace the coil when burnt. Replace the broken parts when available. Replace the valve, contactor, starter, or solenoid-operated device when the parts are not available. *Note:* Determine the fault before installing a new coil when a solenoid has failed due to a burnt or shorted coil. Coils will continue to burn out if the fault is not corrected. Always observe solenoid operation after a solenoid is replaced.

4. Disconnect the solenoid wires from the electrical circuit when no obvious problem is observed.

5. Check the solenoid continuity. Connect the DMM leads to the solenoid wires with all power turned OFF. The DMM should indicate a resistance reading of ±15% of the coil's normal reading.

Tech Fact

Solenoid coils are rated for operation on either 50 Hz or 60 Hz power. The solenoid coil frequency rating should always be checked against the power supply used for a piece of equipment, especially when using imported equipment. A 50 Hz-rated solenoid coil operating on 60 Hz will not develop its full rated force. A 60 Hz-rated solenoid coil operating on 50 Hz will overheat.

TROUBLESHOOTING SOLENOIDS

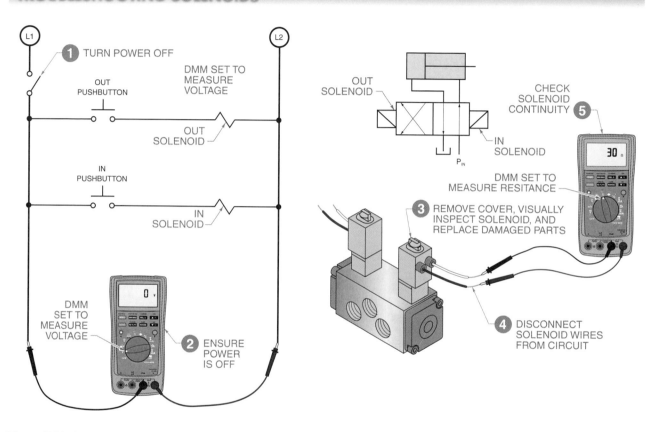

Figure 7-21. A DMM set to measure voltage and resistance is required when troubleshooting a solenoid.

7-5 CHECKPOINT

1. What is the acceptable voltage range (high and low) that should be applied to a solenoid?

2. What device is used to reduce transients produced by solenoids?

3. What is the acceptable resistance range (high and low) that should be measured when testing a solenoid?

Additional Resources

Review and Resources

Access Chapter 7 Review and Resources for *Electrical Motor Controls for Integrated Systems* by scanning the above QR code with your mobile device.

Applying Your Knowledge

Refer to the *Electrical Motor Controls for Integrated Systems* Learner Resources for interactive Applying Your Knowledge questions.

Workbook and Applications Manual

Refer to Chapter 7 in the *Electrical Motor Controls for Integrated Systems Workbook* and the *Applications Manual* for additional exercises.

ENERGY EFFICIENCY PRACTICES

Double Solenoids and Latching Solenoids

Most solenoids require more power to move the plunger or valve spool than they require to hold the plunger or valve spool in position. To save energy, the voltage (power) can be reduced after the solenoid coil has moved the plunger. The voltage required to hold the plunger in place is typically around 50% of the voltage required to move it. In applications that require the solenoid coil to be energized for long periods of time, a voltage control circuit can be used to reduce the applied voltage after the plunger has moved.

A single spring-return solenoid valve requires electrical power be applied to the solenoid at all times to keep the valve switched to the solenoid position. When power is removed from the solenoid, the valve switches to the spring position. When using a double solenoid valve with no springs, power is applied just long enough to switch the valve position. To switch the solenoid to the other position, the other solenoid is energized long enough to shift the position of the valve again.

In addition to double solenoid valves, latching solenoids can also save energy. A latching solenoid changes position when power is applied to the solenoid and remains in position after power is removed. Re-energizing the solenoid changes the solenoid back to the original position.

Some solenoids also include permanent magnets to help save energy. A permanent magnet does not have enough power to move the plunger, though it does have enough power to hold the plunger in place after an electromagnetic coil moves the plunger into position. A second electromagnetic coil and permanent magnet are used to move and hold the plunger into the other position.

Objectives

8-1
- Define relays and explain how they function.
- Define electromechanical relays (EMRs) and solid-state relays (SSRs).

8-2
- Explain general-purpose relays and applications.
- Identify and define single-pole (SP) and double-pole (DP) contacts.
- Identify and define single-throw (ST) and double-throw (DT) contacts.
- Identify and define single-break (SB) contacts and double-break (DB) contacts.
- Explain why a form letter is used to identify relay types and list common form identification letters.
- Explain machine control relay operation and applications.
- Explain reed relay operation and applications.
- Define arcing and explain how to protect relay contacts from electrical damage.
- Explain why silver is alloyed with other metals or gold-flashed when used for relay contacts.
- Explain why relay contacts typically fail and how to help extend relay operating life.

8-3
- Explain why some relays have a manual operating switch.
- Explain how to troubleshoot a relay with a digital mutlimeter (DMM) to verify it is working properly or determine a fault if it is not working properly.

Review and Resources
atplearningresources.com/Quicklinks
Access Code: 362245

Electromechanical Relays

The four main parts to most electrical systems are the power supply, the load, a control (switching) method, and an overload protection device. In some applications, such as controlling the lights in an automobile, a manual switch is used to directly control the lamps. However, when starting the automobile, the manual key switch is used to energize a relay coil that controls the starter motor through the relay contact. Also, relays are used to control loads in appliances (refrigerators, washing machines, etc.), HVAC systems, fluid power systems, transportation systems (ships, aircraft, trucks, heavy machines, etc.), and industrial applications that require moving, mixing, sorting, painting, etc. Understanding where and why relays are used in many applications as well as how to select and troubleshoot them is fundamental when designing or working on many different electrical systems.

8-1 ELECTROMECHANICAL RELAYS

A *relay* is a device that controls one electrical circuit by opening and closing contacts in another circuit. Depending on design, relays normally do not control power-consuming devices directly, except for small loads which draw less than 15 A. Relays have traditionally been used in machine tool control, industrial assembly lines, and commercial equipment. Relays are used to switch starting coils in contactors and motor starters, heating elements, pilot lights, audible alarms, and some small motors (less than ⅛ HP).

A small voltage applied to a relay results in a larger voltage being switched. **See Figure 8-1.** For example, applying 24 V to the relay coils may operate a set of contacts that control a 230 V circuit. In this case, the relay acts as an amplifier of the voltage or current in the control circuit because relay coils require a low current or voltage to switch, but can energize larger currents or voltages.

Another example of a relay providing an amplifying effect is when a single input to the relay results in several other circuits being energized. **See Figure 8-2.** An input may be considered amplified because certain mechanical relays provide eight or more sets of contacts controlled from any one input.

The two major types of relays are the electromechanical relay and the solid-state relay. An *electromechanical relay (EMR)* is a switching device that has sets of contacts that are closed by a magnetic effect. A *solid-state relay (SSR)* is a switching device that has no contacts and switches entirely by electronic means.

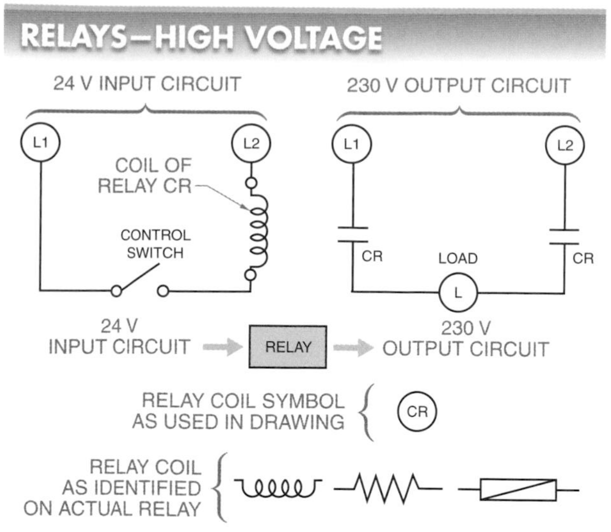

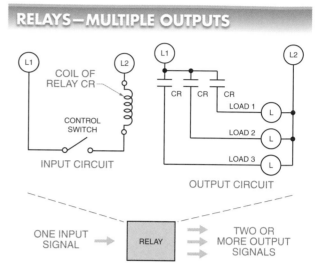

RELAYS—HIGH VOLTAGE

Figure 8-1. Relays may be compared to amplifiers in that small voltage input results in large voltage output.

RELAYS—MULTIPLE OUTPUTS

Figure 8-2. Relays may be compared to amplifiers in that a single input may result in multiple outputs.

8-1 CHECKPOINT

1. Are relays usually used to control higher current loads, such as a 20 A electric motor?
2. Does the load (alarm, motor starter coil, etc.) that the relay contacts control need to be the same voltage rating and type (AC or DC) as the relays coil?
3. Can one relay coil control several different relay contacts?
4. Can a relay with multiple contacts use the different contacts to control different loads, each with a different voltage level, as long as each circuit is isolated from the other circuits?

8-2 TYPES OF ELECTROMECHANICAL RELAYS

EMRs that are common to commercial and industrial applications may be general-purpose, machine control, or reed relays. The major differences between the types of EMRs are their intended use in the circuit, cost, and life expectancy of each device.

General-Purpose Relays

A *general-purpose relay* is a mechanical switch operated by a magnetic coil. **See Figure 8-3.** General-purpose relays are available in AC and DC designs. These relays are available with coils that can open or close contacts ranging from millivolts to several hundred volts. Relays with 6 V, 12 V, 24 V, 48 V, 115 V, and 230 V coils are the most common. General-purpose relays are available that require as little as 4 mA at 5 VDC or 22 mA at 12 VDC. These relays are available in a wide range of switching configurations.

General-purpose relays are EMRs that include several sets (normally two, three, or four) of nonreplaceable NO and NC contacts (normally rated at 5 A to 15 A) that are activated by a coil. **See Figure 8-4.** A general-purpose relay is a good relay for applications that simplify troubleshooting and reduce costs. Special attention must be given to the contact current rating when using general-purpose relays because the contact rating for switching DC is less than the contact rating for switching AC. For example, a 15 A AC-rated contact normally is only rated for 8 A to 10 A DC.

General-purpose relays are designed for commercial and industrial applications where economy and fast replacement are high priorities. Most general-purpose relays have a plug-in feature that makes for quick replacement and simple troubleshooting. Because they are inexpensive, they are thrown away rather than replaced.

GENERAL-PURPOSE RELAYS

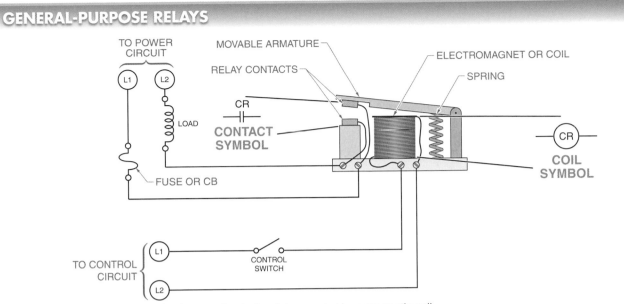

Figure 8-3. A general-purpose relay is a mechanical switch operated by a magnetic coil.

GENERAL-PURPOSE RELAY CONTACTS

Omron Electronics, LLC

Figure 8-4. General-purpose relays are EMRs that include several sets of nonreplaceable NO and NC contacts that are activated by a coil.

Contacts. Contacts are the conducting part of the relay that acts as a switch to make or break a circuit. The most common contacts are the single-pole, double-throw (SPDT); double-pole, double-throw (DPDT); and the three-pole, double-throw (3PDT) contacts. Relay contacts are described by their number of poles, throws, and breaks. **See Figure 8-5.**

A *pole* is the number of completely isolated circuits that a relay can switch. A single-pole contact can carry current through only one circuit at a time. A double-pole contact can carry current through two circuits simultaneously. In a double-pole contact, the two circuits are mechanically connected to open or close simultaneously and are electrically insulated from each other. The mechanical connection is represented by a dashed line connecting the poles. Relays are available with 1 to 12 poles.

A *throw* is the number of closed contact positions per pole. A single-throw contact can control only one circuit. A double-throw contact can control two circuits.

A *break* is the number of separate places on a contact that open or close an electrical circuit. For example, a single-break contact breaks an electrical circuit in one place. A double-break (DB) contact breaks the electrical circuit in two places. All contacts are single break or double break. Single-break (SB) contacts are normally used when switching low-power devices such as indicating lights. Double-break contacts are used when switching high-power devices such as solenoids.

Tech Tip

A relay contact block may contain more contacts than are required for a particular application. Unused relay contacts can be connected in parallel with the used contacts to divide the current flow over both sets of contacts. Contact life can be extended because dividing the current flow over multiple sets of contacts reduces the current flow through any one set.

RELAY CONTACTS

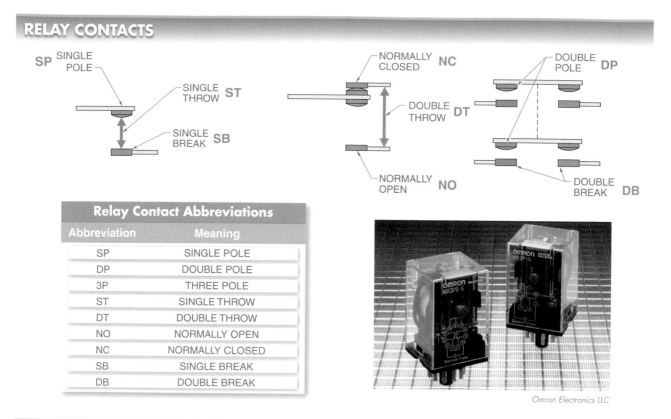

Relay Contact Abbreviations	
Abbreviation	**Meaning**
SP	SINGLE POLE
DP	DOUBLE POLE
3P	THREE POLE
ST	SINGLE THROW
DT	DOUBLE THROW
NO	NORMALLY OPEN
NC	NORMALLY CLOSED
SB	SINGLE BREAK
DB	DOUBLE BREAK

Omron Electronics LLC

Relay Contact Arrangements

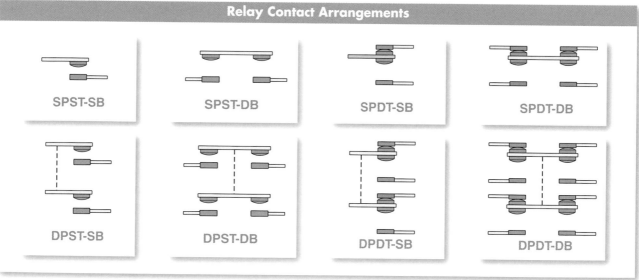

Figure 8-5. Relay contacts are described by their number of poles, throws, and breaks.

Relay manufacturers use a common code to simplify the identification of relays. **See Figure 8-6.** This code uses a form letter to indicate the type of relay. For example, Form A has one contact that is NO and closes (makes) when the coil is energized. Form B has one contact that is NC and breaks (opens) when the coil is energized. Form C has one pole that first breaks one contact and then makes a second contact when the coil is energized. Form C is the most common form.

In some electrical applications, the exact order in which each contact operates (makes or breaks) must be known so the circuit can be designed to reduce arcing. Arcing occurs at any electrical contact that has current flowing through it when the contact is opened.

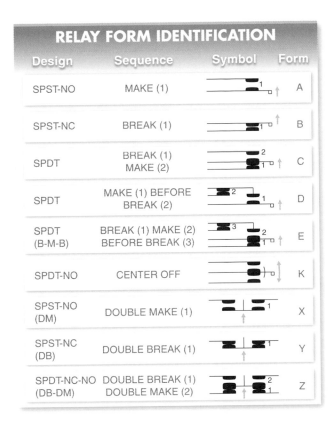

RELAY FORM IDENTIFICATION

Design	Sequence	Symbol	Form
SPST-NO	MAKE (1)		A
SPST-NC	BREAK (1)		B
SPDT	BREAK (1) MAKE (2)		C
SPDT	MAKE (1) BEFORE BREAK (2)		D
SPDT (B-M-B)	BREAK (1) MAKE (2) BEFORE BREAK (3)		E
SPDT-NO	CENTER OFF		K
SPST-NO (DM)	DOUBLE MAKE (1)		X
SPST-NC (DB)	DOUBLE BREAK (1)		Y
SPDT-NC-NO (DB-DM)	DOUBLE BREAK (1) DOUBLE MAKE (2)		Z

Figure 8-6. Relay manufacturers use a common code (form letter) to simplify the identification of relays.

Machine Control Relays

A *machine control relay* is an EMR that includes several sets (usually two to eight) of NO and NC replaceable contacts (typically rated at 10 A to 20 A) that are activated by a coil. Machine control relays are the backbone of electromechanical control circuitry and are expected to have a long life and a minimal amount of problems. Machine control relays are used extensively in machine tools for direct switching of solenoids, contactors, and starters. **See Figure 8-7.** Machine control relays provide easy access for contact maintenance and may provide additional features like time delay, latching, and convertible contacts for maximum circuit flexibility. Convertible contacts are mechanical contacts that can be placed in either an NO or NC position. Machine control relays are also known as heavy-duty or industrial control relays.

In a machine control relay, each contact is a separate removable unit that may be installed to obtain any combination of NO and NC switching. These contacts are also convertible from NO to NC and from NC to NO. **See Figure 8-8.** The unit may be used as either an NO or NC contact by changing the terminal screws and

rotating the unit 180°. Relays of 1 to 12 contact poles are readily assembled from stock parts. Machine control relays may have additional decks (groups of contacts) stacked onto the base unit.

MACHINE CONTROL RELAYS

Sprecher & Schuh

Figure 8-7. Machine control relays are used extensively in machine tools for direct switching of solenoids, contactors, and starters.

MACHINE CONTROL RELAY CONVERSION

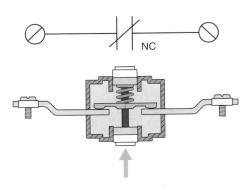

Figure 8-8. Machine control relay contacts are convertible from NO to NC and from NC to NO.

The control coils for machine control relays are easily changed from one control voltage to another and are available in AC or DC standard ratings. Machine control relays have a large number of accessories that may be added to the relay unit. These include indicating lights, transient suppression, latching controls, and time controls.

Reed Relays

A *reed relay* is a fast-operating, single-pole, single-throw switch with normally open (NO) contacts hermetically sealed in a glass envelope. **See Figure 8-9.** During the sealing operation, dry nitrogen is forced into the tube, creating a clean inner atmosphere for the contacts. Because the contacts are sealed, they are unaffected by dust, humidity, and fumes. The life expectancy of reed relay contacts is quite long.

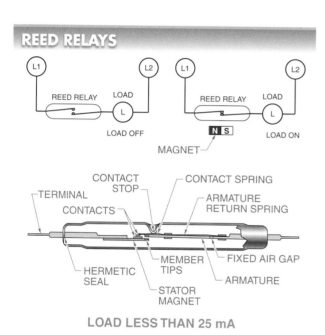

Figure 8-9. A reed relay is a fast-operating, single-pole, single-throw switch that is activated by a magnetic field.

A reed relay includes a very low current-rated contact (less than 400 mA) that is activated by the presence of a magnetic field. Reed relays may be activated in a variety of ways, which allows them to be used in circuit applications where other relay types are inappropriate.

Reed relays are designed to be actuated by an external movable permanent magnet or DC electromagnet. When a magnetic field is brought close to the two reeds, the ferromagnetic (easily magnetized) ends assume opposite

magnetic polarity. If the magnetic field is strong enough, the attracting force of the opposing poles overcomes the stiffness of the reed, drawing the contacts together. Removing the magnetizing force allows the contacts to spring open. AC electromagnets are not suitable for reed relays because the reed relay switches so fast that it would energize and de-energize on alternate half-cycles of a standard 60 Hz line.

Reed Contacts. To obtain a low and consistent contact resistance, the overlapping ends of the contacts may be plated with gold, rhodium, silver alloy, or other low-resistance metal. Contact resistance is often under $0.1\ \Omega$ when closed. Reed contacts have an open contact resistance of several million ohms.

Most reed contacts are not directly capable of switching of industrial solenoids, contactors, and motor starters. Reed relay contact ratings indicate the maximum current, voltage, and volt-amps that may be switched by the relay. Under no circumstances should these values be exceeded.

Tech Tip

Reed relays are operated by a magnetic field. Unshielded relays (the least expensive type) that are placed close to each other can cause false relay operation. Shielded relays should be used when the relays must be located near each other or when magnetic fields from motors and transformers may cause false operation.

Reed Relay Actuation. A permanent magnet is the most common actuator for a reed relay. Permanent-magnet actuation can be arranged in several ways depending on the switching requirement. One of the most commonly used arrangements is proximity motion.

The proximity motion arrangement uses the presence of a magnetic field that is brought within a specific proximity (close distance) to the reed relay to close the contacts. The distance for activating any given relay depends on the sensitivity of the relay and the strength of the magnet. A more sensitive relay or stronger magnet needs less distance for actuation. Methods of proximity motion operation are the pivoted motion, perpendicular motion, parallel motion, and front-to-back motion. **See Figure 8-10.** In each method, either the magnet or relay is moved. In some applications, both the magnet and relay are in motion. The contacts operate quickly, with snap action and little wear. The application and switching requirements determine the best method.

PROXIMITY MOTION

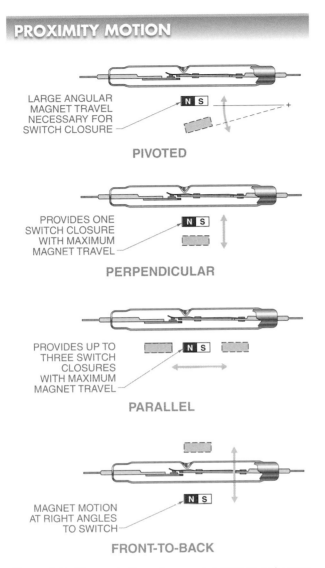

Figure 8-10. The proximity motion arrangement uses the presence of a magnetic field brought within a specific proximity to the reed relay to close the contacts.

Magnetic Level Switch

A *magnetic level switch* is a switch that contains a float, a moving magnet, and a magnetically operated reed switch to detect the level of a liquid. The float moves with the level of the liquid. A permanent magnet inside the float moves up and down with the liquid. The magnet passes alongside a magnetically operated reed switch as it moves with the level of the liquid. The reed switch contacts change position when the magnetic field is present. The reed switch contacts return to their normal position when the magnetic field moves away. An advantage of using a magnetic level switch is that several individual switches may be placed in one housing. **See Figure 8-11.**

MAGNETIC LEVEL SWITCHES

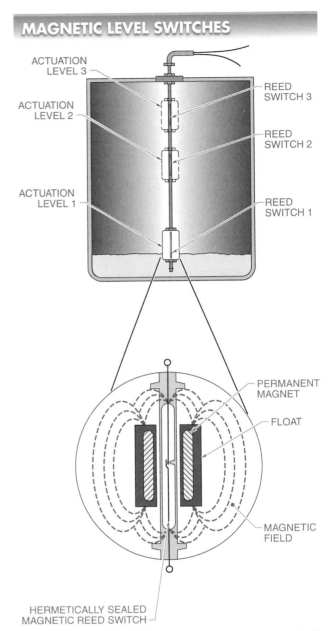

Figure 8-11. A magnetic level switch uses a magnetically operated reed switch and a moving magnet to detect the level of a liquid.

EMR Life

EMR life expectancy is rated in contact life and mechanical life. *Contact life* is the number of times the contacts of a relay switch the load controlled by the relay before malfunctioning. Typical contact life ratings are 100,000 to 500,000 operations. *Mechanical life* is the number of times the mechanical parts of a relay operate before malfunctioning. Typical mechanical life ratings are 1,000,000 to 10,000,000 operations.

Relay contact life expectancy is lower than mechanical life expectancy because the life of a contact depends on the application. The contact rating of a relay is based on the full-rated power of the contact. Contact life is increased when contacts switch loads less than their full-rated power. Contact life is reduced when contacts switch loads that develop destructive arcs. *Arcing* is the discharge of an electric current across a gap, such as when an electric switch is opened. Arcing causes contact burning and temperature rise. **See Figure 8-12.**

ARCING

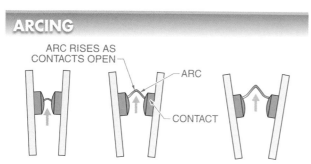

Figure 8-12. Arcing is the discharge of an electric current across a gap, such as when an electric switch is opened.

Arcing is minimized through the use of an arc suppressor and the correct contact material for the application. An arc suppressor is a device that dissipates the energy present across opening contacts. Arc suppression is used in applications that switch arc-producing loads such as solenoids, coils, motors, and other inductive loads.

Arc suppression is also accomplished by using a contact protection circuit. A *contact protection circuit* is a circuit that protects contacts by providing a nondestructive path for generated voltage as a switch is opened. A contact protection circuit may contain a diode or a metal-oxide varistor (MOV).

A diode is used as contact protection in DC circuits. The diode does not conduct electricity when the load is energized. The diode conducts electricity and shorts the generated voltage when the switch is opened. Because the diode shorts the generated voltage, the voltage is dissipated across the diode and not the relay contacts. **See Figure 8-13.**

A snubber is used as contact protection in AC circuits. The capacitor in an RC circuit is a high impedance to the 60 Hz line power and a short circuit to generated high frequencies. The short circuit dissipates generated voltage.

CONTACT PROTECTION CIRCUITS

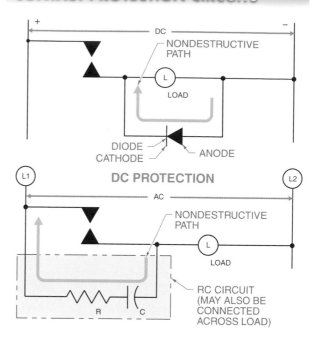

Figure 8-13. A contact protection circuit, such as a diode, protects contacts by providing a nondestructive path for generated voltage as a switch is opened.

Contact Material

All relay contacts are available in fine silver, silver-cadmium oxide, gold-flashed silver, and tungsten. Fine silver has the highest electrical conductivity of all metals. However, fine silver sticks, welds, and is subject to sulfidation when used for many applications. *Sulfidation* is the formation of film on the contact surface. Sulfidation increases the resistance of the contacts. Silver is alloyed with other metals to reduce sulfidation.

Contact Failure

In most applications, a relay fails due to contact failure. In some low-current applications, the relay contacts may look clean but may have a thin film of sulfidation, oxidation, or contaminants on the contact surface. This film increases the resistance to the flow of current through the contact. Normal contact wiping or arcing usually removes the film. In low-power circuits, this action may not take place. In most applications, contacts are oversized for maximum life. However, low-power circuit contacts should not be oversized and should switch no more than 75% of their rating.

Contacts are often subject to high-current surges. High-current surges reduce contact life by accelerating sulfidation and contact burning. For example, a 100 W incandescent lamp has a current rating of about 1 A. The life of the contacts is reduced if a relay with 5 A contacts is used to switch the lamp because the lamp's filament has a low resistance when cold. When first turned ON, the lamp draws 12 A or more. Though the 5 A relay switches the lamp, it will not switch it for the rated life of the relay. Contacts are oversized in applications that have high-current surges.

Silver is alloyed with cadmium to produce a silver-cadmium alloy. Silver-cadmium alloy contacts have good electrical characteristics and low resistance, which helps the contact resist arcing but not sulfidation. Silver or silver-cadmium alloy contacts are used in circuits that switch several amperes at more than 12 V, which burns off the sulfidation.

Sulfidation can damage silver contacts when used in intermittent applications. Gold-flashed silver contacts are used in intermittent applications to minimize sulfidation and provide a good electrical connection. Gold-flashed silver contacts are not used in high-current applications because the gold burns off quickly.

Gold-flashed silver contacts are good for switching loads of 1 A or less.

Tungsten contacts are used in high-voltage applications because tungsten has a high melting temperature and is less affected by arcing. Tungsten contacts are used when high repetitive switching is required.

8-2 CHECKPOINT

1. What does DPST-DB stand for?
2. What type of contact is best to use when switching higher power loads?
3. A form letter is used to simplify relay contacts. What form letter would identify a relay with a SPST-NC contact type?
4. Which relay type comes with convertible contacts that can be placed in either an NO or NC operating position?
5. Are reed relays usually used to switch loads in the 0.1 A to 10 A range?
6. Why are gold-flashed silver contacts used in automobile air bag activation control systems?

8-3 TROUBLESHOOTING ELECTROMECHANICAL RELAYS

When troubleshooting EMRs, the input and output of the relay are checked to determine whether the circuit on the input side of the relay is the problem, the circuit on the output side of the relay is the problem, or the relay itself is the problem. The relay coil and contacts are checked to determine whether the relay is the problem.

Relay Inspection

A relay should be checked for contact sticking or binding if it is not functioning properly. Any loose parts should be tightened. Any broken, bent, or badly worn parts should be replaced. All contacts should be checked for signs of excessive wear and dirt buildup. Contacts are not harmed by discoloration or slight pitting. Contacts should be vacuumed or wiped with a soft cloth to remove dirt.

A contact cleaner should not be used on relay contacts. Contacts require replacement when the silver surface has become badly worn. When severe contact wear is evident on any contact, all contacts should be replaced. Replacing all contacts prevents uneven and unequal contact closing. A contact should never be filed.

Relay coils should be free of cracks and burn marks. The coil should be replaced if there is any evidence of overheating, cracking, melting, or burning. The coil terminals should be checked for the correct voltage level. Overvoltage or undervoltage conditions of more than 10% should be corrected. It is important to only use replacement parts recommended by the manufacturer when replacing parts of a relay. Using nonapproved parts can void the manufacturer's warranty and may transfer product liability from the manufacturer to the owner. Relays are tested by manual operation and by using a digital multimeter (DMM).

Manual Relay Operation

Most relays can be manually operated. Manually operating a relay determines whether the circuit that the relay is controlling (output side) is working correctly. A relay is manually operated by pressing down at a designated area on the relay. This closes the relay contacts. Electromechanical relays may include a push-to-test button. **See Figure 8-14.**

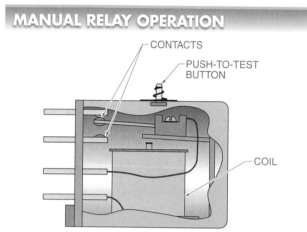

MANUAL RELAY OPERATION

CONTACTS

PUSH-TO-TEST BUTTON

COIL

Figure 8-14. Manually operating a relay determines whether the circuit that the relay is controlling (output side) is working correctly.

WARNING: Use caution when manually operating a relay because loads may start or stop without warning.

When manually operating relay contacts, the circuit controlling the coil is bypassed. Troubleshooting is performed from the relay through the control circuit when the load controlled by the relay operates manually. Troubleshooting is performed on the circuit that the relay is controlling if the load controlled by the relay does not operate when the relay is manually operated.

Digital Multimeter Test

A DMM is also used to test an electromechanical relay. A DMM is connected across the input and output side of a relay. Troubleshooting is performed from the input of the relay through the control circuit when no voltage is present at the input side of the relay. The relay is the problem if the relay is not delivering the correct voltage.

Troubleshooting is performed from the output of the relay through the power circuit when the relay is delivering the correct voltage. The supply voltage measured across an open contact indicates that the DMM is completing the circuit across the contact. The contacts are not closing and the relay is defective if the voltage measured across the contact remains at full voltage when the coil is energized and de-energized. The contacts are welded closed and the relay is defective if the voltage measured across the contacts remains zero (or very low) when the coil is energized and de-energized. **See Figure 8-15.** To troubleshoot an electromechanical relay, the following procedure is applied:

1. Measure the voltage in the circuit containing the control relay coil. The voltage should be within 10% of the voltage rating of the coil. The relay coil cannot energize if the voltage is not present. The coil may not energize properly if the voltage is not at the correct level. Troubleshoot the power supply when the voltage level is incorrect.

2. Measure the voltage across the control relay coil. The voltage across the coil should be within 10% of the coil's rating. Troubleshoot the switch controlling power to the coil when the voltage level is incorrect.

3. Measure the voltage in the circuit containing the control relay contacts. The voltage should be within 10% of the rating of the load. Troubleshoot the power supply if the voltage level is incorrect.

4. Measure the voltage across the control relay contacts. The voltage across the contacts should be less than 1 V when the contacts are closed and nearly equal to the supply voltage when open. The contacts have too much resistance and are in need of service if the voltage is more than 1 V when the contacts are closed. Troubleshoot the load when the voltage is correct at the contacts and the circuit does not work.

Tech Tip

When testing a relay, the voltage across the relay contacts should be measured when the load is on and the relay contacts are closed. The lower the voltage drop, the less heat produced at the contacts and the greater the contact life. The higher the voltage drop, the greater the heat produced at the contacts and the shorter the contact life.

TROUBLESHOOTING ELECTROMECHANICAL RELAYS

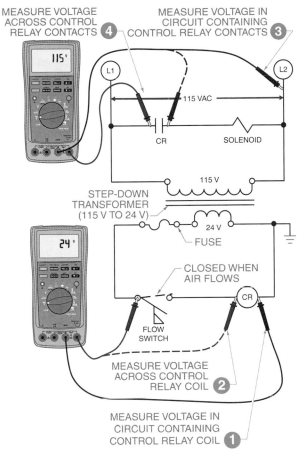

MEASURE VOLTAGE ACROSS CONTROL RELAY CONTACTS **4**

MEASURE VOLTAGE IN CIRCUIT CONTAINING CONTROL RELAY CONTACTS **3**

L1

L2

115 VAC

CR

SOLENOID

115 V

STEP-DOWN TRANSFORMER (115 V TO 24 V)

24 V

FUSE

CLOSED WHEN AIR FLOWS

CR

FLOW SWITCH

MEASURE VOLTAGE ACROSS CONTROL RELAY COIL **2**

MEASURE VOLTAGE IN CIRCUIT CONTAINING CONTROL RELAY COIL **1**

Figure 8-15. A DMM is connected across the input and output side to test an EMR.

Electromechanical relays are used in HVAC systems to control a variety of functions.

8-3 CHECKPOINT

1. Should relay contacts be filed when showing signs of wear?
2. In a 120 V circuit, is it acceptable when 2 V are measured across a closed contact operating a load?
3. In a 120 V circuit, is it acceptable when the full 120 V are measured across an open relay contact?
4. Can reed relays be manually operated?

Additional Resources

Review and Resources

Access Chapter 8 Review and Resources for *Electrical Motor Controls for Integrated Systems* by scanning the above QR code with your mobile device.

Applying Your Knowledge

Refer to the *Electrical Motor Controls for Integrated Systems* Learner Resources for interactive Applying Your Knowledge questions.

Workbook and Applications Manual

Refer to Chapter 8 in the *Electrical Motor Controls for Integrated Systems Workbook* and the *Applications Manual* for additional exercises.

ENERGY EFFICIENCY PRACTICES

Distributed Generator Interconnection Relays

Distributed generation is the exporting of electricity generated by a secondary power source to the utility grid. Distributed generation adds electrical demand capacity to a utility grid and enables consumers to sell excess electricity to the utility. A distributed generator interconnection relay is a specialized relay that monitors a primary power source and a secondary power source for the purpose of paralleling the systems. The interconnection relay includes sensors for voltage, current, frequency, direction of power flow, and other parameters. The interconnection relay only allows the paralleling of connections if the two power sources are in phase and immediately opens the connection if they become out of phase.

When paralleling two power sources, the power sources must be carefully synchronized. In addition, potential equipment damage and safety hazards may occur if the utility fails while connected, back-feeding the secondary power source into the utility grid. Distributed generators must be immediately disconnected from the grid if there is a utility outage. Otherwise, utility workers can be seriously injured by unexpectedly energized equipment.

Sections

Objectives

9-1
- State the operating principle of how a DC generator produces electricity.
- Identify the major parts of a DC generator and state their operating functions.
- Explain the left-hand generator rule.

9-2
- Explain the difference between a series-wound generator and a shunt-wound generator.
- Describe a compound-wound generator.

Review and Resources
atplearningresources.com/Quicklinks
Access Code: 362245

Chapter 9

DC Generators

The two types of electrical power are DC and AC. Both DC and AC are generated, distributed, and used to operate DC loads and AC loads. DC power can be converted into AC using an inverter and AC power can be converted into DC using a converter. Most electrical systems include both DC and AC components and circuits. Each type has advantages and disadvantages in different applications. It is important to understand both DC and AC generation and usages when working in the electrical field.

9-1 DC GENERATOR COMPONENTS

A *generator* is a machine that converts mechanical energy into electrical energy by means of electromagnetic induction. DC generators operate on the principle that when a coil of wire is rotated in a magnetic field, a voltage is induced in the coil. The amount of voltage induced in the coil is determined by the rate at which the coil is rotated in the magnetic field. When a coil is rotated in a magnetic field at a constant rate, the voltage induced in the coil depends on the number of magnetic lines of force in the magnetic field at each given instant of time. DC generators consist of field windings, an armature, a commutator, and brushes. **See Figure 9-1.**

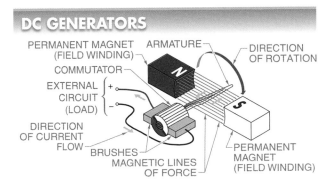

Figure 9-1. DC generators consist of field windings, an armature, a commutator, and brushes. They operate on the principle that when a coil of wire is rotated in a magnetic field, a voltage is induced in the coil.

Field Windings

Field windings are electromagnets used to produce the magnetic field in a generator. The magnetic field used in a generator can be produced by permanent magnets or electromagnets. Permanent magnets are used in very small machines referred to as magnetos. A disadvantage of permanent magnets is that their magnetic lines of force decrease as the age of the magnet increases. Another disadvantage is that the strength of a permanent magnet cannot be varied for control purposes. Most generators use electromagnets, which must be supplied with current. If the current for the field windings is supplied by an outside source (a battery or another generator), the generator is separately excited. If the generator itself supplies current for the field windings, the generator is referred to as self-excited. DC generators are usually self-excited.

Armatures

An *armature* is the movable coil of wire in a generator that rotates through the magnetic field. A DC generator always has a rotating armature and a stationary field (field windings). The rotating armature may consist of many coils. Although increasing the number of coils reduces the ripples (pulsations) in the output voltage, it is impossible to remove the ripples completely.

Commutators

A *commutator* is a ring made of segments that are insulated from one another. Each end of a coil of wire is connected to a segment. A voltage is induced in the coil whenever the coil cuts the magnetic lines of force of a magnetic field. **See Figure 9-2.** The commutator segments reverse the connections to the brushes every half cycle. This maintains a constant polarity of output voltage produced by the generator.

Brushes

A *brush* is the sliding contact that rides against the commutator segments or slip rings and is used to connect the armature to the external circuit. Brushes are made from soft carbon (natural graphite). Brushes are softer than the commutator bars yet strong enough so that the brushes do not chip or break from vibration. One brush makes contact with each segment of the commutator.

A DC generator is designed so that the brushes ride on the different segments of the commutator each time the current is zero. Therefore, the current in the external circuit (load) always flows in one direction; however, its magnitude varies continuously. The action of reversing the connections to the coil (armature) to obtain a direct current is referred to as commutation. The resulting output voltage of a DC generator is a pulsating DC voltage. The pulsations of the output voltage are known as ripples.

A generator is commonly used as an emergency power supply when the primary power supply fails.

Left-Hand Generator Rule

The *left-hand generator rule* is the relationship between the current in a conductor and the magnetic field existing around the conductor. The left-hand generator rule states that with the thumb, index finger, and middle finger of the left hand set at right angles to each other, the index finger points in the direction of the magnetic field, the thumb points in the direction of the motion of the conductor, and the middle finger points in the direction of the induced current. **See Figure 9-3.** When using the left-hand generator rule, it is assumed that the magnetic field is stationary and that the conductor is moving through the field.

INDUCED DC VOLTAGE

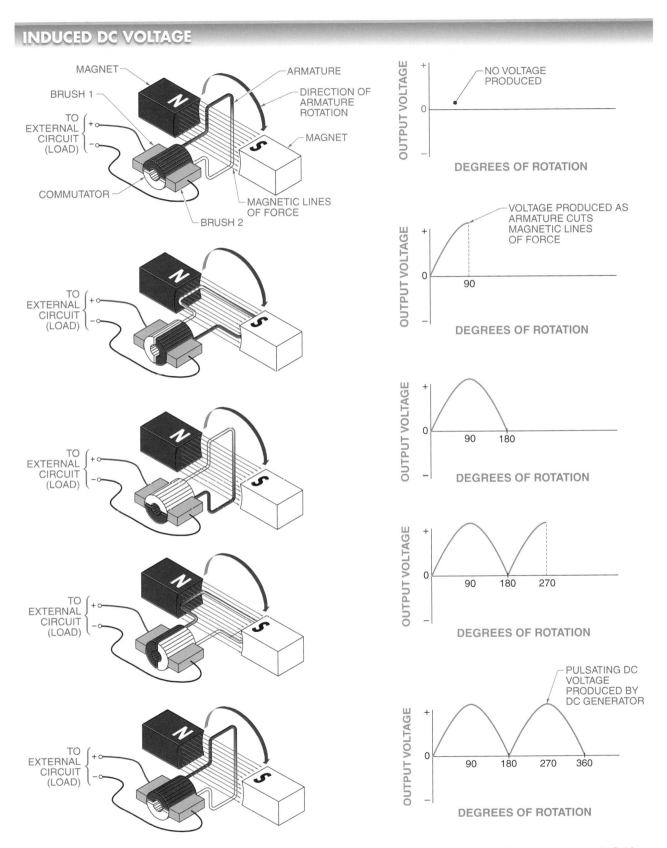

Figure 9-2. A voltage is induced in the coil (armature) of a generator when the coil cuts the lines of force of a magnetic field.

LEFT-HAND GENERATOR RULE

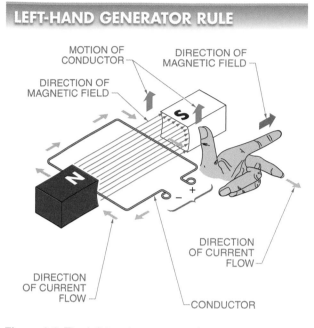

Figure 9-3. The left-hand generator rule expresses the relationship between the conductor, magnetic field, and induced voltage in a generator.

9-1 CHECKPOINT

1. What is the stationary part of a DC generator that produces a magnetic field?
2. What is the rotating part of a DC generator that produces electricity as it rotates through the magnetic field?
3. Do most DC generators use electromagnets or permanent magnets to produce a magnetic field?

9-2 DC GENERATOR TYPES

The three types of DC generators are series-wound, shunt-wound, and compound-wound generators. The difference between the types is based on the relationship of the field windings to the external circuit.

Series-Wound Generators

A *series-wound generator* is a generator that has its field windings connected in series with the armature and the external circuit (load). **See Figure 9-4.** In a series-wound generator, the field windings consist of a few turns of low-resistance wire because the load current flows through them.

The ability of a generator to have a constant voltage output under varying load conditions is referred to as the generator voltage regulation. Series-wound generators have poor voltage regulation. Because of their poor voltage regulation, series-wound DC generators are not used frequently. The output voltage of a series-wound generator may be controlled by a rheostat (variable resistor) connected in parallel with the field windings.

Shunt-Wound Generators

A *shunt-wound generator* is a generator that has its field windings connected in parallel (shunt) with the armature and the external circuit (load). **See Figure 9-5.** Because the field windings are connected in parallel with the load, the current through them is wasted as far as output is concerned. The field windings consist of many turns of high-resistance wire to keep the current flow through them low.

A shunt-wound generator is suitable if the load is constant. However, if the load fluctuates, the voltage also varies. The output voltage of a shunt-wound generator may be controlled by means of a rheostat connected in series with the shunt field.

SERIES-WOUND GENERATORS

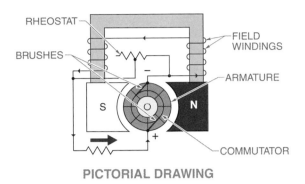

PICTORIAL DRAWING

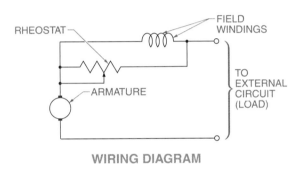

WIRING DIAGRAM

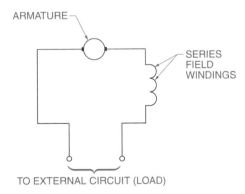

SCHEMATIC DIAGRAM

Figure 9-4. A series-wound generator has its field windings connected in series with the armature and load.

SHUNT-WOUND GENERATORS

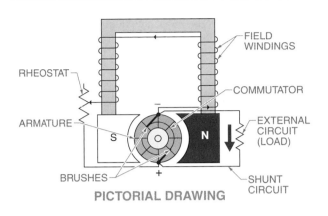

PICTORIAL DRAWING

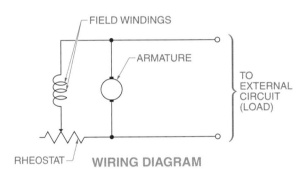

WIRING DIAGRAM

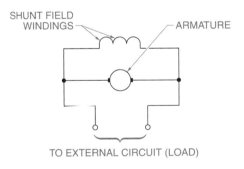

SCHEMATIC DIAGRAM

Figure 9-5. A shunt-wound generator has its field windings connected in parallel (shunt) with the armature and load.

Compound-Wound Generators

A *compound-wound generator* is a generator that includes series and shunt field windings. In a compound-wound generator, the series field windings and shunt field windings are combined in a manner

to take advantage of the characteristics of each. The shunt field is normally the stronger of the two. The series field is used only to compensate for effects that tend to decrease the output voltage. **See Figure 9-6.**

COMPOUND-WOUND GENERATORS

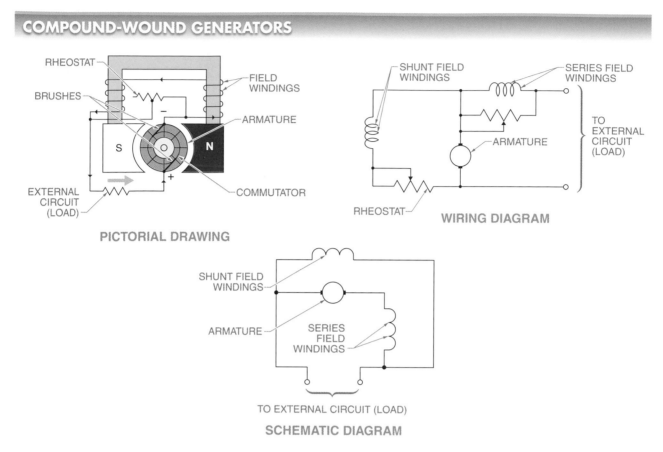

Figure 9-6. A compound-wound generator includes series and shunt field windings.

9-2 CHECKPOINT

1. Does a series-wound generator have good or poor voltage regulation?
2. Does "shunt" mean connected in series or connected in parallel?

3. What type of DC generator uses both series and parallel connected field windings?

Additional Resources

Review and Resources

Access Chapter 9 Review and Resources for *Electrical Motor Controls for Integrated Systems* by scanning the above QR code with your mobile device.

Applying Your Knowledge

Refer to the *Electrical Motor Controls for Integrated Systems* Learner Resources for interactive Applying Your Knowledge questions.

Workbook and Applications Manual

Refer to Chapter 9 in the *Electrical Motor Controls for Integrated Systems Workbook* and the *Applications Manual* for additional exercises.

ENERGY EFFICIENCY PRACTICES

DC Generators

DC generators are used in a variety of applications such as in vehicles and boats. They are also used in off-the-utility-grid locations such as mining operations. DC generators are also used in photovoltaic and wind turbine systems to charge battery backup storage when there is not enough light or wind.

A DC generator is selected based on the power output (both peak and standard) of the generator, voltage level, operating fuel, expected run times, and operating noise level (in dB). An energy-efficient DC generator should operate as quietly as possible. A dB sound rating of 70 dB or less is good.

In addition, the energy efficiency rating of the generator should be considered to reduce energy usage and operating cost. The energy efficiency rating of a DC generator is listed as the amount of power (in kW) the generator can produce in an hour (h) per gallon (gal.) of fuel (kWh/gal.). The higher the kWh/gal. rating, the more energy efficient the generator. For example, a DC generator with an energy efficiency rating of 22 kWh/gal. is about 18% more energy efficient than a DC generator with an 18 kWh/gal. rating. An energy-efficient DC generator should have an energy efficiency rating of 20 kWh/gal. or higher.

Objectives

10-1
- State the purpose of each of the components used in an AC generator.
- Explain how AC voltage generated in the armature is delivered to an external circuit.

10-2
- Explain how a single-phase (1ϕ) generator produces the positive and negative parts of a sine wave as the armature rotates through the magnetic field of the field windings.
- Explain how a three-phase (3ϕ) AC generator produces three simultaneous voltages that are 120° out of phase from each other.
- Describe a 3ϕ AC generator with neutral wye-connected system.
- Describe a 3ϕ AC generator with neutral delta-connected system.

10-3
- Describe the differences between momentary, temporary, and sustained power interruptions.

Sections

10-1 AC Generator Components

10-2 AC Generator Types

10-3 Voltage Changes

Review and Resources
atplearningresources.com/Quicklinks
Access Code: 362245

Chapter
AC Generators

Almost every part of daily life depends on the use of electricity to provide light, produce useful products, allow easy communication, provide a safe and comfortable environment, provide entertainment, and provide transportation. Although electricity can be produced by batteries and photocells, the vast majority of electricity is produced by AC generators. Generators have been used to convert mechanical energy into electrical energy ever since Michael Faraday discovered that moving a wire through a magnetic field produces electricity in 1831. Understanding how generators produce electricity is important to understand the operation of other electrical devices, such as electric motors and transformers.

10-1 AC GENERATOR COMPONENTS

Generators convert mechanical energy into electrical energy by means of electromagnetic induction. AC generators, also referred to as alternators, convert mechanical energy into AC voltage and current. AC generators consist of field windings or permanent magnets, an armature, slip rings, and brushes. **See Figure 10-1.**

Field Windings

A *field winding* is an electromagnet used to produce the stationary magnetic field in a generator. The magnetic field in a generator can be produced by permanent magnets or electromagnets. However, most generators use electromagnets, which must be supplied with current. **See Figure 10-2.**

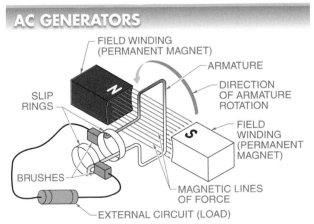

Figure 10-1. AC generators consist of field windings, an armature (coil), slip rings, and brushes.

FIELD WINDINGS

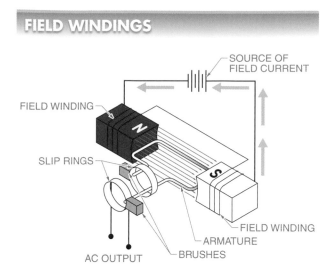

Figure 10-2. Field windings are used to produce the stationary magnetic field in a generator.

Armatures

An *armature* is the movable coil of wire in a generator that rotates through the magnetic field. The armature may consist of many coils. The ends of the coils are connected to slip rings.

Slip Rings

Slip rings are metallic rings connected to the ends of the armature and are used to connect the induced voltage to the brushes. When the armature is rotated in the magnetic field, a voltage is generated in each half of the armature coil.

Brushes

A *brush* is the sliding contact that rides against the commutator segments or slip rings and is used to connect the armature to the external circuit. AC generators are similar in construction and operation to DC generators. The major difference between AC and DC generators is that DC generators contain a commutator that reverses the connections to the brushes every half cycle. This maintains a constant

polarity of output voltage produced by the generator. AC generators use slip rings to connect the armature to the external circuit (load). The slip rings do not reverse the polarity of the output voltage produced by the generator. The result is an alternating sine wave output.

As the armature rotates, each half cuts across the magnetic lines of force at the same speed. Thus, the strength of the voltage induced in one side of the armature is always the same as the strength of the voltage induced in the other side of the armature.

Each half of the armature cuts the magnetic lines of force in a different direction. For example, as the armature rotates in the clockwise direction, the lower half of the coil cuts the magnetic lines of force from the bottom up to the left, while the top half of the coil cuts the magnetic lines of force from the top down to the right. Therefore, the voltage induced in one side of the coil is opposite to the voltage induced in the other side of the coil. The voltage in the lower half of the coil enables current flow in one direction, and the voltage in the upper half enables current flow in the opposite direction.

However, since the two halves of the coil are connected in a closed loop, the voltages add to each other. The result is that the total voltage of a full rotation of the armature is twice the voltage of each coil half. This total voltage is obtained at the brushes connected to the slip rings and may be applied to an external circuit.

10-1 CHECKPOINT

1. What is another name for an AC generator?
2. What produces the stationary magnetic field of a generator?
3. What is the rotating coil of a generator called?
4. Are slip rings used to convert generated AC voltage to DC voltage?

10-2 AC GENERATOR TYPES

The types of AC generators include single-phase (1φ) and three-phase (3φ) AC generators. Single-phase and three-phase AC generators operate similarly; however, single-phase generators have one armature and three-phase generators have three armatures. Three-phase generators provide a more consistent voltage than single-phase generators. Three-phase generators are either wye-connection or delta-connection.

Single-Phase AC Generators

Each complete rotation of the armature in a single-phase (one wire) AC generator produces one complete alternating current cycle. **See Figure 10-3.** In position A, before the armature begins to rotate in a clockwise direction, there is no voltage and no current in the external (load) circuit because the armature is not cutting across any magnetic lines of force (0° of rotation).

SINGLE-PHASE AC GENERATORS

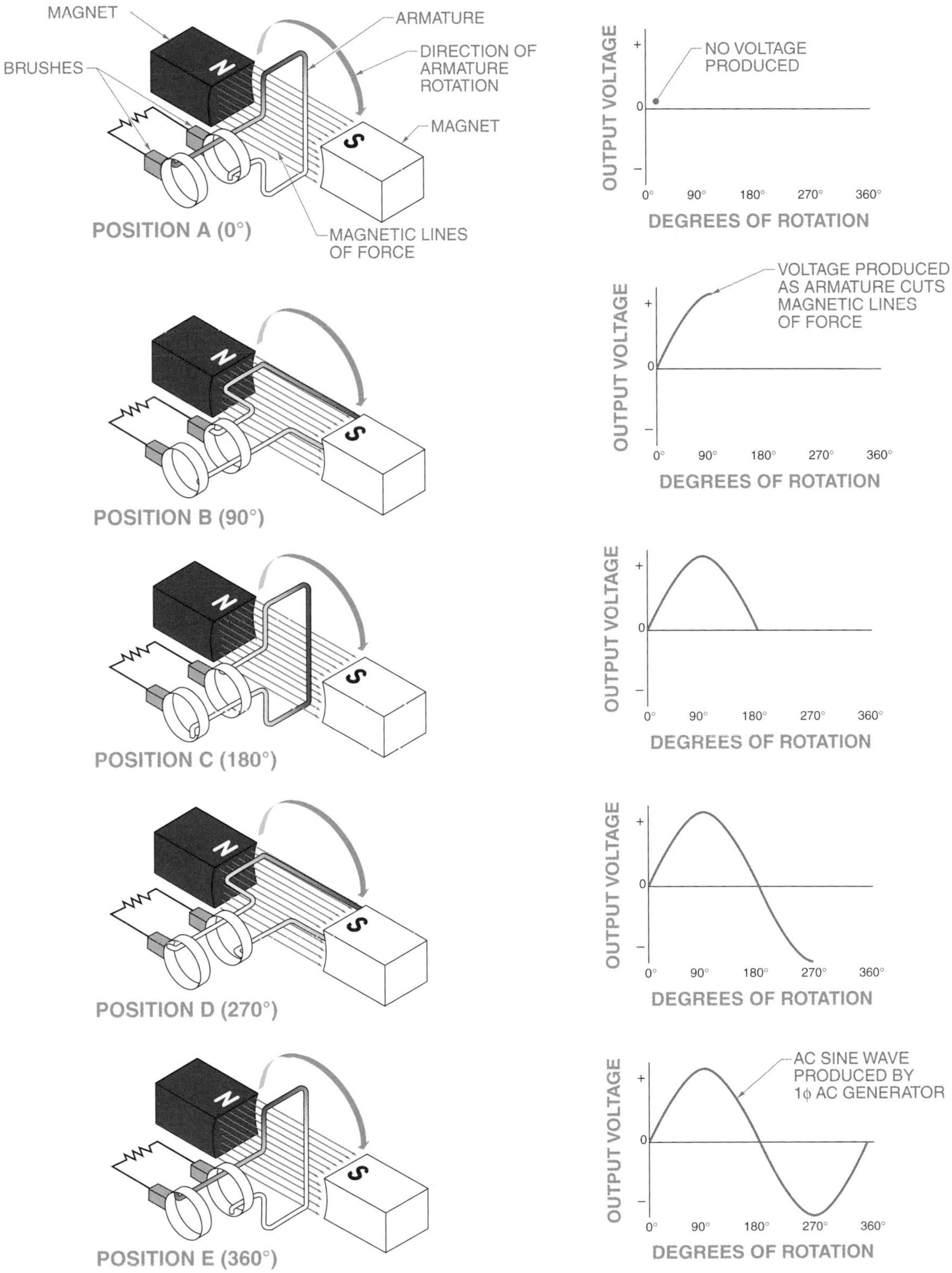

Figure 10-3. In a 1ϕ AC generator, as the armature rotates through 360° of motion, the voltage generated is a continuously changing AC sine wave.

As the armature rotates from position A to position B, each half of the armature cuts across the magnetic lines of force, producing current in the external circuit. The current increases from zero to its maximum value in one direction. This changing value of current is represented by the first quarter (90° of rotation) of the sine wave.

As the armature rotates from position B to position C, current continues in the same direction. The current decreases from its maximum value to zero. This changing value of current is represented by the second quarter (91° to 180° of rotation) of the sine wave.

As the armature continues to rotate to position D, each half of the coil cuts across the magnetic lines of force in the opposite direction. This changes the direction of the current. During this time, the current increases from zero

to its maximum negative value. This changing value of current is represented by the third quarter (181° to 270° of rotation) of the sine wave.

As the armature continues to rotate to position E (or position A again), the current decreases to zero. This completes one 360° cycle of the sine wave.

Three-Phase AC Generators

The principles of a three-phase (3ϕ) AC generator are the same as a 1ϕ AC generator except that there are three equally spaced armature windings 120° out of phase with each other. **See Figure 10-4.** The output of a 3ϕ AC generator results in three output voltages 120° out of phase with each other.

THREE-PHASE AC GENERATORS

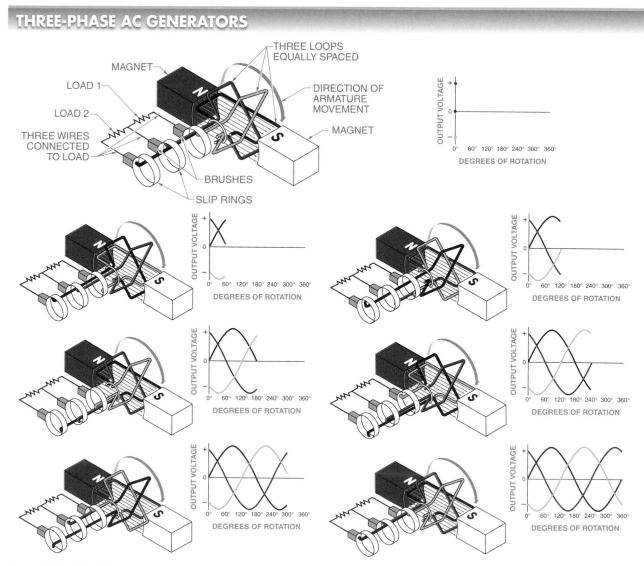

Figure 10-4. A 3ϕ AC generator has three equally spaced armature windings 120° out of phase with each other.

A 3ɸ AC generator has six leads coming from the armature coils. When these leads are brought out from the generator, they are connected so that only three leads appear for connection to the load. Armature coils can be connected in a wye (Y) connection or a delta (Δ) connection. The manner in which the leads are connected determines the electrical characteristics of the generator output. **See Figure 10-5.**

Wye (Y) Connections. A *wye (Y) connection* is a connection that has one end of each coil connected together and the other end of each coil left open for external connections. This circuit can be simplified by connecting the A2, B2, and C2 phase ends together. **See Figure 10-6.**

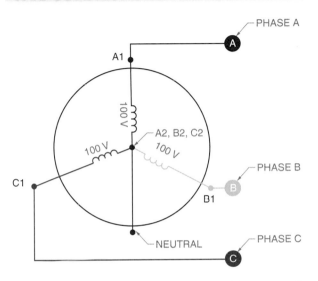

Figure 10-6. A common neutral wire can safely connect the internal leads of a wye-connected alternator to form a common return for lighting loads.

The three ends can be safely connected at the neutral point because no voltage difference exists between them. As phase A is at its maximum, phases B and C are opposite to A. If the equal opposing values of B and C are added vectorially, the opposing force of B and C combined is exactly equal to A. **See Figure 10-7.** For example, if three people are pulling with the same amount of force on ropes tied together at a single point, the resulting forces cancel each other and the resultant force is zero in the center (neutral point).

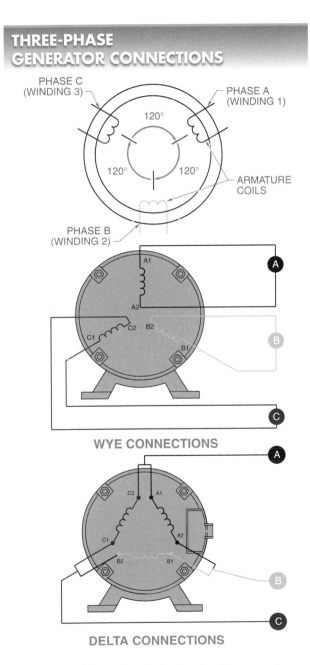

Figure 10-5. When the six leads of a 3ɸ generator are brought out, they are connected in a delta connection or a wye connection.

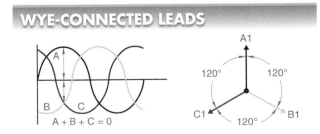

Figure 10-7. The 3ɸ voltages of a wye-connected alternator effectively cancel each other at the neutral point, allowing the three leads of the alternator to be connected.

The net effect is a large voltage (pressure) difference between the A1, B1, and C1 coil ends, but no voltage difference between the A2, B2, and C2 coil ends. The common wire (lead) may or may not be connected from the generator. If it is connected, it is called the neutral. A simplified drawing shows a wye-connected generator with the common wire or neutral not being connected outside the generator. In this configuration each load is connected across two phases in series. Voltage between any two lines in a wye-connected AC generator is 1.73 (√3) multiplied by any of the individual phase voltages. If the 1φ voltage is 100 V, then the voltage between any two lines in the wye configuration will be 173 V (100 × 1.73). Since both coils are in series, the current remains the same throughout the coils.

Another way to look at the output voltage of an AC generator is by using vectors. A vector is used to show magnitude and direction. A vector can be visualized as an arrow drawn in a specific direction, with a length that is equal to the magnitude (voltage or current) drawn in a specific direction. **See Figure 10-8.**

The voltage measured across a single winding or phase is known as the phase voltage. The voltage measured between the lines is known as line-to-line voltage or, simply, the line voltage. When referring to the wiring schematics, terms such as Phase A, Phase 1, and Line 1 are often used interchangeably.

A light can be connected to each separate phase of a wye-connected generator. Each light illuminated from the generated 1φ power is delivered from each phase. The A2, B2, and C2 wires return to the generator together. In a 3φ wye-connected lighting circuit, the 3φ circuit is balanced because the loads are all equal in power consumption. In a balanced circuit, there is no current flow in the neutral wire because the sum of all currents is zero.

All large power distribution systems are designed as 3φ systems with the loads balanced across the phases as closely as possible. The only current that flows in the neutral wire is the unbalanced current. This is normally kept to a minimum because most systems can be kept fairly balanced. The neutral wire is normally connected to a ground such as the earth.

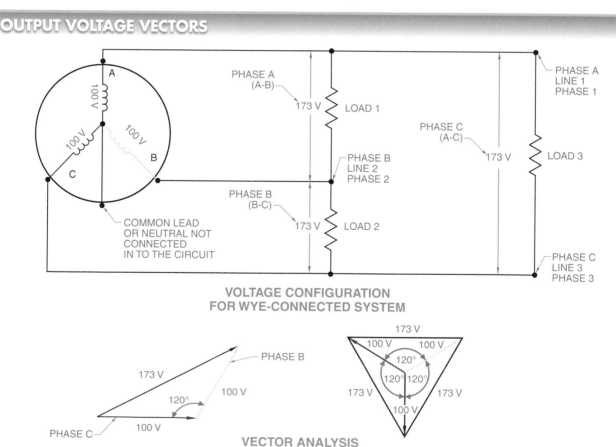

OUTPUT VOLTAGE VECTORS

VOLTAGE CONFIGURATION FOR WYE-CONNECTED SYSTEM

VECTOR ANALYSIS

Figure 10-8. Vectors can be used to illustrate the magnitude and direction of AC generator output voltages.

A wye connection can be used to obtain phase-to-neutral voltage (1ϕ low voltage), phase-to-phase voltage (1ϕ high voltage), and phase-to-phase-to-phase voltage (3ϕ voltage). **See Figure 10-9.** Phase-to-phase voltage is also referred to as line voltage.

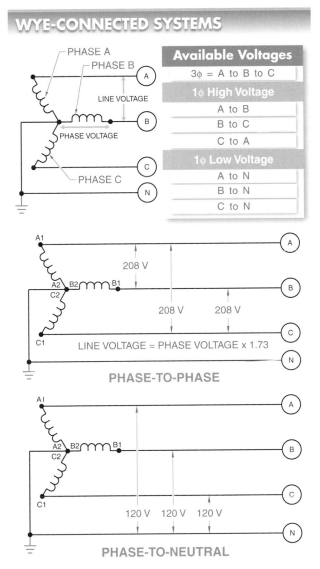

Figure 10-9. In wye-connected systems, the neutral wire is connected to ground and has various available voltages.

In a 3ϕ wye-connected system, the phase-to-neutral voltage is equal to the voltage generated in each coil. For example, if a generator produces 120 V from A1 to A2, the equivalent 120 V is present from B1 to B2 and C1 to C2. Thus, in a 3ϕ wye-connected system, the output voltage of each coil appears between each phase and the neutral.

In a high-voltage, wye-connected generator such as those found in power plants, a phase-to-neutral voltage of 2400 V creates a phase-to-phase voltage of 4152 V, and a phase-to-neutral voltage of 7200 V creates a phase-to-phase voltage of 12,456 V. One of the benefits wye-connected systems bring to a utility company is that even though their generators are rated at 2400 V or 7200 V per coil, they can transmit at these higher phase-to-phase voltages with a reduction in losses. This is because the higher the transmitted voltage, the lower the power loss.

According to the formula $P = E \times I$, power equals voltage multiplied by current. When voltage is higher, current is lower for a given amount of power. According to the formula $P = I^2 \times R$, power (or power loss) equals current squared multiplied by resistance. With a lower current, there is a lower power loss for a power line with a given resistance. Therefore, power lines can carry more power at higher voltages than at lower voltages. The reduction in power losses across transmission lines is especially important for long rural power lines.

Delta (Δ) Connections. A *delta (Δ) connection* is a connection that has each coil end connected end-to-end to form a closed loop. Alternator coil windings of a 3ϕ system can also be connected as a delta connection. **See Figure 10-10.** As in a wye-connected system, the coil windings are spaced 120° apart.

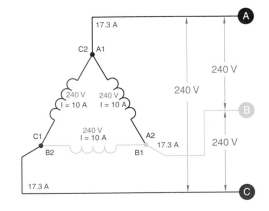

Figure 10-10. A delta connection has each coil end connected end-to-end to form a closed loop.

In a delta-connected system, the voltage measured across any two lines is equal to the voltage generated in the coil winding. This is because the voltage is measured directly across the coil winding. For example, if the generated coil voltage is equal to 240 V, the voltage between any two lines equals 240 V. In the delta configuration, line voltage and phase voltage are the same.

Following any line in a delta-connected system back to the connection point shows that the current supplied to that line is supplied by two coils. Phase A can be traced back to connection point A1, C2. However, as in a wye-connected system, the coils are 120° apart. Therefore, the line current is the vector sum of the two coil currents. In a balanced system, the phase currents are equal. In a balanced 3ϕ delta-connected system, the line current is equal to 1.73 times the current in one of the coils.

For example, it is assumed that each of the phase windings in a delta-connected system has a current flow of 10 A. The current in each of the lines, however, is 17.32 A. The reason for this difference in current is that current flows through different windings at different times in a 3ϕ circuit. During some periods, current will flow between two lines only. At other times, current will flow from two lines to the third line. The delta connection is similar to a parallel connection because there is always more than one path for current flow. Since these currents are 120° out of phase with each other, vector addition must be used when finding the sum of the currents. **See Figure 10-11.**

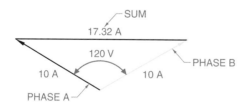

Figure 10-11. Vectors may be added to find the sum of currents and voltages that are out of phase.

To determine the current in a delta connection, the following formulas are used:

$$I_L = I_\phi \times 1.73$$
and
$$I_\phi = I_L \div 1.73$$
where
I_L = line current (in A)
I_ϕ = phase current (in A)

Three-Phase Power

Technicians may become confused when computing power in 3ϕ circuits. One reason for confusion is that there are actually two formulas that can be used. If the line values of voltage and current are known, the power (in watts) of a pure resistive load can be computed using the following formula:

$$P = \sqrt{3} \times V_L \times I_L$$

where
P = power (in W)
V_L = line voltage (in V)
I_L = line current (in A)

If the phase values of voltage and current are known, the apparent power can be computed using the following formula:

$$P_A = \sqrt{3} \times V_\phi \times I_\phi$$

where
P_A = apparent power (in VA)
V_ϕ = phase voltage (in V)
I_ϕ = phase current (in A)

In the first formula, the line values of voltage and current are multiplied by the square root of 3. In the second formula, the phase values of voltage and current are multiplied by the square root of 3. The first formula is used more often because it is generally more convenient to obtain line values of voltage and current, which can be measured with a voltmeter and clamp-on ammeter.

In a delta-connected system, only three wires appear in the system. None of the three wires are normally connected to ground. However, when a delta-connected system is not grounded, it is possible for one phase to accidentally become grounded without anyone being aware of this. This problem is not apparent until another phase also grounds. For this reason, some plants deliberately ground one corner of the delta-connected system so that inadvertent faults on the other two phases cause a fuse or circuit breaker to trip.

A delta-connected system permits different voltage possibilities. Three-phase power (240 V) is available between A, B, and C. Single-phase power (120 V) is available from A to N and from C to N. **See Figure 10-12.** Single-phase power (240 V) is available from A to B, B to C, and C to A. Also available is approximately 195 V, 1ϕ power from B to N.

WARNING: Voltage for B to N is high/dangerous voltage and should be avoided because it could damage equipment. Phase B is also known as the high leg (stinger). In a 120/240 V, 3ϕ delta-connected system, phase B is never used for 120 VAC loads.

Generator types range from small portable units to large permanently installed, pad-mounted units with large fuel-driven engines.

DELTA-CONNECTED VOLTAGES

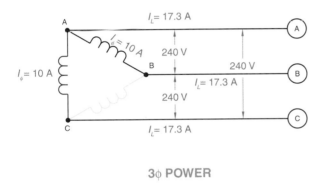

3ϕ POWER

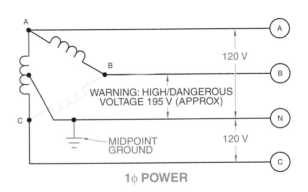

1ϕ POWER

Figure 10-12. A delta-connected system permits different voltage possibilities.

10-2 CHECKPOINT

1. In a two field pole (one north and one south) single-phase AC generator, how many mechanical degrees of rotation does the armature have to rotate to produce one complete AC sine wave?

2. In a three-phase (3ϕ) AC generator, each of the three generated voltages is how many degrees apart from each other?

3. In what type of generator are the ends of each of the three coil windings connected together?

4. In what type of generator are the three coil windings connected end-to-end to form a closed loop?

5. In what type of connection is the voltage between any one of the three phases and the neutral the same?

6. In what type of connection are there two phase-to-neutral legs with the same voltage and a third phase-to-neutral leg that should not be used?

10-3 VOLTAGE CHANGES

AC generators are designed to produce a rated output voltage. In addition, all electrical and electronic equipment is rated for operation at a specific voltage. The rated voltage is a voltage range that was normally ±10%. Today, however, with many components derated to save energy and operating cost, the range is normally +5% to −10%. A voltage range is used because an overvoltage is generally more damaging than an undervoltage. Equipment manufacturers, utility companies, and regulating agencies must routinely compensate for changes in system voltage.

Back-up generators are used to compensate for voltage changes. A back-up generator can be powered by a diesel, gasoline, natural gas, or propane engine connected to the generator. If there is any power interruption in the time period between the loss of main utility power and when the generator starts providing power, the generator is usually classified as a standby (emergency) power supply. Voltage changes in a system may be categorized as momentary, temporary, or sustained. **See Figure 10-13.**

Momentary Power Interruptions

A *momentary power interruption* is a decrease to 0 V on one or more power lines lasting from 0.5 cycles up to 3 sec. All power distribution systems have momentary power interruptions during normal operation. Momentary power interruptions can be caused when lightning strikes nearby, by utility grid switching during a problem (such as a short on one line), or during open circuit transition switching. *Open circuit transition switching* is a process in which power is momentarily disconnected when switching a circuit from one voltage supply (or level) to another.

Tech Fact

Electrical equipment and surge suppressors may be described by the terms "ride-through capability" and "let-through capability." The term "ride-through capability" is used by manufacturers to describe the ability of electrical equipment to withstand a momentary power interruption. The term "let-through capability" is used to describe how much voltage a surge suppressor will allow through to the equipment it protects.

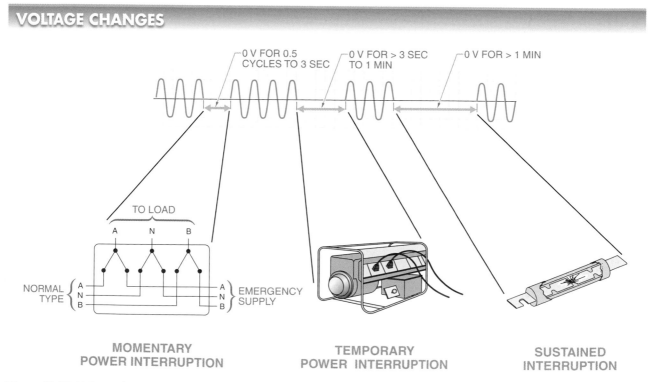

Figure 10-13. Voltage changes in an electrical system may be categorized as momentary, temporary, or sustained.

Temporary Power Interruptions

A *temporary power interruption* is a decrease to 0 V on one or more power lines lasting for more than 3 sec up to 1 min. Automatic circuit breakers and other circuit protection equipment protect all power distribution systems. Circuit protection equipment is designed to remove faults and restore power. An automatic circuit breaker normally takes from 20 cycles to about 5 sec to close. If the power is restored, the power interruption is only temporary. If power is not restored, a temporary power interruption becomes a sustained power interruption. A temporary power interruption can also be caused by a time gap between power interruptions and when a back-up power supply (generator) takes over or if someone accidentally opens the circuit by switching the wrong circuit breaker switch.

Sustained Power Interruptions

A *sustained power interruption* is a decrease to 0 V on all power lines for a period of more than 1 min. All power distribution systems have a complete loss of power at some time. Sustained power interruptions (outages) are commonly the result of storms, tripped circuit breakers, blown fuses, and/or damaged equipment.

The effect of a power interruption on a load depends on the load and the application. If a power interruption could cause equipment, production, and/or security problems that are not acceptable, an uninterruptible power system can be used. An *uninterruptible power system (UPS)* is a power supply that provides constant on-line power when the primary power supply is interrupted. For long-term power interruption protection, a generator/UPS is used. For short-term power interruptions, a static UPS is used.

Transient Voltages

A *transient voltage,* also referred to as a voltage spike, is a temporary, unwanted voltage in an electrical circuit. Transient voltages typically exist for a very short time but are often larger in magnitude than voltage surges and very erratic. Transient voltages occur due to lightning strikes, unfiltered electrical equipment, and power being switched on and off. High transient voltages may reach several thousand volts. A transient voltage on a 120 V power line can reach 1000 V (1 kV) or more.

High transient voltages exist close to a lightning strike or when large (high-current) loads are switched off. For example, when a large motor (100 HP) is turned off, a transient voltage can move down the power distribution system. If a DMM is connected to a point along the system in which the high transient voltage is present, an arc can be created inside the DMM. Once started, the arc can cause a high-current short in the power distribution system even after the original high transient voltage is gone. The high-current short can turn into an arc blast. An *arc blast* is an explosion that occurs when the surrounding air becomes ionized and conductive. **See Figure 10-14.**

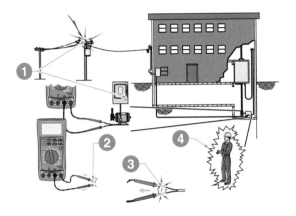

TRANSIENT VOLTAGE

1 LIGHTNING STRIKE OR LARGE LOAD SWITCHING CAUSES A TRANSIENT VOLTAGE ON POWER LINE, CREATING ARC BETWEEN DMM INPUT TERMINALS

2 HIGH CURRENT FLOWS IN CLOSED CIRCUIT; ARC STARTS AT PROBE TIPS

3 WHEN PROBES ARE PULLED IN REACTION TO LOUD NOISE, ARCS ARE DRAWN TO TERMINALS

4 IF ARCS ARE JOINED, RESULTING HIGH-ENERGY ARC CAN CREATE A LIFE-THREATENING SITUATION FOR USER

Figure 10-14. When taking measurements in an electrical circuit, transient voltages can cause electrical shock and/or damage to equipment.

The amount of current drawn and potential damage caused depends on the specific location of the power distribution system. All power distribution systems have current limits set by fuses and circuit breakers along the system. The current rating (size) of fuses and circuit breakers decreases as distance from the main distribution panel increases. The farther away from the main distribution panel, the less likely the high transient voltage is to cause damage.

Transient voltages are normally erratic, large voltages or surges that have a short duration and a short rise time. Computers, electronic circuits, and specialized electrical equipment require protection against transient voltages. Protection methods commonly include proper wiring, grounding, shielding of the power lines, and use of surge suppressors. A *surge suppressor* is an electrical device that provides protection from transient voltages by limiting the level of voltage allowed downstream from the surge suppressor. Surge suppressors can be installed at service entrance panels and at individual loads. **See Figure 10-15.**

SURGE SUPPRESSORS

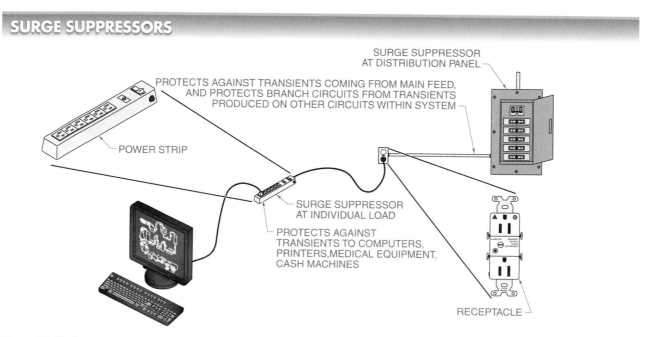

SURGE SUPPRESSOR AT DISTRIBUTION PANEL

PROTECTS AGAINST TRANSIENTS COMING FROM MAIN FEED, AND PROTECTS BRANCH CIRCUITS FROM TRANSIENTS PRODUCED ON OTHER CIRCUITS WITHIN SYSTEM

POWER STRIP

SURGE SUPPRESSOR AT INDIVIDUAL LOAD

PROTECTS AGAINST TRANSIENTS TO COMPUTERS, PRINTERS, MEDICAL EQUIPMENT, CASH MACHINES

RECEPTACLE

Figure 10-15. A surge suppressor is an electrical device that provides protection from high-level transients by limiting the level of voltage allowed downstream from the surge suppressor.

10-3 CHECKPOINT

1. What is the acceptable voltage range applied to electrical and electronic equipment today?

2. What are the three types of power interruptions stated from the shortest to the longest?

3. What electrical device is used to help protect against the damage that may occur from a transient voltage?

4. What device is used to maintain power during a power interruption?

Additional Resources

Review and Resources

Access Chapter 10 Review and Resources for *Electrical Motor Controls for Integrated Systems* by scanning the above QR code with your mobile device.

Applying Your Knowledge

Refer to the *Electrical Motor Controls for Integrated Systems* Learner Resources for interactive Applying Your Knowledge questions.

Workbook and Applications Manual

Refer to Chapter 10 in the *Electrical Motor Controls for Integrated Systems Workbook* and the *Applications Manual* for additional exercises.

ENERGY EFFICIENCY PRACTICES

Generator Efficiency

All generators produce electrical energy by converting some other form of energy, such as wind, falling water, coal, or oil, to electrical energy. How efficiently a generator produces electrical energy varies greatly based on the type of generator and its condition, the type and cost of power used to drive the generator, the load on the generator, and the environmental conditions. In addition, the entire generator system must be factored into how efficiently electricity is being generated. For example, a generator driven by falling water has high costs associated with building a dam. Similarly, a nuclear plant has very high costs associated with building the nuclear plant, operating the plant safely, and meeting regulation requirements.

The efficiency rating of electricity produced by a generator system can be estimated and compared to other types of systems by taking into account the fuel cost, installation cost, generator operation and maintenance requirements, and average available power output. Average efficiency ratings for different types of generator systems include the following:

- Hydroelectric generator systems are 90% to 95% efficient.
- Tidal generator systems are 85% to 90% efficient.
- Coal-fired generator systems are 45% to 50% efficient.
- Nuclear generator systems are 45% to 55% efficient.
- Wind generator systems are 45% to 50% efficient.
- Gas-fired generator systems are 35% to 40% efficient.

Generator system efficiency varies greatly based on the system load, power source reliability, fuel cost, operating temperature, and required dependents.

- System load—Even an efficient generator is not efficient if it is only required to deliver 10% to 50% of its rated output. It is more efficient to operate a generator at 75% to 85% of its rating.
- Power source reliability—Generator efficiency is based partly on the reliability of its power source. For example, if there is no wind, a wind generator is 0% efficient. If wind is blowing at optimum operating speed, a wind generator is nearly as efficient as a hydroelectric generator. However, since wind speed varies and water pressure behind a dam is constant, a hydroelectric generator system is rated much higher than a wind generator system.
- Fuel cost—The higher the cost of coal, gas, or oil, the less efficient the generator system, even if the generator itself is very efficient.
- Operating temperature—All generators are less efficient at higher operating temperatures. This includes the ambient temperature and the heat produced within the generator.
- Required dependents—The required dependents for the generated power and the available power delivery system must also be considered when determining the efficiency of a generator system. For example, there are many places where wind blows constantly, large rivers flow, and fossil fuels are abundant, but few (or no) customers need power or there is no distribution system to deliver the power to the customers.

Objectives

11-1
- Define transformer and describe the primary and secondary windings.
- Explain turns ratio.
- Explain how resistance, eddy currents, and hysteresis in the iron core of a transformer produce heat loss in the transformer.

11-2
- Explain how residential electrical service uses single-phase (1ϕ) transformer connections.
- Explain how a transformer secondary tap is used.
- Explain how to obtain a control voltage of 120 VAC from a line voltage of 240 VAC.
- Explain how to obtain a control voltage of 120 VAC from a line voltage of 480 VAC.

11-3
- Explain how to select a transformer.

11-4
- Describe how to troubleshoot a transformer by taking resistance and voltage measurements.
- Explain how to troubleshoot control transformers.

Review and Resources
atplearningresources.com/Quicklinks
Access Code: 362245

Transformers

Transformers are used today in almost every electrical system to either step up, step down, or isolate one voltage level from another. The extensive use of transformers started after what is known in the electrical field as the "War of Currents." Thomas Edison believed that the U.S. power distribution system should be DC with DC generators delivering power over a DC distribution system to DC loads. George Westinghouse believed an AC distribution system would be much better because AC can be distributed over long distances using step-up and step-down transformers. Westinghouse's system of AC generators and AC transformers proved to be the best and is now the standard distribution system used around the world. Because transformers are so important in an electrical system, learning their operating principle and usages is important when designing, installing, or troubleshooting an electrical system.

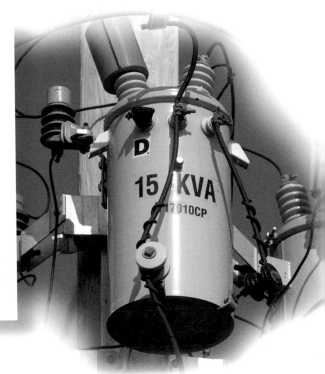

11-1 TRANSFORMERS

A *transformer* is an electric device that uses electromagnetism to change voltage from one level to another or to isolate one voltage from another. Transformers are used in electrical distribution systems to increase or decrease the voltage and current safely and efficiently. For example, transformers are used to increase generated voltage to a high level for transmission across the country and then decrease it to a low level for use by electrical loads. **See Figure 11-1.**

Transformers allow power companies to distribute large amounts of power at a reasonable cost. Large transformers are used for power distribution along city streets and in large manufacturing or commercial buildings. Large transformers are normally maintained by a power company or by workers who have been

specifically trained in high-voltage transformer operation and maintenance.

Technicians often work with small control transformers. Control transformers isolate the power circuit from the control circuit, providing additional safety for the circuit operator. Transformers are also used in the power supplies of most electronic equipment to step the power line voltage up or down to provide the required operating voltage for the equipment.

A transformer has a primary winding and a secondary winding wound around an iron core. **See Figure 11-2.** The *primary winding* is the coil of a transformer that draws power from the source. The *secondary winding* is the coil of a transformer that delivers the energy at the transformed or changed voltage to the load.

CONTROL TRANSFORMERS

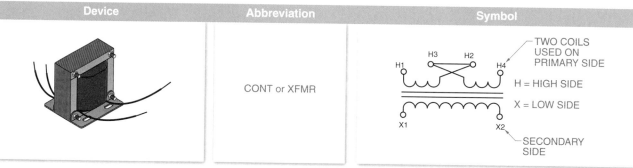

ABB Power T&D Company Inc.

TRANSMISSION TRANSFORMER

General Electric Company

DISTRIBUTION TRANSFORMER

Device	Abbreviation	Symbol
	CONT or XFMR	H3 H2 H1 ○———○———○ H4 X1 ○ ○ X2 TWO COILS USED ON PRIMARY SIDE H = HIGH SIDE X = LOW SIDE SECONDARY SIDE

Figure 11-1. Transformers are used to increase voltage to a high level for transmission across the country and then decrease it to a low level for use by electrical loads.

TRANSFORMER WINDINGS

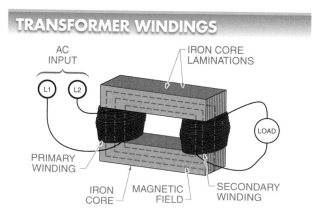

Figure 11-2. A transformer has a primary winding and a secondary winding wound around an iron core.

Transformer Operation

A transformer transfers AC energy from one circuit to another. The energy transfer is made magnetically through the iron core. A magnetic field builds up around a wire when AC is passed through the wire. The magnetic field builds up and collapses each half cycle because the wire is carrying AC. **See Figure 11-3.**

The primary coil of the transformer supplies the magnetic field for the iron core. The secondary coil supplies the load with an induced voltage proportional to the number of turns of a conductor cut by the magnetic field of the core. A transformer is either a step-up or step-down transformer depending on the ratio between the number of turns of the conductor in the primary and secondary sides of the transformer. **See Figure 11-4.**

The *turns ratio* is the ratio of the number of turns in the primary winding to the number of turns in the secondary winding of a transformer. If twice as many turns are on the secondary, twice the voltage is induced on the secondary. The ratio of primary to secondary is 1:2, making the transformer a step-up transformer.

However, if only half as many turns are on the secondary, only half the voltage is induced on the secondary. In this case, the ratio of primary to secondary is 2:1, making the transformer a step-down transformer.

MAGNETIC FIELDS

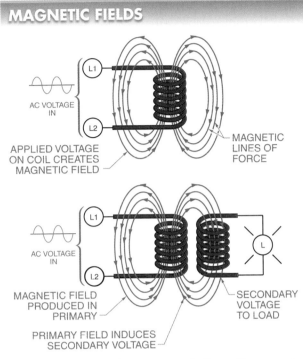

Figure 11-3. In a transformer, magnetic lines of force created by one coil induce a voltage in a second coil.

In a step-up transformer, a ratio of 1:2 doubles the voltage. This may seem like a gain or a multiplication of voltage without any sacrifice. However, the amount of power transferred in a transformer is equal on both the primary and the secondary, excluding small losses within the transformer.

Because power is equal to voltage times current ($P = E \times I$) and power is always equal on both sides of a transformer, the voltage cannot change without changing the current. For example, when voltage is stepped down from 240 V to 120 V in a 2:1 ratio, the current increases from 1 A to 2 A, keeping the power equal on each side of the transformer. By contrast, when the voltage is stepped up from 120 V to 240 V in a 1:2 ratio, the current is reduced from 2 A to 1 A to maintain the power balance. In other words, voltage and current may be changed for particular reasons, but power is constant.

One advantage of increasing voltage and reducing current is that power may be transmitted through smaller gauge wire, thus reducing the cost of power lines.

For this reason, the generated voltages are stepped up very high for distribution across large distances and then stepped back down to meet consumer needs. Although both the voltage and current can be stepped up or down, the terms step up and step down, when used with transformers, always apply to voltage.

TRANSFORMER RATIOS

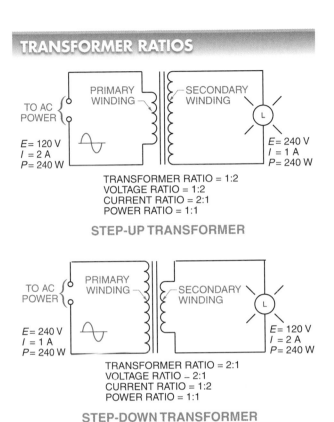

Figure 11-4. Voltage and current change from the primary to secondary winding in step-up and step-down transformers.

Transformer Losses

Although transformers are very efficient, they are not perfect. Not all of the energy delivered to the primary side by the source is transferred to the secondary load circuit. The majority of the energy lost is lost as heat in the transformer. The three types of losses in an iron core transformer are resistive, eddy current, and hysteresis losses.

All of these losses make the typical iron core transformer hot when operating under full load. A transformer may be too hot to touch during normal operation, but there should be no odor of burning insulation or varnish or signs of discoloration or smoke. Any one of these indicates to the technician that the transformer is overloaded or defective.

Resistive Losses. Resistive losses come from the resistance of the coil winding. When current passes through a winding, the winding heats up and loses energy that could have been transferred to the secondary.

Eddy Current Losses. Because iron is a fair conductor of electricity, the varying magnetic field that induces a voltage in the secondary of a transformer also induces small voltages in the iron core of the transformer. These small voltages produce eddy currents, which in turn produce heat. This heat also represents a loss because it does no useful work.

Eddy currents are minimized either by making the core out of thin sheets (laminations) that are insulated from each other or by using powdered-iron cores instead of solid blocks of iron. The insulation between the laminations of a laminated core breaks up the current paths within the core and reduces the eddy currents. This is the same technique used to reduce eddy currents in solenoids.

Hysteresis Losses. Each time the magnetizing force produced by the primary side of a transformer changes, the atoms of the core realign themselves in the direction of the force. The energy required to realign the iron atoms must be supplied by the input power and is not transferred to the secondary load circuit. The realignment of the iron atoms does not follow the magnetizing force instantaneously but instead lags slightly behind it. This lagging action is called hysteresis. The degree of hysteresis is a measure of the amount of energy required to realign the iron atoms in the core; the energy lost to do so is called hysteresis loss.

Hysteresis results in heating of the iron core. Because of this and other similarities to mechanical friction, hysteresis is sometimes referred to as magnetic friction. Hysteresis losses are minimized by using high-silicon steel and other alloys in the core.

Transformer Efficiency

In an ideal transformer, energy is transferred from the primary circuit to the secondary circuit and there is no power loss. However, all transformers have some power loss. Most transformers operate with little power loss (normally 0.5% to 8%). The less the power loss, the more efficient the transformer. The efficiency of a transformer is expressed as a percentage. To calculate the efficiency of a transformer, the following formula is applied:

$$Eff = \frac{P_S}{P_P} \times 100$$

where

Eff = efficiency (in %)

P_S = power of secondary circuit (in W)

P_P = power of primary circuit (in W)

For example, what is the efficiency of a transformer that uses 1200 W of primary power to deliver 1110 W of secondary power?

$$Eff = \frac{P_S}{P_P} \times 100$$

$$Eff = \frac{1110}{1200} \times 100$$

$$Eff = 0.925 \times 100$$

$$Eff = \textbf{92.5\%}$$

Tech Fact

In a typical heavy industrial facility, electricity may be delivered directly from a transmission substation to an outside transformer vault. Service-entrance conductors are routed from the outside transformer vault through an outdoor busway to a metered switchboard. Power is then fed through circuit breakers in the panelboard and routed through busways to power distribution panels and busways with plug-in sections to the points of use. Depending on customer needs, the power distribution system delivers power at standard voltage levels and fixed current ratings to set points such as receptacles.

11-1 CHECKPOINT

1. What are the two windings of a transformer called?

2. If a transformer has a turns ratio of 1:4, what is the transformer's voltage ratio?

3. If a transformer has a turns ratio of 1:4, what is the transformer's current ratio?

4. If a transformer has a turns ratio of 1:4, what is the transformer's power ratio?

5. If a transformer has a turns ratio of 1:4 and 115 V is applied to the primary, how many volts are on the secondary?

6. Why are transformer iron cores laminated?

11-2 TRANSFORMER CONNECTIONS

Transformers may be connected in various configurations depending on the application. Configurations consist of single-phase and three-phase connections. Single-phase connections are typically found in residential applications, while three-phase connections are found in commercial and industrial applications. High voltage transformer windings are marked H1, H2, etc., and low voltage transformer windings are marked X1, X2, etc.

Single-Phase Transformer Connections

Electricity is used in residential applications (one-family, two-family, and multifamily dwellings) to provide energy for lighting, heating, cooling, cooking, etc. The electrical service to dwellings is normally 1φ, 120/240 V. The low voltage (120 V) is used for general-purpose receptacles and general lighting. The high voltage (240 V) is used for heating, cooling, cooking, etc.

Residential electrical service may be overhead or lateral. *Overhead service* is an electrical service in which service-entrance conductors are run through the air from the utility pole to the building. *Service lateral* is an electrical service in which service-entrance conductors are run underground from the utility system to the service point. **See Figure 11-5.**

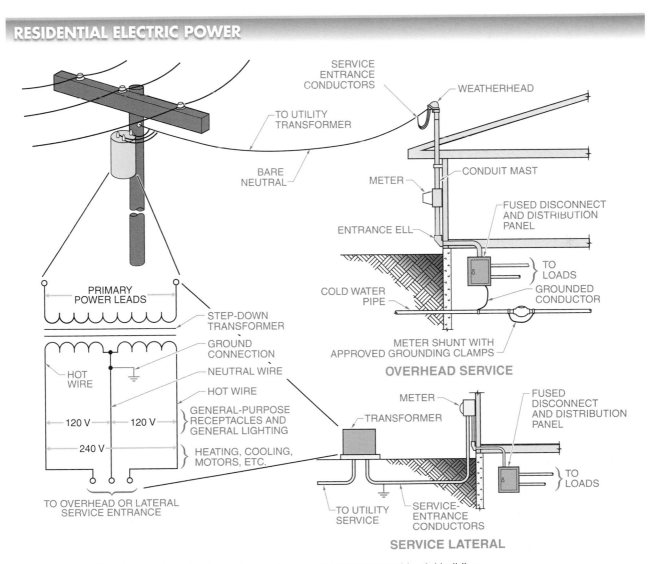

Figure 11-5. Overhead or service lateral may be used to supply power to a residential building.

Three-Phase Transformer Connections

Three 1ϕ transformers are connected to develop 3ϕ voltage. The three transformers may be connected in a wye or delta connection. In a wye connection, the end of each coil is connected to the incoming power lines (primary side) or used to supply power to the load or loads (secondary side). The other ends of each coil are connected together. In a delta connection, each transformer coil is connected end-to-end to form a closed loop. Each connecting point in a delta connection is attached to the incoming power lines or used to supply power to the load or loads. The voltage output and type available for the load or loads is determined by whether the transformer is connected in a wye or delta connection. **See Figure 11-6.**

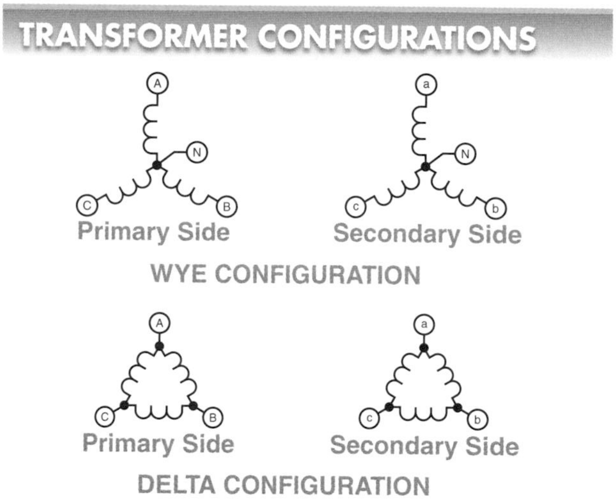

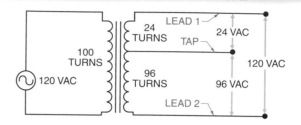

Figure 11-7. Taps allow different output voltages to be obtained from a transformer.

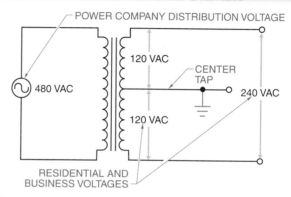

Figure 11-8. A center-tapped transformer is used to change the high voltage of power company distribution lines to the common 240/120 VAC supply of residences and businesses.

TRANSFORMER CONFIGURATIONS

Primary Side | Secondary Side
WYE CONFIGURATION

Primary Side | Secondary Side
DELTA CONFIGURATION

Figure 11-6. Three-phase transformers may be connected in a wye or delta configuration.

Transformer Secondary Taps

Many transformers have a secondary coil that has an extra lead (tap) attached to it. A *tap* is a connection brought out of a winding at a point between its endpoints to allow the voltage or current ratio to be changed. Taps allow different output voltages to be obtained from a transformer. **See Figure 11-7.** For example, the output voltage between leads 1 and 2 is 120 VAC because the turns ratio is 1:1 (100 to 100). The output between the tap and lead 1 is 24 VAC because the turns ratio is approximately 4.17:1 (100 to 24).

A tap that splits a secondary in half is referred to as a center tap. A common application of a transformer with a center tap is a distribution transformer. A distribution transformer is used in residences and businesses to change the high voltage of power company distribution lines to the common 240/120 VAC supply of residences and businesses. **See Figure 11-8.**

The center tap is connected to earth ground and becomes a common conductor. The voltage across the output lines is 240 VAC. However, the voltage measured between either output line and the center tap is 120 VAC.

This circuit is a typical circuit used by a power company to deliver power to a residence. The 240 VAC power is used to supply devices in the residence that require a large amount of operating power, such as a central air conditioner, water heater, clothes dryer, and cooking range. These high-power devices run on 240 VAC to allow smaller conductor wires to deliver power to them. The 120 VAC power is wired to the electrical outlets and lighting system. This provides a much safer level of voltage, which can be used on smaller electrical devices.

Control Transformers

A *control transformer* is a transformer that is used to step down the voltage to the control circuit of a system or machine. The most common control transformers have two primary coils and one secondary coil. **See Figure 11-9.**

CONTROL TRANSFORMERS

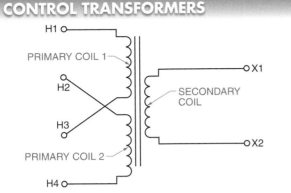

Figure 11-9. The most common control transformers have two primary coils and one secondary coil.

The primary coils of a control transformer are crossed so that metal links can be used to connect the primaries for either 240 VAC or 480 VAC operation. In most applications, a control transformer is used to reduce the main or line voltage of 240 VAC or 480 VAC to a control voltage of 120 VAC.

240 V Primary. To obtain a control voltage of 120 VAC from a line voltage of 240 VAC, the two primary coils must be connected in parallel. **See Figure 11-10.** If the primary coils are connected in parallel, the effective turns of the two primary coils is 200, the same as if there were only one primary coil. If the secondary has 100 turns, the turns ratio is 2:1. This means an input voltage of 240 VAC produces an output voltage of 120 VAC.

CONTROL TRANSFORMERS— 240 V PRIMARY

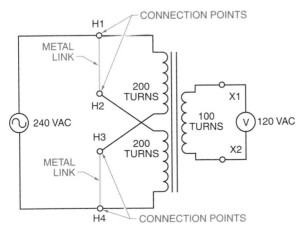

Figure 11-10. To obtain a control voltage of 120 VAC from a line voltage of 240 VAC, the two primary coils must be connected in parallel.

480 V Primary. To obtain a control voltage of 120 VAC from a line voltage of 480 VAC, the two primary coils must be connected in series. **See Figure 11-11.** If the primary coils are connected in series, the effective turns of the two primary coils is 400, making the turns ratio 4:1. This means an input voltage of 480 VAC produces an output voltage of 120 VAC.

CONTROL TRANSFORMERS— 480 V PRIMARY

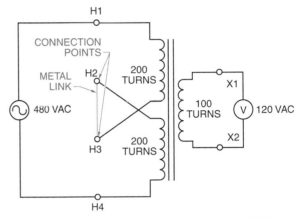

Figure 11-11. To obtain a control voltage of 120 VAC from a line voltage of 480 VAC, the two primary coils must be connected in series.

11-2 CHECKPOINT

1. What type of electrical service is the most common for residential dwellings?
2. Why do control transformers usually have two primary coils?
3. What do "X" markings on a transformer represent?
4. What do "H" markings on a transformer represent?
5. When connecting the primary of a control transformer to high voltage, are the coils connected in series or parallel?
6. When connecting the primary of a control transformer to low voltage, are the coils connected in series or parallel?

11-3 TRANSFORMER SELECTION

A transformer that is selected for an application must have a higher volt/amperage (VA) rating than necessary for the application. However, a transformer that has a VA rating excessively higher than the required rating should not be selected because the transformer will be less efficient in that application. A transformer should be selected that has a rating above but close to the required value. The information needed to size a transformer includes input voltage available, output voltage desired, and output current required (both inrush and steady-state). With this information, a catalog specification sheet can be used to select the proper transformer. **See Figure 11-12.**

The most important guideline when sizing a transformer is to select a transformer that safely and efficiently provides the maximum current that can be drawn by a load. A common instance of sizing a transformer occurs when selecting a transformer to operate a machine. Many machines require a transformer to step down the line voltage (480 VAC or 240 VAC) to the operating voltage of 120 VAC. Machines with motors or other high inrush devices draw their maximum current when the devices are first started. For these machines, this inrush current is the critical value that must be considered when selecting a transformer.

If a machine does not have devices with high inrush characteristics, inrush current is not as much of a consideration. In this case, the steady-state current is more important. Most machines list both their maximum inrush and steady-state current requirements.

Transformer specification sheets normally list a steady-state volt/amperage (VA). The steady-state VA is the secondary voltage multiplied by the secondary current ($V_s \times I_s$) load or loads during steady-state current. Another listed value is the maximum inrush VA. This is the secondary voltage multiplied by the secondary current during the inrush period.

Tech Fact

The K rating should be considered when a power transformer is selected for an application. The K rating of a transformer is a measurement of the ability of a transformer to operate properly when connected to nonlinear loads that cause problems such as harmonics. The higher the K rating, the better the transformer is at handling nonlinear loads. For example, K-1 rated transformers can be used for motors and incandescent lamps, K-4 rated transformers can be used for high-intensity discharge lamps and welders, K-13 rated transformers can be used for computer, and K-20 rated transformers can be used for motor drives and data processing rooms.

TRANSFORMER ELECTRICAL SPECIFICATIONS AND ORDERING DATA (SUPPLY VOLTAGE 220 VAC)

VA*	Maximum Inrush VA†	Temperature Rise‡	Dimensions§			Model J201
			H	L	W	
**110 V To 120 V Secondary Voltage Rating**
50	180	55	3 5/16	3 3/8	2 1/2	1111
75	218	55	3 9/16	3 3/8	2 7/8	1121
100	273	55	3 3/4	3 3/8	2 7/8	1131
150	660	55	4 5/16	4 1/2	3 13/16	1141
250	1360	55	5	4 1/2	3 13/16	1161
500	1964	115	5 1/2	4 1/2	3 3/4	1191
1000	4014	115	6 3/4	5 1/4	4 3/8	1211
**22 V To 24 V Secondary Voltage Rating**
50	180	55	3 5/16	3	2 1/2	1111-824
100	273	55	3 3/4	3 3/8	2 7/8	1131-824
150	660	55	4 5/16	4 1/2	3 13/16	1141-824

* Terminal type
† Capability VA. Refers to maximum inrush VA after calculations are made
‡ 0° C
§ in in.

Figure 11-12. Transformer specification sheets are used to obtain required information when selecting the proper transformer for an application.

11-3 CHECKPOINT

1. Is a transformer's VA rating based on the primary or secondary of the transformer?

2. What information is needed to size a transformer?

11-4 TROUBLESHOOTING TRANSFORMERS

After a transformer is installed in a circuit, it may operate without failure for a long time. One reason for this is that transformers have no moving parts. If a transformer does fail, it appears either as a short circuit or an open circuit in one of the coils. The two methods that can be used to determine whether a transformer has failed are to measure the input and output voltages and to check the transformer resistance.

Measuring Input and Output Voltages

If a transformer is connected in a circuit, the transformer can be tested by measuring the input and output voltages. The transformer is good if the input and output voltages are reasonably close to the theoretical values. The current levels are tested if the voltage does not stay constant. Although the initial voltage may appear normal, it may not hold up when the transformer is fully loaded.

Checking Transformer Resistance

A DMM set to measure resistance can be used to check for open circuits in the coils, short circuits between the primary and secondary coils, or coils shorted to the core without power applied to the transformer. **See Figure 11-13.**

- Open circuits in the coils—The resistance of each coil is checked with a DMM. The winding is open and the transformer is bad if any of the coils show an infinite resistance reading. It should be noted that very low resistance readings do not indicate a short, just the resistance of the wire.

- Short circuits between the primary and secondary coils—A check for short circuits should be made between the primary and secondary coils of the transformer. A DMM should show an infinite resistance reading between the primary and secondary coils.

- Coils shorted to the core—A resistance check is made from each transformer coil to the core of the transformer. All coils should show an infinite resistance reading. The transformer should not be used if a resistance is shown between any coil and the core.

Testing Control Transformers

A control transformer is used in a circuit to step down the supply voltage to provide a safe voltage level for the control circuit. A control transformer should be checked when there is a problem in a control circuit that may be related to the power supply. **See Figure 11-14.**

All transformers are capable of delivering a limited current output at a given voltage. The power limit of a transformer equals the current times the voltage. This power limit is listed on the nameplate of the transformer as its kilovolt-ampere (kVA) rating. This rating indicates the apparent power the transformer can deliver. If this limit is exceeded, the transformer overheats and the control circuit does not function properly. The transformer comes closer to reaching its limit when loads are added. To test a control transformer, apply the following procedure:

1. Check the input and output voltages of the transformer with the power supply energized. The input and output voltages should be within 5% of the transformer nameplate rating (10% max). The transformer is good if the voltage is within the rating or proportionally low.

2. Measure the current drawn by the transformer with a clamp-on ammeter. The apparent power drawn by the control circuit is determined by multiplying the current reading by the voltage reading. A larger transformer is required if the volt-amperes (VA) drawn are more than the rating of the transformer.

3. Check the transformer ground. A ground test should be performed on new transformer installations or when a ground problem is suspected. Connect one lead of a DMM set to measure voltage to the metal frame of the transformer. Do not connect it to a painted or varnished surface. Connect the second lead to each lead of the transformer on the secondary. Under normal circumstances, if X2 is grounded, the DMM displays a voltage when connected to X1. Under normal circumstances, the DMM will also read a voltage when connected to H1 (if hot side of primary) and no voltage on H2 (if neutral side of primary).

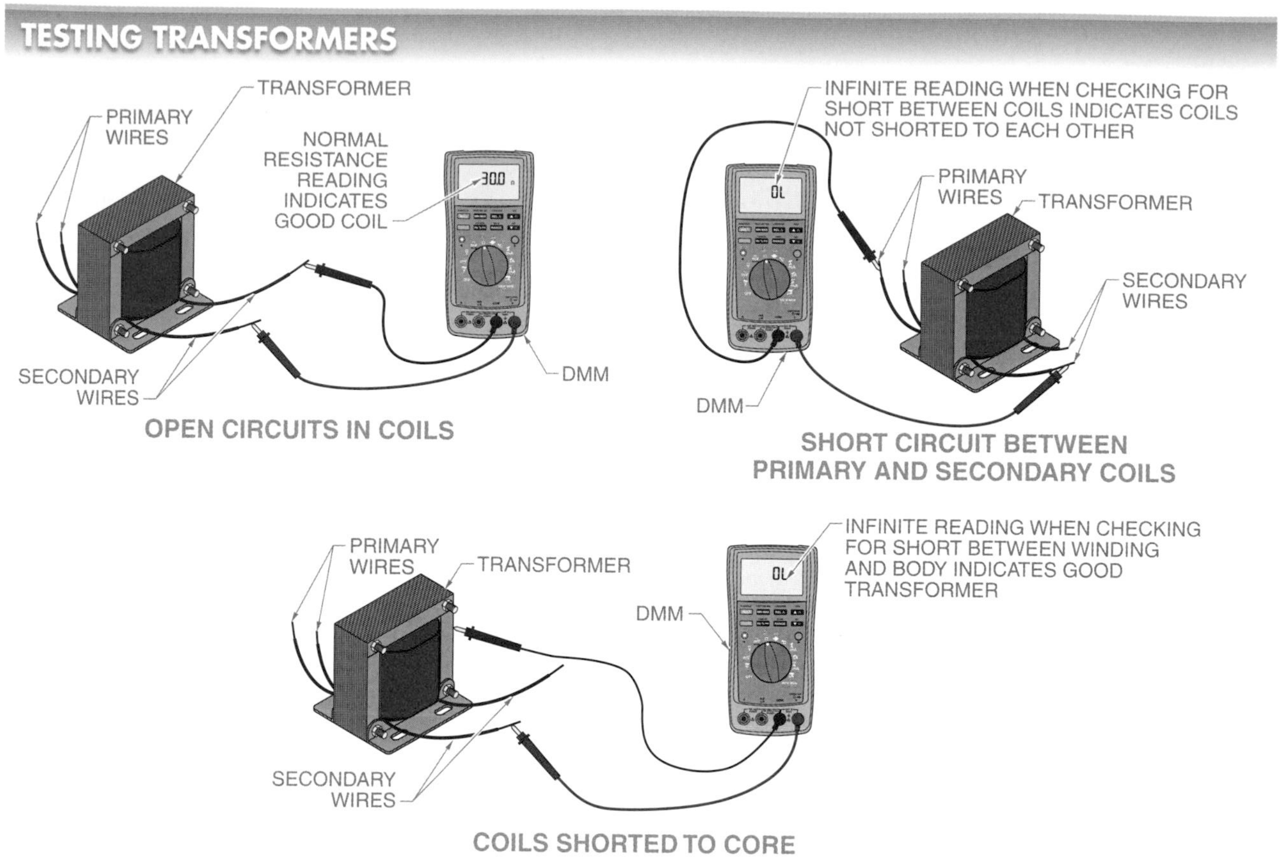

TESTING TRANSFORMERS

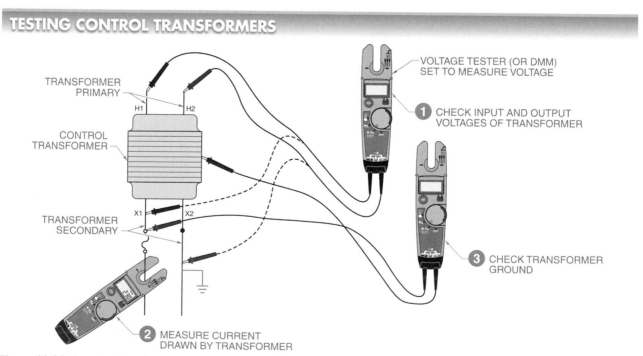

PRIMARY WIRES
TRANSFORMER
NORMAL RESISTANCE READING INDICATES GOOD COIL
30.0 Ω
DMM
SECONDARY WIRES

OPEN CIRCUITS IN COILS

INFINITE READING WHEN CHECKING FOR SHORT BETWEEN COILS INDICATES COILS NOT SHORTED TO EACH OTHER
OL
PRIMARY WIRES
TRANSFORMER
SECONDARY WIRES
DMM

SHORT CIRCUIT BETWEEN PRIMARY AND SECONDARY COILS

PRIMARY WIRES
TRANSFORMER
DMM
INFINITE READING WHEN CHECKING FOR SHORT BETWEEN WINDING AND BODY INDICATES GOOD TRANSFORMER
OL
SECONDARY WIRES

COILS SHORTED TO CORE

Figure 11-13. Transformers are tested by checking for open circuits in the coils, short circuits between the primary and secondary coils, and coils shorted to the core.

TESTING CONTROL TRANSFORMERS

TRANSFORMER PRIMARY
H1   H2
CONTROL TRANSFORMER
X1   X2
TRANSFORMER SECONDARY

VOLTAGE TESTER (OR DMM) SET TO MEASURE VOLTAGE

1 CHECK INPUT AND OUTPUT VOLTAGES OF TRANSFORMER

3 CHECK TRANSFORMER GROUND

2 MEASURE CURRENT DRAWN BY TRANSFORMER

Figure 11-14. A control transformer should be checked when there is a problem in a control circuit that may be related to the power supply.

11-4 CHECKPOINT

1. If a 240/480 V to 120 V, 360 VA rated transformer is tested using a ammeter to measure the secondary current, what is the maximum current that should be measured before the transformer is overloaded when the transformer primary is connected to 240 V?

2. If a 240/480 V to 120 V, 360 VA rated transformer is tested using a ammeter to measure the secondary current, what is the maximum current that should be measured before the transformer is overloaded when the transformer primary is connected to 480 V?

Additional Resources

Review and Resources

Access Chapter 11 Review and Resources for *Electrical Motor Controls for Integrated Systems* by scanning the above QR code with your mobile device.

Applying Your Knowledge

Refer to the *Electrical Motor Controls for Integrated Systems* Learner Resources for interactive Applying Your Knowledge questions.

Workbook and Applications Manual

Refer to Chapter 11 in the *Electrical Motor Controls for Integrated Systems Workbook* and the *Applications Manual* for additional exercises.

ENERGY EFFICIENCY PRACTICES

Energy-Efficient Transformers

According to the Environmental Protection Agency (EPA), 61 billion kWh of electricity are wasted each year in transformer losses. This finding led to the 1992 Energy Act, which mandated that the Department of Energy (DOE) evaluate distribution transformers and work closely with the National Electrical Manufacturers Association (NEMA) to help increase transformer efficiency.

Transformers normally deliver more than 90% of their input power to the load. However, even small changes in efficiency can result in large energy savings. Transformers are composed of a core made of magnetically permeable material and windings normally made of a low-resistance material such as aluminum or copper. Energy losses in transformers arise from both of these components. Transformer loss data is readily available from most manufacturers.

Core (no-load) loss is the amount of power required to energize the core of a transformer. The biggest contributor to core loss is hysteresis loss. Hysteresis is the resistance of the molecules in the core laminations of a transformer to being magnetized and demagnetized by an alternating magnetic field. This resistance causes friction resulting in heat. Hysteresis loss can be reduced by choosing the correct size and type of transformer core. Energy-efficient transformers have cores made of silicon steel or amorphous steel.

Manufacturers also reduce core loss by using thin laminations in the core and by using step-lapped joints, which increase the amount of steel that bridges the joint gap. This reduces the resistance between the laminations and thus reduces eddy current loss. Since most transformers are energized continuously, core loss is present at all times, regardless of whether a load is connected to the transformer. When lightly loaded, core loss represents the greatest portion of the total transformer loss.

Winding (resistive) loss results from the resistance in the windings of a transformer when there is a load on the transformer. Because winding loss is a function of the square of the load current, it increases quickly as the transformer is loaded. Copper windings have lower resistance per cross-sectional area than aluminum windings. Thus, copper windings require smaller cores that produce lower winding losses and offer greater reliability. When heavily loaded, winding loss represents the greatest portion of the total transformer loss.

Objectives

12-1
- Explain why knife switches were discontinued as a means of controlling motors and how they were improved.

12-2
- Define manual contactor and explain why one should not be used as a motor starter.
- Describe double-break contacts and explain how they are used.
- Describe how to draw a wiring diagram for manual contactors.

12-3
- Explain the difference between a manual starter and a manual contactor.
- Explain the difference between how a fuse or circuit breaker protects a circuit and how overload relays protect running motors.
- Explain how overload heater coils operate to automatically turn off an overloaded motor.
- Describe how to select the proper AC manual starter.

12-4
- Describe magnetic contactors and explain how they are used.
- Explain how to design a two-wire control circuit that can be used to control a magnetic contactor.
- Explain how to design a three-wire control circuit that can be used to control a magnetic contactor.
- Explain why opening a DC circuit causes more of an arc contact problem than when opening an AC circuit.
- Define arc chute and explain its usage.
- Explain how DC magnetic blowout coils work.
- Explain how to choose a magnetic contactor.

12-5
- Define magnetic motor starter and describe their different means of overload protection.
- Describe the characteristics that must be considered when selecting an overload heater.
- Explain how to select the correct overload heater for a given motor using a manufacturer selection chart.
- Define inherent motor protector and describe the different types.
- Explain how to troubleshoot circuit breakers.

12-6
- Explain what devices may be added to basic contactors or magnetic motor starters.

12-7
- State the procedure for troubleshooting a motor starter.

Sections

12-1 Manual Switching

12-2 Manual Contactors

12-3 Manual Starters

12-4 Magnetic Contactors

12-5 Magnetic Motor Starters

12-6 Contactor and Magnetic Motor Starter Modifications

12-7 Troubleshooting Contactors and Motor Starters

Review and Resources
atplearningresources.com/Quicklinks
Access Code: 362245

Contactors and Magnetic Motor Starters

Motors use more total power than any other type of electrical load. Large lighting banks and high-power electric heaters also use large amounts of power. Motors, lighting banks, and heating elements must be controlled through some type of controller. Motor starters are used to control and provide running protection for motors. Lighting contactors are used to control banks of lights, and heating contactors are used to control high-power heating elements.

Motor starters and contactors are the interface between the high-power motor, lighting, and heating loads and the low-power control devices, such as pushbuttons, photoelectric switches, and PLCs, that are used to control when the loads are ON and OFF. By using a motor starter or contactor, it is possible for a switch that is rated at a few volts and milliamps to control a load that is rated for hundreds of horsepower and/or amps. Understanding motor starters and contactors is important when designing, installing, or troubleshooting an electrical circuit that includes motors, lighting banks, or high-powered heating elements.

12-1 MANUAL SWITCHING

In the late 1800s, when electric motors were introduced, a method had to be found to start and stop them. This was accomplished through the use of knife switches. **See Figure 12-1.** Knife switches were eventually discontinued as a means of controlling motors for three basic reasons. First, the open knife switch had exposed (live) parts that presented an extreme electrical hazard. In addition, any applications where dirt or moisture were present made the open switch vulnerable to problems. Second, the speed of opening and closing contacts was determined by the operator. Considerable arcing and pitting of the contacts led to rapid wear if the operator did not open or close the switch quickly. Finally, most knife switches were made of soft copper, which required replacement after repeated arcing, heat generation, and mechanical fatigue.

KNIFE SWITCHES

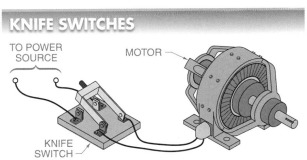

Figure 12-1. Knife switches were the first devices used to start and stop electric motors.

Mechanical Improvements

As industry demanded more electric motors, improvements were made to knife switches to make them more acceptable as control devices. First, the knife switch was enclosed in a steel housing to protect the switch. **See Figure 12-2.** An insulated external handle was added to protect the operator. Also, an operating spring was attached to the handle to ensure quick opening and closing of the knife blade. The switch handle was designed so that once the handle was moved a certain distance, the tension on the spring forced the contacts to open or close at the same continuous speed each time it was operated.

Even with these improvements, the blade and jaw mechanism of a knife switch had a short mechanical life when the knife switch was used as a direct control device. The knife switch mechanism was discontinued as a means of direct control for motors because of its short life. Knife switches are currently used as electrical disconnects. A *disconnect* is a device used only periodically to remove electrical circuits from their supply source. The mechanical life of the knife switch mechanism is not of major concern because a disconnect is used infrequently. Lockout/tagout procedures should always be followed when using a disconnect.

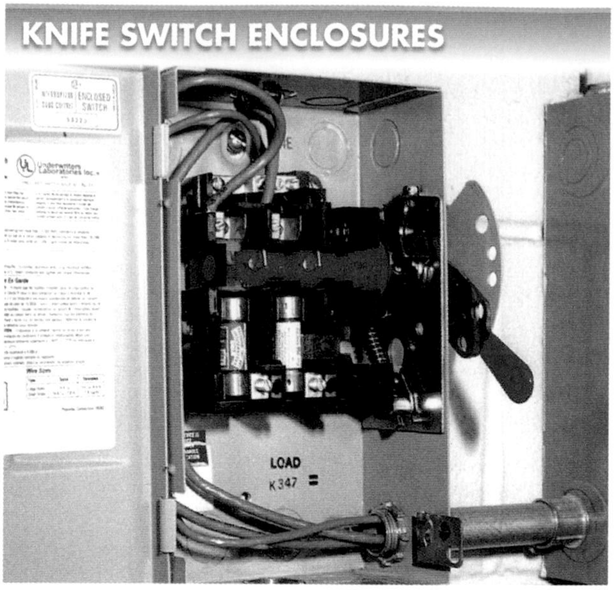

KNIFE SWITCH ENCLOSURES

Figure 12-2. A knife switch is enclosed in a steel housing, has an insulated external handle, and includes an operating spring for improved operation and safety.

12-1 CHECKPOINT

1. Why is it important to open (break) an electrical circuit as fast as possible?

2. Despite being eliminated for direct control of motors, knife switch operation is still used in what type of electrical equipment?

12-2 MANUAL CONTACTORS

A *manual contactor* is a control device that uses pushbuttons to energize or de-energize the load connected to it. **See Figure 12-3.** A manual contactor manually opens and closes contacts in an electrical circuit. Manual contactors cannot be used to start and stop motors because they have no overload protection built into them. Manual contactors are normally used with lighting circuits and resistive loads such as heaters or large lamp loads. A fuse or circuit breaker is normally included in the same enclosure with a manual contactor.

Double-Break Contacts

Double-break contacts can act as a direct controller. *Double-break contacts* are contacts that break an electrical circuit in two places. Double-break contacts are used in pushbuttons. **See Figure 12-4.**

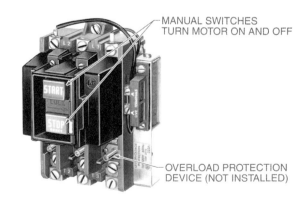

MANUAL CONTACTORS

MANUAL SWITCHES TURN MOTOR ON AND OFF

OVERLOAD PROTECTION DEVICE (NOT INSTALLED)

Rockwell Automation, Allen-Bradley Company, Inc.

Figure 12-3. A manual contactor uses pushbuttons to energize or de-energize the load connected to it.

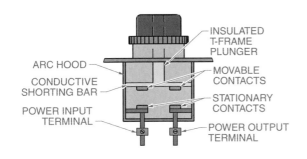

CONTACTS OPEN

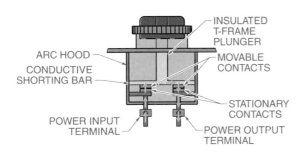

CONTACTS CLOSED

Figure 12-4. Double-break contacts break an electrical circuit in two places.

Double-break contacts allow devices to be designed with a higher contact rating (current rating) in a smaller space than devices designed with single-break contacts. With double-break contacts, the movable contacts are forced against the two stationary contacts to complete the electrical circuit when a set of normally open (NO) double-break contacts are energized. The movable contacts are pulled away from the stationary contacts and the circuit is opened when the manual contactor is de-energized. The procedure is reversed when normally closed (NC) double-break contacts are used.

A 3ϕ manual contactor has three sets of NO double-break contacts. One set of NO double-break contacts is used to open and close each phase in the circuit. The movable contacts are located on an insulated T-frame and are provided with springs to soften their impact. The T-frame is activated by a pushbutton mechanism. Similar to a disconnect, the mechanical linkage consistently and quickly makes or breaks the circuits. **See Figure 12-5.**

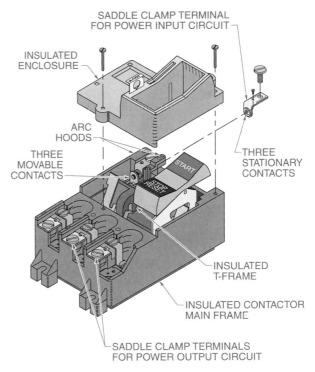

Figure 12-5. A 3ϕ manual contactor has three sets of normally open (NO) double-break contacts.

The movable contacts have no physical connection to external electrical wires. The movable contacts move into arc hoods and bridge the gap between a set of fixed contacts to make or break the circuit. All physical electrical connections are made indirectly to the fixed contacts, normally through saddle clamps.

Contact Construction

In the past, a major problem with knife switches was that they were constructed from soft copper. Today, most contacts are made of a low-resistance silver alloy. Silver is alloyed (mixed) with cadmium or cadmium oxide to make an arc-resistant material that has good conductivity (low resistance). In addition, the silver alloy has good mechanical strength, enabling it to endure the continual wear encountered by many openings and closings. Another advantage of silver alloy contacts is that the oxide (rust) that forms on the metal is an excellent conductor of electricity. Even when the contacts appear dull or tarnished, they are still capable of operating normally. **See Figure 12-6.**

SILVER ALLOY CONTACT OXIDATION

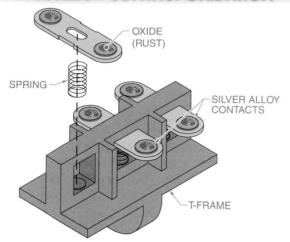

Figure 12-6. The oxide (rust) that forms on silver alloy contacts is an excellent conductor of electricity.

Manual contactors directly control power circuits. Power circuit wiring is shown on a wiring diagram. An understanding of wiring diagrams is required because an electrician may be required to make changes in power circuits as well as in control circuits. **See Figure 12-7.**

POWER CIRCUIT WIRING

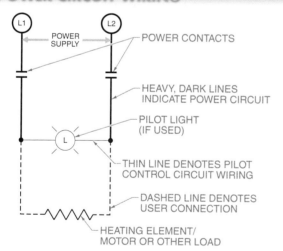

Figure 12-7. A wiring diagram shows the connection of an installation or its component devices or parts.

The wiring diagram for a double-pole manual contactor and pilot light shows the power contacts and their connection to the load. As in a line diagram, the power circuit is indicated through heavy, dark lines and the control circuit is indicated by thin lines. In this circuit, current passes from L1 through the pilot light to L2,

causing the pilot light to glow when the power contacts in L1 and L2 close. At the same time, current passes from L1 through the heating element to L2, causing the heating element to be activated. The pilot light and heating element are connected in parallel with each other.

Wiring diagrams may be complex. For example, the wiring diagram for a dual-element heater with pilot lights contains various circuit paths. In this circuit, the low-heat heating element is operated when the low contacts in L1 and L2 are closed so that a connection is made to the low and common terminals of the heater. This allows the low-heat heating element to be energized. **See Figure 12-8.**

DUAL-ELEMENT HEATER— LOW-HEAT HEATING ELEMENT

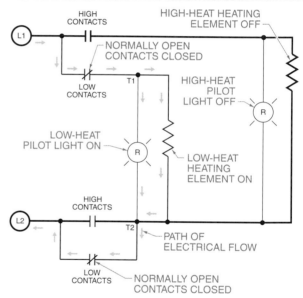

Figure 12-8. In the wiring diagram for a dual-element heater with pilot lights, the low-heat heating element is operated when the low contacts in L1 and L2 are closed so that a connection is made to the low and common terminals of the heater.

To operate the high-heat heating element, the high contacts in L1 and L2 are closed so that a connection is made to the high and common terminals of the heater. This allows the high-heat heating element to be energized. A low-heat pilot light and high-heat pilot light turn on to indicate each condition because each pilot light is in parallel with the appropriate heating element. **See Figure 12-9.**

One problem that may arise with a dual-element start is that someone may try to energize both sets of elements at the same time. This causes serious damage to the heater. To prevent this problem from occurring, most manual contactors are equipped with a mechanical interlock.

A *mechanical interlock* is the arrangement of contacts in such a way that both sets of contacts cannot be closed at the same time. Mechanical interlocking can be established by a mechanism that forces open one set of contacts while the other contacts are being closed. Another method is to provide a blocking bar or holding mechanism that does not allow the first set of contacts to close until the second set of contacts opens. An electrician can determine whether a device is mechanically interlocked by consulting the wiring diagram information provided by the manufacturer. This information is either normally packaged with the equipment when it is delivered or attached to the inside of the enclosure.

Tech Fact

An understanding of interlocking is required because a lack of interlocking can cause shorted power lines, injury to personnel, and damage to machines, motors, chains, and belts. An understanding of interlocking is also required to ensure it is being used correctly and can be tested for proper operation.

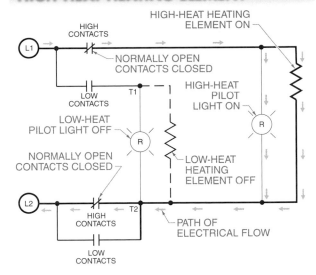

Figure 12-9. In the wiring diagram for a dual-element heater with pilot lights, the high-heat heating element is operated when the high contacts in L1 and L2 are closed so that a connection is made to the high and common terminals of the heater.

12-2 CHECKPOINT

1. Why are manual contactors not used to control motors?
2. What is the advantage of using double-break contacts instead of single-break contacts?
3. Why are silver alloy contacts better then copper contacts?
4. Why are mechanical interlocks included on dual-element heating contactors?

12-3 MANUAL STARTERS

A *manual starter* is a contactor with an added overload protective device. Manual starters are used only in electrical motor circuits. The primary difference between a manual contactor and a manual starter is the addition of an overload protective device. **See Figure 12-10.**

The overload protective device must be added because the National Electrical Code® (NEC®) requires that a control device shall not only turn a motor on and off, but it shall also protect the motor from destroying itself under an overloaded situation, such as a locked rotor. A *locked rotor* is a condition when a motor is loaded so heavily that the motor shaft cannot turn. A motor with a locked rotor draws excessive current and its windings and other components will burn up if the motor is not disconnected

from the line voltage. To protect the motor, the overload device senses the excessive current and opens the circuit.

Motor Overload Protection

A motor goes through three stages during normal operation: resting, starting, and operating under load. **See Figure 12-11.** A motor at rest requires no current because the circuit is open. A motor that is starting draws a tremendous inrush current (normally six to eight times the running current) when the circuit is closed. Fuses or circuit breakers must have a sufficiently high ampere rating to prevent the immediate opening of the circuit caused by the large inrush current required for a motor when starting.

MANUAL STARTERS

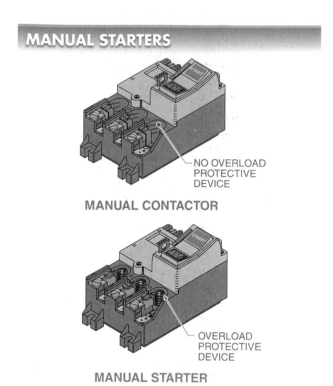

MANUAL CONTACTOR

— NO OVERLOAD
PROTECTIVE
DEVICE

— OVERLOAD
PROTECTIVE
DEVICE

MANUAL STARTER

Figure 12-10. A manual starter is a contactor with an added overload protective device.

MOTOR OPERATING STAGES

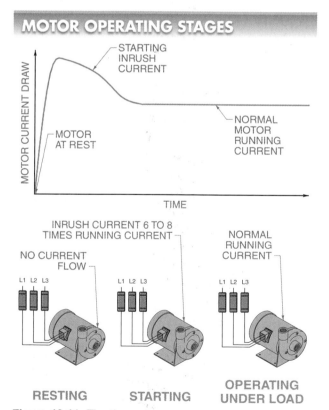

Figure 12-11. The three stages a motor goes through during normal operation include resting, starting, and operating under load.

A motor may encounter an overload while running. While it may not draw enough current to blow the fuses or trip the circuit breakers, it is large enough to produce sufficient heat to burn up the windings and other components in the motor. The intense heat concentration generated by excessive current in the windings causes the insulation to fail and burns the motor. It is estimated that every 1°C (1.8°F) rise over normal ambient temperature ratings for insulation can reduce the life expectancy of a motor by almost a year. *Ambient temperature* is the temperature of the air surrounding a motor. The normal rating for many motors is about 40°C (104°F).

Fuses or circuit breakers must protect the circuit against the very high current of a short circuit or a ground fault. An overload relay that does not open the circuit while the motor is starting but does open the circuit if the motor gets overloaded and the fuses do not blow is required. **See Figure 12-12.**

OVERLOAD RELAYS

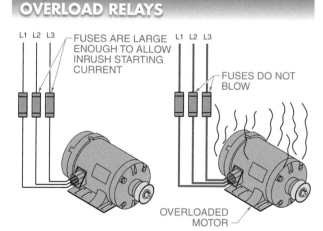

Figure 12-12. An overload relay is required that does not open a circuit while a motor is starting, but opens the circuit if the motor gets overloaded and the fuses do not blow.

To meet motor protection needs, overload relays are designed to have a time delay to allow harmless, temporary overloads without disrupting the circuit. Overload relays must also have a trip capability to open the circuit if mildly dangerous currents that could result in motor damage continue over a period of time. All overload relays have some means of resetting the circuit once the overload is removed.

Melting Alloy Overloads. Heat is the end product that destroys a motor. To be effective, an overload relay must measure the temperature of the motor by monitoring the amount of current being drawn. The overload relay must indirectly monitor the temperature conditions of the motor because the overload relay is normally located at some distance from the motor. One of the most popular methods of providing overload protection is to use a melting alloy overload relay.

A *heater coil* is a sensing device used to monitor the heat generated by excessive current and the heat created through ambient temperature rise. Many different types of heater coils are available. The operating principle of each is the same. A heater coil converts the excess current drawn by a motor into heat, which is used to determine whether the motor is in danger. **See Figure 12-13.**

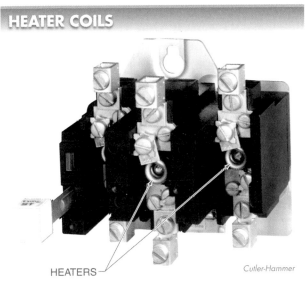

HEATER COILS

HEATERS

Cutler-Hammer

Figure 12-13. A heater coil is a sensing device used to monitor the heat generated by excessive current and the heat created through ambient temperature rise.

Most manufacturers rely on a eutectic alloy in conjunction with a mechanical mechanism to activate a tripping device when an overload occurs. A *eutectic alloy* is a metal that has a fixed temperature at which it changes directly from a solid to a liquid state. This temperature never changes and is not affected by repeated melting and resetting.

Most manufacturers use a ratchet wheel and eutectic alloy tube combination to activate a trip mechanism when an overload occurs. The eutectic alloy tube consists of an outer tube and an inner shaft connected to a ratchet wheel. The ratchet wheel is held firmly in the

tube by the solid eutectic alloy. The inner shaft and ratchet wheel are locked into position by a pawl (locking mechanism) so that the wheel cannot turn when the alloy is cool. **See Figure 12-14.** Excessive current applied to the heater coil melts the eutectic alloy. This allows the ratchet wheel to turn freely.

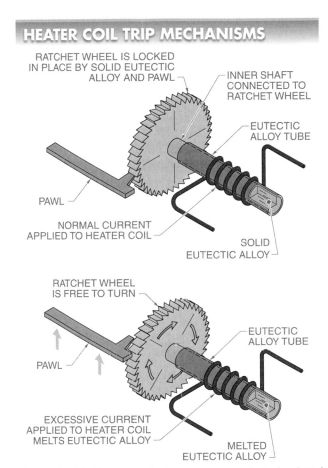

HEATER COIL TRIP MECHANISMS

RATCHET WHEEL IS LOCKED IN PLACE BY SOLID EUTECTIC ALLOY AND PAWL

INNER SHAFT CONNECTED TO RATCHET WHEEL

EUTECTIC ALLOY TUBE

PAWL

NORMAL CURRENT APPLIED TO HEATER COIL

SOLID EUTECTIC ALLOY

RATCHET WHEEL IS FREE TO TURN

EUTECTIC ALLOY TUBE

PAWL

EXCESSIVE CURRENT APPLIED TO HEATER COIL MELTS EUTECTIC ALLOY

MELTED EUTECTIC ALLOY

Figure 12-14. Most manufacturers use a ratchet wheel and eutectic alloy combination to activate a trip mechanism when an overload occurs.

The main device in an overload relay is the eutectic alloy tube. The compressed spring tries to push the NC overload contacts open when motor current conditions are normal. The pawl is caught in the ratchet wheel and does not let the spring push up to open the contacts. **See Figure 12-15.**

The heater coil heats the eutectic alloy tube when an overload occurs. The heat melts the alloy, which allows the ratchet wheel to turn. The spring pushes the reset button up, which opens the contacts to the voltage coil of the contactor. The contactor opens the circuit to the motor, which stops the current flow through the heater coil. The heater coil cools, which solidifies the eutectic alloy tube.

EUTECTIC ALLOY TUBES

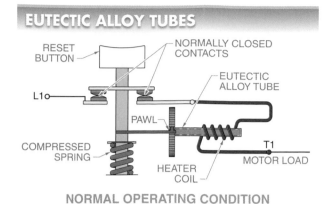

NORMAL OPERATING CONDITION

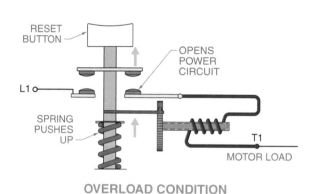

OVERLOAD CONDITION

Figure 12-15. In a manual starter overload relay, the compressed spring tries to push the normally closed (NC) contacts open under normal operating conditions.

RESETTING OVERLOAD DEVICES

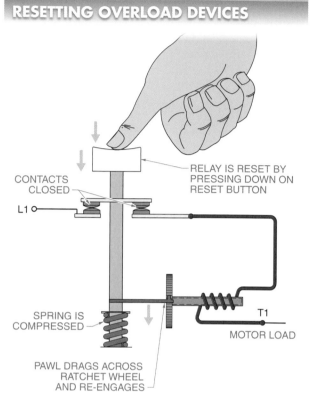

Figure 12-16. The overload relay is reset by pressing the reset button, which forces the pawl across the ratchet wheel until the contacts are closed and the spring and ratchet wheel are returned to their original condition.

Only the NC overload contacts open during an overload condition. The NC overload contacts can be manually reset to the closed position. The actual heating elements (heaters) installed in the motor starter do not open during an overload. The heaters are only used to produce heat. The higher the current draw of the motor, the more heat produced.

Resetting Overload Devices. The cause of an overload must be found before resetting an overload relay. A relay trips on resetting if the overload is not removed. Once the overload is removed, the device can be reset. The reset button is pushed, which forces the pawl across the ratchet wheel until the contacts are closed and the spring and ratchet wheel are returned to their original condition. The start pushbutton can then be pressed to start the motor. **See Figure 12-16.**

Nothing requires replacement or repair when an overload device trips because the heaters do not open like a fuse would open. Once the cause of the overload is removed, the reset button may be pressed. Normally, a few minutes should be allowed for the eutectic alloy to cool.

The same basic overload relay is used with all sizes of motors. The only difference is that the heater coil size is changed. For small horsepower motors, a small heater coil is used. For large horsepower motors, a large heater coil is used. The NEC® should be consulted for selecting appropriate overload heater sizes.

Selecting AC Manual Starters

Electricians are often required to select AC manual starters for new installations or replace ones that have been severely damaged due to an electrical fire or explosion. In either case, the electrician must specify certain characteristics of the starter to obtain the proper replacement. **See Figure 12-17.** AC manual starters are selected based on phasing, number of poles, voltage, starter size, and enclosure type. Starter sizes are given in general motor protection tables. General motor protection tables indicate motor protective device sizes based on motor horsepower, current, fuse classification, and wire size. **See Appendix.**

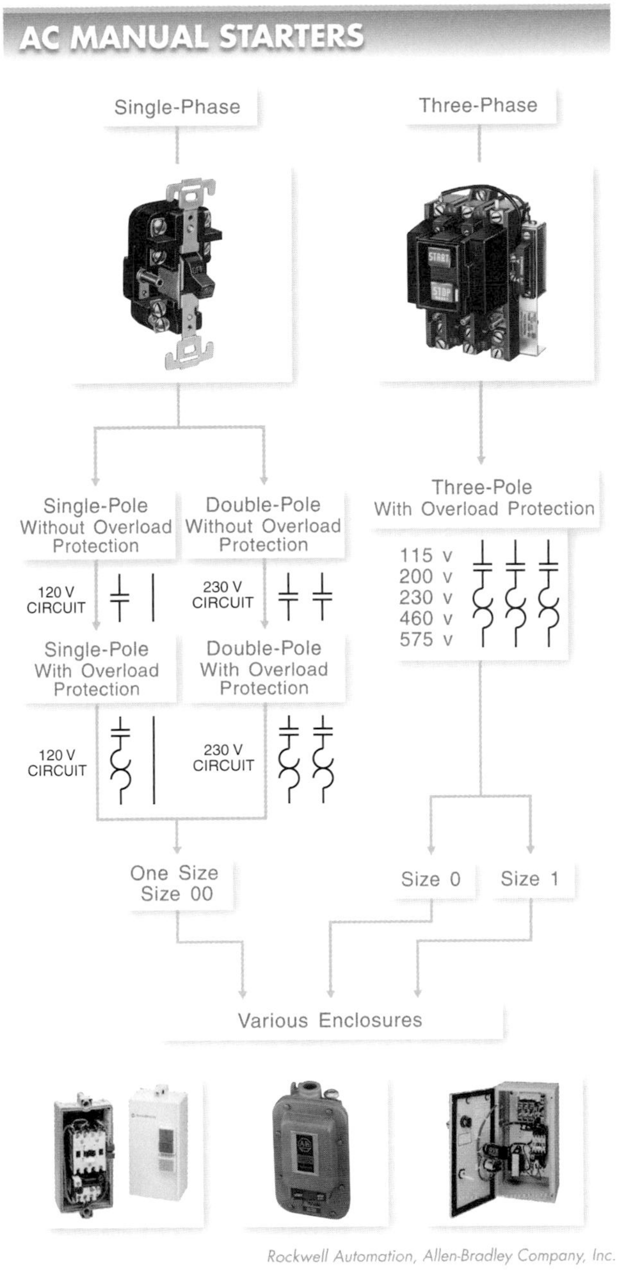

AC MANUAL STARTERS

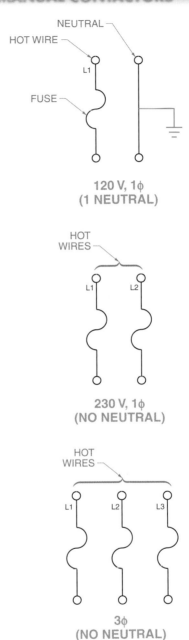

Single-Phase

Three-Phase

Single-Pole
Without Overload
Protection

Double-Pole
Without Overload
Protection

Three-Pole
With Overload Protection

120 V
CIRCUIT

230 V
CIRCUIT

115 v
200 v
230 v
460 v
575 v

Single-Pole
With Overload
Protection

Double-Pole
With Overload
Protection

120 V
CIRCUIT

230 V
CIRCUIT

One Size
Size 00

Size 0   Size 1

Various Enclosures

Rockwell Automation, Allen-Bradley Company, Inc.

Figure 12-17. AC manual starters are selected based on phasing, number of poles, voltage, starter size, and enclosure type.

AC MANUAL CONTACTORS

NEUTRAL

HOT WIRE

L1

FUSE

**120 V, 1φ
(1 NEUTRAL)**

HOT
WIRES

L1   L2

**230 V, 1φ
(NO NEUTRAL)**

HOT
WIRES

L1   L2   L3

**3φ
(NO NEUTRAL)**

Figure 12-18. AC manual contactors can be divided into 1φ and 3φ contactors.

Phasing. AC manual starters/contactors can be divided into 1φ and 3φ contactors. **See Figure 12-18.** A 120 V, 1φ power source has one hot wire (ungrounded conductor) and one neutral wire (grounded conductor). A 230 V, 1φ power source has two hot wires, L1 and L2 (ungrounded conductors), and no neutral. A 3φ power source has three hot wires, L1, L2, and L3, and no neutral.

Single-phase manual starters are available as single-pole and double-pole devices because the NEC® requires that each ungrounded conductor (hot wire) be open when disconnecting a device. A single-pole device is used on 120 V circuits and a double-pole device is used on 230 V circuits.

Single-phase manual starters have limited horsepower ratings because of their physical size and are normally used as starters for motors of 1 HP or less. Single-phase manual starters are often available in only one size for all motors rated at 1 HP or less. The size established for 1φ starters is classified as NEMA size 00. IEC manual starters/contactors are horsepower rated. Single-phase manual contactors and starters are normally used for 1φ, 1 HP or less motors where low-voltage protection is not needed. They are also used for 1φ motors that do not require a high frequency of operation.

Three-phase manual starters are physically larger than 1φ manual starters and may be used for motors of 10 HP or less. Three-phase manual contactors are normally pushbutton-operated instead of toggle-operated like 1φ starters.

Motor circuits require a manual starter that has overloads. Contactors, however, can be used in certain applications, such as in lighting circuits, without overload devices. In those cases, the fuse or circuit breaker in the main disconnect provides the overload protection.

Three-phase devices are designed with three-pole switching because 3φ devices have three hot wires that must be disconnected. Similar to 1φ devices, 3φ devices use contacts and have quick-make and quick-break mechanisms. Three-phase contactors and starters are normally designed to be used on circuits from 115 V up to and including 575 V.

Three-phase starters are normally used for 3φ, 7.5 HP and less motors operating at 208/230 V or 3φ, 10 HP and less motors operating at 380/575 V. Three-phase starters are also used for 3φ motors where low-voltage protection is not needed, for motors that do not require a high frequency of operation, and for motors that do not need remote operation by pushbuttons or limit switches.

Enclosures. Enclosures provide mechanical and electrical protection for the operator and the starter. **See Appendix.** Although the enclosures are designed to provide protection against a variety of contaminants such as water, dust, and oil, as well as contaminants from hazardous locations, the internal electrical wiring and physical construction of the starter remain the same.

The NEC® and local codes should be consulted to determine the proper selection of an enclosure for a particular application. For example, NEMA Type 1 enclosures are intended for indoor use primarily to provide a degree of protection against human contact with the enclosed equipment in locations where unusual service conditions do not exist.

Manual Starter Applications

Manual motor starters are used in applications such as conveyor systems and drill presses. **See Figure 12-19.** In most applications, the manual starter provides the means of turning on and off the device while providing motor overload protection.

MANUAL MOTOR STARTER APPLICATIONS

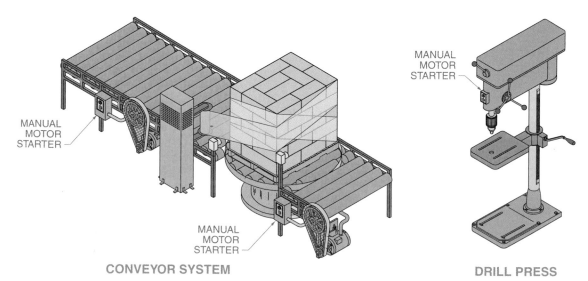

MANUAL MOTOR STARTER

MANUAL MOTOR STARTER

MANUAL MOTOR STARTER

CONVEYOR SYSTEM

DRILL PRESS

Figure 12-19. Manual motor starters are used in applications such as conveyor systems and drill presses.

12-3 CHECKPOINT

1. What is the primary difference between a manual contactor and a manual starter?
2. When does a motor controlled by a motor starter draw the highest amount of current?
3. If a fuse or circuit breaker directly opens a circuit when the current limit is reached, do overload heaters open and break the circuit when their current limit is reached?
4. When an overload device trips, does it have to be replaced like a fuse?
5. In a 1ϕ motor starter, does the starter have to open both the ungrounded (hot) conductor and the neutral conductor?
6. Do all ungrounded (hot) conductors have to be opened when controlling loads?

12-4 MAGNETIC CONTACTORS

Contactors may be operated manually or magnetically. Contactors are devices for repeatedly establishing and interrupting an electrical power circuit. Contactors are used to make and break the electrical power circuit to loads such as lights, heaters, transformers, and capacitors. **See Figure 12-20.**

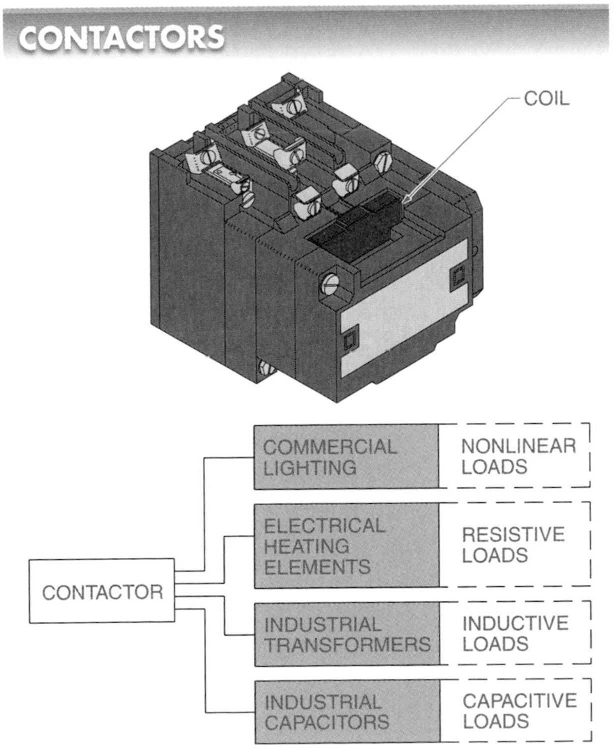

Figure 12-20. Contactors are used to make and break the electrical power circuit to lights, heaters, transformers, and capacitors.

Magnetic Contactor Construction

Solenoid action is the principal operating mechanism for magnetic contactors. The linear action of a solenoid is used to open and close sets of contacts instead of pushing and pulling levers and valves. **See Figure 12-21.** The use of solenoid action rather than manual input is an advantage of a magnetic contactor over a manual contactor. Remote control and automation, which are impossible with manual contactors, can be designed into a system using magnetic contactors.

Tech Fact

To help understand the difference between a relay, contactor, and motor starter, it should be remembered that they are each designed to switch current using contacts operated by a coil. Relays are used to switch low currents that are usually less than 15 A. Contactors are basically the same as relays but are used to switch higher currents that are usually up to hundreds of amperes for large lighting or heating loads. Motor starters are contactors that have an additional overload section to protect a running motor.

Magnetic Contactor Wiring

Control circuits are often referred to by the number of conductors used in the control circuit, such as two-wire and three-wire control. Two-wire control involves two conductors to complete the circuit. Three-wire control involves three conductors to complete the circuit.

SOLENOID ACTION

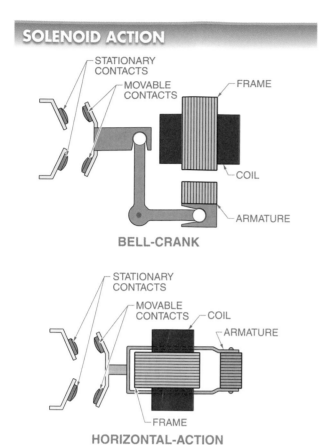

BELL-CRANK

HORIZONTAL-ACTION

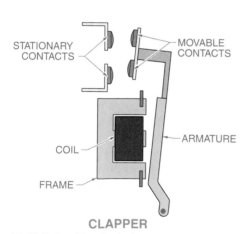

CLAPPER

Figure 12-21. Solenoid action is the principal operating mechanism for magnetic contactors.

Two-Wire Control. Two-wire control has two wires leading from the control device to the contactor or starter. **See Figure 12-22.** The control device could be a thermostat, float switch, or other contact device. When the contacts of the control device close, they complete the coil circuit of the contactor, causing it to energize. This connects the load to the line through the power contacts. The contactor coil is de-energized when the

contacts of the control device open. This de-energizes coil C, which opens the contacts that control the load. The contactor functions automatically in response to the condition of the control device without the attention of an operator.

TWO-WIRE CONTROL

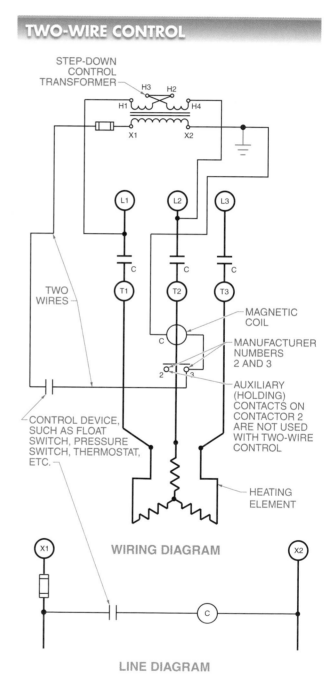

Figure 12-22. In two-wire control, two wires lead from the control device to the contactor or starter.

A two-wire control circuit provides low-voltage release but not low-voltage protection. In the event of a power loss in the control circuit, the contactor de-energizes (low-voltage release), but it also re-energizes it if the control device remains closed when the circuit has power restored. Low-voltage protection cannot be provided in this circuit because there is no way for the operator to be protected from the circuit once it has been re-energized.

Caution must be exercised in the use and service of two-wire control circuits because of the lack of low-voltage protection. Two-wire control is normally used for remote or inaccessible installations, such as pumping stations, water or sewage treatment, air conditioning or refrigeration systems, and process line pumps where an immediate return to service after a power failure is required.

Two-wire control circuits are used with motor loads and nonmotor loads. Motor overload protection must be added to a contactor that is used to control a motor load. When motor overload protection is included as part of the contactor assembly, the unit is referred to as a motor starter. Contactors are not used to control motors unless the motor is a small horsepower motor (normally fractional HP) that includes internal protection, or the contactor is used with a separate motor overload protection unit. With nonmotor loads, the contactor is used to directly control the power applied to the load.

Three-Wire Control. Three-wire control has three wires leading from the control device to the starter or contactor. **See Figure 12-23.** The circuit uses a momentary contact OFF pushbutton (NC) wired in series with a momentary contact ON pushbutton (NO) wired in parallel to a set of contacts that form a holding circuit interlock (memory).

When the normally open ON pushbutton is pressed, current flows through the normally closed OFF pushbutton, through the momentarily closed ON pushbutton, through magnetic coil C, and on to L2. This causes the magnetic coil to energize. When energized, the auxiliary holding circuit interlock contacts (memory) close, sealing the path through to the coil circuit even if the start pushbutton is released.

Pressing the OFF pushbutton (NC) opens the circuit to the magnetic coil, causing the contactor to de-energize. A power failure also de-energizes the contactor. The interlock contacts (memory) reopen when the contactor de-energizes. This opens both current paths to the coil through the ON pushbutton and the interlock.

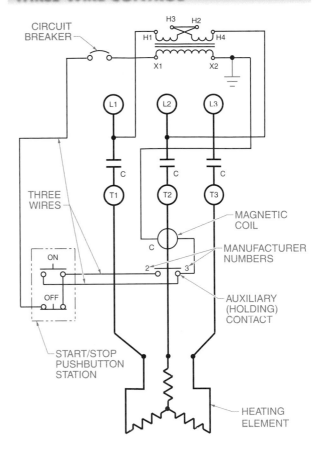

WIRING DIAGRAM

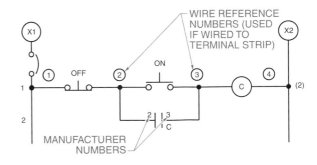

LINE DIAGRAM

Figure 12-23. In three-wire control, three wires lead from the control device to the starter or contactor.

Three-wire control provides low-voltage release and low-voltage protection. The coil drops out at low or no voltage and cannot be reset unless the voltage returns and the operator presses the start pushbutton.

Control Circuit Voltage

Pushbuttons, limit switches, pressure switches, temperature switches, etc. are used to control the flow of power to the contactor (or motor starter) magnetic coil in the control circuit. When the control circuit is connected to the same voltage level as the load (lamps, heating elements, or motors), the control circuit must be rated for the same voltage.

In most circuits in which the load is rated higher than 115 V (normally 208 V, 230 V, 240 V, 460 V, and 480 V), the control circuit is operated at a lower voltage level than the load. A step-down control transformer is used to step down the voltage to the level required in the control circuit. Normally, the secondary of the transformer is rated for 12 V, 24 V, or 120 V. The voltage of a control circuit can be any voltage (AC or DC), but it is commonly less than 120 V. **See Figure 12-24.**

AC and DC Contactors

AC contactor assemblies may have several sets of contacts. DC contactor assemblies typically have only one set of contacts. **See Figure 12-25.** In 3φ AC contactors, all three power lines must be broken. This creates the need for several sets of contacts. For multiple contact control, a T-bar assembly allows several sets of contacts to be activated simultaneously. In a DC contactor, it is necessary to break only one power line.

AC contactor assemblies are made of laminated steel, while DC assemblies are solid. Laminations are unnecessary in a DC coil because the current travels in one direction at a continuous rate and does not create eddy current problems. The other major differences between AC and DC contactors are the electrical and mechanical requirements necessary for suppressing the arcs created in opening and closing contacts under load.

STEP-DOWN CONTROL TRANSFORMERS

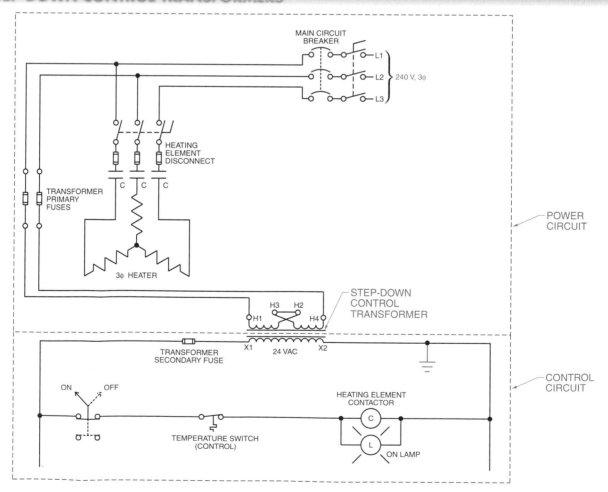

Figure 12-24. A step-down control transformer is used to step down the voltage to the level required in the control circuit.

AC AND DC CONTACTORS

General Electric Company

AC CONTACTOR

DC COIL

AC OR DC CONTACTS

Rockwell Automation, Allen-Bradley Company, Inc.

DC CONTACTOR

Figure 12-25. Contactors have either an AC coil or a DC coil, but they may have either AC or DC contacts.

Arc Suppression

Arc suppression is required on contactors and motor starters. An *arc suppressor* is a device that dissipates the energy present across opening contacts. Without arc suppression, contactors and motors may require premature maintenance that results in excessive downtime.

Opening Contact Arc. A short period of time (a few thousandths of a second) exists when a set of contacts is opened under load. During this time, the contacts are neither fully in touch with each other nor completely separated. **See Figure 12-26.**

As the contacts continue to separate, the contact surface area decreases, which increases the electrical resistance. With full-load current passing through the increasing resistance, a substantial temperature rise is created on the surface of the contacts. This temperature rise is often high enough to cause the contact surfaces to become molten and emit ions of vaporized metal into the gap between the contacts. This hot ionized vapor permits the current to continue to flow in the form of an arc, even though the contacts are completely separated. The arcs produce additional heat, which, if continued, can damage the contact surfaces. The sooner the arc is extinguished, the longer the life expectancy of the contacts.

CONTACT ARCING

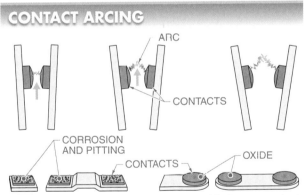

ARC

CONTACTS

CORROSION AND PITTING

CONTACTS

OXIDE

Figure 12-26. An electrical arc is created between contacts as they are opened. Prolonged arcing may result in damage to contact surfaces.

DC Arc Suppression. DC arcs are considered the most difficult to extinguish because the continuous DC supply causes current to flow constantly and with great stability across a much wider gap than does an AC supply of equal voltage. To reduce arcing in DC circuits, the switching mechanism must be such that the contacts separate rapidly and with enough of an air gap to extinguish the arc as soon as possible on opening. DC contactors are larger than AC contactors to allow for the additional air gap. In addition, the operating characteristics of DC contactors are faster than AC contactors.

When closing DC contacts, it is necessary to move the contacts together as quickly as possible to prevent some of the same problems encountered in opening them. One disadvantage to rapidly closing DC contactors is that the contacts must be buffered to eliminate contact bounce due to excessive closing force. Contact bounce may be minimized through the use of certain types of solenoid action and springs attached under the contacts to absorb some of the shock.

AC Arc Suppression. An AC arc is self-extinguishing when a set of contacts is opened. In contrast to a DC supply of constant voltage, an AC supply has a voltage that reverses its polarity 120 times a second when operated on a 60 hertz (Hz) line frequency. The alternation allows the arc to have a maximum duration of no more than a half-cycle. During any half-cycle, the maximum arcing current is reached only once in that half-cycle. **See Figure 12-27.**

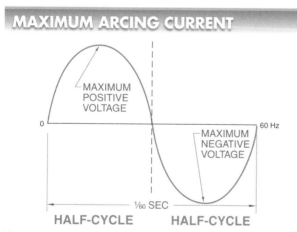

MAXIMUM ARCING CURRENT

Figure 12-27. The maximum arcing current is reached only once during any half-cycle of AC voltage.

The contacts can be separated more slowly and the gap length may be shortened because an AC arc is self-extinguishing. This short gap keeps the voltage across the gap and the arc energy low. With low gap energy, ionizing gases cool more rapidly, extinguishing the arc and making it difficult to restart. AC contactors need less room to operate and run cooler, which increases contact life.

Arcs at Closing. Arcing may also occur on AC and DC contactors when they are closing. The most common arcing occurs when the contacts come close enough that an arc is able to bridge the open space between the contacts.

Arcing also occurs if a whisker or rough edge of the contact touches first and melts, causing an ionized path that allows current to flow. In either case, the arc lasts until the contact surfaces are fully closed. Contactor design is quite similar for both AC and DC devices. The contactor should be designed so that the contacts close as rapidly as possible, without bouncing, to minimize the arc at each closing.

Arc Chutes. An *arc chute* is a device that confines, divides, and extinguishes arcs drawn between contacts opened under load. **See Figure 12-28.** Arc chutes are used to contain large arcs and the gases created by them. Arc chutes employ the de-ion principle, which confines, divides, and extinguishes the arc for each set of contacts.

Arcs may also be extinguished by using special arc traps and arc-quenching compounds. This method of extinguishing arcs is a circuit breaker technique that attracts, splits, and quickly cools arcs as well as vents ionized gases. Vertical barriers between each set of contacts, as well as arc covers, confine arcs to separate chambers and quickly quench them.

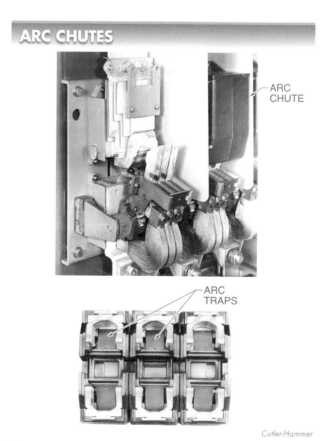

ARC CHUTES

Cutler-Hammer

Figure 12-28. Arc chutes and arc traps are used to confine, divide, and extinguish arcs drawn between contacts opened under load.

DC Magnetic Blowout Coils. When a DC circuit carrying large amounts of current is interrupted, the collapsing magnetic field of the circuit current may induce a voltage that helps sustain the arc. Action must be taken to quickly limit the damaging effect of the heavy current arcs because a sustained electrical arc may melt the contacts, weld them together, or severely damage them.

One way to stop the arc quickly is to move the contacts some distance from each other as quickly as possible. The problem is that the contactor has to be large enough to accommodate such a large air gap.

Magnetic blowout coils are used to reduce the distance required and yet quench arcs quickly. Magnetic blowout coils provide a magnetic field that blows out the arc similarly to blowing out a match.

A magnetic field is created around the current flow whenever a current flows through a conductive medium (in this case ionized air). The direction of the magnetic field around the conductor is determined by wrapping the right or left hand around the conductor. When the thumb on the right hand points in the direction of conventional current flow, the wrapping fingers point in the direction of the resulting magnetic field. When the thumb on the left hand points in the direction of electron current flow, the wrapping fingers point in the direction of the resulting magnetic field. **See Figure 12-29.**

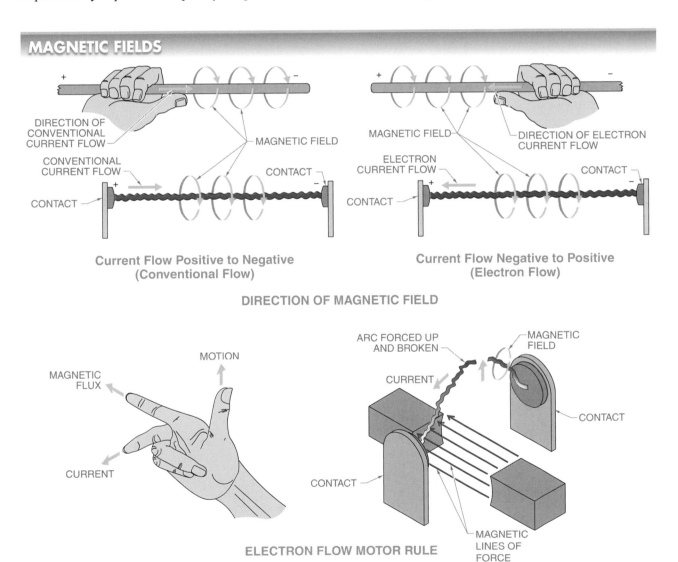

Figure 12-29. The direction of the magnetic field around the conductor is determined by wrapping the right or left hand around the conductor. The electron flow motor rule indicates the motion of an arc cutting through magnetic lines of force.

The electron flow motor rule states that when a current-carrying conductor (represented by the middle finger) is placed in a parallel magnetic field (represented by the index finger), the resulting force or movement is in the direction of the thumb. This action occurs because the magnetic field around the current flow opposes the parallel magnetic field above the current flow. This makes the magnetic field above the current flow weaker, while aiding the magnetic field below the current flow, making the magnetic field stronger. The net result is an upward push that quickly elongates the arc current so that it breaks (blows out). An electromagnetic blowout coil is often referred to as a puffer because of its blowout ability. **See Figure 12-30.**

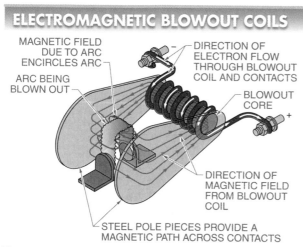

Figure 12-30. Electromagnetic blowout coils rapidly extinguish DC arcs.

Contact Construction

Contact design and materials depend on the size, current rating, and application of the contactor. Double-break contacts are normally made of a silver-cadmium alloy. Single-break contacts in large contactors are frequently made of copper because of the low cost.

Single-break copper contacts are designed with a wiping action to remove the copper oxide film that forms on the copper tips of the contacts. The wiping action is necessary because copper oxide formed on the contacts when not in use is an insulator and must be eliminated for good circuit conductivity.

In most cases, the slight rubbing action and burning that occur during normal operation keep the contact surfaces clean for proper operation. Copper contacts that seldom open or close, or those being replaced, should be cleaned to reduce contact resistance. High contact resistance often causes serious heating of the contacts.

General-Purpose AC/DC Contactor Sizes and Ratings

Magnetic contactors, like manual contactors, are rated according to the size and type of load by the National Electrical Manufacturers Association (NEMA). Tables are used to indicate the number/size designations and establish the current load carried by each contact in a contactor. **See Figure 12-31.** The rating is for each contact individually, not for the entire contactor. For example, a size 0, three-pole contactor rated at 18 A is capable of, and rated for, switching three separate 18 A loads simultaneously.

60 Hz AC CONTACTOR STANDARD NEMA RATINGS

Size	Rating*	Power Rating† 3φ 200 V	230 V	230/460 V	1φ 115 V	230 V
00	9	1½	1½	2	⅓	1
0	18	3	3	5	1	2
1	27	7½	7½	10	2	3
2	45	10	15	25	3	7½
3	90	25	30	50	—	—
4	135	40	50	100	—	—
5	270	75	100	200	—	—
6	540	150	200	400	—	—
7	810	—	300	600	—	—
8	1215	—	450	900	—	—
9	2250	—	800	1600	—	—

* in A
† in HP

DC CONTACTOR STANDARD NEMA RATINGS

Size	Rating*	Power Rating† 115 V	230 V	550 V
1	25	3	5	—
2	50	5	10	20
3	100	10	25	50
4	15	20	40	75
5	300	40	75	150
6	600	75	150	300
7	900	110	225	450
8	1350	175	350	700
9	2500	300	600	1200

* in A
† in HP

Figure 12-31. Tables indicate the number/size designations and establish the current load carried by each contact in a contactor.

Contactor dimensions vary greatly, ranging from inches to several feet in length. Contactors are selected based on type, size, and voltage available. **See Figure 12-32.**

Contactors are also available in a variety of enclosures. The enclosures offer protection ranging from the most basic protection to high levels of protection required in hazardous locations where any spark caused by the closing or opening of the contact could cause an explosion.

AC/DC MAGNETIC CONTACTORS

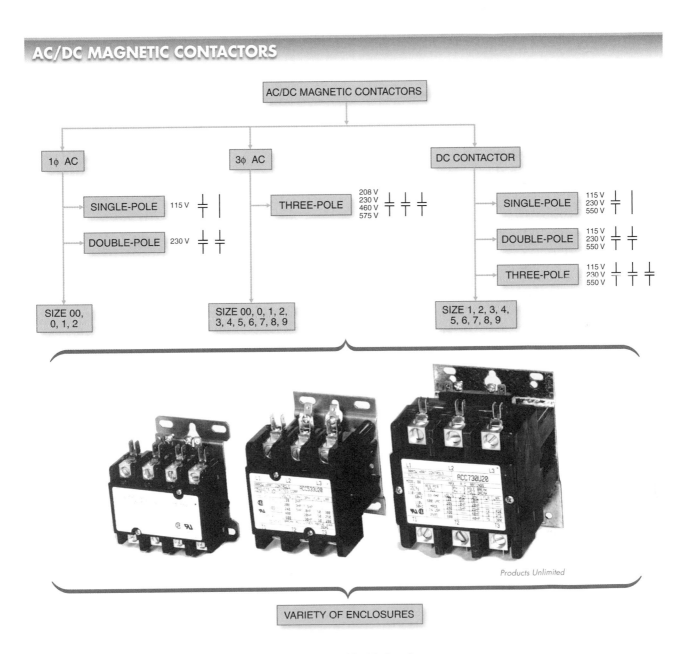

Products Unlimited

VARIETY OF ENCLOSURES

Figure 12-32. Contactor dimensions vary from inches to several feet in length.

12-4 CHECKPOINT

1. What happens to a load that was energized when a two-wire control device is used and power is lost and then reapplied?
2. What happens to a load that was energized when a three-wire control device is used and power is lost and then reapplied?
3. Does opening an AC or DC circuit cause the most problems in arc suppression?
4. What happens to electrical contacts if their resistance increases?
5. Why are arc chutes used?
6. As conductor size numbers increase (AWG 12 to 14), they can carry less current and power. What happens as NEMA contactor size number (size 1 to 2) increases?

12-5 MAGNETIC MOTOR STARTERS

A *magnetic motor starter* is an electrically operated switch (contactor) that includes motor overload protection. Magnetic motor starters include overload relays that detect excessive current passing through a motor and are used to switch all types and sizes of motors. Magnetic motor starters are available in sizes that can switch loads of a few amperes to several hundred amperes. **See Figure 12-33.**

MAGNETIC MOTOR STARTERS

Furnas Electric Co.

Figure 12-33. A magnetic motor starter is a contactor that includes overload protection added.

Overload Protection

The main difference between the sensing device for a manual motor starter and a magnetic motor starter is that on a manual motor starter a manual overload opens the power contacts on the starter. The overload device on a magnetic motor starter opens a set of contacts to the magnetic coil, de-energizing the coil and disconnecting the power. Overload devices include melting alloy, magnetic, and bimetallic overload relays. The overload unit (heater) does not open as a fuse or CB does, but it produces the heat required to open the overload contacts.

Melting Alloy Overload Relays. The melting alloy overload relays used in magnetic motor starters are similar to the melting alloy overload relays used in manual motor starters. They consist of a heater coil, eutectic alloy, and mechanical mechanism to activate a tripping device when an overload occurs.

Magnetic Overload Relays. Magnetic overload relays provide another means of monitoring the amount of current drawn by a motor. A magnetic overload relay operates through the use of a current coil. At a specified overcurrent value, the current coil acts as a solenoid, causing a set of normally closed contacts to open. This causes the circuit to open and protect the motor by disconnecting it from power. **See Figure 12-34.**

Magnetic overload relays are used in special applications such as steel mill processing lines or other heavy-duty industrial applications where holding a specified level of motor current is required. A magnetic overload relay is also ideal for special applications such as slow-acceleration motors, high-inrush-current motors, or any use where normal time/current curves of thermal overload relays do not provide satisfactory operation. This flexibility is made possible because the magnetic unit may be set for either instantaneous or inverse time-tripping characteristics. The device may also offer independent adjustable trip time and trip current.

of the bimetallic strip. A bimetallic strip is made of two pieces of dissimilar metal that are permanently joined by lamination. Heating the bimetallic strip causes it to warp because the dissimilar metals expand and contract at different rates. The warping effect of the bimetallic strip is used as a means of separating contacts. **See Figure 12-35.**

MAGNETIC OVERLOAD RELAYS

Furnas Electric Co.

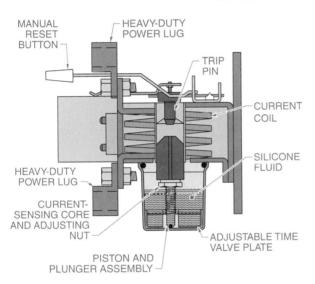

Figure 12-34. Magnetic overload relays use a current coil that, at a specific overcurrent value, acts like a solenoid and causes a set of normally closed contacts to open.

Magnetic overload relays are extremely quick to reset because they do not require a cooling-off period before being reset. Magnetic overload relays are much more expensive than thermal overload relays.

Bimetallic Overload Relays. In certain applications such as walk-in meat coolers, remote pumping stations, and some chemical process equipment, overload relays that reset automatically to keep the unit operating up to the last possible moment may be required. A *bimetallic overload relay* is an overload relay that resets automatically. Bimetallic overload relays operate on the principle

BIMETALLIC OVERLOAD RELAYS

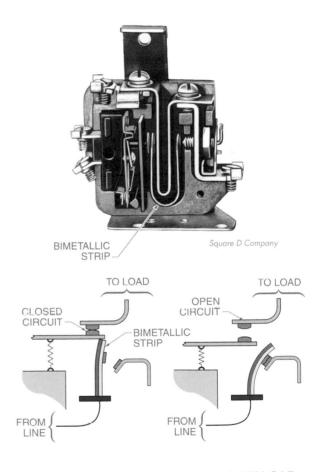

Square D Company

Figure 12-35. The warping effect of a bimetallic strip is used as a means for separating contacts.

Once the tripping action has taken place, the bimetallic strip cools and reshapes itself. In certain devices, such as circuit breakers, a trip lever needs to be reset to make the circuit operate again. In other devices, such as bimetallic overload relays, the device automatically resets the circuit when the bimetallic strip cools and reshapes itself.

The motor restarts even when the overload has not been cleared and trips and resets itself again at given intervals. Care must be exercised in the selection of a bimetallic overload relay because repeated cycling eventually burns out the motor. The bimetallic strip may be shaped in the form of a U. The U-shape provides a uniform temperature response.

Trip Indicators. Many overload devices have a trip indicator built into the unit to indicate to the operator that an overload has taken place within the device. **See Figure 12-36.** A red metal indicator appears in a window located above the reset button when the overload relay has tripped. The red indicator informs the operator or electrician why the unit is not operating and that it is potentially capable of restarting with an automatic reset.

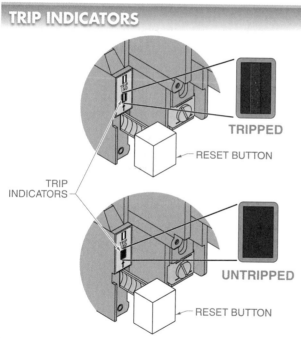

TRIP INDICATORS

TRIPPED

RESET BUTTON

TRIP INDICATORS

UNTRIPPED

RESET BUTTON

Figure 12-36. Trip indicators indicate that an overload has taken place within the device.

Overload Current Transformers. Large-horsepower motors have currents that exceed the values of standard overload relays. To make the overload relays larger would greatly increase their physical size, which would create a space problem in relation to the magnetic motor starter. To prevent such a conflict, current transformers are used to reduce the current in a fixed ratio. **See Figure 12-37.** A current transformer is used to change the amount of current flowing to a motor but reduces the current to a lower value for the overload relay. For example, if 50 A were flowing to

a motor, only 5 A would flow to the overload relay through the use of the current transformer. Standard current transformers are normally rated in primary and secondary rated current such as 50/5 or 100/5.

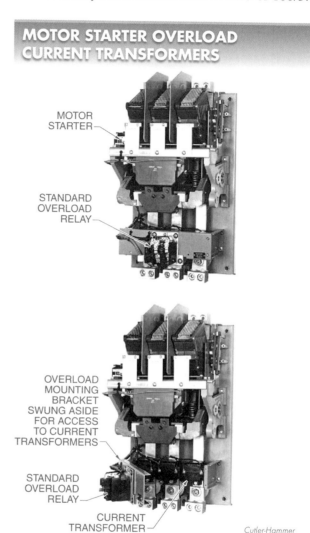

MOTOR STARTER OVERLOAD CURRENT TRANSFORMERS

MOTOR STARTER

STANDARD OVERLOAD RELAY

OVERLOAD MOUNTING BRACKET SWUNG ASIDE FOR ACCESS TO CURRENT TRANSFORMERS

STANDARD OVERLOAD RELAY

CURRENT TRANSFORMER

Cutler-Hammer

Figure 12-37. Standard overload relays may be used on very large starters by using current transformers with specific reduction ratios.

Because the ratio is always the same, an increase in the current to a motor also increases the current to the overload relay. If the correct current transformer and overload relay combination is selected, the same overload protection can be provided to a motor as if the overload relay were actually in the load circuit. The overload relay contacts open and the coil to the magnetic motor starter is de-energized when excessive current is sensed. This shuts the motor off. Several different current transformer ratios are available to make this type of overload protection easy to provide.

Overload Heater Sizes

Each motor must be sized according to its own unique operating characteristics and applications. Thermal overload heaters are selected based on the full-load current (FLC) rating, service factor (SF), and ambient temperature (surrounding air temperature) of the motor when it is operating.

Full-Load Current Rating. Selection of thermal overload heaters is based on the FLC rating shown on the motor nameplate or in the motor manufacturer specification sheet. The current value reflects the current to be expected when the motor is running at specified voltages, specified speeds, and normal torque operating characteristics. Heater manufacturers develop current charts indicating the heater that should be used with each full-load current.

Service Factor. In most motor applications, there are times when the motor must produce more than its rated horsepower for a short period of time without damage. A *service factor (SF)* is a number designation that represents the percentage of extra demand that can be placed on a motor for short intervals without damaging the motor. Common SFs range from 1.00 to 1.25, indicating that the motor can produce 0% to 25% extra demand over that for which it is normally rated. A 1.00 SF indicates that the motor cannot produce more power than it is rated for and to do so would result in damage. A 1.25 SF indicates that the motor can produce up to 25% more power than it is rated for, but only for short periods of time.

The excessive current that can be safely handled by a given motor for short periods of time is approximated by multiplying the SF by the FLC rating. For example, if a motor has an FLC rating of 10 A with an SF of 1.15, the excess short-term current equals 11.5 A ($10 \times 1.15 = 11.5$ A). The motor could handle an additional 1.5 A for a short period of time.

Ambient Temperature. A thermal overload relay operates on the principle of heat. When an overload takes place, sufficient heat is generated by the excessive current to melt a metal alloy, produce movement in a current coil, or warp a bimetallic strip and allow the device to trip. The temperature surrounding a thermal overload relay must be considered because the relay is sensitive to heat from any source. The ambient temperature is a factor when considering moving a thermal overload relay from a refrigerated meat packing plant to a location near a blast furnace.

Overload relay devices are normally rated to trip at a specific current when surrounded by an ambient temperature of 40°C (104°F). This standard ambient temperature is acceptable for most control applications. Compensation must be provided for higher or lower ambient temperatures.

Overload Heater Selection

Overload heater coils for continuous-duty motors are selected from manufacturer tables based on the motor nameplate full-load current for maximum motor protection and compliance with Section 430.32 of the NEC®. The class, type, and size information of a magnetic motor starter are found on the nameplate on the face of the starter. **See Figure 12-38.** The phase, service factor, and full-load current of the motor are determined from the motor nameplate. Common applications use 40°C as the ambient temperature. Questionable ambient temperatures should be measured at the job site or determined by some other method.

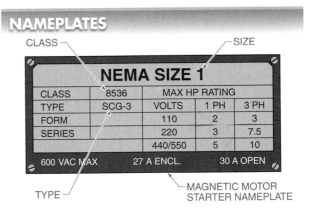

Figure 12-38. The nameplate of a magnetic motor starter includes the class, type, and size of the starter.

It is important to always refer to the manufacturer instructions on thermal overload relay selection to see if any restrictions are placed on the class of starter required. **See Figure 12-39.** For example, unless a class 8198 starter is used, motors with service factors of 1.15 to 1.25 may use 100% of the motor full-load current for thermal overload selection.

Manufacturer Heater Selection Charts. Manufacturers provide charts for use in selecting proper thermal overload heaters. The correct heater selection chart must be used for the appropriate size starter. **See Figure 12-40.** This information is also found within the enclosure of many motor starters. Each motor starter manufacturer has a chart that applies to their specific brand.

MANUFACTURER INSTRUCTIONS

CLASS RESTRICTIONS —

Motor and controller in *same ambient temperature:*

a. All starter classes, except Class 8198

    1. For 1.15 to 1.25 service factor motors use 100% of motor full-load current for thermal unit selection.

    2. For 1.0 service factor motors use 90% of motor full-load current for thermal unit selection.

b. Class 8198 only:

    1. For 1.0 service factor motors use 100% of motor full-load current for thermal unit selection.

    2. For 1.15 to 1.25 service factor motors use 110% of motor full-load current for thermal unit selection.

Figure 12-39. Manufacturer instructions on thermal overload relay selection detail restrictions that are placed on classes of starters.

HEATER SELECTION CHARTS—THERMAL UNIT CURRENT RATINGS

Motor Full-Load Current (Amps)			Thermal Unit Number
1 Unit (Heater)	2 Units (Heaters)	3 Units (Heaters)	
0.29 – 0.31	0.29 – 0.31	0.28 – 0.30	B0.44
0.32 – 0.34	0.32 – 0.34	0.31 – 0.34	B0.51
0.35 – 0.38	0.35 – 0.38	0.35 – 0.37	B0.51
0.39 – 0.45	0.39 – 0.45	0.38 – 0.44	B0.63
0.46 – 0.54	0.46 – 0.54	0.45 – 0.53	B0.71
0.55 – 0.61	0.55 – 0.61	0.54 – 0.59	B0.81
0.62 – 0.66	0.62 – 0.66	0.60 – 0.64	B0.92
0.67 – 0.73	0.67 – 0.73	0.65 – 0.72	B1.03
0.74 – 0.81	0.74 – 0.81	0.73 – 0.80	B1.16
0.82 – 0.94	0.82 – 0.94	0.81 – 0.90	B1.30
0.95 – 1.05	0.95 – 1.05	0.91 – 1.03	B1.45
1.06 – 1.22	1.06 – 1.22	1.04 – 1.14	B1.67
1.23 – 1.34	1.23 – 1.34	1.15 – 1.27	B1.88
1.35 – 1.51	1.35 – 1.51	1.28 – 1.43	B2.10
1.52 – 1.71	1.52 – 1.71	1.44 – 1.62	B2.40
			B2.65

Figure 12-40. Manufacturers provide charts to use for selecting proper overload heaters.

For example, a thermal unit number B2.40 is the correct overload heater for controlling a 3ϕ motor with an FLC rating of 1.50 A. Column three in the heater selection chart is used because all three phases of the 3ϕ motor must have thermal overload protection. The heater must provide protection of approximately 1.5 A (1.44 to 1.62) based on the motor full-load current. Manufacturers have different numbers that relate to their specific heaters, but the selection procedure is similar.

Checking Selections. Section 430.32 of the NEC® indicates that a motor must be protected up to 125% of its FLC rating. Because the minimum full-load current of a B2.40 overload device is 1.44 A, the device trips at 125% of this value or 1.8 A (1.44 × 1.25 = 1.8 A). Dividing the minimum trip current (1.8 A) by the full-load current of the motor (1.5 A) and multiplying by 100% determines if this range is acceptable (1.8 / 1.5 × 100% = 120%). The heater selection is correct because the trip current is less than the NEC® limit of 125%.

Ambient Temperature Compensation. As ambient temperature increases, less current is needed to trip overload devices. As ambient temperature decreases, more current is needed to trip overload devices. Most heater manufacturers provide special overload heater selection tables that provide multipliers to compensate for temperature changes above or below the standard temperature of 40°C. The multipliers ensure that the increase or decrease in temperature does not affect the proper protection provided by the overload relay. **See Figure 12-41.**

THERMAL UNIT SELECTION

		Melting Alloy and Noncompensated Bimetallic Relays		
		Ambient Temperature of Motor		
Controller Class	Continuous Duty Motor Service Factor	*	†	‡
		Full-Load Current Multiplier		
All classes except 8198	1.15 – 1.25	1.0	0.9	1.05
	1.0	0.9	0.8	0.95
Class 8198	1.15 – 1.25	1.1	1.0	1.15
	1.0	1.0	0.9	1.05

* same as controller ambient
† constant 10°C (18°F) higher than controller ambient
‡ constant 10°C (18°F) lower than controller ambient

Figure 12-41. Special overload heater selection tables provide multipliers to compensate for ambient temperatures above or below the standard temperature of 40°C.

For example, a multiplier of 0.9 is required for an ambient temperature increase of 10°C to 50°C. Multiplying the motor full-load current (1.5 A) by the correction factor (0.9) determines the compensated overload heater current rating of 1.35 A (1.5 A × 0.9 = 1.35 A). Using the heater selection chart, the acceptable current range is 1.28 A to 1.43 A. A B2.10 heater is required based on the increase in ambient temperature. This is one size smaller than the heater required (B2.40) at a 40°C ambient temperature.

The temperature surrounding an overload heater is 30°C if the ambient temperature is decreased 10°C. The correction multiplier is 1.05 for a 10°C decrease in ambient temperature. The corrected current is 1.575 A using a full-load current of 1.5 A (1.5 A × 1.05 = 1.575 A). Using the heater selection chart, the acceptable current range is 1.44 to 1.62 A. In this case, the same size heater could be used. Manufacturer specifications and tables should always be consulted for proper heater sizing.

In rare instances, such as older installations or severely damaged equipment, it may be impossible to determine a motor full-load current from its nameplate. Manufacturers provide charts listing approximate full-load currents based on average motor full-load currents. **See Figure 12-42.**

AMPERE RATINGS OF 3φ, 60 Hz AC INDUCTION MOTORS

HP	Speed (RPM)	200 V	230 V	380 V	460 V	575 V	220 V
¼	1800	1.09	0.95	0.55	0.48	0.38	—
	1200	1.61	1.40	0.81	0.70	0.56	
	900	1.84	1.60	0.93	0.80	0.64	
⅓	1800	1.37	1.19	0.69	0.60	0.48	—
	1200	1.83	1.59	0.92	0.80	0.64	
	900	2.07	1.80	1.04	0.90	0.72	
½	1800	1.98	1.72	0.99	0.86	0.69	—
	1200	2.47	2.15	1.24	1.08	0.86	
	900	2.74	2.38	1.38	1.19	0.95	
¾	1800	2.83	2.46	1.42	1.23	0.98	—
	1200	3.36	2.92	1.69	1.46	1.17	
	900	3.75	3.26	1.88	1.63	1.30	
1	3600	3.22	2.80	1.70	1.40	1.12	—
	1800	4.09	3.56	2.06	1.78	1.42	
	1200	4.32	3.76	2.28	1.88	1.50	
	900	4.95	4.30	2.60	2.15	1.72	
1½	3600	5.01	4.36	2.64	2.18	1.74	—
	1800		4.86				
10			29.2	15..	12.7	10.1	
	1800	30.8	26.8	16.3	13.4	10.7	
	1200	32.2	28.0	16.9	14.0	11.2	
	900	35.1	30.5	18.5	15.2	12.2	
15	3600	41.9	36.4	22.0	18.2	14.5	—
	1800	45.1	39.2	23.7	19.6	15.7	
	1200	47.6	41.4	25.0	20.7	16.5	
	900	51.2	44.5	26.9	22.2	17.8	
20	3600	58.0	50.4	30.5	25.2	20.1	—
	1800	58.9	51.2	31.0	25.6	20.5	
	1200	60.7	52.8	31.9	26.4	21.1	
	900	63.1	54.9	33.2	27.4	21.9	
25	3600	69.9	60.8	36.8	30.4	24.3	—
	1800	74.5	64.8	39.2	32.4	25.9	
	1200	75.4	65.6	39.6	32.8	26.2	
	900	77.4	67.3	40.7	33.7	27.0	
30	3600	84.4	73.7	44.4	36.8	29.4	—
	1800	86.9	75.6	45.7	37.8	30.2	
	1200	90.6	78.8	47.6	39.4	31.5	
	900	94.1	81.8	49.5	40.9	32.7	
40	3600	111.0	96.4	58.2	48.2	38.5	—
	1800	116.0	101.0	61.0	50.4	40.3	
	1200	117.0	102.0	61.2	50.6	40.4	
	900	121.0	105.0	63.2	52.2	41.7	
50	3600	138.0	120.0	72.9	60.1	48.2	—
	1800	143.0	124.0	75.2	62.2	49.7	
	1200	145.0	126.0	76.2	63.0	50.4	
	900	150.0	130.0	78.5	65.0	52.0	
60	3600	164.0	143.0	86.8	71.1	57.3	—
	1800	171.0	149.0	90.0	74.5	59.4	
	1200	173.0	150.0	91.0	75.0	60.0	
	900	177.0	154.0	93.1	77.0	61.5	
75	3600	206.0	179.0	108.0	89.6	71.7	—
	1800	210.?	183.0	..1.0	91.	73.2	
	1800	56..	343.0	284.0	227.0	57.5	
	1200	573.0	345.0	287.0	229.0	58.5	
	900	600.0	347.0	300.0	240.0	60.5	
300	1800	—	678.0	392.0	339.0	271.0	69.0
	1200		684.0	395.0	342.0	274.0	70.0
400	1800	—	896.0	518.0	448.0	358.0	91.8
500	1800	—	1110.0	642.0	555.0	444.0	116.0

Figure 12-42. Most manufacturers provide charts for approximating full-load current when motor nameplate information is not available.

Note: These charts should be used only as a last resort. This technique is not suggested as a standard procedure because the average rating could be higher or lower for a specific motor and, therefore, selection on this basis always involves risk. For fully reliable motor protection, heat coils should be selected based on the motor full-load current rating shown on the motor nameplate. The full-load current of a motor stated on charts should be used in the selection of a heater using the same procedure as if it were the motor nameplate information. These charts provide approximately the same information that may be found on the motor nameplate, but they should be used only if motor nameplate information is not available.

Magnetic Motor Starter Control Circuit

Magnetic motor starters are used to turn motors on and off and provide overload protection. A magnetic motor starter also provides additional control contacts (NO and NC auxiliary contacts) in addition to power contacts (L1/T1, L2/T2, L3/T3) used to switch the motor on and off. Magnetic motor starters work well in applications that require basic ON and OFF motor control.

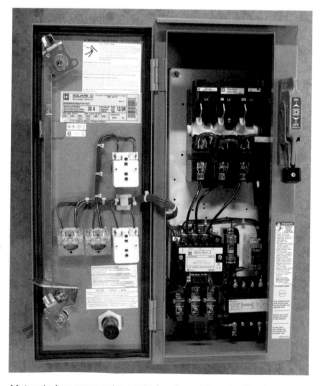

Motor starters are used to control and provide protection for motors.

When using a magnetic motor starter to control a motor, the starter is wired following a standard line diagram. For example, a pump motor is controlled by a magnetic motor starter that uses a control circuit. The control circuit includes a three-position selector switch (HAND/OFF/AUTO) and a liquid level switch to control the motor starter. **See Figure 12-43.** When the selector switch is placed in the HAND position, the motor is ON. When the selector switch is placed in the AUTO position, the motor is ON only when the liquid level switch contacts are closed (liquid at or above switch level). In the OFF position, the motor is OFF regardless of the position of the liquid level switch contacts.

The magnetic motor starter can turn the motor on and off, but it cannot set the acceleration and deceleration times for the motor. Also, the motor starter cannot be used to set the motor speed, provide circuit condition readouts, or display circuit or motor faults. A magnetic motor starter does provide overload protection, and some models may provide phase-loss detection that turns the motor off if one of the three phases is lost.

Inherent Motor Protectors

An *inherent motor protector* is an overload device located directly on or in a motor to provide overload protection. Certain inherent motor protectors base their sensing element on the amount of heat generated or the amount of current consumed by a motor. Inherent motor protectors directly or indirectly (using contactors) trip a circuit that disconnects the motor from the power circuit based on what the motor protector senses. Bimetallic thermodiscs and thermistor overload devices are inherent motor protectors.

Bimetallic Thermodiscs. A bimetallic thermodisc operates on the same principle as a bimetallic strip. The differences between these devices are the shapes of the devices and their locations. A thermodisc has the shape of a miniature dinner plate and is located within the frame of a motor. **See Figure 12-44.** A bimetallic thermodisc warps and opens the circuit when a motor is overloaded. Bimetallic thermodiscs are normally used on small-horsepower motors to disconnect the motor directly from the power circuit. Bimetallic thermodiscs may be tied into the control circuit of a magnetic contactor coil where they can be used as indirect control devices.

It is important to always ensure power to the motor is turned off before resetting a manual-reset thermodisc. This prevents a potential hazard when the motor restarts.

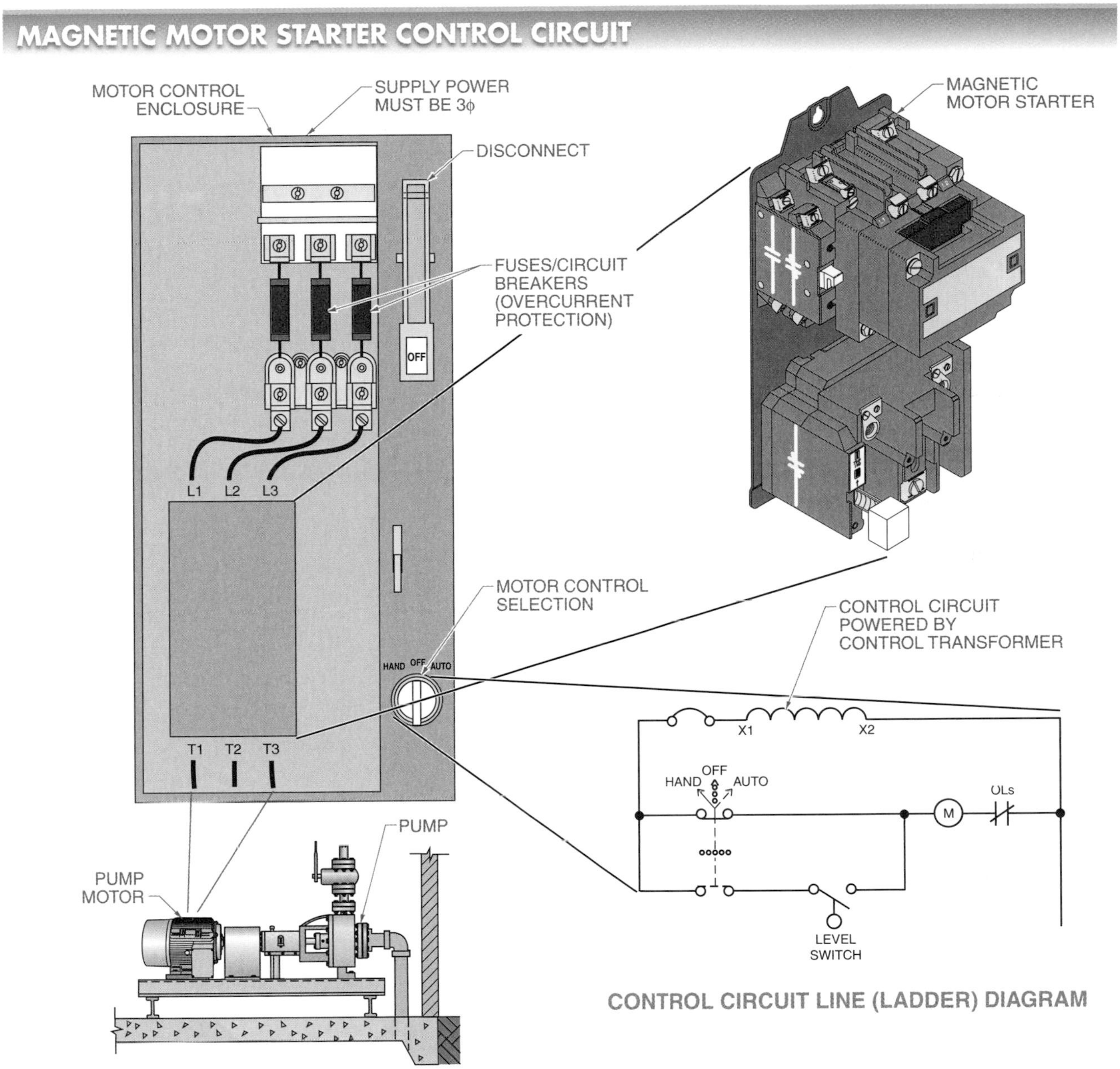

MAGNETIC MOTOR STARTER CONTROL CIRCUIT

MOTOR CONTROL ENCLOSURE

SUPPLY POWER MUST BE 3φ

DISCONNECT

FUSES/CIRCUIT BREAKERS (OVERCURRENT PROTECTION)

MAGNETIC MOTOR STARTER

OFF

L1 L2 L3

MOTOR CONTROL SELECTION

HAND OFF AUTO

T1 T2 T3

CONTROL CIRCUIT POWERED BY CONTROL TRANSFORMER

X1   X2

OFF
HAND   AUTO

OLs

M

LEVEL SWITCH

PUMP

PUMP MOTOR

CONTROL CIRCUIT LINE (LADDER) DIAGRAM

Figure 12-43. A magnetic motor starter can be used to control a pump motor.

BIMETALLIC THERMODISCS

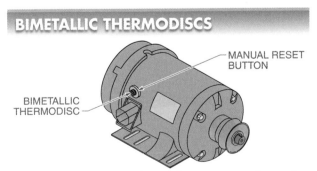

MANUAL RESET BUTTON

BIMETALLIC THERMODISC

Figure 12-44. Bimetallic thermodiscs are normally used on small-horsepower motors to directly disconnect the motor from the power circuit.

Thermistor Overload Devices. A thermistor-based overload is a sophisticated form of inherent motor protection. A thermistor overload device combines a thermistor, solid-state relay, and contactor into a custom-built overload protector. **See Figure 12-45.**

Tech Fact

Overcurrent protective devices are designed for fast operation when protecting a circuit from short circuits and overcurrent. Overload protective devices are designed to protect a circuit from an overcurrent condition that exists for a relatively long time, such as during motor acceleration.

THERMISTORS

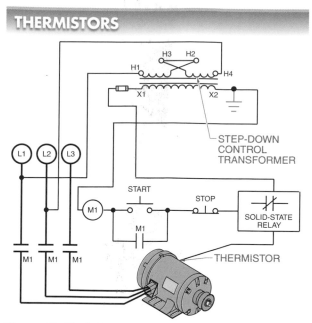

Figure 12-45. A thermistor overload device combines a thermistor, solid-state relay, and contactor into a custom-built overload protector.

A thermistor is similar to a resistor in that its resistance changes with the amount of heat applied to it. As the temperature increases, the resistance of the thermistor decreases and the amount of current passing through the thermistor increases. The changing signal must be amplified before it can do any work, such as triggering a relay, because the thermistor is a low-power device (normally in the thousandths of an ampere range). When a thermistor overload device is amplified, a relay may open a set of contacts in the control circuit of a magnetic motor starter, de-energizing the power circuit of the motor.

The major drawback to thermistor overload devices is that they require a close coordination between the user and the manufacturer to customize the design. Custom-designed overload protectors are more costly than standard, off-the-shelf overload protectors. With the exception of special and high-priced motors requiring extensive protection, custom-designed overload protectors are uneconomical and are not recommended.

Troubleshooting Magnetic Breakers

In addition to troubleshooting motor overload protective devices, the power source must be checked. The power source may have one or more damaged circuit breakers that can cause problems for motors and motor starters.

A *circuit breaker (CB)* is an overcurrent protective device with a mechanism that automatically opens the circuit when an overload condition or short circuit occurs. CBs are connected in series with the circuit and operate much like a relay. They protect a circuit from overcurrents or short circuits. CBs are magnetically operated and are reset after an overload. A DMM set to measure voltage is used to test CBs. CBs perform the same function as fuses and are tested the same way. **See Figure 12-46.** To troubleshoot CBs, the following procedure is applied:

1. Turn the handle of the safety switch or combination starter to the OFF position.

2. Open the door of the safety switch or combination starter. The operating handle must be capable of opening the switch. Replace the operating handle if it does not open the switch.

3. Check the enclosure and interior parts for deformation, displacement of parts, and burns.

4. Check the incoming voltage between each pair of power leads. Incoming voltage should be within 10% of the voltage rating of the motor.

5. Test the enclosure for grounding if voltage is present and at the correct level.

6. Examine the CB. It will be in one of three positions: ON, TRIPPED, or OFF.

7. Reset the CB. If no evidence of damage is present, reset the CB by moving the handle to the OFF position and then back to the ON position. CBs are designed so they cannot be held in the ON position if an overload or short is present. Check the voltage of the reset CB if resetting it does not restore power. Replace all faulty CBs. Never try to service a faulty CB.

A digital multimeter (DMM) set to measure voltage is used to test circuit breakers.

TROUBLESHOOTING CIRCUIT BREAKERS

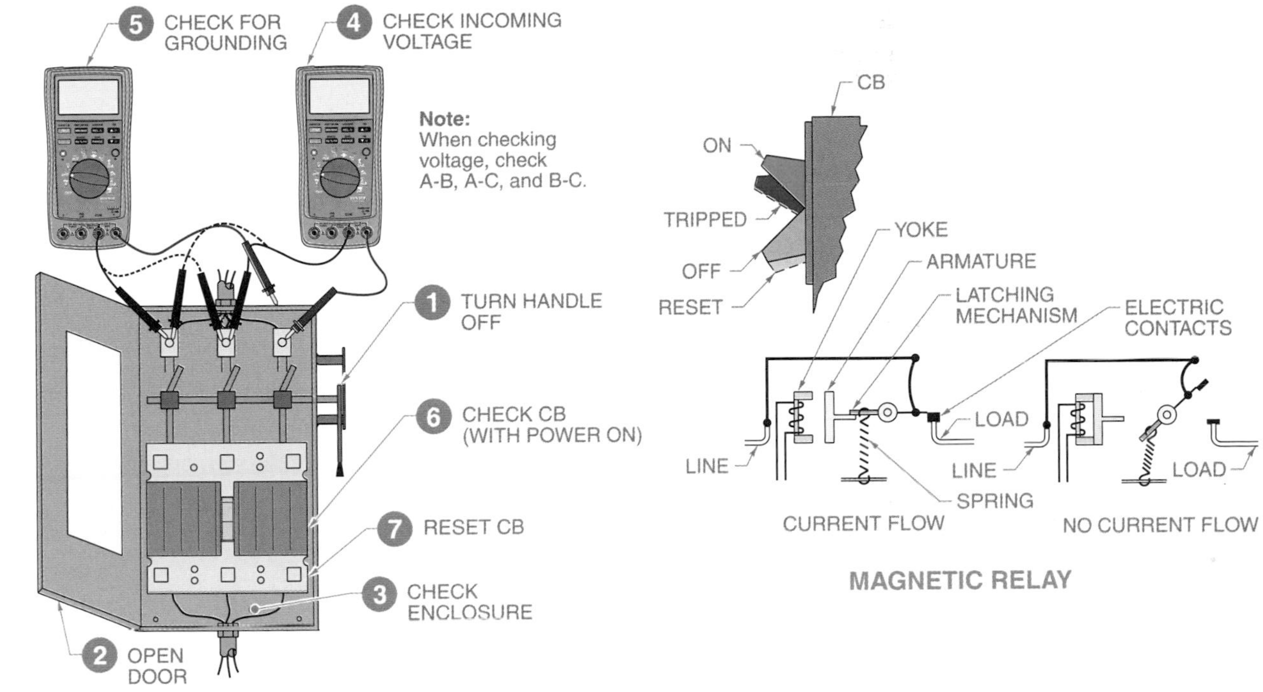

Figure 12-46. Circuit breakers (CBs) perform the same function as fuses and are tested in the same manner.

12-5 CHECKPOINT

1. Is the current rating listed on a motor nameplate the full-load current (FLC) rating, the starting current rating, or the no-load rating?
2. If a motor has an 8 A FLC rating and a service factor of 1.25, how much current is the motor designed to handle for short periods of time?
3. Will an increase in the ambient temperature increase or decrease the amount of current required to trip overloads?
4. How are inherent motor protective devices reset once they are tripped?

12-6 CONTACTOR AND MAGNETIC MOTOR STARTER MODIFICATIONS

Certain devices may be added to basic contactors or motor starters to expand their capability. These devices include additional electrical contacts, power poles, pneumatic timers, transient suppression modules, and control circuit fuse holders. **See Figure 12-47.**

Additional Electrical Contacts

Most contactors and motor starters have the ability to control several additional electrical contacts if the additional contacts are added to existing auxiliary contacts. The additional contacts may be used as extra auxiliary contacts. Both NO and/or NC contacts may be wired to

control additional loads. NC contacts are used to turn on additional loads anytime the contactor or starter is OFF, as well as to provide electrical interlocking. NO contacts are used to turn on additional loads anytime a contactor or starter is ON.

Power Poles

In certain cases, additional power poles (contacts capable of carrying a load) may be added to a contactor. The power poles are available with NO or NC contacts. Normally, only one power pole unit with one or two contacts is added per contactor or motor starter.

ADDITIONAL CONTACTOR/MOTOR STARTER DEVICES

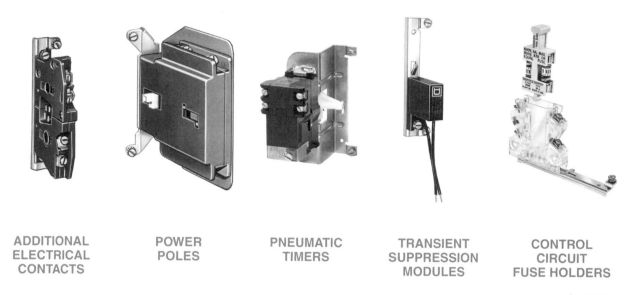

ADDITIONAL
ELECTRICAL
CONTACTS

POWER
POLES

PNEUMATIC
TIMERS

TRANSIENT
SUPPRESSION
MODULES

CONTROL
CIRCUIT
FUSE HOLDERS

Square D Company

Figure 12-47. The devices that may be added to basic contactors or magnetic motor starters to expand their capability include additional electrical contacts, power poles, pneumatic timers, transient suppression modules, and control circuit fuse holders.

In certain cases with large-sized contactors or motor starters, it may be necessary to replace the coil to handle the additional load created when energizing the additional poles. Most power poles are factory or field installed.

Pneumatic Timers

A mechanically operated pneumatic timer can be mounted on some sizes of contactors and motor starters for applications requiring the simultaneous operation of a timer and a contactor. The use of mechanically operated timers results in considerable savings in panel space over a separately mounted timer. Available in time delay after de-energization (off-delay) or time delay after energization (on-delay), the timer attachment has an adjustable timing period over a specified range. Most manufacturers provide units that are field convertible from on-delay to off-delay (or vice versa) without additional parts. The pneumatic timers are ordered either fixed or variable. Most pneumatic timers mount on the side of the contactor and are secured firmly. One single-pole, double-throw contact is provided.

Transient Suppression Modules

Transient suppression modules are designed to be added where the transient voltage generated when opening the coil circuit interferes with the power operation of nearby components and solid-state control circuits. Transient suppression modules normally consist of resistance/capacitance (RC) circuits and are designed to suppress the voltage transients to approximately 200% of peak coil supply voltage.

In certain cases, a voltage transient is generated when switching the integral control transformer that powers the coil control circuit. A transient suppression module, when used with devices wired for common control, is connected across the 120 V transformer secondary. The transient suppression module is not connected across the control coil.

Control Circuit Fuse Holders

Control circuit fuse holders can be attached to contactors or starters when either one or two control circuit fuses may be required. The fuse holder helps satisfy the NEC® requirements in Section 430.72.

12-6 CHECKPOINT

1. Is a transient suppression module used to help reduce transient voltages in the control circuit or motor power circuit?

2. Is the time delay adjustable when a pneumatic timer is added to a motor starter?

12-7 TROUBLESHOOTING CONTACTORS AND MOTOR STARTERS

Contactors or motor starters are the first devices checked when troubleshooting a circuit that does not work or has a problem. Contactors or motor starters are checked first because they are the point where the incoming power, load, and control circuit are connected. Basic voltage and current readings are taken at a contactor or motor starter to determine where the problem lies. The same basic procedure used to troubleshoot a motor starter works for contactors because a motor starter is a contactor with added overload protection.

The tightness of all terminals and busbar connections is checked when troubleshooting control devices. Loose connections in the power circuit of contactors and motor starters cause overheating. Overheating leads to equipment malfunction or failure. Loose connections in the control circuit cause control malfunctions. Loose connections of grounding terminals lead to electrical shock and cause electromagnetic-generated interference.

The power circuit and the control circuit are checked if the control circuit does not correctly operate a motor. The two circuits are dependent on each other but are considered two separate circuits because they are normally at different voltage levels and always at different current levels. **See Figure 12-48.** To troubleshoot a motor starter, the following procedure is applied:

1. Inspect the motor starter and overload assembly. Service or replace motor starters that show heat damage, arcing, or wear. Replace motor starters that show burns. Check the motor and driven load for signs of an overload or other problem.

2. Reset the overload relay if there is no visual indication of damage. Replace the overload relay if there is visual indication of damage.

3. Observe the motor starter for several minutes if the motor starts after resetting the overload relay. The overload relay will continue to open if an overload problem continues to exist.

4. Check the voltage into the starter if resetting the overload relay does not start the motor. Check circuit voltage ahead of the starter if the voltage reading is 0 V. The voltage is acceptable if the voltage reading is within 10% of the motor voltage rating. The voltage is unacceptable if the voltage reading is not within 10% of the motor voltage rating.

5. Energize the starter and check the starter contacts if the voltage into the starter is present and at the correct level. The starter contacts are good if the voltage reading is acceptable. Open the starter, turn the power off, and replace the contacts if there is no voltage reading.

6. Check the overload relay if voltage is coming out of the starter contacts. Turn the power off and replace the overload relay if the voltage reading is 0 V. The problem is downstream from the starter if the voltage reading is acceptable and the motor is not operating.

Since troubleshooting is typically performed while circuits are energized, technicians must protect themselves with PPE rated for potential hazards.

TROUBLESHOOTING MOTOR STARTERS

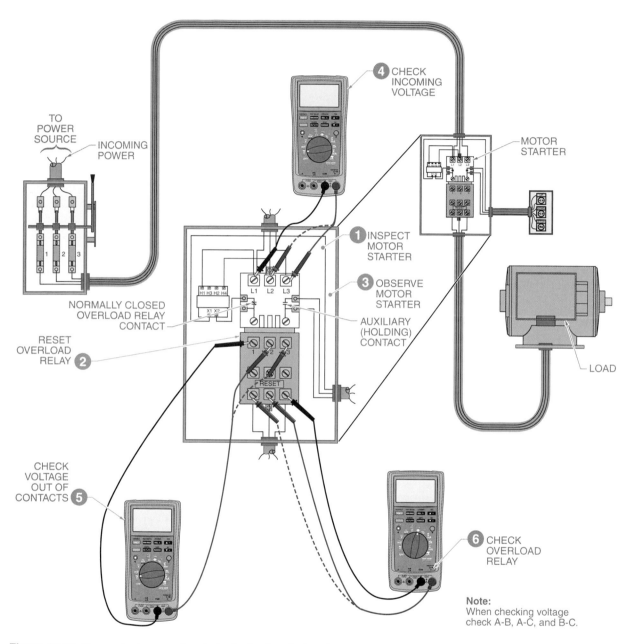

Figure 12-48. The contactor or motor starter is the first device checked when troubleshooting a circuit that does not work or has a problem.

Troubleshooting Guides

A troubleshooting guide is used when troubleshooting contactors and motor starters. Troubleshooting guides state a problem, its possible cause(s), and corrective action(s) that may be taken. **See Figure 12-49.**

Tech Fact

When a contactor or motor starter is checked, voltage and current measurements should always be taken on each load power line. The voltage values should be within 3% of each line-to-line value and the current should be within 10% of each line value. The highest current value should be less than 80% of the device rating.

CONTACTOR AND MOTOR STARTER TROUBLESHOOTING GUIDE

Problem	Possible Cause	Corrective Action
Humming noise	Magnetic pole faces misaligned	Realign; replace magnet assembly if realignment is not possible
	Too low voltage at coil	Measure voltage at coil; check voltage rating of coil; correct any voltage that is 10% less than coil rating
	Pole face obstructed by foreign object, dirt, or rust	Remove any foreign object and clean as necessary; never file pole faces
Loud buzz noise	Pole face obstructed by foreign object, dirt, or rust	Replace coil assembly
Controller fails to drop out	Voltage to coil not being removed	Measure voltage at coil; trace voltage from coil to supply; search for shorted switch or contact if voltage is present
POWER IN	Worn or rusted parts causing binding	Replace worn parts; clean rusted parts
	Pole face obstructed by foreign object, dirt, or rust	Pole face obstructed by foreign object, dirt, or rust
POWER OUT	Pole face obstructed by foreign object, dirt, or rust	Pole face obstructed by foreign object, dirt, or rust
Controller fails to pull in	No coil voltage	Measure voltage at coil terminals; trace voltage loss from coil to supply voltage if voltage is not present
POWER IN	Too low voltage	Measure voltage at coil terminals; correct voltage level if voltage is less than 10% of rated coil voltage; check for a voltage drop as large loads are energized
	Coil open	Measure voltage at coil; remove coil if voltage is present and correct but coil does not pull in; measure coil resistance for open circuit; replace if open
POWER OUT	Coil shorted	Shorted coil may show signs of burning; the fuse or breakers should trip if coil is shorted; disconnect one side of coil and reset if tripped; remove coil and check resistance for short if protective device does not trip; a shorted coil has zero or very low resistance; replace shorted coil; replace any coil that is burned
	Mechanical obstruction	Remove any obstructions
Contacts badly burned or welded	Too high inrush current	Measure inrush current; check load for problem of higher-than-rated load current; change to larger controller if load current is correct but excessive for controller
	Too fast load cycling	Change to larger controller if load cycles ON and OFF repeatedly
	Too large overcurrent protective device	Size overcurrent protective device to load and controller
	Short circuit	Check fuses or circuit breakers; clear any short circuit
	Insufficient contact pressure	Check to ensure contacts are making good connection
Nuisance tripping	Incorrect overload size	Check size of overload against rated load current; adjust size as permissible per NEC®
	Lack of temperature compensation	Check setting of overload if controller and load are at different ambient temperatures
	Loose connections	Check for loose terminal connection

Figure 12-49. A troubleshooting guide used when troubleshooting contactors and motor starters states a problem, its possible cause or causes, and corrective actions that may be taken.

12-7 CHECKPOINT

1. When troubleshooting a motor control circuit, what measuring points should have the higher voltage reading, L1 and L2 or X1 and X2?

2. When troubleshooting a motor control circuit, if the correct voltage is measured between T1 and T2, and not between T1 and T3 or between T2 and T3, what line (L1, L2, or L3) has the problem?

Additional Resources

Review and Resources

Access Chapter 12 Review and Resources for *Electrical Motor Controls for Integrated Systems* by scanning the above QR code with your mobile device.

Applying Your Knowledge

Refer to the *Electrical Motor Controls for Integrated Systems* Learner Resources for interactive Applying Your Knowledge questions.

Workbook and Applications Manual

Refer to Chapter 12 in the *Electrical Motor Controls for Integrated Systems Workbook* and the *Applications Manual* for additional exercises.

ENERGY EFFICIENCY PRACTICES

Contactors and Lamps

Contactors are control devices used for repeatedly establishing and interrupting an electrical power circuit. Contactors may be operated manually or magnetically and are used to make and break the electrical power circuit of loads such as lamps, heating elements, transformers, and capacitors.

Contactors are not dependent on a continuous control voltage to maintain the position of their contacts. Energy-efficient contactors use less energy than electrically held contactors because they do not have their coils energized continuously. Electrically held contactors require a constant supply of control voltage to maintain the position of their contacts. Energy-efficient contactors can be used in low-noise environments because the coil hum and continuous coil energy consumption is eliminated, which reduces background noise and saves energy.

A common use of contactors is in switching energy-efficient lamps. Depending on the contactor size and application, the contactor is normally used to switch several lamps simultaneously. Applications in which contactors are used to simultaneously switch several lamps include lighting for auditoriums, theaters, halls, churches, factories, sporting fields, showrooms, and roads.

Compact fluorescent lamps (CFLs) may be switched using contactors. CFLs are designed to replace incandescent lamps and fit into existing incandescent lamp fixtures. CFLs last, on average, ten times longer than incandescent lamps and use approximately 25% of the energy used by incandescent lamps.

Objectives

13-1
- Define motor and explain the electron flow motor rule.
- Explain torque and magnetic force rotation.

13-2
- List and describe the parts of a DC motor.
- State how a commutator and brushes deliver voltage to the armature.

13-3
- List the basic types of DC motors.
- Describe a DC series motor.
- Describe a DC shunt motor.
- Describe a DC compound motor.
- Describe a DC permanent-magnet motor.

13-4
- Define stepper motor and state how it operates to produce small incremental steps of the motor shaft.
- Describe how switches are used with stepper motors.
- Explain the function of an encoder in a stepper motor application.

13-5
- Define work and state how it is calculated.
- Define torque and horsepower.
- Explain the relationship between speed, torque, and horsepower.

13-6
- Explain how to troubleshoot brushes.
- Explain how to troubleshoot commutators.
- Explain how to troubleshoot for grounded, open, or short circuits.
- Explain DC voltage variations.

Sections

Review and Resources
atplearningresources.com/Quicklinks
Access Code: 362245

DC Motors

Direct current (DC) motors, like all types of motors, have advantages and disadvantages. Their main advantages are that they can produce higher torque than AC motors of the same physical size and can operate on batteries, making them ideal for portable tools and equipment. The disadvantage of DC motors is that most types have comparatively high maintenance requirements and/or short operating life because their moving parts (brushes and commutator) are in contact with each other and will wear.

Types of DC motors range from large-horsepower motors used to move heavy loads, such as trains and mining equipment, to very small stepper motors used to control the precise movement of a motor shaft to a few degrees, such as in applications with robotic arms and laser-guided equipment. Learning the different types of DC motors and their wide range of uses is important for anyone designing, installing, or servicing a system that includes a DC motor.

13-1 DC MOTOR OPERATION

A *motor* is a machine that converts electrical energy into mechanical energy by means of electromagnetic induction. Motors operate on the principle that when a current-carrying conductor is placed in a magnetic field, a force that tends to move the conductor out of the field is exerted on the conductor. The conductor tends to move at right angles to the field. **See Figure 13-1.**

The direction of the movement of the conductor depends on the direction of the current and the magnetic field. The electron flow motor rule is used to determine the direction of motion of a current-carrying conductor in a magnetic field. **See Figure 13-2.** The electron flow motor rule states that with the thumb, index finger, and middle finger of the right hand set at right angles to each other, the index finger points in the direction of the magnetic field (N to S), the thumb points in the direction of the induced conductor motion, and the middle finger points in the direction of the electron current flow in the conductor.

DC MOTORS

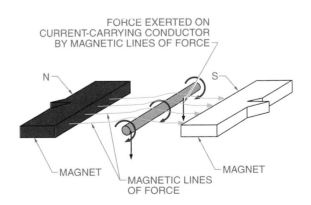

FORCE EXERTED ON
CURRENT-CARRYING CONDUCTOR
BY MAGNETIC LINES OF FORCE

N

S

MAGNET

MAGNETIC LINES
OF FORCE

MAGNET

Figure 13-1. When a current-carrying conductor is placed in a magnetic field, a force that tends to move the conductor out of the field is exerted on the conductor.

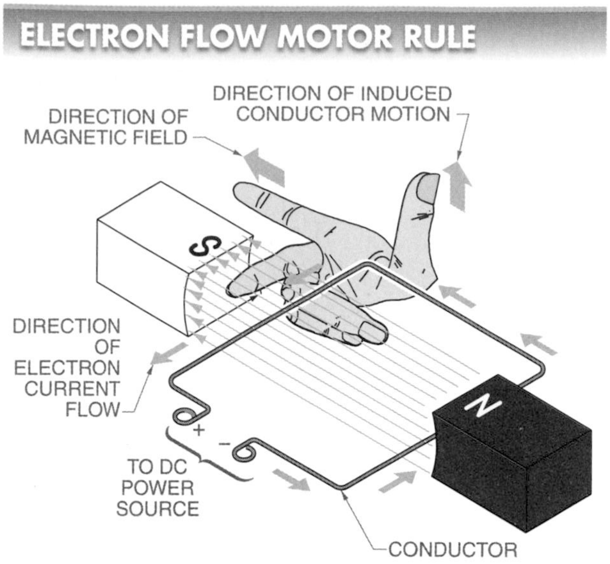

ELECTRON FLOW MOTOR RULE

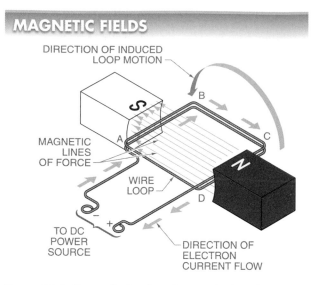

Figure 13-2. The electron flow motor rule is used to determine the direction of motion of a current-carrying conductor in a magnetic field.

The current in the conductor flows at right angles to the magnetic lines of force of the magnetic field. The force on the conductor is at right angles to both the current in the conductor and the magnetic lines of force. The amount of force on the conductor depends on the intensity of the magnetic field, the current through the conductor, and the length of the conductor. The intensity of the magnetic field and the amount of current in the conductor are normally changed to increase the force on the conductor. However, the amount of force can be increased by increasing any of these three factors.

Torque is developed on a wire loop in a magnetic field. **See Figure 13-3.** Electron current flow must be at a right angle to the magnetic field. This is required for induced motion because no force is exerted on a conductor if the direction of electron current flow through the conductor and direction of the magnetic lines of force are the same (parallel).

Both sections of loop AB and CD have a force exerted on them because the direction of electron current flow in these segments is at right angles to the magnetic lines of force. The exertion of force on AB and CD is opposite in direction because the current flow is opposite in each section of the wire loop.

The result of the two magnetic fields intersecting creates a rotating force, referred to as torque, on the loop. The magnetic lines of force cause the loop to rotate when they straighten. The left side of the conductor is forced downward and the right side of the conductor is forced upward, causing a counterclockwise rotation. **See Figure 13-4.**

MAGNETIC FIELDS

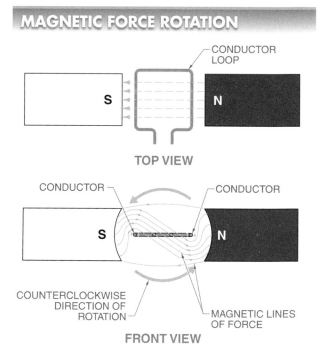

Figure 13-3. Torque is developed on a wire loop in a magnetic field.

MAGNETIC FORCE ROTATION

Figure 13-4. The distortion of the magnetic lines of force causes the conductor loop to rotate in a counterclockwise direction.

13-1 CHECKPOINT

1. What three factors determine the amount of force exerted on a conductor in a magnetic field?
2. What is the rotating force of a motor shaft called?

13-2 DC MOTOR CONSTRUCTION

A *direct current (DC) motor* is a motor that uses direct current connected to the field and armature to produce shaft rotation. A DC motor consists of field windings, an armature, a commutator, and brushes. **See Figure 13-5.** *Field windings* are the stationary windings or magnets of a DC motor. An *armature* is the rotating part of a DC motor. A magnetic field is produced in the armature by current flowing through the armature coils. The armature magnetic field interacts with the field windings magnetic field as a result of the current flow through the coils. The interaction of the magnetic fields causes the armature to rotate.

A *commutator* is a ring made of segments that are insulated from one another. The commutator connects each armature winding to the brushes using copper bars (segments) that are insulated from each other with pieces of mica. The commutator is mounted on the same shaft as the armature and rotates with the shaft.

A *brush* is the sliding contact that rides against the commutator segments and is used to connect the armature to the external circuit. Brushes are made of carbon or graphite material and are held in place by brush holders. A pigtail connects a brush to the external circuit (power supply). A *pigtail* is an extended, flexible connection or a braided copper conductor. Brushes are free to move up and down in the brush holder. This freedom allows the brush to follow irregularities in the surface of the commutator. A spring placed behind the brush forces the brush to make contact with the commutator. Normally, the spring pressure is adjustable, as is the entire brush holder assembly. The brushes make contact with the successive copper bars of the commutator as the shaft, armature, and commutator rotate.

DC power is delivered to the armature coils through the brushes and commutator segments. The armature coils, commutator, and brushes are arranged so that the flow of current is in one direction in the loop on one side of the armature, and the flow of current is in the opposite direction in the loop on the other side of the armature. For example, brush 2 breaks contact with side B of the commutator and makes contact with side A.

The flow of current through the commutator reverses because the flow of current is at the same polarity on the brushes at all times. This allows the commutator to rotate another 180°. After the additional 180° rotation, brush 1 breaks contact with side B of the commutator and makes contact with side A. Likewise, brush 2 breaks contact with side A of the commutator and makes contact with side B. This reverses the direction of current in the commutator again and allows for another 180° of rotation. The armature continues to rotate as long as the commutator winding is supplied with current and there is a magnetic field.

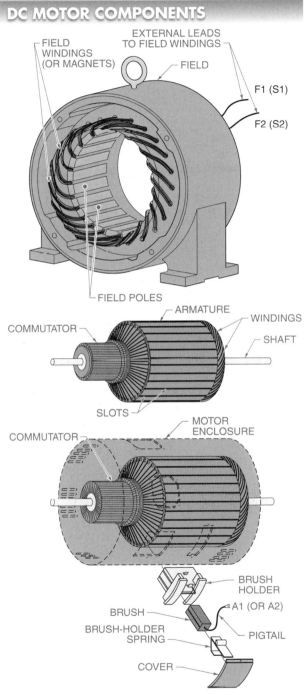

Figure 13-5. DC motors consist of field windings, an armature, commutator, and brushes that make contact with the successive copper bars of the commutator as the shaft, armature, and commutator rotate.

Torque is exerted on the armature when it is positioned so that the plane of the armature loop is parallel to the field, and the armature loop sides are at right angles to the magnetic field. **See Figure 13-6.** No movement takes place if the armature loop is stopped in the vertical (neutral) position. In this position, no further torque is produced because the forces acting on the armature are upward on the top side of the loop and downward on the lower side of the loop.

The armature does not stop because of inertia. The armature continues to rotate for a short distance. As it rotates, the magnetic field in the armature is opposite that of the field. This pushes the conductor back in the direction it came, stopping the rotating motion. A method is required to reverse the current in the armature every one-half rotation so that the magnetic fields work together. Brushes and a commutator are added to maintain this positive rotation.

Connecting voltage directly to the field and armature of a DC motor allows the motor to produce higher torque in a smaller frame than AC motors. DC motors provide excellent speed control for acceleration and deceleration with effective and simple torque control. DC motors perform better than AC motors in most traction equipment applications. DC motors require more maintenance than AC motors because they have brushes that wear. DC motors are used as the drive motor in mobile equipment such as golf carts, quarry and mining equipment, and locomotives.

ARMATURE POSITIONS

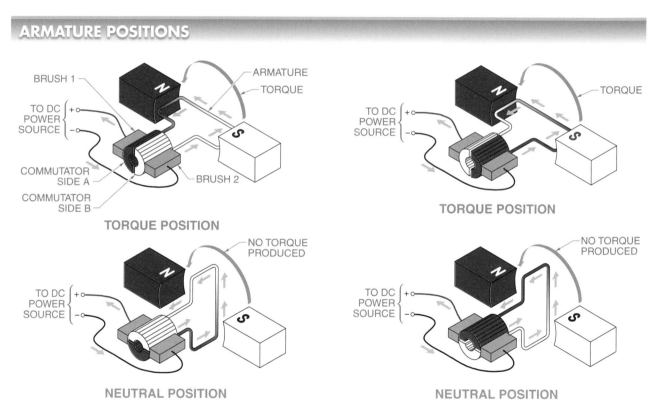

Figure 13-6. A rotating force (torque) is exerted on the armature when it is positioned so that the plane of the armature loop is parallel to the field and the armature loop sides are at a right angle to the magnetic field.

13-2 CHECKPOINT

1. What are the stationary windings of a DC motor called?

2. What are the rotating windings of a DC motor called?

3. What is the function of brushes?

13-3 DC MOTOR TYPES

The four basic types of DC motors are DC series motors, DC shunt motors, DC compound motors, and DC permanent-magnet motors. **See Figure 13-7.** These DC motors have similar external appearances but are different in their internal construction and output performance.

Tech Fact

DC motor armature leads are labeled A1 A2, shunt field leads are labeled F1 and F2, and series field leads are labeled S1 and S2. When DC motor leads are connected to DC power, A1, F1, and S1 are typically connected closest to the positive power side and A2, F2, and S2 are connected closest to the negative power side.

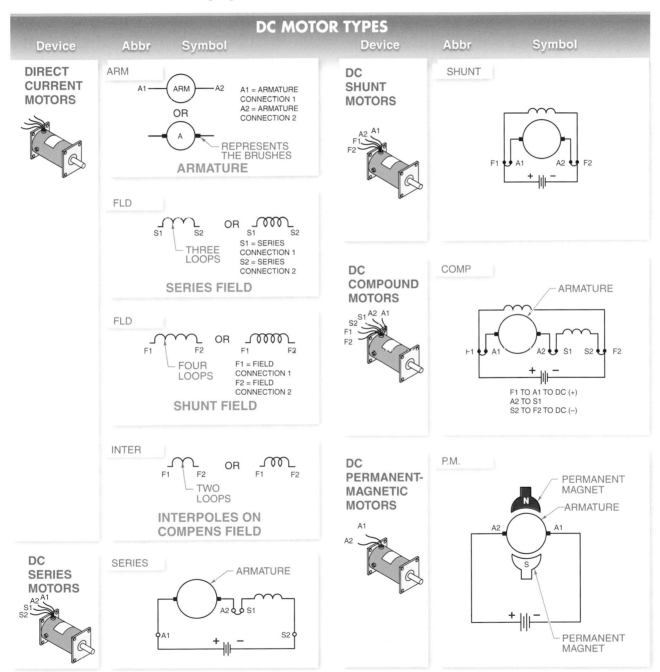

Figure 13-7. The four basic types of DC motors are DC series motors, DC shunt motors, DC compound motors, and DC permanent-magnet motors.

DC Series Motors

A *DC series motor* is a DC motor that has the series field connected in series with the armature. The field must carry the load current passing through the armature. The field has relatively few turns of heavy-gauge wire. The wires extending from the series coil are marked S1 and S2. The wires extending from the armature are marked A1 and A2. **See Figure 13-8.**

DC SERIES MOTORS

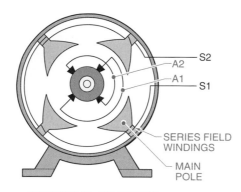

PICTORIAL DRAWING

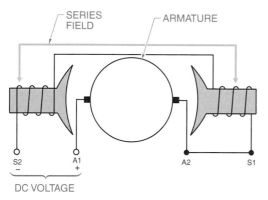

WIRING DIAGRAM

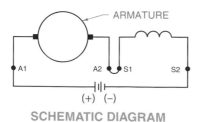

SCHEMATIC DIAGRAM

Figure 13-8. A DC series motor is a motor with the field connected in series with the armature.

DC Series Motor Torque. A DC series motor produces high starting torque. **See Figure 13-9.** The field coil (series field) of the motor is connected in series with the armature. Although speed control is poor, a DC series motor produces very high starting torque and is ideal for applications in which the starting load is large. Applications include cranes, hoists, electric buses, streetcars, railroads, and other heavy-traction applications.

SERIES MOTOR CHARACTERISTICS

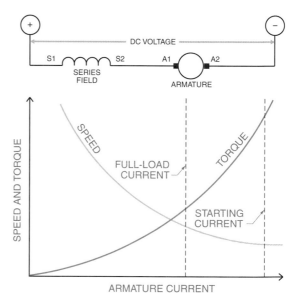

Figure 13-9. A DC series motor produces high starting torque.

The torque that is produced by a motor depends on the strength of the magnetic field in the motor. The strength of the magnetic field depends on the amount of current that flows through the series field. The amount of current that flows through a motor depends on the size of the load. The larger the load, the greater the current flow. Any increase in load increases current in both the armature and series field because the armature and field are connected in series. This increased current flow is what gives a DC series motor a high torque output.

In DC series motors, speed changes rapidly when torque changes. When torque is high, speed is low; and when speed is high, torque is low. This occurs because there is a large flux increase as the increased current (created by the load) flows through the series field. This increased flux produces a large counter electromotive force, which greatly decreases the speed of the motor. As the load is removed, the motor rapidly increases speed. Without a load, the motor would gain speed uncontrollably. In certain cases, the speed may become great enough to damage the motor. For this reason, a DC series motor should always be connected directly to the load and not through belts, chains, etc.

The speed of a DC series motor is controlled by varying the applied voltage. Although the speed control of a series motor is not as good as the speed control of a shunt motor, not all applications require good speed regulation. The advantage of a high torque output outweighs good speed control in certain applications, such as the starter motor in automobiles.

DC Shunt Motors

A *DC shunt motor* is a DC motor that has the field connected in shunt (parallel) with the armature. The wires extending from the shunt field of a DC shunt motor are marked F1 and F2. The armature windings are marked A1 and A2. **See Figure 13-10.**

The field has numerous turns of wire, and the current in the field is independent of the armature, providing the DC shunt motor with excellent speed control. The shunt field may be connected to the same power supply as the armature or may be connected to another power supply. A *self-excited shunt field* is a shunt field connected to the same power supply as the armature. A *separately excited shunt field* is a shunt field connected to a different power supply than the armature.

DC shunt motors are used where constant or adjustable speed is required and starting conditions are moderate. Typical applications include fans, blowers, centrifugal pumps, conveyors, elevators, woodworking machinery, and metalworking machinery.

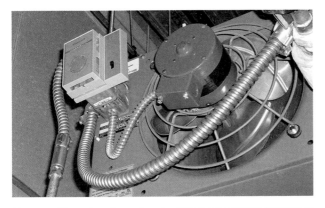

Fans used to cool equipment or circulate air are powered by motors.

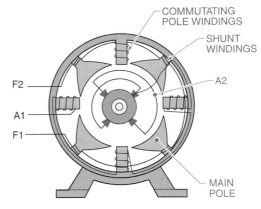

PICTORIAL DRAWING

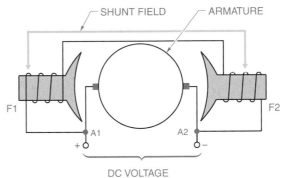

WIRING DIAGRAM

SCHEMATIC DIAGRAM

Figure 13-10. A DC shunt motor is a motor with the field connected in shunt (parallel) with the armature.

In a DC shunt motor, if the voltage to the armature is reduced, the speed is also reduced. If the strength of the magnetic field is reduced, the motor speeds up. DC shunt motors speed up with a reduction in shunt field strength because with less field strength, there is less counter electromotive force developed in the armature. When the counter electromotive force is lowered, the armature current increases, producing increased torque and speed. To control the speed of a DC shunt motor, the voltage to the armature is varied or the shunt field current is varied. **See Figure 13-11.**

SHUNT MOTOR CHARACTERISTICS

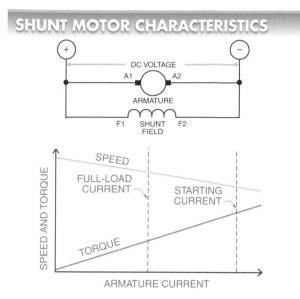

Figure 13-11. To control the speed of a DC shunt motor, the voltage to the armature is varied as the shunt field current is varied.

A field rheostat or armature rheostat is used to adjust the speed of a DC shunt motor. **See Figure 13-12.** The rheostat is used to increase or decrease the strength of the field or armature. Once the strength of the field is set, it remains constant regardless of changes in armature current. As the load is increased on the armature, the armature current and torque of the motor increase. This slows the armature, but the reduction of counter electromotive force (CEMF) simultaneously allows a further increase in armature current and thus returns the motor to the set speed. The motor runs at a fairly constant speed at any control setting.

RHEOSTATS

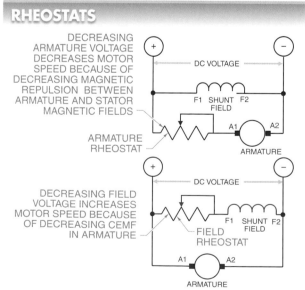

Figure 13-12. A field rheostat or armature rheostat is used to adjust the speed of a DC shunt motor.

A DC shunt motor has relatively high torque at any speed. The motor torque is directly proportional to the armature current. As armature current is increased, so is motor torque, with only a slight drop in motor speed.

DC Compound Motors

A *DC compound motor* is a DC motor with the field connected in both series and shunt with the armature. The field coil is a combination of the series field (S1 and S2) and the shunt field (F1 and F2). **See Figure 13-13.**

The series field is connected in series with the armature. The shunt field is connected in parallel with the series field and armature combination. This arrangement gives the motor the advantages of the DC series motor (high torque) and the DC shunt motor (constant speed). **See Figure 13-14.**

DC compound motors are used when high starting torque and constant speed are required. Typical applications include punch presses, shears, bending machines, and hoists.

Speed control is obtained in a DC compound motor by changing the shunt field current strength or changing the voltage applied to the armature. This is accomplished by using a controller that uses resistors to reduce the applied voltage or by using a variable voltage supply.

Harrington Hoists, Inc.

Motorized hoists are used in storage and warehouse facilities to lift and move materials.

DC COMPOUND MOTORS

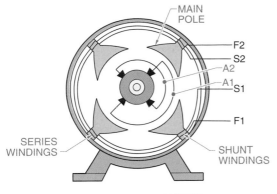

PICTORIAL DRAWING

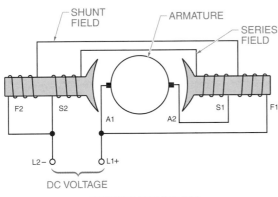

WIRING DIAGRAM

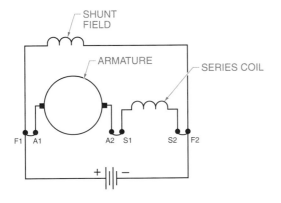

| A1 TO F1 TO DC+ |
| A2 TO S1 |
| S2 TO F2 TO DC- |

SCHEMATIC DIAGRAM

Figure 13-13. A DC compound motor is a motor with the field connected in both series and shunt with the armature.

COMPOUND MOTOR CHARACTERISTICS

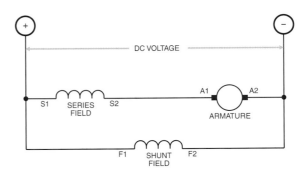

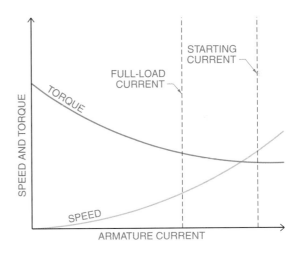

Figure 13-14. A DC compound motor combines the operating characteristics of series and shunt motors.

DC Permanent-Magnet Motors

A *DC permanent-magnet motor* is a motor that uses magnets, not a winding, for the field poles. DC permanent-magnet motors have molded magnets mounted into a steel shell. The permanent magnets are the field coils. DC power is supplied only to the armature. **See Figure 13-15.**

DC permanent-magnet motors are used in automobiles to control power seats, power windows, and windshield wipers. DC permanent-magnet motors produce relatively high torque at low speeds and provide some self-braking when removed from power. Not all DC permanent-magnet motors are designed to run continuously because they overheat rapidly. Overheating destroys the permanent magnets.

DC PERMANENT-MAGNET MOTORS

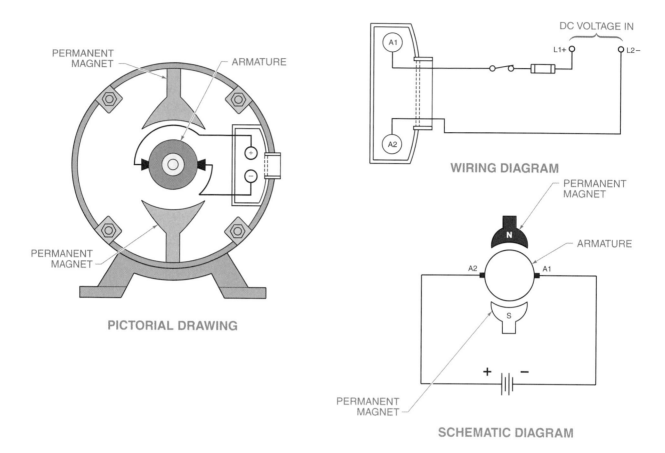

Figure 13-15. A DC permanent-magnet motor uses magnets, not a winding, for the field poles.

13-3 CHECKPOINT

1. What are the labels of the four wires coming out of a DC series motor?

2. What are the labels of the four wires coming out of a DC shunt motor?

3. What are the labels of the two wires coming out of a DC permanent-magnet motor?

4. Which DC motor type has the highest torque but poor speed regulation?

13-4 STEPPER MOTORS

A *stepper motor* is a motor that divides shaft rotation into discrete distances (steps). Unlike other motors that rotate their shafts when power is applied, stepper motors only move their shafts in small controlled increments called steps.

Stepper motors are used in applications that require precise control of the position of the motor shaft. Typical applications include pen positioning, rotary and indexing table control, robotic arm movement control, laser positioning, fax machine operation, and printer control.

The shaft of a stepper motor rotates at fixed angles when it receives an electrical pulse. The electrical pulse magnetizes the motor's stationary field, called the stator field. The magnetic stator field moves the permanent magnet. The permanent magnet is called the rotor or armature. The rotor steps forward to align with the stator's magnetic field (N to S and S to N) and stops movement until another stator field is energized. **See Figure 13-16.**

To increase the number of steps the rotor can take per 360° rotation, many individual stator and rotor segments are used. The total number of segments a motor has determines the number of individual steps the motor can take in one revolution. Stepper motors can have 4, 8, 20, or hundreds of steps per revolution. **See Figure 13-17.**

REVERSING DIRECTION OF SHAFT MOVEMENT

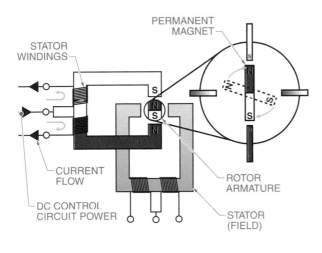

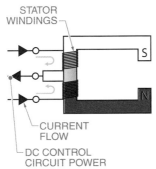

Figure 13-16. Stepper motors rotate at fixed angles when they receive electrical impulses.

The direction of shaft rotation is controlled by the polarity of the produced magnetic field in the stator coil. The direction of the magnetic field determines the positive and negative side of the coil. The direction can be reversed by reversing the direction of current through the coil.

STEPPER MOTORS

Typical Stepping Increments		
Steps*	Step Angle†	Maximum Run Rate‡
4	90	100
8	45	260
20	18	500
24	15	600
48	7.5	1000
72	5.0	1000
96	3.75	1000
144	2.5	1000
180	2.0	1500
200	1.8	2000
500	0.72	1000

* per revolution
† in degrees
‡ in steps per second (typical)

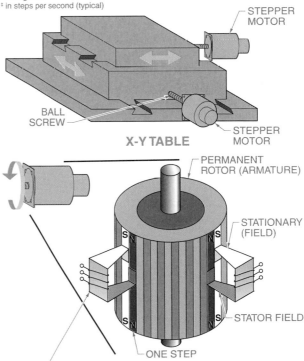

Figure 13-17. The total number of segments a motor has determines the number of individual steps a motor can take in one revolution.

Each input rotor pulse produces shaft rotation through the rated step angle of the stepper motor. For example, if 50 pulses are applied to a 1.8° stepper motor, the shaft rotates exactly 90° (50 × 1.8 = 90); and if 200 pulses are applied, the shaft rotates 360° (200 × 1.8 = 360).

Although stepper motors can provide precise positioning, they cannot operate large loads because their torque output is low compared to other motor types. Torque also decreases at higher operating speeds. The operating characteristics of a stepper motor must be closely checked and compared with the operating characteristic of the application for proper operation.

Stepper motors include important information on their nameplates and specification sheets. Typical stepper motor specifications include the following:

• shaft length/diameter (2.48″/1.5″, etc.)
• rated voltage (3 VDC, 12 VDC, etc.)
• rated current (32 mA, etc.)
• rated impedance per phase (150 Ω, 380 Ω, etc.)
• step angle (1.5°, 2.5°, etc.)
• torque (5 oz/in, 100 oz/in, etc.)
• steps per revolution (24, 200, etc.)
• noise (40 dB, etc.)

Switches that turn on and off (pulse) the voltage to the coil control when the individual stator coils are energized to produce the magnetic field. The switches provide the required direct current needed to magnetize the stationary field. The order (clockwise or counterclockwise) in which the coils are turned on and off determines the direction of rotation of the rotor and shaft of a stepper motor. The control switches do not remain on. They are turned on to move the rotor and then turned off (pulsed). **See Figure 13-18.**

Because the stator coils are turned on and off rapidly, solid-state transistors switches are used instead of mechanical switches. The transistor switches are controlled by an electronic control circuit that sends signals to the transistor to start or stop the flow of current through the coil. The electronic circuit can be from a stand-alone stepper controller, a programmable logic controller (PLC), or another electronic circuit. **See Figure 13-19.**

All coils produce high current (inductive surge) when their magnetic fields collapse, which can damage switch contacts. This damage can include burnt contacts in mechanical switches and component failure in electronic switches such as transistors, silicon-controlled rectifiers (SCRs), and triacs. Transistors and SCRs are used to switch DC, and triacs are used to switch AC. To help reduce damage, a snubber circuit is used to protect the switch. For example, a diode may be used in a DC circuit.

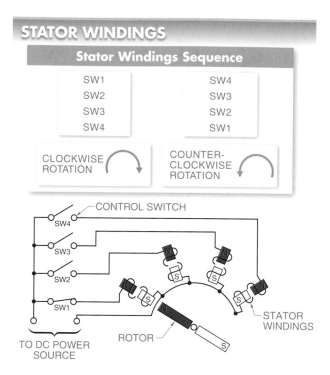

Figure 13-18. The order in which the stator windings are turned on and off determines the direction of rotation of the rotor and shaft of a stepper motor.

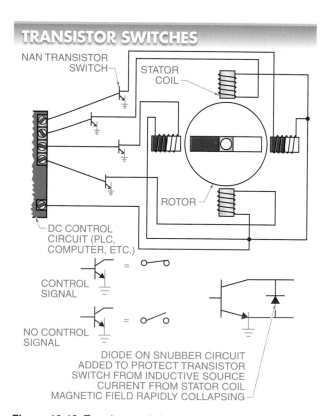

Figure 13-19. Transistor switches are used to rapidly turn on and off the stator coils to move the rotor.

The electronic control circuit outputs are connected so that each output performs a specific task within the system. One stepper motor might control the x-axis and another stepper motor might control the y-axis on a machine table. When using stepper motors to work together in an application, such as when controlling an x-axis and a y-axis, it is best to use the same stepper motor type for both.

Stepper motor circuits can be open-loop or closed-loop types. In the open-loop type, the control signals are sent to the motor and it is assumed that the motor shaft moved as required. However, the shaft may not have moved as instructed. This happens when the motor is underpowered, the load has increased above design, or there is an obstruction.

In some stepper motor applications, feedback is important since the motor may not have moved the actual steps the controller told it to. To provide feedback to the control circuit about the position of the motor shaft, an optical encoder with a disk on the motor shaft is typically used. An *encoder* is a sensor (transducer) that produces discrete electrical pulses during each increment of shaft rotation. **See Figure 13-20.**

The encoder disk has open sections in which light beams shine through. The pulses of the light beams are sent to the electronic control circuit to compare the instructed movement to the actual movement. Any differences can be compensated for by the design of the control circuit.

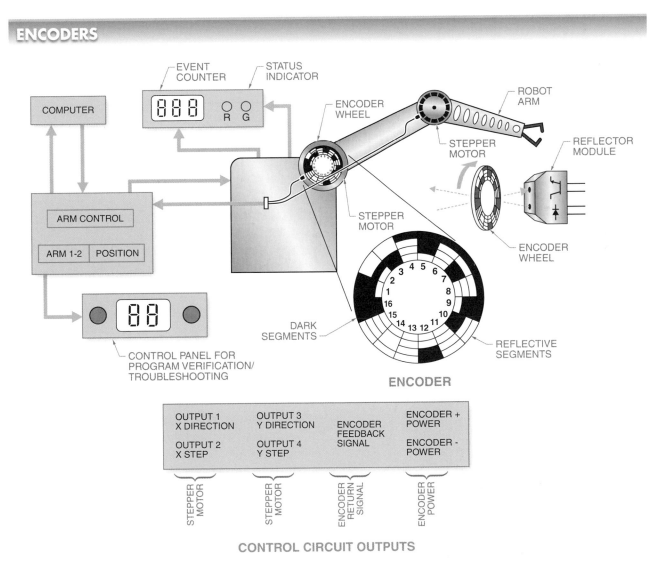

Figure 13-20. An encoder is used to produce discrete electrical pulses during each increment of shaft rotation of a stepper motor.

13-4 CHECKPOINT

1. What are the two names used for the rotating part of a stepper motor?
2. What are the two names used for the stationary part of a stepper motor?

3. What causes the high current that can damage the switching contacts/device?
4. What device is used to provide feedback about the position of the motor shaft to the control circuit?

13-5 DC MOTOR LOAD REQUIREMENTS

The loads that are connected to and controlled by motors vary considerably. Each motor has its own ability to control different loads at different speeds. For example, certain motors are rated at high starting torque with low running torque and others are rated at low starting torque with high running torque. Load requirements must be determined to select the correct motor for a given application. This is especially true in applications that require speed control. The requirements a motor must meet in controlling a load include work, force, torque, and horsepower in relation to speed.

Work

Work is the application of force over a distance. *Force* is any cause that changes the position, motion, direction, or shape of an object. Work is done when a force overcomes a resistance. *Resistance* is any force that tends to hinder the movement of an object. If an applied force does not cause motion, no work is produced. **See Figure 13-21.**

The amount of work (*W*) produced is determined by multiplying the force (*F*) that must be overcome by the distance (*D*) through which it acts. Thus, work is measured in pound-feet (lb-ft). To calculate the amount of work produced, the following formula is applied:

$W = F \times D$

where

W = work (in lb-ft)

F = force (in lb)

D = distance (in ft)

Example: Calculating Work

How much work is required to carry a 25 lb bag of groceries vertically from street level to the fourth floor of a building 30′ above street level?

$W = F \times D$

$W = 25 \times 30$

$W = 750$ lb-ft

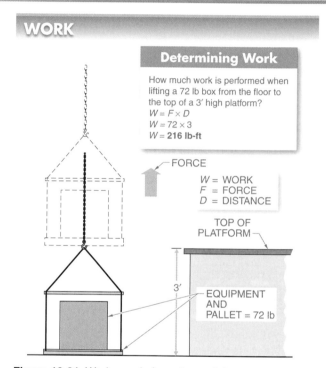

WORK

Determining Work

How much work is performed when lifting a 72 lb box from the floor to the top of a 3′ high platform?

$W = F \times D$

$W = 72 \times 3$

$W = $ **216 lb-ft**

FORCE

W = WORK
F = FORCE
D = DISTANCE

TOP OF PLATFORM

3′

EQUIPMENT AND PALLET = 72 lb

Figure 13-21. Work equals force times distance.

Torque

Torque is the force that produces rotation. Torque causes an object to rotate. Torque (*T*) consists of a force (*F*) acting on a radius (*r*). **See Figure 13-22.** Torque, like work, is measured in pound-feet (lb-ft). However, torque, unlike work, may exist even though no movement occurs. To calculate torque, the following formula is applied:

$T = F \times r$

where

T = torque (in lb-ft)

F = force (in lb)

r = radius (in ft)

TORQUE

T = TORQUE
F = FORCE
r = RADIUS

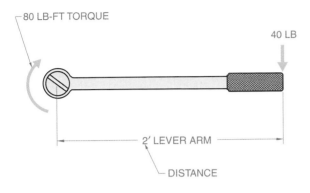

80 LB-FT TORQUE

40 LB

2' LEVER ARM

DISTANCE

Determining Torque

How much torque is produced by a 40 lb force pushing on a 2' lever arm?
$T = F \times r$
$T = 40 \times 2$
$T = \textbf{80 lb-ft}$

Figure 13-22. Torque is the force that produces or tends to produce rotation.

Example: Calculating Torque

What is the torque produced by a 60 lb force pushing on a 3' lever arm?

$$T = F \times r$$
$$T = 60 \times 3$$
$$T = \textbf{180 lb-ft}$$

Horsepower

Electrical power is rated in horsepower or watts. *Horsepower (HP)* is a unit of power equal to 746 W or 33,000 lb-ft per min (550 lb-ft per sec). A *watt (W)* is a unit of measure equal to the power produced by a current of 1 A across a potential difference of 1 V. A watt is ¹⁄₇₄₆ of 1 HP. The watt is the base unit of electrical power. Motor power is rated in horsepower and watts. **See Figure 13-23.**

Horsepower is used to measure the energy produced by an electric motor while doing work. To calculate the horsepower of a motor when current, efficiency, and voltage are known, the following formula is applied:

$$HP = \frac{I \times E \times Eff}{746}$$

where
HP = horsepower
I = current (in A)
E = voltage (in V)
Eff = efficiency (in %)
746 = constant

MOTOR POWER

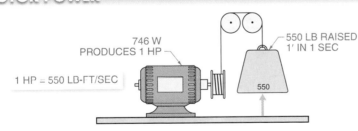

746 W PRODUCES 1 HP

1 HP = 550 LB-FT/SEC

550 LB RAISED 1' IN 1 SEC

550

Mechanical Energy	Electrical Energy
½ HP	373 W
1 HP	746 W
2 HP	1492 W
5 HP	3730 W
100 HP	74,600 W

Determining HP: Voltage and Current Given

What is the horsepower of a 120 V motor pulling 7.2 A and having 88% efficiency?

$HP = \dfrac{I \times E \times Eff}{746}$

$HP = \dfrac{7.2 \times 120 \times 0.88}{746}$

$HP = \dfrac{760.32}{746}$

$HP = \textbf{1 HP}$

7.2 A at 88% Eff

120 V MOTOR

Determining HP: Speed and Torque Given

What is the horsepower of a 1180 rpm motor with an FLT of 2.25 lb-ft?

$HP = \dfrac{rpm \times T}{5252}$

$HP = \dfrac{1180 \times 2.25}{5252}$

$HP = \dfrac{2655}{5252}$

$HP = \textbf{½ HP}$

1180 rpm

FLT = 2.25 lb-ft

Figure 13-23. Motor power is rated in horsepower and watts.

Example: Calculating Horsepower Using Voltage, Current, and Efficiency

What is the horsepower of a 230 V motor pulling 4 A and having 82% efficiency?

$$HP = \frac{I \times E \times Eff}{746}$$

$$HP = \frac{4 \times 230 \times 0.82}{746}$$

$$HP = \frac{754.4}{746}$$

$$HP = \textbf{1 HP} \text{ (rounded)}$$

The horsepower of a motor determines the size of a load that the motor can operate and the speed at which the load turns. To calculate the horsepower of a motor when the speed and torque are known, the following formula is applied:

$$HP = \frac{rpm \times T}{5252}$$

where

HP = horsepower

rpm = revolutions per minute

T = torque (lb-ft)

5252 = constant $\left(\frac{33,000}{2\pi} = 5252 \right)$

Example: Calculating Horsepower Using Speed and Torque

What is the horsepower of a 1725 rpm motor with a full-load torque of 3.1 lb-ft?

$$HP = \frac{rpm \times T}{5252}$$

$$HP = \frac{1725 \times 3.1}{5252}$$

$$HP = \frac{5347.5}{5252}$$

$$HP = \textbf{1 HP} \text{ (rounded)}$$

Formulas for determining torque and horsepower are for theoretical values. When these formulas are applied to specific applications, an additional 15% to 40% capability may be required to start a given load. Loads that are more difficult to start require the higher rating. To increase the rating, multiply the calculated theoretical value by 1.15 (115%) to 1.4 (140%). For example, what is the horsepower of a 1725 rpm motor with a full-load torque of 3.1 lb-ft with an added 25% output capability?

$$HP = \frac{rpm \times T}{5252} \times \%$$

$$HP = \frac{1725 \times 3.1}{5252} \times 1.25$$

$$HP = \frac{5347.5}{5252} \times 1.25$$

$$HP = 1 \times 1.25$$

$$HP = \textbf{1.25 HP}$$

Relationship between Speed, Torque, and Horsepower

The operating speed, torque, and horsepower rating of a motor determine the work that the motor can produce. These three factors are interrelated when applied to driving a load. **See Figure 13-24.**

If the torque remains constant, speed and horsepower are proportional. (A) If the speed increases, the horsepower must increase to maintain a constant torque. (B) If the speed decreases, the horsepower must decrease to maintain a constant torque.

If speed remains constant, torque and horsepower are proportional. (C) If the torque increases, the horsepower must increase to maintain a constant speed. (D) If the torque decreases, the horsepower must decrease to maintain a constant speed.

If torque and speed vary simultaneously but in opposite directions, the horsepower remains constant. (E) If the torque increases and the speed decreases, the horsepower remains constant. (F) If the torque decreases and the speed increases, the horsepower remains constant.

Milwaukee Tool Corporation
Cordless hand drills typically contain a brushed DC motor that is powered by a rechargeable battery.

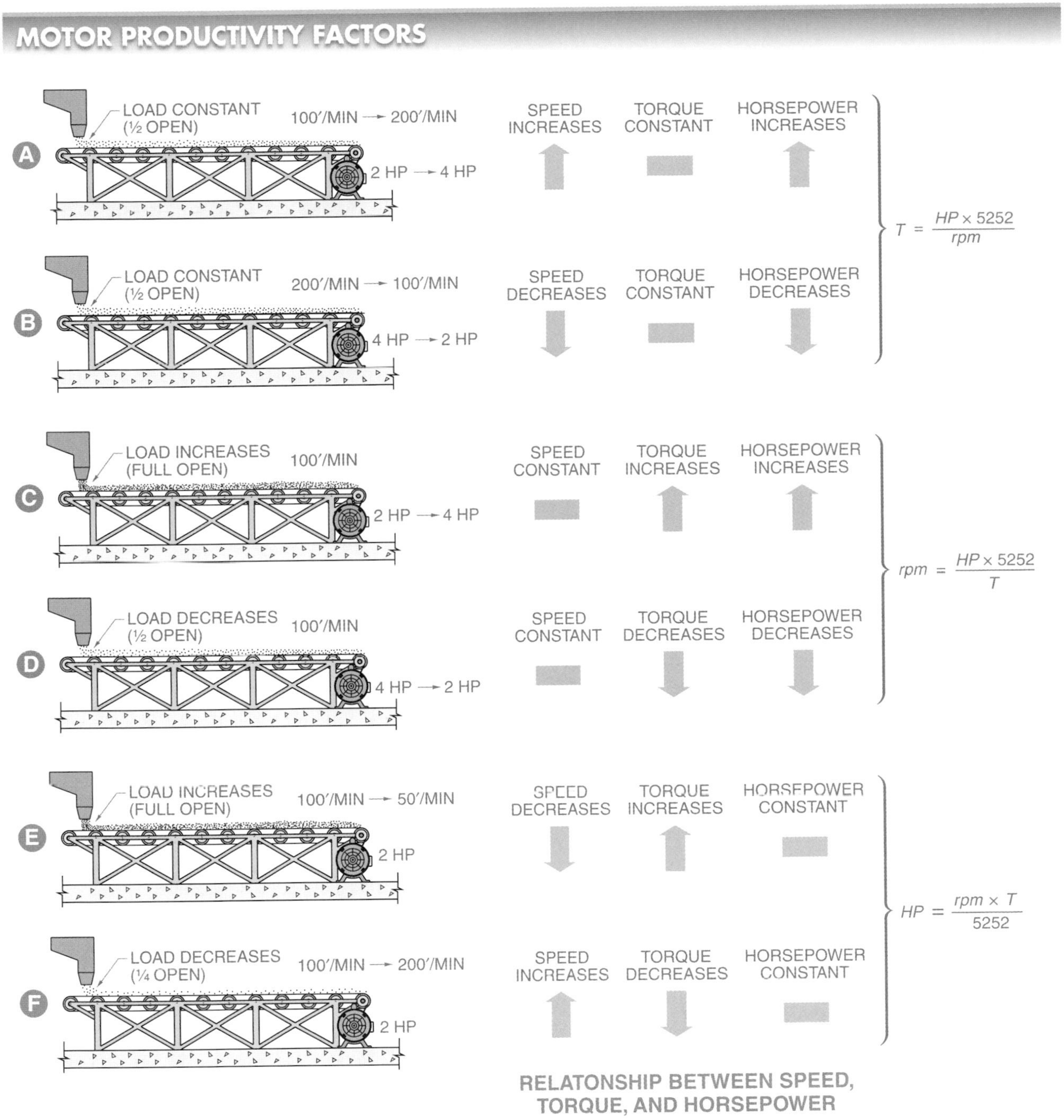

MOTOR PRODUCTIVITY FACTORS

A LOAD CONSTANT (½ OPEN)   100'/MIN → 200'/MIN   2 HP → 4 HP
SPEED INCREASES   TORQUE CONSTANT   HORSEPOWER INCREASES

B LOAD CONSTANT (½ OPEN)   200'/MIN → 100'/MIN   4 HP → 2 HP
SPEED DECREASES   TORQUE CONSTANT   HORSEPOWER DECREASES

$$T = \frac{HP \times 5252}{rpm}$$

C LOAD INCREASES (FULL OPEN)   100'/MIN   2 HP → 4 HP
SPEED CONSTANT   TORQUE INCREASES   HORSEPOWER INCREASES

D LOAD DECREASES (½ OPEN)   100'/MIN   4 HP → 2 HP
SPEED CONSTANT   TORQUE DECREASES   HORSEPOWER DECREASES

$$rpm = \frac{HP \times 5252}{T}$$

E LOAD INCREASES (FULL OPEN)   100'/MIN → 50'/MIN   2 HP
SPEED DECREASES   TORQUE INCREASES   HORSEPOWER CONSTANT

F LOAD DECREASES (¼ OPEN)   100'/MIN → 200'/MIN   2 HP
SPEED INCREASES   TORQUE DECREASES   HORSEPOWER CONSTANT

$$HP = \frac{rpm \times T}{5252}$$

RELATIONSHIP BETWEEN SPEED, TORQUE, AND HORSEPOWER

Figure 13-24. The operating speed, torque, and horsepower rating of a motor determine the work that the motor can produce.

13-5 CHECKPOINT

1. Can there be work if no movement occurs?
2. Can there be torque if no movement occurs?
3. How many watts does a 2 HP motor use?
4. If speed must increase, what must happen to horsepower to maintain constant torque?

13-6 TROUBLESHOOTING DC MOTORS

DC motors require considerable troubleshooting because of their brushes. The brushes and commutator of a DC motor are subject to wear. The brushes are designed to wear as the motor ages. Most DC motors are designed so that the brushes and the commutator can be inspected without disassembling the motor. Some motors require disassembly for close inspection of the brushes and commutator. Troubleshooting DC motors also includes troubleshooting for grounded, open, or short circuits.

Troubleshooting Brushes

Brushes wear faster than any other component of a DC motor. The brushes ride on the fast-moving commutator. **See Figure 13-25.** Typically, bearings and lubrication are used to reduce friction when two moving surfaces touch. However, no lubrication is used between the moving brushes and the commutator because the brushes must carry current from the armature. Sparking occurs as the current passes from the commutator to the brushes. Sparking causes heat, burns, and wear of electric parts.

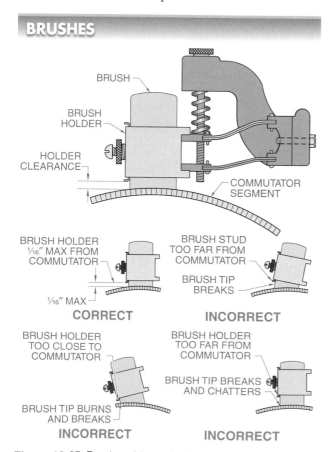

BRUSHES

BRUSH

BRUSH HOLDER

HOLDER CLEARANCE

COMMUTATOR SEGMENT

BRUSH HOLDER
1/16" MAX FROM COMMUTATOR

1/16" MAX

CORRECT

BRUSH STUD TOO FAR FROM COMMUTATOR

BRUSH TIP BREAKS

INCORRECT

BRUSH HOLDER TOO CLOSE TO COMMUTATOR

BRUSH TIP BURNS AND BREAKS

INCORRECT

BRUSH HOLDER TOO FAR FROM COMMUTATOR

BRUSH TIP BREAKS AND CHATTERS

INCORRECT

Figure 13-25. Brushes ride on the fast-moving commutator.

Replacing worn brushes is easier and less expensive than servicing or replacing a worn commutator. When troubleshooting brushes, the brushes are observed as the motor operates. The brushes should ride smoothly on the commutator with little or no sparking. There should be no brush noise, such as chattering. Brush sparking, chattering, or a rough commutator indicates service is required. Brushes must be positioned correctly for proper contact with the commutator. **See Figure 13-26.** To troubleshoot brushes, the following procedure is applied:

1. Turn the handle of the safety switch or combination starter to the OFF position. Lock out and tag out the starting mechanism per company policy.

2. Measure the voltage at the motor terminals to ensure that the power is OFF.

3. Check the brush movement and tension. Remove the brushes. The brushes should move freely in the brush holder. The spring tension should be approximately the same on each brush.

4. Check the length of the brushes. Brushes should be replaced when they have worn down to about half of their original size. Replace all brushes if any brush is less than half its original length. Never replace only one brush. Always replace brushes with brushes of the same composition. Check manufacturer recommendations for proper brush position and brush pressure.

Troubleshooting Commutators

Brushes wear faster than the commutator. After the brushes have been changed once or twice, the commutator usually needs to be serviced. Any markings on the commutator, such as grooves or ruts, or discolorations other than a polished brown color where the brushes ride, indicate a problem. **See Figure 13-27.** To troubleshoot commutators, the following procedure is applied:

1. Visually check the commutator. The commutator should be smooth and concentric. A uniform, dark, copper oxide-carbon film should be present on the surface of the commutator. This naturally occurring film acts like a lubricant by prolonging the life of the brushes and reducing wear on the commutator surface.

2. Check the mica insulation between the commutator segments. The mica insulation separates and insulates each commutator segment. The mica insulation should be undercut (lowered below the surface) approximately ¹⁄₃₂″ to ¹⁄₁₆″, depending on the size of the motor. The larger the motor, the deeper the undercut. Replace or service the commutator if the mica is raised.

Troubleshooting for Grounded, Open, or Short Circuits

A DC motor is tested for a grounded, open, or short circuit by using a test light. A *grounded circuit* is a circuit in which current leaves its normal path and travels to the frame of the motor. A grounded circuit is caused when insulation breaks down or is damaged, which allows circuit wiring to come in contact with the metal frame of the motor.

TROUBLESHOOTING BRUSHES

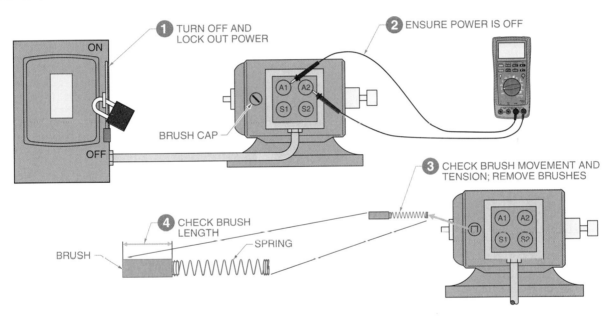

Figure 13-26. Brushes must be positioned correctly to maintain proper contact with the commutator.

TROUBLESHOOTING COMMUTATORS

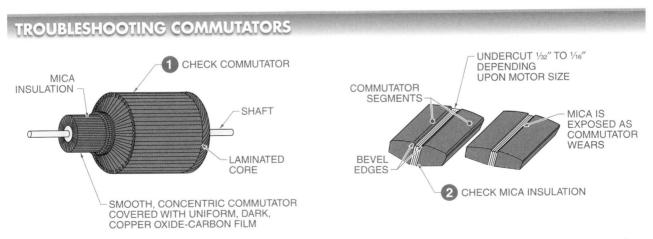

Figure 13-27. Any markings on the commutator, such as grooves or ruts, or discolorations other than a polished brown color where the brushes ride, indicate a problem.

An *open circuit* is an electrical circuit that has an incomplete path that prevents current flow. An open circuit is caused when a conductor or connection has physically moved apart from another conductor or connection.

A *short circuit* is a circuit in which current takes a shortcut around the normal path of current flow. A short circuit is caused when the insulation of two conductors fails, which allows different parts of a circuit to come in contact with one another. Short circuits are usually a result of insulation breakdown. Insulation will break down after extended periods of vibration, friction, or abrasion.

A continuity tester can be used for a quick check of a motor. A continuity tester can detect basic problems (shorts to ground). **See Figure 13-28.** An insulation tester (megohmmeter) should be used for more accurate testing of insulation. To troubleshoot for a grounded circuit, open circuit, or short circuit, the following procedure is applied:

1. Check for a grounded circuit. Connect one lead of the continuity tester to the frame of the motor. Touch the other lead from one motor lead to the other. A grounded circuit is present if the continuity tester beeps. Service and repair the motor.

2. Check for an open circuit. Connect the two test leads to the motor field and armature circuits as follows:
   • series motors: A1 to A2 and S1 to S2
   • shunt motors: A1 to A2 and F1 to F2
   • compound motors: A1 to A2, F1 to F2, and S1 to S2

   The circuits are complete if the continuity tester beeps. The circuits are open if the continuity tester does not beep. Service and repair the motor.

3. Check for a short between windings. Connect the two leads to the motor field and armature circuits as follows:
   • series motors: A1 to S1, A1 to S2, A2 to S1, and A2 to S2
   • shunt motors: A1 to F1, A1 to F2, A2 to F1, and A2 to F2
   • compound motors: A1 to F1, A2 to F2; A1 to F2, A2 to S1; A1 to S1, A2 to S2; A1 to S2, F1 to S1; and A2 to F1, F1 to S2

   The circuit is shorted if the continuity tester beeps. The circuit is not shorted if the continuity tester does not beep. Service and repair the motor.

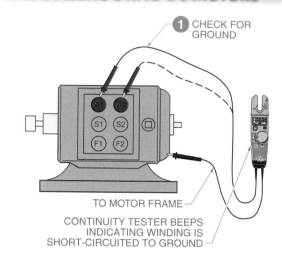

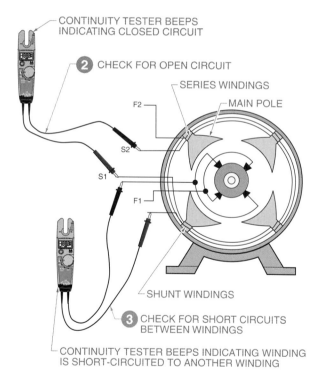

Figure 13-28. A DC motor is tested for a grounded, open, or short circuit by using a continuity tester.

Re-Marking DC Motor Connections

All three types may have the same armature and frame but differ in the way the field coil and armature are connected. For all DC motors, terminal markings A1 and A2 always indicate the armature leads. Terminal markings S1 and S2 always indicate the series field leads. Terminal markings F1 and F2 always indicate the shunt field leads.

DC motor terminals can be re-marked using a DMM by measuring the resistance of each pair of wires. A pair of wires must have a resistance reading or they are not a pair.

The field reading can be compared to the armature reading because each DC motor must have an armature. The series field normally has a reading less than the armature. The shunt field has a reading considerably larger than the armature. The armature can be easily identified by rotating the shaft of the motor when taking the readings. The armature varies the DMM reading as it makes and breaks different windings. One final check can be made by lifting one of the brushes or placing a piece of paper under the brush. The DMM moves to the infinity reading.

From this information, a motor is either a DC series or DC shunt motor if it has two pairs of leads (four wires) coming out. A coil is the series field if the reading of the coil is less than the armature coil resistance. A coil is the shunt field if the reading is considerably larger than the armature resistance.

DC Voltage Variations

DC motors should be operated on pure DC power. Pure DC power is power obtained from a battery or DC generator. DC power is also obtained from rectified AC power. Most industrial DC motors obtain power from a rectified AC power supply. DC power obtained from a rectified AC power supply varies from almost pure DC power to half-wave DC power. **See Figure 13-29.**

Half-wave rectified power is obtained by placing a diode in one of the AC power lines. Full-wave rectified power is obtained by placing a bridge rectifier (four diodes) in an AC power line. Rectified DC power is filtered by connecting a capacitor in parallel with the output of the rectifier circuit.

Computer disc drives and other small pieces of office equipment are commonly driven by DC motors.

DC motor operation is affected by a change in voltage. The change may be intentional, as in a speed-control application, or the change may be caused by variations in the power supply. The power supply voltage normally should not vary by more than 10% of a motor's rated voltage. Motor speed, current, torque, and temperature are affected if the DC voltage varies from the motor rating. **See Figure 13-30.**

Tech Fact

DC motors are designed to operate at the voltage listed on the nameplate (90 VDC, 180 VDC, etc.). Any variation in voltage will also change the amount of current draw and power output. A DMM with a MIN/MAX recording mode can be used to record measurements over the complete cycle of the motor usage to get an accurate depiction of any voltage variations.

DC POWER TYPES

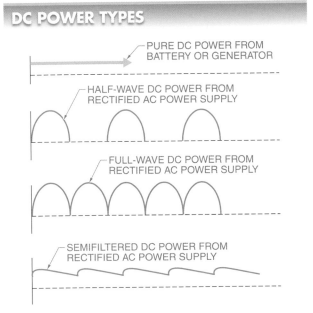

PURE DC POWER FROM BATTERY OR GENERATOR

HALF-WAVE DC POWER FROM RECTIFIED AC POWER SUPPLY

FULL-WAVE DC POWER FROM RECTIFIED AC POWER SUPPLY

SEMIFILTERED DC POWER FROM RECTIFIED AC POWER SUPPLY

Figure 13-29. DC power obtained from a rectified AC power supply varies from almost pure DC power to half-wave DC power.

DC MOTOR PERFORMANCE CHARACTERISTICS

Performance Characteristics	Voltage 10% Below Rated Voltage		Voltage 10% Above Rated Voltage	
	Shunt	Compound	Shunt	Compound
Starting current	−15%	−15%	−15%	+15%
Full-load current	−5%	−6%	+5%	+6%
Motor torque	+12%	+12%	−8%	−8%
Motor efficiency	Decreases	Decreases	Increases	Increases
Speed	Increases	Increases	Decreases	Decreases
Temperature rise	Increases	Increases	Decreases	Decreases

Figure 13-30. Motor speed, current, torque, and temperature are affected if the DC voltage varies from the motor rating.

13-6 CHECKPOINT

1. What usually wears out first on a DC motor and requires occasional service?

2. What usually wears out second on a DC motor and requires occasional service?

3. Which motor winding usually has a higher resistance than the motor's armature winding?

4. Which motor winding usually has a lower resistance than the motor's armature winding?

Additional Resources

Review and Resources

Access Chapter 13 Review and Resources for *Electrical Motor Controls for Integrated Systems* by scanning the above QR code with your mobile device.

Applying Your Knowledge

Refer to the *Electrical Motor Controls for Integrated Systems* Learner Resources for interactive Applying Your Knowledge questions.

Workbook and Applications Manual

Refer to Chapter 13 in the *Electrical Motor Controls for Integrated Systems Workbook* and the *Applications Manual* for additional exercises.

ENERGY EFFICIENCY PRACTICES

Determining Motor Operating Costs

Motors perform work by converting electrical energy to mechanical energy. Depending on the size and design, motors normally convert between 75% and 95% of supplied electrical energy into usable mechanical energy. The mechanical energy is used to produce work. The balance of the electrical energy is lost. Lost energy adds to the cost of electricity but performs no work.

When the power consumed and cost of electrical power are known, the operating cost per hour of a motor can be found by multiplying the power consumed per hour ($P_{/hr}$) by the operating cost per hour ($C_{/kWh}$) and dividing by 1000. For a given motor, the power consumed per hour is found by multiplying the motor horsepower by 746 and dividing by the motor efficiency. For example, the operating cost per hour for a standard 50 HP motor with an 82% efficiency rating at a cost of $0.10/kWh is $4.55.

$$P_{/hr} = HP \times \frac{746}{Eff}$$

$$P_{/hr} = 50 \times \frac{746}{0.82}$$

$$P_{/hr} = \frac{37,300}{0.82}$$

$$P_{/hr} = \mathbf{45,487.80 \text{ W}}$$

$$C_{/hr} = \frac{P_{/hr} \times C_{/kWh}}{1000}$$

$$C_{/hr} = \frac{45,487.80 \times 0.10}{1000}$$

$$C_{/hr} = \frac{4548.78}{1000}$$

$$C_{/hr} = \mathbf{\$4.55}$$

The operating cost per week, based on a 40 hr week, is found by multiplying the operating cost per hour times 40. The operating cost per year is found by multiplying the operating cost per week times 52, as there are 52 weeks per year. For example, the operating cost per week for a standard 50 HP motor with an 82% efficiency rating is $182.00 ($4.55 × 40 = $182.00). The operating cost per year is $9,464.00 ($182.00 × 52 = $9,464.00).

Energy-efficient motors cost more to purchase and less to operate than standard motors. Energy-efficient motors normally cost 20% more than standard motors of the same rating and type. If a motor is not operated often, a standard motor is more cost efficient. If a motor is operated often, an energy-efficient motor is more cost efficient. To calculate the annual savings of operating an energy-efficient motor compared to a standard motor, the operating cost per year for an energy-efficient motor is subtracted from the operating cost per year for a standard motor.

For example, if the operating cost per year of a standard motor is $9,464.00 and the operating cost per year for an energy-efficient motor is $8174.40, the annual savings of operating the energy-efficient motor is $1,289.60 ($9,464.00 – $8,174.40 = $1,289.60).

To find the payback period of using an energy-efficient motor instead of a standard motor, the price premium for the energy-efficient motor is divided by the annual savings. For example, a standard 50 HP, 3ϕ, open-enclosure motor normally costs $1,700.00. An energy-efficient 50 HP, 3ϕ, open-enclosure motor normally costs $2,040.00. The difference (price premium) in the purchase price is $340.00 ($2,040.00 – $1,700.00 = $340.00). The payback period for the energy-efficient motor is 0.2636 years or 96.2 days ($340.00 ÷ $1,289.60 = 0.2636 years or 96.2 days).

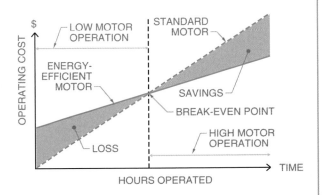

COST SAVINGS USING ENERGY-EFFICIENT MOTORS

Objectives

14-1
- Define AC motor and identify the main parts of an AC motor and state their operating functions.
- Explain how a shaded-pole motor operates and give examples of its usage.
- Explain how a split-phase motor operates and give examples of its usage.
- Explain how a capacitor-start motor, capacitor-run motor, and capacitor start-and-run motor operate and give examples of their usages.
- Explain how a three-phase (3ϕ) motor operates and why no extra starting methods are needed.
- Describe a single-voltage, 3ϕ motor.
- Describe a dual-voltage, 3ϕ motor.

14-2
- Describe motor nameplates and enclosures.
- Explain the effects of long motor starting times and overcycling.
- Explain how excessive heat affects a motor.
- Describe an overload.
- Explain how altitude affects a motor.

14-3
- Explain how to troubleshoot a shaded-pole motor.
- Explain how to troubleshoot a split-phase motor.
- Explain how to troubleshoot a capacitor motor.
- Explain how to troubleshoot a 3ϕ motor.
- Explain how a short circuit can occur in a motor.
- Explain how to determine wye or delta connections.
- Explain how to re-mark a dual-voltage, wye-connected motor.
- Explain how to re-mark a dual-voltage, delta-connected motor.

Alternating current (AC) motors are the most common type of motor used to produce work. AC motors range in size from fractional horsepower (HP) to thousands of HP. Fractional HP AC motors are used in residential buildings to drive refrigerators, washing machines, and dryers; circulate air; and operate appliances. Larger AC motors are used in commercial buildings to drive large appliances such as HVAC systems, commercial washers and dryers, elevators/escalators, and baggage carousels at airports. AC motors of all sizes are used in industrial applications to produce food and other consumable products; operate mining equipment; pump water, sewage, and petroleum; and provide instantaneous work for any application that requires a safe and efficient power source. Since all AC motors operate on the same basic principles, understanding their operation, abilities, and limits is important when designing, installing, servicing, and troubleshooting any system that includes a motor.

14-1 AC MOTOR TYPES

An *alternating current (AC) motor* is a motor that uses alternating current to produce rotation. AC motors have several advantages over DC motors. One advantage is that AC motors have only two bearings that can wear. Secondly, there are no brushes to wear because the motor does not have a commutator. For these reasons, maintenance is minimal. Also, no sparks are generated to create a hazard in the presence of flammable materials. The main parts of an AC motor are the rotor and stator. A *rotor* is the rotating part of an AC motor. A *stator* is the stationary part of an AC motor. **See Figure 14-1.** AC motors are either single-phase (1ϕ) or three-phase (3ϕ).

Single-Phase Motors

Single-phase motors are used in residential applications for AC motor-driven appliances such as furnaces, air conditioners, washing machines, etc. Single-phase motors include shaded-pole, split-phase, and capacitor motors.

Shaded-Pole Motors. A *shaded-pole motor* is a 1ϕ AC motor that uses a shaded stator pole for starting. Shading the stator pole is the simplest method used to start a 1ϕ motor. Shaded-pole motors are commonly ½₀ HP or less and have low starting torque. Common applications of shaded-pole motors include small cooling fans found in computers and home entertainment centers. The shaded pole is normally a solid single turn of copper wire placed around a portion of the main pole laminations. **See Figure 14-2.**

The shaded pole delays the magnetic field in the area of the pole that is shaded. Shading causes the magnetic field at the pole area to be positioned approximately 90° from the magnetic field of the main stator pole. The offset magnetic field causes the rotor to move from the main pole toward the shaded pole. This movement determines the starting direction of a shaded-pole motor.

AC MOTORS

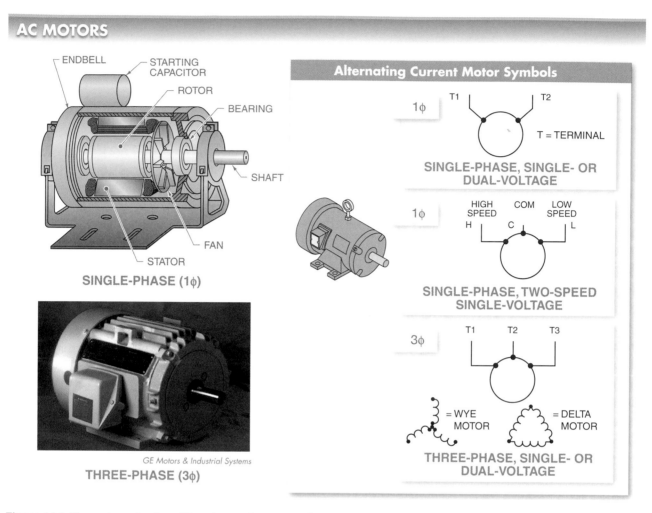

Figure 14-1. The main parts of an AC motor are the rotor and stator.

SHADED-POLE MOTORS

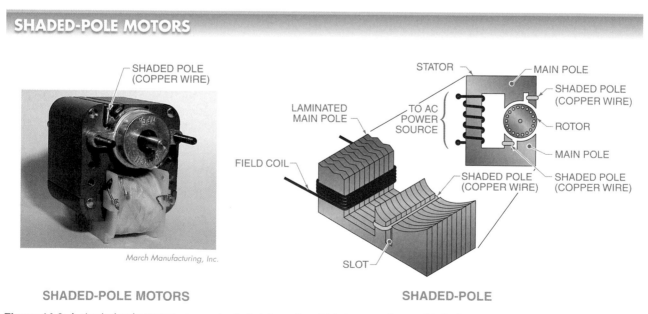

Figure 14-2. A shaded-pole motor uses a shaded stator pole, which is normally a solid single turn of copper wire.

Split-Phase Motors. A *split-phase motor* is a 1ϕ AC motor that includes a running winding (main winding) and a starting winding (auxiliary winding). Split-phase motors are AC motors of fractional horsepower, usually ⅟₂₀ HP to ⅓ HP. Split-phase motors are commonly used to operate washing machines, oil burners, and small pumps and blowers.

A split-phase motor has a rotating part (rotor), a stationary part consisting of the running winding and starting winding (stator), and a centrifugal switch that is located inside the motor to disconnect the starting winding at approximately 60% to 80% of full-load speed. **See Figure 14-3.**

SPLIT-PHASE MOTORS

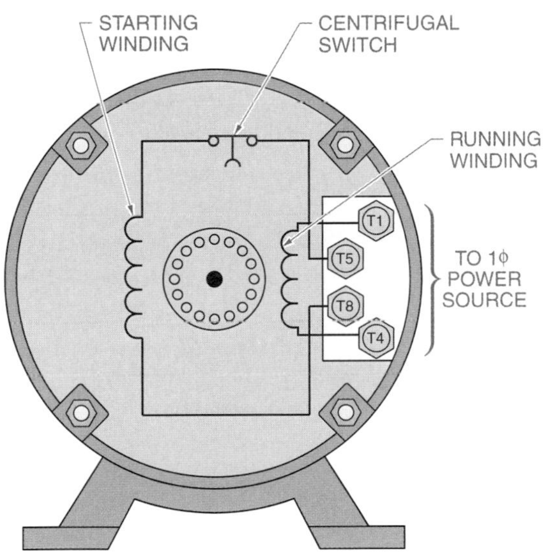

Figure 14-3. A split-phase motor includes a running winding, a starting winding, and a centrifugal switch.

When starting, both the running windings and the starting windings are connected in parallel. The running winding is normally made up of heavy insulated copper wire, and the starting winding is made of fine insulated copper wire. When the motor reaches approximately 75% of full speed, the centrifugal switch opens, disconnecting the starting winding from the circuit. This allows the motor to operate on the running winding only. When the motor is turned off (power removed), the centrifugal switch recloses at approximately 40% of full-load speed.

The running winding is made of larger wire and has a greater number of turns than the starting winding. When the motor is first connected to power, the reactance of the running winding is higher and the resistance is lower than the starting winding. *Reactance is the opposition to the flow of alternating current in a circuit due to inductance.*

The starting winding is made of relatively small wire and has fewer turns than the running winding. When the motor is first connected to power, the reactance of the starting winding is lower and the resistance is higher than the running winding.

When power is first applied, both the running winding and the starting winding are energized. The running winding current lags the starting winding current because of its different reactance. This produces a phase difference between the starting and running windings. A 90° phase difference is required to produce maximum starting torque, but the phase difference is commonly much less. A rotating magnetic field is produced because the two windings are out of phase.

The rotating magnetic field starts the rotor rotating. With the running and starting windings out of phase, the current changes in magnitude and direction, and the magnetic field moves around the stator. This movement forces the rotor to rotate with the rotating magnetic field.

Capacitor Motors. A *capacitor motor* is a 1ϕ AC motor that includes a capacitor in addition to the running and starting windings. Capacitor motor sizes range from ⅛ HP to 10 HP. Capacitor motors are used to operate refrigerators, compressors, washing machines, and air conditioners. The construction of a capacitor motor is similar to that of a split-phase motor, except that in a capacitor motor, a capacitor is connected in series with the starting winding. The addition of a capacitor in the starting winding gives a capacitor motor more torque than a split-phase motor. The three types of capacitor motors are capacitor-start, capacitor-run, and capacitor start-and-run motors.

A capacitor-start motor operates much the same as a split-phase motor in that it uses a centrifugal switch that opens at approximately 60% to 80% of full-load speed. In a capacitor-start motor, the starting winding and the capacitor are removed when the centrifugal switch opens. The capacitor used in the starting winding gives a capacitor-start motor high starting torque. **See Figure 14-4.**

CAPACITOR-START MOTORS

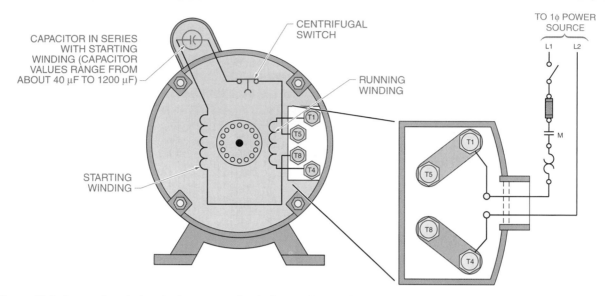

Figure 14-4. A capacitor-start motor has a capacitor in the starting winding, which gives the motor a high starting torque.

A capacitor-run motor has the starting winding and capacitor connected in series at all times. A lower-value capacitor is used in a capacitor-run motor than in a capacitor-start motor because the capacitor remains in the circuit at full-load speed. This gives a capacitor-run motor medium starting torque and somewhat higher running torque than a capacitor-start motor. **See Figure 14-5.**

A capacitor start-and-run motor uses two capacitors. A capacitor start-and-run motor starts with one value capacitor in series with the starting winding and runs with a different value capacitor in series with the starting winding. Capacitor start-and-run motors are also known as dual-capacitor motors. **See Figure 14-6.**

CAPACITOR-RUN MOTORS

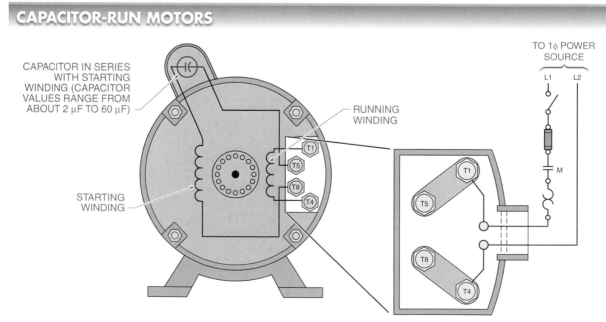

Figure 14-5. A capacitor-run motor has the starting winding and capacitor connected in series at all times.

CAPACITOR START-AND-RUN MOTORS

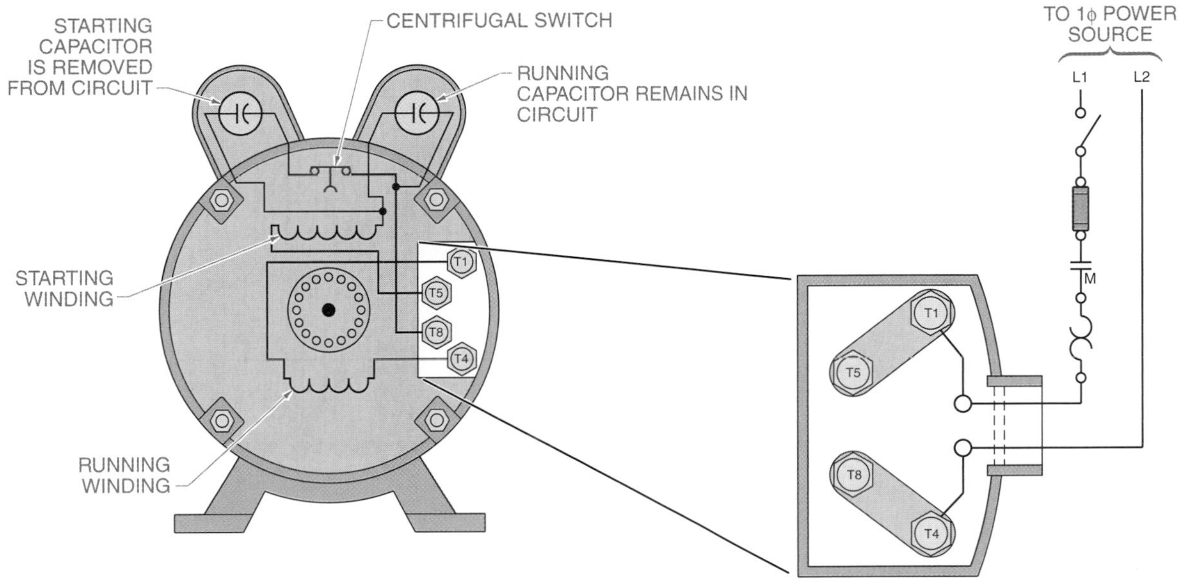

Figure 14-6. In a capacitor start-and-run motor, the starting capacitor is removed when the motor reaches full-load speed, but the running capacitor remains in the circuit.

A capacitor start-and-run motor has the same starting torque as a capacitor-start motor. A capacitor start-and-run motor has more running torque than a capacitor-start motor or capacitor-run motor because the capacitance is better matched for starting and running.

In a typical capacitor start-and-run motor, one capacitor is used for starting the motor and the other capacitor remains in the circuit while the motor is running. A large-value capacitor is used for starting and a small-value capacitor is used for running. Capacitor start-and-run motors are used to run refrigerators and air compressors.

Three-Phase Motors

Three-phase motors are the most common motors used in industrial applications. Three-phase motors are used in applications ranging from fractional horsepower to over 500 HP. Three-phase motors are used in most applications because they are simple in construction, require little maintenance, and cost less to operate than 1ϕ or DC motors.

The most common 3ϕ motor used in most applications is the induction motor. An *induction motor* is a motor that has no physical electrical connection to the rotor. Induction motors have no brushes that wear or require maintenance. Current in the rotor is induced by the rotating magnetic field of the stator.

In a 3ϕ motor, a rotating magnetic field is set up automatically in the stator when the motor is connected to 3ϕ power. The coils in the stator are connected to form three separate windings (phases). Each phase contains one-third of the total number of individual coils in the motor. These composite windings or phases are the A phase, B phase, and C phase. **See Figure 14-7.**

Each phase is placed in the motor so that it is 120° from the other phases. A rotating magnetic field is produced in the stator because each phase reaches its peak magnetic strength 120° away from the other phases. Three-phase motors are self-starting and do not require an additional starting method because of the rotating magnetic field in the motor.

To develop a rotating magnetic field in a motor, the stator windings must be connected to the proper voltage level. This voltage level is determined by the manufacturer and stamped on the motor nameplate. Three-phase motors are designed as either single-voltage motors or dual-voltage motors.

Tech Fact

Three-phase motors can be used where only single-phase power exists because motor drives are available to deliver a three-phase (typically 230 VAC) output up to about 2 HP when connected to a 120 or 230 VAC single-phase power source.

THREE-PHASE MOTORS

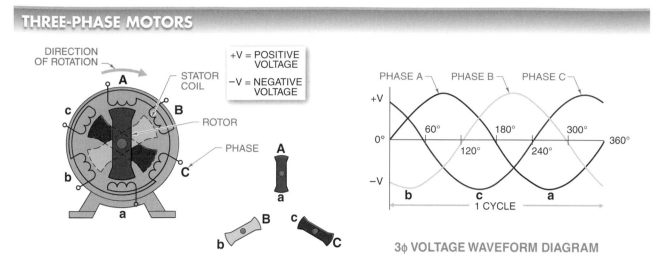

Figure 14-7. The coils in the stator of a 3ϕ motor are connected to form three separate windings (phases).

Single-Voltage, 3ϕ Motors. A *single-voltage motor* is a motor that operates at only one voltage level. Single-voltage motors are less expensive to manufacture than dual-voltage motors, but they are limited to locations having the same voltage as the motor. Common single-voltage, three-phase motor ratings are 230 V, 460 V, and 575 V. Other single-voltage, three-phase motor ratings are 200 V, 208 V, and 220 V. All three-phase motors are wired so that the phases are connected together in either a wye (Y) or delta (Δ) configuration.

In a single-voltage, wye-connected, 3ϕ motor, one end of each of the three phases is internally connected to the other phases. **See Figure 14-8.** The remaining end of each phase is brought out externally and connected to a power line. The leads that are brought out externally are labeled terminal one (T1), terminal two (T2), and terminal three (T3). When connected, terminals T1, T2, and T3 are matched to the 3ϕ power lines labeled line one (L1), line two (L2), and line three (L3). For the motor to operate properly, the 3ϕ lines supplying power to the wye motor must have the same voltage and frequency as the motor.

SINGLE-VOLTAGE, WYE-CONNECTED, 3ϕ MOTORS

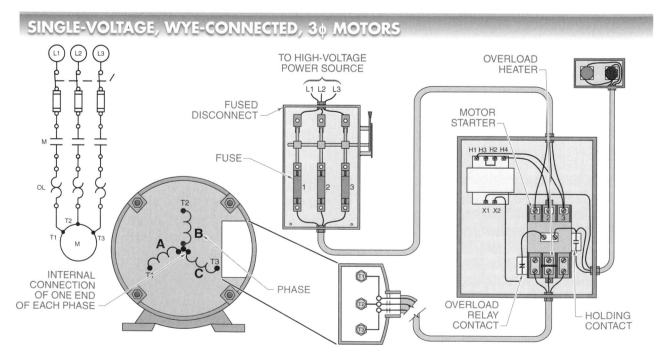

Figure 14-8. In a single-voltage, wye-connected, 3ϕ motor, one end of each phase is internally connected to the other phases.

In a single-voltage, delta-connected, 3φ motor, each winding is wired end-to-end to form a completely closed loop circuit. **See Figure 14-9.** At each point where the phases are connected, leads are brought out externally and labeled terminal one (T1), terminal two (T2), and terminal three (T3). These terminals, like those of a wye-connected motor, are attached to power lines one (L1), two (L2), and three (L3). The 3φ lines supplying power to the delta motor must have the same voltage and frequency rating as the motor.

Dual-Voltage, 3φ Motors. Most 3φ motors are made so that they may be connected for either of two voltages. Making motors for two voltages enables the same motor to be used with two different power line voltages. The normal dual-voltage rating of industrial motors is 230/460 V. The motor nameplate should be reviewed for proper voltage ratings.

The higher voltage is preferred when a choice between voltages is available. The motor uses the same amount of power and gives the same horsepower output for either high or low voltage, but as the voltage is doubled (230 V to 460 V), the current is cut in half. Using a reduced current enables the use of a smaller wire size, which reduces the cost of installation. Dual-voltage 3φ motors are wired so that the phases are connected in either a wye or delta configuration.

A wiring diagram is used to show the terminal numbering system for a dual-voltage, wye-connected, 3φ motor. **See Figure 14-10.** Nine leads are brought out of the motor. These leads are marked T1 through T9 and may be externally connected for either of the two voltages. The terminal connections for high and low voltage are normally provided on the motor nameplate.

The nine leads are connected in either series (high voltage) or parallel (low voltage). To connect a wye-connected motor for high voltage, L1 is connected to T1, L2 to T2, and L3 to T3; T4 is tied to T7, T5 to T8, and T6 to T9. This connects the individual coils in phases A, B, and C in series, each coil receiving 50% of the line-to-neutral point voltage. The neutral point equals the internal connecting point of all three phases.

To connect a wye-connected motor for low voltage, L1 is connected to T1 and T7, L2 to T2 and T8, and L3 to T3 and T9; T4 is tied to T5 and T6. This connects the individual coils in phases A, B, and C in parallel so that each coil receives 100% of the line-to-neutral point voltage.

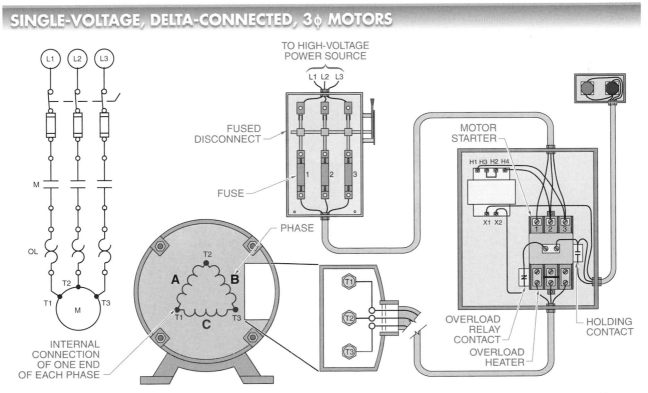

Figure 14-9. In a single-voltage, delta-connected, 3φ motor, each phase is wired end-to-end to form a completely closed loop.

DUAL-VOLTAGE, WYE-CONNECTED, 3ϕ MOTORS

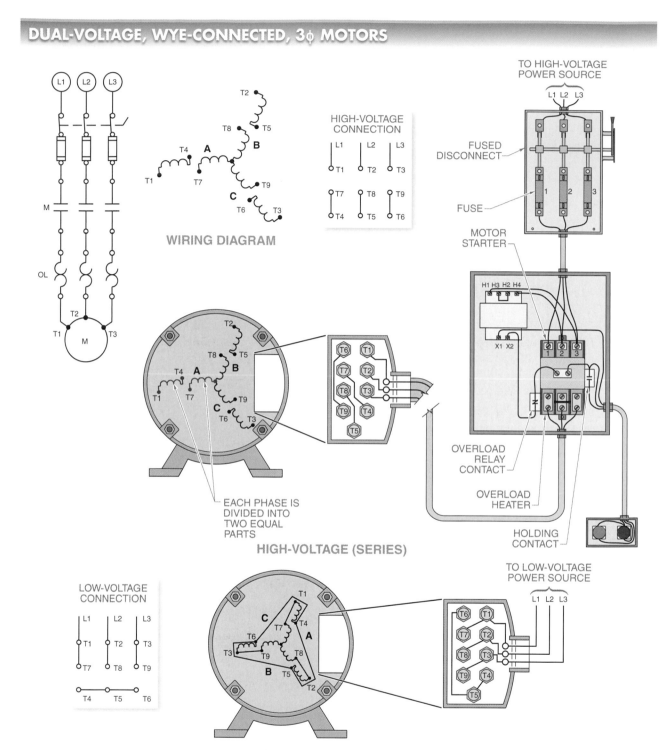

Figure 14-10. In a dual-voltage, wye-connected, 3ϕ motor, each phase coil is divided into two equal parts, and a wiring diagram is used to show the terminal numbering system.

A wiring diagram is used to show the terminal numbering system for a dual-voltage, delta-connected, 3ϕ motor. **See Figure 14-11.** The leads are marked T1 through T9 and a terminal connection chart is provided for wiring high- and low-voltage operations.

DUAL-VOLTAGE, DELTA-CONNECTED, 3φ MOTORS

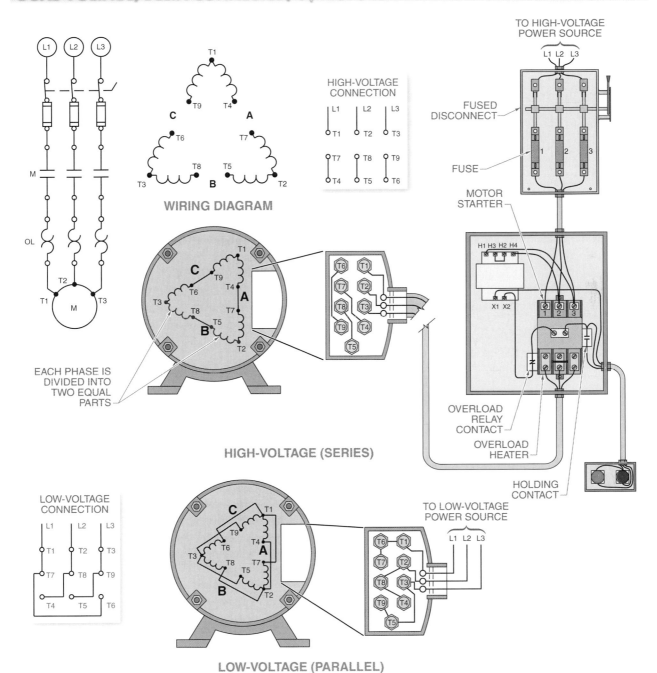

Figure 14-11. In a dual-voltage, delta-connected, 3φ motor, each phase coil is divided into two equal parts, and a wiring diagram is used to show the terminal numbering system.

The nine leads are connected in either series or parallel for high or low voltage. In the high-voltage configuration, the coils are wired in series. In the low-voltage configuration, the coils are wired in parallel to distribute the voltage to the individual coil ratings.

Tech Fact

NEMA motor leads are T1, T2, and T3 where the power lines are connected. IEC motor leads are labeled U, V, and W where the power lines are connected. Likewise, NEMA power lines are labeled L1, L2, and L3 and IEC power lines are labeled R, S, and T.

14-1 CHECKPOINT

1. What is the stationary coil of an AC motor called?
2. What is the rotating coil of an AC motor called?
3. Can a shaded-pole motor be reversed?
4. What is the function of the capacitor in 1ϕ motors?

5. When connecting a dual-voltage, 3ϕ motor for high voltage, are the windings connected in series or parallel?
6. When connecting a dual-voltage, 3ϕ motor for low voltage, are the windings connected in series or parallel?

14-2 AC MOTOR MAINTENANCE

Electric motors are very dependable machines. An electric motor gives good service under all operating conditions for which it is designed. For the safest service possible, the information given on the motor nameplate should be checked before putting a motor into operation. **See Figure 14-12.** The nameplate should be checked to ensure that the proper voltage and current are being used.

motor should be unplugged immediately. This prevents the windings from becoming seriously overheated. To prevent an ordinary motor from becoming overheated, the air openings on its frame should be kept clear at all times. When oiling a motor, oil should be applied to the bearings only. Excessive oil used on a motor damages the winding insulation and causes the motor to collect an excessive amount of dirt and dust.

MOTOR NAMEPLATE INFORMATION

MOTOR FLC RATING

MODEL	425	TYPE		PHASE	3
H.P.	7.5	HERTZ (CYCLES)	60	AMPS	17.0
VOLTS	460		RPM	1725	
TEMP. RISE	40° C		GEI	1058	
DUTY RATING	CONTINUOUS		SERIAL #	211576	
CODE		S.F.	1.15	FR	56

MOTOR NAMEPLATE   SERVICE FACTOR

Figure 14-12. Motor nameplate information should be checked to ensure that the proper voltage and current are being used.

A standard motor should not be operated in very damp locations or where water may enter the motor frame. Motors are designed for use in specific locations. Typical enclosures available are open motor enclosures and totally enclosed motor enclosures. **See Figure 14-13.** When replacing a motor, it is important to ensure that the motor enclosure meets the proper specifications.

The frame of a motor should be grounded, especially if the motor is used in a damp location. If a motor shaft does not rotate after the switch has been turned on, the

MOTOR ENCLOSURES

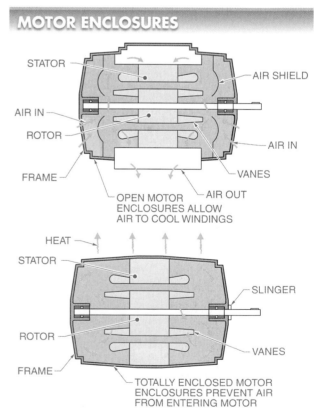

Figure 14-13. An open motor enclosure allows air to flow through the motor to cool the windings. A totally enclosed motor enclosure prevents air from entering the motor.

Allowable Motor Starting Time

A motor must accelerate to its rated speed within a limited time period. **See Figure 14-14.** The longer a motor takes to accelerate, the higher the temperature rise in the motor. The larger the load that is placed on the motor, the longer the acceleration time. The maximum recommended acceleration time depends on the motor frame size. Large motor frames dissipate heat faster than small motor frames.

MAXIMUM ACCELERATION TIME

Frame Number	Maximum Acceleration Time (in sec)
48 and 56	8
143 to 286	10
324 to 326	12
364 to 505	15

Figure 14-14. A motor must accelerate to its rated speed within a limited time period.

Overcycling

Overcycling is the process of turning a motor on and off repeatedly. Motor starting current is usually six to eight times the full-load running current of a motor. Most motors are not designed to start more than 10 times per hour. Overcycling occurs when a motor is at its operating temperature and still cycles on and off. This further increases the temperature of the motor, destroying the motor insulation. **See Figure 14-15.**

Open, dripproof motors rated for continuous duty are the best choice for applications that include overcycling. When a motor application requires a motor to be cycled often, the following steps should be taken:

* Use a motor with a 50°C rise instead of the standard 40°C.
* Use a motor with a 1.25 or 1.35 service factor instead of a 1.00 or 1.15 service factor.
* Provide additional cooling by forcing air over the motor.

Heat Problems

Excessive heat is a major cause of motor failure and other motor problems. Heat destroys motor insulation, which short-circuits the windings. The motor is not functional when motor insulation is destroyed.

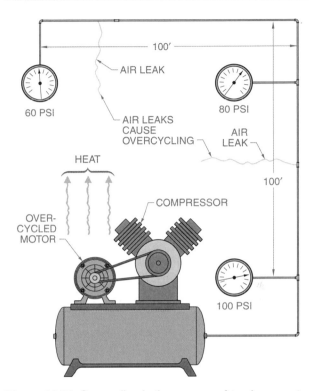

Figure 14-15. Overcycling is the process of turning a motor on and off repeatedly, increasing the temperature of the motor and destroying the motor insulation.

The life of motor insulation is shortened as the heat in a motor increases beyond the temperature rating of the insulation. The higher the temperature, the sooner the insulation fails. The temperature rating of motor insulation is listed as the insulation class. **See Figure 14-16.**

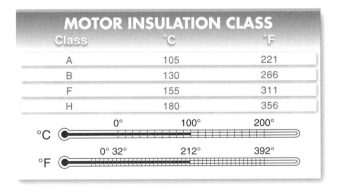

MOTOR INSULATION CLASS

Class	°C	°F
A	105	221
B	130	266
F	155	311
H	180	356

Figure 14-16. The temperature rating of motor insulation is listed as the insulation class.

The insulation class is given in Celsius (°C) and/or Fahrenheit (°F). A motor nameplate normally lists the insulation class of the motor. Heat buildup in a motor can be caused by the following conditions:
- incorrect motor type or size for the application
- improper cooling (normally from dirt buildup)
- excessive load (normally from improper use)
- excessive friction (normally from misalignment or vibration)
- electrical problem (normally voltage unbalance, phase loss, or a voltage surge)

All motors produce heat as they convert electrical energy to mechanical energy. This heat must be removed to prevent destruction of motor insulation. Motors are designed with air passages that permit a free flow of air over and through the motor. Airflow removes heat from a motor. Anything that restricts airflow through a motor causes the motor to operate at a higher than designed temperature. Airflow through a motor may be restricted by the accumulation of dirt, dust, lint, grass, pests, rust, etc. Airflow is restricted much faster if a motor becomes coated with oil from leaking seals or from overlubrication. **See Figure 14-17.**

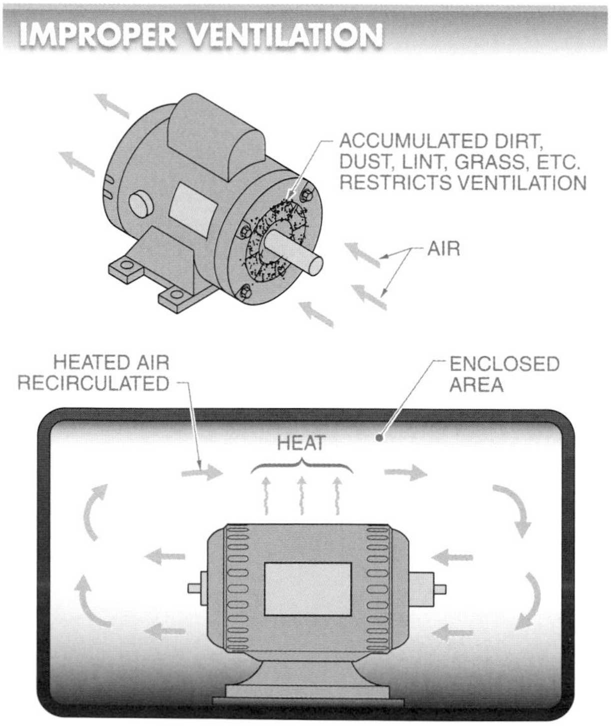

IMPROPER VENTILATION

ACCUMULATED DIRT, DUST, LINT, GRASS, ETC. RESTRICTS VENTILATION

AIR

HEATED AIR RECIRCULATED

ENCLOSED AREA

HEAT

Figure 14-17. Anything that restricts airflow through a motor causes the motor to operate at a higher than designed temperature.

Overheating can also occur if a motor is placed in an enclosed area. A motor overheats due to the recirculation of heated air when a motor is installed in a location that does not permit the heated air to escape. Vents added at the top and bottom of the enclosed area allow a natural flow of heated air.

An *overload* is the application of excessive load to a motor. Motors attempt to drive the connected load when the power is ON. The larger the load, the more power required. All motors have a limit to the load they can drive. For example, a 5 HP, 460 V, 3φ motor should draw no more than 7.6 A. Further information is available in NEC® Table 430.150.

Overloads should not harm a properly protected motor. Any overload present longer than the delay time built into the protection device is detected and removed. Properly sized heaters in the motor starter ensure that an overload is removed before any damage is done.

An electrician can observe the even blackening of all motor windings that occurs when a motor has failed due to overloading. **See Figure 14-18.** The even blackening is caused by the slow destruction of the motor over a long period of time. No obvious damage or isolated areas of damage to the insulation are visible.

Current readings are taken at a motor to determine an overload problem. **See Figure 14-19.** A motor is working to its maximum if it is drawing rated current. A motor is overloaded if it is drawing more than rated current. The motor size may be increased or the load on the motor decreased if overloads are a problem.

Tech Fact

Heat is produced evenly in the motor windings anytime the motor is operated. This heat should be dissipated through the frame of the motor. Phase imbalance, phase loss, dirt, and poor ventilation around or through the motor can cause uneven heat. Over heated motor insulation at any one point can cause total motor failure. Infrared (IR) or thermal imaging test tools can be used to check heat in motor windings as well as heat in motor bearings and couplings.

Altitude Correction

Temperature rise of motors is based on motor operation at altitudes of 3300′ or less. A motor with a service factor of 1.0 is derated when it operates at altitudes above 3300′. A motor with a service factor above 1.0 is derated based on the altitude and service factor. **See Figure 14-20.**

MOTOR OVERLOADING

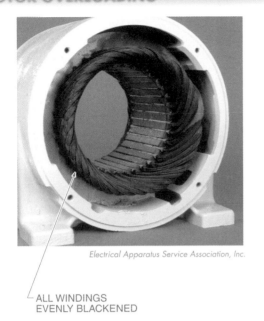

Electrical Apparatus Service Association, Inc.

⌐ ALL WINDINGS
  EVENLY BLACKENED

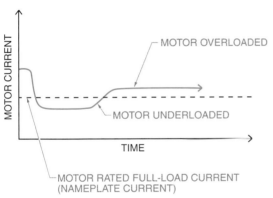

Figure 14-18. Overloading causes an even blackening of all motor windings.

MOTOR CURRENT READINGS

Rated Current of Motor	Meter Reading		
	Motor Underloaded	Motor Fully Loaded	Motor Overloaded
20 A	12 A	20 A	22 A

| NAMEPLATE LISTED VALUE | 0% TO 95% OF LISTED VALUE | 95% TO 105% OF LISTED VALUE | 105% + OF LISTED VALUE |

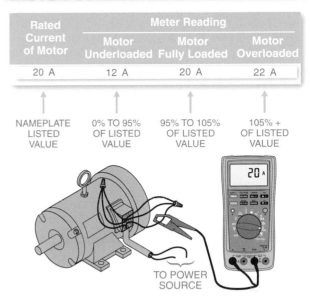

TO POWER SOURCE

Figure 14-19. Current readings are taken at a motor to determine an overload problem.

MOTOR ALTITUDE DERATINGS

Altitude Range (in ft)	Service Factor			
	1.0	1.15	1.25	1.35
3300 to 9000	93%	100%	100%	100%
9000 to 9900	91%	98%	100%	100%
9900 to 13,200	86%	92%	98%	100%
13,200 to 16,500	79%	85%	91%	94%
Over 16,500	Consult manufacturer			

Figure 14-20. A motor with a service factor of 1.0 is derated when it operates at altitudes above 3300′.

14-2 CHECKPOINT

1. What does the maximum rated acceleration time of a motor depend on?
2. When a motor application requires a motor to be cycled often, what two rated specifications on the motor nameplate should be increased from the normal rating?
3. Which motor insulation class has the highest temperature rating?
4. How is a motor rating affected when operating at altitudes above 3300′?

14-3 TROUBLESHOOTING AC MOTORS

Most problems with 1ϕ AC motors involve the centrifugal switch, thermal switch, or capacitor(s). A motor is usually serviced and repaired if the problem is in the centrifugal switch, thermal switch, or capacitor(s). However, if the motor is less than ⅛ HP, it is almost always replaced. Three-phase motors usually operate for many years without any problems because they have fewer components that may malfunction than other motors.

Troubleshooting Shaded-Pole Motors

Shaded-pole motors that fail are usually replaced. However, the reason for the motor failure should be investigated. Simply replacing a motor because it has failed due to a jammed load does not solve the problem. **See Figure 14-21.** To troubleshoot a shaded-pole motor, the following procedure is applied:

1. Visually inspect the motor after turning power to motor off. Replace the motor if it is burned, if the shaft is jammed, or if there is any sign of damage.
2. Check the stator winding. The stator winding is the only electrical circuit that may be tested without taking the motor apart. Measure the resistance of the stator winding. Set a DMM to the lowest resistance scale for taking the reading. The winding is open if the DMM indicates an infinity reading. Replace the motor. The winding is short-circuited if the DMM indicates a zero reading. Replace the motor. The winding may still be good if the DMM indicates a low resistance reading. Check the winding with a megohmmeter before replacing the motor.

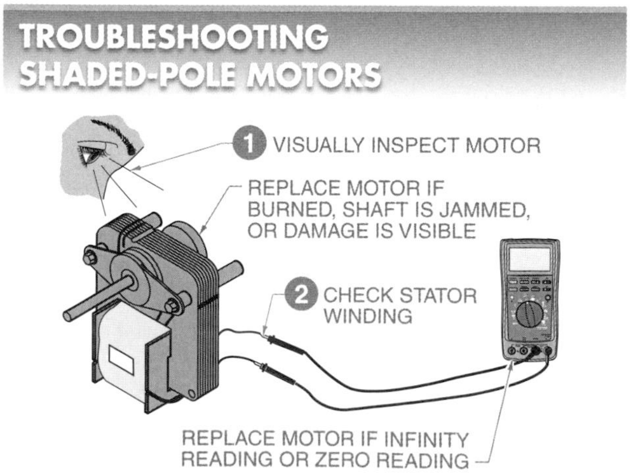

TROUBLESHOOTING SHADED-POLE MOTORS

1 VISUALLY INSPECT MOTOR

REPLACE MOTOR IF BURNED, SHAFT IS JAMMED, OR DAMAGE IS VISIBLE

2 CHECK STATOR WINDING

REPLACE MOTOR IF INFINITY READING OR ZERO READING

Figure 14-21. Shaded-pole motors that fail are usually replaced; however, the reason for motor failure should be investigated.

Troubleshooting Split-Phase Motors

Some split-phase motors include a thermal switch that automatically turns the motor off when it overheats. Thermal switches may have a manual reset or an automatic reset. Caution should be taken with any motor that has an automatic reset because the motor may automatically restart at any time. **See Figure 14-22.** To troubleshoot a split-phase motor, the following procedure is applied:

1. Visually inspect the motor after turning the power to the motor off. Replace the motor if it is burned, if the shaft is jammed, or if there is any sign of damage.
2. Check to determine if the motor is controlled by a thermal switch. Reset the thermal switch and turn the motor on if the thermal switch is manual.
3. Check for voltage at the motor terminals using a DMM set to measure voltage if the motor does not start. The voltage should be within 10% of the motor listed voltage. Troubleshoot the circuit leading to the motor if the voltage is not correct. If the voltage is correct, turn the power to the motor off so the motor may be tested.
4. Turn off and lock out and tag out the starting mechanism per company policy.
5. With power turned off, connect a DMM set to measure resistance to the same motor terminals from which the incoming power leads were disconnected. The DMM reads the resistance of the starting and running windings. Their combined resistance is less than the resistance of either winding alone because the windings are connected in parallel. A short circuit is present if the DMM reads zero. An open circuit is present if the DMM reads infinity. Replace the motor. *Note:* Split-phase motors are normally too small for a repair to be cost efficient.
6. Check the centrifugal switch for signs of burning or broken springs. Service or replace the switch if any obvious signs of problems are present. Check the switch using a DMM set to measure resistance if no obvious signs of problems are present. Manually operate the centrifugal switch. The endbell on the switch side may have to be removed. The resistance on the DMM decreases if the motor is good. A problem exists if the resistance does not change.

TROUBLESHOOTING SPLIT-PHASE MOTORS

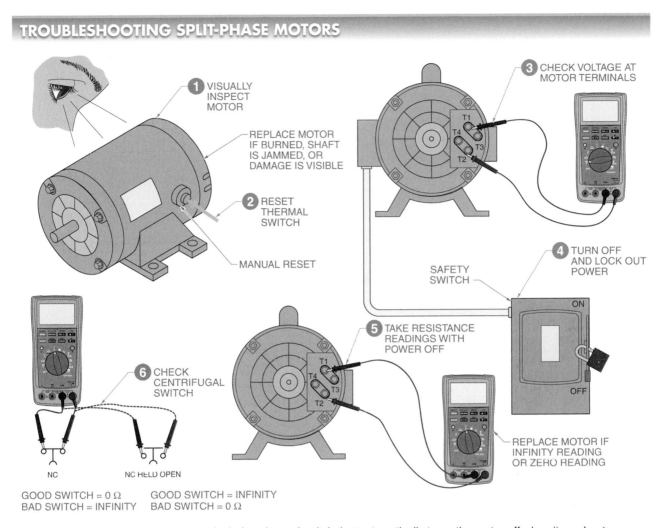

1 VISUALLY INSPECT MOTOR

REPLACE MOTOR IF BURNED, SHAFT IS JAMMED, OR DAMAGE IS VISIBLE

2 RESET THERMAL SWITCH

MANUAL RESET

3 CHECK VOLTAGE AT MOTOR TERMINALS

4 TURN OFF AND LOCK OUT POWER

SAFETY SWITCH

ON

OFF

5 TAKE RESISTANCE READINGS WITH POWER OFF

REPLACE MOTOR IF INFINITY READING OR ZERO READING

6 CHECK CENTRIFUGAL SWITCH

NC   NC HELD OPEN

GOOD SWITCH = 0 Ω   GOOD SWITCH = INFINITY
BAD SWITCH = INFINITY   BAD SWITCH = 0 Ω

Figure 14-22. Some split-phase motors include a thermal switch that automatically turns the motor off when it overheats.

Troubleshooting Capacitor Motors

Troubleshooting capacitor motors is similar to troubleshooting split-phase motors. The only additional device to be tested is the capacitor. Capacitors have a limited life and are often the problem in capacitor motors. Capacitors may have a short circuit or an open circuit, or they may deteriorate to the point where they must be replaced.

Deterioration may also change the value of a capacitor, which causes additional problems. When a capacitor short-circuits, the winding in the motor may be damaged if the OCPD does not quickly open the circuit. When a capacitor opens, there is no current flow in the starting winding and no phase shift. Poor starting torque may prevent the motor from starting, which usually trips the overloads.

All capacitors are made with two conducting surfaces separated by dielectric material. *Dielectric material* is a medium in which an electric field is maintained with

little or no outside energy supply. Dielectric material is used to insulate the conducting surfaces of a capacitor. Capacitors are either oil or electrolytic. Oil capacitors are filled with oil and sealed in a metal container. The oil serves as the dielectric material.

More motors use electrolytic capacitors than oil capacitors. Electrolytic capacitors are formed by winding two sheets of aluminum foil separated by pieces of thin paper impregnated with an electrolyte. An *electrolyte* is a conducting medium in which the current flow occurs by ion migration. The electrolyte is used as the dielectric material. The aluminum foil and electrolyte are encased in a cardboard or aluminum cover. A vent hole is provided to prevent a possible explosion in the event the capacitor is shorted or overheated. AC capacitors are used with capacitor motors. Capacitors that are designed to be connected to AC have no polarity. **See Figure 14-23.**

TROUBLESHOOTING CAPACITOR MOTORS

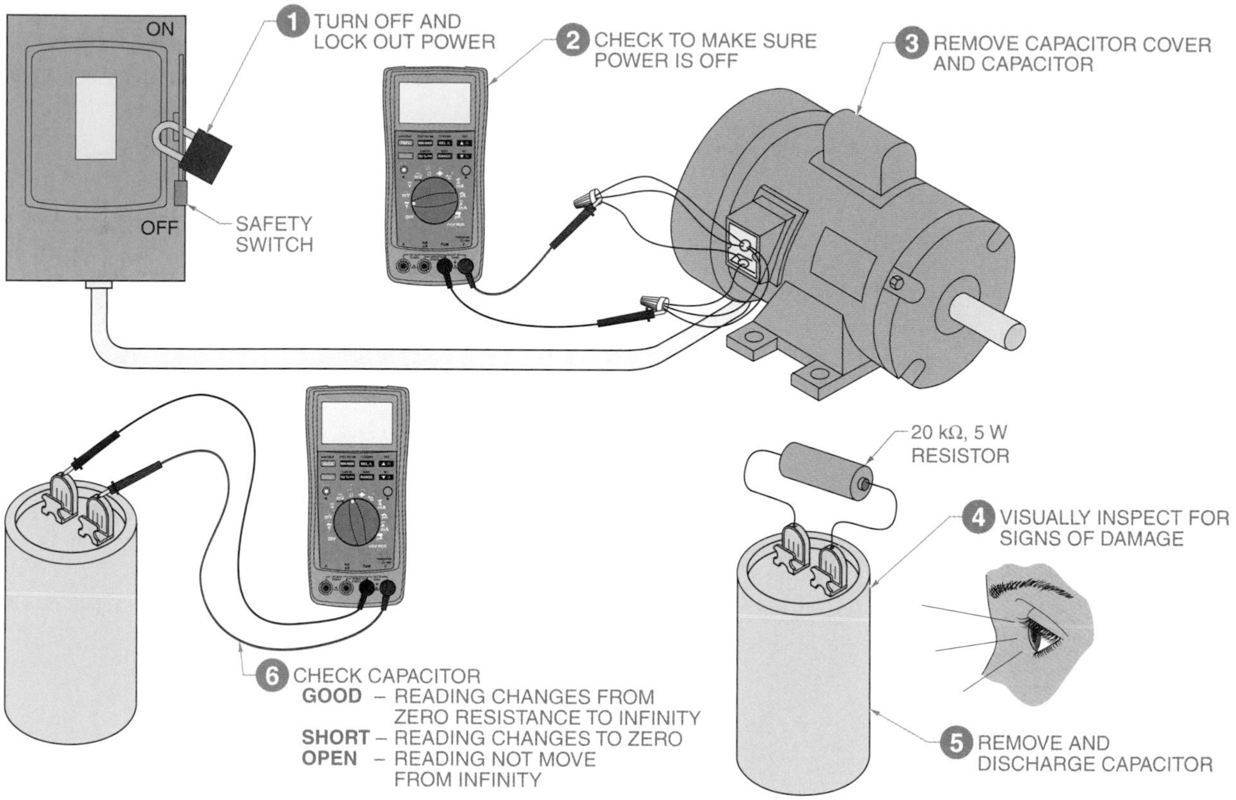

Figure 14-23. Capacitors have a limited life and are often the problem in capacitor motors.

To troubleshoot a capacitor motor, the following procedure is applied:

1. Turn the handle of the safety switch or combination starter to the OFF position. Lock out and tag out the starting mechanism per company policy.

2. Use a DMM set to measure voltage at the motor terminals to ensure the power is OFF.

3. Remove the cover of the capacitor. Capacitors are located on the outside frame of a motor. **CAUTION:** A good capacitor will hold a charge even when power is removed.

4. Visually inspect the capacitor for leakage, cracks, or bulges. If these are present, replace the capacitor.

5. Remove the capacitor from the circuit and discharge it. To safely discharge a capacitor, place a 20,000 Ω, 5 W resistor across the terminals for 5 sec or place a voltmeter across the terminals and watch it until the potential falls to 0. *Note:* Use the analog scale on a DMM or use a capacitor tester (preferred) when testing capacitors.

6. After a capacitor is discharged, connect the leads of a DMM set to measure resistance to the capacitor terminals. The DMM indicates the general condition of the capacitor. A capacitor is either good, shorted, or open.

• Good capacitor — The reading changes from zero resistance to infinity. When the reading reaches the halfway point, remove one of the leads and wait 30 sec. When the lead is reconnected, the reading should change back to the halfway point and continue to infinity. This demonstrates that the capacitor can hold a charge. The capacitor cannot hold a charge and must be replaced when the reading changes back to zero resistance.

• Short capacitor — The reading changes to zero and does not move. The capacitor is bad and must be replaced.

• Open capacitor — The reading does not change from infinity. The capacitor is bad and must be replaced.

Troubleshooting 3φ Motors

The extent of troubleshooting a 3φ motor depends on the motor application. Testing is normally limited to checking the voltage and current at the motor if it is used in an application that is critical to an operation or production.

The motor is assumed to be the problem if the voltage is present and correct. Unless it is very large, the motor is normally replaced at this time so production may continue. Further tests may be made to determine the exact problem if time is not a critical factor. **See Figure 14-24.**

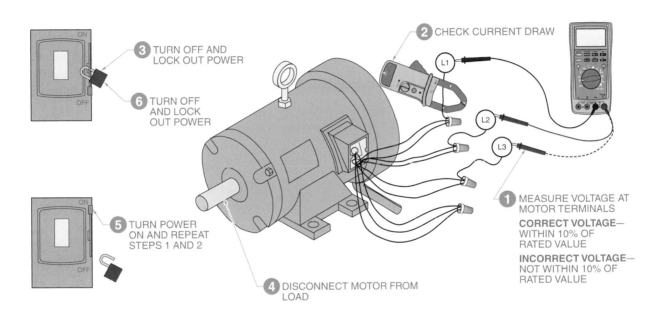

TROUBLESHOOTING 3φ MOTORS

3 TURN OFF AND LOCK OUT POWER

6 TURN OFF AND LOCK OUT POWER

5 TURN POWER ON AND REPEAT STEPS 1 AND 2

2 CHECK CURRENT DRAW

1 MEASURE VOLTAGE AT MOTOR TERMINALS

CORRECT VOLTAGE— WITHIN 10% OF RATED VALUE

INCORRECT VOLTAGE— NOT WITHIN 10% OF RATED VALUE

4 DISCONNECT MOTOR FROM LOAD

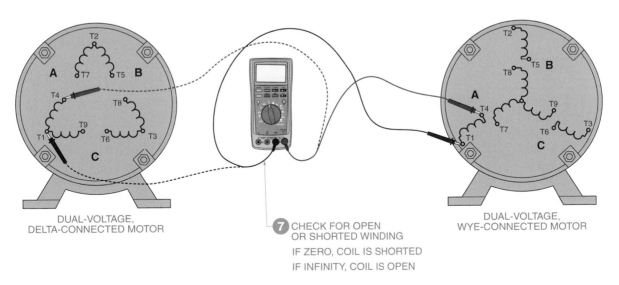

DUAL-VOLTAGE, DELTA-CONNECTED MOTOR

7 CHECK FOR OPEN OR SHORTED WINDING

IF ZERO, COIL IS SHORTED

IF INFINITY, COIL IS OPEN

DUAL-VOLTAGE, WYE-CONNECTED MOTOR

Figure 14-24. Testing of 3φ motors is normally limited to checking the voltage and current at the motor if it is used in an application that is critical to an operation or production.

To troubleshoot a 3φ motor, the following procedure is applied:

1. Measure the voltage at the motor terminals using a DMM set to measure voltage. The motor must be checked if the voltage is present and at the correct level on all three phases. The incoming power supply must be checked if the voltage is not present on all three phases.

2. Check the current draw on each motor lead using a clamp-on ammeter and compare the current to the motor nameplate current rating. If the voltage and current at the motor are not within nameplate ratings, additional tests are required.

3. Turn the handle of the safety switch or combination starter to the OFF position if voltage is present but the motor is not operating. Lock out and tag out the starting mechanism per company policy.

4. Disconnect the motor from the load to verify the load is not the problem.

5. Turn power on and repeat steps 1 and 2. If the current is within listed nameplate ratings, the motor load may be the problem.

6. Turn power off and lock out the power.

7. Check the motor windings with a DMM set to measure resistance for any opens or shorts. Take a resistance reading of the T1-T4 coil. The coil must have a resistance reading. The coil is shorted when the reading is zero. The coil is open when the reading is infinity. The resistance is low because the coil winding is made of wire only. However, there is resistance on a good coil winding. The larger the motor, the smaller the resistance reading.

After the resistance of one coil has been found, the basic laws of series and parallel circuits are applied. When measuring the resistance of two coils in series, the total resistance is twice the resistance of one coil. When measuring the resistance of two coils in parallel, the total resistance is one-half the resistance of one coil.

Tech Fact

Motor overloads should not harm a properly protected motor. However, when overload protection is improperly sized or applied, the motor may draw excessive current and overheat. When troubleshooting a motor, a digital multimeter set to measure current can be used to determine whether the motor is overloaded. Current readings should be below the nameplate rating for best motor performance.

Motor Short Circuits

A short circuit occurs any time current takes a shortcut around the normal path of current flow. In a motor, a short circuit can occur due to the following:

- The insulation on the motor winding breaks down due to overheating, which occurs when the motor winding must carry higher currents or the point of the short is the weakest point of the winding insulation.

- The insulation is nicked (or removed) because of a foreign object entering the motor housing (file shaving, etc.).

- There is a manufacturer fault that occurred when the insulation was placed on the winding and the fault only showed up after the motor was operated (subjected to vibration, heat, etc.).

If motor insulation breaks down between two windings, there is a phase-to-phase short circuit. If motor insulation breaks down between a winding and the ground wire, there is a phase-to-ground short circuit. **See Figure 14-25.**

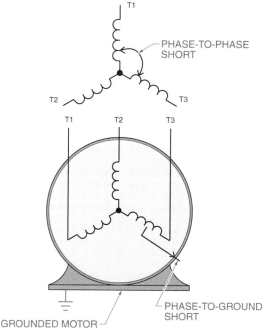

MOTOR SHORT CIRCUITS

Note: Current will not be the same on each line (T1, T2, and T3) when there is a phase-to-phase or phase-to-ground short.

Figure 14-25. A phase-to-phase short circuit occurs when motor insulation breaks down between two windings, while a phase-to-ground short circuit occurs when motor insulation breaks down between a winding and the ground wire.

Re-marking 3φ Induction Motor Connections

Three-phase induction motors are the most common motors used in industrial applications. Three-phase induction motors operate for many years with little or no required maintenance. It is not uncommon to find 3φ induction motors that have been in operation for 10 to 20 years in certain applications. The length of time a motor is in operation may cause the markings of the external leads to become defaced. This may also happen to a new or rebuilt motor that has been in the maintenance shop for some time. To ensure proper operation, each motor lead must be re-marked before troubleshooting and reconnecting the motor to a power source.

The two most common 3φ motors are the single-voltage, 3φ, three-lead motor and the dual-voltage, 3φ, nine-lead motor. Both may be internally connected in a wye or delta configuration.

The three leads of a single-voltage, 3φ, three-lead motor can be marked T1, T2, and T3 in any order. The motor can be connected to the rated voltage and allowed to run. Any two leads may be interchanged if the rotation is in the wrong direction. The industry standard is to interchange T1 and T3.

Determining Wye or Delta Connections

A standard dual-voltage motor has nine leads extending from it and may be internally connected with a wye or delta configuration. The internal connections must be determined when re-marking the motor leads. A DMM is used to measure resistance or a continuity tester is used to determine whether a dual-voltage motor is internally connected in a wye or delta configuration.

A dual-voltage, wye-connected motor has four separate circuits. A dual-voltage, delta-connected motor has three separate circuits. See Figure 14-26. A wye-connected motor has three circuits of two leads each (T1-T4, T2-T5, and T3-T6) and one circuit of three leads (T7-T8-T9). A delta-connected motor has three circuits of three leads each (T1-T4-T9, T2-T5-T7, and T3-T6-T8).

A DMM is used to determine the winding circuits (T1-T4, T2-T5, etc.) on an unmarked motor by connecting one meter lead to any motor lead and temporarily connecting the other meter lead to each remaining motor lead. See Figure 14-27. *Note:* Ensure that the motor is disconnected from the power supply. A resistance reading other than infinity indicates a complete circuit.

A continuity tester may also be used to determine the winding circuits on an unmarked motor by connecting one test lead to any motor lead and temporarily connecting the other test lead to each remaining motor lead. See Figure 14-28.

WYE AND DELTA CONNECTIONS

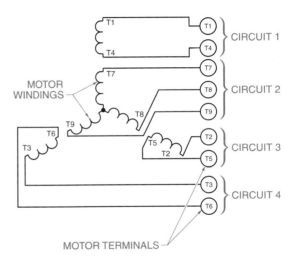

WYE-CONNECTED MOTOR

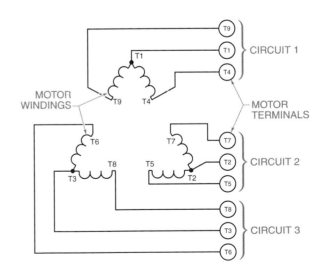

DELTA-CONNECTED MOTOR

Figure 14-26. The internal connections of a delta- or wye-connected motor must be determined when re-marking the motor leads.

DETERMINING WINDING CIRCUITS—DMMs

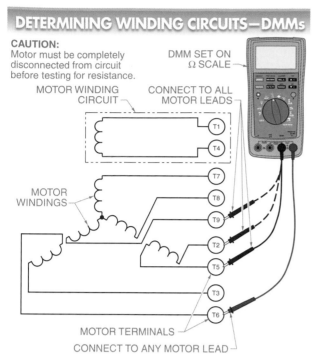

CAUTION:
Motor must be completely disconnected from circuit before testing for resistance.

DMM SET ON Ω SCALE

MOTOR WINDING CIRCUIT

CONNECT TO ALL MOTOR LEADS

T1
T4
T7
MOTOR WINDINGS
T8
T9
T2
T5
T3
T6

MOTOR TERMINALS

CONNECT TO ANY MOTOR LEAD

Figure 14-27. A DMM is used to determine the winding circuits on an unmarked motor by connecting one meter lead to any motor lead and temporarily connecting the other meter lead to each of the remaining motor leads.

DETERMINING WINDING CIRCUITS—CONTINUITY TESTERS

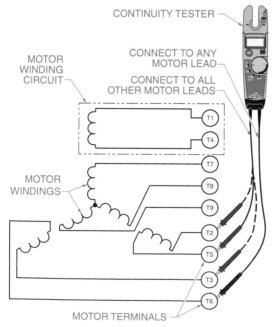

CONTINUITY TESTER

MOTOR WINDING CIRCUIT

CONNECT TO ANY MOTOR LEAD
CONNECT TO ALL OTHER MOTOR LEADS

T1
T4
T7
MOTOR WINDINGS
T8
T9
T2
T5
T3
T6

MOTOR TERMINALS

Figure 14-28. A continuity tester is used to determine a winding circuit on an unmarked motor by connecting one test lead to any motor lead and temporarily connecting the other test lead to each of the remaining motor leads.

The continuity tester indicates a complete circuit by an audible beep. Each connection that indicates a complete circuit is marked by taping or pairing the leads together. All pairs of leads are checked with all the remaining motor leads to determine if the circuit is a two- or three-lead circuit. The motor is a wye-connected motor if three circuits of two leads and one circuit of three leads are found. The motor is a delta-connected motor if three circuits of three leads are found.

Re-marking Dual-Voltage, Wye-Connected Motors

To re-mark a dual-voltage, wye-connected motor with no power or load conductors connected, the following procedure is applied:

1. Determine the winding circuits using a DMM or continuity tester. **See Figure 14-29.**

2. Mark the leads of the one three-lead circuit T7, T8, and T9 in any order. Separate the other motor leads into pairs, making sure none of the wires touch.

3. While wearing proper PPE, connect the motor to the correct supply voltage. Connect T7 to L1, T8 to L2, and T9 to L3. The correct supply voltage is the lowest voltage rating of the dual-voltage rating given on the motor nameplate. The low voltage is normally 220 V because the standard dual-voltage motor operates on 220/440 V. For any other voltage, all test voltages should be changed in proportion to the motor rating.

4. Turn on the supply voltage and let the motor run. The motor should run with no apparent noise or problems. Although the motor is running on only half of the windings (T7, T8, and T9), the motor should operate since it is not connected to any load. The starting voltage should be reduced through a reduced-voltage starter if the motor is too large to be started by connecting it directly to the supply voltage.

5. While wearing proper PPE, measure the voltage across each of the three open circuits while the motor is running, using a DMM set on at least the 440 VAC measuring scale. Care must be taken when measuring the high voltage of a running motor. Insulated test leads must be used. Connect only one test lead at a time. The voltage measured should be about 127 V or slightly less and should be the same on all three circuits.

The voltage is read on all circuits even though the two-wire circuits are not connected to the power lines because the voltage applied to the three-lead circuit induces a voltage in the two-wire circuits.

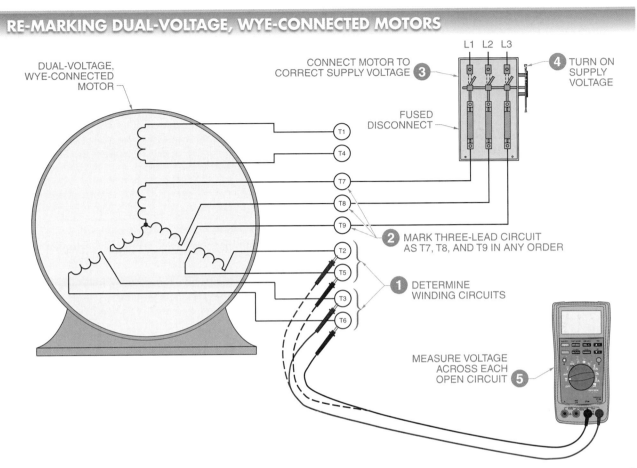

RE-MARKING DUAL-VOLTAGE, WYE-CONNECTED MOTORS

DUAL-VOLTAGE,
WYE-CONNECTED
MOTOR

CONNECT MOTOR TO
CORRECT SUPPLY VOLTAGE **3**

L1 L2 L3

4 TURN ON
SUPPLY
VOLTAGE

FUSED
DISCONNECT

T1
T4
T7
T8
T9

2 MARK THREE-LEAD CIRCUIT
AS T7, T8, AND T9 IN ANY ORDER

T2
T5
T3
T6

1 DETERMINE
WINDING CIRCUITS

MEASURE VOLTAGE
ACROSS EACH
OPEN CIRCUIT **5**

Figure 14-29. When re-marking a dual-voltage, wye-connected motor, the three-lead circuit is connected to the correct supply voltage and the voltage across each of the three open circuits is measured.

This occurs because the operating coils (T7, T8, and T9) are acting as a transformer primary inducing a voltage into the T1-T4, T2-T5, and T3-T6 windings, which are now the transformer secondary windings. Because the motor is acting like a transformer, voltage measurements are taken to determine just which motor windings line up with each other. For example, motor winding T1-T4 needs to line up with motor winding T7.

To determine the remaining T1-T4, T2-T5, and T3-T6 motor windings, apply the following procedure:

1. While wearing proper PPE, connect one lead of any two-wire circuit to T7 and connect the other lead of the circuit to one side of a DMM. Temporarily mark the lead connected to T7 as T4 and the lead connected to the DMM as T1. **See Figure 14-30.**

2. Connect the other lead of the DMM to T8 and then to T9. Mark T1 and T4 permanently if the two voltages are the same and are approximately 335 V.

3. Perform the same procedure on another two-wire circuit if the voltages are unequal. Mark the new terminals T1 and T4 if the new circuit gives the correct voltage (335 V). T1, T7, and T4 are found by this first test.

4. Connect one lead of the two remaining unmarked two-wire circuits to T8 and the other lead to one side of the DMM. Temporarily mark the lead connected to T8 as T5 and the lead connected to the DMM as T2.

5. Connect the other side of the DMM to T7 and T9 and measure the voltage. Measurements and changes should be made until a position is found at which both voltages are the same and approximately 335 V. T2, T5, and T8 are found by this second test.

6. Check the third circuit in the same way until a position is found at which both voltages are the same and approximately 335 V. T3, T6, and T9 are found by this third test.

7. After each motor lead is found and marked, turn off the motor and connect L1 to T1 and T7, L2 to T2 and T8, and L3 to T3 and T9. Connect T4, T5, and T6 together.

8. Start the motor and let it run. Check the current on each power line with a clamp-on ammeter. The markings are correct and may be marked permanently if the current is approximately equal on all of the three power lines. The measured current should be less than the motor's rated nameplate current since the motor is not connected to any load.

9. Take voltage readings between each line (T1-T2, T2-T3, and T1-T3). The voltage should be equal between each line.

AC motors are commonly used in manufacturing facilities to drive conveyor belts.

Re-marking Dual-Voltage, Delta-Connected Motors

A dual-voltage, delta-connected motor has nine leads grouped into three separate circuits. Each circuit has three motor leads connected, which makes the circuits T1-T4-T9, T2-T5-T7, and T3-T6-T8. **See Figure 14-31.** To re-mark a dual-voltage, delta-connected motor with no load, the following procedure is applied:

1. Determine the winding circuits using a DMM or continuity tester.

2. Measure the resistance of each circuit to find the center terminal. The resistance from the center terminal to the other two terminals is one-half the resistance between the other two terminals. Separate the three circuits and mark the center terminal for each circuit as T1, T2, and T3. Temporarily mark the two leads in the T1 group as T4 and T9, the two leads in the T2 group as T5 and T7, and the two leads in the T3 group as T6 and T8. Disconnect the DMM.

3. Connect the group marked T1, T4, and T9 to L1, L2, and L3 of a 220 V power supply. This should be the low-voltage rating on the nameplate of the motor. The other six leads should be left disconnected and must not touch because a voltage is induced in these leads even though these leads are not connected to power.

4. Turn the motor on and let it run with the power applied to T1, T4, and T9.

DUAL-VOLTAGE, WYE-CONNECTED MOTOR WIRING DIAGRAMS

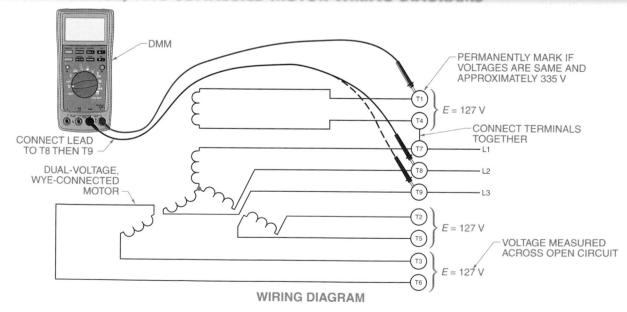

Figure 14-30. A wiring diagram is drawn when re-marking a dual-voltage, wye-connected motor to clarify the internal winding circuits.

5. Connect T4 (which is also connected to L2) to T7 and measure the voltage between T1 and T2. Set the DMM on at least a 460 VAC range. Use insulated test leads and connect one meter lead at a time. The lead markings for T4 and T9, and T7 and T5, are correct if the measured voltage is approximately 440 V. Interchange T5 with T7 or T4 with T9 if the measured voltage is approximately 380 V. Interchange both T5 with T7, and T4 with T9 if the new measured voltage is approximately 220 V. T4, T9, T7, and T5 may be permanently marked if the voltage is approximately 440 V.

To correctly identify T6 and T8, connect T6 and T8 and measure the voltage from T1 and T3. The measured voltage should be approximately 440 V. Interchange leads T6 and T8 if the voltage does not equal 440 V. T6 and T8 may be permanently marked if the voltage is approximately 440 V.

Turn off the motor and reconnect the motor to a second set of motor leads. Connect L1 to T2, L2 to T5, and L3 to T7. Restart the motor and observe the direction of rotation. The motor should rotate in the same direction as with the previous connection. Turn off the motor and reconnect the motor to the third set of motor leads (L1 to T3, L2 to T6, and L3 to T8) after the motor has run and the direction is determined.

Restart the motor and observe the direction of rotation. The motor should rotate in the same direction as the first two connections. Start over carefully, re-marking each lead if the motor does not rotate in the same direction for any set of leads.

Turn off the motor and reconnect the motor for the low-voltage connection. Connect L1 to T1-T6-T7, L2 to T2-T4-T8, and L3 to T3-T5-T9. Restart the motor and take current readings on L1, L2, and L3 with a clamp-on ammeter. The markings are correct if the motor current is approximately equal on each line.

RE-MARKING DUAL-VOLTAGE, DELTA-CONNECTED MOTORS

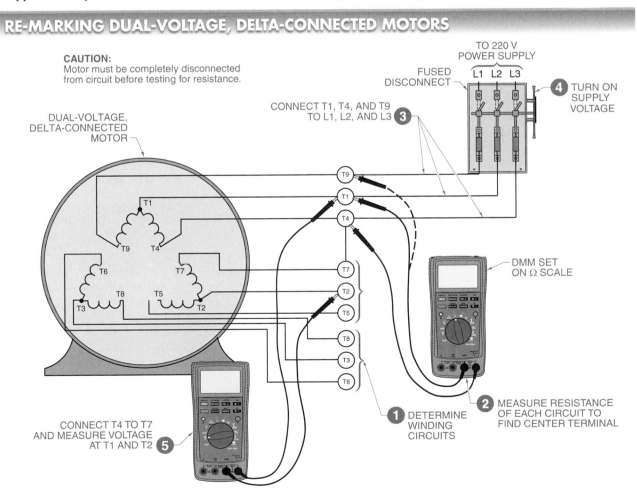

Figure 14-31. A dual-voltage, delta-connected motor has nine leads grouped into three separate circuits.

14-3 CHECKPOINT

1. What is the most likely part of a split-phase motor to fail over time with a properly sized and installed motor?

2. Does a voltage measurement at the motor indicate whether the motor is under loaded, fully loaded, or overloaded?

3. How is a motor tested to determine the load of a motor?

4. If T2 and T5 measure 20 Ω when testing the windings of a 3ϕ, delta-connected, dual-voltage motor with an ohmmeter, what should the reading be between T5 and T7?

Additional Resources

Review and Resources

Access Chapter 14 Review and Resources for *Electrical Motor Controls for Integrated Systems* by scanning the above QR code with your mobile device.

Applying Your Knowledge

Refer to the *Electrical Motor Controls for Integrated Systems* Learner Resources for interactive Applying Your Knowledge questions.

Workbook and Applications Manual

Refer to Chapter 14 in the *Electrical Motor Controls for Integrated Systems Workbook* and the *Applications Manual* for additional exercises.

ENERGY EFFICIENCY PRACTICES

Energy-Efficient Motor Facts

Although it is true that an energy-efficient motor uses less electrical energy than a standard motor of the same horsepower, it is not always true that the energy-efficient motor will always save money. Several facts must be considered when determining whether an energy-efficient motor can save operating cost and save energy.

- An energy-efficient motor costs about 20% more than a standard motor. The time it takes to recover the extra cost depends upon the operating time of the motor, the size of the motor load, and the cost of electrical energy.

- Energy-efficient motors are designed to recover the additional cost in about two years. However, this actually depends upon how often the motor operates. The more time the motor operates, the shorter the payback time. Manufacturers typically list a two year payback time, but this is based on the motor operating an average of 16 hours (two shifts) every day. Thus, if a motor does not operate often, there may be little or no payback. However, if a motor operates for long periods of time, the extra initial cost may be paid back over time.

- A motor that is properly sized to the driven load can save more energy than an energy-efficient motor. Motors are most efficient in converting electrical energy to mechanical energy when they are loaded to 70% to 100% of their rating. An oversized energy-efficient motor that is loaded less than 70% will waste more energy than a properly sized standard motor because the energy-efficient motor will draw more power than actually needed.

- The cost of energy is a major factor in determining the total operating cost of an energy-efficient motor over a long period of time. The higher the cost of energy, the faster the payback time for the initial purchase and installation costs as well as the lower the operating cost per year. High energy costs can reduce the initial payback period by months and save monthly operating costs over the entire life of the motor.

- Increasing the efficiency of a motor is only one part in increasing the efficiency of an entire electrical system. For the highest system efficiency, all parts must be considered. The incoming power lines should be about 98% efficient, the motor drive should be about 96% efficient, the motor should be about 84% (small motors) to about 90% (large motors) efficient, and the motor drive mechanical components (gears, chains, etc.) and load should be about 75% efficient.

Objectives

15-1
- Describe a manual starter and explain how a mechanical interlock works.
- Explain how to reverse three-phase (3φ) motors using manual starters.
- Explain how to reverse single-phase (1φ) motors using manual starters.
- Explain how to reverse DC motors using manual starters.

15-2
- Define drum switch and explain how they are used to reverse motors.

15-3
- Explain the difference between a magnetic reversing starter and a manual reversing starter.
- Explain how auxiliary contact interlocking works.
- Explain how pushbutton interlocking works.
- Describe a power circuit and a control circuit.

15-4
- Explain how the circuit for starting and stopping in forward and reverse with indicator lights works.
- Explain how the circuit for starting and stopping in forward and reverse with limit switches controlling reversing works.
- Explain how the circuit for starting and stopping in forward and reverse with limit switches as safety stops works.
- Explain how the circuit for a selector switch used to determine direction of motor travel works.
- Explain how the circuit for starting, stopping, and jogging in forward and reverse with jogging controlled through a selector switch works.

15-5
- Explain direct hardwiring.
- Describe hardwiring using terminal strips.

15-6
- Explain how to troubleshoot a power circuit.
- Explain how to troubleshoot a control circuit.

Sections

Review and Resources
atplearningresources.com/Quicklinks
Access Code: 362245

Reversing Motors

All electric motors are used to convert electrical energy into a rotating mechanical force that performs work. The amount of required work depends upon the application, which also determines the motor size, type, and control requirements. Some motor applications require the motor to operate in only one direction and other applications require the motor to be reversible.

When motors are used to operate loads such as pumps, escalators, and timers, the motor must operate in only one direction. If the motor is reversed, the loads and driven components can be damaged. When motors are used to operate loads such as overhead doors, elevators, cranes, tooling equipment, electric automobiles, electric boats, and electric trains, the motor must be designed to be reversible by an external control device or circuit.

All motor types, except the shaded-pole motor, are reversible or can be designed to be reversible by bringing the required motor internal wires outside the motor. Three-phase and DC motors are reversible without bringing out additional internal motor wires. To be reversed, split-phase and capacitor-type motors must have their starting winding leads brought outside of the motor. Learning how each motor type is reversed and knowing the reversing component or circuit requirements is important when designing, installing, servicing, or troubleshooting any application that requires the motor to operate in both directions.

15-1 REVERSING MOTORS USING MANUAL STARTERS

A manual starter is a contactor with an added overload protective device. Two manual starters are connected together to create a manual reversing starter. Manual starters are used in pairs to change the direction of rotation of 3ϕ, 1ϕ, and DC motors. **See Figure 15-1.**

Manual reversing starters are used to operate low-horsepower motors, such as those found in fans, small machines, pumps, and blowers, in forward and reverse directions. Typically, individual manual starters are marked start/stop instead of forward/stop or reverse/stop. Because these markings are common when two manual starters are placed in the same enclosure to make up a manual reversing starter, the electrician must correctly label the unit once it is properly wired.

MANUAL STARTERS

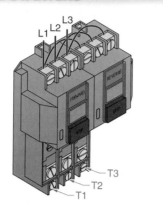

Figure 15-1. Manual starters are used in pairs to reverse 3ϕ, 1ϕ, and DC motors.

Since a motor cannot run in forward and reverse simultaneously, some means must be included to prevent both starters from energizing at the same time. A manual reversing starter uses a mechanical interlock to separate the starter contacts. A *mechanical interlock* is the arrangement of contacts in such a way that both sets of contacts cannot be closed at the same time. Mechanical interlock devices are inserted between the two starters to ensure that both switching mechanisms cannot be energized at the same time. Crossing dashed lines are used between the manual starters in the wiring diagram to indicate a mechanical interlock. **See Figure 15-2.** An electrician must ensure that the interlock is provided if the unit is not preassembled.

MECHANICAL INTERLOCKS— MANUAL STARTERS

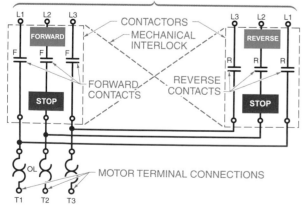

Figure 15-2. Mechanical interlock devices are inserted between the two starters to ensure that both switching mechanisms cannot be energized at the same time.

Reversing 3ϕ Motors Using Manual Starters

A wiring diagram illustrates the electrical connections necessary to properly reverse a 3ϕ motor using a manual reversing starter. **See Figure 15-3.** Only one set of overloads is required.

ELECTRICAL CONNECTIONS FOR REVERSING 3ϕ MOTORS USING MANUAL STARTERS

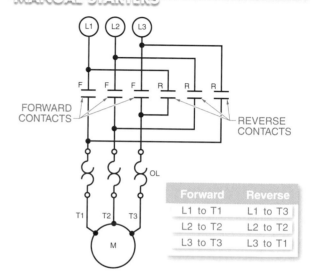

Forward	Reverse
L1 to T1	L1 to T3
L2 to T2	L2 to T2
L3 to T3	L3 to T1

Figure 15-3. A wiring diagram illustrates the electrical connections necessary to properly reverse a 3ϕ motor using a manual reversing starter.

Reversing the direction of rotation of 3ϕ motors is accomplished by interchanging any two of the three main power lines to the motor. Although any two lines may be interchanged, the industry standard is to interchange L1 and L3. This standard is true for all 3ϕ motors including three-, six-, and nine-lead, wye- and delta-connected motors.

When using a manual reversing starter to reverse a 3ϕ motor, regardless of the type, L1 is connected to T1, L2 is connected to T2, and L3 is connected to T3 for forward rotation (when the forward contacts close). **See Figure 15-4.** L1 is connected to T3, L2 is connected to T2, and L3 is connected to T1 for reverse rotation (when the reverse contacts close and the forward contacts open). The motor changes direction each time forward or reverse is pressed because it is necessary to interchange only two leads on a 3ϕ motor to reverse rotation. If a 3ϕ motor has more than three leads coming out, these leads are connected according to the motor wiring diagram.

REVERSING 3ϕ MOTORS USING MANUAL STARTERS

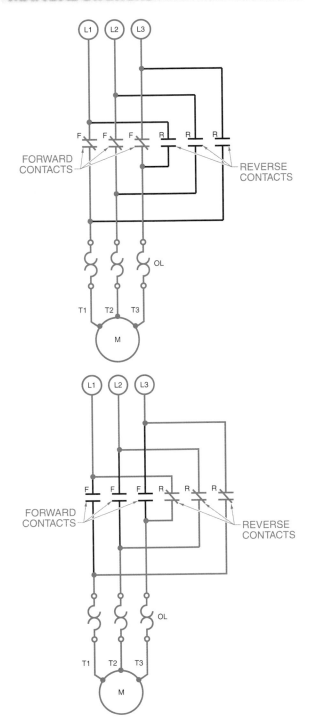

Figure 15-4. When using a manual reversing starter to reverse a 3ϕ motor, L1 is connected to T1, L2 is connected to T2, and L3 is connected to T3 for forward rotation. L1 is connected to T3, L2 is connected to T2, and L3 is connected to T1 for reverse rotation.

Interchanging L1 and L3 is standard for safety reasons. When first connecting a motor, the direction of rotation is not usually known until the motor is started. A motor may be temporarily connected to determine the direction of rotation before making permanent connections. Motor lead temporary connections are not permanently taped. By always interchanging L1 and L3, L2 can be permanently connected to T2, creating an insulated barrier between L1 and L3.

Reversing 1ϕ Motors Using Manual Starters

Reversing the rotation of 1ϕ motors is accomplished by interchanging the leads of the starting or running windings. The manufacturer wiring diagram is used to determine the exact wires to interchange to properly reverse a 1ϕ motor using a manual reversing starter. **See Figure 15-5.** *Note: Always check the manufacturer wiring diagrams for the proper reversal of 1ϕ motors. An electrician can measure the resistance of the starting winding and running winding to determine which leads are connected to which windings if the manufacturer information is not available. The running winding is made of a heavier gauge wire than the starting winding, so the running winding has a much lower resistance than the starting winding.*

ELECTRICAL CONNECTIONS FOR REVERSING 1ϕ MOTORS USING MANUAL STARTERS

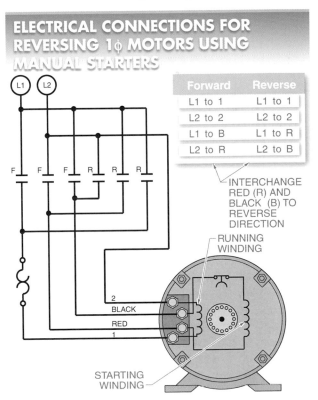

Forward	Reverse
L1 to 1	L1 to 1
L2 to 2	L2 to 2
L1 to B	L1 to R
L2 to R	L2 to B

Figure 15-5. A wiring diagram illustrates the electrical connections necessary to properly reverse a 1ϕ motor using a manual reversing starter.

When the forward contacts are closed, L1 is connected to the black lead of the starting winding and side 1 of the running winding, and L2 is connected to the red lead of the starting winding and side 2 of the running winding. **See Figure 15-6.**

When the reverse contacts are closed and the forward contacts open, L1 is connected to the red lead of the starting winding and side 1 of the running winding, and L2 is connected to the black lead of the starting winding and side 2 of the running winding. The starting windings are interchanged while the running windings remain the same.

The motor changes direction each time forward or reverse is pressed because it is necessary to interchange only the starting windings on a 1ϕ motor to reverse rotation. *Note:* Always check the manufacturer wiring diagram when reversing 1ϕ motors to determine which leads are connected to the starting winding. The red and black wires are normally used for reversal.

The direction of rotation for a capacitor motor can be changed by reversing the connections to the starting or running windings. Whenever possible, the manufacturer wiring diagram should be checked for the exact wires to interchange.

Reversing DC Motors Using Manual Starters

A manual starter can be used to reverse the direction of current flow through the armature of all DC motors. The motor is wired to the starter so that the polarity of the applied DC voltage on the field remains the same in either direction, but the polarity on the armature is opposite for each direction. The direction of rotation for DC series, shunt, and compound motors may be reversed by reversing the direction of the current through the field without changing the direction of the current through the armature or by reversing the direction of the current through the armature, but not both. The industry standard is to reverse the current through the armature.

REVERSING 1ϕ MOTORS USING MANUAL STARTERS

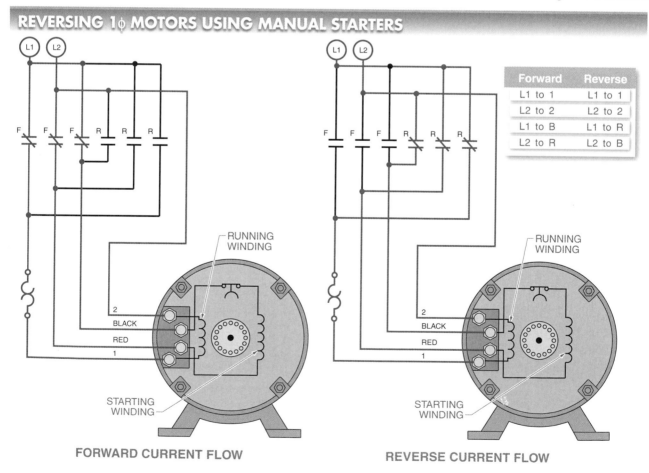

Forward	Reverse
L1 to 1	L1 to 1
L2 to 2	L2 to 2
L1 to B	L1 to R
L2 to R	L2 to B

FORWARD CURRENT FLOW

REVERSE CURRENT FLOW

Figure 15-6. To reverse the direction of a 1ϕ motor, the direction of current through the starting winding is reversed.

A wiring diagram is used to properly wire a DC series motor for reversing. A DC series motor is wired to the starter so that A2 is positive and A1 is negative when the forward contacts are closed and A2 is negative and A1 is positive when the reverse contacts are closed. **See Figure 15-7.** Regardless of whether the forward contacts or reverse contacts are closed, S2 is always positive and S1 is always negative. The motor reverses direction for each position of the starter because only the polarity of the armature reverses direction.

ELECTRICAL CONNECTIONS FOR REVERSING DC MOTORS USING MANUAL STARTERS

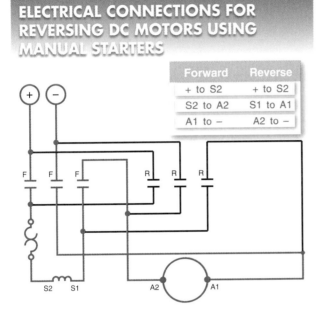

Forward	Reverse
+ to S2	+ to S2
S2 to A2	S1 to A1
A1 to −	A2 to −

Figure 15-7. A DC series motor is wired to a starter so that A2 is positive and A1 is negative when the forward contacts are closed and A2 is negative and A1 is positive when the reverse contacts are closed.

A wiring diagram is used to properly wire a DC shunt motor for reversing. A DC shunt motor is wired to the starter so that A2 is positive and A1 is negative when the forward contacts are closed and A2 is negative and A1 is positive when the reverse contacts are closed. **See Figure 15-8.** Regardless of whether the forward contacts or reverse contacts are closed, F2 is always positive and F1 is always negative. The motor reverses direction for each position of the starter because only the polarity of the armature reverses direction.

A wiring diagram is used for properly wiring a DC compound motor for reversing. **See Figure 15-9.** A DC compound motor is wired to the starter so that A2 is positive and A1 is negative when the forward contacts are closed and A2 is negative and A1 is positive when

the reverse contacts are closed. Regardless of whether the forward contacts or reverse contacts are closed, S2 and F2 are always positive and S1 and A1 are always negative. The motor reverses direction for each position of the starter because only the polarity of the armature reverses direction.

REVERSING DC SHUNT MOTORS USING MANUAL STARTERS

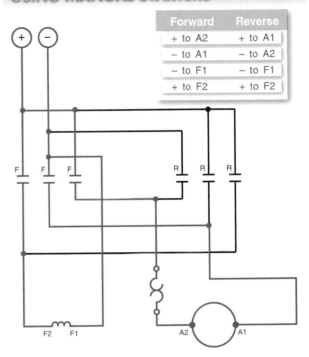

Forward	Reverse
+ to A2	+ to A1
− to A1	− to A2
− to F1	− to F1
+ to F2	+ to F2

Figure 15-8. A wiring diagram is used to properly wire a DC shunt motor for reversing.

Tech Fact

DC motors can instantly reverse direction, which can damage the equipment driven by the motor. In applications where it is best for a motor to first come to a full stop before reversing, a timer can be used in the control circuit to set a time delay before the motor reverses direction.

In a DC compound motor, the series and shunt field relationship to the armature must be left unchanged. The shunt field must be connected in parallel with the armature and the series field must be connected in series with the armature. Reversal is accomplished by reversing the armature connections only. If the motor has commutating pole windings, these windings are considered a part of the armature circuit and the current through them must be reversed when the current through the armature is reversed.

REVERSING DC COMPOUND MOTORS USING MANUAL STARTERS

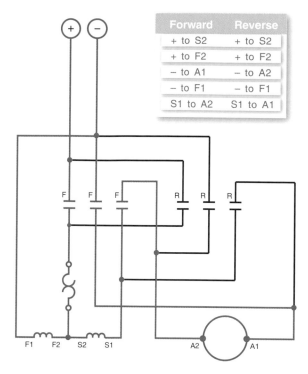

Forward	Reverse
+ to S2	+ to S2
+ to F2	+ to F2
– to A1	– to A2
– to F1	– to F1
S1 to A2	S1 to A1

Figure 15-9. A DC compound motor is wired to a starter so that A2 is positive and A1 is negative when the forward contacts are closed and A2 is negative and A1 is positive when the reverse contacts are closed.

A wiring diagram is used for properly wiring a DC permanent-magnet motor for reversing. **See Figure 15-10.** A DC permanent-magnet motor is wired to the starter so that A2 is positive and A1 is negative when the forward contacts are closed and A2 is negative and A1 is positive when the reverse contacts are closed. A permanent-magnet field never reverses its direction of polarity regardless of the polarity to which the armature is connected. The direction of rotation of a DC permanent-magnet motor is reversed by reversing the direction of the current through the armature only since there are no field connections available.

REVERSING DC PERMANENT-MAGNET MOTORS USING MANUAL STARTERS

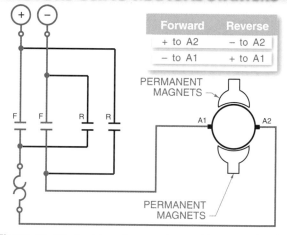

Forward	Reverse
+ to A2	– to A2
– to A1	+ to A1

Figure 15-10. A wiring diagram is used to properly wire a DC permanent-magnet motor for reversing.

15-1 CHECKPOINT

1. How does a reversing manual motor starter prevent the motor from being told to operate simultaneously in both directions?

2. How is a 3ϕ motor reversed using the industry standard?

3. Does the running winding or starting winding of a 1ϕ motor have higher resistance?

4. When reversing a DC motor, is the standard to reverse polarity on the armature or on the field?

15-2 REVERSING MOTORS USING DRUM SWITCHES

A *drum switch* is a manual switch made up of moving contacts mounted on an insulated rotating shaft. **See Figure 15-11.** The moving contacts make and break contact with stationary contacts within the switch as the shaft is rotated.

Drum switches are totally enclosed and an insulated handle provides the means for moving the contacts from point to point. Drum switches are available in several sizes and can have different numbers of poles and positions. Drum switches are usually used where an operator's eyes must remain on a particular operation such as when a crane is raising and lowering a load.

DRUM SWITCHES

Figure 15-11. A drum switch is a manual switch with moving contacts mounted on an insulated rotating shaft.

A drum switch may be purchased with maintained contacts or spring-return contacts. In either case, when the motor is not running in forward or reverse, the handle is in the center (OFF) position. To reverse a running motor, the handle must first be moved to the center position until the motor stops and then moved to the reverse position.

Drum switches are not motor starters because they do not contain protective overloads. Separate overload protection is normally provided by placing a nonreversing starter in line before the drum switch. This provides the required overload protection and acts as a second disconnecting means. A drum switch is used only as a means of controlling the direction of a motor by switching the leads of the motor.

Reversing 3ϕ Motors Using Drum Switches

A 3ϕ motor may be connected to the contacts of a drum switch to change the direction of rotation from forward to reverse. **See Figure 15-12.** Charts are used to show the internal operation of a drum switch and the resulting motor connections for forward and reverse. L1 and L3 are interchanged as the drum switch is moved from the forward to the reverse position. The motor changes direction each time the drum switch is moved to forward or reverse because only the two leads on a 3ϕ motor must be interchanged to reverse rotation.

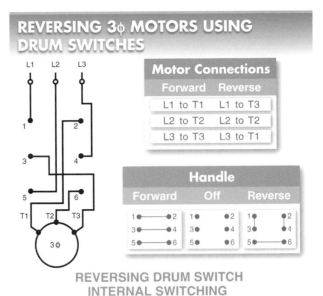

Figure 15-12. A 3ϕ motor may be connected to the contacts of a drum switch to change the direction of rotation from forward to reverse.

Reversing 1ϕ Motors Using Drum Switches

A 1ϕ motor may be connected to the contacts of a drum switch to change the direction of rotation from forward to reverse. **See Figure 15-13.** Charts are used to show the internal operation of a drum switch. Always consult the manufacturer wiring diagram to ensure proper wiring.

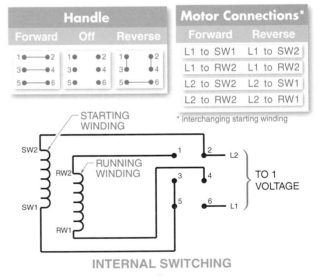

Figure 15-13. A 1ϕ motor may be connected to the contacts of a drum switch to change the direction of rotation from forward to reverse.

The motor changes direction each time the drum switch is moved to forward or reverse. This occurs because only the starting windings must be interchanged on a 1ϕ motor to reverse rotation.

Reversing DC Motors Using Drum Switches

The direction of rotation of any DC series, shunt, compound, or permanent-magnet motor may be reversed by reversing the direction of the current through the fields without changing the direction of the current through the armature. The direction of rotation may also be reversed by reversing the direction of the current through the armature without changing the direction of the current through the fields. The industry standard is to reverse the direction of current through the armature.

A drum switch may be connected to change the direction of rotation of any DC series, shunt, compound, or permanent-magnet motor. **See Figure 15-14.** In each circuit, the current through the armature is changed. Some DC motors have commutating windings (interpoles) that are used to prevent sparking at the brushes in the motor. For this reason, the armature circuit (armature and commutating windings) must be reversed on all DC motors with commutating windings to reverse the direction of motor rotation.

Tech Fact

Standard drum switches have three positions (FOR/OFF/REV). Drum switches are designed to either maintain the selected position or spring back to the center position. In applications where it is best for the operator keep a hand on the controls as the motor operates, a drum switch that springs back to center should be used.

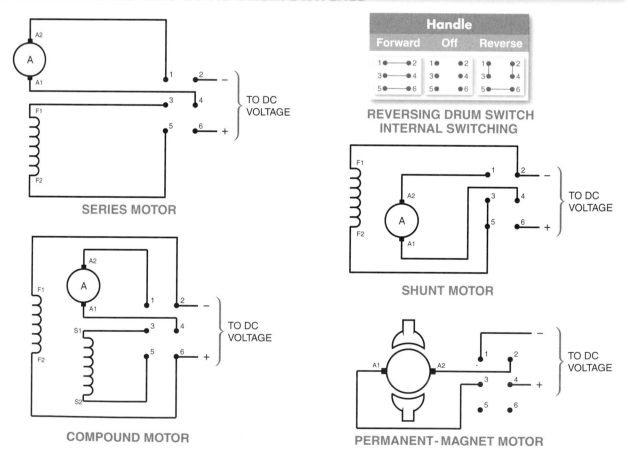

Figure 15-14. A drum switch may be connected to change the direction of rotation of any DC series, shunt, compound, or permanent-magnet motor.

15-2 CHECKPOINT

1. What is the advantage of using a drum switch to reverse a motor instead of a pushbutton-operated manual motor starter?

2. Why are drum switches not considered motor starters?

15-3 REVERSING MOTORS USING MAGNETIC STARTERS

A magnetic reversing starter performs the same function as a manual reversing starter. The only difference between manual and magnetic reversing starters is the addition of forward and reverse coils and the use of auxiliary contacts. **See Figure 15-15.** The forward and reverse coils replace the pushbuttons of a manual starter and the auxiliary contacts provide additional electrical protection and circuit flexibility. The reversing circuit is the same for both manual and magnetic starters.

ELECTRICAL CONNECTIONS FOR REVERSING MOTORS USING MAGNETIC STARTERS

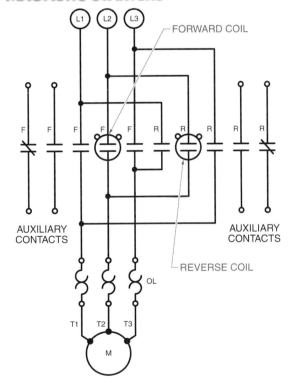

Figure 15-15. A magnetic reversing starter has forward and reverse coils, which replace the pushbuttons of a manual starter, and auxiliary contacts that provide additional electrical protection and circuit flexibility.

Mechanical Interlocking

A magnetic reversing starter may be controlled by forward and reverse pushbuttons. **See Figure 15-16.** *Note:* A line diagram does not show the power contacts. The power contacts are found in the wiring diagram. The broken lines running from the forward coil to the reverse coil indicate that the coils are mechanically interlocked like those of a manual reversing starter. This mechanical interlock is normally factory-installed by the manufacturer.

MECHANICAL INTERLOCKS— MAGNETIC STARTERS

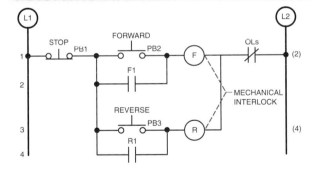

Figure 15-16. A magnetic reversing starter may be controlled by forward and reverse pushbuttons.

In this circuit, pressing forward pushbutton PB2 completes the forward coil circuit from L1 to L2, energizing coil F. Coil F energizes auxiliary contacts F1, providing memory. Mechanical interlocking keeps the reversing circuit from closing. Pressing stop pushbutton PB1 opens the forward coil circuit, causing coil F to de-energize and contacts F1 to return to their normally open (NO) position. Pressing reverse pushbutton PB3 completes the reverse coil circuit from L1 to L2, energizing coil R. Coil R energizes auxiliary contacts R1, providing memory. Mechanical interlocking keeps the forward circuit from closing. Pressing stop pushbutton

PB1 opens the reverse coil circuit, causing coil R to de-energize and contacts R1 to return to their NO position. Overload protection is provided in forward and reverse by the same set of overloads.

Auxiliary Contact Interlocking

Although most magnetic reversing starters provide mechanical interlock protection, some circuits are provided with a secondary backup or safety backup system that uses auxiliary contacts to provide electrical interlocking. **See Figure 15-17.**

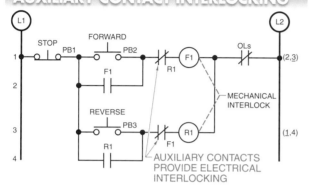

AUXILIARY CONTACT INTERLOCKING

Figure 15-17. Although most magnetic reversing starters provide mechanical interlock protection, some circuits are provided with a secondary backup system that uses auxiliary contacts to provide electrical interlocking.

In this circuit, one normally open (NO) set and one normally closed (NC) set of contacts are activated when the forward coil circuit is energized. The NO contacts close, providing memory, and the NC contacts open, providing electrical isolation in the reverse coil circuit. When the forward coil circuit is energized, the reverse coil circuit is automatically opened or isolated from the control voltage. Even if the reverse pushbutton is closed, no electrical path is available in the reverse circuit. For the reverse circuit to operate, the stop pushbutton must be pressed so that the forward circuit de-energizes and returns the NC contacts to their normal position. Pressing the reverse pushbutton provides the same electrical interlock for the reverse circuit when the forward contacts are in their normal position.

Pushbutton Interlocking

Pushbutton interlocking may be used with either or both mechanical and auxiliary interlocking. Pushbutton interlocking uses both NO and NC contacts mechanically connected on each pushbutton. **See Figure 15-18.**

In this circuit, when NO contacts on the forward pushbutton close to energize the F coil circuit, the NC contacts wired into the R coil circuit open and provide electrical isolation. Conversely, when the NO contacts on the reverse pushbutton close to energize the R coil circuit, the NC contacts wired into the F coil circuit open and provide electrical isolation. Mechanical and auxiliary contact electrical interlocking is also provided in the circuit.

CAUTION: In many cases, motors or the equipment they are powering cannot withstand a rapid reversal of direction. Care must be exercised to determine the equipment that can be safely reversed under load. Also consider the braking that must be provided to slow the machine to a safe speed before reversal.

Motors used in crane operations are designed to operate in two directions.

Reversing Power and Control Circuits

A power circuit and a control circuit are required when using motor starters or motor drives. The power circuit includes the incoming circuit main power, the motor starter (or drive), and the motor. The control circuit includes the required circuit inputs (pushbuttons, limit switches, etc.), motor starter coils, motor starter auxiliary contacts, overload contacts, timers, counters, and any other device designed to operate in the control circuit.

The control circuit is normally operated at a lower voltage than the power circuit. The low control circuit voltage is obtained by using a step-down transformer. When used with motor control circuits, this transformer is often referred to as the control transformer.

Although the power circuit and control circuit operate together to control the motor, they are electrically isolated from each other through the transformer. This electrical isolation allows individual control circuits to control different motor types (1φ motors, DC motors, and 3φ motors). **See Figure 15-19.**

PUSHBUTTON INTERLOCKING

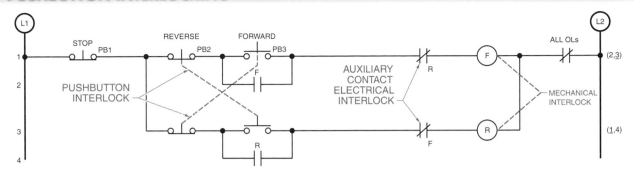

Figure 15-18. Pushbutton interlocking uses both NO and NC contacts mechanically connected on each pushbutton.

REVERSING POWER AND CONTROL CIRCUITS

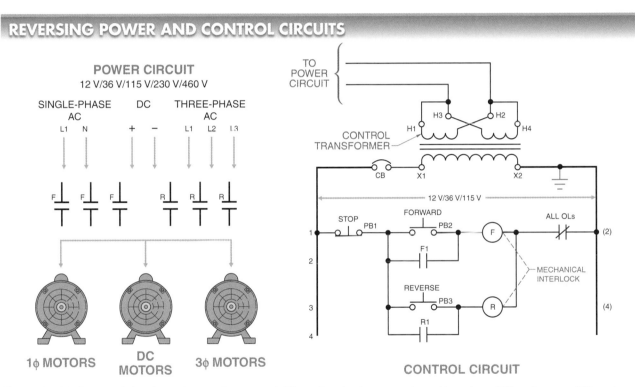

Figure 15-19. A control circuit can be used to control different motor types, such as 1φ motors, DC motors, and 3φ motors.

15-3 CHECKPOINT

1. How are mechanical interlocks in the wiring diagram shown in the line diagram control circuit?

2. How do auxiliary contacts prevent a motor from being reversed when the motor is already running in one direction?

3. How does pushbutton interlocking prevent a motor from being reversed when the motor is already running in one direction and the pushbutton for the other direction is pressed?

4. When using magnetic motor starters to reverse a 3φ motor, is the control circuit usually at the same voltage level as the power circuit?

15-4 MAGNETIC REVERSING STARTER APPLICATIONS

Many applications can be built around a basic magnetic reversing starter because magnetic reversing starters are controlled electrically. These circuits include the functions of starting and stopping motors in forward and reverse and controlling the motors with various control devices.

Starting and Stopping in Forward and Reverse with Indicator Lights

Operators are often required to know the direction of rotation of a motor at a given moment. A start/stop/forward/reverse circuit with indicator lights enables an operator to know the direction of rotation of a motor at any time. **See Figure 15-20.** An example is a motor controlling a crane that raises and lowers a load. The line diagram shows how lights can indicate the direction the motor is operating. If an electrician adds nameplates, these lights could indicate up and down directions of the hoist.

In this circuit, pressing the momentary contact forward pushbutton causes the NO and NC contacts to move simultaneously. The NO contacts close, energizing coil F while the pushbutton is pressed. Coil F causes the memory contacts F to close and the NC electrical interlock to open, isolating the reversing circuit. The forward pilot light turns on when holding contacts F are closed. For the period of time the pushbutton is pressed, the NC contacts of the forward pushbutton open and isolate reversing coil R.

Pressing the momentary contact reverse pushbutton causes the NO and NC contacts to move simultaneously. The opening of the NC contacts de-energizes coil F. With coil F de-energized, the memory contacts F open and the electrical interlock F closes. The closing of the NO contacts energizes coil R. Coil R causes the holding contacts R to close and the NC electrical interlock to open, isolating the forward circuit. The reverse pilot light turns on when the memory contacts R close. Pressing the stop pushbutton with the motor running in either direction stops the motor and causes the circuit to return to its normal state.

Overload protection for the circuit is provided by heater coils. Operation of the overload contacts breaks the circuit, opening the overload contacts. The motor cannot be restarted until the overloads are reset and the forward or reverse pushbutton is pressed.

This circuit provides protection against low voltage or a power failure. A loss of voltage de-energizes the circuit, and holding contacts F or R open. This design prevents the motor from starting automatically after the power returns.

Tech Fact

When magnetic motor starters are used to control the direction of a motor, two motor starters are required with an external interlocking control circuit connected between them. The external interlocking control circuit prevents both starters from energizing at the same time, which would cause a short circuit and equipment damage. When a motor drive is used, control of motor direction is accomplished by the same drive that has automatic interlocking already built into it.

ELECTRICAL CONNECTION FOR INDICATOR LIGHTS

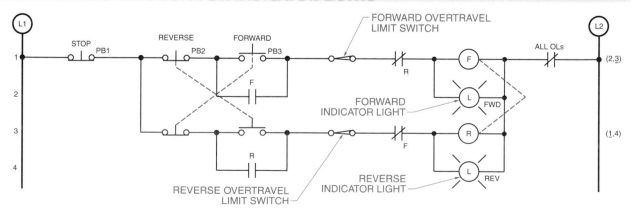

Figure 15-20. A start/stop/forward/reverse circuit with indicator lights enables an operator to know the direction of rotation of a motor at any time.

Starting and Stopping in Forward and Reverse with Limit Switches Controlling Reversing

Limit switches may be used to provide automatic control of reversing circuits. **See Figure 15-21.** This circuit uses limit switches and a control relay to automatically reverse the direction of a machine at predetermined points. For example, this circuit could control the table of an automatic grinding machine where the operation must be periodically reversed.

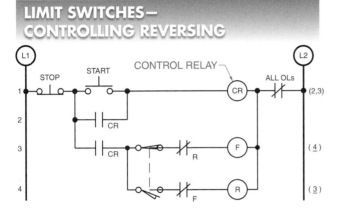

LIMIT SWITCHES— CONTROLLING REVERSING

Figure 15-21. Limit switches may be used to provide automatic control of reversing circuits.

In this circuit, pressing the start pushbutton causes control relay CR to become energized. The auxiliary CR contacts close when control relay CR is energized. One set of contacts form the holding circuit, and the other contacts connect the limit switch circuit. The motor runs when the limit switch circuit is activated. The motor runs in the forward direction if the forward limit switch is closed. The motor runs in the opposite direction if the reverse limit switch is closed.

Overload protection for the circuit is provided by heater coils. Operation of the overload contacts breaks the circuit. The motor cannot be restarted until the overloads are reset and the start pushbutton is pressed.

This circuit provides protection against low voltage or a power failure. A loss of voltage de-energizes the circuit and the holding contacts CR open. This prevents the motor from starting automatically after the power returns.

Starting and Stopping in Forward and Reverse with Limit Switch as Safety Stop in Either Direction

For safety reasons, it may be necessary to ensure that a load controlled by a reversing motor does not go beyond

certain operating points in the system. For example, a hydraulic lift should not rise too high. Limit switches are incorporated to shut the operation down if a load travels far enough to be unsafe. **See Figure 15-22.** The circuit provides overtravel protection through the use of limit switches.

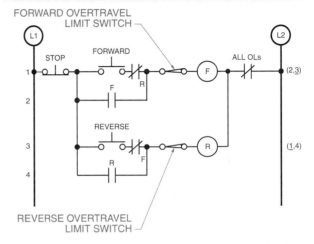

LIMIT SWITCHES— OVERTRAVEL PROTECTION

Figure 15-22. Limit switches may be used in a circuit to provide overtravel protection.

In this circuit, pressing the forward pushbutton activates coil F. Coil F pulls in the holding contacts F and opens the electrical interlock F, isolating the reversing circuit. The motor runs in the forward direction until either the stop pushbutton is pressed or the limit switch is activated. The circuit is broken and the holding contacts and electrical interlock return to their normal state if either control is activated.

Pressing the reverse pushbutton activates coil R. Coil R pulls in the holding contacts R and opens the electrical interlock R, isolating the forward circuit. The motor runs in the reverse direction until either the stop pushbutton is pressed or the limit switch is activated. The circuit is broken and the holding contacts and electrical interlock return to their normal state if either control is activated. The circuit may still be reversed to clear a jam or undesirable situation if either limit switch is opened. This allows the operator to operate the motor in the direction opposite the tripped overtravel limit switch to reset the tripped limit switch.

Overload protection for the circuit is provided by heater coils. Operation of the overload contacts breaks the circuit. The motor cannot be restarted until the overloads are reset and the forward pushbutton is pressed.

This circuit provides protection against low voltage or a power failure in that a loss of voltage de-energizes the circuit and holding contacts F or R open. This prevents the motor from starting automatically after the power returns.

Selector Switch Used to Determine Direction of Motor Travel

A selector switch and a basic start/stop station can be used to reverse a motor. **See Figure 15-23.** The motor can be run in either direction, but the desired direction must be set by the selector switch before starting the motor.

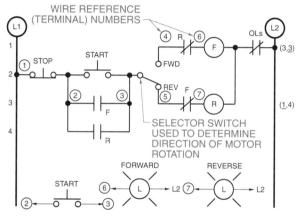

Figure 15-23. A selector switch and a basic start/stop station can be used to reverse a motor.

In this circuit, pressing the start pushbutton with the selector switch in the forward position energizes coil F. Coil F closes the holding contacts F and opens the electrical interlock F, isolating the reversing circuit. Pressing the stop pushbutton de-energizes coil F, which releases the holding contacts and the electrical interlock. Pressing the start pushbutton with the selector switch in the reverse position energizes coil R. Coil R closes the holding contacts R and opens the electrical interlock R, isolating the forward circuit.

Overload protection for the circuit is provided by heater coils. Operation of the overload contacts breaks the circuit. The motor cannot be restarted until the overloads are reset and the start pushbutton is pressed.

This circuit provides protection against low voltage or a power failure in that a loss of voltage de-energizes the circuit and holding contacts F or R open. This prevents the motor from starting automatically after the power returns.

This circuit also illustrates the proper connections for adding forward and reverse indicator lights. The forward indicator light is connected to wire 6 and L2. The reverse indicator light is connected to wire 7 and L2. Additional start pushbuttons are connected to wires 2 and 3. It is standard industrial practice to mark the NO memory contacts 2 and 3. It is also standard industrial practice to mark the wire coming from the forward coil and leading to the NC reverse contact (used for interlocking) as wire 6. Likewise, the wire coming from the reverse coil and leading to the NC forward contact is marked as wire 7. These numbers are usually printed on the magnetic starters to help in wiring the circuit.

Starting, Stopping, and Jogging in Forward and Reverse with Jogging Controlled through a Selector Switch

In certain industrial operations, it may be necessary to reposition equipment a little at a time for small adjustments. A jogging circuit allows the operator to start a motor for short times without memory. **See Figure 15-24.** *Jogging* is the frequent starting and stopping of a motor for short periods of time.

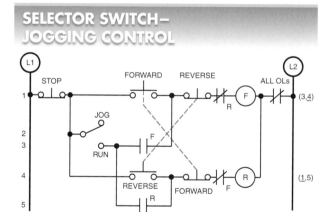

Figure 15-24. A jogging circuit allows the operator to start a motor for short times without memory.

In this circuit, small adjustments may be made in forward and reverse motor rotation or in continuous operation, depending on the position of the selector switch. Pressing the forward pushbutton with the selector switch in the run position activates coil F. Coil F pulls in the NO holding contacts F and opens the NC electrical interlock F, isolating the reversing circuit. The motor starts and continues to run. Pressing the reverse pushbutton with the selector switch in the run position activates coil R. Coil R pulls in the NO holding contacts R and opens the NC electrical interlock R, isolating the forward circuit. The motor starts in the reverse direction and continues to run.

Pressing the stop pushbutton in either direction breaks the circuit and returns the circuit contacts to their normal positions. Pressing the forward pushbutton with the selector switch in the jog position activates coil F and the motor only for the period of time that the forward pushbutton is pressed. In addition, the NC electrical interlock F opens and isolates the reversing circuit. Pressing the reverse pushbutton with the selector switch in the jog position activates coil R and the motor only for the period of time that the reverse pushbutton is pressed. In addition, the NC electrical interlock R opens and isolates the forward circuit.

Overload protection is provided by heater coils. Operation of the overload contacts breaks the circuit. The motor cannot be restarted until the overloads are reset and the start pushbutton is pressed.

This circuit provides protection against low voltage or a power failure in that a loss of voltage de-energizes the circuit and holding contacts F or R open. This prevents the motor from starting automatically after the power returns.

15-4 CHECKPOINT

1. When a limit switch, or any other type of control switch, is used to automatically stop a motor operating in one direction, are the switches used in the control circuit NO or NC contacts?

2. How are indicating lights that visually show motor direction added into the control circuit?

15-5 MOTOR CONTROL WIRING METHODS

A motor must have a method of control in order to operate safely and efficiently. Motor control circuits vary from simple to complex. Reversing motor control circuits, similar to nonreversing motor control circuits, can be wired using manual controls (manual starters, drum switches), magnetic controls (magnetic starters), motor drives, or PLCs to control the operation of a motor.

Several different methods of wiring a motor and motor control circuit are available. These methods can be used individually or in combination to control the operation of a motor. Each motor control wiring method has advantages and disadvantages. The four basic methods of motor control wiring are direct hardwiring, hardwiring using terminal strips, PLC wiring, and electric motor drive wiring.

Direct Hardwiring

Direct hardwiring is the oldest and most straightforward motor control wiring method used. In direct hardwiring, the power circuit and the control circuit are wired point-to-point. **See Figure 15-25.** *Point-to-point wiring* is wiring in which each component in a circuit is connected (wired) directly to the next component as speci-

fied on the wiring and line diagrams. For example, the transformer X1 terminal is connected directly to the fuse, the fuse is connected directly to the stop pushbutton, the stop pushbutton is connected directly to the reverse pushbutton, the reverse pushbutton is connected directly to the forward pushbutton, and so on until the final connection from the overload (OL) contact is made back to the transformer X2 terminal.

A direct hardwired circuit may operate properly for a period of time. The disadvantage of a direct hardwired circuit is that circuit troubleshooting and circuit modification are time consuming.

For example, when a problem occurs in a direct hardwired circuit, the circuit operation must be understood, measurements taken, and the problem identified. Circuit operation can be understood from a wiring diagram. Without a wiring diagram, the circuit wiring can be determined by tracing each wire throughout the circuit. The circuit problem can eventually be found; however, tracing each wire in a circuit to find the wire with a problem is time consuming. Time is saved as experience is gained from working on a circuit several times and understanding its operation and components.

ELECTRICAL CONNECTIONS FOR DIRECT HARDWIRED CIRCUITS

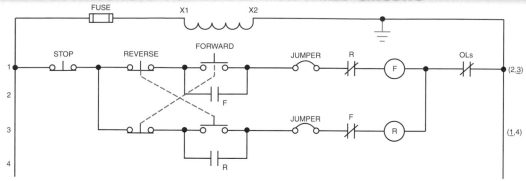

REVERSING CONTROL CIRCUIT LINE DIAGRAM

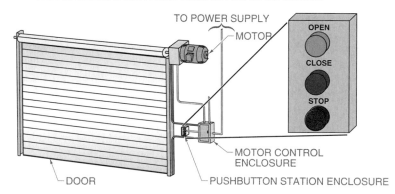

INDUSTRIAL DOOR APPLICATION

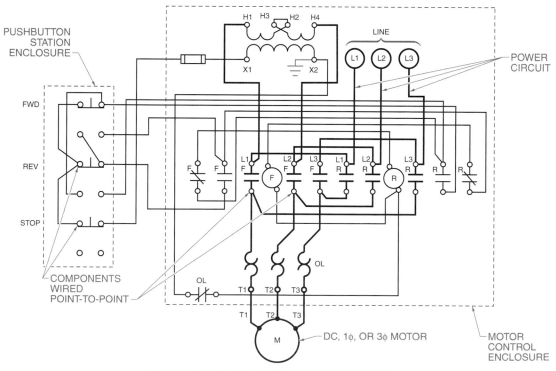

REVERSING CIRCUIT WIRING DIAGRAM

Figure 15-25. In direct hardwiring, the power circuit and the control circuit are wired point-to-point.

A direct hardwired circuit is difficult to modify. For example, if a forward indicator lamp and a reverse indicator lamp are to be added to a motor control circuit, their exact connection points must be found. Once the exact connection points are found, the lamps can be wired into the control enclosure. Even when the exact connection points are found, problems may arise when making the actual connection (such as there not being enough room under the terminal screw, etc.).

Some circuit modifications, such as adding forward and reverse indicator lamps, may not be a problem because they only require adding new wires. In these modifications, old wires do not need to be moved or removed. Some circuit modifications, such as adding limit switches, are more difficult. For example, if forward and reverse limit switches are to be added to a circuit, some wiring must be removed from the circuit and/or the new wiring for the limit switches must be added. **See Figure 15-26.**

CIRCUIT MODIFICATIONS FOR DIRECT HARDWIRED CIRCUITS

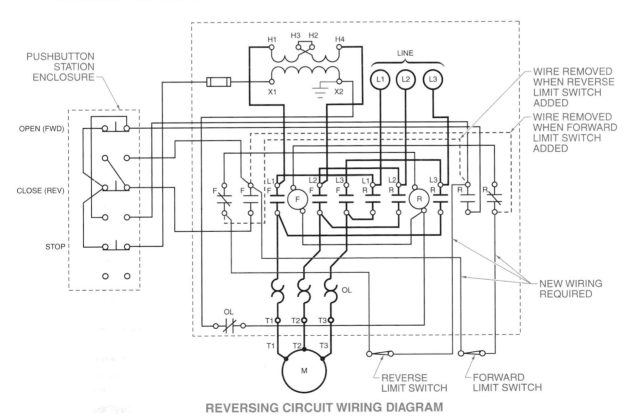

REVERSING CONTROL CIRCUIT LINE DIAGRAM

INDUSTRIAL DOOR APPLICATION

REVERSING CIRCUIT WIRING DIAGRAM

Figure 15-26. In direct hardwired circuits, circuit modifications may require the removal and/or addition of circuit wiring.

In this circuit, before the limit switches are added, the wires connecting the NC interlock contacts of the forward and reverse coils to the pushbuttons must be removed (or opened) and the limit switch wired in the opening. To do this, the technician making the circuit modification must have a wiring diagram of the circuit (or understand the circuit from past experience) in order to know which wires to open and where to locate the limit switches.

Hardwiring Using Terminal Strips

Hardwiring to a terminal strip allows for easy circuit modification and simplifies circuit troubleshooting. When wiring using a terminal strip, each wire in the control circuit is assigned a reference point on the line diagram to identify the different wires that connect the components in the circuit. Each reference point is assigned a wire reference number. **See Figure 15-27.**

HARDWIRING USING TERMINAL STRIPS

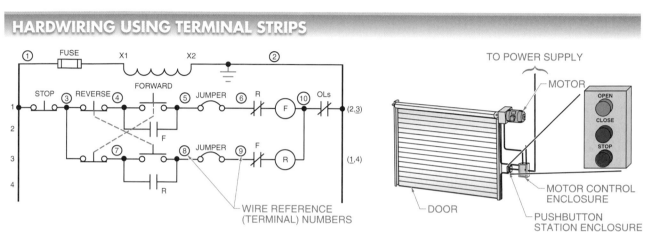

REVERSING CONTROL CIRCUIT LINE DIAGRAM

INDUSTRIAL DOOR APPLICATION

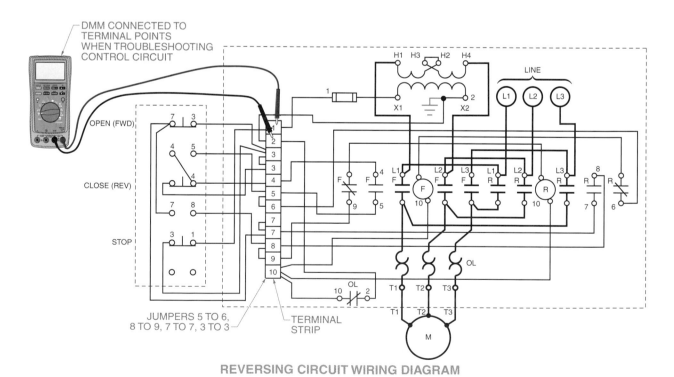

REVERSING CIRCUIT WIRING DIAGRAM

Figure 15-27. When hardwiring a circuit using a terminal strip, each wire in the control circuit is assigned a reference point on the line diagram to identify the different wires that connect the components in the circuit.

Wire reference numbers were commonly assigned from the top left to the bottom right. However, in most new diagrams, the power line on the left (usually L1 or X1) is assigned the number 1, and the power line on the right (usually L2 or X2) is assigned the number 2. This way the control circuit voltage can always be found at terminal 1 and at terminal 2. This aids a technician when he or she is troubleshooting a circuit. If several connections of a given number are required, jumpers can be added to the terminal strip to provide multiple connection points to one given terminal number.

When troubleshooting a circuit with a terminal strip, the technician can go directly to the terminal strip and take measurements to help isolate the problem. The DMM is first placed on terminals 1 and 2. If the voltage is not correct at that point, the problem is located on the primary side of the transformer. If the voltage is correct at terminals 1 and 2, one DMM lead is left on terminal 2 and the other lead is moved to different terminals until the problem is located.

In addition to the terminal strip and wire reference numbers being an aid when troubleshooting, they also make circuit modification easier. This is because most, if not all, of the wires required to make the change are disconnected and reconnected at the terminal strip. **See Figure 15-28.**

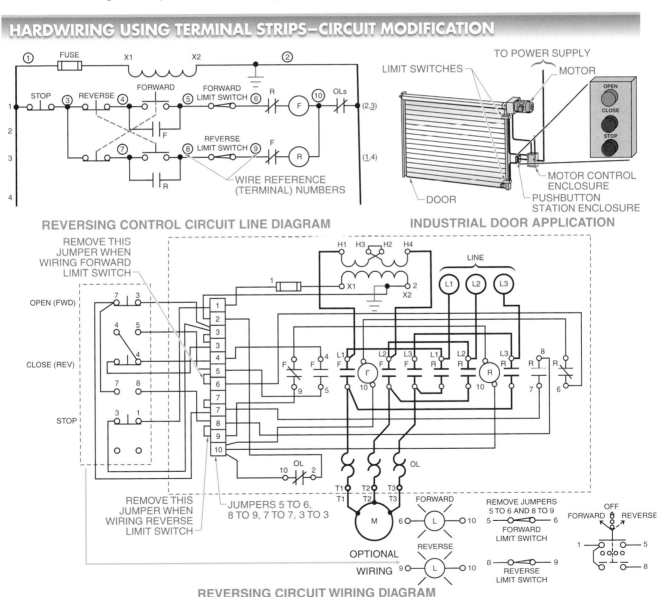

Figure 15-28. Terminal strips and wire reference numbers enable easy circuit modification because most wires required to make a change are disconnected and reconnected at the terminal strip.

15-5 CHECKPOINT

1. What are two disadvantages of using a direct hardwiring method?

2. What is the usual order in which wire reference numbers are assigned to a control circuit?

15-6 TROUBLESHOOTING REVERSING POWER CIRCUITS

A *power circuit* is the part of an electrical circuit that connects the loads to the main power lines. Troubleshooting reversing power circuits normally involves determining the point in the system where power is lost. **See Figure 15-29.** To troubleshoot a reversing power circuit, the following procedure is applied:

1. Measure the incoming voltage between each pair of power leads. Incoming voltage must be within 10% of the voltage rating of the motor. Measure the voltage at the main power panel feeding the control cabinet if no voltage is present or if the voltage is not at the correct level.

2. Measure the voltage out of each fuse or circuit breaker. The fuse or circuit breaker is open if no voltage reading is obtained. Replace any blown fuse or tripped circuit breaker.

WARNING: Use caution when manually operating starter contacts because loads may start or stop without warning.

3. Measure the voltage out of the motor starter. The voltage should be present when either the forward power contacts or reverse power contacts are closed. The contacts can be closed manually at most motor starters if the power contacts cannot be closed by using the control circuit pushbuttons. Disconnect the incoming power and check the motor starter contacts for burning or wear if the voltage is not at the correct level.

4. Measure the voltage at the motor terminals. The voltage must be within 10% of the motor rating and equal on each power line. There is a problem with the motor or mechanical connection if the voltage is correct and the motor does not operate.

TROUBLESHOOTING POWER CIRCUITS

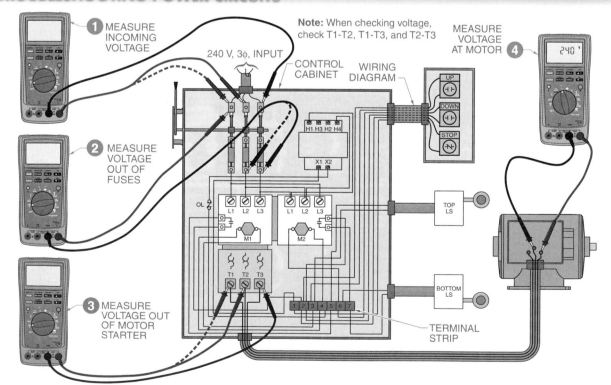

Figure 15-29. Troubleshooting reversing power circuits normally involves determining the point in the system where power is lost.

Troubleshooting Reversing Control Circuits

When troubleshooting reversing control circuits, a line diagram is used to illustrate circuit logic and a wiring diagram is used to locate the actual test points at which a DMM is connected to the circuit. **See Figure 15-30.** To troubleshoot a reversing control circuit, the following procedure is applied:

1. Measure the supply voltage of the control circuit by connecting a DMM set to measure voltage between line 1 (hot conductor) and line 2 (neutral conductor). The voltage must be within 10% of the control circuit rating. Test the power circuit if the voltage is not correct. The control circuit voltage rating is determined by the voltage rating of the loads used in the control circuit (motor starter coils, etc.).

2. Measure the voltage out of the overload contacts to ensure the contacts are closed. The contacts are tripped or faulty if no voltage is present. Reset the overloads if they are tripped. Overloads are installed to protect the motor during operation. The control circuit does not operate when the overloads are tripped.

3. Measure the voltage into and out of the control switch or contacts. NC switches (stop pushbuttons, etc.) should have a voltage output before they are activated. NO switches (start pushbuttons, memory contacts, etc.) should have a voltage output only after they are activated.

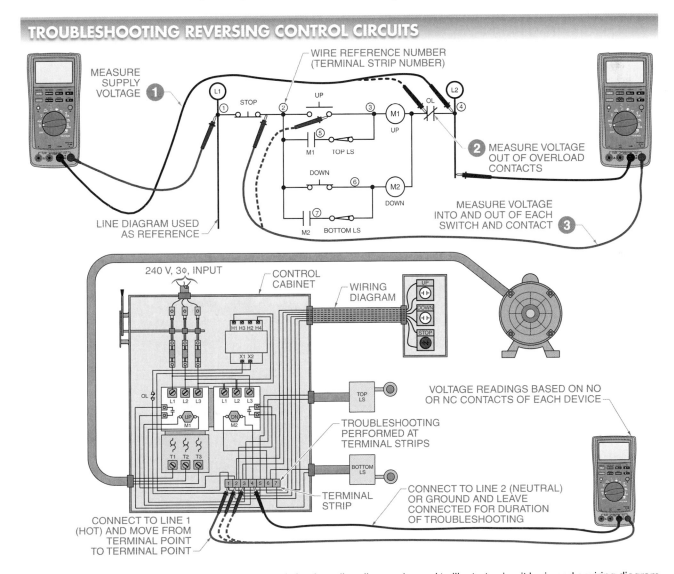

TROUBLESHOOTING REVERSING CONTROL CIRCUITS

Figure 15-30. When troubleshooting reversing control circuits, a line diagram is used to illustrate circuit logic and a wiring diagram is used to locate the actual test points at which a DMM is connected.

15-6 CHECKPOINT

1. Can the power line fuses be checked in the power circuit, even if the control circuit is not working?

2. If the forward part of the control circuit is working properly, can the overloads or the control circuit fuse be the problem if the reversing part of the circuit is not working?

Additional Resources

Review and Resources

Access Chapter 15 Review and Resources for *Electrical Motor Controls for Integrated Systems* by scanning the above QR code with your mobile device.

Applying Your Knowledge

Refer to the *Electrical Motor Controls for Integrated Systems* Learner Resources for interactive Applying Your Knowledge questions.

Workbook and Applications Manual

Refer to Chapter 15 in the *Electrical Motor Controls for Integrated Systems Workbook* and the *Applications Manual* for additional exercises.

ENERGY EFFICIENCY PRACTICES

Reversing Motor Circuits

The most efficient means of reversing the direction of rotation of industrial motors is by using PLCs or electric motor drives. The use of PLCs in reversing motor control circuits allows for greater flexibility and the ability to monitor circuits. The advantage of using PLCs in reversing motor circuits is that PLCs simplify the circuit by eliminating much of the wiring and required components. For example, a PLC can eliminate the need for normally closed (NC) auxiliary contacts on the starter that are used for interlocking. A magnetic motor starter that is replaced with an electric motor drive saves energy and enables the load to be operated at different speeds. In addition, motor drives allow 1ϕ power (115/230 VAC) to operate 3ϕ motors in applications of 3 HP or less. Additional energy savings can be realized by using an energy-efficient motor. Energy-efficient motors save approximately 7% of the energy cost as compared to standard motors.

Objectives

Sections

16-1
- Describe the differences between how a dashpot timer, a synchronous clock timer, a solid-state timer, and a solid-state programmable timer produce a time delay.

16-2
- Describe how an on-delay timer operates and give an example of its usage.
- Describe how an off-delay timer operates and give an example of its usage.
- Explain how plugging can be accomplished by using timing relays.
- Describe how a one-shot timer operates and give an example of its usage.
- Describe how a recycle timer operates and give an example of its usage.
- Describe multiple-function timers.

16-3
- Describe multiple-contact timer wiring diagrams.
- Give an example of the usage of a multiple-contact timer.
- Explain the difference between supply voltage-controlled timers, contact-controlled timers, and sensor-controlled timers.

16-4
- Explain how to troubleshoot timing circuits.
- Describe transistor-controlled timers.
- Describe 555 and 556 timers.
- Describe solid-state programmable timers.
- Describe programmable timers.

16-5
- Define totalizer and counter.
- Describe up counters and up/down counters.

Review and Resources
atplearningresources.com/Quicklinks
Access Code: 362245

Chapter
Timing and Counting Functions
16

Many aspects of everyday life are controlled by time, such as when the work or school day begins, how long it takes for a stoplight to change, and when events or programs start and stop. Likewise, many aspects of everyday life are also controlled by count, such as the number of products purchased for a given amount, the number of products in a package, and the number of seats or spaces allowed.

Most time- and count-controlled events are electronically controlled by timers and counters. Likewise, commercial and industrial manufacturing and processing are all time- and count-controlled by electrical and electronic circuits and systems. It is important for anyone working with time- or count-controlled circuits to understand the many different types of timers, timing functions, and counters and how each one is set or programmed for the given application requirements.

16-1 TIMERS

The four major categories of timers are dashpot, synchronous clock, solid-state, and solid-state programmable. **See Figure 16-1.** Dashpot, synchronous clock, and solid-state timers are stand-alone timers. Stand-alone timers are physically connected between the input device (limit switch, etc.) and the output device (solenoid, etc.) controlled by the timer. Solid-state programmable timers are timing functions that are included in electrical control devices such as programmable logic relays (PLRs).

Dashpot timers are the oldest type of industrial timers. Dashpot timers can be found on old equipment but are rarely used in new installations. Synchronous clock timers have been installed in millions of control applications and are still specified in some new applications. Solid-state timers are the most common stand-alone timer used in control applications today. However, in most electrical systems that include PLRs, the internal solid-state

programmable timers of the PLRs can be used to replace any stand-alone timer. Each timer device accomplishes its task in a different way, but all timers have the common ability to introduce some degree of time delay into a control circuit.

Dashpot Timers

A *dashpot timer* is a timer that provides time delay by controlling how rapidly air or liquid is allowed to pass into or out of a container through an orifice (opening) that is either fixed or variable in diameter. **See Figure 16-2.** For example, if the piston of a hand-operated tire pump is forced down, the piston moves down rapidly if the valve opening is unrestricted. However, if the valve opening is restricted, the travel time of the piston increases. The smaller the opening is, the longer the travel time.

TIMERS

Rockwell Automation, Allen-Bradley Company, Inc.

DASHPOT

Eagle Signal Industrial Controls

SYNCHRONOUS CLOCK

Carlo Gavazzi Inc. Electromatic Business Unit

SOLID-STATE

Teco

SOLID-STATE PROGRAMMABLE

Figure 16-1. The four major categories of timers are dashpot, synchronous clock, solid-state, and solid-state programmable.

DASHPOT TIMERS

Figure 16-2. A dashpot timer provides time delay by controlling how rapidly air or liquid is allowed to pass into or out of a container.

Synchronous Clock Timers

A *synchronous clock timer* is a timer that opens and closes a circuit depending on the position of the hands of a clock. **See Figure 16-3.** Synchronous clock timers may have one or more contacts through which the circuit may be opened or closed.

SYNCHRONOUS CLOCK TIMERS

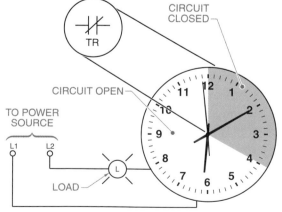

Figure 16-3. A synchronous clock timer opens and closes a circuit depending on the position of the hands of a clock.

The time delay is provided by the speed at which the clock hands move around the perimeter of the face of the clock. For example, the contacts can close once every 12 hr. A synchronous clock motor operates the timer. Synchronous clock motors are AC-operated and maintain their speed based on the frequency of the AC power line that feeds them. Synchronous clock timers are accurate timers because power companies regulate the line frequency within strict tolerances.

Solid-State Timers

A *solid-state timer* is a timer with a time delay that is provided by solid-state electronic devices enclosed within the timing device. **See Figure 16-4.** A solid-state timing circuit provides a very accurate timing function at the most economical cost. Solid-state timers can control timing functions ranging from a fraction of a second to hundreds of hours. Most solid-state timers are designed as plug-in modules for quick replacement.

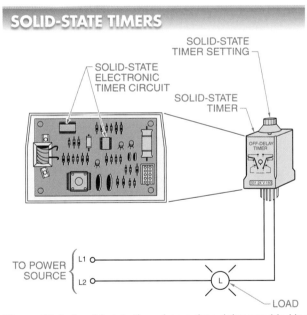

Figure 16-4. A solid-state timer has a time delay provided by solid-state electronic devices enclosed within the timing device.

Solid-state timers can replace dashpot and synchronous timers in most applications. Solid-state timers are less susceptible to outside environmental conditions because they, like relay coils, are often encapsulated in epoxy resin for protection. However, because they are encapsulated and therefore impossible to repair, they are normally discarded when they fail since they cost less than other timers.

Solid-State Programmable Timers

A *solid-state programmable timer* is a timer that is programmed within a programmable logic relay (PLR) or other programmable logic device (PLD). A *programmable logic relay (PLR)* is a solid-state control device that includes internal relays, timers, counters, and other control functions that can be programmed and reprogrammed to automatically control small residential, commercial, and industrial circuits. A PLR includes an input, output, and programming section.

The programming section of a PLR is used to program internal relays, timers, and counters that can be programmed into a control circuit without the need for external components like timers and counters. PLRs typically include at least 12 timers. Each timer can be programmed for a specific function and time setting based on the application requirements. PLRs often also include real-time clocks that allow the timer to be set on specific days and times to operate the circuit loads.

The input section of a PLR is where the circuit's digital (ON/OFF, OPEN/CLOSED, etc.) switches are connected. The input section often includes an analog (0 VDC to 10 VDC, 4 mA to 20 mA, etc.) input section. The output section is where the circuit's loads (lamps, solenoids, etc.) are connected. The output section provides either mechanical switches or solid-state output contacts. **See Figure 16-5.**

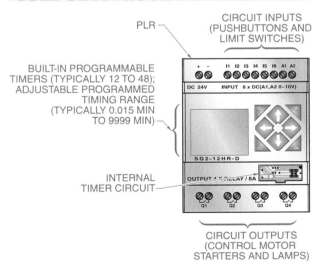

Figure 16-5. A programmable logic relay can be programmed for specific functions and time settings based on the application.

16-1 CHECKPOINT

1. Which type of timer can provide a time delay between activation and contact actuation, but its time setting cannot be precisely set and cannot be set for long periods of time?
2. Which type of timer includes multiple separate timers that can be set individually for different times?

3. What is meant by a digital input controlling a timer?
4. What is meant by an analog input controlling a timer?

16-2 TIMING FUNCTIONS

Several different timing functions are available to meet the many different requirements of time-based circuits and applications. On-delay and off-delay timing functions were the only two timing functions available when dashpot and synchronous timers were the only timers being used.

When solid-state timers became available, they offered on-delay, off-delay, one-shot, and recycle timing functions. Solid-state timers today offer dozens of special timing functions in addition to the four basic timing functions because solid-state timing circuits can be easily modified. Normally several of the special timing functions are combined into one multiple-function timer.

On-Delay Timers

An *on-delay (delay-on-operate) timer* is a device that has a preset time period that must pass after the timer has been energized before any action occurs on the timer contacts. Once activated, the timer may be used to turn a load on or off, depending on the way the timer contacts are connected into the circuit. The load energizes after the preset time delay when a normally open (NO) timer contact is used. The load de-energizes after the preset time delay when a normally closed (NC) timer contact is used.

An on-delay timer can be designed to open or close a circuit after a preset time delay. On-delay timer contacts do not change position until the set time period passes after the timer receives power. **See Figure 16-6.** After the preset time has passed, the timer contacts change position.

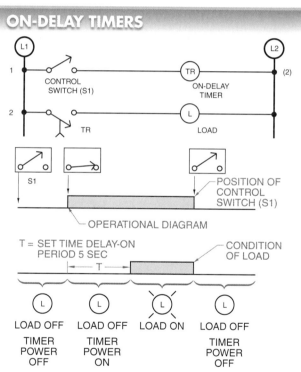

Figure 16-6. On-delay timer contacts do not change position until the set time period passes after the timer receives power.

In the on-delay timer circuit, the NO contacts close and energize the load. The load remains energized as long as the control switch remains closed. The load de-energizes the second the control switch is opened. An operational diagram is used to show timer operation. In the operational diagram, the top line shows the position of the control switch and the bottom line shows the condition of the load.

On-Delay (Timed-Closed). An on-delay (timed-closed) function may be illustrated using two balloons. **See Figure 16-7.** The solenoid plunger forces air out of balloon A, through orifice B, and into balloon C when control switch S1 is closed. Contacts TR1 close, energizing the circuit to the load after balloon C is filled. This energizes the load. The on-delay function will take 5 sec if it takes 5 sec for balloon C to fill.

One-half of an arrow is used to indicate the direction of time delay of the NO timing contacts in on-delay timers. The half arrow points in the direction of on delay. The operational diagram should be used if an arrow is not used with an on-delay timer.

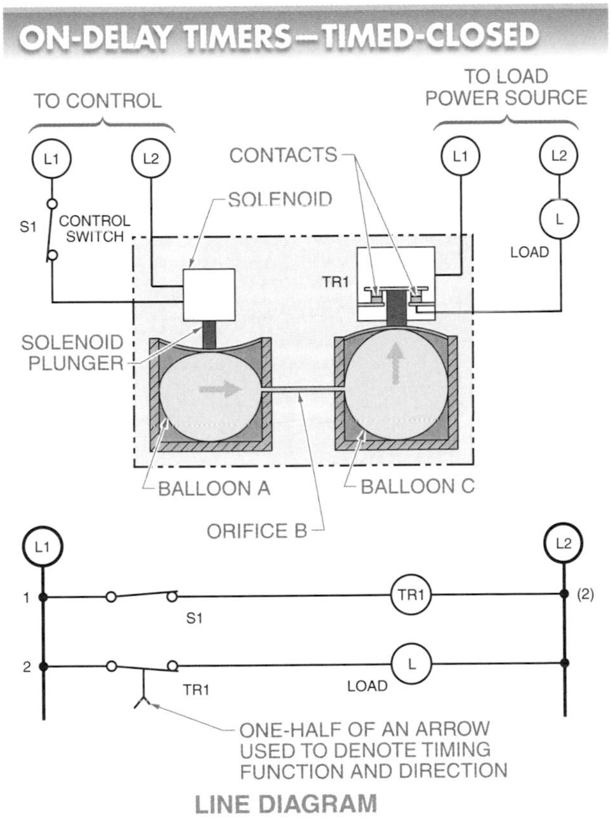

that the NC contacts open after the on-delay function has taken place. This pneumatically operated timing function is the way dashpot timers operate. A synchronous clock timer or solid-state timer could be substituted for the pneumatic timer. A pneumatic timer is the easiest to understand in terms of mechanical and timing operation.

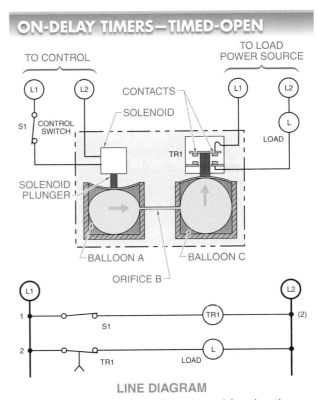

Figure 16-8. With an on-delay (timed-open) function, the contacts open after the timing cycle is complete.

On-Delay (Timed-Open). An on-delay (timed-open) function also may be illustrated using balloons to show how the contacts are forced open after the timing cycle is complete. **See Figure 16-8.** With control switch S1 closed, the solenoid plunger forces air from balloon A through orifice B and into balloon C. After 5 sec, contacts TR1 open the circuit to the load and the load is de-energized. One-half of an arrow is shown in the line diagram. The arrow indicates

On-Delay Timer Applications

On-delay timers are the most common type of timer in use. For example, an on-delay timer is often used to monitor a medical patient's breathing. **See Figure 16-9.** In this application, the timer is used to sound an alarm if a patient does not take a breath within 10 sec. The circuit includes a low-pressure switch built into a patient monitoring system. Pressure switches are available that can activate electrical contacts at pressures less than 1 psi. The circuit is turned on by the ON/OFF switch once the patient is connected to the monitor. If the patient does not take a breath, the timer starts timing and continues timing until the patient takes a breath (which resets the timer) or the timer times out. If the timer times out, the timer contacts close, sounding a warning.

ON-DELAY TIMERS—TIMED-CLOSED

TO CONTROL

TO LOAD POWER SOURCE

CONTACTS

SOLENOID

S1   CONTROL SWITCH

L1   L2

L1   L2

L

LOAD

TR1

SOLENOID PLUNGER

BALLOON A   BALLOON C

ORIFICE B

L1   L2

1   S1   TR1   (2)

2   TR1   LOAD   L

ONE-HALF OF AN ARROW USED TO DENOTE TIMING FUNCTION AND DIRECTION

LINE DIAGRAM

Figure 16-7. With an on-delay (timed-closed) function, the contacts close after the timing cycle is complete.

ON-DELAY TIMER APPLICATIONS

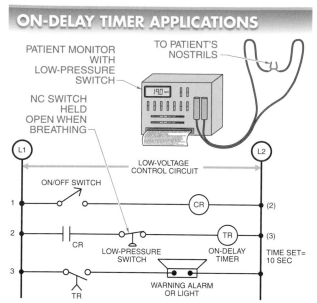

Figure 16-9. On-delay timers may be used to monitor a medical patient's breathing.

Off-Delay Timers

An *off-delay (delay-on-release) timer* is a device that does not start its timing function until the power is removed from the timer. **See Figure 16-10.** In this circuit, a control switch is used to apply power to the timer. The timer contacts change immediately and the load energizes when power is first applied to the timer.

OFF-DELAY TIMERS

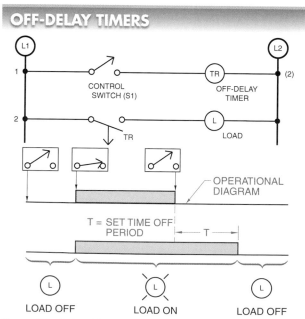

Figure 16-10. An off-delay (delay-on-release) timer is a device that does not start its timing function until the power is removed from the timer.

The timer contacts remain in the changed position and the time period starts when power is removed from the timer. The timer contacts return to their normal position and the load is de-energized when the set time period expires.

Off-Delay (Timed-Open). An off-delay (timed-open) contact circuit may be used to continue to provide cooling in a projector once the bulb has been turned off but has not had time to cool down. **See Figure 16-11.**

OFF-DELAY TIMERS—TIMED-OPEN

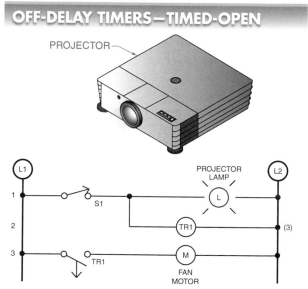

Figure 16-11. An off-delay (timed-open) contact circuit may be used to continue to provide cooling in a projector once the bulb has been turned off but has not had time to cool down.

In this circuit, closing switch S1 turns on the projector bulb and activates timer coil TR1. With timer TR1 energized, NO contacts TR1 immediately close, energizing the fan motor, which controls the cooling of the projector.

The projector bulb and the cooling fan remain on as long as switch S1 stays closed. When switch S1 is opened, the projector bulb turns off and power is removed from the timer. Contacts TR1 remain closed for a preset off-delay and then open, causing the cooling fan to turn off. This off-delay, timed-open circuit is generally set to adequately cool the projector equipment before it shuts off. This circuit can also be used for large cooling fan motors when the fan motor in the control circuit is replaced with a motor starter. The motor starter could be used to control any size motor.

Off-Delay (Timed-Closed). An off-delay (timed-closed) contact circuit may be used to provide a pumping system with backspin protection and surge protection on stopping. **See Figure 16-12.**

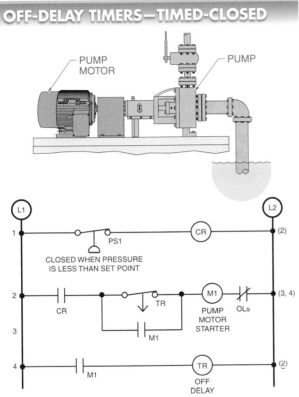

OFF-DELAY TIMERS—TIMED-CLOSED

Figure 16-12. An off-delay (timed-closed) contact circuit may be used to provide a pumping system with backspin protection and surge protection on stopping.

Surge protection is often necessary when a pump is turned off and a high column of water is stopped by a check valve. The force of the sudden stop may cause surges that operate the pressure switch contacts, subjecting the starter to chatter (open and close). Backspin is the backward turning of a centrifugal pump when the head of water runs back through the pump just after it has been turned off. Starting the pump during backspin may damage the pump motor.

To minimize any damage resulting from surge and backspin, the pressure switch PS1 closes on low pressure and energizes the control relay CR. The CR contacts close and energize the pump motor starter M1 through the NC contacts TR. For M1, one set of NO contacts energize the timer and the other set of NO contacts keep the motor energized even after the NC timer contacts open. When PS1 opens at the set system pressure, M1 cannot be re-energized until the off-delay contacts TR are allowed to time out and

reclose the NC contacts, regardless of the number of times the PS1 contacts chatter during backspin pressures.

Off-Delay Timer Applications

Off-delay timers are used in applications that require a load to remain energized even after the input control has been removed, such as in emergency industrial showers. **See Figure 16-13.** In this circuit, the off-delay timer is used to keep the water flowing for 1 min after the pushbutton for the emergency industrial shower is pressed and released.

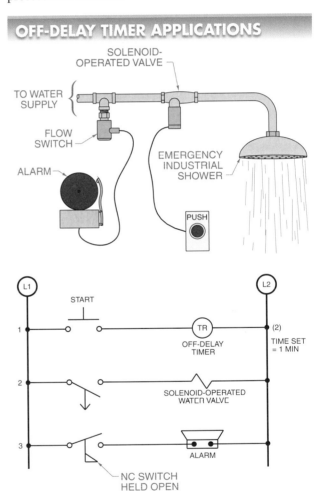

OFF-DELAY TIMER APPLICATIONS

Figure 16-13. Off-delay timers are used in applications that require a load to remain energized even after the input control has been removed.

Tech Fact

A stand-alone off-delay timer is called an off-delay or delay-on-release timer. However, when an off-delay timer is programmed using a PLC software program, it is programmed as a TOF timer. Likewise, a stand-alone on-delay timer is called an on-delay or delay-on-operate timer. But when an on-delay timer is programmed using a PLC software program, it is programmed as a TON timer.

After the pushbutton is pressed, the timer contacts close and the solenoid-operated valve starts the flow of water. The water flows even if the pushbutton is released. A flow switch is used to indicate when water is flowing. The flow switch sounds an alarm that can be used to bring help. The flow switch also sounds the alarm if there is a break in the flow of water at any point downstream from the switch.

Plugging Using Timing Relays

Plugging can also be accomplished by using a timing relay. Normally, the advantage of using a timing relay is a lower cost since the timer is inexpensive and does not have to be connected mechanically to the motor shaft or driven machine. The disadvantage is that, unlike a plugging switch, the timer does not compensate for a change in the load condition (which affects stopping time) once the timer is preset.

An off-delay timer may be used in applications where the time needed to decelerate a motor is constant and known. **See Figure 16-14.** In this circuit, the NO contacts of the timer are connected into the circuit in the same manner as a plugging switch. The coil of the timer is connected in parallel with the forward starter.

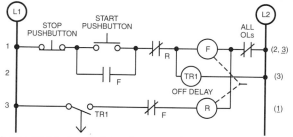

Figure 16-14. An off-delay timer may be used in applications where the time needed to decelerate the motor is constant.

The motor is started and memory is added to the circuit when the start pushbutton is pressed. In addition to energizing the forward starter, the off-delay timer is also energized. The energizing of the off-delay timer immediately closes the NO timer contacts. The closing of these contacts does not energize the reverse contacts due to the interlocks.

The forward starter and timer coil are de-energized when the stop pushbutton is pressed. The NO timing contact remains held closed for the setting of the timer. The reversing starter is energized for the

period of time set on the timer when the timing contact is held closed. This plugs the motor to a stop. The timer's contact must reopen before the motor is actually reversed. The motor reverses direction if the time setting is too long.

On-delay and off-delay timers can be used to control motors that operate conveyor lines.

An off-delay timer may also be used for plugging a motor to a stop during emergency stops. **See Figure 16-15.** In this circuit, the timer's contacts are connected in the same manner as the plugging switch. The motor is started and memory is added to the circuit when the start pushbutton is pressed. The forward starter and timer are de-energized if the stop pushbutton is pressed. Although the timer's NO contacts are held closed for the time period set on the timer, the reversing starter is not energized. This is because no power is applied to the reversing starter from L1.

If the emergency stop pushbutton is pressed, the forward starter and timer are de-energized and the reversing starter is energized. The energizing of the reversing starter adds memory to the circuit and stops the motor. The opening of the timing contacts de-energizes the reversing starter and removes the memory.

A comparison chart may be used to compare the operation of on-delay and off-delay timing functions and contacts. **See Figure 16-16.** To help compare timing functions, instantaneous relay contacts are also included. Some manufacturers also use abbreviations in their catalogs to describe the type of contacts used.

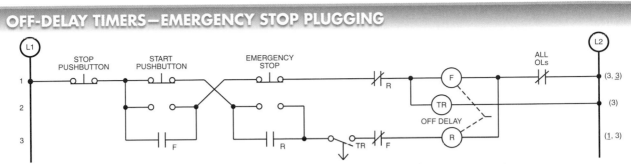

OFF-DELAY TIMERS—EMERGENCY STOP PLUGGING

Figure 16-15. An off-delay timer may also be used for plugging a motor to a stop during emergency stops.

TIMER COMPARISON			
Abbreviation	**Meaning**	**Function**	**Symbols**
NOTC	Normally open, timed-closed	On-delay (timed-closed) contact – Timer contact normally open: timed-closed on timer energization; opens immediately on timer de-energization	
NCTO	Normally closed, timed-open	On-delay (timed-open) contact – Timer contact normally closed: timed-open on timer energization; closes immediately on timer de-energization	
NOTO	Normally open, timed-open	Off-delay (timed-open) contact – Timer contact normally open: closes immediately on timer energization; timer contact times open on timer de-energization	
NCTC	Normally closed, timed-closed	Off-delay (timed-closed) contact – Timer contact normally closed: opens immediately on timer energization; timer contact times closed on timer de-energization	
NO	Normally open	Instantaneous contact – Normally open: contact closes immediately on relay energization; opens immediately on relay de-energization	
NC	Normally closed	Instantaneous contact – Normally closed: contact opens immediately on relay energization; closes immediately on relay de-energization	

Figure 16-16. A comparison chart may be used to compare the operation of on-delay and off-delay timing functions and contacts.

One-Shot Timers

A *one-shot (interval) timer* is a device in which the contacts change position immediately and remain changed for the set period of time after the timer has received power. **See Figure 16-17.** After the set period of time has passed, the contacts return to their normal position.

One-shot timers are used in applications in which a load is ON for only a set period of time. One-shot timer applications include coin-operated games, dryers, car washes, and other machines. One-shot timing functions have not been available as long as on-delay and off-delay timing functions because the one-shot timing function became available only when solid-state timers became available. For this reason, and the fact that so many other timing functions are now available, no standard symbol was established for any other timer contacts except on-delay and off-delay. Today, the symbol for basic NO and NC contacts, along with the timer type and/or operational diagram, are used with all timers that are not on-delay or off-delay.

ONE-SHOT TIMERS

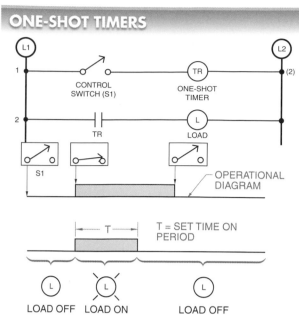

Figure 16-17. A one-shot (interval) timer has contacts that change position and remain changed for the set period of time after the timer has received power.

One-Shot Timer Applications

One-shot timers are used in applications that require a fixed-time output for a set period of time. For example, a one-shot timer can be used to control the amount of time that plastic wrap is wound around a pallet of cartons. **See Figure 16-18.**

In this application, a photoelectric switch detects a pallet entering the plastic wrap machine. The photoelectric switch energizes the one-shot timer. The one-shot timer contacts close, starting the wrapping process. The wrapping process continues for the setting of the timer. A second photoelectric switch could be used to detect that the plastic wrap is actually being applied. This can help indicate a tear in the plastic or an empty roll.

Recycle Timers

A *recycle timer* is a device in which the contacts cycle open and closed repeatedly once the timer has received power. The cycling of the contacts continues until power is removed from the timer. **See Figure 16-19.** In a recycle timer circuit, the closing of the control switch starts the cycling function. The load continues to turn on and off at regular time intervals as long as the control switch is closed. The cycling function stops when the control switch is opened.

ONE-SHOT TIMER APPLICATIONS

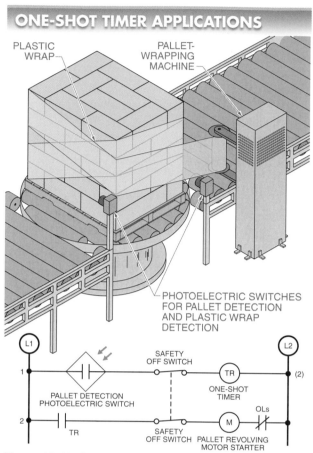

Figure 16-18. One-shot timers are used in applications that require a fixed-time output for a set period of time.

RECYCLE TIMERS

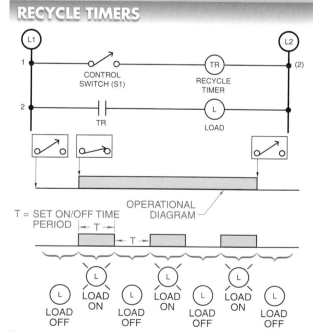

Figure 16-19. A recycle timer is a device in which the contacts cycle open and closed repeatedly once the timer has received power.

Recycle timers may be symmetrical or asymmetrical. A *symmetrical recycle timer* is a timer that operates with equal on and off time periods. An *asymmetrical recycle timer* is a timer that has independent adjustments for the on and off time periods. Asymmetrical timers always have two different time adjustments.

Recycle Timer Applications

Recycle timers are used in applications that require a fixed on and off time period. For example, a recycle timer can be used to automatically keep a product mixed. **See Figure 16-20.** In this application, power is applied to the timer when the three-position selector switch is placed in the automatic position. The timer starts recycling for as long as the selector switch is in the automatic position. The recycle timer turns the mixing motor on and off at the set time. An asymmetrical timer works best for this application. The on time period (mixer motor is ON) is set less than the off time period (mixer motor is OFF). For example, the timer may be set to mix the product for 5 min every 2 hr.

Multiple-Function Timers

On-delay, off-delay, one-shot, and recycle timers are considered mono-function timers. That is, they perform only one timing function, such as on-delay or off-delay. Multiple-function timers are solid-state timers that can perform many different timing functions. Multiple-function timers are normally programmed for different timing functions by the placement of dual in-line package (DIP) switches located on the timer. **See Figure 16-21.**

MULTIPLE-FUNCTION TIMERS

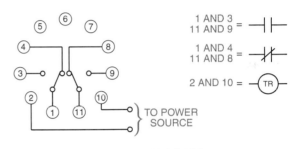

WIRING DIAGRAM

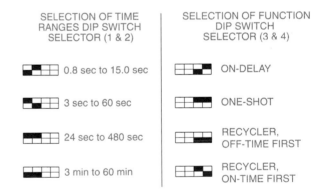

TIME/FUNCTION SETTINGS

RECYCLE TIMER APPLICATIONS

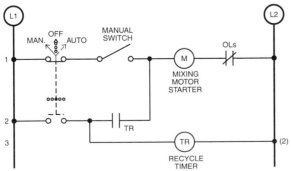

Figure 16-20. Recycle timers are used in applications that require a fixed on and off time period.

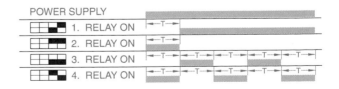

OPERATIONAL DIAGRAM

Figure 16-21. A multiple-function timer may use the placement of DIP switches to determine the type of timing function and timer setting.

In this timer, four DIP switches are used to set the timer range and function. The first two DIP switches set the time range from 0.8 sec to 60 min. The last two DIP switches set the timer function. The timer can be set for an on-delay, one-shot, or recycle timer function. The recycle timer function can be set to start with the off time period occurring first or the on time period occurring first.

Machines used in commercial and industrial manufacturing and processing are time- and count-controlled by electrical and electronic circuits and systems.

In addition to some (or all) of the basic timing functions, many multiple-function timers can also be programmed for special timing functions. **See Figure 16-22.** In this multiple-function timer, many timing functions can be programmed with a time range from 0.15 sec to 220 hr. This timer includes standard timing functions such as an on-delay (program setting 1) and special timing functions such as a combination of both on-delay and off-delay (program setting 5).

MULTIPLE-FUNCTION TIMERS—SPECIAL TIMING FUNCTIONS

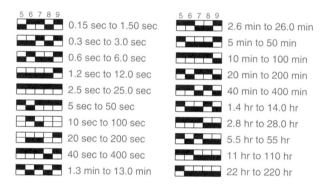

OPERATIONAL DIAGRAM

SELECTION OF TIME RANGE
DIP SWITCH SELECTOR (5 to 9)

5 6 7 8 9		5 6 7 8 9	
	0.15 sec to 1.50 sec		2.6 min to 26.0 min
	0.3 sec to 3.0 sec		5 min to 50 min
	0.6 sec to 6.0 sec		10 min to 100 min
	1.2 sec to 12.0 sec		20 min to 200 min
	2.5 sec to 25.0 sec		40 min to 400 min
	5 sec to 50 sec		1.4 hr to 14.0 hr
	10 sec to 100 sec		2.8 hr to 28.0 hr
	20 sec to 200 sec		5.5 hr to 55 hr
	40 sec to 400 sec		11 hr to 110 hr
	1.3 min to 13.0 min		22 hr to 220 hr

TIME SETTING

Figure 16-22. Multiple-function timers can be programmed for many timing functions with various time ranges.

16-2 CHECKPOINT

1. What type of timer is used to control the fan motor in a furnace so that the fan continues to operate for a time after the heating source is turned off in order to distribute the remaining heat from the furnace?

2. What type of timer is used to control the light in a garage door opener so that the light turns on when the opener is activated and remains on for a set period of time after the opener button is released?

3. What is a symmetrical recycle timer?

4. What is an asymmetrical recycle timer?

16-3 MULTIPLE-CONTACT TIMERS

In the past, manufacturers of synchronous clock timers included instantaneous and time delay contacts on their timers to meet the many different requirements of timer applications. These timers were used for numerous applications where both types of contacts were included. They are often still used and have been updated to include solid-state timing circuits and can be converted from on-delay to off-delay timing functions.

Multiple-Contact Timer Wiring Diagrams

Wiring diagrams are required on multiple-contact timers because several different connection points must be located and wired to the timer for it to perform properly. The wiring diagram for a timer is normally located on the back of the timer. By quickly surveying the diagram, the timer clutch coil, the synchronous motor (or solid-state circuit), and the timer pilot light can be located. **See Figure 16-23.**

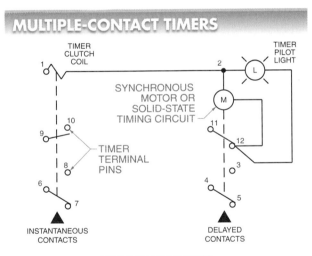

Figure 16-23. Wiring diagrams simplify the process of locating connection points and wiring multiple contact timers.

In this timer, the timer clutch coil engages and disengages the motor. This is similar to what is found in an automobile. The timer clutch coil engages and disengages contacts 9 and 10 as well as 6, 7, and 8. When the clutch is engaged, contacts 9 and 10 and 6 and 8 close instantaneously. Contacts 6 and 7 open simultaneously. In other words, the timer clutch controls two NO contacts and one NC contact instantaneously. The timing motor controls contacts 11 and 12 as well as 3, 4, and 5 through a time delay. When the motor times out, contacts 4 and 5 and 11 and 12 open and contacts 3 and 4 close. Contacts 11 and

12 open slightly later than contacts 4 and 5 after the motor times out. The timer pilot light, wired in parallel with the motor, indicates when the motor is timing. Contacts 4 and 5 and 11 and 12 close and contacts 3 and 4 open when the timer is reset.

Wiring Motor-Driven Timers. A motor-driven timer can be wired into L1 and L2 to provide power to the circuit. **See Figure 16-24.** In this circuit, current flows from L1, through the timer clutch coil, and on to L2. Current also flows from terminal 1 through the closed contacts 11 and 12, feeding the parallel circuit provided by the timer motor and timer pilot light, and then on to L2.

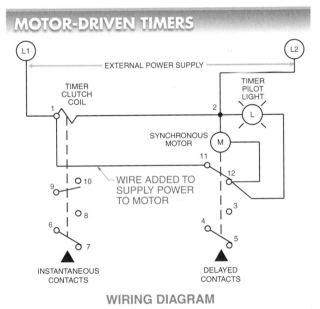

Figure 16-24. A motor-driven timer can be wired into L1 and L2 of a multiple-contact timer to provide power.

Operation of Motor-Driven Timers. A timer that is activated by a limit switch can affect various loads wired into the circuit. **See Figure 16-25.** The timer can be controlled by a manual, mechanical, or automatic input. The timer may be used to achieve control using a sustained mechanical input. This input (limit switch) must remain closed to energize the timer clutch coil and power the timer motor. The timer is connected for on-delay, requiring input power to close the clutch and start the timer. The contacts, both instantaneous and time delay, are connected to four loads marked A, B, C, and D. These loads may be any load, such as a solenoid, magnetic motor starter, or light.

MOTOR-DRIVEN TIMER OPERATIONS

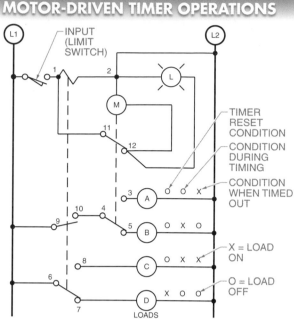

Figure 16-25. Multiple-contact timers may be used to achieve control using a sustained mechanical input.

A code is added above each load to illustrate the sequence during reset, during timing, and when timed out. The code is used to indicate the condition (ON or OFF) of each load during the three stages of the timer. The three stages include the reset condition (no power applied to timer), the timing condition (time at which the timer is timing but not timed out), and the timed-out condition. An "O" indicates when the load is de-energized and an "X" indicates when the load is energized.

Loads C and D are relay-type responses, using only the instantaneous contacts. Loads A and B use the combined action of the instantaneous and delay contacts to achieve the desired sequence. In this circuit, the timing motor is wired through delay contacts 11 and 12 to ensure motor cutoff after the timer times out. This is required because the limit switch remains closed after timing out. The limit switch otherwise would have to be opened to reset the timer. This also means that a loss of plant power resets the timer because the clutch opens when power is lost.

Multiple-Contact Timer Applications

Multiple-contact timers are used in applications where a sustained input is used to control a circuit. **See Figure 16-26.** In this circuit, the cartons coming down the conveyor belt are to be filled with detergent. Each carton must be filled with the same amount of detergent

and the process must be automatic. To accomplish this, the timer circuit is used to control the time it takes to fill one carton. A limit switch with sustained input is used to detect the carton. A motor drives the conveyor belt and a solenoid opens or closes the hopper full of detergent. The limit switch could be replaced with a photoelectric or proximity switch.

MULTIPLE-CONTACT TIMER APPLICATIONS

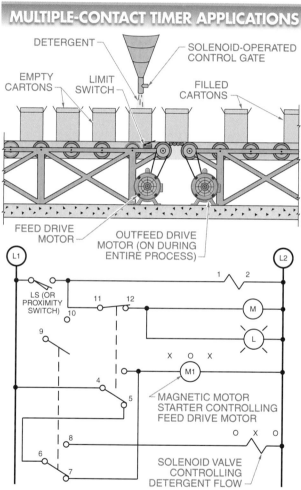

Figure 16-26. Multiple-contact timers are used in applications where a sustained input is used to control a circuit.

As the cartons are coming down the conveyor, the feed drive motor is ON and the solenoid valve is OFF (no detergent fill). As a carton contacts the limit switch, the feed drive motor shuts off. This stops the conveyor and energizes the solenoid. The solenoid opens the control gate on the hopper of detergent to fill the carton. After the timer times out, the feed drive motor turns on and the solenoid valve closes. This removes the filled carton, which opens the limit switch, resetting the timer. The code for each load is added above the respective load. The code illustrates the desired sequence of operation of the loads.

Memory could be added to a circuit if the circuit requires a momentary input, such as a pushbutton, to initiate timing. **See Figure 16-27.** In this circuit, a pushbutton is used as the input signal. As with any memory circuit, a NO instantaneous contact must be connected in parallel with the pushbutton if memory is to be added into the circuit. To accomplish this, NO instantaneous contacts 9 and 10 are connected in parallel with the pushbutton so that power is maintained when the control switch is released. A separate reset (stop) switch is used to reset the timer. After the timer has timed out, the timer resets only if the reset pushbutton is pressed. The sequence of each load is illustrated by the code above each load.

Supply Voltage-Controlled Timers

A *supply voltage-controlled timer* is a timer that requires the control switch to be connected so that it controls power to the timer coil. **See Figure 16-28.** In this circuit, the control switch is connected in series with the timer coil. The advantage of this control method is that it is exactly the same as most electrical control circuits used over the years. The disadvantage is that if a standard 115 VAC timer is used, the control switch has to switch the 115 VAC. This means that the control switch has to be installed using standard AWG 14 copper wire and properly enclosed for safety.

MULTIPLE-CONTACT TIMER APPLICATIONS—ADDING MEMORY

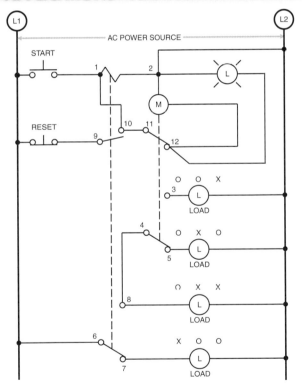

Figure 16-27. Memory could be added to a circuit if the circuit requires a momentary input, such as a pushbutton, to initiate timing.

SUPPLY VOLTAGE-CONTROLLED TIMERS

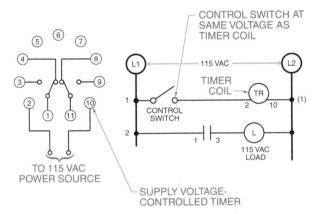

Figure 16-28. A supply voltage-controlled timer requires the control switch to be connected so that it controls power to the timer coil.

Wiring diagrams are not true line diagrams. Line diagrams are standards used by all manufacturers. Wiring diagrams are the actual diagrams matching the logic of the line diagram to the manufacturer's product that is designed to perform that logic. A wiring diagram is used because it is the actual diagram found on the timer.

Programmable logic controllers (PLCs) and programmable logic relays (PLRs) have internal timers that must be programmed into a circuit.

Contact-Controlled Timers

A *contact-controlled timer* is a timer that does not require the control switch to be connected in line with the timer coil. **See Figure 16-29.** In a contact-controlled timer, the timer supplies the voltage to the circuit in which the control switch is placed. The voltage of the control circuit is normally less than 24 VDC.

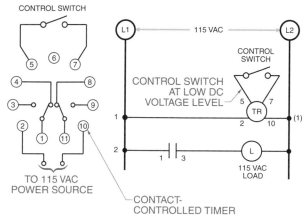

Figure 16-29. A contact-controlled timer does not require the control switch to be connected in line with the timer coil.

The advantage of this control method is that the control switch can be wired using low-voltage wire (AWG 16, 18, or 20). The control switch contacts can be small and normally require less than a 100 mA rating because the timer control circuit requires little current to pass through the control switch. The disadvantage of this control method is that many electricians are unfamiliar with how to connect the control switch outside a standard 115 VAC control circuit. The timer low-voltage circuit (pins 5 and 7) is often connected to the timer 115 VAC circuit (pins 2 and 10). This results in the destruction of the timer.

Sensor-Controlled Timers

A sensor-controlled timer is like a contact-controlled timer or transistor-controlled timer that includes an additional output from the timer. A *sensor-controlled timer* is a timer controlled by an external sensor in which the timer supplies the power required to operate the sensor. **See Figure 16-30.** The advantage of a sensor-controlled timer is that, by supplying power out of the timer itself, no external power supply is required to operate the control sensor. Sensor-controlled timers are generally used with photoelectric and proximity controls, but they may be used with any control that meets the specifications of the timer.

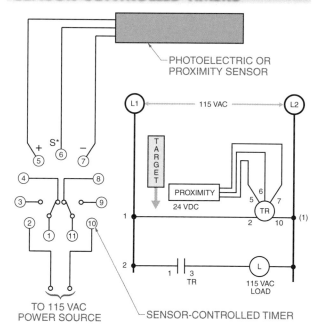

Figure 16-30. A sensor-controlled timer is controlled by an external sensor in which the timer supplies the power required to operate the sensor.

16-3 CHECKPOINT

1. When adding a timing code to a load controlled by a timer, what does an "X" indicate?

2. When adding a timing code to a load controlled by a timer, what does an "O" indicate?

3. When adding an XOX timing code to a load, is the load ON during the timing period?

4. When connecting the input switch that controls a timer, what is the advantage of using a contact-controlled timer instead of a supply voltage-controlled timer?

16-4 TROUBLESHOOTING CONTACT-BASED TIMING CIRCUITS

Troubleshooting timing circuits is a matter of checking power to the timer and checking for proper timer contact operation. **See Figure 16-31.** The timer is replaced if there is a problem with any part of the timer. The most common problem is contact failure. The current at the contacts is measured if the contacts failed prematurely. The current must not exceed the rating of the contacts. To troubleshoot timers, the following procedure is applied:

1. Measure the voltage of the control circuit. The voltage must be within the specification range of the timer. Correct the voltage problem if the voltage is not within the specification range of the timer. A common problem is an overloaded control transformer that is delivering a low-voltage output.

2. Measure the voltage at the timer coil. The voltage should be the same as the voltage of the control circuit. Check the control switch if the voltage is not the same. Check the voltage at the timer contacts if the voltage is correct.

3. Measure the voltage into the timer contacts. The voltage must be within the range of the load the timer is controlling. Correct the voltage problem if the voltage is not within this range.

4. Measure the voltage out of the timer contacts if the voltage into the timer contacts is correct. The voltage should be the same as the voltage of the control circuit. Check the timer contact connection points and wiring for a bad connection or corrosion if the voltage is not the same.

Transistor-Controlled Timers

Transistor-controlled timers are similar to contact-controlled timers. A *transistor-controlled timer* is a timer that is controlled by an external transistor from a separately powered electronic circuit. **See Figure 16-32.** Modern industrial control circuits are often connected to DC electronic circuits that include a transistor as their output. Such devices include solid-state temperature controls, counters, computers, and other control devices. A timer that uses a transistor as the control input can be connected to almost any electronic circuit that includes a transistor output.

The advantages and disadvantages of a transistor-controlled timer are the same as a contact-controlled timer. Any transistor-controlled timer can use a contact as the control device. However, not all contact-controlled timers can use a transistor as the control device. Manufacturer specification sheets should be checked when any unfamiliar timer is used.

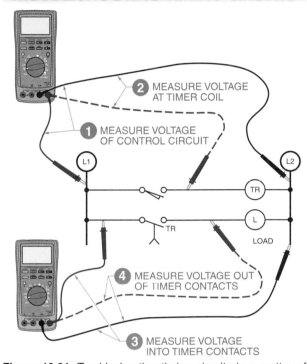

Figure 16-31. Troubleshooting timing circuits is a matter of checking power to the timer and checking for proper timer contact operation.

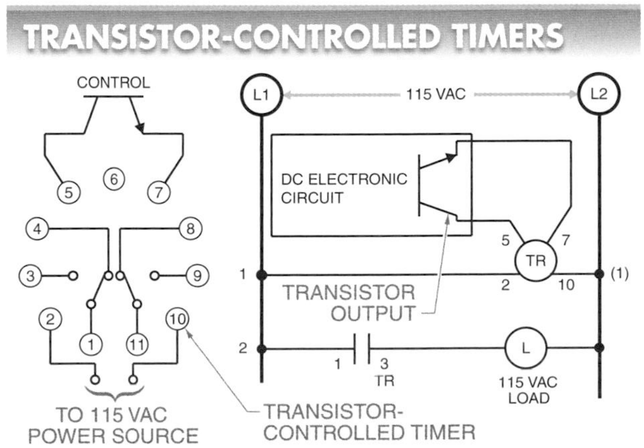

Figure 16-32. A transistor-controlled timer is controlled by an external transistor from a separately powered electronic circuit.

555/556 Timers

A 555 timer is an integrated circuit (IC) that can function as either a timer or an oscillator. A 556 timer includes two 555 timers housed in a 14-pin package. With a 556 timer, the two timers share the same power supply pins. **See Figure 16-33.** A 556 timer can also be used as a single 555 timer by using half of the IC.

IC TIMERS

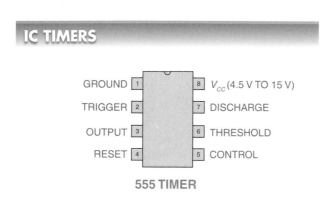

555 TIMER

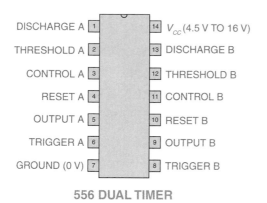

556 DUAL TIMER

Figure 16-33. Integrated circuit timers are available with different pin arrangements.

The circuit symbol for a 555 timer and 556 timer is a box with pins arranged the circuit diagram. For example, pin 8 is on the top right for the common-collector voltage (V_{CC}) and pin 4 is on the lower left. Typically, pins are represented by numbers and are not labeled by function. The 555 timer and 556 timer can be used with a V_{CC} of 4.5 V to 16 V.

555 Timer Operation

A 555 timer consists of a voltage-divider network (R1, R2, and R3), two comparators (Comp 1 and Comp 2),

a flip-flop, two control transistors (Q1 and Q2), and a power-output amplifier. *Note:* A flip-flop is the electronic equivalent of a toggle switch. It has one high output and one low output. When one is high, the other is low.

The comparators compare the input voltages to the internal reference voltages created by the voltage divider, which consists of resistors R1, R2, and R3. Since the resistors are of equal value, the reference voltage provided by two of the resistors is two-thirds of the common-collector (supply) voltage (V_{CC}). The other resistor provides one-third of V_{CC}. The value of V_{CC} may change (for example, from 9 V to 12 V to 15 V) from IC to IC. However, the ratios remain the same. When the input voltage to either one of the comparators is higher than the reference voltage, the comparator goes into saturation and produces a signal that will trigger the flip-flop. In the IC, the flip-flop has two inputs: S and R.

Note: The two comparators feed signals into the flip-flop. Comp 1 is called the threshold comparator, and Comp 2 is called the trigger comparator. Comp 1 is connected to the S input of the flip-flop. Comp 2 is connected to the R input of the flip-flop.

Whenever the voltage at S is positive and the voltage at R is zero, the output of the flip-flop is high. Whenever the voltage at S is zero and the voltage at R is positive, the output of the flip-flop is low. The output from the flip-flop at point Q is then applied to transistors Q1 and Q2 and to the output amplifier simultaneously. If the signal is high, Q1 will turn on such that pin 7 (the discharge pin) will be grounded through the emitter-collector circuit. Then, Q2 will be in a position to turn on pin 7 to ground through the emitter-collector circuit.

The flip-flop signal is also applied to Q2. A signal to pin 4 can be used to reset the flip-flop. Pin 4 can be activated when a low-level voltage signal is applied. Once applied, this signal will override the output signal from the flip-flop. The reset pin (pin 4) will force the output of the flip-flop to go low, regardless of the state of the other inputs to the flip-flop.

Finally, the flip-flop signal is applied to the power-output amplifier. The power-output amplifier boosts the signal, and the 555 timer delivers up to 200 mA of current when operated at 15 V. The output can be used to drive other transistor circuits and even devices such as a small audio speaker. The output will always be an inverted signal compared to the input. If the input to the power-output amplifier is high, the output will be low. If the input is low, the output will be high.

An external time-delay circuit is needed to operate the 555 timer. A variable resistor and a capacitor are used to determine the amount of time delay. **See Figure 16-34.** The capacitor is usually held in the discharge state by transistor Q1, which shorts the capacitor to ground. The timing cycle begins when a negative pulse is applied to the trigger input (pin 2). The negative pulse forces a flip-flop to go low on the output, which in turn removes the base bias to the discharge transistor (Q1). With transistor Q1 turned off, the short circuit (ground) is removed from the capacitor. Therefore, the capacitor is allowed to start charging with a time constant established by the values of the resistor and capacitor.

When the voltage across the capacitor reaches two-thirds of V_{CC}, as determined by the voltage divider, Comp 1 resets the flip-flop, returning it to its original high state. This change causes Q1 to turn on, which discharges the external capacitor. With the capacitor discharging, the charging cycle stops. The resetting of the flip-flop drives the power-output amplifier, causing the output to return to its normal low operating condition.

The overall timing sequence can be seen by comparing input and output waveforms to the charging rate of the resistor–capacitor (RC) network. **See Figure 16-35.** A short input pulse at pin 2 produces a relatively long output pulse at pin 3.

EXTERNAL TIME-DELAY CIRCUITS

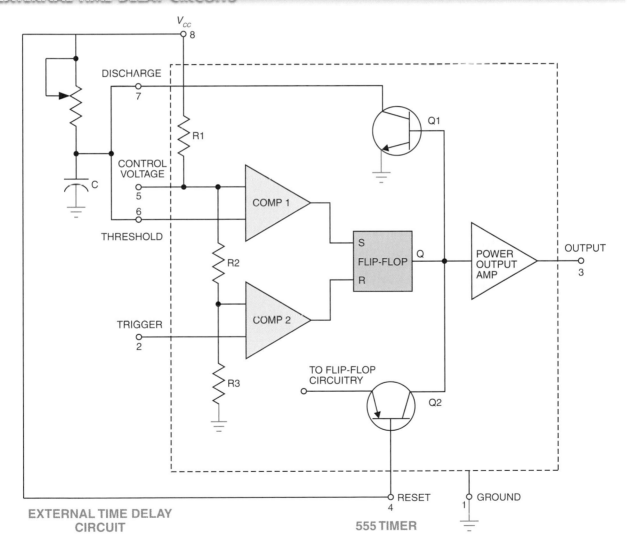

Figure 16-34. An external time-delay circuit uses a resistor and capacitor to determine the amount of time delay for a 555 timer.

TIMING SEQUENCE OF 555 TIMERS

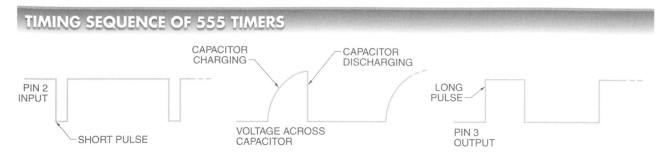

Figure 16-35. The overall timing sequence of a 555 timer is shown by comparing input and output waveforms to the charging of the RC network.

Note: The leading edge of the output pulse starts with the leading edge of the input pulse. The width of the output pulse is determined by the voltage produced across the capacitor, which is determined by the RC time constant.

The 555 timer can be connected to time on and off repeatedly by itself. It can also time out when the circuit receives an outside trigger signal. For the 555 timer to be fully functional, certain external components must be connected to the IC. The actual timing is accomplished with a RC time-delay network.

The accuracy of the timing sequence may vary as much as ±20% of the stated delay. Other precision timers may vary ±1% or less of the stated delay. The time it takes for a capacitor to charge or discharge is determined by the value of the resistance and the size of the capacitor.

Solid-State Programmable Timers

Solid-state programmable timers are devices included in many control circuits to provide timing functions. Programmable logic relays (PLRs) have internal solid-state timers that are programmed into the control circuit rather than hardwired. The internal timers may be programmed with line (ladder) diagram or function block diagram format. Internal timers must be programmed with a specific mode, time base, and preset value. Timer modes normally include at least several timing functions, such as on-delay, off-delay, one-shot, recycle, and special timing functions. **See Figure 16-36.**

Time base is used to set the precision of a timer. Preset value is used to set the amount of time that a timer counts before completing an operation. For example, a preset value of 1234 with a time base setting of 0.01 sec is equal to 12.34 sec. A preset value of 1234

with a time base of 1 sec equals 1234 sec. The time base is normally set to 0.01 sec, 0.1 sec, 1 sec, 1 min, or 1 hr. The mode is set for on-delay with a preset value of 1000 and a time base of 0.01 sec, resulting in a timing function of 10 sec duration.

SOLID-STATE PROGRAMMABLE TIMERS—PLRs

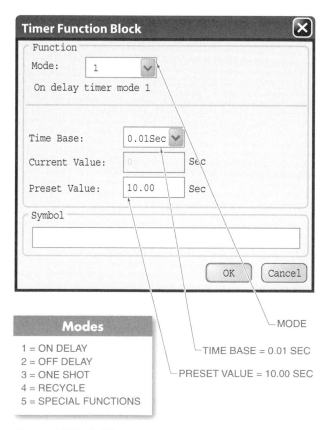

Figure 16-36. Solid-state programmable timers may be programmed using a PLR rather than hardwiring them into a circuit.

In applications that require multiple relays or circuit changes, PLRs are used. The advantages of using a PLR include the following:

- relatively low cost
- can include as many contacts as needed because the contacts are programmed (not physical contacts as on stand-alone relays)
- additional control-circuit devices such as timers, counters, and internal relays
- changing, saving, and printing of programs allowed
- can be programmed using multiple languages
- includes a simulation function that allows circuits to be tested before they are implemented

A timer is programmed and set in the same manner regardless of whether the circuit is programmed in ladder logic format or function block format. In addition, circuit inputs (pushbuttons and limit switches) and circuit outputs (solenoids and motor starters) are wired to the PLR in the same way regardless of how the PLR is programmed. **See Figure 16-37.**

Real-Time Clocks. A real-time clock allows for outputs to be turned on and off at preset times of the day and days of the week. For example, in an outdoor automatic lawn sprinkler system, the water valve solenoid (output) can be set to turn on at 6:30 AM and turn off at 7:30 AM for one hour of watering each day. The real-time clock can also be set to turn on and off the sprinkler system on set days. For example, the sprinkler system may be turned on only on Sundays or Monday through Friday. **See Figure 16-38.**

A real-time clock for a PLR is based on a 24 hr (military time) clock. The main difference between a 12 hr and 24 hr clock is how the hours are expressed. The 12 hr clock uses numbers 1 through 12 to identify each hour in a day. The first 12 hr set (12:00 AM through 11:59 AM) uses the "AM" designation, and the second 12 hr set (12:00 PM through 11:59 PM) uses the "PM" designation. The 24 hr clock uses the numbers 0 through 23 to identify each hour in a day. Under this system, 12:00 AM is 00:00, 1:00 AM is 01:00, 1 PM is 13:00, and 11:00 PM is 23:00.

SOLID-STATE PROGRAMMABLE TIMERS—PLR DIAGRAMS

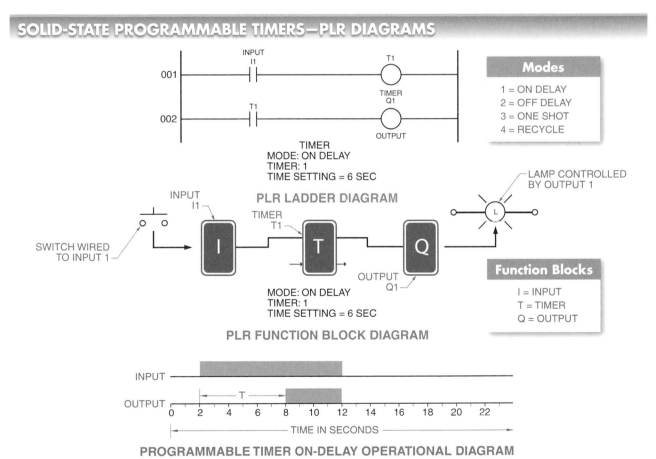

Figure 16-37. Circuit inputs and outputs are wired in the same manner regardless of how a PLR programmable timer is programmed.

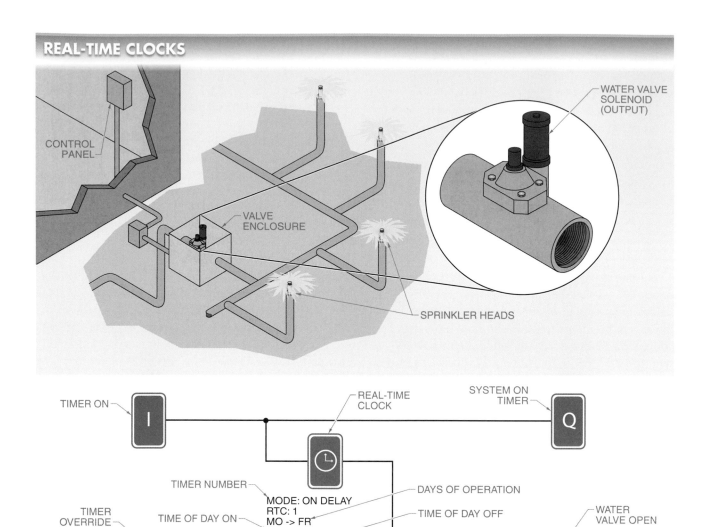

Figure 16-38. A real-time clock can be used to control an automatic sprinkler system.

Minutes and seconds are expressed the same way for both 12 hr clocks and 24 hr clocks. When converting from the 12 hr to 24 hr format and vice versa, the minutes and seconds remain the same.

Retentive and Nonretentive Timers

Programmable timers include retentive and nonretentive timers. A *retentive timer* is a timer that maintains its current accumulated time value when its control input signal is interrupted or power to the timer is removed. A *nonretentive timer* is a timer that does not maintain its current accumulated time value when its control input signal is interrupted or power to the timer is removed. **See Figure 16-39.**

When a retentive timer is used, the timer retains all accumulated time values if it is interrupted during a procedure. Retentive timers are used in applications where the timer value is to continue where it left off after an interruption. When a nonretentive timer is used, the timer does not retain any accumulated time values if it is interrupted during a procedure. Nonretentive timers reset to the values they had before they began timing. Retentive and nonretentive timers retain all preset values during a power failure.

When using a PLR for a control circuit, no relay is actually wired into the circuit. The control circuit is programmed into the PLR using standard line (ladder) or function block diagram format. **See Figure 16-40.**

Only the inputs (off, on, pressure switch, and push-to-test switch) of the circuit are wired to the PLR input section. In addition, the outputs (solenoids 1 and 2, red lamp, and green lamp) of the circuit are wired to the PLR output section.

Tech Fact

When programming timers using a PLC software program, a retentive timer is programmed by selecting an RTO (retentive timer operation) timer. An RTO timer requires a separate reset input to reset accumulated time value back to zero.

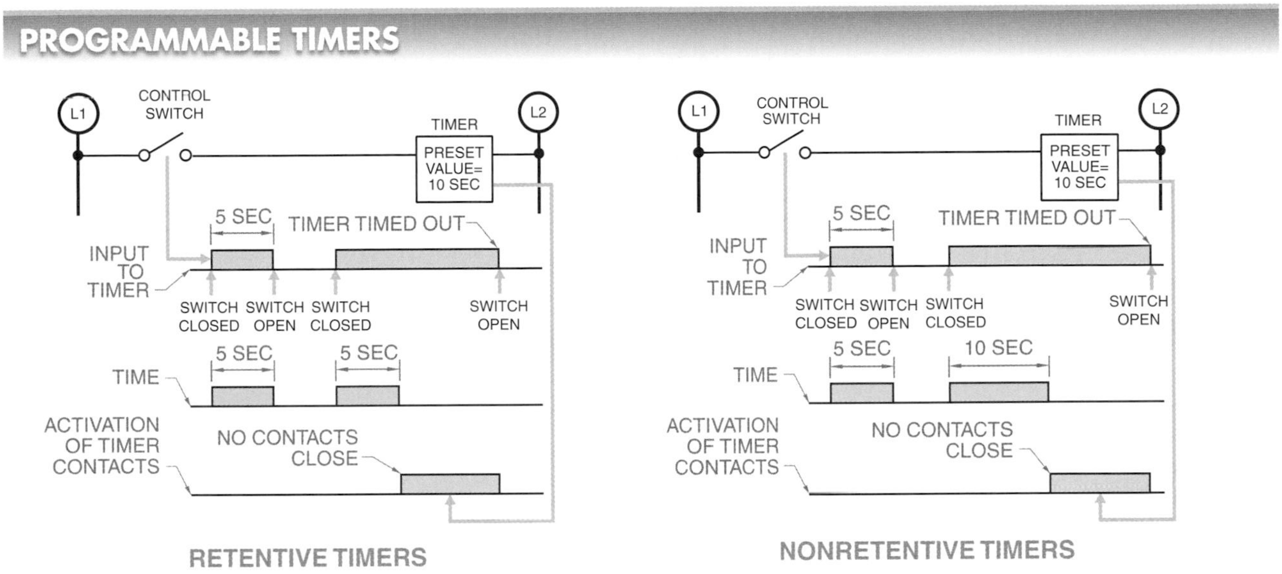

PROGRAMMABLE TIMERS

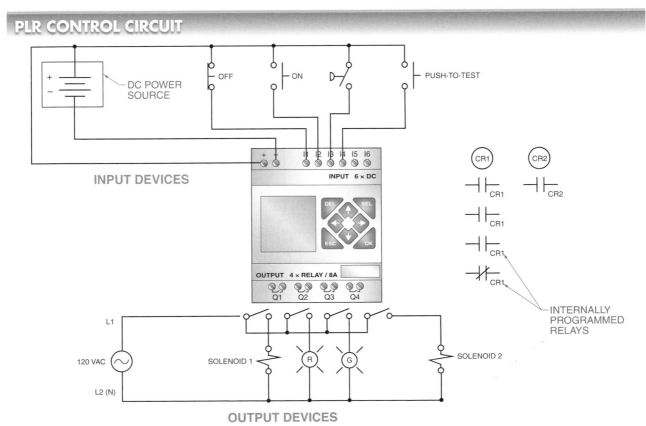

RETENTIVE TIMERS **NONRETENTIVE TIMERS**

Figure 16-39. Programmable timers include retentive and nonretentive timers.

PLR CONTROL CIRCUIT

Figure 16-40. PLRs are versatile and normally used for control circuits that require numerous relays and/or frequent changes.

16-4 CHECKPOINT

1. In a 555 timing circuit that uses an external RC time-delay circuit, what components are used to determine the amount of time delay?
2. What is the name for an electronic switch that has two outputs in which one is high anytime the other is low?
3. If a programmable timer is programed for a time setting of 1020 and the time base is programmed for 0.1 sec, how long is the timer programmed for?
4. If a programmable timer is programed for a time setting of 1020 and the time base is programmed for 1 sec, how long is the timer programmed for?
5. When programming a real-time clock to operate its contacts from 4:30 PM to 6:00 PM, what is the time setting in military time?
6. What type of timer still retains its accumulated time value after a power loss?

16-5 COUNTERS AND TOTALIZERS

In most applications it is necessary to account for the number of events within the system. This may include counting the number of products made, the number of products required to fill a carton, the number of rejected parts, the number of gallons flowing through a pipe, etc. A totalizer can be used when only the total number of units or events is required to be known. A *totalizer* is a counting device that keeps track of the total number of units or events and displays the total counted value.

Counters are used to count inputs and provide an output. A *counter* is a counting device that accounts for the total number of inputs entering into it and can provide an output (mechanical or solid-state contacts) at predetermined counts in addition to displaying the counted value. The two basic types of counters are up counters and up/down counters.

Up Counters

An *up counter* is a device used to count inputs and provide an output (contacts) after the preset count value is reached. An up counter has either one count input or one count input and one reset input. Removing power to the counter resets it with one count input. A counter with two inputs uses one input to add a count into the counter each time it is activated and the second input to reset the counter to zero counts when it is activated. **See Figure 16-41.**

The count input adds one count each time it is closed and opened. When the preset count value is reached, the counter contacts are activated (closing NO contacts and opening NC contacts). If a counter has a total count display in addition to a preset count setting, any additional counts sent into the counter are displayed as part of the total.

UP COUNTERS

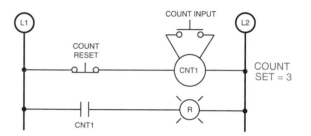

SINGLE-INPUT COUNTER

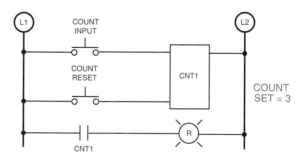

DUAL-INPUT COUNTER

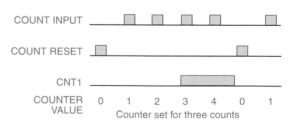

Figure 16-41. Up counters are used in applications that require an output after a fixed number of counts.

Up/Down Counters

An *up/down counter* is a device used to count input from two different inputs, one input that adds a count and another input that subtracts a count. Up/down counters provide an output (contacts) after the preset count value is reached. A third input is used as the reset input, which resets the counter to zero counts when it is activated. **See Figure 16-42.**

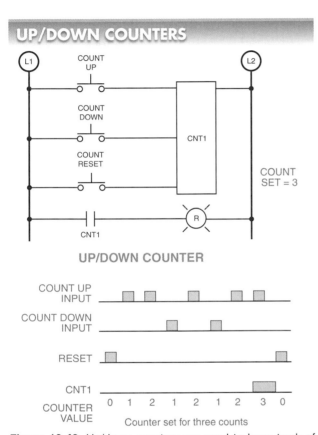

Figure 16-42. Up/down counters are used to keep track of the number of counts when counts are both added to and subtracted from an application.

Up/down counters are used in applications such as counters for parking garages. Each time a car enters the garage, one count is added to the count. Each time a car leaves the garage, one count is subtracted from the count. When the garage is full (preset count value reached), the counter contact is used to turn on a "Lot Full" sign. If one car leaves, then the sign turns off.

Totalizers and counters are commonly used in product production applications. **See Figure 16-43.** For example, in a shrink-wrap application, totalizers and counters can be used to count the number of times a shrink-wrap roller motor has turned on and off.

The motor is turned on and off once for the wrapping of each product, which is detected by a proximity or photoelectric switch. There are only a set number of times the shrink-wrap roller motor can be turned on and off before a new roll of shrink wrap must be placed in the machine. The counter can indicate when a new roll is required by using the counter output contacts to turn on an alarm or lamp.

In this application, the products coming off the production line can be sent to storage area 1 or 2. A counter can be used to count the number of products coming out of the heat machine and send a fixed number (such as 6, 12, 36, etc.) first to storage area 1 and then a new group to storage area 2, thereby allowing distribution control of the produced item.

An inspection station could be added to the output of the heat machine that inspects the product and rejects any that do not meet predetermined requirements such as proper weight, size, labels, etc. Items that pass inspection can be sent to storage area 1 and items that fail inspection can be sent to storage area 2. A counter may be used to track the number of products going to storage area 1 and a second counter may be used to track the number of products going to storage area 2. An up/down counter can also be used to count the number of products coming out of production and moving to storage area 1 and subtract the rejected products moving to storage area 2. Finally, a counter can be used to automatically stop the production line after a predetermined count.

Sensors can be used with counters to track items moving along a conveyor line.

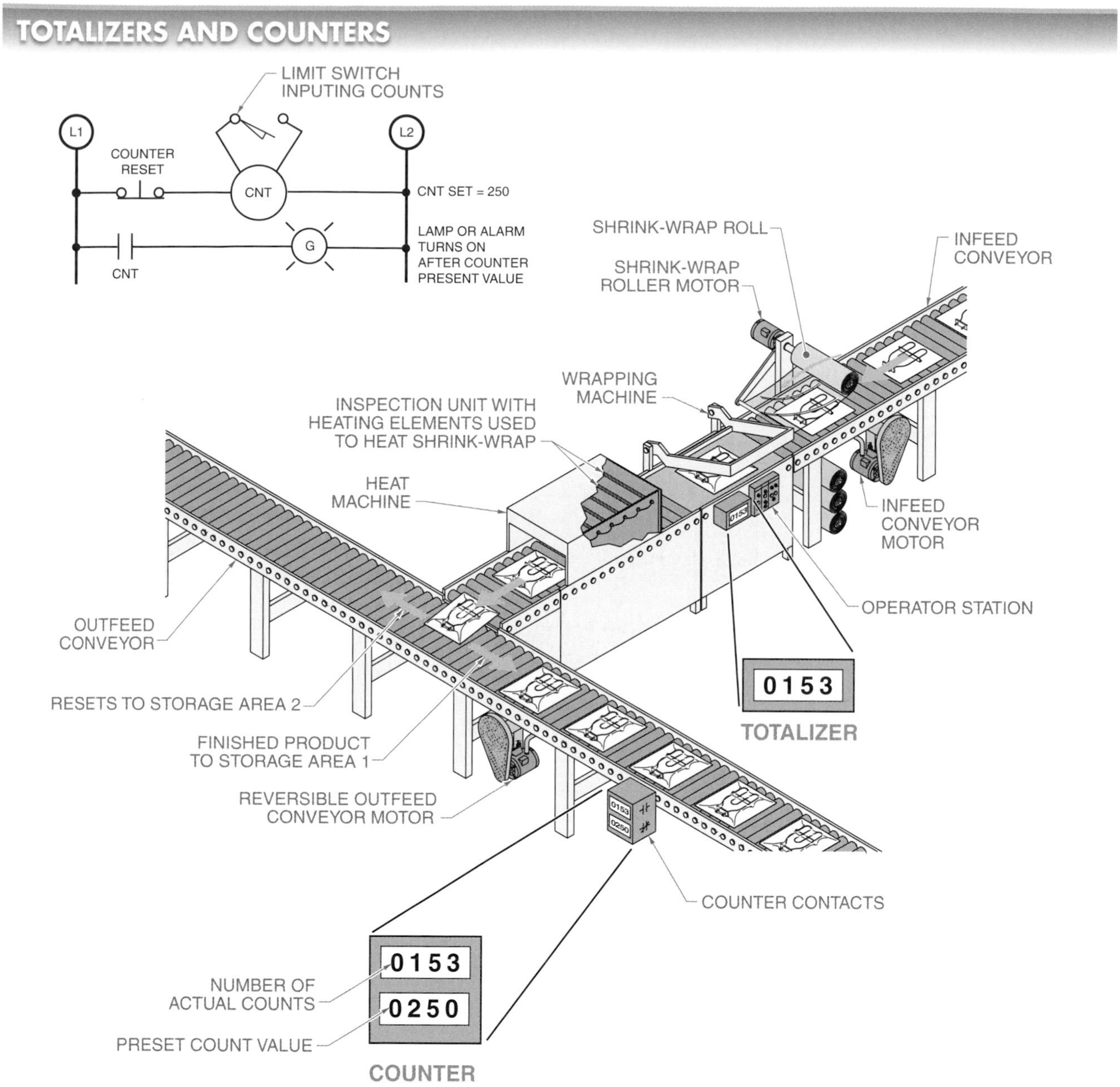

TOTALIZERS AND COUNTERS

LIMIT SWITCH
INPUTING COUNTS

L1

COUNTER
RESET

CNT

L2

CNT SET = 250

G

LAMP OR ALARM
TURNS ON
AFTER COUNTER
PRESENT VALUE

CNT

SHRINK-WRAP ROLL

SHRINK-WRAP
ROLLER MOTOR

INFEED
CONVEYOR

WRAPPING
MACHINE

INSPECTION UNIT WITH
HEATING ELEMENTS USED
TO HEAT SHRINK-WRAP

HEAT
MACHINE

INFEED
CONVEYOR
MOTOR

OPERATOR STATION

OUTFEED
CONVEYOR

0153

TOTALIZER

RESETS TO STORAGE AREA 2

FINISHED PRODUCT
TO STORAGE AREA 1

REVERSIBLE OUTFEED
CONVEYOR MOTOR

COUNTER CONTACTS

NUMBER OF
ACTUAL COUNTS

0153

0250

PRESET COUNT VALUE

COUNTER

Figure 16-43. Totalizers and counters are commonly used in applications such as counting for shrinkable plastic wrap on large quantities of finished goods.

16-5 CHECKPOINT

1. If a totalizer and counter both keep track of the number of counts, what is difference between the two?

2. If the timer of an up counter uses an input switch (such as a limit or photoelectric switch, etc.) to add counts, how are the counter's accumulated counts reset back to zero?

3. What is the minimum number of individual inputs required to operate an up/down counter?

Additional Resources

Review and Resources

Access Chapter 16 Review and Resources for *Electrical Motor Controls for Integrated Systems* by scanning the above QR code with your mobile device.

Applying Your Knowledge

Refer to the *Electrical Motor Controls for Integrated Systems* Learner Resources for interactive Applying Your Knowledge questions.

Workbook and Applications Manual

Refer to Chapter 16 in the *Electrical Motor Controls for Integrated Systems Workbook* and the *Applications Manual* for additional exercises.

ENERGY EFFICIENCY PRACTICES

Time-Based Control

Time-based control strategies use the time of day to determine the desired operation of energy-consuming loads in a building. Time-based control strategies involve turning a load on and off at specific times. Early time-based control functions were performed by electromechanical time clocks. However, electromechanical time clocks do not allow for control flexibility. In addition, electromechanical time clocks are unpredictable and are less accurate than other control strategies. Time-based control strategies consist of adjusting time schedules for loads in various areas of a building and include seven-day programming, daily multiple-time-period scheduling, and timed overrides.

Seven-day programming allows technicians to individually program a system's on and off time functions for each day of the week. Daily multiple-time-period scheduling allows system software to accommodate unusual building occupancy. The HVAC, lighting, and other loads can be scheduled to operate during multiple independent time periods. A timed override is a control strategy in which the occupants of a building can change a zone from an unoccupied mode to an occupied mode for temporary occupancy. A timed override can be activated by a pushbutton switch that is configured as a digital input and wired to the building control system. A timed override can also be activated by a personal computer.

Objectives

17-1
- Describe friction brakes.

17-2
- Define and describe plugging.

17-3
- Define and describe electric braking.

17-4
- Define and describe dynamic braking.

Sections

Review and Resources
atplearningresources.com/Quicklinks
Access Code: 362245

Motor Stopping Methods

Electric motors produce rotary motion that is used to produce work in applications that range in requirements, size, and environments, as well as other conditions. All electric motors require power to get them started and to keep them running. Once running, all motors will stop rotating when power is removed. For most applications, removing power to stop motor rotation meets the needs of the application.

However, in some applications, consideration must be given to stopping a motor by some other method than just removing power and allowing the motor to come to a stop based on the motor's operating speed and load applied to the motor shaft. In applications that require a faster than normal stop or a controlled stop time, some type of braking method must be applied. There are several different methods used to stop a motor; usually one method is best to meet the requirements of the application. Learning how motors can be stopped is important when designing or working on circuits in which faster motor stopping can provide a safer operating condition for the driven parts and persons working around the equipment.

17-1 FRICTION BRAKES

The length of time required for a motor to come to rest depends on the inertia of the moving parts (motor and motor load) and friction. *Inertia* is the property of matter by which a mass persists in its state of rest or motion until acted upon by an external force. *Friction* is the resistance to motion that occurs when two surfaces slide against each other.

Friction brakes normally consist of two friction surfaces (shoes or pads) that come in contact with a rotor (wheel) mounted on the motor shaft. Spring tension holds the shoes on the rotor and braking occurs as a result of the friction between the shoes and the rotor. Friction brakes (magnetic or mechanical) are the oldest motor stopping method. Friction brakes are similar to the brakes on automobiles. **See Figure 17-1.**

FRICTION BRAKES

Heidelberg Harris, Inc.

Figure 17-1. Friction brakes normally consist of two friction surfaces that come in contact with a rotor mounted on the motor shaft.

Solenoid Operation

Friction brakes are normally controlled by a solenoid, which activates the brake shoes. The solenoid is energized when the motor is running. This keeps the brake shoes from touching the drum mounted on the motor shaft. The solenoid is de-energized and the brake shoes are applied through spring tension when the motor is turned off.

Two methods are used to connect the solenoid into the circuit so that it activates the brake whenever the motor is turned on and off. **See Figure 17-2.** The first method is used if the solenoid has a voltage rating equal to the motor voltage rating. The second method is used if the solenoid has a voltage rating equal to the voltage between L1 and the neutral. The brake solenoid should always be connected directly into the motor circuit, not into the control circuit. This eliminates improper activation of the brake.

Brake Shoes

In friction braking, the braking action is applied to a rotor mounted on the shaft of the motor rather than directly to the shaft. The rotor provides a much larger braking surface than could be obtained from the shaft alone. This permits the use of large brake shoe linings and low shoe pressure. Low shoe pressure equally distributed over a large area results in even wear and braking torque. The braking torque developed is directly proportional to surface area and spring pressure. The spring pressure is adjustable on nearly all friction brakes.

Advantages and Disadvantages of Friction Brakes

The advantages of using friction brakes are lower initial cost and simplified maintenance. Friction brakes are less expensive to install than other braking methods because fewer expensive electrical components are required. Maintenance is simplified because it is easy to see whether the shoes are worn and if the brake is working. Friction brakes are available in both AC and DC designs to meet almost any application requirements. The disadvantage of friction brakes is that they require more maintenance than other braking methods. Maintenance consists of replacing the shoes and pads regularly. Shoe replacement depends on the number of times the motor is stopped. A motor that is stopped often needs more maintenance than a motor that is rarely stopped. Friction brake applications include printing presses, cranes, overhead doors, hoisting equipment, and machine tool control.

All friction brakes produce heat when stopping a motor. Excessive braking can cause high temperatures on the brakes and may cause a fire in areas containing flammable materials. In these locations, friction brakes that include a built-in thermal protective device (thermistor) can be used to monitor brake temperature and provide feedback to the control/safety circuit.

SOLENOID CIRCUIT CONNECTIONS FOR BRAKING

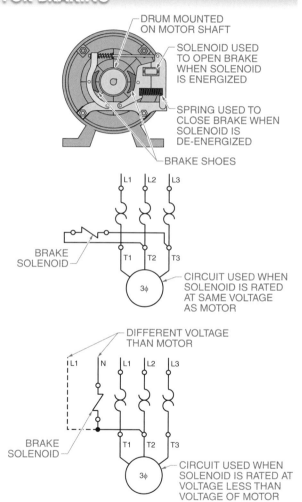

Figure 17-2. A friction brake may be connected to full-line voltage equal to that produced between L1 and the neutral.

17-1 CHECKPOINT

1. Would a motor automatically apply the brakes if the brake solenoid lost power?
2. Is the braking force still applied to a motor shaft when the circuit has no power?

17-2 PLUGGING

Plugging is a method of motor braking in which the motor connections are reversed so that the motor develops a countertorque that acts as a braking force. The countertorque is accomplished by reversing the motor at full speed with the reversed motor torque opposing the forward inertia torque of the motor and its mechanical load. Plugging a motor allows for very rapid stopping. Although manual and electromechanical controls can be used to reverse the direction of a motor, a plugging switch is normally used in plugging applications. **See Figure 17-3.**

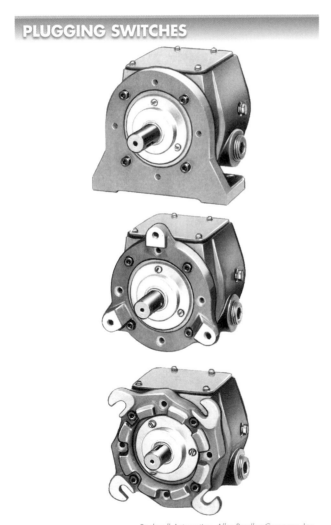

PLUGGING SWITCHES

Rockwell Automation, Allen-Bradley Company, Inc.

Figure 17-3. Plugging switches prevent the reversal of the controlled load after the load has stopped.

A plugging switch is connected mechanically to the shaft of the motor or driven machinery. The rotating motion of the motor is transmitted to the plugging switch contacts either by a centrifugal mechanism or by a magnetic induction arrangement. The contacts on the plugging switch are normally open (NO), normally closed (NC), or both and actuate at a given speed. The primary function of a plugging switch is to prevent the reversal of the load once the countertorque action of plugging has brought the load to a standstill. The motor and load would start to run in the opposite direction without stopping if the plugging switch were not present.

Plugging Switch Operation

Plugging switches are designed to open and close sets of contacts as the shaft speed on the switch varies. As the shaft speed increases, the contacts are set to change at a given rpm. As the shaft speed decreases, the contacts return to their normal condition. As the shaft speed increases, the contact setpoint (point at which the contacts operate) reaches a higher rpm than the point at which the contacts reset (return to their normal position) on decreasing speed. The difference in these contact operating values is the differential speed or rpm.

In plugging operations, the continuous running speed must be many times the speed at which the contacts are required to operate. This provides high contact holding force and reduces possible contact chatter or false operation of the switch.

Continuous Plugging

A plugging switch may be used to plug a motor to a stop each time the stop pushbutton is pressed. **See Figure 17-4.** In the circuit, the NO contacts of the plugging switch are connected to the reversing starter through an interlock contact. Pushing the start pushbutton energizes the forward starter, starting the motor in forward and adding memory to the control circuit. As the motor accelerates, the NO plugging contacts close. The closing of the NO plugging contacts does not energize the reversing starter because of the interlocks. Pushing the stop pushbutton drops out the forward starter and interlocks. This allows the reversing starter to immediately energize through the plugging switch and the NC forward interlock. The motor is reversed and the motor brakes to a stop. After the motor is stopped, the plugging switch opens to disconnect the reversing starter before the motor is actually reversed.

CONTINUOUS PLUGGING

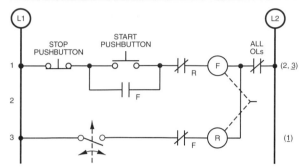

Figure 17-4. A plugging switch may be used to plug a motor to a stop each time the stop pushbutton is pressed.

Plugging for Emergency Stops

A plugging switch may be used in a circuit where plugging is required only in an emergency. **See Figure 17-5.** In this circuit, the motor is started in the forward direction by pushing the run pushbutton. This starts the motor and adds memory to the control circuit. As the motor accelerates, the NO plugging contacts close. Pushing the stop pushbutton de-energizes the forward starter but does not energize the reversing starter. This is because there is no path for the L1 power to reach the reversing starter, so the motor coasts to a stop.

Pushing the emergency stop pushbutton de-energizes the forward starter and simultaneously energizes the reversing starter. Energizing the reversing starter adds memory in the control circuit and plugs the motor to a stop. When the motor is stopped, the plugging switch opens to disconnect the reversing starter before the motor is actually reversed. The de-energizing of the reversing starter also removes the memory from the circuit.

Limitations of Plugging

Plugging may not be applied to all motors and/or applications. Braking a motor to a stop using plugging requires that the motor be a reversible motor and that the motor can be reversed at full speed. Even if the motor can be reversed at full speed, the damage that plugging may do outweighs its advantages.

Reversing. A motor cannot be used for plugging if it cannot be reversed at full speed. For example, a 1ϕ shaded-pole motor cannot be reversed at any speed. Thus, a 1ϕ shaded-pole motor cannot be used in a plugging circuit. Likewise, most 1ϕ split-phase and capacitor-start motors cannot be plugged because their centrifugal switches remove the starting windings when the motor accelerates. Without the starting winding in the circuit, the motor cannot be reversed.

Heat. Many motor types can be used for plugging. However, high current and heat result from plugging a motor to a stop. A motor is connected in reverse at full speed when plugging a motor. The current may be three or more times higher during plugging than during normal starting. For this reason, a motor designated for plugging or a motor with a high service factor should be used in all cases except emergency stops. The service factor (SF) should be 1.35 or more for plugging applications.

Tech Fact

Plugging is used to bring a motor to a fast stop. However, plugging a motor to a stop produces high temperatures in the rotor due to the reversal of the magnetic field during stopping. These temperatures are higher than the heat produced during locked rotor current. Plugging is an effective way to bring a motor to a fast stop, but it must not be used to frequently stop the motor. Frequent stops may damage the motor. Fewer stops or the use of a higher-rated HP motor and proper cooling will help reduce heat buildup in the motor.

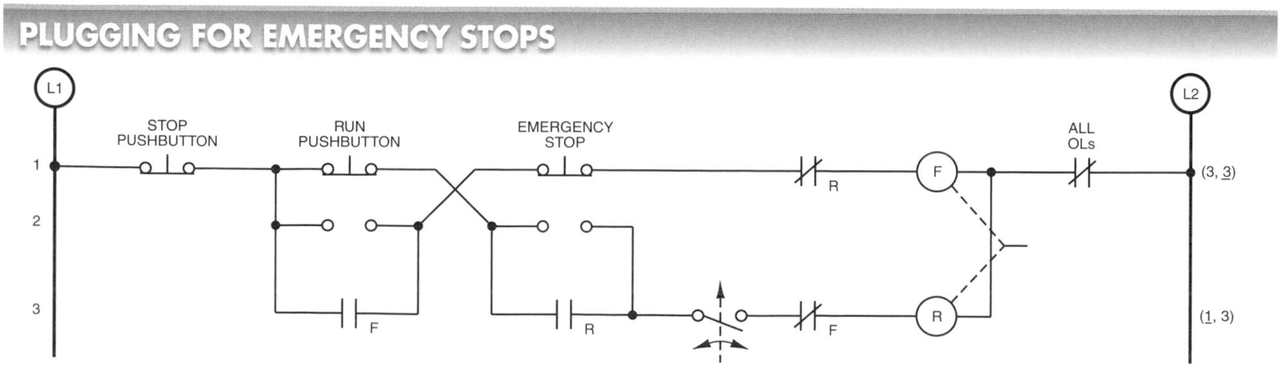

PLUGGING FOR EMERGENCY STOPS

Figure 17-5. A plugging switch may be used in a circuit where plugging is required only in an emergency.

17-2 CHECKPOINT

1. Can any type of electric motor be used with a plugging switch to apply a braking force?

2. When the plugging switch connects a motor for the reverse direction, why does the motor never actually reverse direction?

17-3 ELECTRIC BRAKING

Electric braking is a method of braking in which a DC voltage is applied to the stationary windings of a motor after the AC voltage is removed. Electric braking is also known as DC injection braking. **See Figure 17-6.** Electric braking is an efficient and effective method of braking most AC motors. Electric braking provides a quick and smooth braking action on all types of loads, including high-speed and high-inertia loads. Maintenance is minimal because there are no parts that come in physical contact during braking.

Electric Braking Operating Principles

The principle stating that unlike magnetic poles attract each other and like magnetic poles repel each other explains why a motor shaft rotates. However, the method by which the magnetic fields are created changes from one type of motor to another.

In AC induction motors, the opposing magnetic fields are induced from the stator windings into the rotor windings by transformer action. The motor continues to rotate as long as the AC voltage is applied. The motor coasts to a standstill over a period of time when the AC voltage is removed because there is no induced field to keep it rotating.

Electric braking can be used to provide an immediate stop if the coasting time is unacceptable, particularly in an emergency situation. Electric braking is accomplished by applying a DC voltage to the stationary windings once the AC voltage is removed. The DC voltage creates a magnetic field in the stator that does not change polarity.

ELECTRIC BRAKING

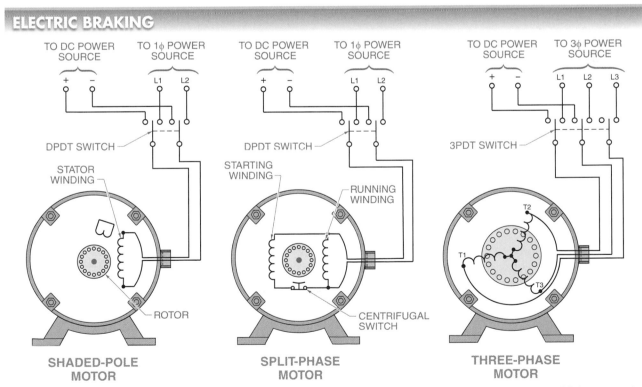

Figure 17-6. Electric braking is achieved by applying DC voltage to the stationary windings of a motor after the AC is removed.

The constant magnetic field in the stator creates a magnetic field in the rotor. Because the magnetic field of the stator does not change in polarity, it attempts to stop the rotor when the magnetic fields are aligned (N to S and S to N). **See Figure 17-7.** The only force that can keep the rotor from stopping with the first alignment is the rotational inertia of the load connected to the motor shaft. However, because the braking action of the stator is present at all times, the motor brakes quickly and smoothly to a standstill.

ELECTRIC BRAKING OPERATIONS

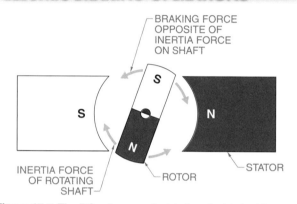

Figure 17-7. The DC voltage applied during electric braking creates a magnetic field in the stator that does not change polarity.

DC Electric Braking Circuits

DC is applied after the AC is removed to bring the motor to a stop quickly. **See Figure 17-8.** This circuit, like most DC braking circuits, uses a bridge rectifier circuit to change the AC into DC. In this circuit, a 3φ AC motor is connected to 3φ power by a magnetic motor starter.

The magnetic motor starter is controlled by a standard stop/start pushbutton station with memory. An off-delay timer is connected in parallel with the magnetic motor starter. The off-delay timer controls a NO contact that is used to apply power to the braking contactor for a short period of time after the stop pushbutton is pressed. The timing contact is adjusted to remain closed until the motor comes to a stop.

The braking contactor connects two motor leads to the DC supply. A transformer with tapped windings is used to adjust the amount of braking torque applied to the motor. Current-limiting resistors could be used for the same purpose. This allows for a low- or high-braking action, depending on the application. The larger the applied DC voltage, the greater the braking force.

DC ELECTRIC BRAKING CIRCUITS

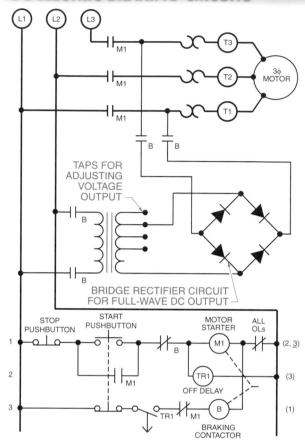

Figure 17-8. DC is applied after AC is removed to bring the motor to a stop quickly.

The interlock system in the control circuit prevents the motor starter and braking contactor from being energized at the same time. This is required because the AC and DC power supplies must never be connected to the motor simultaneously. Total interlocking should always be used on electrical braking circuits. Total interlocking is the use of mechanical, electrical, and pushbutton interlocking. A standard forward and reversing motor starter can be used in this circuit, as it can with most electric braking circuits.

17-3 CHECKPOINT

1. What determines the amount of braking force applied to the motor when using electric braking?
2. What determines the amount of time that a braking force is applied to the motor during stopping?

17-4 DYNAMIC BRAKING

Dynamic braking is a method of motor braking in which a motor is reconnected to act as a generator immediately after it is turned off. Connecting the motor in this way makes the motor act as a loaded generator that develops a retarding torque, which rapidly stops the motor. The generator action converts the mechanical energy of rotation to electrical energy that can be dissipated as heat in a resistor with a high power rating. **See Figure 17-9.**

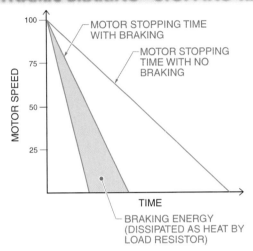

Figure 17-9. Dynamic braking significantly reduces motor stopping time when a motor is reconnected to act as a generator immediately after it is turned off.

Dynamic braking is normally applied to DC motors because there must be access to the armature via the brushes to take advantage of the generator action. Access is accomplished through the brushes on DC motors. **See Figure 17-10.** Dynamic braking of a DC motor may be needed because DC motors are often used for lifting and moving heavy loads that may be difficult to stop.

In this circuit, the armature terminals of the DC motor are disconnected from the power supply and immediately connected across a resistor that acts as a load. The smaller the resistance of the resistor, the greater the rate of energy dissipation and the faster the motor comes to rest. The field windings of the DC motor are left connected to the power supply. The armature generates a counter electromotive force (CEMF). The CEMF causes current to flow through the resistor and armature.

The current causes heat to be dissipated in the resistor in the form of electrical watts. This removes energy from the system and slows the motor rotation.

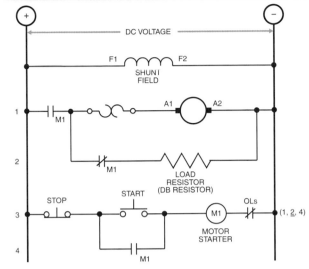

Figure 17-10. Dynamic braking is normally applied to DC motors because there must be access to the armature via the brushes to take advantage of the generator action.

The generated CEMF decreases as the speed of the motor decreases. As the motor speed approaches 0 rpm, the generated voltage also approaches 0 V. The braking action lessens as the speed of the motor decreases. As a result, a motor cannot be braked to a complete stop using dynamic braking. Dynamic braking also cannot hold a load once it is stopped because there is no braking action.

Electromechanical friction brakes are often used along with dynamic braking in applications that require the load to be held. A combination of dynamic braking and friction braking can also be used in applications where a large, heavy load is to be stopped. In these applications, the force of the load wears the friction brake shoes excessively; so then dynamic braking can be used to slow the load before the friction brakes are applied. This is similar to using a parachute to slow a racecar before applying the brakes.

Dynamic Braking Electric Motor Drives

A motor drive can control the time it takes for a motor to stop. The stopping time is programmed by setting the deceleration parameter. The deceleration parameter can be set for one second (or less) to several minutes. However, for fast stops (especially with high inertia loads), a braking resistor is added. **See Figure 17-11.**

DYNAMIC BRAKING ELECTRIC MOTOR DRIVES

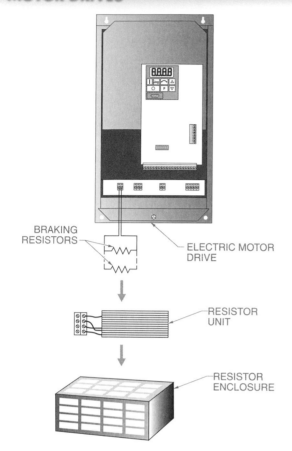

BRAKING RESISTORS

ELECTRIC MOTOR DRIVE

RESISTOR UNIT

RESISTOR ENCLOSURE

Sample Manufacturer Braking Resistance Chart			
HP Rating	Resistor (in W)	Wattage	Model Number
½	180	120	A .5
1	100	150	A 1
3	50	175	A 3
5	35	200	A 5
7.5	17	300	A 7.5
10	15	400	A 10

Figure 17-11. A braking resistor can be added to a motor drive to control fast stops on high-inertia loads.

The braking resistor also helps control motor torque during repeated on/off motor cycling. The size (wattage rating) and resistance (in ohms) of the braking resistor are typically selected using the drive manufacturer braking resistor chart. The resistor wattage must be high enough to dissipate the heat delivered at the resistor. The resistor resistance value must be low enough to allow current flow at a level high enough to convert the current to heat, but not so low as to cause excessive current flow. The braking resistor(s) can be housed in an enclosure such as a NEMA Type 1 enclosure.

Motor stopping methods are commonly used in applications where a controlled stop time is required.

17-4 CHECKPOINT

1. Does the braking force applied to a motor increase, decrease, or remain the same at all deceleration speeds during dynamic braking?
2. Can dynamic braking be used to hold a load after the motor comes to a stop?

Additional Resources

Review and Resources

Access Chapter 17 Review and Resources for *Electrical Motor Controls for Integrated Systems* by scanning the above QR code with your mobile device.

Applying Your Knowledge

Refer to the *Electrical Motor Controls for Integrated Systems* Learner Resources for interactive Applying Your Knowledge questions.

Workbook and Applications Manual

Refer to Chapter 17 in the *Electrical Motor Controls for Integrated Systems Workbook* and the *Applications Manual* for additional exercises.

ENERGY EFFICIENCY PRACTICES

ENERGY SAVINGS THROUGH MOTOR SPEED REDUCTION

Traveling at a high rate of speed requires a large amount of energy. The higher the rate of speed, the more energy consumed over the same distance traveled, regardless of whether the machine is an internal combustion engine or electric motor. Traveling at a low rate of speed requires a reduced amount of energy. For example, a decrease in speed from 70 mph to 60 mph in an automobile can reduce energy consumption by 5% to 10%.

Likewise, reducing the speed of airplanes also saves fuel. For example, one airline reduced the average speed of its airplanes from 542 mph to 532 mph on its flights from Paris to Minneapolis. This added 8 min to the flight time and saved approximately 162 gal. of fuel. Another airline estimated that it would save $600,000 in fuel costs for 2008 by adding only 4 min to its flights to Hawaii.

A reduction in energy consumption can also be realized by reducing the speed of electric motors. The greatest energy savings can be realized in applications that do not require a motor to run at full speed. In the past, almost all electric motors were operated at full speed. If any reduction in output speed was required, gears, pulleys, or other means were used. Today the use of electric motor drives enables precise control of motor speed. By reducing the speed of electric motors, a reduction in energy consumption can be realized, because motors account for approximately 62% of all electricity consumed. For example, in fan and pump applications, energy consumption can be reduced by reducing the speed at which the fan or pump motor operates. As the motor speed is reduced, the electrical power consumed is reduced by a greater percentage.

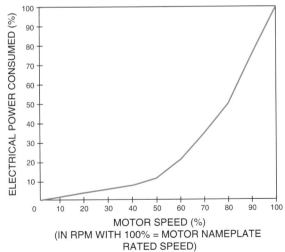

POWER CONSUMPTION AND MOTOR SPEED

Objectives

18-1
- Explain why the type of load and torque requirements for a motor must be understood.

18-2
- Define voltage fluctuation and describe different voltage fluctuation problems.
- Explain how to test for transient voltages.

18-3
- Describe the different classifications of motor loads.

18-4
- Define and describe the types of motor torque.

18-5
- Explain the difference between open circuit transition and closed circuit transition.

Sections

Review and Resources
atplearningresources.com/Quicklinks
Access Code: 362245

Motor Load, Torque, and Power Quality Requirements

All motors are designed to produce work. The applications in which motors are used vary widely between the type of application, operating environments, changing load conditions, service requirements, and expected operating life. In order to select the best motor type and size for an application, the type of load that the motor is required to drive must be known, as well as the types of torque that are required by the motor to properly operate the load under all operating conditions. Since motors require different amounts of power under different operating conditions and can draw large amounts of currents from the power supply, their effects on the system must be understood and considered to help prevent any power quality problems within the system.

18-1 UNDERSTANDING MOTOR LOADS AND TORQUE REQUIREMENTS

Since all motors produce torque that can be used to operate loads, all motors are capable of producing work. The amount of work a motor produces depends upon its size, which can range from fractional HP to hundreds of HP, and the type of load it must drive. In the past, a motor was connected to a motor starter (manual or magnetic), and if the motor produced enough torque to start and operate the load, the exact type of load and required motor torque did not always need to be fully understood. However, in today's motor control systems that use programmable motor drives, the type of load and operating torque requirements of the application must be understood. This is because the type of load and any changes in required operating torque (such as torque boost) are programmed into the drive before the system is operated and when any load requirement changes are made. **See Figure 18-1.**

Motor Load Power Quality Problems

Motors, as a group, consume more power than all other types of loads (lamps, heating elements, electronic devices, etc.) combined. Motors also draw large amounts of current when running and even more current when starting. Starting and stopping motors can produce power quality problems such as voltage sags and transients within the distribution system, which can cause other loads in the system to be damaged or fail to operate properly. For example, the starting of a large motor can cause a voltage sag that causes relays, motor starters, computers, etc. to drop out or reset. Also, a motor turning off can produce transient (high-voltage spike) that can damage equipment connected to the system.

MOTOR LOAD TYPES AND TORQUE REQUIREMENTS

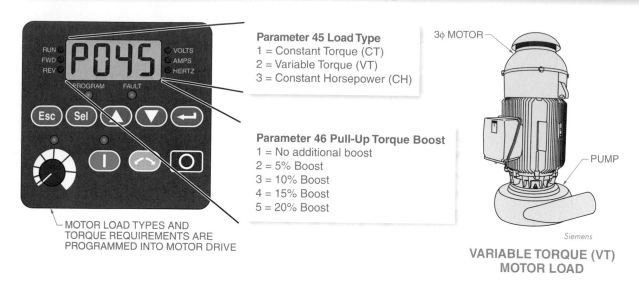

Parameter 45 Load Type
1 = Constant Torque (CT)
2 = Variable Torque (VT)
3 = Constant Horsepower (CH)

Parameter 46 Pull-Up Torque Boost
1 = No additional boost
2 = 5% Boost
3 = 10% Boost
4 = 15% Boost
5 = 20% Boost

MOTOR LOAD TYPES AND TORQUE REQUIREMENTS ARE PROGRAMMED INTO MOTOR DRIVE

3ϕ MOTOR

PUMP

Siemens

VARIABLE TORQUE (VT) MOTOR LOAD

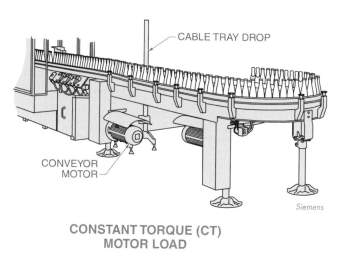

CABLE TRAY DROP

CONVEYOR MOTOR

Siemens

CONSTANT TORQUE (CT) MOTOR LOAD

Figure 18-1. Motor load types and torque requirements must be understood for both mechanical (magnetic motor starter) and solid-state (motor drive) applications.

Understanding what effects motors have on the distribution system is important to protect the motor loads (including other motors) connected to the system and the components within the system. Understanding the terminology used when describing power quality problems, such as sags, swells, and transients, is important to understand utility penalties that apply when motors cause problems, select corrective and preventive equipment (capacitors and filters), and troubleshoot problems at the motor or in the systems.

18-1 CHECKPOINT

1. When a motor drive is used to control a motor, can the drive be programmed to add an additional torque boost if the application requires it?
2. When does a motor draw the highest amount of current from the power supply?

18-2 POWER QUALITY

The quality of incoming power to a load is important because all electrical and electronic equipment is rated for operation at a specific voltage. The rated voltage is actually a voltage range. Typically, this range is ±10%. However, with many newer components derated to save energy and operating cost, the operating voltage range is now often between +5% and –10%. This range is used because an overvoltage is generally more damaging than an undervoltage.

The amount and duration of voltage change can be classified into commonly used industrial terms. Voltage changes range from total power interruption to very short variations that may last only a few milliseconds (transients). Power interruptions are classified as momentary, temporary, or sustained power interruption. The different classifications are based on the time power is lost.

A *momentary power interruption* is a decrease to 0 V on one or more power lines lasting from 0.5 cycles up to 3 sec. A *temporary power interruption* is a decrease to 0 V on one or more power lines lasting for more than 3 sec up to 1 min. A *sustained power interruption* is a decrease in voltage to 0 V on all power lines for a period of more than 1 min. Voltage changes other than power interruptions occur much more often and can be a continued change that causes both short term and long term problems. **See Figure 18-2.**

VOLTAGE CHANGES

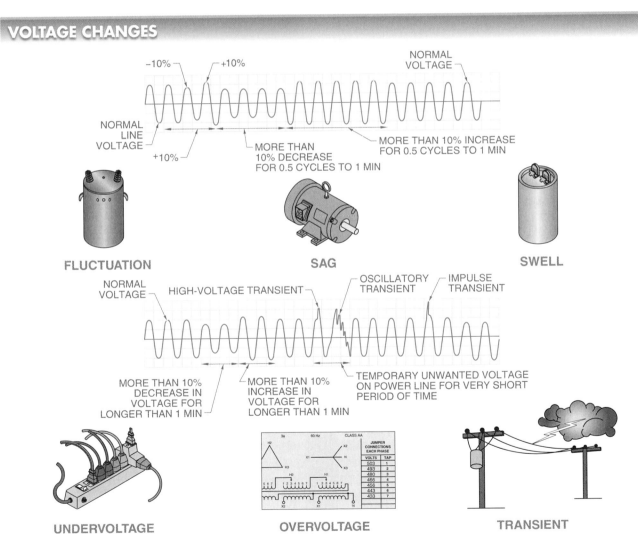

Figure 18-2. Voltage changes, such as voltage sags and swells, undervoltage, and overvoltage, are difficult to detect because voltage changes are often intermitent.

Voltage Fluctuations

A *voltage fluctuation* is an increase or decrease in the normal line voltage within the range of +5% to −10%. Voltage fluctuations are commonly caused by overloaded transformers, unbalanced transformer loading, and/or high impedance caused by long circuit runs, undersized conductors, or poor or loose electrical connections in the system.

Voltage Sags

A *voltage sag* is a voltage drop of more than 10% (but not to 0 V) below the normal rated line voltage that lasts from 0.5 cycles up to 1 min. Voltage sags commonly occur when high-current loads, such as motors, are turned on. When very large motors are turned on, a voltage sag can be followed by a voltage swell as voltage regulators overcompensate during the voltage sag.

A voltmeter can be used to measure the voltage at the power source and at the load to determine if there is a problem with bad connections, too long of run, or undersized conductors. Voltmeters with a MIN/MAX recording mode are used to measure voltage over time and can be used to determine if there is a voltage sag or other voltage fluctuation problem. **See Figure 18-3.**

Voltage Swells

A *voltage swell* is a voltage increase of more than 10% above the normal rated line voltage that lasts from 0.5 cycles up to 1 min. Voltage swells commonly occur when large loads, such as motors, are turned off and voltage on the power line increases above the normal voltage fluctuation (+10%) for a short period of time. Voltage swells are not as common as voltage sags. However, voltage swells are more destructive than voltage sags because voltage swells damage electrical equipment in very short periods of time.

Undervoltages

Undervoltage is a drop in voltage of more than 10% (but not to 0 V) below the normal rated line voltage for a period of time longer than 1 min. Undervoltage (low voltage) is more common than overvoltage (high voltage) on power lines. Undervoltages are commonly caused by overloaded transformers, undersized conductors, conductor runs that are too long, too many loads on a circuit, or brownouts. A *brownout* is a reduction of the voltage level by a power company to conserve power during times of peak usage or excessive loading of the power distribution system.

Overvoltages

Overvoltage is an increase of voltage of more than 10% above the normal rated line voltage for a period of time longer than 1 min. Overvoltages are caused when loads are near the beginning of a power distribution system or when taps on a transformer are not wired correctly. A *tap* is a connection brought out of a winding at a point between its end points to allow the voltage or current ratio to be changed. Transformer taps are commonly provided at 2.5% increments.

Taps on a transformer must be wired correctly to prevent overvoltages.

Transient Voltages

Research by the American Institute of Electrical Engineers (AIEE), service technicians, and equipment manufacturers clearly indicates that solid-state circuits and devices do not tolerate momentary voltage surges exceeding twice the normal operating voltage. Voltage surges are produced in all electrical systems from outside sources (lightning) and within the system when loads that include coils (solenoids, magnetic motor starters, and motors) are turned off. The voltage surge produced is called a transient voltage.

A *transient voltage (voltage spike)* is a temporary, undesirable voltage in an electrical circuit. Transient voltages range from a few volts to several thousand volts and last from a few microseconds up to a few milliseconds. Two types of transient voltages are oscillatory transient voltages and impulse transient voltages. An *oscillatory transient voltage* is a transient voltage commonly caused by turning off high inductive loads and by switching off large utility power factor correction capacitors. An *impulse transient voltage* is a transient voltage commonly caused by lightning strikes and when loads with coils (motor starters and motors) are turned off.

DETERMINING VOLTAGE PROBLEMS

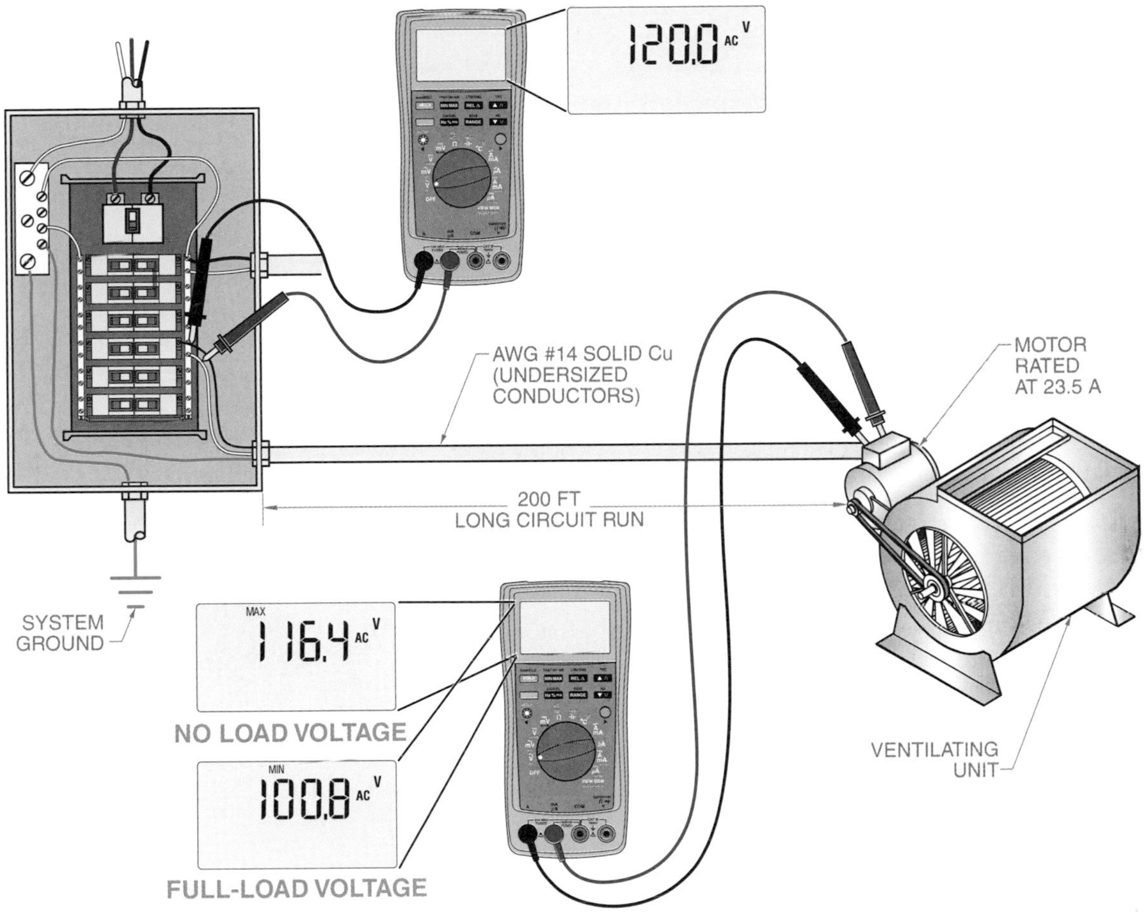

Figure 18-3. A voltmeter with a MIN/MAX recording mode is used to determine the type of voltage problem.

Transient Voltage Measurement Procedures. Power analyzer meters are typically used to monitor transient voltages. The size, duration, and time of transient voltages can be displayed at a later time when using power analyzer meters. When transient voltages are identified as a problem within a facility, surge protection devices (voltage surge suppressors) must be used. A *surge protection device* is a device that limits the intensity of voltage surges that occur on the power lines of a power distribution system.

Transient voltage measurements must be taken any time equipment prematurely fails and a transient voltage is suspected. Transient voltage measurements are taken to verify that surge protection devices are working. A transient voltage measurement is different from a standard voltage measurement because a standard voltage measurement displays the RMS voltage value of a circuit and

a transient voltage measurement displays the peak voltage value of a circuit. For example, a standard voltmeter that displays 115 VAC is displaying the RMS voltage measurement of the circuit. The peak voltage of a 115 VAC RMS circuit is about 162 V (RMS voltage multiplied by 1.414 equals peak voltage). **See Figure 18-4.**

Tech Fact

Transients can damage any electrical or electronic equipment. Transients can also cause problems in electronic circuits without damaging them by producing false signals. The high, short-duration voltage spikes can also cause errors in digital-based electronic circuits because they operate on low-level digital signals. Transients should be checked for at the power line input of electronic equipment and transient protection devices (snubbers) should be used to reduce transient problems. Transient voltage damage can be reduced when electronic devices are plugged into a good surge suppressor with a rating of at least 1200 Joules (J).

TRANSIENT VOLTAGE MEASUREMENT PROCEDURES

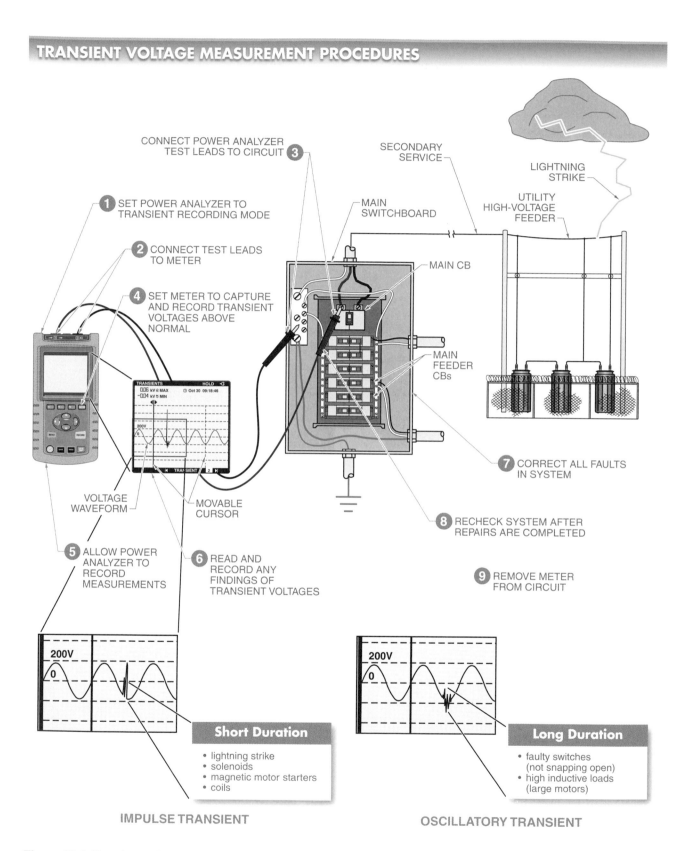

1 SET POWER ANALYZER TO TRANSIENT RECORDING MODE

2 CONNECT TEST LEADS TO METER

3 CONNECT POWER ANALYZER TEST LEADS TO CIRCUIT

4 SET METER TO CAPTURE AND RECORD TRANSIENT VOLTAGES ABOVE NORMAL

5 ALLOW POWER ANALYZER TO RECORD MEASUREMENTS

6 READ AND RECORD ANY FINDINGS OF TRANSIENT VOLTAGES

7 CORRECT ALL FAULTS IN SYSTEM

8 RECHECK SYSTEM AFTER REPAIRS ARE COMPLETED

9 REMOVE METER FROM CIRCUIT

SECONDARY SERVICE

LIGHTNING STRIKE

UTILITY HIGH-VOLTAGE FEEDER

MAIN SWITCHBOARD

MAIN CB

MAIN FEEDER CBs

VOLTAGE WAVEFORM

MOVABLE CURSOR

TRANSIENTS   HOLD
0.06 kV ≃ MAX   Oct 30 09:18:46
−0.04 kV ≃ MIN

200V
0

Short Duration
- lightning strike
- solenoids
- magnetic motor starters
- coils

IMPULSE TRANSIENT

200V
0

Long Duration
- faulty switches (not snapping open)
- high inductive loads (large motors)

OSCILLATORY TRANSIENT

Figure 18-4. Transient voltage measurements must be taken any time equipment prematurely fails and transient voltages are suspected.

Before taking any transient voltage measurements using a power analyzer meter, it is important to ensure that the meter is designed to take measurements on the system being tested. The operating manual of the test instrument should be referred to for all measuring precautions, limitations, and procedures. The required personal protective equipment (PPE) should always be worn and all safety rules should be followed when taking the measurement. To test for transient voltages, the following procedure is applied:

1. Set the power analyzer meter to transient voltage recording mode.

2. Connect the test leads of the power analyzer meter to the meter jacks as required.

3. Connect the power analyzer meter voltage test leads to the system being tested. When using a power analyzer meter to record transient voltages, the test leads of the meter must be connected to the powered system before recording starts. The test leads must be connected first because a power analyzer meter will be looking for a voltage higher than the setting of the meter (50%, 100%, 150%, etc.). When the test leads are not connected before the meter starts recording, the meter will record 50% (or whatever the meter is set on) of nothing (0 V).

4. Set the transient voltage recording mode to record transient voltages at a set level (50%, 100%, 200%), or above normal voltage. Normal voltage is the voltage applied to the test leads before the meter is set to start recording transients. When a meter is connected to a standard 120 V circuit and is set to record transient voltages greater than 100%, the meter records any voltage over 340 V peak (120 V RMS multiplied by 1.414 equals 170 V peak, and the meter is set at 100%, so 170 V multiplied by 2 equals 340 V peak).

5. Allow the meter to record as loads are switched on and off.

6. Read and record any findings of recorded transient voltages.

7. Correct for system transients by using surge suppressors to filter any transient voltages from the system.

8. Retest the system after repairs are completed.

9. Remove the power analyzer meter from the system.

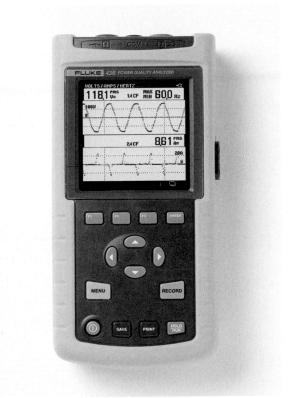

Fluke Corporation

A power analyzer meter is used to measure transient voltages to verify that surge protection devices are working.

18-2 CHECKPOINT

1. What is the acceptable operating voltage range (percentage) for newer electrical and electronic equipment?

2. Are voltage sags or voltage swells a more common problem?

3. What is a brownout?

4. What are the two types of transients voltages produced on a power distribution system?

18-3 MOTOR LOADS

Motor loads may require constant torque, variable torque, constant horsepower, or variable horsepower when operating at different speeds. Each type of motor has its own ability to control different loads at different speeds. The best type of motor to use for a given application depends on the type of load the motor must drive. Loads are generally classified as constant torque/variable horsepower (CT/VH), constant horsepower/variable torque (CH/VT), or variable torque/variable horsepower (VT/VH). **See Figure 18-5.**

Constant Torque/Variable Horsepower Loads

A *constant torque/variable horsepower (CT/VH) load* is a load in which the torque requirement remains constant. Any change in operating speed requires a change in horsepower. **See Figure 18-6.** CT/VH loads include loads that produce friction. Examples of these loads include conveyors, gear pumps and machines, metal-cutting tools, load-lifting equipment, and other loads that operate at different speeds.

MOTOR LOADS				
Load	Motor Torque*		Classification	NEMA Motor Design
	LRT	PUT		
Ball Mill (mining)	125 to 150	175 to 200	CT/VH	C-D
Band Saws				
Production	50 to 80	175 to 225	CT/VH	C
Small	40	250 to 300	CT/VH	B
Car Pullers				
Automobile	150	200 to 225	CT/VH	C
Railroad	175	250 to 300	CT/VH	D
Chipper	60	225	CT/VH	B
Compressor (air)	60	150	VT/VH	B
Conveyors				
Unloaded at start	50	125 to 150	CT/VH	B
Loaded at start	125 to 175	200 to 250	CT/VH	C
Screw	100 to 125	50 to 175	CT/VH	C
Crushers				
Unloaded at start	75 to 100	150 to 175	CT/VH	B
With flywheel	125 to 150	175 to 200	CT/VH	D
Dryer, Industrial (loaded rotary drum)	150 to 175	175 to 225	CT/VH	D
Fan and Blower	40	150	VT/VH	B
Machine Tools				
Drilling	40	150	CT/VH	B
Lathe	75	150	CT/VH	B
Press (with flywheel)	50 to 100	250 to 350	CH/VT	D
Pumps				
Centrifugal	50	150	VT/VH	B
Positive displacement	60	175	CT/VH	B
Propeller	50	150	VT/VH	B
Vacuum	60	150	CT/VH	B-C

* in % of FLT

Figure 18-5. Loads are generally classified as constant torque/variable horsepower (CT/VH), constant horsepower/variable torque (CH/VT), or variable torque/variable horsepower (VT/VH).

CONSTANT TORQUE/VARIABLE HORSEPOWER LOADS

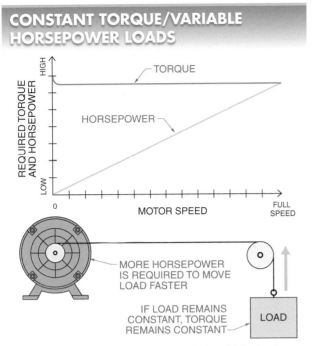

Figure 18-6. CT/VH loads are loads in which the torque requirement remains constant.

Although the operating speed may change, a CT/VH load requires the same torque at low speeds as at high speeds. Since the torque requirement remains constant, an increase in speed requires an increase in horsepower.

Constant Horsepower/Variable Torque Loads

A *constant horsepower/variable torque (CH/VT) load* is a load that requires high torque at low speeds and low torque at high speeds. Since the torque requirement decreases as speed increases, the horsepower remains constant. Speed and torque are inversely proportional in CH/VT loads. **See Figure 18-7.**

An example of a CH/VT load is a center-driven winder used to roll and unroll materials such as paper or metal. Since the work is performed with a varying diameter, but with tension and linear speed of the material constant, horsepower must also be constant. Although the speed of the material is kept constant, the motor speed is not. The diameter of the material on the roll that is driven by the motor is constantly changing as material is added. At the start, the motor must run at high speed to maintain the correct material speed while torque is kept at a minimum. As material is added to the roll, the motor must deliver more torque at a lower speed. As the material is rolled, both torque and speed are constantly changing while the motor horsepower remains the same.

CONSTANT HORSEPOWER/VARIABLE TORQUE LOADS

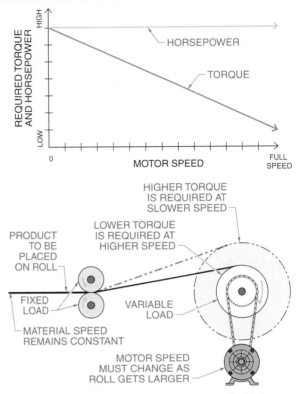

Figure 18-7. CH/VT loads are loads that require high torque at low speeds and low torque at high speeds.

Variable Torque/Variable Horsepower Loads

A *variable torque/variable horsepower (VT/VH) load* is a load that requires a varying torque and horsepower at different speeds. **See Figure 18-8.** With this type of load, a motor must work harder to deliver more output at a faster speed. Both torque and horsepower are increased with increased speed. Examples of VT/VH loads include fans, blowers, centrifugal pumps, mixers, and agitators.

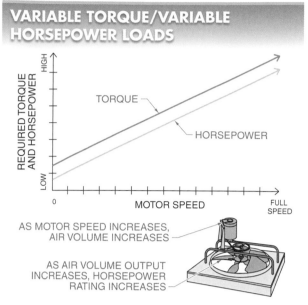

VARIABLE TORQUE/VARIABLE HORSEPOWER LOADS

AS MOTOR SPEED INCREASES, AIR VOLUME INCREASES

AS AIR VOLUME OUTPUT INCREASES, HORSEPOWER RATING INCREASES

Figure 18-8. VT/VH loads are loads that require different torque and horsepower at different speeds.

A fan is an example of a variable torque/variable horsepower (VT/VH) load that requires varying torque and horsepower at different speeds.

18-3 CHECKPOINT

1. If a drive is used to operate the spool of a paper roller, what setting (CT, VT, or CH) should be used?

2. If a drive is used to operate a variable-speed industrial food-mixing machine, what setting (CT, VT, or CH) should be used?

18-4 MOTOR TORQUE TYPES

A motor must produce enough torque to start the load and keep it moving for the motor to operate the load connected to it. A motor connected to a load produces four types of torque. The four types of torque are locked rotor torque (LRT), pull-up torque (PUT), breakdown torque (BDT), and full-load torque (FLT). **See Figure 18-9.**

Locked Rotor Torque

Locked rotor torque (LRT) is the torque a motor produces when its rotor is stationary and full power is applied to the motor. **See Figure 18-10.** All motors can safely produce a higher torque output than the rated full-load torque for short periods of time. Since many loads require a higher torque to start them moving than

to keep them moving, a motor must produce a higher torque when starting the load. LRT is also referred to as breakaway or starting torque. Starting torque is the torque required to start a motor and is normally expressed as a percentage of full-load torque.

Pull-Up Torque

Pull-up torque (PUT) is the torque required to bring a load up to its rated speed. **See Figure 18-11.** If a motor is properly sized to the load, PUT is brief. If a motor does not have sufficient PUT, the locked rotor torque may start the load turning but the PUT cannot bring it up to rated speed. Once the motor is up to rated speed, full-load torque keeps the load turning. PUT is also referred to as accelerating torque.

MOTOR TORQUE

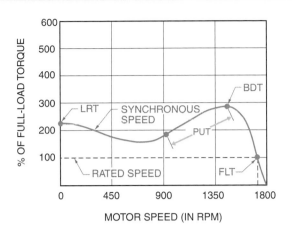

LRT = Locked rotor torque—produced when rotor is stationary and full power is applied to motor

PUT = Pull-up torque—required to bring motor up to correct speed

BDT = Breakdown torque—maximum torque motor can provide without reduction in motor speed

BDT = Full-load torque—produces rated power at full speed of motor

Figure 18-9. A motor connected to a load produces four types of torque: locked rotor torque (LRT), pull-up torque (PUT), breakdown torque (BDT), and full-load torque (FLT).

LOCKED ROTOR TORQUE (LRT)

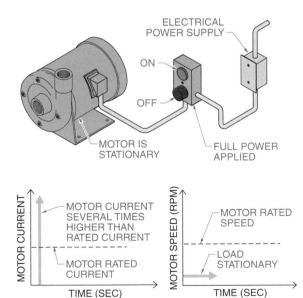

Figure 18-10. Locked rotor torque (LRT) is the torque a motor produces when its rotor is stationary and full power is applied to the motor.

PULL-UP TORQUE (PUT)

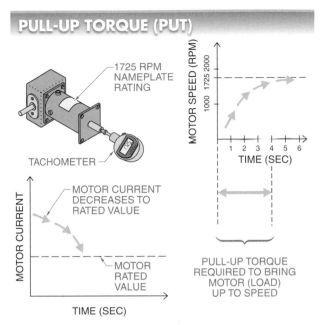

Figure 18-11. Pull-up torque (PUT) is the torque required to bring a load up to its rated speed.

Breakdown Torque

Breakdown torque (BDT) is the maximum torque a motor can provide without an abrupt reduction in motor speed. **See Figure 18-12.** As the load on a motor shaft increases, the motor produces more torque. As the load continues to increase, the point at which the motor stalls is reached. This point is the breakdown torque.

BREAKDOWN TORQUE (BDT)

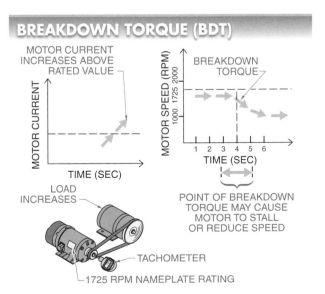

Figure 18-12. Breakdown torque (BDT) is the maximum torque a motor can provide without an abrupt reduction in motor speed.

Full-Load Torque

Full-load torque (FLT) is the torque required to produce the rated power at the full speed of the motor. **See Figure 18-13.** The amount of torque a motor produces at rated power and full speed (full-load torque) can be found by using a horsepower-to-torque conversion chart. To calculate motor FLT, the following formula is applied:

$$T = \frac{HP \times 5252}{rpm}$$

where

T = torque (in lb-ft)
HP = horsepower
5252 = constant
rpm = revolutions per minute

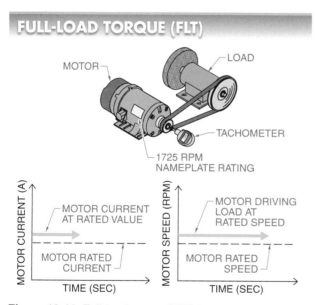

Figure 18-13. Full-load torque (FLT) is the torque required to produce the rated power at full speed of the motor.

Example: Calculating Full-Load Torque

What is the FLT of a 30 HP motor operating at 1725 rpm?

$$T = \frac{HP \times 5252}{rpm}$$

$$T = \frac{30 \times 5252}{1725}$$

$$T = \frac{157,560}{1725}$$

$$T = \textbf{91.34 lb-ft}$$

If a motor is fully loaded, it produces FLT. If a motor is underloaded, it produces less than FLT. If a motor is overloaded, it must produce more than FLT to keep the load operating at the motor's rated speed. **See Figure 18-14.**

For example, a 30 HP motor operating at 1725 rpm can develop 91.34 lb-ft of torque at full speed. If the load requires 91.34 lb-ft at 1725 rpm, the 30 HP motor produces an output of 30 HP. However, if the load to which the motor is connected requires only half as much torque (45.67 lb-ft) at 1725 rpm, the 30 HP motor produces an output of 15 HP. The 30 HP motor draws less current (and power) from the power lines and operates at a lower temperature when producing 15 HP.

The torque required to operate a load from initial startup to final shutdown is considered when determining the type and size of motor required for the application.

However, if the 30 HP motor is connected to a load that requires twice as much torque (182.68 lb-ft) at 1725 rpm, the motor must produce an output of 60 HP. The 30 HP motor draws more current (and power) from the power lines and operates at a higher temperature. If the overload protection device is sized correctly, the 30 HP motor automatically disconnects from the power line before any permanent damage is done to the motor.

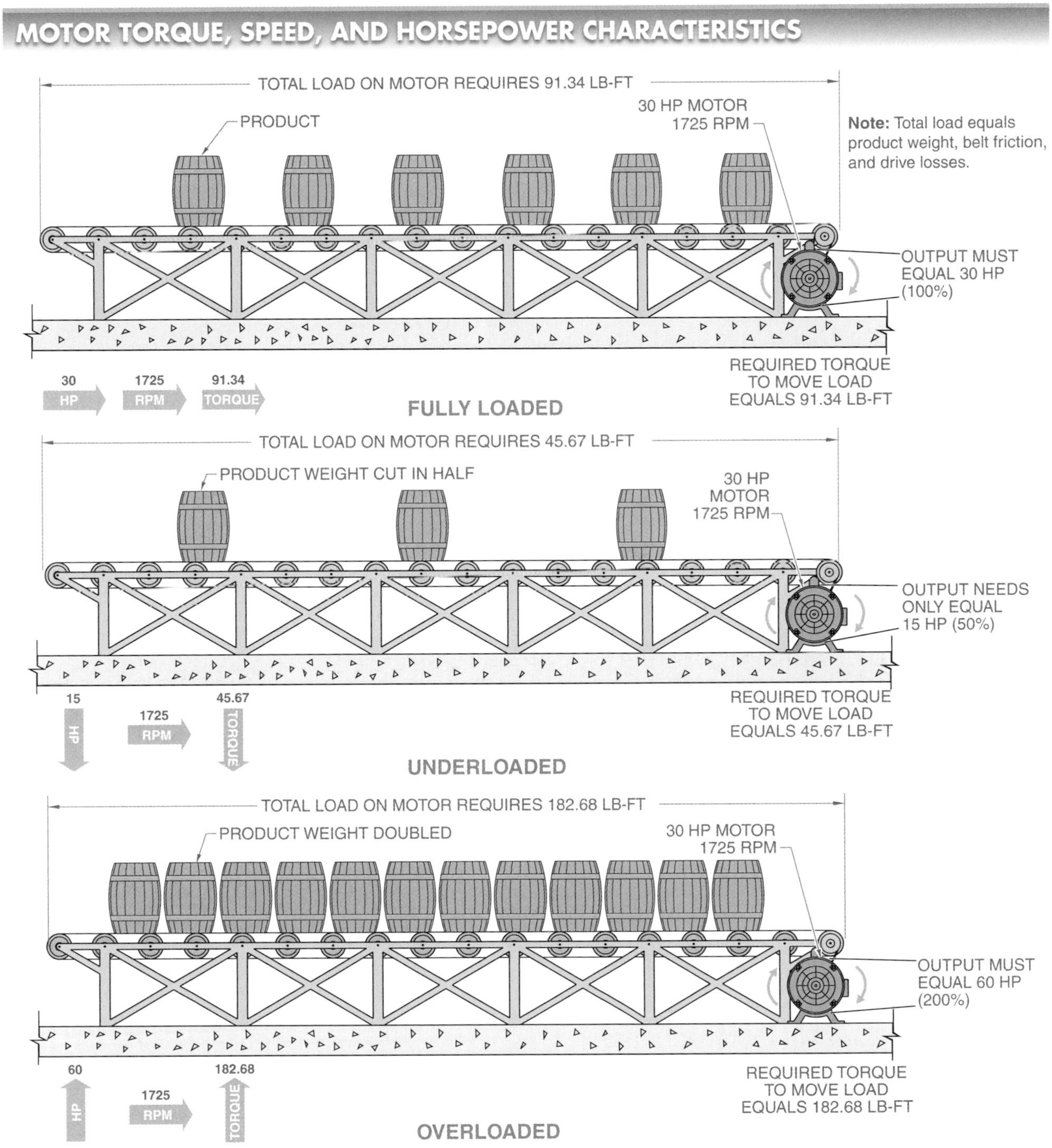

MOTOR TORQUE, SPEED, AND HORSEPOWER CHARACTERISTICS

TOTAL LOAD ON MOTOR REQUIRES 91.34 LB-FT

PRODUCT

30 HP MOTOR
1725 RPM

Note: Total load equals product weight, belt friction, and drive losses.

OUTPUT MUST EQUAL 30 HP (100%)

| 30 HP | 1725 RPM | 91.34 TORQUE |

REQUIRED TORQUE
TO MOVE LOAD
EQUALS 91.34 LB-FT

FULLY LOADED

TOTAL LOAD ON MOTOR REQUIRES 45.67 LB-FT

PRODUCT WEIGHT CUT IN HALF

30 HP
MOTOR
1725 RPM

OUTPUT NEEDS ONLY EQUAL 15 HP (50%)

| 15 HP | 1725 RPM | 45.67 TORQUE |

REQUIRED TORQUE
TO MOVE LOAD
EQUALS 45.67 LB-FT

UNDERLOADED

TOTAL LOAD ON MOTOR REQUIRES 182.68 LB-FT

PRODUCT WEIGHT DOUBLED

30 HP MOTOR
1725 RPM

OUTPUT MUST EQUAL 60 HP (200%)

| 60 HP | 1725 RPM | 182.68 TORQUE |

REQUIRED TORQUE
TO MOVE LOAD
EQUALS 182.68 LB-FT

OVERLOADED

Figure 18-14. A motor may be fully loaded, underloaded, or overloaded.

18-4 CHECKPOINT

1. What type of torque is required to produce the rated power for a motor at full speed?

2. What type of torque is required to start turning a motor shaft?

18-5 OPEN AND CLOSED CIRCUIT TRANSITION

Motors that are started at reduced voltage must be switched to line voltage before reaching full speed. The two methods used to switch motors from starting voltage to line voltage include open circuit transition and closed circuit transition. In open circuit transition, a motor is temporarily disconnected from the voltage source when switching from a reduced starting voltage level to a running voltage level before reaching full motor speed. In closed circuit transition, a motor remains connected to the voltage source when switching from a reduced starting voltage level to a running voltage level before reaching full motor speed.

Closed circuit transition is preferable to open circuit transition because closed circuit transition does not cause a high-current transition surge. However, closed circuit transition is the more expensive circuit transition method. Open circuit transition produces a higher current surge than closed circuit transition at the transition point because the motor is momentarily disconnected from the voltage source. **See Figure 18-15.**

The high-current surge during open circuit transition is based on the motor speed at the time of transition. Transfer from the low starting voltage to the high line voltage should occur as close to full motor speed as possible. If the transition occurs when the motor is at a low speed, a surge current even higher than the starting current can occur. **See Figure 18-16.**

CIRCUIT TRANSITIONS—OPEN/CLOSED

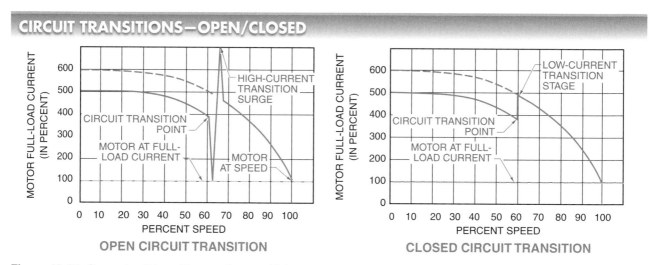

Figure 18-15. Open circuit transition produces a higher current surge than closed circuit transition at the transition point because the motor is momentarily disconnected from the voltage source.

CIRCUIT TRANSITIONS—LOW/FULL MOTOR SPEED

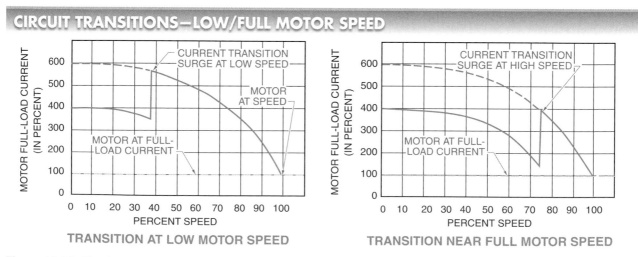

Figure 18-16. Circuit transition should occur at or near full motor speed to minimize current surges.

18-5 CHECKPOINT

1. What circuit transition method is the least expensive type that can be used?

2. Why is a closed circuit transition preferred when switching a motor from reduced voltage to full voltage?

Additional Resources

Review and Resources

Access Chapter 18 Review and Resources for *Electrical Motor Controls for Integrated Systems* by scanning the above QR code with your mobile device.

Applying Your Knowledge

Refer to the *Electrical Motor Controls for Integrated Systems* Learner Resources for interactive Applying Your Knowledge questions.

Workbook and Applications Manual

Refer to Chapter 18 in the *Electrical Motor Controls for Integrated Systems Workbook* and the *Applications Manual* for additional exercises.

ENERGY EFFICIENCY PRACTICES

Power Correction Capacitors Energy and Operating Cost Savings

Power factor (PF) is the ratio of true power used in a circuit to apparent power delivered to the circuit. Almost all electrical systems require that the apparent power be higher than the true power because no system has a perfect power factor of 100%. Most electrical systems have a power factor of 70% to 90%. The lower the power factor, the higher the utility cost of operating the system and the more utility-generated power is wasted.

To save energy and reduce utility cost, power correction capacitors can be added into an electrical system. Power correction capacitors improve the power factor of the system. The addition of power correction capacitors to the system improves the system in the following ways:

- There is more true power to operate more loads without the need to increase the apparent power of the system. This means that transformers, switchgears, panelboards, conductors, and other system components will not need to increase in size.

- The utility cost of operating loads will decrease. A decrease in cost can be significant in areas where utilities impose penalties for poor power factor.

- Loads such as motors will operate better because the capacitors will help reduce voltage sags (power line voltage dips).

- Utilities generate less power, which means that less fossil fuel is burned and fewer power plants or wind turbines are needed.

There are two places within an electrical system that power correction capacitors can be added to improve the power factor. The best place to add power correction capacitors is at the motor loads that cause most of the poor power factor in a system. The addition of these capacitors at the motor requires fewer capacitor sizing calculations because the capacitor size can be selected from a chart based on the rated HP size of the motor. Since the capacitors are switched on only when the motor is ON, their operating life and serviceability is improved.

Power correction capacitors can also be added at the main distribution points. The addition of these capacitors to transformers, switchgears, and panelboards improves the distribution system at that point. By using a power quality meter that measures power factor, true power, and apparent power, the improvements can be measured and recorded for each amount of added capacitance.

Objectives

19-1
- Explain why a reduced-voltage starting method may be used instead of full-voltage starting.

19-2
- Explain how a reduced-voltage starting method reduces the amount of voltage and current to a DC motor during starting.

19-3
- Explain why reduced-voltage starting is used for three-phase (3ϕ) induction motors.

19-4
- Define primary resistor starting and explain how it reduces the amount of voltage and current to an AC motor during starting.

19-5
- Explain how an autotransformer reduced-voltage starting method reduces the amount of voltage and current to an AC motor during starting.

19-6
- Define part-winding starting and explain how it reduces the amount of voltage and current to an AC motor during starting.

19-7
- Explain how a wye-delta reduced-voltage starting method reduces the amount of voltage and current to an AC motor during starting.

19-8
- Describe the considerations for selecting a starting method and describe the advantages and disadvantages of each.

19-9
- Explain how to troubleshoot a reduced-voltage starting circuit.

Sections

Review and Resources
atplearningresources.com/Quicklinks
Access Code: 362245

Chapter

Reduced-Voltage Starting Circuits

19

Electric motors are used in applications that require power ranging from fractional horsepower to thousands of horsepower. The higher the horsepower rating that the motor has, the higher the motor's starting and running current draw. Since motors draw several times more starting current than running current, the electrical system supplying power to the motor must supply enough current to start the motor without causing any problems within the system. Supplying enough starting current can be accomplished by increasing the available current. The available current is increased by increasing the system's conductor size, fuse/circuit breaker size, and transformer size.

However, if the starting current drawn by a motor can be reduced, the size of the system will not need to be increased. The starting current of a motor can be reduced several different ways based on the motor type, application requirements, and motor size. Control circuit methods used to reduce the starting current of a motor are called reduced-voltage starting circuits. It is important to learn the different reduced-voltage starting methods for any system that uses large motors.

19-1 REDUCED-VOLTAGE STARTING

Full-voltage starting is the least expensive and most efficient means of starting a motor for applications involving small-horsepower motors. Many applications involve large-horsepower DC and AC motors that require reduced-voltage starting because full-voltage starting may create interference with other systems. Reduced-voltage starting reduces interference in the power source.

Power Source

Reduced-voltage starting is used to reduce the large current drawn from the electric utility lines by an across-the-line start of a large motor. An induction motor acts much like a short circuit in the secondary of a transformer when it is started. The current drawn by a motor during starting is typically about two to six times the current rating found on the motor nameplate. This sudden demand for a large amount of current can reflect back into the power lines and create problems. Reduced-voltage starting reduces the amount of starting current a motor draws when starting. **See Figure 19-1.**

Electric utilities normally limit the inrush current drawn from their lines to a maximum amount for a specified period of time. Such limitations are necessary for smooth, steady power regulation and for eliminating objectionable voltage disturbances such as annoying light flicker.

REDUCED-VOLTAGE STARTING CURRENT

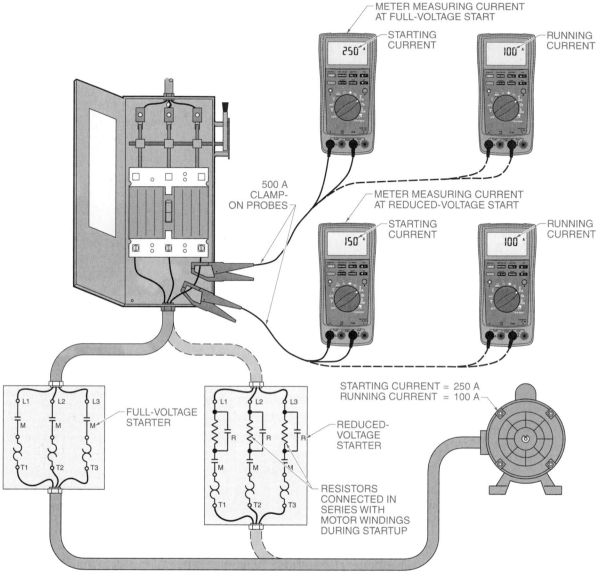

Figure 19-1. Reduced-voltage starting reduces the amount of current a motor draws when starting.

In these cases, the electric utility is not limiting the total maximum amount of current that can be drawn, but rather dividing the amount of current into steps. This permits an incremental start that allows the utility voltage regulators sufficient time to compensate for the large current draw. *Increment current* is the maximum current permitted by the utility in any one step of an incremental start. The increment current may be determined by checking with the local electric utility company. Reduced-voltage starting provides incremental current draw over a longer period of time.

19-1 CHECKPOINT

1. What is the main goal of using a reduced-voltage starting with electric motors?

2. Why does an electric utility need to set limits on the amount of current drawn at any one short period of time?

19-2 DC MOTOR REDUCED-VOLTAGE STARTING

A DC motor is used to convert electrical energy into a rotating mechanical force. Although AC and DC motors operate on the same fundamental principles of magnetism, they differ in the way the conversion of the electrical power to mechanical power is accomplished. This difference gives each motor its own operating characteristics. The two fundamental operating characteristics of DC motors that make them the preferred choice for some applications are high torque output and good speed control.

Another factor in using DC motors is the available source of power. For applications such as an automobile starter, a DC motor is compatible with the power source (a battery), which delivers only DC power. DC motors run by batteries are also used in industrial applications using portable power equipment such as forklift trucks, dollies, and small locomotives that move materials and supplies. In applications where the motor is to be connected to a power source other than a battery, the available power source may be either AC or DC.

Baldor Electric Co.

Reduced-voltage starting is used with large industrial motors to reduce the damaging effect of a large starting current.

All DC motors are supplied with current to the armature and field windings. During startup, current is limited only by the higher resistance of the wire in the armature and the field windings when current is connected directly to the motor. The larger the horsepower (HP) size of the motor, the less the resistance and the higher the current. In large DC motors, the higher starting current can damage the motor. To prevent this damage, the motor must be connected to a reduced-voltage device such as a DC motor drive.

DC drives are now used for starting DC motors with a reduced voltage and for operating the motor under more controlled conditions than were possible with older reduced-voltage starting circuits. When any motor is started, high inrush current (starting current) is produced. A decrease in current results as a motor accelerates because the motor generates a voltage that is opposite to the applied voltage.

This opposing voltage, known as counter electromotive force (CEMF), depends on the speed of the motor. CEMF is zero at standstill and increases with motor speed. Ohm's law is used to calculate motor starting current. To calculate motor starting current, the following formula is applied:

$$I = \frac{E}{R}$$

where

I = starting current (in A)

E = applied voltage (in V)

R = resistance (in Ω)

Example: Calculating Motor Starting Current

What is the starting current of a DC motor with an armature resistance of 0.1 Ω that is connected to a 200 V supply?

$$I = \frac{E}{R}$$

$$I = \frac{200}{0.1}$$

$$I = \textbf{2000 A}$$

As the motor accelerates, CEMF is generated. The CEMF reduces the current in the motor. To calculate the current drawn by a motor during starting, the following formula is applied:

$$I = \frac{E - C_{EMF}}{R}$$

where

I = starting current (in A)

E = applied voltage (in V)

C_{EMF} = generated counter electromotive force (in V)

R = resistance (in Ω)

Example: Calculating Current During Starting

What is the current during the starting of a DC motor with an armature resistance of 0.1 Ω that is connected to a 200 V supply and is generating 100 V of CEMF?

$$I = \frac{E - C_{EMF}}{R}$$

$$I = \frac{200 - 100}{0.1}$$

$$I = \frac{100}{0.1}$$

$$I = \textbf{1000 A}$$

A motor at full speed generates an even higher CEMF. The higher CEMF further reduces the current in the motor. To prevent damage, it is the 200 A (or starting current of any large DC motor) that the motor must be protected against.

Example: Calculating Motor Running Current

What is the running current of a DC motor with an armature resistance of 0.1 Ω that is connected to a 200 V supply and is generating 180 V of CEMF?

$$I = \frac{E - C_{EMF}}{R}$$

$$I = \frac{200 - 180}{0.1}$$

$$I = \frac{20}{0.1}$$

$$I = \textbf{200 A}$$

A starting rheostat or solid-state circuit is used for DC motor reduced-voltage starting, although many have recently been replaced by DC drives. A starting rheostat is connected in series with the incoming power line (typically the positive DC line) and the motor. The starting rheostat reduces the voltage applied to the motor during starting by placing a high resistance in series with the motor. The resistance is decreased as the rheostat is moved to the run position. **See Figure 19-2.**

Tech Fact

Starting motor current is several times higher than the running current. If fuses or circuit breakers trip on starting, check to make sure that time delay fuses (TDFs) or inverse time breakers (ITBs) are used and not the instantaneous trip type of circuit breakers. If the fuses and breakers trip after startup while the motor is running, measure the running current and compare it to the nameplate rated current to make sure the motor is not overloaded. Also check the size of the motor starter overload (heater) ratings.

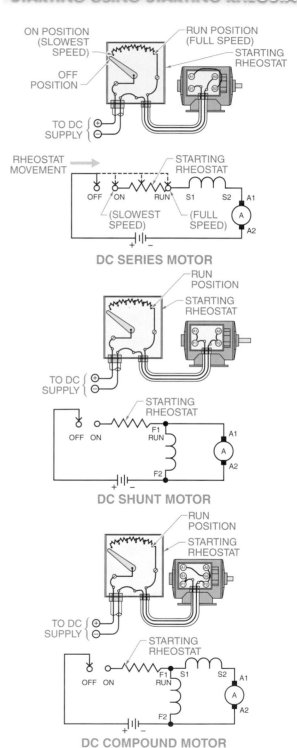

DC MOTOR REDUCED-VOLTAGE STARTING USING STARTING RHEOSTAT

Figure 19-2. A starting rheostat reduces the voltage applied to a motor during starting by placing a high resistance in series with the motor.

The starting rheostat is controlled manually, which means that the operator determines the exact starting time. Although a starting rheostat can also be used to control motor speed (the speed of a DC motor varies with the applied voltage), the purpose of a starting rheostat is to reduce the voltage (and thus current and torque) during starting. After the motor is started, a different circuit can be used to control motor speed.

A solid-state circuit can also be used to reduce the voltage applied to a motor. A silicon-controlled rectifier (SCR) is typically used to control the applied voltage. The voltage is adjusted through a control circuit that controls the voltage applied to the gate of the SCR. The higher the applied gate voltage, the higher the output voltage applied to the motor. **See Figure 19-3.**

DC MOTOR REDUCED-VOLTAGE STARTING USING SOLID-STATE CIRCUIT

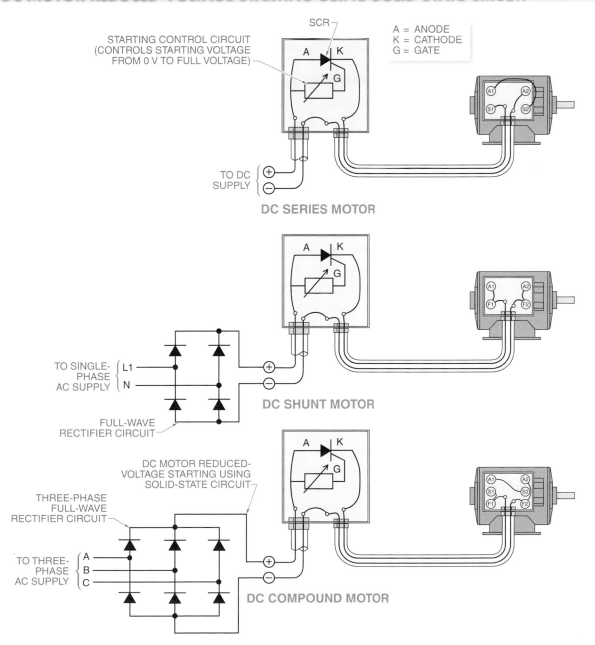

Figure 19-3. A solid-state circuit, such as a silicon-controlled rectifier (SCR), can be used to reduce the voltage to a motor.

19-2 CHECKPOINT

1. Does motor resistance increase or decrease as the horsepower (HP) size of the motor increases?

2. What specific electronic component is used to perform the same function as a rheostat in reducing the starting voltage to a DC motor?

19-3 REDUCED-VOLTAGE STARTING FOR 3φ INDUCTION MOTORS

The majority of industrial applications normally use 3φ induction motors. Three-phase induction motors are chosen over other types of motors because of their simplicity, ruggedness, and reliability. Both 1φ and 3φ induction motors have become the standard for all-purpose, constant-speed AC motor applications. Reduced-voltage starting is applied to 3φ motors because 1φ motors are small (typically 5 HP and less).

AC Motor Reduced-Voltage Starting

A heavy current is drawn from the power lines when an induction motor is started. This sudden demand for large current can reflect back into the power lines and create problems such as voltage sags (temporary low voltage) on the power lines.

The revolving field of the stator induces a large current in the short-circuited rotor bars. The current is highest when the rotor is at a standstill and decreases as the motor speed increases. The current drawn by a motor when starting is excessive due to a lack of CEMF at the instant of starting. Once rotation begins, CEMF is built up in proportion to speed, and the current decreases. **See Figure 19-4.**

The percent of full-load current (FLC) is marked on the vertical scale and the percent of motor speed is marked on the horizontal scale. The starting current is quite high compared to the running current. The starting current remains fairly constant at this high value as the speed of the motor increases, but then it drops sharply during the last few percentages up to 100%. This illustrates that the heating rate is quite high during acceleration because the heating rate is a function of current. A motor may be considered in the locked condition during nearly all of the accelerating period.

Locked rotor current (LRC) is the steady-state current taken from the power line with the rotor locked (stopped) and with the voltage applied. LRC and the resulting torque produced in the motor shaft (in addition to load requirements) determine whether the motor can be connected across the line or whether the current must be reduced through reduced-voltage starting.

Full-load current (FLC) is the current required by a motor to produce full-load torque at the motor's rated speed. FLC is the current given on the motor nameplate. The load current is less than what is given on the nameplate if the motor is not required to deliver full torque. This information is required when testing a motor (running a motor without a load). The only torque a motor must produce when not connected to a load is the torque that is needed to overcome its own internal friction and winding losses. For this reason, the current is less than the value given on the motor nameplate.

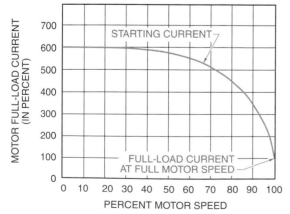

Figure 19-4. The current drawn by a motor when starting is excessive because of a lack of CEMF at the instant of starting.

19-3 CHECKPOINT

1. Why is reduced-voltage starting not applied to 1φ AC motors?

2. Is reduced-voltage starting used to reduce locked rotor current (LRC), full-load current (FLC), or both?

19-4 PRIMARY RESISTOR STARTING

Primary resistor starting is a reduced-voltage starting method that uses a resistor connected in each motor line (in one line for a 1φ starter) to produce a voltage drop. This reduces the motor starting current as it passes through the resistor. A timer is provided in the control circuit to short the resistors after the motor accelerates to a specified point. **See Figure 19-5.** The motor is started at reduced voltage but operates at full line voltage.

contactor coil C. Coil M controls the motor starter, which energizes the motor and provides overload protection. The timer provides a delay from the point where coil M energizes until contacts C close, shorting resistors R1, R2, and R3. Coil C energizes the contactor, which provides a short circuit across the resistors.

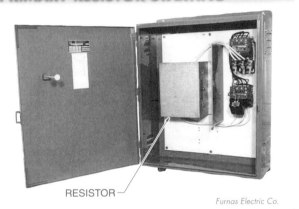

RESISTOR

Furnas Electric Co.

Figure 19-5. Primary resistor starting uses a resistor connected in each motor line and a timer to short the resistors after the motor accelerates to a specified point.

Baldor Electric

It is more feasible to use methods to reduce the starting current of a motor than to increase the available current.

Primary resistor starters provide extremely smooth starting due to increasing voltage across the motor terminals as the motor accelerates. Standard primary resistor starters provide two-point acceleration (one step of resistance) with approximately 70% of line voltage at the motor terminals at the instant of motor starting. Multiple-step starting is possible by using additional contacts and resistors when extra-smooth starting and acceleration are needed. Multiple-step starting may be required in paper or fabric applications where even a small jolt during starting may tear the paper or snap the fabric.

Primary Resistor Starting Circuits

In a primary resistor starting circuit, external resistance is added to and taken away from the motor circuit. **See Figure 19-6.** The control circuit consists of the motor starter coil M, on-delay timer TR1, and

Pressing start pushbutton PB2 energizes motor starter coil M and the on-delay timer coil TR1. Motor starter coil M closes contacts M to create memory. On-delay timer coil TR1 causes contacts TR1 to remain open during reset, stay open during timing, and close after timing out. Once timed out, contactor coil C energizes, causing contacts C to close and the resistors to short.

This circuit is a common reduced-voltage starting circuit. Changes are often made in the values of resistance and wattage to accommodate motors of different horsepower ratings.

PRIMARY RESISTOR STARTING CIRCUITS

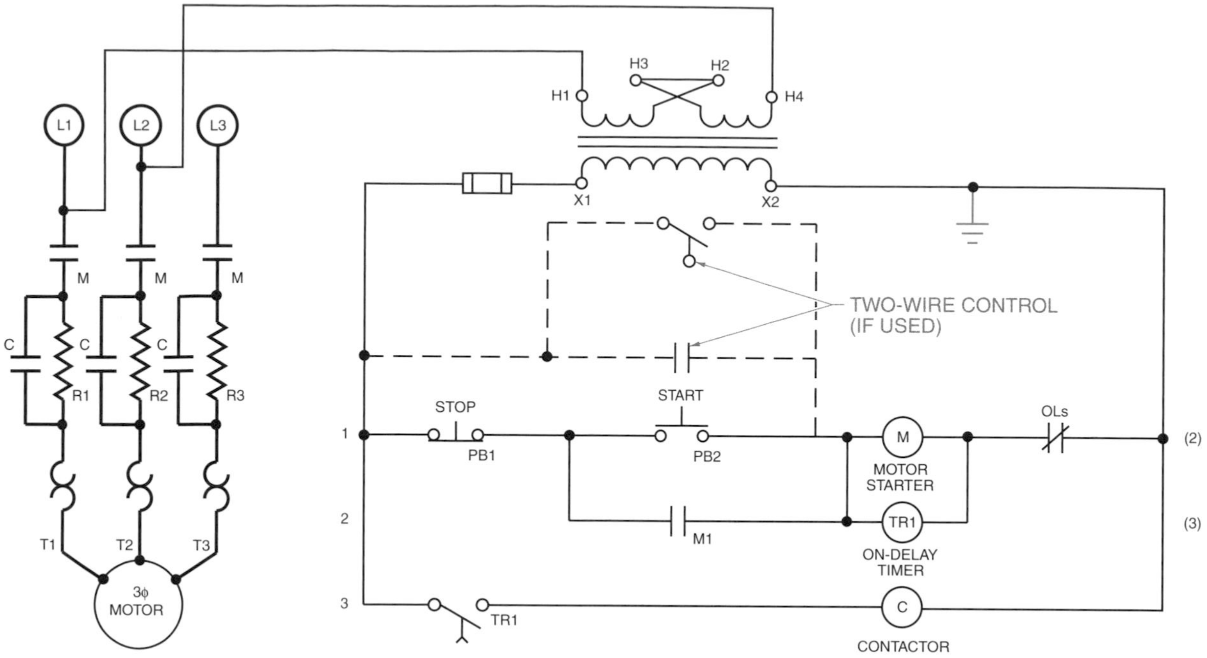

Figure 19-6. In a primary resistor starting circuit, external resistance is added to and taken away from the motor circuit.

19-4 CHECKPOINT

1. What does the control circuit of a primary resistor starting circuit consist of?

2. In the primary resistor starting circuit, what determines the setting of the on-delay timer?

19-5 AUTOTRANSFORMER STARTING

Autotransformer starting uses a tapped 3ϕ autotransformer to provide reduced-voltage starting. **See Figure 19-7.** Autotransformer starting is one of the most effective methods of reduced-voltage starting. Autotransformer starting is preferred over primary resistor starting when the starting current drawn from the line must be held to a minimum value but the maximum starting torque per line ampere is required.

In autotransformer starting, the motor terminal voltage does not depend on the load current. The current to the motor may change because of the changing characteristics of the motor, but the voltage to the motor remains relatively constant.

Autotransformer starting may use its turns ratio advantage to provide more current on the load side of the transformer than on the line side. In autotransformer starting, transformer motor current and line current are not equal as they are in primary resistor starting.

For example, a motor has a full-voltage starting torque of 120% and a full-voltage starting current of 600%. The electric utility has set a limitation of 400% current draw from the power line. This limitation is set only for the line side of the controller. Because the transformer has a step-down ratio, the motor current on the transformer secondary is larger than the line current as long as the primary of the transformer does not exceed 400%.

In this example, with the line current limited to 400%, 80% voltage can be applied to the motor, generating 80% motor current. The motor draws only 64% line current ($0.8 \times 80 = 64\%$) due to the 1:0.8 turns ratio of the transformer. The advantage is that the starting torque is 77% (0.8×80 of 120%) instead of the 51% obtained in primary resistor starting. This additional percentage may be sufficient accelerating energy to start a load that may be otherwise difficult to start.

Autotransformer Starting Circuits

In an autotransformer starting circuit, the various windings of the transformer are added to and taken away from the motor circuit to provide reduced voltage when starting. **See Figure 19-8.** The control circuit consists of an on-delay timer TR1 and contactor coils C1, C2, and C3.

AUTOTRANSFORMER STARTING

Furnas Electric Co.

└TRANSFORMER

Figure 19-7. Autotransformer starting uses a tapped 3ϕ auto-transformer to provide reduced-voltage starting.

AUTOTRANSFORMER STARTING CIRCUITS

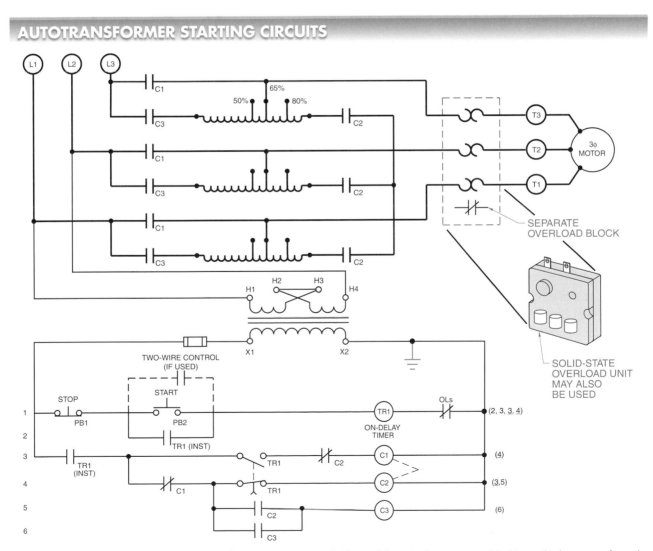

Figure 19-8. In an autotransformer starting circuit, the various windings of the transformer are added to and taken away from the motor circuit to provide reduced voltage when starting.

Pressing start pushbutton PB2 energizes the timer, causing instantaneous contacts TR1 in line 2 and 3 of the line diagram to close. Closing the normally open (NO) timer contacts in line 2 provides memory for timer TR1, while closing NO timer contacts in line 3 completes an electrical path through line 4 and energizing contactor coil C2. The energizing of coil C2 causes NO contacts C2 in line 5 to close, energizing contactor coil C3.

The normally closed (NC) contacts in line 3 also provide electrical interlocking for coil C1 so that they may not be energized together. The NO contacts of contactor C2 close, connecting the ends of the autotransformers together when coil C2 energizes. When coil C3 energizes, the NO contacts of contactor C3 close and connect the motor through the transformer taps to the power line, starting the motor at reduced inrush current and starting torque. Memory is also provided to coil C3 by contacts C3 in line 6.

After a predetermined time, the on-delay timer times out and the NC timer contacts TR1 open in line 4, de-energizing contactor coil C2. Also, NO timer contacts TR1 close in line 3, energizing coil C1. In addition, NC contacts C1 provide electrical interlock in line 4 and NC contacts C2 in line 3 return to their NC position. The net result of de-energizing C2 and energizing C1 is the connecting of the motor to full line voltage.

During the transition from starting to full line voltage, the motor was not disconnected from the circuit, which indicates closed circuit transition. As long as the motor is running in the full-voltage condition, timer TR1 and contactor C1 remain energized. Only an overload or pressing the stop pushbutton will stop the motor and reset the circuit. Overload protection is provided by a separate overload block.

In this circuit, pushbuttons are used to control the motor. However, any NO and/or NC device may be used to control the motor. Thus, in an air conditioning system, the pushbuttons would be replaced with a temperature switch and the circuit would be connected for two-wire control.

19-5 CHECKPOINT

1. In the autotransformer starting circuit in Figure 19-8, when contactors C2 and C3 are initially energized to connect the motor through the autotransformer's 65% taps, are the transformers connected into a wye or delta configuration?
2. In the autotransformer starting circuit in Figure 19-8, what other special feature must the on-delay timer have in addition to TR1 time contacts?

19-6 PART-WINDING STARTING

Part-winding starting is a method of starting a motor by first applying power to part of the motor coil windings for starting and then applying power to the remaining coil windings for normal running. The motor stator windings must be divided into two or more equal parts for a motor to be started using part-winding starting. Each equal part must also have its terminal available for external connection to power. In most applications, a wye-connected motor is used, but a delta-connected motor can also be started using part-winding starting.

Wye-Connected Motors

Part-winding starting requires the use of a part-winding motor. A part-winding motor has two sets of identical windings, which are intended to be used in parallel. **See Figure 19-9.** These windings produce reduced starting current and reduced starting torque when energized in sequence. Most dual-voltage, 230/460 V motors are suitable for part-winding starting at 230 V.

Part-winding starters are available in either two- or three-step construction. The more common two-step starter is designed so that when the control circuit is energized, one winding of the motor is connected directly to the line. This winding draws about 65% of normal LRC and develops approximately 45% of normal motor torque. After about 1 sec, the second winding is connected in parallel with the first winding in such a way that the motor is electrically complete across the line and develops its normal torque.

Part-winding starting is not a true reduced-voltage starting method. Part-winding starting is usually classified as reduced-voltage starting because of the resulting reduced current and torque.

PART-WINDING, WYE-CONNECTED, 3φ MOTORS

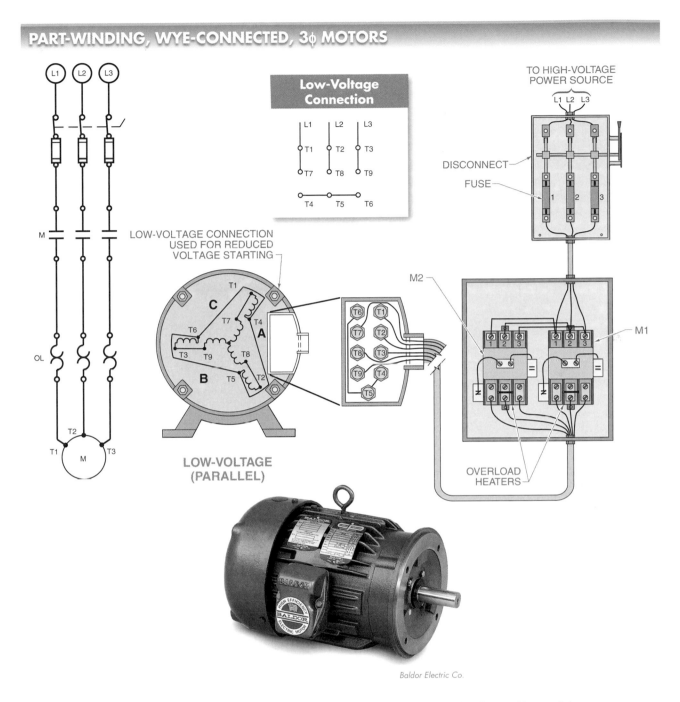

Low-Voltage Connection

LOW-VOLTAGE CONNECTION USED FOR REDUCED VOLTAGE STARTING

LOW-VOLTAGE (PARALLEL)

TO HIGH-VOLTAGE POWER SOURCE

DISCONNECT

FUSE

OVERLOAD HEATERS

Baldor Electric Co.

Figure 19-9. A part-winding motor has two sets of identical windings, which are intended to be used in parallel.

Delta-Connected Motors

When a dual-voltage, delta-connected motor is operated at 230 V from a part-winding starter having a three-pole starting contactor and a three-pole running contactor, an unequal current division occurs during normal operation. This results in overloading the starting contactor.

To overcome this problem, some part-winding starters are furnished with a four-pole starting contactor and a two-pole running contactor. This arrangement eliminates the unequal current division obtained with a delta-wound motor and enables wye-connected part-winding motors to be given either a one-half or two-thirds part-winding start.

Advantages and Disadvantages of Part-Winding Starting

Part-winding starting is less expensive than most other starting methods because it does not require voltage-reducing components, such as transformers or resistors, and it uses only two half-size contactors. Also, its transition is inherently closed circuit.

Part-winding starting has poor starting torque because the starting torque is fixed. In addition, the starter is almost always an incremental start device. Not all motors should be part-winding started. The manufacturer specifications should be consulted before applying part-winding starting to a motor. Some motors are wound sectionally with part-winding starting in mind. Indiscriminate application to any dual-voltage motor can lead to excessive noise and vibration during starting, overheating, and extremely high transient currents when switching.

The fuses in a part-winding starter must be sized to protect the small contactors and overload devices allowed because of the low-current requirements in part-winding starters. Dual-element fuses are normally required.

Part-Winding Starter Circuits

Part-winding reduced-voltage starting is less expensive than other starting methods and produces less starting torque. **See Figure 19-10.** The control circuit consists of motor starter coil M1, on-delay timer TR1, and motor starter coil M2.

Pressing start pushbutton PB2 energizes starter M1 and timer TR1. M1 energizes the motor and closes contacts M1 in line 2 to provide memory. With the motor starter M1 energized, L1 is connected to T1, L2 is connected to T2, and L3 is connected to T3. This starts the motor at reduced current and torque through one-half of the wye windings.

The on-delay NO contacts of on-delay timer TR1 in line 2 remain open during timing and close after timing out, energizing coil M2. When M2 energizes, L1 is connected to T7, L2 is connected to T8, and L3 is connected to T9. This applies voltage to the second set of wye windings. The motor now has both sets of windings connected to the supply voltage for full current and torque. The motor may normally be stopped by pressing stop pushbutton PB1 or by an overload in any line. Each magnetic motor starter only needs to be half-size because each one controls only one-half of the winding. Overloads must be sized accordingly.

PART-WINDING STARTER CIRCUITS

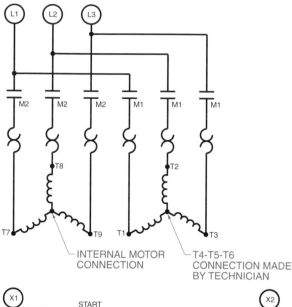

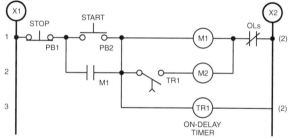

Figure 19-10. Part-winding reduced-voltage starting is less expensive than other starting methods and produces less starting torque.

Tech Fact

Part-winding starting is the least expensive method of reduced starting and meets the local utility requirement that motors over a specified horsepower must be started using reduced voltage. However, this starting method only produces half the starting torque of a full-voltage start. In some applications it may be difficult for the motor to start the load. A horsepower motor that is rated higher than the load requirement can be used or a higher starting torque starting method can be selected.

19-6 CHECKPOINT

1. Why is part-winding starting less expensive than resistor or autotransformer starting?
2. What does the control circuit of a part-winding starting circuit consist of?

19-7 WYE-DELTA STARTING

Wye-delta starting accomplishes reduced-voltage starting by first connecting the motor leads in a wye configuration for starting. A motor started in the wye configuration receives approximately 58% of the normal voltage and develops approximately 33% of the normal torque.

Wye-delta motors are specially wound with six leads extending from the motor to enable the windings to be connected in either a wye or delta configuration. When a wye-delta starter is energized, two contactors close. One contactor connects the windings in a wye configuration and the second contactor connects the motor to line voltage. After a time delay, the wye contactor opens (momentarily de-energizing the motor) and the third contactor closes to reconnect the motor to the power lines with the windings connected in a delta configuration. A wye-delta starter is inherently an open transition system because the leads of the motor are disconnected and then reconnected to the power supply.

This starting method does not require any accessory voltage-reducing equipment such as resistors and transformers. Wye-delta starting gives a higher starting torque per line ampere than part-winding starting, with considerably less noise and vibration.

However, wye-delta starters have the disadvantage of being open transition. Closed transition versions are available at additional cost. In wye-delta starters that are closed transition, the motor windings are kept energized for the few cycles required to transfer the motor windings from wye to delta. Such starters are provided with one additional contactor plus a resistor bank.

Wye-Delta Motors

The windings of a wye-delta motor may be joined to form a wye or delta configuration. **See Figure 19-11.** There are no internal connections on this motor as there are on standard wye and delta motors. This allows a technician to connect the motor leads into a wye-connected motor or into a delta-connected motor.

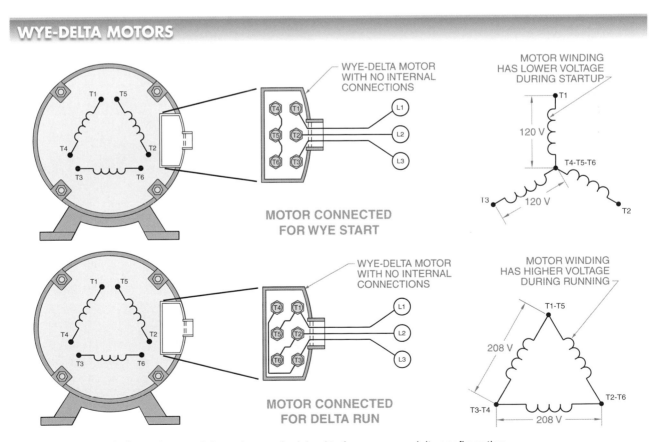

WYE-DELTA MOTORS

WYE-DELTA MOTOR WITH NO INTERNAL CONNECTIONS

MOTOR WINDING HAS LOWER VOLTAGE DURING STARTUP

MOTOR CONNECTED FOR WYE START

WYE-DELTA MOTOR WITH NO INTERNAL CONNECTIONS

MOTOR WINDING HAS HIGHER VOLTAGE DURING RUNNING

MOTOR CONNECTED FOR DELTA RUN

Figure 19-11. The windings of a wye-delta motor may be joined to form a wye or delta configuration.

Each coil winding in the motor receives 208 V if a delta-connected motor is connected across a 208 V, 3ϕ power line. This is because each coil winding in the motor is connected directly across two power lines. **See Figure 19-12.**

DELTA-CONNECTED MOTOR

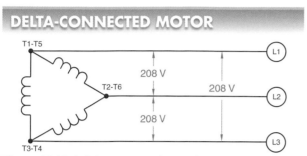

Figure 19-12. A delta-connected motor has each coil winding directly connected across two power lines so each winding receives the entire source voltage of 208 V.

Each coil winding in a wye-connected motor receives 120 V if it is connected across a 208 V, 3ϕ power line. This is because there are two coils connected in series across any pair of power lines. **See Figure 19-13.**

WYE-CONNECTED MOTOR

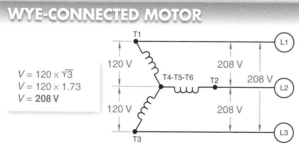

Figure 19-13. A wye-connected motor has two power lines connected across two sets of windings.

When calculating the voltage in the coil for a wye-connected circuit, the voltage is equal to the line voltage divided by the square root of 3 (1.73). The coil voltage is equal to 120 V ($^{208}/_{1.73}$) because the line voltage is equal to 208 V. A wye-delta motor connected to a line voltage of 208 V starts with 120 V (wye) and runs with 208 V (delta) across the motor windings, thus reducing starting voltage.

Wye-Delta Starting Circuits

The control circuit of a typical wye-delta starting circuit consists of motor starter coils M1 and M2, contactor C1,

and on-delay timer TR1. **See Figure 19-14.** Pressing start pushbutton PB2 energizes coil M1. This provides memory in line 2 and connects the power lines L1 to T1, L2 to T2, and L3 to T3. Contactor coil C1 in line 3 is energized, providing electrical interlock in line 2 and connecting motor terminals T4 and T5 to T6 so the motor starts in a wye configuration. TR1 in line 3 is also energized. After a preset time, the on-delay timer times out, which causes the NO TR1 contacts in line 2 to close and the NC TR1 contacts in line 3 to open. The opening of the NC contacts in line 3 disconnects contactor C1, and an instant later the NO contacts in line 2 energize the second motor starter through coil M2.

Motor starters and control transformers are often placed in the same enclosure.

The short time delay between M2 and C1 is necessary to prevent a short circuit in the power lines and is provided through the NC auxiliary contacts of C1 in line 2. With contactor C1 de-energized, terminals T5, T6, and T4 are connected to power lines T1, T2, and T3 because L1, L2, and L3 are still connected to run in a delta configuration. The circuit can normally be stopped only by an overload in any line or by pressing the stop pushbutton PB1.

WYE-DELTA STARTING CIRCUITS

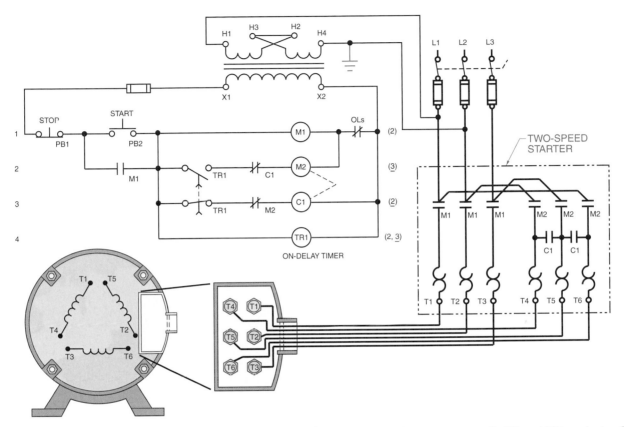

Figure 19-14. The control circuit of a typical wye-delta starting circuit consists of motor starter coils M1 and M2, contactor C1, and on-delay timer TR1.

19-7 CHECKPOINT

1. Can a wye-delta start motor have any internally connected motor windings?

2. What does the control circuit of a wye-delta starting circuit consist of?

19-8 STARTING METHOD COMPARISON

Several starting methods are available when an industrial application calls for using reduced-voltage starting. The amount of reduced current, the amount of reduced torque, and the cost of each starting method must be considered when selecting the appropriate starting method. The selection is not simply a matter of selecting the starting method that reduces the current the most. For example, the motor will not start

and the motor overloads will trip if the starting torque is reduced too much.

A general comparison can be made of the amount of reduced current for each type of starting method compared to across-the-line starting. **See Figure 19-15.** Some primary resistor starters are adjustable, while others are not adjustable. Part-winding and wye-delta starters are not adjustable.

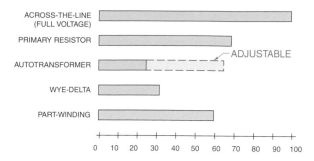

PERCENTAGE OF FULL-VOLTAGE LINE STARTING CURRENT

Figure 19-15. The different methods of reduced-voltage starting produce different percentages of full-voltage current.

A general comparison can be made of the amount of reduced torque for each type of starting method compared to across-the-line starting. **See Figure 19-16.** The amount of reduced torque is adjustable when using the autotransformer starting method, which has taps. The motor overloads trip if the load requires more torque than the motor can deliver. The torque requirements of the load must be taken into consideration when selecting a starting method.

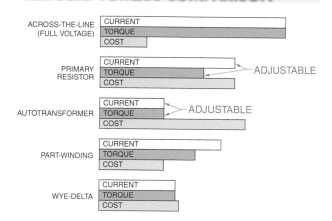

REDUCED-TORQUE COMPARISON

Figure 19-16. The different methods of reduced-voltage starting produce different amounts of reduced torque.

A general comparison can also be made of the costs for each type of starting method compared to across-the-line starting. Although reducing the amount of starting current or starting torque in comparison to the load requirements is the primary consideration for selecting a starting method, cost may also have to be considered. The costs vary for each starting method. **See Figure 19-17.**

The primary resistor starting method is used when it is necessary to restrict inrush current to predetermined increments. Primary resistors can be built to meet almost any inrush current limitation. Primary resistors also provide smooth acceleration and can be used where it is necessary to control starting torque. Primary resistor starting may be used with any motor.

The autotransformer starting method provides the highest possible starting torque per ampere of line current and is the most effective means of motor starting for applications where the inrush current must be reduced with a minimum sacrifice of starting torque. Three taps are provided on the transformers, thereby making it field adjustable. Autotransformer starting can be used with any motor. However, cost must be considered because autotransformers are the most expensive type of transformer.

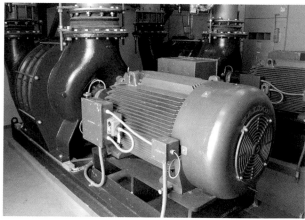

Baldor Electric Co.

The occurrence of voltage sags and swells may indicate a weak power distribution system. In such a system, voltage will change dramatically when a large motor is switched on or off.

The part-winding starting method is simple in construction and economical in cost. Part-winding starting provides a simple method of accelerating fans, blowers, and other loads involving low starting torque. The part-winding starting method requires a nine-lead wye motor. The cost is less because no external resistors or transformers are required.

The wye-delta starting method is particularly suitable for applications involving long accelerating times or frequent starts. Wye-delta starting is commonly used for high-inertia loads such as centrifugal air conditioning units, although it can be used in applications where low starting torque is necessary or where low starting current and low starting torque are permissible. The wye-delta starting method requires a special six-lead motor.

MOTOR STARTING METHOD COMPARISON

Starter Type	Starting Characteristics			Standard Motor	Transition	Extra Acceleration Steps Available	Installation Cost	Advantages	Disadvantages	Applications
	Volts at Motor	Line Current	Starting Torque							
Across-the-Line (Full Voltage)	100%	100%	100%	Yes	None	None	Lowest	Inexpensive, readily available, simple to maintain, maximum starting torque	High inrush, high starting torque	Many and various
Primary Resistor	65%	65%	42%	Yes	Closed	Yes	High	Smooth acceleration, high power factor during start, less expensive than autotransformer starter in low HPs, available with as many as 5 accelerating points	Low torque efficiency, resistors give off heat, starting time in excess of 5 sec, requires expensive resistors, difficult to change starting torque under varying conditions	Belt and gear drives, conveyors, textile machines
Autotransformer	80% 65% 50%	64% 42% 25%	64% 42% 25%	Yes	Closed	No	High	Provides highest torque per ampere of line current, 3 different staring torques available through autotransformer taps, suitable for relatively long starting periods, motor current is greater than line current during starting	Is most expensive design in lower HP ratings, low power factor, large physical size	Blowers, pumps, compressors, conveyors
Part-Winding	100%	65%	48%	*	Closed	Yes†	Low	Least expensive reduced-voltage starter, most dual-voltage motors can be started part-winding on lower voltage, small physical size	Unsuited for high-inertia, long-starting loads, requires special motor design for voltage higher than 230 V, motor does not start when torque demanded by load exceeds that developed by motor when first half of motor is energized, first step of acceleration must not exceed 5 sec or motor overheats	Reciprocating compressors, pumps, blowers, fans
Wye-Delta	100%	33%	33%	No	Open‡	No	Medium	Suitable for high-inertia, long acceleration loads, high torque efficiency, ideal for especially stringent inrush restrictions, ideal for frequent starts	Requires special motor, low starting torque, momentary inrush occurs during open transition when delta contactor is closed	Centrifugal compressors, centrifuges

* Standard dual-voltage 230/460 V motor can be used on 230 V systems
† Very uncommon
‡ Closed transition available for average of 30% more cost

Figure 19-17. Factors considered when selecting a reduced-voltage starting method include voltage at the motor, line current, starting torque, and installation costs.

19-8 CHECKPOINT

1. Which starting method(s) do not allow for any adjustments in the amount of reduced starting current and torque?

2. Which starting method would be the most expensive to add to a system?

19-9 TROUBLESHOOTING REDUCED-VOLTAGE STARTING CIRCUITS

The two main sections that must be considered when troubleshooting reduced-voltage starting circuits, as with all motor circuits, are the power circuit and control circuit. The power circuit connects the motor to the main power supply. In addition to including the main switching contacts and overload detection device (which can be heaters or solid-state), the power circuit also includes the power resistors (in the case of primary resistor starting) or autotransformer (in the case of autotransformer starting).

The control circuit determines when and how the motor starts. The control circuit includes the motor starter (mechanical or solid-state), overload contacts, and timing circuit. To troubleshoot the control circuit, the same troubleshooting procedure is used as when troubleshooting any other motor control circuit.

Voltage and current readings are taken when troubleshooting power circuits. Current readings can be taken at the incoming power leads or the motor leads. The current reading during starting should be less than the current reading when starting without reduced-voltage starting. The amount of starting current varies by the starting method.

When troubleshooting the power circuit, voltage and current readings are taken. **See Figure 19-18.** Current readings can be taken at the incoming power leads or the motor leads, since the current draw is the same at either point. With each starting method, there should be a reduction in starting current, as compared to a full-voltage start. When troubleshooting a reduced-voltage power circuit, the following procedure is applied:

1. Visually inspect the motor starter. Look for loose wires, damaged components, and signs of overheating (discoloration).

2. Measure the incoming voltage coming into the power circuit. The voltage should be within 10% of the voltage rating listed on the motor nameplate. If the voltage is not within 10%, the problem is upstream from the reduced-voltage power circuit.

3. Measure the voltage delivered to the motor from the reduced-voltage power circuit during starting and running. For primary resistor starting, the voltage during starting should be 10% to 50% less than the incoming measured voltage. The exact amount depends on the resistance added into the circuit. The resistance is set by using the resistor taps or adding resistors in series/parallel.

For autotransformer starting, the voltage during starting should be 50%, 65%, or 80% of the incoming measured voltage. The exact amount depends on which tap connection is used on the autotransformer. For part-winding starting, the voltage during starting should be equal to the incoming measured voltage.

For wye-delta starting, the voltage during starting should equal the incoming measured voltage. For solid-state starting, the voltage during starting should be 15% to 50% less than the incoming measured voltage. The exact amount depends on the setting of the solid-state starting control switch.

The voltage measured after the motor is started should equal the incoming voltage with each method of reduced-voltage starting. There is a problem in the power circuit or control circuit if the voltage out of the starting circuit is not correct.

4. Measure the motor current draw during starting and after the motor is running. In each method of reduced-voltage starting, the starting current should be less than the current that the motor draws when connected for full-voltage starting. The current should normally be about 40% to 80% less. After the motor is running, the current should equal the normal running current of the motor. This current value should be less than or equal to the current rating listed on the motor nameplate.

TROUBLESHOOTING REDUCED-VOLTAGE STARTING CIRCUITS

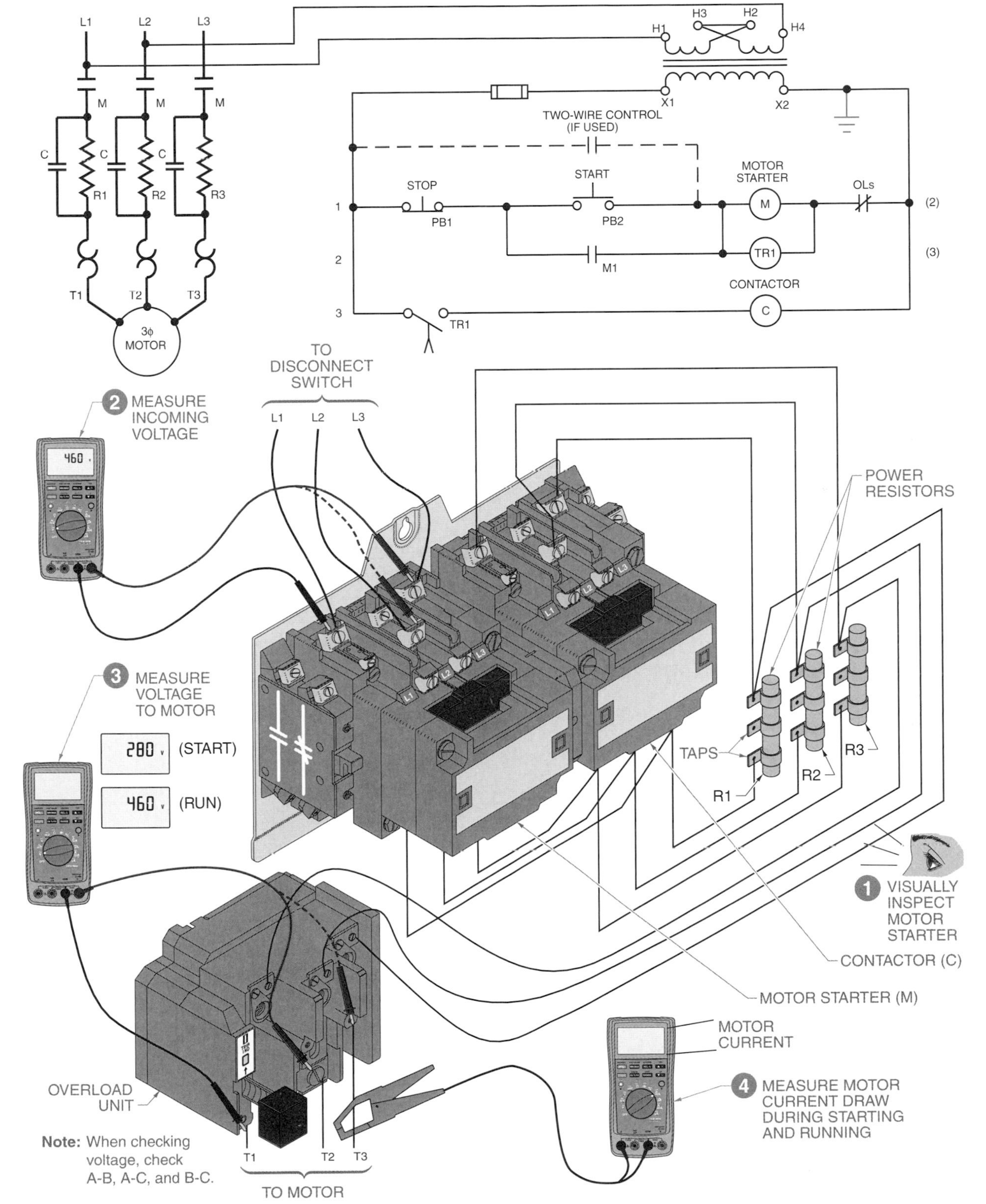

Figure 19-18. Voltage and current readings are taken when troubleshooting reduced-voltage power circuits.

19-9 CHECKPOINT

1. When measuring the current at a motor controlled by a reduced-voltage starting method, should the starting current always be less than it would be with full-voltage starting?

2. In the troubleshooting circuit in Figure 19-18, if the motor does not switch from a reduced-voltage start to a full-voltage run condition after the motor is started, what is the most likely control circuit component that may not be operating properly?

Additional Resources

Review and Resources

Access Chapter 19 Review and Resources for *Electrical Motor Controls for Integrated Systems* by scanning the above QR code with your mobile device.

Applying Your Knowledge

Refer to the *Electrical Motor Controls for Integrated Systems* Learner Resources for interactive Applying Your Knowledge questions.

Workbook and Applications Manual

Refer to Chapter 19 in the *Electrical Motor Controls for Integrated Systems Workbook* and the *Applications Manual* for additional exercises.

ENERGY EFFICIENCY PRACTICES

Large Motor Energy Cost Reduction

Reduced-voltage starting is used for large motors because large motors cannot generally be connected to full voltage when starting (across-the-line starting) because they would draw much more power than the utility system is normally designed to handle. Because large power-consuming motors are used with reduced-voltage starting methods, there is an opportunity to reduce the amount of electrical power consumption.

The cost of electrical power is based on the number of kilowatt-hours (kWh) of electricity consumed. For this reason, the operating cost of a motor is dependent on the cost of electricity. Power companies use different rates that change the cost of electrical power based on factors such as summer/winter rates, peak usage, power factor, and type of customer (commercial, industrial, government, or military). A power factor (demand) charge is normally applied to the utility bill of industrial customers. The power factor charge normally ranges from 0.5% to 2.5% of the total utility bill. The power factor charge is applied because motors cause the current to lag behind the voltage on the power lines. This lag causes power inefficiency in the power distribution system.

In addition to reduced-voltage starting methods, the use of energy-efficient motors instead of standard motors can lead to reduced electrical power costs. Motor efficiency is the measure of the effectiveness with which a motor converts electrical energy to mechanical energy. Improvements in motor efficiency can be achieved only by reducing power losses in the motor. Power losses in a motor are a result of energy losses in the stator core, stator windings, bearings, and rotor. Power losses are considered the cost of converting electrical energy into mechanical energy. Power losses are always present to some degree. Energy-efficient motors have greater efficiency and cost less to operate than a standard motor of the same rating and type.

Objectives

20-1
- Define DC power supply and describe the different types of rectifiers used in single-phase DC power supplies.

20-2
- Define filter and describe the different types of filters.

20-3
- Explain how a shunt regulator regulates the output voltage in a DC power supply.

20-4
- Define fuel cell and explain how fuel cells are connected for higher voltage or current.

20-5
- Explain photovoltaic cell operation.
- Explain photovoltaic cell output.

20-6
- Explain how to troubleshoot a half-wave rectifier.
- Explain how to troubleshoot a full-wave rectifier.
- Explain how to troubleshoot a zener diode voltage regulator.

20-1 Rectifiers

20-2 DC Power Supply Filters

20-3 Voltage Regulators

20-4 Fuel Cells

20-5 Photovoltaic Cells

20-6 Troubleshooting DC Power Supplies

Review and Resources
atplearningresources.com/Quicklinks
Access Code: 362245

Chapter

DC Power Sources

20

Since AC can easily be stepped up and down using transformers, it is used in power distribution systems to transmit power from utility power plants to customer service panels and points of usage. Although common loads such as appliances, computers, printers, entertainment systems, and most electronic devices are connected to an AC receptacle, the devices depend upon DC circuits for operation. Industrial devices such as programmable logic controllers, motor drives, timers, counters, photoelectric and proximity sensors, and digital displays all include DC circuits for operation as well. Industrial devices such as motor drives take incoming AC, convert the AC into DC (DC bus), and invert the DC back into variable-frequency AC to operate motors at controlled speeds. Understanding how AC is converted into DC is important when servicing and troubleshooting devices and circuits that include DC circuits.

20-1 RECTIFIERS

When a source of DC power is needed, a DC power supply is used. A *DC power supply* is a device that converts alternating current (AC) to regulated direct current (DC) for use in electrical circuits. A DC power supply consists of a rectifier, filter, voltage regulator, and voltage divider. **See Figure 20-1.**

A DC power supply may also include a transformer and a voltage multiplier. A *transformer* is an electric device that uses electromagnetism to change voltage from one level to another or to isolate one voltage from another. Depending on the cost and application, a DC power supply may contain all or some of these components.

DC POWER SUPPLY COMPONENTS

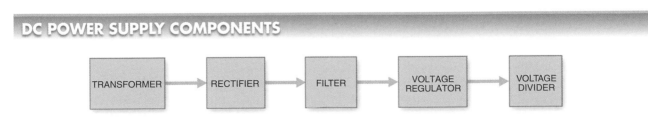

Figure 20-1. A block diagram shows the basic components of a DC power supply.

A *rectifier* is an electrical circuit that changes AC into DC. Rectifiers typically consist of one or more diodes used to control the flow of current in a circuit. Three basic types of rectifiers used in single-phase DC power supplies are half-wave, full-wave, and full-wave bridge rectifiers.

Half-Wave Rectifiers

A *half-wave rectifier* is an electrical circuit containing an AC source, a load resistor (R_L), and a diode that permits only the positive half cycles of the AC sine wave to pass, which creates pulsating DC. **See Figure 20-2.** Half-wave rectification is accomplished because current is allowed to flow only when the anode terminal of diode D1 is positive with respect to the cathode. Electrons are not allowed to flow through the rectifier when the cathode is positive with respect to the anode.

The output voltage of the half-wave rectifier is considered pulsating DC when half of the AC sine wave is cut off. *Pulsating DC* is direct current that varies in amplitude but does not change polarity. The rectifier can pass either the positive or negative half of the input AC cycle depending on the polarity of the diode in the circuit. Half-wave rectifiers are considered inefficient for many applications because one-half of the input cycle is not used.

HALF-WAVE RECTIFIERS

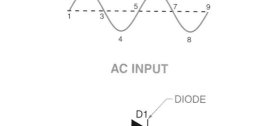

AC INPUT

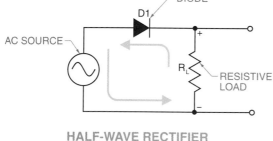

HALF-WAVE RECTIFIER

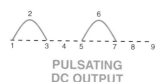

PULSATING
DC OUTPUT

Note: The numbers on the sine wave are for reference only.

Figure 20-2. A half-wave rectifier is used to convert AC to pulsating DC.

Full-Wave Rectifiers

A *full-wave rectifier* is an electrical circuit containing two diodes and a center-tapped transformer used to produce pulsating DC. **See Figure 20-3.** The center-tapped transformer supplies out-of-phase voltages to the two diodes. When voltage is induced in the secondary from point A to B, point A is positive with respect to point N. Current flows from N to A, through the load R_L, and through diode D1. Diode D1 is forward-biased and allows electrons to flow. Diode D2 is reverse-biased and blocks current flow because point B is negative (−) and point A is positive (+).

When the voltage across the secondary reverses during the negative half cycle of the AC sine wave, point B is positive with respect to point N. Current then flows from N to B, through the load, and through diode D2. Diode D2 is forward-biased and allows electrons to flow. Diode D1 is reverse-biased and blocks current because B is positive with respect to N. This is repeated every cycle of the AC sine wave, producing a full-wave DC output.

The output voltage of a full-wave rectifier has no off cycle. Electrons flow through the load during both half cycles. The constant electron flow results in a complete output signal. A full-wave rectifier is more efficient and has a smoother output than a half-wave rectifier.

FULL-WAVE RECTIFIERS

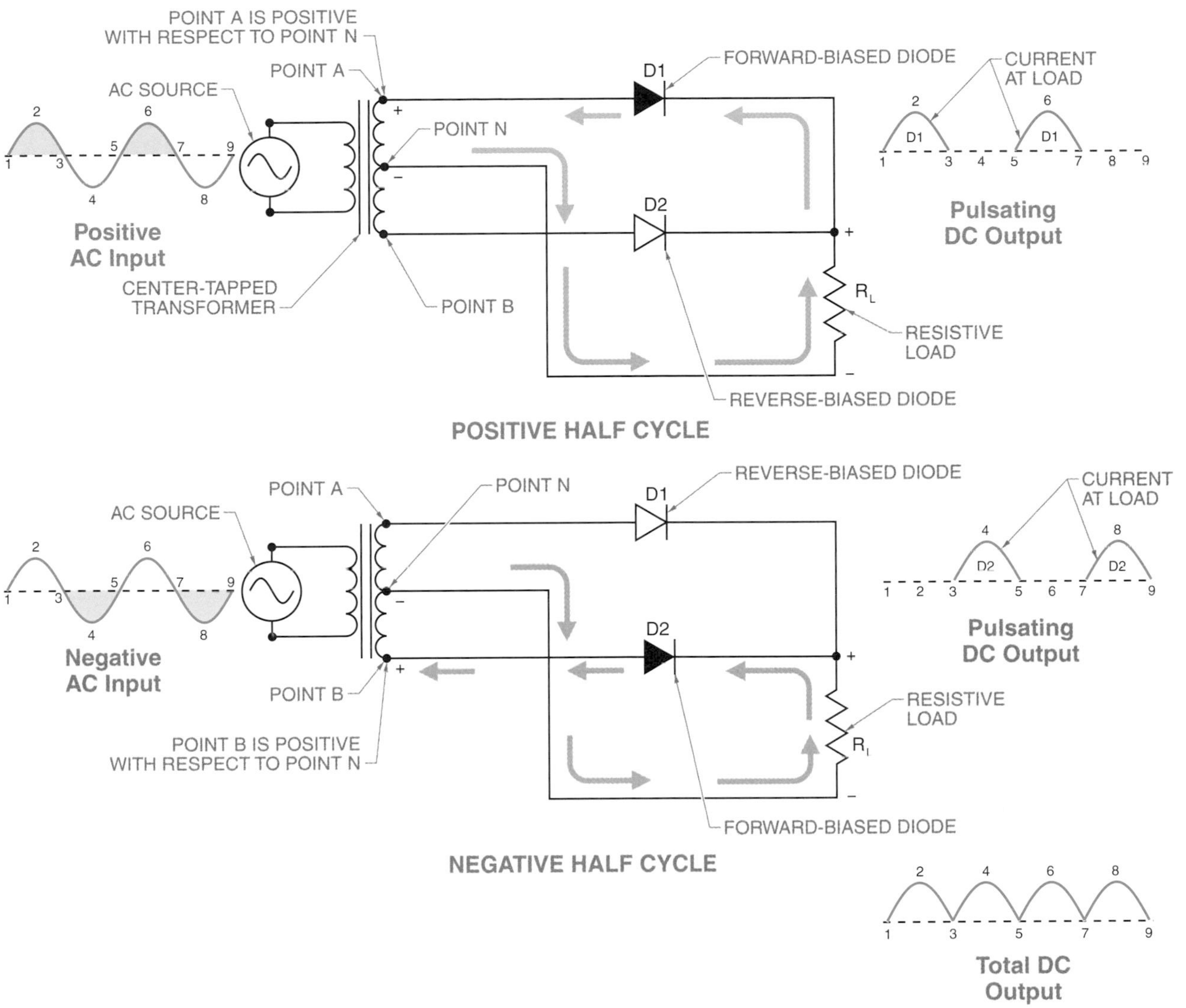

Figure 20-3. A full-wave rectifier uses two diodes and a center-tapped transformer to produce pulsating DC.

Full-Wave Bridge Rectifiers

A *full-wave bridge rectifier* is an electrical circuit containing four diodes that allow both halves of a sine wave to be changed into pulsating DC. A full-wave bridge rectifier does not require a center-tapped transformer. It uses lower voltage diodes than the center-tapped circuit of a full-wave rectifier. The bridge diodes only need to block half as much reverse voltage as center-tapped transformer diodes for the same output voltage. **See Figure 20-4.**

When the voltage is positive at point A and negative at point B, electrons flow from point B through diode D2,

load R_L, and diode D1 to point A. When the AC supply voltage is negative at point A and positive at point B, electrons flow from point A through diode D4, load R_L, and diode D3 to point B.

One disadvantage of a full-wave bridge rectifier is that on each alternation, the DC in the circuit must flow through two series-connected diodes. The forward DC voltage drop across the two rectifiers is therefore greater than the drop across a single rectifying diode. However, the voltage drop across silicon diodes is small (0.6 V), and the loss typically can be tolerated.

FULL-WAVE BRIDGE RECTIFIERS

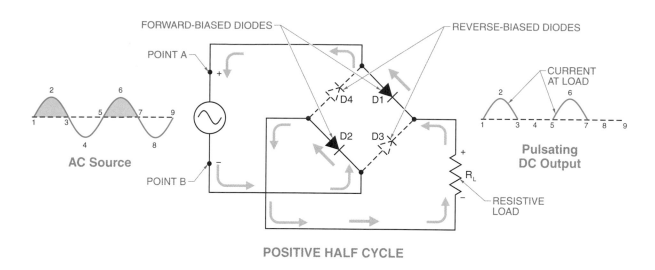

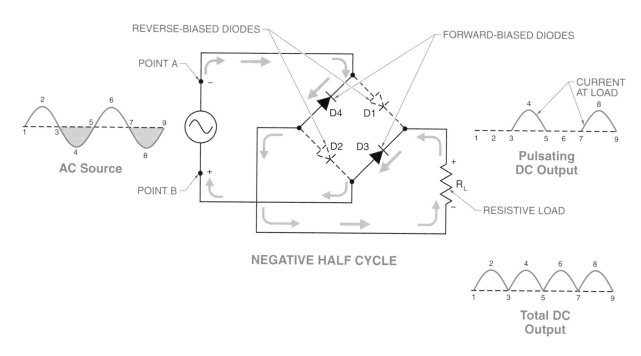

Figure 20-4. A full-wave bridge rectifier requires four diodes but does not require a center-tapped transformer.

Rectifier Voltage Measurements

Measuring voltages in a rectifier requires knowledge of some of the terminology, math, and instruments necessary to take accurate measurements. A rectifier uses both AC and DC. Care must be exercised when measuring AC and DC and when converting or comparing any AC value to a DC value. Measurements in a rectifier are typically taken with either a DMM or an oscilloscope. An oscilloscope can show the waveforms in addition to the voltage value. **See Figure 20-5.**

RECTIFIER MEASUREMENTS

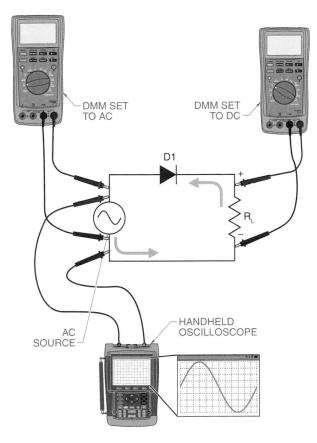

Figure 20-5. Either a DMM or an oscilloscope can be used to take measurements in a rectifier circuit.

Rectifiers can be small solid-state components that have no moving parts.

20-1 CHECKPOINT

1. What are the three basic types of rectifiers?
2. Which type of rectifier requires a center-tapped transformer?
3. Why can a full-wave bridge rectifier use lower voltage diodes than a full-wave rectifier?
4. What would be the advantage of using an oscilloscope to troubleshoot a rectifier circuit instead of a DMM?

20-2 DC POWER SUPPLY FILTERS

A *filter* is a circuit in a power supply section that smooths the pulsating DC to make it more consistent. A filter minimizes or removes ripple voltage from a rectified output by opposing changes in voltage and current. The filtering process is accomplished by connecting parallel capacitors and series resistors or inductors to the output of the rectifier. A capacitor smoothes voltage, while an inductor smoothes current.

Capacitive Filters

A *capacitive filter* is a circuit consisting of a capacitor and resistor connected in parallel. The capacitive filter provides maximum voltage output to a load. Since a large capacitor is needed, an electrolytic capacitor is typically used.

As pulsating DC voltage from a rectifier is applied across capacitor C1, it charges to the peak voltage. **See Figure 20-6.** Between peaks, the capacitor discharges through the resistive load R_L, and the voltage gradually drops. *Ripple voltage* is the amount of varying voltage present in a DC power supply. In a capacitive filter, ripple voltage is the voltage drop before the capacitor begins to charge again. The amount of discharge between voltage peaks is controlled by the resistor-capacitor (RC) time constant of the capacitor and the load resistance. If the load resistance is large and the capacitance is large, the ripple voltage will be small, resulting in a smooth output. Ripple voltage increases when the load increases on the capacitive filter.

CAPACITIVE FILTERS

INPUT FROM FULL-WAVE RECTIFIER

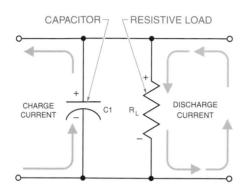

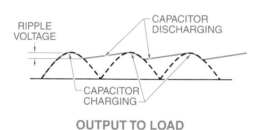

OUTPUT TO LOAD

Figure 20-6. A capacitive filter consists of a capacitor connected in parallel with a load resistor.

L-Section Resistive Filters

An *L-section resistive filter* is a filter that reduces or eliminates the amount of DC ripple at the output of a circuit by using a resistor and capacitor as an RC time constant. An L-section resistive filter reduces surge currents by using a current-limiting resistor (R1). **See Figure 20-7.** R1 controls surge currents by limiting the current flow to slow the charging of the capacitor. R1 should always be used in series with the rectifier and the input capacitor of the filter system. This protects the rectifier from the high surge of charging current that flows through the rectifier from the input capacitor C1 when the circuit is first energized. A low-value resistor of about 50 Ω or less is typically used in the application. The filtering of the resistor is not as good as other filters, but it is less expensive.

L-SECTION RESISTIVE FILTERS

INPUT FROM FULL-WAVE RECTIFIER

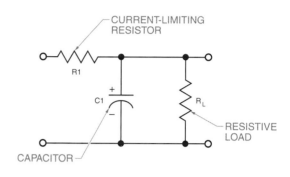

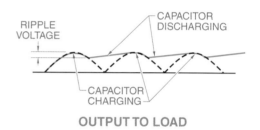

OUTPUT TO LOAD

Figure 20-7. An L-section resistive filter reduces surge currents by using a current-limiting resistor (R1).

L-Section Inductive Filters

An *L-section inductive filter* is a filter that reduces surge currents by using a current-limiting inductor and a capacitor. **See Figure 20-8.** An inductor (L1) in series opposes a change in current by creating a counter electromotive force (CEMF) or countervoltage. Surge current is greatly reduced and the capacitor charges slowly. The inductor also aids the filtering effectiveness of the capacitor since the CEMF of the inductor tends to cancel out the effects of the ripple voltage.

The operation of an L-section inductive filter can also be seen through the effect that inductive reactance has on the circuit. When the pulsating DC voltage is applied to the inductor, the changing voltage produces a high inductive reactance. Therefore, the inductor tends to block the pulsating DC voltage. The DC portion of the signal is allowed to pass through the inductor. The pulses not blocked by the inductor are bypassed by the capacitor.

L-SECTION INDUCTIVE FILTERS

INPUT FROM FULL-WAVE RECTIFIER

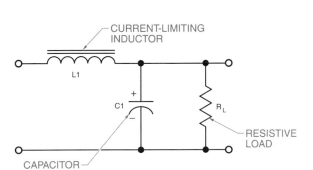

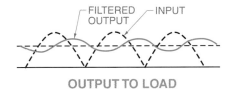

OUTPUT TO LOAD

Figure 20-8. An L-section inductive filter reduces surge currents by using a current-limiting inductor (L1).

Pi-Section Filters

A *pi-section filter* is a filter made with two capacitors and an inductor or resistor to smooth out the AC ripple in a rectified waveform. Pi-section filters get their name from the Greek letter pi (π) because the filter configuration resembles the symbol for pi. The two types of pi-section filters are inductive and resistive.

A pi-section filter consists of three elements. In a pi-section inductive filter, there is a shunt input capacitor, C1; a series inductor (choke), L1; and a shunt output capacitor, C2. **See Figure 20-9.** As the input voltage reaches the first capacitor (C1), the capacitor shunts most of the AC ripple current to ground. This presents a much smoother current to L1. Since L1 presents a high inductive reactance to the remaining AC ripple, L1 tends to block the AC ripple much better than a resistor in a resistive pi-section filter. Finally, C2 shunts to ground any remaining AC ripple. The result is a relatively smooth DC voltage.

PI-SECTION FILTERS

INPUT FROM FULL-WAVE RECTIFIER

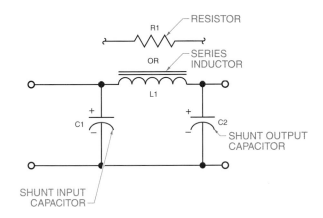

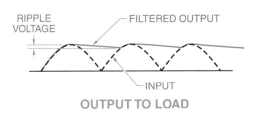

OUTPUT TO LOAD

Figure 20-9. A pi-section filter consists of two capacitors and an inductor, which may be substituted for a resistor.

Tech Fact

Full-wave rectifiers are used to produce unfiltered DC voltage. Filtering helps produce a more pure DC voltage, but a small amount of fluctuation, called ripple, can still remain. A 12 VDC rated power supply that fluctuates between 11.8 V and 12.2 V has a 0.4 V ripple. The lower the rated power supply ripple, the better.

20-2 CHECKPOINT

1. What component in a filter circuit is used to help limit a change in current?
2. What component in a filter circuit is used to help limit a change in voltage?

20-3 VOLTAGE REGULATORS

A *voltage regulator* is an electrical circuit that is used to maintain a relatively constant value of output voltage over a wide range of operating situations. DC power supplies are either regulated or unregulated, depending on whether the final output voltage must be constant or if it can fluctuate over certain limits.

A *regulated power supply* is a power supply that maintains a constant voltage across an output even when loads vary. An *unregulated power supply* is a power supply with an output that varies depending on changes of line voltage or load. Many electronic devices require regulated power supplies. These include motor controls, computers, and critical timing equipment. Voltage regulators that are used to regulate power supplies include shunt regulators.

Shunt Regulators

A shunt regulator combined with the resistance of a DC power supply or with an additional resistor forms a voltage divider to help regulate the output voltage. **See Figure 20-10.** As shunt resistor R2 increases resistance, more voltage appears across it as an output to the load. As R2 decreases resistance, less voltage appears across it.

The resistance of a shunt regulator needs to increase when the output voltage decreases. Alternately, the resistance of a shunt regulator needs to decrease when the output voltage increases. Therefore, the shunt regulator returns the output voltage to normal.

A zener diode can be used in a shunt regulator. In a simple power supply with direct zener regulation, the "raw" DC output of the power supply must exceed the zener diode regulated voltage. **See Figure 20-11.** Resistor R1 supplies current electron flow to the zener diode and load. The zener diode is regulated by drawing more or less current through R1 as the voltage across it tries to increase or decrease. As a result, the voltage drop across R1 increases or decreases.

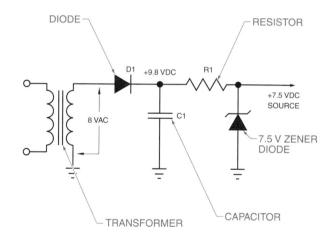

Figure 20-11. A zener diode can be used as a voltage regulator.

SHUNT REGULATORS

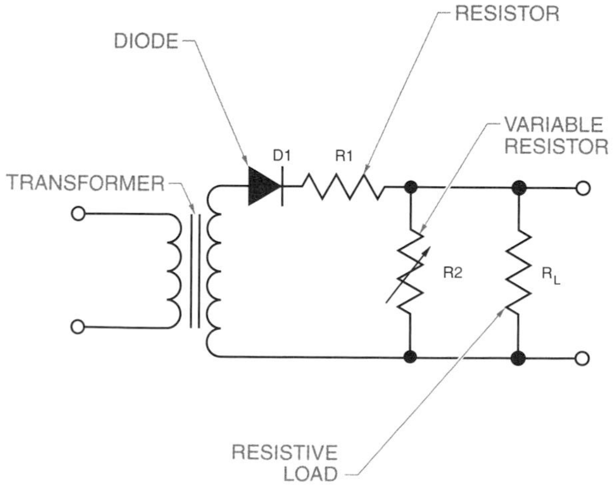

Note: Variable resistor R1 acts as a voltage regulator.

Figure 20-10. The voltage regulator of a shunt-regulated power supply is a variable resistor in parallel. It automatically changes as the output voltage changes.

20-3 CHECKPOINT

1. What type of power supply is used when the output voltage must stay constant, even with a changing load?
2. What is the function of the shunt resistor in a power supply?

20-4 FUEL CELLS

A *fuel cell* is an energy source that transforms the chemical energy from fuel into electrical energy. All fuel cells operate quite similarly. Each fuel cell has one positive electrode called the cathode and one negative electrode called the anode. The chemical reaction that creates electricity takes place at the electrodes. All fuel cells also have an electrolyte that carries charged particles from one electrode to another. A catalyst, which increases the reaction of the electrodes, is also usually present. **See Figure 20-12.**

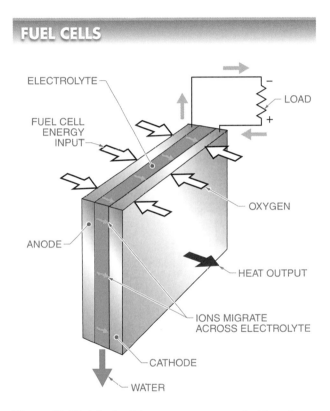

Figure 20-12. A fuel cell is an energy source that transforms chemical energy from fuel into electrical energy.

For example, a fuel cell consists of two electrodes separated by an electrolyte. Hydrogen fuel is fed into the anode of the fuel cell. Oxygen from the air enters the fuel cell at the cathode. Using a catalyst, the hydrogen splits into protons and electrons. The protons pass directly through the electrolyte. The electrons create a separate current which can be used before returning to the cathode. Once returned to the cathode, the electrons reunite with the hydrogen and oxygen to form a molecule of water.

A fuel cell may only generate a small amount of DC electricity. To be useful, many fuel cells are assembled into stacks. For higher voltages, fuel cells are connected in series. For higher current, fuel cells are connected in parallel.

One of the main advantages of fuel cells is that they generate electricity with very little pollution. This is possible because most of the hydrogen and oxygen used as fuel recombine to form a by-product (waste) that is pure water. In addition to hydrogen fuel cells, there are hydrocarbon fuel cells and chemical hybrids. The by-products of these types of cells are carbon dioxide and water. Because fuel cells produce very little pollution, they are considered environmentally friendly. Some parts of the United States have exempted power plants operating on fuel cells from some of the more stringent air permit requirements.

Fuel cells are rapidly replacing batteries in applications in which a continuous source of electricity must be maintained. For example, the fuel cells in video equipment last longer than batteries and can be replaced or refurbished without having to shut off the equipment during a film shooting sequence.

The high reliability of fuel cells provides certain businesses a level of security during power outages. Organizations such as hospitals, security companies, and 911 emergency centers can use fuel cells during power outages. By using fuel cells, a facility can continue to operate during a power outage because an independent system continues to power the facility.

Fuel cells can be made in a variety of sizes and with varying power outputs. Stationary fuel-cell applications provide power for buildings such as residential and commercial property. Transportation applications range from utility vehicles, golf carts, cars, boats, and buses. Portable applications include laptop computers, cellular phones, and digital cameras.

20-4 CHECKPOINT

1. What is the positive electrode of a fuel cell called?
2. What is the negative electrode of a fuel cell called?
3. How are fuel cells connected to increase current?
4. How are fuel cells connected to increase voltage?

20-5 PHOTOVOLTAIC CELLS

A photovoltaic cell (solar cell) is a device that converts solar energy to electrical energy. A photovoltaic cell is sensitive to light and produces a voltage without an external source. **See Figure 20-13.** Several different photovoltaic cells are available.

a potential difference between a pair of terminals when exposed to sunlight. The voltage potential is caused by the absorption of photons across the PN junction of the semiconductor material. The *photovoltaic effect* is the production of electrical energy due to the absorption of light photons in a semiconductor material.

PHOTOVOLTAIC CELLS

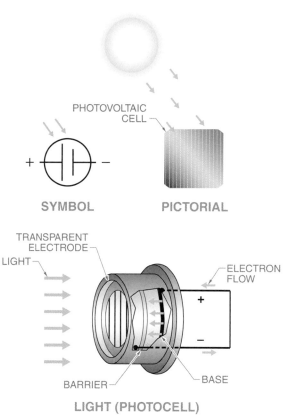

Figure 20-13. A photovoltaic cell is sensitive to light and produces voltage without an external power source.

The use of photovoltaic cells as a remote power source is becoming more popular. Many manufacturers are designing photovoltaic cells into their products in individual and multi-cell applications. For example, most handheld calculators are powered by photovoltaic cells and do not require batteries.

Photovoltaic Cell Operation

Photovoltaic cells generate energy by using a PN junction to convert sunlight (solar energy) into electrical energy. **See Figure 20-14.** The photovoltaic cell produces

PHOTOVOLTAIC CELL VOLTAGE MEASUREMENTS

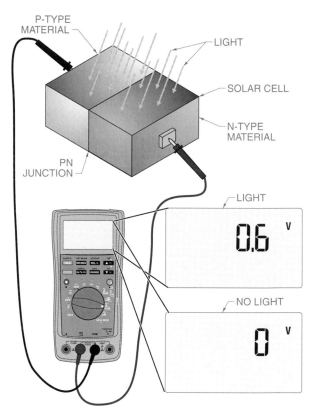

Figure 20-14. A photovoltaic cell generates energy by using a PN junction to convert sunlight into electrical energy.

Photons contain various amounts of energy depending on their wavelength. Higher energies are associated with shorter wavelengths (higher frequencies). The photons from sunlight are absorbed by the semiconductor material in photovoltaic cells, which cause electrons to flow through the semiconductor material and produce electricity. Certain materials, such as cesium, selenium, and cadmium, emit electrons when exposed to light.

Photovoltaic Cell Output

Photovoltaic cells are rated by the amount of energy they convert. Most manufacturers rate photovoltaic cell output in terms of voltage (V) and current (mA). Photovoltaic cells produce a limited amount of voltage and current. For example, each photovoltaic cell may produce up to 0.6 V. To increase the voltage output, cells are connected in series. In addition to the maximum voltage, each photovoltaic cell may produce up to 40 mA of current. To increase the current output, cells are connected in parallel. To increase both voltage and current, the individual cells are connected both in series and parallel. **See Figure 20-15.**

INCREASING CELL OUTPUT

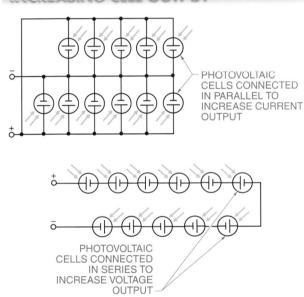

PHOTOVOLTAIC CELLS CONNECTED IN PARALLEL TO INCREASE CURRENT OUTPUT

PHOTOVOLTAIC CELLS CONNECTED IN SERIES TO INCREASE VOLTAGE OUTPUT

Figure 20-15. Photovoltaic cells are connected in parallel to increase current output or connected in series to increase voltage output.

The photovoltaic effect is measured using a high-impedance voltage-measuring device such as a DMM. In the dark, there is no open-circuit voltage present. When sunlight strikes the cell, the light is absorbed and, if the photon energy is large enough, it frees hole-electron pairs. **See Figure 20-16.** A voltage potential now exists between the cell terminals.

At the PN junction, some recombination of the electrons and holes occurs, but the junction itself acts as a barrier between the two charges. The electrical field at the PN junction maintains the negative charges in the N-type material and the positive charges or holes in the P-type material.

PHOTOVOLTAIC EFFECT

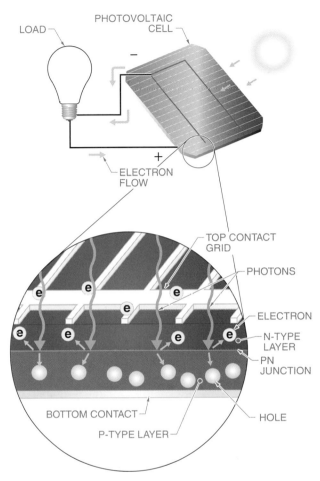

LOAD

PHOTOVOLTAIC CELL

ELECTRON FLOW

TOP CONTACT GRID

PHOTONS

ELECTRON

N-TYPE LAYER

PN JUNCTION

BOTTOM CONTACT

P-TYPE LAYER

HOLE

Figure 20-16. The photovoltaic effect produces free electrons that must travel through conductors in order to recombine with electron holes.

When a load is connected across the PN junction, current flows. When current flows through the load, electron-hole pairs formed by sunlight recombine and return to the normal condition prior to the application of sunlight. Consequently, there is no loss of electrons from or addition of electrons to the semiconductor material during the process of converting absorbed sunlight into electrical energy.

20-5 CHECKPOINT

1. How are photovoltaic cells connected to increase current?
2. How are photovoltaic cells connected to increase voltage?

20-6 TROUBLESHOOTING DC POWER SUPPLIES

The operations of several types of basic DC supplies, including half-wave, full-wave, and full-wave bridge rectifiers, all have certain traits in common. For example, all power supplies may exhibit similar problems and can be tested using a DMM or an oscilloscope. However, the tests used for problems in one power supply may not necessarily be the same tests used for another power supply.

Troubleshooting Half-Wave Rectifiers

In a simple half-wave rectifier, AC power is provided by the power transformer. Diode D1 allows current in one direction only during the positive half cycle of the AC source. The output of the diode is half of a sine wave, as shown on the oscilloscope screen. The output of the power supply is affected by capacitor C1. **See Figure 20-17.** Capacitor C1 is charged from the output of diode D1 and then supplies power to the circuit when diode D1 is reverse-biased. This results in a smoothing process, as shown in the oscilloscope waveforms with a load and no load. *Note:* When capacitor C1 is open there is no smoothing of the waveform.

Some problems found in many types of power supplies can be explained by troubleshooting the half-wave rectifier. Common problems with half-wave rectifiers include blown fuses and tripped circuit breakers, no output voltage, and low power supply voltage.

Blown Fuses and Tripped Circuit Breakers. One of the most common problems in half-wave rectifiers is a shorted diode. A shorted diode will cause a fuse to blow or a circuit breaker to trip. This happens because with diode D1 shorted, capacitor C1 is effectively across the transformer secondary and has excess current. This excess current is reflected into the primary of the transformer and the fuse blows or the circuit breaker trips. If the circuit is not overload protected, the transformer burns out.

Another possible cause of a blown fuse or tripped circuit breaker is a short in the load. For example, when a transistor or integrated circuit is shorted, it draws excess current from the power supply, causing a fuse to blow or circuit breaker to trip. As with a diode, this problem can typically be isolated with a DMM.

No Output Voltage. No output voltage, or a reading of 0 VDC, occurs when there is no AC input voltage. The lack of AC input voltage occurs when a fuse or circuit breaker is open, a switch does not close, the power transformer winding is open, or there is a loose connection such as a loose plug or fuse holder. All of these possible problems can be checked with a DMM. Maximum voltage is read across an open circuit, provided that it is the only open in the circuit.

Low Power Supply Voltage. The most common cause of low power supply voltage is an open input-filter capacitor. A large increase in ripple can be observed on an oscilloscope when an input-filter capacitor is open.

A simple way to check a capacitor is to temporarily shunt another capacitor of the same value across it. It is important to make sure that the polarity of the capacitor is correct. If the voltage increases dramatically, the capacitor is bad. Any type of bypass jumping with a shunt should be used only as a temporary test.

Troubleshooting Full-Wave Rectifiers

When troubleshooting full-wave rectifiers, tracing the AC signal through the power supply is the most accurate method of troubleshooting problems that cannot be found by a visual check. The AC voltage is traced from the transformer where the AC sine wave changes to pulsating DC at the rectifier output. Then, the DC pulses are smoothed out by the filter. The point where the signal stops or becomes distorted is where the problem can be found. If there is no DC output voltage, there may be an open or short in the signal tracing process. If there is a low DC voltage, there may be a defective part.

An unregulated power supply converts and filters a power signal. **See Figure 20-18.** AC voltage is brought in from the power line by a line cord. AC voltage is connected to the primary of the transformer through switch S1. At the secondary winding of the transformer (points 1 and 2), the oscilloscope shows the stepped-up voltage developed across each half of the secondary winding as a complete sine wave. Each of the two stepped-up voltages is connected between a ground (point 3) and one of two diodes (points 4 and 5).

Tech Fact

DMMs are used for taking voltage measurements on AC and DC circuits. Power supplies take AC and rectify it into DC. Thus it is common to measure both AC and DC voltages when testing a rectifier's input (AC) and output (DC). When the exact voltage type is not known at the measuring point, the DMM is first set to the DC setting. The voltage is DC if a negative sign appears when the meter leads are reversed.

TROUBLESHOOTING HALF-WAVE RECTIFIERS

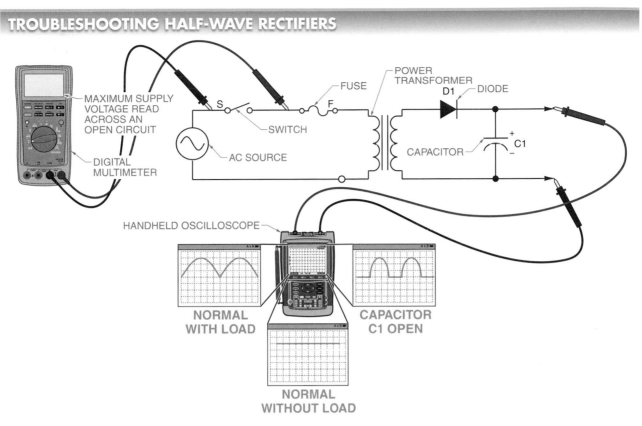

Figure 20-17. The output of the half-wave rectifier circuit is affected by the capacitor.

TROUBLESHOOTING FULL-WAVE RECTIFIERS

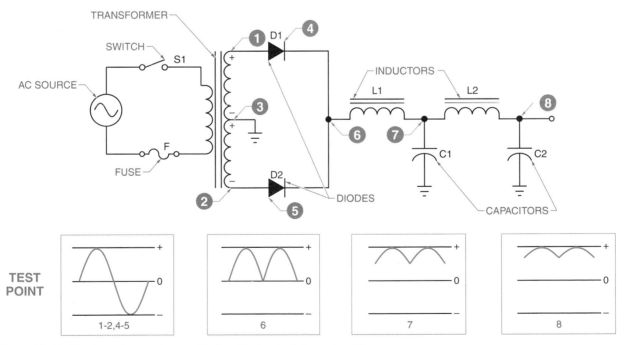

Figure 20-18. Circuit tracing can be used to troubleshoot a full-wave rectifier.

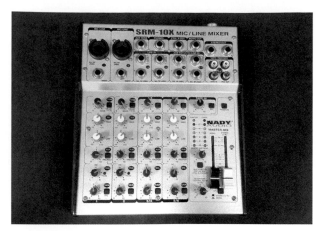

Because rectifiers convert AC power to DC power, they are commonly used inside the power supply of electronic equipment.

After passing through the diodes (point 6) the waveform is changed to pulsating DC. Pulsating DC is fed through inductor L1 and filter capacitor C1, which removes a large part of the ripple (point 7). Finally, the DC voltage is fed through inductor L2 and filter capacitor C2, which removes most of the remaining ripples (point 8).

Full-wave rectifiers may experience some problems. The problems that may occur in full-wave rectifiers include the following:

• an open fuse or circuit breaker in the primary or secondary, causing a loss of voltage

• a shorted or open transformer, which does not allow voltage across the circuit

• an open or shorted diode (D1 or D2), which causes the output voltage to drop by one-half

• both diodes open, indicated by 0 V at the load

• both diodes shorted, which shorts the entire transformer secondary

Troubleshooting Zener Diode Voltage Regulators

Zener diodes that are not working properly may have too little or too much voltage across them. Too little voltage across the zener diode may be caused by a bad diode, low DC input voltage, increased value of a resistor, or a load that is too large. Too much voltage across the zener diode indicates an open.

A zener diode can be checked in its circuit using a DMM. However, a DMM only indicates whether a zener diode is open or shorted. One lead of the diode must be disconnected from the circuit to get an accurate reading. The test does not tell whether the breakdown voltage of a zener diode is correct. For example, due to leakage, a 15 V zener diode may hold the voltage to 12 V. A DMM may not indicate any leakage because it only uses low test voltages. Like all diodes, the zener diode should have a low resistance in the forward-bias direction and a high resistance in the reverse-bias direction.

A zener diode must be safely checked to confirm it is working properly. A DC source can be used to check a zener diode as long as it has a somewhat higher voltage than the zener diode to be checked. The limiting resistor value is not critical but should be large enough to limit the expected zener diode current to approximately 10 mA for zener diodes rated at 24 V or to 5 mA for higher voltage units. If a zener diode is rated at more than 1 W, higher test currents may be used. If an adjustable DC power supply is used, a 1000 Ω resistor is used. The supply voltage is increased until the voltage reading of the DMM across the zener diode does not increase. The DMM should indicate the zener diode rating.

Occasionally a zener diode may appear to fail only in certain situations. To check for intermittent failures, a zener diode must be tested while in operation. An oscilloscope is used for testing the characteristics of a zener diode in an operating situation. An oscilloscope displays the dynamic operating characteristics of the zener diode. **See Figure 20-19.**

TESTING ZENER DIODES

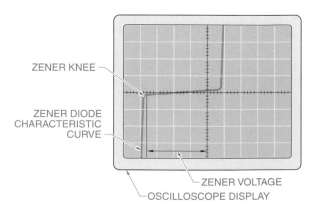

ZENER KNEE

ZENER DIODE CHARACTERISTIC CURVE

ZENER VOLTAGE

OSCILLOSCOPE DISPLAY

Figure 20-19. An oscilloscope test display indicates whether a zener diode is good.

20-6 CHECKPOINT

1. What are two causes of a blown fuse in a half-wave rectifier?

2. What is the primary reason for low power supply voltage for a half-wave rectifier?

Additional Resources

Review and Resources

Access Chapter 20 Review and Resources for *Electrical Motor Controls for Integrated Systems* by scanning the above QR code with your mobile device.

Applying Your Knowledge

Refer to the *Electrical Motor Controls for Integrated Systems* Learner Resources for interactive Applying Your Knowledge questions.

Workbook and Applications Manual

Refer to Chapter 20 in the *Electrical Motor Controls for Integrated Systems Workbook* and the *Applications Manual* for additional exercises.

ENERGY EFFICIENCY PRACTICES

Rectifier Efficiency

Even though many devices such as computers, printers, TVs, monitors, and chargers are plugged into an AC receptacle, these devices do not operate on AC power. In fact, AC is converted to DC to operate those devices and others. Even an AC motor drive takes the incoming AC and converts it to DC (DC bus voltage) to operate the drive before converting the DC back into a variable AC that operates the motor. The conversion of AC into DC uses energy and produces heat.

To reduce the amount of energy and heat, some systems are designed to reduce the amount of required rectification to operate equipment that operates on DC power. For example, data center equipment operates on DC power. Traditionally, each piece of equipment included a rectifier section to convert AC to DC. To reduce the amount of energy, cost, and space, some data centers have switched to equipment that operates on a connected DC power source. This eliminates the need for a rectifier in every piece of equipment, reduces the size of the equipment, and lowers the cost of equipment.

In such a system, the incoming utility-supplied AC power is rectified at one central place and the DC power is distributed to each piece of equipment that requires DC. One main DC uninterruptible power supply (UPS) can be used and power quality can be controlled and monitored more easily with one central rectifier. It is estimated that in such a system about 10% less power is used, space is reduced by about 25%, and equipment cost is reduced by about 20%.

Objectives

21-1
- Define thermistor and describe the classes of thermistors.
- Explain how to test thermistors.

21-2
- Describe photoconductive cells (photocells) and give examples of how they are used.
- Explain how to test photocells.

21-3
- Define photoconductive diode (photodiode) and explain how it operates.

21-4
- Define and describe pressure sensors.
- Explain how to test a pressure sensor.

21-5
- Define and describe flow detection sensors.

21-6
- Define Hall effect sensor and explain the Hall effect.
- Explain how Hall effect sensors operate.
- Explain how Hall effect sensors may be actuated.
- Describe Hall effect sensor applications.

21-7
- Define proximity sensor and explain how it operates.
- Explain the difference between an inductive proximity sensor and a capacitive proximity sensor.

21-8
- Define ultrasonic sensor and explain how it operates.
- Explain the difference between a direct mode ultrasonic sensor and a diffused mode ultrasonic sensor.

Sections

Review and Resources
atplearningresources.com/Quicklinks
Access Code: 362245

Chapter

Semiconductor
Input Devices

21

Semiconductor input devices are often referred to as transducers. Transducers provide a link in integrating industrial control systems because they convert mechanical, magnetic, thermal, electrical, optical, and chemical variations into electrical voltage and current signals. These voltage and current signals are used directly or indirectly to control devices within a system. Because of the variety of solid-state devices available, many types of electromechanical transducers are being replaced with solid-state transducers. Solid-state transducers include thermistors, photoconductive cells (photocells), photoconductive diodes (photodiodes), pressure sensors, flow detection sensors, Hall effect sensors, proximity sensors, and ultrasonic sensors.

21-1 THERMISTORS

A *thermistor* is a temperature-sensitive resistor whose resistance changes with a change in temperature. **See Figure 21-1.** Thermistors are popular because of their small size, which allows them to be mounted in places that are inaccessible to other temperature-sensing devices.

The operation of a thermistor is based on the electron-hole theory. As the temperature of the semiconductor increases, the generation of electron-hole pairs increases due to thermal agitation. Increased electron-hole pairs cause a drop in resistance.

Thermistors may be directly heated or indirectly heated. Directly heated thermistors are used in voltage regulators, vacuum gauges, and electronic time-delay circuits. Indirectly heated thermistors are used for precision temperature measurement and temperature compensation. Each type of thermistor is represented by a separate schematic symbol.

Thermistors are commonly used in small appliances such as coffee makers to control temperature.

THERMISTORS

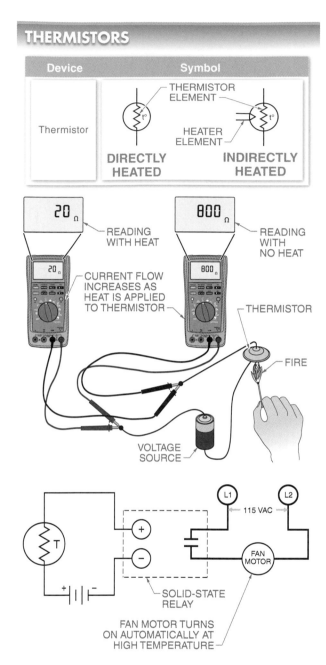

Figure 21-1. A thermistor is a temperature-sensitive resistor whose resistance changes with a change in temperature.

A typical application of a thermistor is to control a fan motor. As the thermistor is heated, its resistance decreases and more current flows through the circuit. When enough current flows through the circuit, a solid-state relay turns on. The solid-state relay is used to switch on a fan motor at high temperatures. Such a circuit can be used to automatically reduce heat in attics or to circulate warm air.

PTC and NTC Thermistors

Two classes of thermistors are positive temperature coefficient (PTC) and negative temperature coefficient (NTC). Although the most commonly used thermistors are NTC thermistors, some applications require PTC thermistors. With a PTC thermistor, an increase in temperature causes the resistance of the thermistor to increase. With an NTC thermistor, an increase in temperature causes the resistance of the thermistor to decrease. The resistance of each thermistor returns to its original state (resistance value) when the heat is removed.

Resistance refers to the operating resistance of a thermistor at extreme temperatures. Cold resistance is measured at 25°C (room temperature). However, some manufacturers specify lower temperatures. The specification sheet should always be checked for the correct temperature specification. Hot resistance is the resistance of a heated thermistor. In a directly heated thermistor, heat is generated from the ambient temperature, the current, and the heating element of the thermistor.

Thermistor Applications

A fire alarm circuit is a common application of an NTC thermistor. **See Figure 21-2.** The purpose of this circuit is to detect a fire and activate an alarm. In normal operating environments, the resistance of the thermistor is high because ambient temperatures are relatively low. The high resistance keeps the current to the control circuit low. The alarm remains OFF. However, in the presence of a fire, the increased ambient temperature lowers the resistance of the thermistor. The lower resistance allows current flow, activating the alarm.

THERMISTOR APPLICATIONS

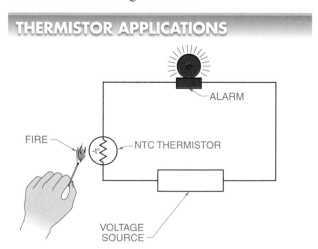

Figure 21-2. In the presence of fire, the increase in temperature lowers the resistance of an NTC thermistor, which increases current and activates an alarm.

Testing Thermistors

A thermistor must be properly connected to an electronic circuit. Loose or corroded connections create a high resistance in series with the thermistor resistance. The control circuit may sense the additional resistance as a false temperature reading. The hot and cold resistance of a thermistor can be checked with a DMM. **See Figure 21-3.** To test the hot and cold resistance of a thermistor, the following procedure is applied:

1. Remove the thermistor from the circuit.

2. Connect the DMM leads to the thermistor leads and place the thermistor and a thermometer in ice water. Record the temperature and resistance readings.

3. Place the thermistor and thermometer in hot water (not boiling). Record the temperature and resistance readings. Compare the hot and cold readings with the manufacturer specification sheet or with a similar thermistor that is known to be good.

Tech Fact

Thermistors have a base resistance rating such as 3 kΩ, 10 kΩ, 30 kΩ, 50 kΩ, and 1 MΩ. The base resistance rating is typically specified as the resistance at a set temperature, usually around room temperature (72°F to 77°F). When testing a thermistor, the resistance is measured after the thermistor has been at room temperature for several minutes to ensure a consistent measurement that can be compared to the rating. As a PTC thermistor is heated, its resistance will increase. As an NTC thermistor is heated, its resistance will decrease.

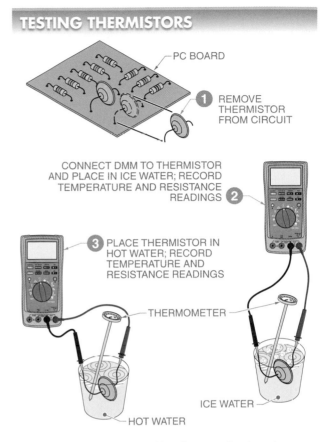

TESTING THERMISTORS

PC BOARD

1 REMOVE THERMISTOR FROM CIRCUIT

CONNECT DMM TO THERMISTOR AND PLACE IN ICE WATER; RECORD TEMPERATURE AND RESISTANCE READINGS 2

3 PLACE THERMISTOR IN HOT WATER; RECORD TEMPERATURE AND RESISTANCE READINGS

THERMOMETER

ICE WATER

HOT WATER

Figure 21-3. The hot and cold resistance of a thermistor can be checked with a DMM.

21-1 CHECKPOINT

1. What happens to the resistance of a PTC thermistor when it is heated?

2. What happens to the resistance of an NTC thermistor when it is heated?

3. What is the most commonly used thermistor type?

21-2 PHOTOCONDUCTIVE CELLS

A *photoconductive cell (photocell)* is a device that conducts current when energized by light. Current increases with the intensity of light because resistance decreases. A photocell is, in effect, a variable resistor. A photocell is formed with a thin layer of semiconductor material, such as cadmium sulfide (CdS) or cadmium selenide (CdSe), deposited on a suitable insulator. Leads are attached to the semiconductor material and the entire assembly is hermetically sealed in glass. The transparency of the glass allows light to reach the semiconductor material. For maximum current-carrying capacity, the photocell is manufactured with a short conduction path having a large cross-sectional area.

Photocell Applications

Photocells are used when time response is not critical. Photocells are not used when several thousand responses per second are needed to transmit accurate data. Applications where photocells can be used efficiently include slow-responding electromechanical equipment such as pilot lights and street lights.

A photocell can be used to control the pilot light in a gas furnace. **See Figure 21-4.** In this application, the light level determines if the pilot light (flame) in the furnace is ON or OFF. When a pilot light is present, the light from the flame reduces the resistance of the photocell. Current is allowed to pass through the cell and activate a control relay. The control relay allows the main gas valve to be energized when the thermostat calls for heat. The same procedure would be used for similar applications such as gas-powered water heaters, clothes dryers, and ovens (commercial or residential), as well as similar electrical applications.

PHOTOCELLS

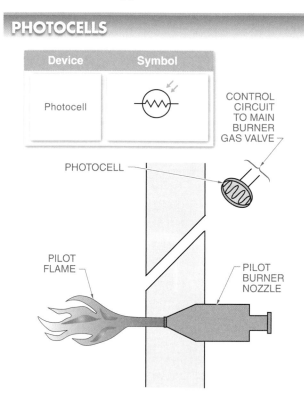

Figure 21-4. A photocell can be used to determine if the pilot light on a gas furnace is ON or OFF.

A photocell can also be used in a streetlight circuit. **See Figure 21-5.** In this circuit, an increase of light at the photocell results in a decrease in resistance and current flow through the solid-state relay. The increased current

in the relay causes the normally closed (NC) contacts to open, and the light turns off. The resistance increases with darkness, causing the NC contacts to return to their original position and turning on the light.

PHOTOCELL APPLICATIONS

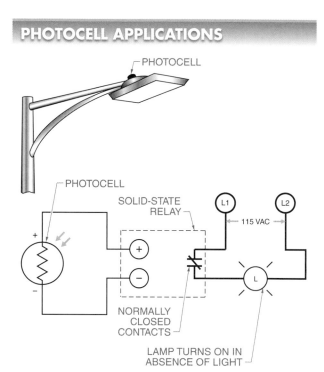

Figure 21-5. A photocell can be used to determine when a streetlight should turn on or off.

Testing Photocells

Humidity and contamination are the primary causes of photocell failure. **See Figure 21-6.** The use of quality components that are hermetically sealed is essential for long life and proper operation. Some plastic units are less rugged and more susceptible to temperature changes than glass units. To test the resistance of a photocell, the following procedure is applied:

1. Disconnect the photocell from the circuit.
2. Connect the DMM leads to the photocell.
3. Cover the photocell and record dark resistance.
4. Shine a light on the photocell and record light resistance.
5. Compare the resistance readings with manufacturer specification sheets. Use a similar photocell that is known to be good when specification sheets are not available. All connections should be tight and corrosion free.

TESTING PHOTOCELLS

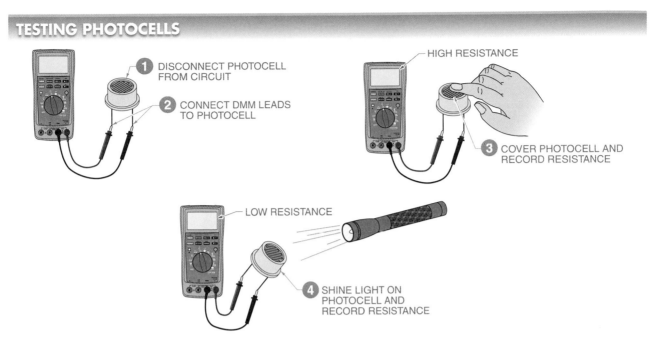

1 DISCONNECT PHOTOCELL FROM CIRCUIT

2 CONNECT DMM LEADS TO PHOTOCELL

HIGH RESISTANCE

3 COVER PHOTOCELL AND RECORD RESISTANCE

LOW RESISTANCE

4 SHINE LIGHT ON PHOTOCELL AND RECORD RESISTANCE

Figure 21-6. Humidity and contamination are the primary causes of photocell failure.

21-2 CHECKPOINT

1. What happens to the resistance of a photoconductive cell as light on it increases?

2. What happens to the current flowing through a photoconductive cell as light on it increases?

21-3 PHOTOCONDUCTIVE DIODES

A *photoconductive diode (photodiode)* is a diode that is switched on and off by light. A photodiode is similar internally to a regular diode. The primary difference is that a photodiode has a lens in its housing for focusing light on the PN junction. **See Figure 21-7.**

Photoconductive Diode Operation

In a photodiode, the conductive properties change when light strikes the surface of the PN junction. Without light, the resistance of the photodiode is high. The resistance is reduced proportionately as the photodiode is exposed to light. The current flowing through the photodiode increases as the resistance decreases.

Photoconductive Diode Applications

Photodiodes respond much faster than photocells. Also, they are usually more rugged than photocells. Photodiodes are found in movie equipment, conveyor systems, and other equipment requiring a rapid response time. Photodiodes can be used for positioning an object and turning functions on and off, such as in a filling machine.

Tech Fact

When a photoconductive diode is used in certain applications, its lens of may become covered with debris and must be cleaned. For example, the sensors used to detect an object under a closing garage door are susceptible to dirt accumulation that may prevent them from operating properly.

PHOTODIODES

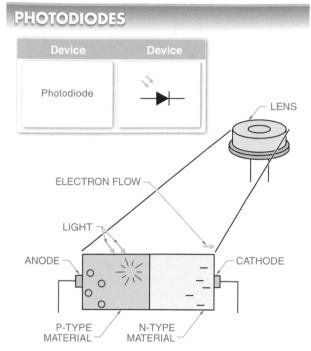

Figure 21-7. A photodiode is a diode that is switched on and off by a light.

In a filling machine, a constant light source is placed across the conveyor from the photodiode so that cartons can move between the light source and the photodiode. **See Figure 21-8.** The photodiode is energized as long as there are no cartons in the way to prevent light from passing. The photodiode de-energizes when a carton passes between the light source and the photodiode. The programmable logic controller (PLC) records the response and stops the conveyor, placing the carton in the correct position and filling the carton. This arrangement eliminates the need for slow mechanical equipment.

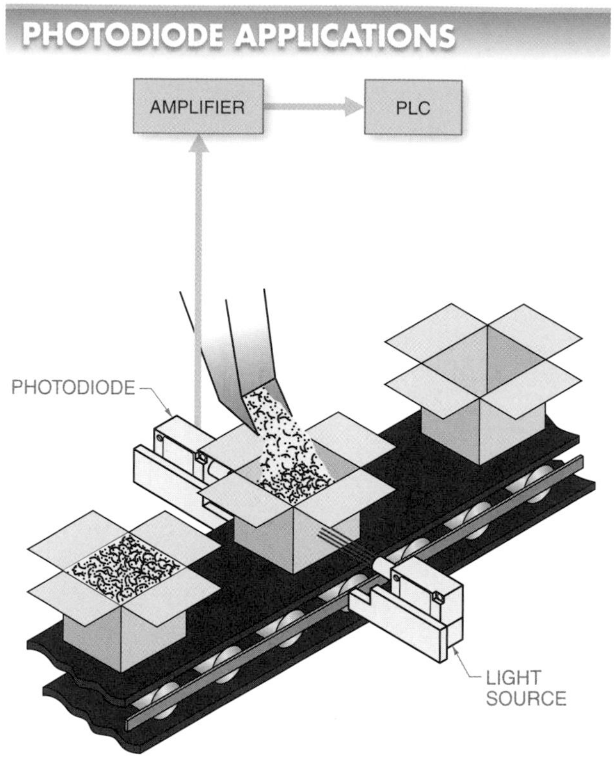

PHOTODIODE APPLICATIONS

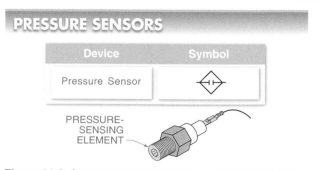

Figure 21-8. Photodiodes are used to position objects and turn machine functions on and off.

21-3 CHECKPOINT

1. What happens to the resistance of a photodiode when light is shining on it?
2. What happens to the current flowing through a photodiode when light is shining on it?

21-4 PRESSURE SENSORS

A *pressure sensor* is a transducer that outputs a voltage or current with a corresponding change in pressure. **See Figure 21-9.** Pressure sensors come in a wide range of pressure ranges. A voltage-output pressure sensor typically outputs 0 VDC to 10 VDC proportional to the rating of the pressure sensor. A current-output pressure sensor typically outputs 4 mA to 20 mA proportional to the rating of the pressure sensor. A pressure sensor can include a switching output (contact or solid-state) that is designed to activate or deactivate when the sensor reaches a predetermined value.

PRESSURE SENSORS

Device	Symbol
Pressure Sensor	◇

PRESSURE-
SENSING
ELEMENT

Figure 21-9. A pressure sensor is a transducer that changes resistance with a corresponding change in pressure.

A pressure sensor is used for high- or low-pressure control, depending on the switching circuit design. Pressure sensors are suited for a wide variety of pressure measurements on compressors, pumps, and other similar equipment. A pressure sensor can detect low pressure or high pressure, or it can trigger a relief valve. Pressure sensors are also used to measure compression in various types of engines because they are extremely rugged.

Testing Pressure Sensors

Pressure sensors are tested by checking the voltage or current output and then comparing the value to manufacturer specification sheets. **See Figure 21-10.** To test a pressure sensor, the following procedure is applied:

1. Disconnect the pressure sensor from the circuit.

2. Connect the DMM leads to the pressure sensor.

3. Activate the device being monitored (compressor, air tank, etc.) until pressure builds up. Record the voltage or current (depending on output type) of the pressure sensor at the high-pressure setting.

4. Open the relief or exhaust valve and reduce the pressure on the sensor. Record the voltage or current of the pressure sensor at the low-pressure setting.

5. Compare the high and low voltage or current readings with manufacturer specification sheets. Use a replacement pressure sensor that is known to be good when manufacturer specification sheets are not available.

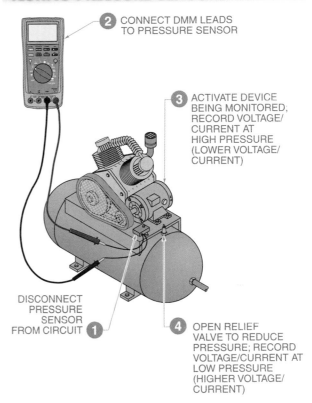

TESTING PRESSURE SENSORS

2 CONNECT DMM LEADS TO PRESSURE SENSOR

3 ACTIVATE DEVICE BEING MONITORED; RECORD VOLTAGE/ CURRENT AT HIGH PRESSURE (LOWER VOLTAGE/ CURRENT)

DISCONNECT PRESSURE SENSOR FROM CIRCUIT 1

4 OPEN RELIEF VALVE TO REDUCE PRESSURE; RECORD VOLTAGE/CURRENT AT LOW PRESSURE (HIGHER VOLTAGE/ CURRENT)

Figure 21-10. Pressure sensors are tested by checking the voltage or current output and then comparing the value to the manufacturer specification sheets.

21-4 CHECKPOINT

1. If a pressure sensor that is rated to output 0 VDC to 10 VDC with a specified pressure operating range of 0 psi to 500 psi outputs 2 VDC, is it working properly according to manufacturer specifications?

2. If a pressure sensor that is rated to output 4 mA to 20 mA DC with a specified pressure operating range of 0 psi to 500 psi outputs 2.5 mA, is the pressure sensor working properly according to manufacturer specifications?

21-5 FLOW DETECTION SENSORS

A *flow detection sensor* is a sensor that detects the movement (flow) of liquid or gas using a solid-state device. Because a flow detection sensor is a solid-state device, there are no moving mechanical parts that can become damaged due to corrosion or product deposits.

A flow detection sensor operates on the principle of thermal conductivity. The flow detection sensor head, which is in contact with the medium (liquid or air) to be detected, is heated to a temperature that is a few degrees higher than the medium to be detected.

When the medium is flowing, the heat produced at the sensor head is conducted away from the sensor, cooling the sensor head. When the medium stops flowing, the heat produced by the sensor head is not conducted away from the sensor head. A thermistor in the sensor head converts the heat not conducted away into a stronger electrical signal than what is produced when the heat is conducted away. This electrical signal is used to operate the sensor's output (contacts, transistor, etc.). **See Figure 21-11.**

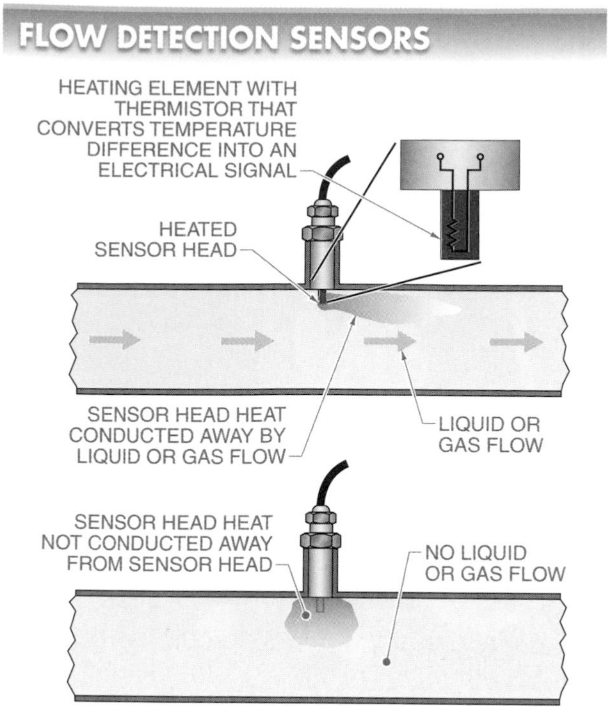

FLOW DETECTION SENSORS

HEATING ELEMENT WITH THERMISTOR THAT CONVERTS TEMPERATURE DIFFERENCE INTO AN ELECTRICAL SIGNAL

HEATED SENSOR HEAD

SENSOR HEAD HEAT CONDUCTED AWAY BY LIQUID OR GAS FLOW

LIQUID OR GAS FLOW

SENSOR HEAD HEAT NOT CONDUCTED AWAY FROM SENSOR HEAD

NO LIQUID OR GAS FLOW

Figure 21-11. A solid-state flow detection sensor operates on the principle of thermal conductivity.

Because there is a small delay (usually less than 30 sec) before the sensor heats enough to signal no flow, the flow detection sensor acts like a motor starter overload, which allows for a short time delay to pass before signaling a problem. This short time delay helps prevent false alarms. The sensor also includes adjustments for setting the sensor to detect different product types and flow rates (speed past the sensor). Different color light-emitting diodes (LEDs) are usually included on the sensor for indicating different operating conditions.

When used to monitor the flow of a liquid, a flow detection sensor is mounted within a pipe in which there should be flow during normal operation. The sensor can be mounted in a vertical or horizontal pipe and the product can flow in either direction. **See Figure 21-12.**

In this circuit, the motor starter contacts in line 3 close each time the pump motor is operating. If there is product flow, an NC (held open during flow) switch is used to prevent the alarm from sounding. The alarm sounds if the flow stops and the motor starter is still ON.

When used to monitor the flow of a gas (usually air), a flow detection sensor can be mounted within a duct in which there should be flow during normal operation. However, if the flow is critical for a safe work environment, several flow sensors can be used to ensure the flow is moving in all parts of the exhaust system. **See Figure 21-13.**

LIQUID FLOW DETECTION SENSORS

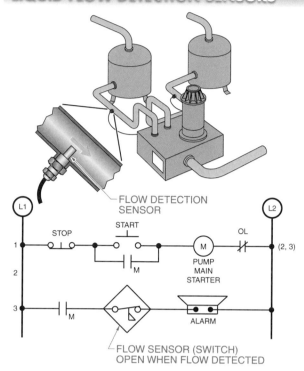

Figure 21-12. A flow detection sensor can be used to monitor product flow in a pipe.

GAS FLOW DETECTION SENSORS

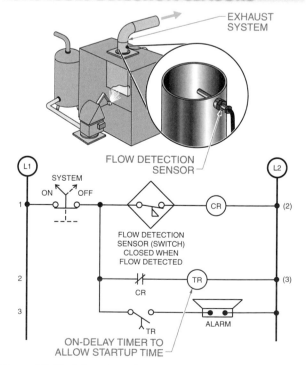

Figure 21-13. A flow detection sensor can be used to monitor airflow in painting or welding exhaust system applications.

In this circuit, an NO (held closed during flow) switch is used to operate a control relay. The control relay operates an on-delay timer. The on-delay timer sounds an alarm if flow stops for longer than the timer's set time period. The time delay prevents the alarm from sounding during the first few seconds of system startup. To eliminate nuisance sounding of the alarm, the time delay can be set for 10 seconds, or longer for less critical applications.

21-5 CHECKPOINT

1. In a thermal-type flow detection sensor, does the sensor's thermistor produce higher or lower electrical signal when there is no flow?
2. Is a thermal-type flow detection sensor a fast-acting or slow-acting change detection type?

21-6 HALL EFFECT SENSORS

A *Hall effect sensor* is a sensor that detects the proximity of a magnetic field. The Hall effect principle was discovered in 1879 by Edward H. Hall at Johns Hopkins University. Hall found that when a magnet is placed in a position where its field is perpendicular to one face of a thin rectangle of gold through which current was flowing, a difference of potential appeared at the opposite edges. He found that this voltage is proportional to the current flowing through the conductor and that the magnetic induction is perpendicular to the conductor.

Today, semiconductors are used for the sensing element (Hall generator) in Hall effect sensors. Hall voltages obtained with semiconductors are much higher than those obtained with gold and are also less expensive.

Theory of Operation

A *Hall generator* is a thin strip of semiconductor material through which a constant control current is passed. **See Figure 21-14.** When a magnet is brought near the Hall generator with its field directed at right angles to the face of the semiconductor, a small voltage (Hall voltage) appears at the contacts placed across the narrow dimension of the Hall generator.

The voltage varies depending on how close the magnet is to the Hall generator, which acts as an analog signal. When the magnet is removed, the Hall voltage drops to zero. The Hall voltage is dependent on the presence of the magnetic field and on the current flowing in the Hall generator. The output of the Hall generator is zero if either the current or the magnetic field is removed. In most Hall effect devices, the control current is held constant and the magnetic induction is changed by movement of a permanent magnet. *Note:* The Hall generator must be combined with associated electronics to form a Hall effect sensor.

Sensor Packaging

To meet many different application requirements, Hall effect sensors are packaged in a number of different configurations. Typical configurations include cylinder, proximity, vane, and plunger. **See Figure 21-15.** Cylinder and proximity Hall effect sensors are used to detect the presence of a magnet. Vane Hall effect sensors include a sensor on one side and a magnet on the other, and are used to detect an object passing through the opening. Plunger Hall effect sensors include a magnet that is moved by an external force acting against a lever.

Hall effect sensors are sometimes used in printers to detect open covers or missing paper.

HALL GENERATORS

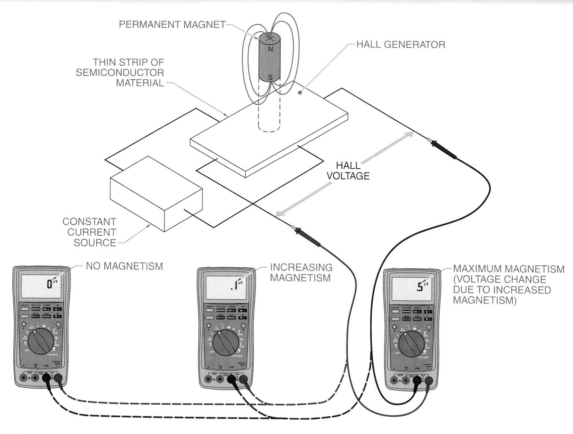

PERMANENT MAGNET

HALL GENERATOR

THIN STRIP OF
SEMICONDUCTOR
MATERIAL

N
S

HALL
VOLTAGE

CONSTANT
CURRENT
SOURCE

NO MAGNETISM

INCREASING
MAGNETISM

MAXIMUM MAGNETISM
(VOLTAGE CHANGE
DUE TO INCREASED
MAGNETISM)

Figure 21-14. A Hall generator is a thin strip of semiconductor material through which a constant control current is passed.

HALL EFFECT SENSORS

Honeywell

Figure 21-15. Hall effect sensors are available in a variety of packages for different applications.

Hall Effect Sensor Actuation

Hall effect sensors may be activated by head-on, slide-by, pendulum, rotary, vane, ferrous proximity shunt, and electromagnetic actuation. The actuation method depends on the application.

Head-on Actuation. *Head-on actuation* is an active method of Hall effect sensor activation in which a magnet is oriented perpendicular to the surface of the sensor and is usually centered over the point of maximum sensitivity. **See Figure 21-16.** The direction of movement is directly toward and away from the Hall effect sensor. The actuator and Hall effect sensor are positioned so the south (S) pole of the magnet approaches the sensitive face of the sensor.

Slide-by Actuation. *Slide-by actuation* is an active method of Hall effect sensor activation in which a magnet is moved across the face of a sensor at a constant distance (gap). **See Figure 21-17.** The primary advantage of slide-by actuation over head-on actuation is that less actuator travel is needed to produce a signal large enough to cycle the device between operate and release.

HEAD-ON ACTUATION

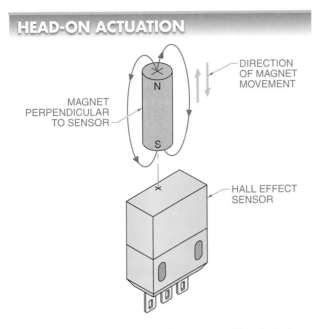

Figure 21-16. In head-on actuation, a magnet is oriented perpendicular to the surface of the sensor and is usually centered over the point of maximum sensitivity.

SLIDE-BY ACTUATION

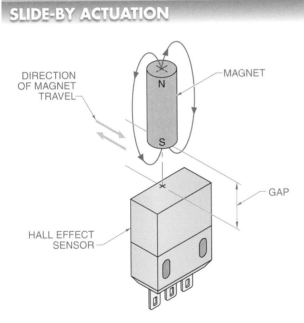

Figure 21-17. In slide-by actuation, a magnet is moved across the face of a Hall effect sensor at a constant distance (gap).

Pendulum Actuation. *Pendulum actuation* is a method of Hall effect sensor activation that is a combination of the head-on and slide-by actuation methods. **See Figure 21-18.** The two methods of pendulum actuation are single-pole and multiple-pole. Single or multiple signals are generated by one actuator.

PENDULUM ACTUATION

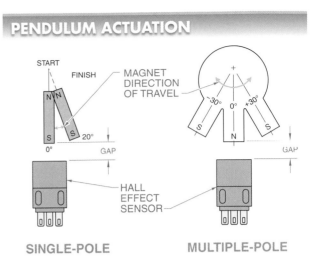

SINGLE-POLE   MULTIPLE-POLE

Figure 21-18. Pendulum actuation is a combination of the head-on and slide-by actuation methods.

Vane Actuation. *Vane actuation* is a passive method of Hall effect sensor activation in which an iron vane shunts or redirects the magnetic field in the air gap away from the sensor. **See Figure 21-19.** When the iron vane is moved through the air gap between the Hall effect sensor and the magnet, the sensor is turned on and off sequentially at any speed due to the shunting effect. The same effect is achieved with a rotary-operated vane.

Hall Effect Sensor Applications

Hall effect sensors are used in a wide range of applications requiring the detection of the presence (proximity) of objects. Hall effect sensors are used in slow-moving and fast-moving applications to detect movement. They are also used to replace mechanical limit switches in applications that require the detection of an object's position. Hall effect sensors are also used to provide a solid-state output in applications that normally use a mechanical output, such as a standard pushbutton on a level switch.

Conveyor Belts. A Hall effect sensor may be used for monitoring a remote conveyor operation. **See Figure 21-20.** In this application, a cylindrical Hall effect sensor is mounted to the frame of the conveyor. A magnet mounted on the tail pulley revolves past the sensor to cause an intermittent visual or audible signal at a remote location to ensure that the conveyor is running. Any shutdown of the conveyor interferes with the normal signal and alerts the operator. Maintenance is minimal because the sensor makes no physical contact and has no levers or linkages to break.

VANE ACTUATION

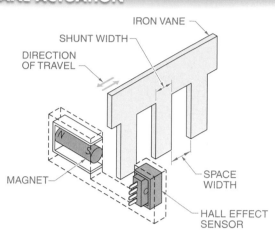

LINEAR VANE SHUNTING

ROTARY VANE SHUNTING

Figure 21-19. In vane actuation, an iron vane shunts or re-directs the magnetic field in the air gap away from the Hall effect sensor.

REMOTE CONVEYOR MONITORING

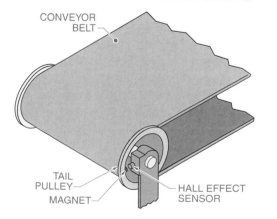

Figure 21-20. A Hall effect sensor may be used for monitoring a remote conveyor operation.

Sensing the speed of a shaft is one of the most common applications of a Hall effect sensor. The magnetic field required to operate the sensor may be furnished by individual magnets mounted on the shaft or hub or by a ring magnet. Each change in polarity results in an output signal. **See Figure 21-21.**

SHAFT SPEED SENSING

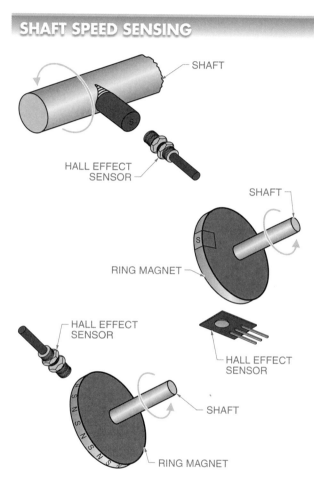

Figure 21-21. Each change in polarity results in an output from a Hall effect sensor used in a shaft speed sensor application.

Liquid Level Monitoring. Another application of a Hall effect sensor is as a low-liquid warning sensor. A low-liquid warning sensor measures and responds to the level of a liquid in a tank. One method used to determine the liquid level in a tank uses a notched tube with a cork floater inserted into the tank. The magnet is mounted in the cork floater assembly, which is forced to move in one plane (vertically). As the liquid level goes down, the magnet passes the digital output sensor Hall effect sensor. The liquid level is indicated when the sensor is actuated. **See Figure 21-22.**

LIQUID LEVEL MONITORING

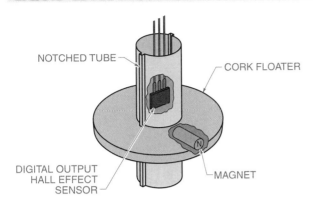

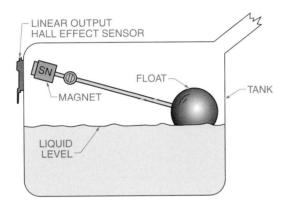

Figure 21-22. A Hall effect sensor can be used to monitor the level of liquid in a tank.

A linear output Hall effect sensor may also be used to indicate the liquid level in a tank. As the liquid level falls, the magnet moves closer to the sensor, causing an increase in output voltage. This method allows measurement of liquid levels without any electrical connections in the interior of the tank. Common applications include fuel tanks, transmission fluid reservoirs, and stationary tanks used in food and chemical processing plants.

Security Systems. A door-interlock security system can be designed using a Hall effect sensor, a magnetic card, and associated electronic circuitry. **See Figure 21-23.** In this circuit, the magnetic card slides by the sensor and produces an analog output signal. This analog signal is converted to a digital signal to provide a crisp signal to energize the door latch relay. When the solenoid of the relay pulls in, the door is unlocked. For systems that require additional security measures, a series of magnets may be molded into the card.

SECURITY SYSTEMS

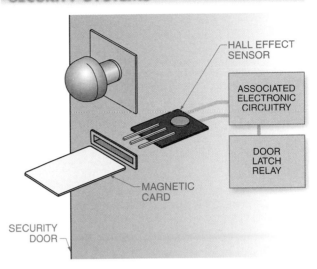

Figure 21-23. A door-interlock system can be designed using a Hall effect sensor, a magnetic card, and associated electronic circuitry.

Beverage Guns. Hall effect sensors are used in beverage gun applications because of their small size, sealed construction, and reliability. **See Figure 21-24.** The small size of the Hall effect sensor allows seven sensors to be installed in a hand-held device. Hall effect sensors cannot be contaminated by syrups, liquids, or foodstuffs because they are completely enclosed in the beverage gun. The beverage gun is also completely submersible in water for easy cleaning and only requires low maintenance.

BEVERAGE GUNS

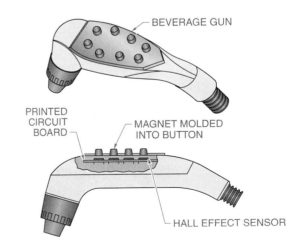

Figure 21-24. Hall effect sensors are used in beverage gun applications because of their small size, sealed construction, and reliability.

Length Measurement. Length measurement can be accomplished by mounting a disk with two notches on the extension of a motor drive shaft. **See Figure 21-25.** In this circuit, a vane Hall effect sensor is mounted so that the disk passes through the gap. Each notch represents a fixed length of material and can be used to measure tape, fabric, wire, rope, thread, aluminum foil, plastic bags, etc.

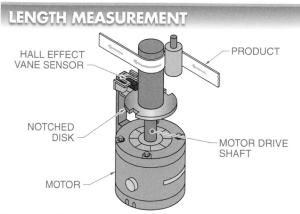

LENGTH MEASUREMENT

HALL EFFECT VANE SENSOR

PRODUCT

NOTCHED DISK

MOTOR DRIVE SHAFT

MOTOR

Figure 21-25. Length measurement can be accomplished by mounting a disk with two notches on the extension of a motor drive shaft.

Level/Degree of Tilt. Hall effect sensors may be installed in the base of a machine to indicate the level or degree of tilt. **See Figure 21-26.** Magnets are installed above the Hall effect sensors in a pendulum fashion. The machine is level as long as the magnet remains directly over the sensor. A change in state of output (when a magnet swings away from a sensor) is indication that the machine is not level. The sensor and magnet combination may also be installed in such a manner as to indicate the degree of tilt.

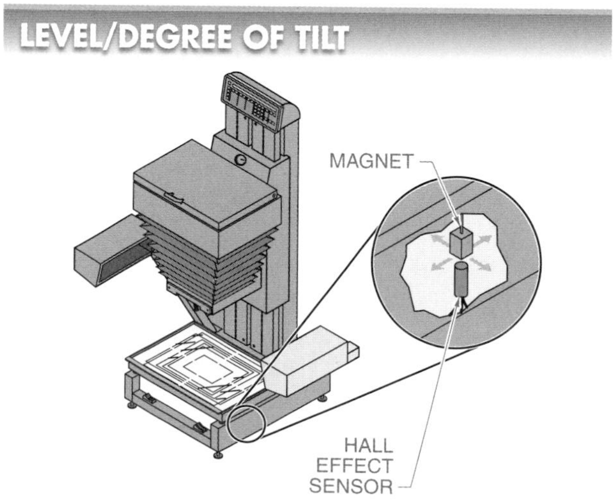

LEVEL/DEGREE OF TILT

MAGNET

HALL EFFECT SENSOR

Figure 21-26. Hall effect sensors may be installed in the base of a machine to indicate the level or degree of tilt.

Joysticks. Hall effect sensors may be used in a joystick application. **See Figure 21-27.** In this application, the Hall effect sensors inside the joystick housing are actuated by a magnet on the joystick. The proximity of the magnet to the sensor controls the activation of different outputs used to control cranes, operators, motor control circuits, wheelchairs, etc. Use of an analog device also achieves degree of movement measurements such as speed.

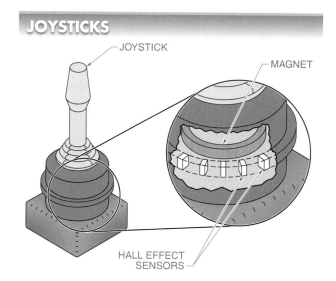

JOYSTICKS

JOYSTICK

MAGNET

HALL EFFECT SENSORS

Figure 21-27. Hall effect sensors may be used in a joystick application.

21-6 CHECKPOINT

1. Is the output of a Hall effect sensor of the digital (ON/OFF) type or analog (varying) type as a magnet moves closer to the sensor?
2. Can a magnet actuate a Hall effect sensor by moving in a sideways or straight manner?

21-7 PROXIMITY SENSORS

A *proximity sensor (proximity switch)* is a solid-state sensor that detects the presence of an object by means of an electronic sensing field. A proximity sensor does not come into physical contact with the object. Proximity sensors can detect the presence or absence of almost any solid or liquid. Proximity sensors are extremely versatile, safe, and reliable. Proximity sensors may be used in applications where limit switches and mechanical level switches cannot be used.

Proximity sensors can detect very small objects, such as microchips, and very large objects, such as automobile bodies. All proximity sensors have encapsulated solid-state circuits that may be used in high-vibration areas, wet locations, and fast-switching applications. Proximity sensors are available in an assortment of sizes and shapes to meet as many application requirements as possible. **See Figure 21-28.**

PROXIMITY SENSORS

Banner Engineering Corp.

Figure 21-28. Proximity sensors are available in an assortment of sizes and shapes to meet as many application requirements as possible.

The two basic proximity sensors are the inductive proximity sensor and the capacitive proximity sensor. An *inductive proximity sensor* is a sensor that detects only conductive substances. Inductive proximity sensors detect only metallic targets. A *capacitive proximity sensor* is a sensor that detects either conductive or nonconductive substances. Capacitive proximity sensors detect solid, fluid, or granulated targets, whether conductive or nonconductive. The proximity sensor used depends on the type and material of the target.

Inductive Proximity Sensors

Inductive proximity sensors operate on the eddy current killed oscillator (ECKO) principle. The ECKO principle states that an oscillator produces an alternating magnetic field that varies in strength depending on whether or not a metallic target is present. The generated alternating field operates at a radio frequency (RF). **See Figure 21-29.**

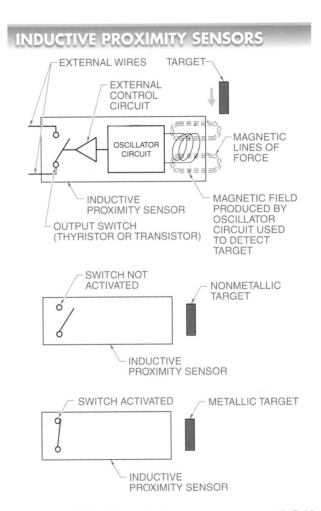

INDUCTIVE PROXIMITY SENSORS

Figure 21-29. Inductive proximity sensors use a magnetic field to detect the presence of a metallic target.

When a metallic target is in front of an inductive proximity sensor, the RF field causes eddy currents to be set up on the surface of the target material. These eddy currents upset the AC inductance of the sensor oscillator circuit, causing the oscillations to be reduced. When the oscillations are reduced to a certain level, the sensor triggers, which indicates the presence of a metallic object. Inductive proximity sensors detect ferrous materials (containing iron, nickel, or cobalt) more readily than nonferrous materials (all other metals, such as aluminum, brass, etc.).

Nominal sensing distances range from 0.5 mm to about 40 mm. Sensitivity varies depending on the size of the object and the type of metal. Iron may be sensed at about 40 mm. Aluminum may be sensed at approximately 20 mm. Applications of inductive proximity sensors include positioning of tools and parts, metal detection, drill bit breakage detection, and solid-state replacement of mechanical limit switches.

Capacitive Proximity Sensors

A capacitive proximity sensor measures a change in capacitance that is caused by the approach of an object to the electrical field of a capacitor. A capacitive proximity sensor detects all materials that are good conductors in addition to insulators that have a relatively high dielectric constant. A *dielectric* is a nonconductor of direct electric current. Capacitive sensors can detect materials such as plastic, glass, water, moist wood, etc. **See Figure 21-30.**

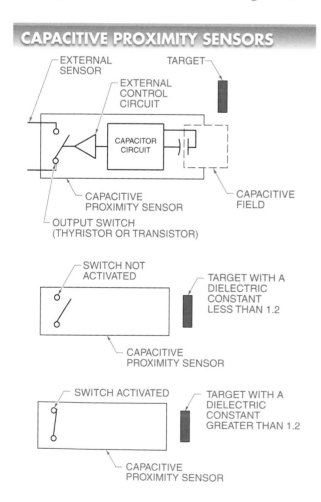

Figure 21-30. Capacitive proximity sensors use a capacitive field to detect the presence of a target.

Two small plates that form a capacitor are located directly behind the front of the sensor. When an object approaches the sensor, the dielectric constant of the capacitor changes, thus changing the oscillator frequency, which activates the sensor output. Nominal sensing distances range from 3 mm to about 15 mm. The maximum sensing distance depends on the physical and electrical characteristics (dielectric) of the object to be detected. The higher the dielectric constant, the easier it is for a capacitive sensor to detect the material. Generally, any material with a dielectric constant greater than 1.2 may be detected. **See Figure 21-31.**

DIELECTRIC CONSTANT	
Material	**Number**
Acetone	20.7
Ammonia (liquid)	15 to 24
Carbon dioxide	1.000985
Epoxy resin	3.3 to 3.7
Gasoline	2.0
Glass	3.7 to 10
Lime marble	2.2 to 2.5
Paper (dry)	8.5
Petroleum jelly	2.0
Porcelain	2.2 to 2.9
Powdered milk	6 to 8
Rubber	3.0
Salt	3 to 15
Sugar	3
Turpentine	2.2
Water	80 to 88
Wood, dry	2 to 6
Wood, wet	10 to 30

Figure 21-31. Capacitive sensors work based on the dielectric constant of the material to be sensed.

21-7 CHECKPOINT

1. What type of proximity switch detects metallic objects?
2. What type of proximity switch detects any object that has a high dielectric constant?

21-8 ULTRASONIC SENSORS

An *ultrasonic sensor* is a solid-state sensor that can detect the presence of an object by emitting and receiving high-frequency sound waves. Ultrasonic sensors can provide an analog voltage output or a digital voltage output (switched output). The high-frequency sound waves are typically in the 200 kHz range. Ultrasonic sensors are used to detect solid and liquid targets (objects) at a distance of up to approximately 1 m (3.3′).

Ultrasonic sensors can be used to monitor the level in a tank, detect metallic and nonmetallic objects, and detect other objects that easily reflect sound waves. Soft materials such as foam, fabric, and rubber are difficult for ultrasonic sensors to detect and are better detected by photoelectric or proximity sensors.

Ultrasonic sensors are used to detect clear objects (glass and plastic), which are difficult to detect with photoelectric sensors. For this reason, ultrasonic sensors are ideal for applications in the food and beverage industry, or for any application that uses clear glass or plastic containers.

Ultrasonic Sensor Operating Modes

The two basic operating modes of ultrasonic sensors are the direct mode and the diffused mode. In the direct mode, an ultrasonic sensor operates like a direct-scan photoelectric sensor. In the diffused mode, an ultrasonic sensor operates like a scan diffuse photoelectric sensor. **See Figure 21-32.**

Direct mode is a method of ultrasonic sensor operation in which the emitter and receiver are placed opposite each other so that the sound waves from the emitter are received directly by the receiver. Ultrasonic sensors used in the direct mode usually include an output that is activated when a target is detected. The output is normally a transistor (PNP or NPN) that can be used to switch a DC circuit. Outputs are available in NO and NC switching modes. Ultrasonic sensors include an adjustment for adjusting (tuning) the sensor sensing distance. Tuning the receiver to the emitter minimizes interference from ambient noise sources that may be present in the area.

Diffused mode is a method of ultrasonic sensor operation in which the emitter and receiver are housed in the same enclosure. In the diffused mode, the emitter sends out a sound wave and the receiver listens for the sound wave echo bouncing back off an object. Ultrasonic sensors used in the diffused mode may include a digital output or an analog output. The analog output provides an output voltage that varies linearly with the target's distance from the sensor. **See Figure 21-33.** The sensor typically includes a light-emitting diode (LED) that glows with intensity proportional to the strength of the echo. The analog output sensor includes an adjustable background suppression feature that allows the sensor to better detect only the intended target and not background objects.

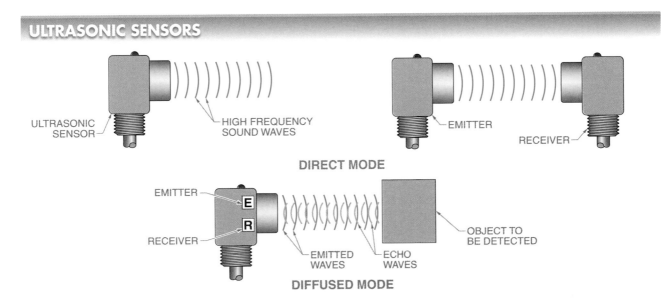

ULTRASONIC SENSORS

ULTRASONIC SENSOR   HIGH FREQUENCY SOUND WAVES   EMITTER   RECEIVER

DIRECT MODE

EMITTER   E   R   RECEIVER   EMITTED WAVES   ECHO WAVES   OBJECT TO BE DETECTED

DIFFUSED MODE

Figure 21-32. Ultrasonic sensors detect objects by bouncing high-frequency sound waves off the objects.

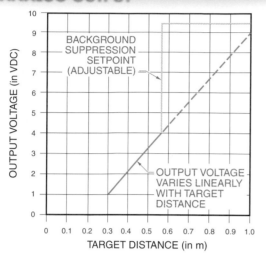

ULTRASONIC SENSOR ANALOG OUTPUT

Figure 21-33. An ultrasonic sensor used in the diffused mode can provide an analog output that varies linearly with the target's distance from the sensor.

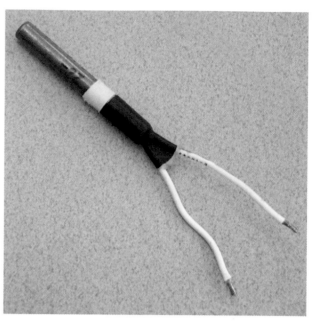

An outdoor air temperature (OAT) sensor is a semiconductor output device used in vehicle and aircraft applications.

21-8 CHECKPOINT

1. What is the operating mode of an ultrasonic sensor called when the emitter sound waves travel in only one direction to the receiver?

2. What is the operating mode of an ultrasonic sensor called when the emitter sound waves travel in one direction to the detected object and bounces back in the opposite direction to the receiver?

Additional Resources

Review and Resources

Access Chapter 21 Review and Resources for *Electrical Motor Controls for Integrated Systems* by scanning the above QR code with your mobile device.

Applying Your Knowledge

Refer to the *Electrical Motor Controls for Integrated Systems* Learner Resources for interactive Applying Your Knowledge questions.

Workbook and Applications Manual

Refer to Chapter 21 in the *Electrical Motor Controls for Integrated Systems Workbook* and the *Applications Manual* for additional exercises.

ENERGY EFFICIENCY PRACTICES

Energy-Efficient Lighting Controls

After motors, lighting is the next highest consumer of energy in buildings, accounting for approximately 20% of the total energy consumed worldwide. Proper building lighting improves the productivity and safety of building occupants. Lighting-system control saves energy and can substantially reduce the operating costs of a building. For example, lighting in an area of a facility can be turned on automatically when an individual enters the area. After a set time period of not sensing an individual, the lights can be set to turn off automatically.

Automated lighting control allows a system to operate without input and saves time and energy. The flexibility of automated lighting systems also accommodates changes to lighting requirements. For example, changes to schedules, tasks, or workplace configurations can be easily implemented while maintaining lighting control goals.

An automated lighting control system may include ultrasonic sensors that activate when sensing changes in the reflected sound waves caused by an individual moving within its detection area. The sensor emits low-power, high-frequency sound waves and monitors for a phase shift in the reflected sound returning to the sensor, which indicates a moving object. The reliance on motion for sensor activation can also cause false deactivations when occupants are still for long periods of time.

Objectives

22-1
- Define amplification and explain gain.
- Define and describe transistors.
- Describe transistor bias and current flow.

22-2
- Define operational amplifier (op amp) and describe its symbol.
- Explain how an op amp operates.
- Describe how op amps are used.

22-3
- Define field-effect transistor (FET) and list the different types.
- Define and describe a junction field-effect transistor (JFET).
- Define and describe a metal-oxide semiconductor field-effect transistor (MOSFET).
- Define and describe an insulated gate bipolar transistor (IGBT).

22-4
- Explain how to troubleshoot an insulated gate bipolar transistor (IGBT).

Review and Resources
atplearningresources.com/Quicklinks
Access Code: 362245

Chapter

Semiconductor
Amplification and Switching

Transistors were among the first solid-state devices used for switching and amplification. As transistor technology for use in circuits expanded, other similar devices were developed that could also be used for switching and amplification. Each device has advantages and disadvantages based on the actual application the device is used in. When used as a switch, the switching speed and the amount of required switching voltage, current, power, and frequency of the device are considered for the selection of the best device for a given application. When used as an amplifier, the device's gain (ratio of output to input), type of input signal the device can amplify, and the type and size of output load the amplified output can control are considered. Since many input devices that measure or detect pressure, temperature, flow, light, sound, etc. require their very low power (voltage/current) signals to be amplified for a practical usage, a basic understanding of the devices used to amplify or switch the input signal helps in the understanding of the total system when servicing or troubleshooting systems that include semiconductors used for switching or amplification.

22-1 AMPLIFICATION

Amplification is the process of taking a small signal and making it larger. In control systems, amplifiers are used to increase small signal currents and voltages so they can do useful work. Amplification is accomplished by using a small input signal to control the energy output from a larger source such as a power supply. Amplification is accomplished by transistors and operational amplifiers (op amps). Amplifiers are essential in increasing the effect of a small change in voltage. Current and resistance is produced by input sensors.

Amplifier Gain

The primary objective of an amplifier is to produce gain. *Gain* is a ratio of the amplitude of an output signal to the amplitude of an input signal. In determining gain, the amplifier can be thought of as a black box. A signal applied to the input of the black box results in output from the box. Mathematically, gain can be found by dividing output by input:

$$Gain = \frac{Output}{Input}$$

Often, a single amplifier does not provide enough gain to increase the output signal to the amplitude needed. In such a case, two or more amplifiers can be used to obtain the gain required. **See Figure 22-1.** For example, amplifier A has a gain of 10 and amplifier B has a gain of 10. The total gain of the two amplifiers is 100 ($10 \times 10 = 100$). If the gains of the amplifiers were 8 and 9, respectively, the total gain would be 72 ($8 \times 9 = 72$). Amplifiers connected in this manner are called cascaded amplifiers. For many amplifiers, gain is in the hundreds or even thousands.

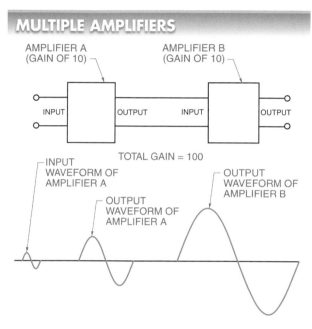

Figure 22-1. Two or more amplifiers can be used to obtain the required gain if a single amplifier does not provide enough gain to increase the output signal to the amplitude needed.

Note: Gain is a ratio of output to input and has no unit of measure, such as volts or amps, attached to it. Therefore, the term gain is used to describe current gain, voltage gain, and power gain. In each case, the output is merely being compared to the input.

Transistors as AC Amplifiers

Transistors may be used as AC amplification devices. A *transistor* is a three-terminal device that controls current through the device depending on the amount of voltage applied to the base. Transistors are bipolar devices. A *bipolar device* is a device in which both holes and electrons are used as internal carriers for maintaining current flow.

Transistors may be PNP or NPN transistors. A PNP transistor is formed by sandwiching a thin layer of N-type material between two layers of P-type material. An NPN transistor is formed by sandwiching a thin layer of P-type material between two layers of N-type material. **See Figure 22-2.**

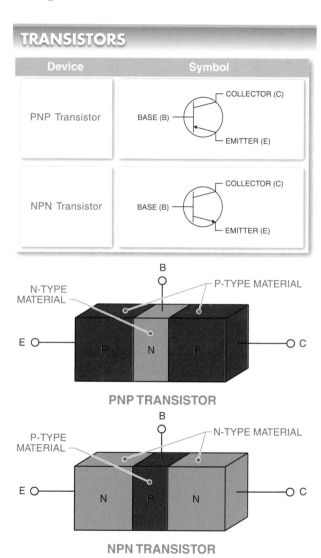

Figure 22-2. A PNP transistor is formed by sandwiching a thin layer of N-type material between two layers of P-type material. An NPN transistor is formed by sandwiching a thin layer of P-type material between two layers of N-type material.

Transistor terminals are the emitter (E), base (B), and collector (C). The symbols for PNP and NPN transistors show the emitter, base, and collector in the same places. The difference in the terminals is the direction in which the emitter arrow points. In both cases, the arrow points from the P-type material toward the N-type material.

The three basic transistor amplifiers are the common-emitter, common-base, and common-collector. **See Figure 22-3.** Each amplifier is named after the transistor connection that is common to both the input and the load. For example, the input of a common-emitter circuit is across the base and emitter, while the load is across the collector and emitter. Thus, the emitter is common to the input and load.

TRANSISTOR AMPLIFIERS

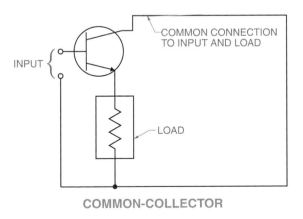

COMMON-EMITTER

COMMON-BASE

COMMON-COLLECTOR

Figure 22-3. The three basic transistor amplifiers are the common-emitter, common-base, and common-collector.

Transistors are manufactured with two or three leads extending from their case. **See Figure 22-4.** These packages are accepted industry-wide regardless of manufacturer. When a transistor with a specific shape must be used, a transistor outline (TO) number is used for reference.

TRANSISTOR PACKAGES

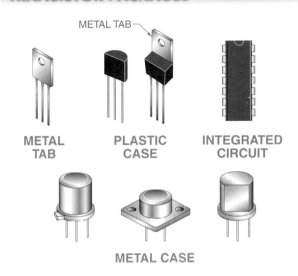

METAL TAB

METAL TAB **PLASTIC CASE** **INTEGRATED CIRCUIT**

METAL CASE

Figure 22-4. Transistors have either two or three leads extending from their case.

A *transistor outline (TO) number* is a number determined by the manufacturer that represents the shape and configuration of a transistor. **See Figure 22-5.** Transistor outline numbers are determined by individual manufacturers. *Note:* The bottom view of transistor TO-3 shows only two leads (terminals). Typically, transistors use the metal case as the collector-pin lead.

TRANSISTOR OUTLINE NUMBERS

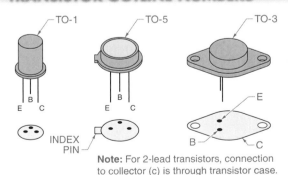

TO-1   TO-5   TO-3

B   B   E
E   C   E   C
INDEX PIN   B   C

Note: For 2-lead transistors, connection to collector (c) is through transistor case.

Figure 22-5. A transistor outline (TO) number is a number determined by the manufacturer that represents the shape and configuration of a transistor.

Biasing Transistor Junctions

In any transistor circuit, the base/emitter junction must always be forward biased and the base/collector junction must always be reverse biased. **See Figure 22-6.** The external voltage (bias voltage) is connected so that the positive terminal connects to the P-type material (base) and the negative terminal connects to the N-type material (emitter). This arrangement forward biases the base/emitter junction. Current flows from the emitter to the base. The action that takes place is the same as the action that occurs for a forward-biased semiconductor diode.

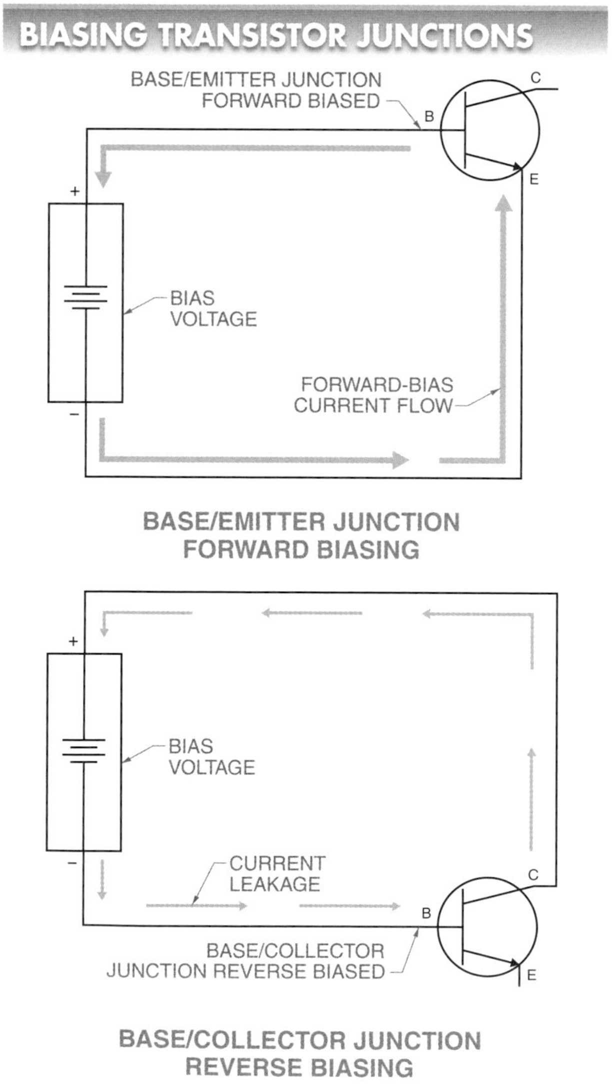

In any transistor circuit, the base/collector junction must always be reverse biased. The external voltage is connected so that the negative terminal connects to the P-type material (base) and the positive terminal connects to the N-type material (collector). This arrangement reverse biases the base/collector junction. Only a very small current (leakage current) flows in the external circuit. The action that takes place is the same as the action that occurs for a semiconductor diode with reverse bias applied.

Transistor Current Flow

Individual PN junctions can be used in combination with two bias arrangements. The base/emitter junction is forward biased while the base/collector junction is reverse biased. This circuit arrangement results in an entirely different current path than the path that occurs when the individual circuits are biased separately. **See Figure 22-7.**

BIASING TRANSISTOR JUNCTIONS

BASE/EMITTER JUNCTION
FORWARD BIASED

BIAS VOLTAGE

FORWARD-BIAS CURRENT FLOW

**BASE/EMITTER JUNCTION
FORWARD BIASING**

BIAS VOLTAGE

CURRENT LEAKAGE

BASE/COLLECTOR
JUNCTION REVERSE BIASED

**BASE/COLLECTOR JUNCTION
REVERSE BIASING**

Figure 22-6. In a transistor circuit, the base/emitter junction must always be forward biased and the base/collector junction must always be reverse biased.

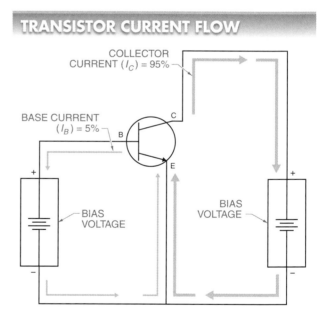

TRANSISTOR CURRENT FLOW

COLLECTOR
CURRENT (I_C) = 95%

BASE CURRENT
(I_B) = 5%

BIAS VOLTAGE

BIAS VOLTAGE

Figure 22-7. In a transistor, an entirely different current path is created when both junctions are biased simultaneously than when each junction is biased separately.

The forward bias of the base/emitter circuit causes the emitter to inject electrons into the depletion region between the emitter and the base. Because the base is less than 0.001″ thick for most transistors, the more positive potential of the collector pulls the electrons through the thin base. As a result, the greater percentage (95%) of the available free electrons from the emitter pass directly through the base (I_C) into the N-type material, which is the collector of the transistor.

22-1 CHECKPOINT

1. What is the total gain of three cascaded amplifiers if amplifier A has a gain of 10, B has a gain of 5, and C has a gain of 10?

2. In a common-emitter amplifier, the emitter is common to what two parts of the circuit?

3. When a transistor uses its metal case as one of its three leads, which lead is the metal base?

4. When the base/emitter junction of a transistor is forward biased and the base/collector junction is reversed biased, is the current flowing through the junctions from positive to negative or negative to positive?

22-2 OPERATIONAL AMPLIFIERS

An operational amplifier (op amp) is one of the most widely used integrated circuits (ICs). An *operational amplifier (op amp)* is a high-gain, directly coupled amplifier that uses external feedback to control response characteristics. **See Figure 22-8.**

An example of feedback control is gain. The gain of an op amp can be controlled externally by connecting a feedback resistor between the output and input. A number of different amplifier applications can be achieved by selecting different feedback components and combinations. With the right component combinations, gains in the thousands are common.

An op amp is very versatile. An op amp can be made to perform math functions, such as addition, subtraction, multiplication, and division, by connecting it to a few components. Today, the main purpose of an op amp is to amplify small signals to levels that can be used for the control of another device. An op amp is often used in digital circuits that require the analog amplification of a weak signal.

Schematic Symbols

The schematic symbol of an operational amplifier may be shown as a triangle with the two inputs of the op amp as inverting (–) and noninverting (+). **See Figure 22-9.** The two inputs are usually drawn with the inverting input at the top. The exception to the inverting input being at the top is when it would complicate the schematic symbol. In either case, the two inputs should be clearly identified on the schematic symbol by polarity symbols.

The advantages of an operational amplifier include the following:

• high input impedance (resistance)—an operational amplifier does not draw much power from the input source

• low output impedance—a high output impedance would reduce the amplified output

• high gain—one operational amplifier can replace many individual transistors

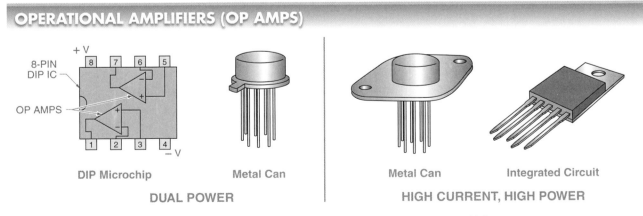

OPERATIONAL AMPLIFIERS (OP AMPS)

8-PIN DIP IC

OP AMPS

DIP Microchip   Metal Can

DUAL POWER

Metal Can   Integrated Circuit

HIGH CURRENT, HIGH POWER

Figure 22-8. Types of operational amplifiers (op amps) include dual-power and high-current, high-power op amps.

OP AMP SCHEMATIC SYMBOL

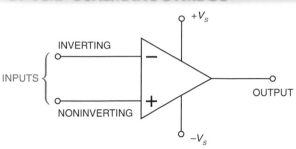

Figure 22-9. The schematic symbol for an op amp includes inverting (–) and noninverting (+) inputs.

Voltage Sources for Op Amps

Like other solid state devices, ICs need DC operating voltages. The DC voltage pins on some ICs are labeled common-collector voltage (V_{CC}). Other pins are labeled input voltage (V_{in}) or supply voltage (V_S). DC voltage ratings in IC data books are labeled all three ways. The voltage source for op amps can be single supply. However, they are usually bipolar or dual supply. **See Figure 22-10.**

OP AMP VOLTAGE SOURCES

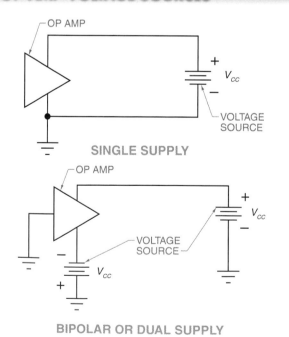

Figure 22-10. While voltage sources for op amps can be single supply, they are usually bipolar or dual supply.

Internal Op Amp Operation

Internally, an op amp has three major sections. It consists of a high-impedance differential amplifier, a high-gain stage, and a low-output impedance power-output stage. The differential amplifier provides the wide bandwidth and the high impedance. The high-gain stage boosts the signal. The power-output stage isolates the gain stage from the load and provides power. **See Figure 22-11.**

OP AMP SECTIONS

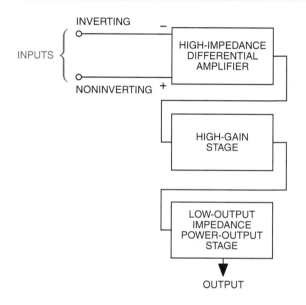

Figure 22-11. Internally, an op amp consists of a high-impedance differential amplifier, a high-gain stage, and a low-output impedance power-output stage.

Gain. Op amps may have gains of 500,000 or more, no gain (unity gain), or controlled gain. Gain in an op amp is controlled by external resistors that provide closed-loop feedback.

In the high-gain mode, a very small change in voltage on either input results in a large change in the output voltage. These high-gain circuits can be far too sensitive and unstable for most applications. The gain is normally reduced to a much lower level due to this instability. The circuit is stabilized by feeding back some of the output signal to one of the inputs through a resistor.

Voltage Followers (Unity-Gain). A voltage follower, or source follower, is a noninverting amplifier. The output voltage (V_{out}) is an exact reproduction of the input voltage (V_{in}). **See Figure 22-12.** The function of the voltage follower is identical to that of the source follower created by a bipolar transistor and FET. The circuit is used to impedance-match an input signal to its load. With the voltage follower, the input impedance is high and the output impedance is low. It should be noted that the voltage follower has no input or feedback components. Because no feedback components are used in this type of circuit, the amplifier is operating in a unity-gain condition.

UNITY-GAIN CONDITION

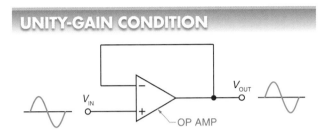

Figure 22-12. An amplifier is operating in a unity-gain condition when the output voltage is an exact reproduction of the input voltage.

Open-Loop Control. Open-loop systems are used almost exclusively for manual control operations. The two variations of the open-loop system are full control and partial control. Full control operation simply turns a system on or off. For example, in an electrical circuit, current flow stops when the circuit path is opened. Switches, circuit breakers, fuses, and relays are used for full control. Partial control operation alters system operations rather than causing them to start or stop. Resistors, inductors, transformers, capacitors, semiconductor devices, and ICs are commonly used to achieve partial control.

Closed-Loop Control. To achieve automatic control, interaction between the control unit and the controlled element must occur. In a closed-loop system this interaction is called feedback. Feedback can be activated by electrical, thermal, light, chemical, or mechanical energy. Both full and partial control can be achieved through a closed-loop system. **See Figure 22-13.**

Many of the automated systems used in industry today are of the closed-loop type. A closed-loop system may have automatic correction control. In the system, energy goes to the control unit and the controlled element. Feedback from the controlled element is directed to a comparator, which compares the feedback signal to a reference signal or standard.

CLOSED-LOOP SYSTEMS

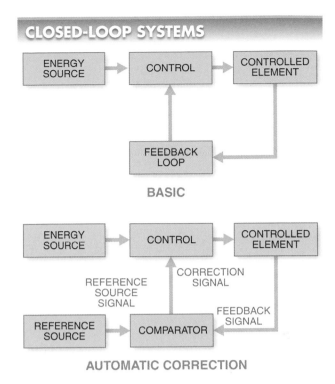

Figure 22-13. Closed-loop systems may be depicted by block diagrams.

A correction signal is developed by the comparator and sent to the control unit. This signal alters the system so that it conforms to the data from the reference source. Systems of this type maintain a specified operating level regardless of external variations or disturbances.

Automated control has gone through many changes in recent years with the addition of control devices that are not obvious to the casual observer. These include devices that change the amplitude, frequency, waveform, time, or phase of signals passing through the system.

Feedback Inverting Amplifiers (Closed-Loop). A feedback inverting amplifier produces a 180° phase inversion from input (V_{in}) to output (V_{out}). **See Figure 22-14.** When a positive-going voltage is applied to the input, a negative-going voltage will be produced at the output. The input signal is applied to the op amp inverting input through R1, while resistor R2 serves as the feedback element.

The voltage gain of the feedback inverting amplifier can be less than, equal to, or greater than 1.0. Its value depends on the values of resistors R1 and R2. Because a feedback component is used in this type of circuit, the amplifier is operating in a closed-loop condition. The closed-loop gain of the feedback inverting amplifier can be controlled by switching in different feedback resistors.

FEEDBACK INVERTING AMPLIFIERS (CLOSED-LOOP)

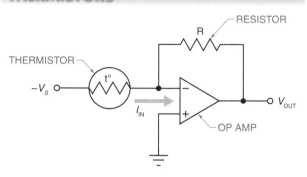

Figure 22-14. A feedback inverting amplifier produces a 180° phase inversion from input to output.

Offset Error and Nulling. Although extreme care is taken in fabricating an op amp, a slight mismatch may still occur between the internal components. *Offset error* is a slight mismatch between internal components. Offset error creates a problem when using an op amp in a DC circuit. The mismatch prevents the amplifier from having a zero output for a zero input.

Even with proper bias, the feedback inverting amplifier has an input bias current (I_B) through the input and feedback resistors with no signal applied. The additional current flow through these resistors produces a voltage drop, which appears as DC input voltage. The op amp then amplifies this DC input voltage, compounding the offset error. To correct offset error, the nulling technique is often used through a nulling resistance network. With this network, the nulling variable resistor is adjusted for zero output with zero input.

Op Amp Applications

Op amps are used for a variety of amplification applications. For example, they are used in audio amplifiers and video amplifiers. They are also ideal for a variety of industrial and commercial control systems.

Current-to-Voltage Converters. A current-to-voltage op amp uses the current sensitivity of the op amp to measure very small currents. The circuit can provide 1 V at the output for 1 µA at the input. This basic current-to-voltage converter is essentially an inverting amplifier without an input resistor. The input current is applied directly to the inverting input of the op amp.

A thermistor varies the amount of current (I_{in}) entering the inverting input of the op amp. **See Figure 22-15.** As the ambient temperature around a thermistor increases, the resistance of the thermistor decreases, current to the input of the op amp increases, and voltage at the output increases. As the temperature decreases, the resistance increases, current decreases, and voltage output decreases. Since the output of the circuit is now voltage, the op amp voltage can be used to drive an output device.

THERMISTORS

Figure 22-15. As the ambient temperature around a thermistor changes, the resistance of the thermistor changes.

Voltage-to-Current Converters. In certain circuits, such as a circuit in an air compressor, a change in voltage becomes the reference for the change in the circuit. When this is the case, a voltage-to-current op amp is used. **See Figure 22-16.** The output voltage of the bridge circuit is a function of the degree of imbalance present in the input bridge. For the bridge circuit, an imbalance can be created by changing the pressure on the pressure sensor (transducer). The potentiometer determines the pressure-set limits.

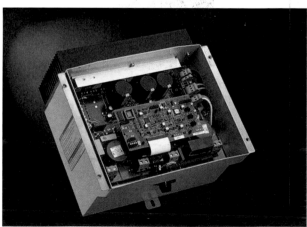

Solar power electronics include semiconductor switching devices.

VOLTAGE-TO-CURRENT OP AMPS

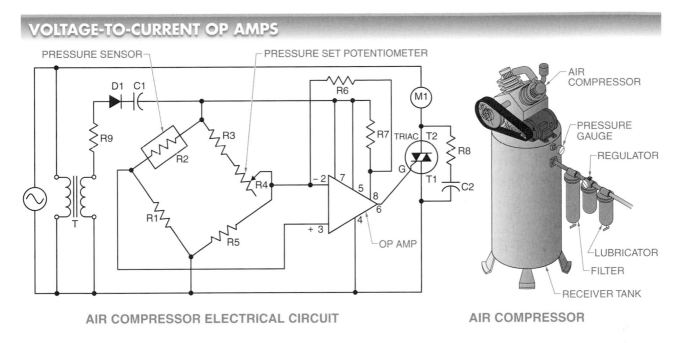

Figure 22-16. A voltage-to-current op amp can be used to control an air compressor.

The unit can operate directly from the AC supply since it includes a step-down transformer and single-phase rectifier. When the pressure drops, the resistance of the pressure sensor decreases. Terminal 3 of the op amp becomes more positive than terminal 2. Under this condition, the output current at pin 6 causes the triac to conduct. With the triac conducting, power is applied to the coil, which turns on the air compressor. When the pressure in the tank is brought up to the preset limit, the value of the pressure sensor increases, balancing the bridge, and the air compressor shuts off.

Thermostat Control. An op amp can be used as a differential amplifier. A differential op amp is a type of op amp used in differential thermostat controls for solar water heaters. **See Figure 22-17.** Small signal changes can be detected easily in the circuit and are amplified for signaling or control purposes. Because the output of an op amp has very little power, an interface device, such as control relay, must be used to control higher power loads, such as a water pump in a hot water system.

When the op amp has signals of equal amplitude and polarity applied to each of its inputs simultaneously, the output will be zero. For the circuit, the bridge must be balanced for the op amp output to be zero. The bridge is formed by a tank sensor; collector sensor; and resistors R1, R2, and R3.

If the tank senses a reduction in temperature, the bridge will become unbalanced, with the difference being presented to the input of the op amp. The signal is amplified to the output. The output is then connected to a relay that starts the pump. When enough hot water passes the collector and tank and heats the tank water, the bridge is again balanced and the relay is turned off.

Troubleshooting Integrated Circuits

An IC that malfunctions cannot be repaired. The only duty of the troubleshooter is to test the IC to determine whether the IC is good or bad. If the IC is bad, the IC must be replaced.

An IC is tested in-circuit using the input/output method. If the input signal is normal, the power sources to the IC are tested. Any external components used with the IC, such as coils or capacitors (resistors are usually built in), are tested next. If the output is still abnormal, a new IC is substituted. Abnormal input usually means that the IC is not the problematic part, although an IC may sometimes distort the input.

Before replacing an IC, the power source must be turned off. Otherwise, the IC may be damaged by power surges if all terminals are not connected to the circuit at the same time.

THERMOSTAT CONTROL CIRCUITS

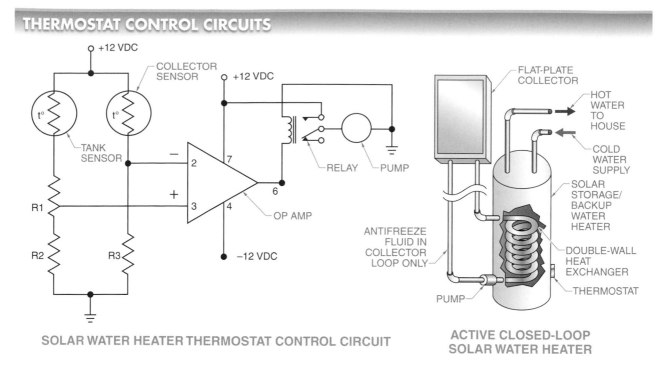

SOLAR WATER HEATER THERMOSTAT CONTROL CIRCUIT

ACTIVE CLOSED-LOOP SOLAR WATER HEATER

Figure 22-17. A differential op amp can be used in a solar water heater thermostat control circuit.

Furthermore, an energized or "hot" circuit should not be soldered. Soldering may create problems that require a lot of time to identify, isolate, and correct. A small drop of solder can destroy a handful of parts instantly. Soldering also presents a safety hazard if there is high voltage.

Tech Fact

A digital logic probe has a TTL/CMOS switch. When a digital logic probe is set to test TTL circuits, a low signal is shown by a green LED at 0 V to 0.8 V and a high signal is shown by a red LED at 2.2 V to 5 V. When a digital logic probe is set to test CMOS circuits, a low signal is show by a green LED at 0% to 30% of the circuit's voltage and a high signal is shown by a red LED at 70% to 100% of the circuit's voltage. No light indicates that the voltage is within the undefined level and can cause circuit errors.

22-2 CHECKPOINT

1. Why is an op amp better for amplifying a very small signal to a larger signal than using a transistor as the amplifier?
2. Why do op amps draw very little power from the input source they are going to amplify?
3. What type of op amp is used to take a very small amount of input current and produce a much larger voltage output?
4. What is an example of an interface device that can be connected between the output of an op amp and a pump motor?

22-3 FIELD-EFFECT TRANSISTORS

A *field-effect transistor (FET)* is a three- or four-terminal device in which output current is controlled by an input voltage. FET terminals include a gate, drain, and source. Two of the most common types of FETs include the junction field-effect transistor (JFET) and the metal-oxide semiconductor field-effect transistor (MOSFET). The newest type of FET is the insulated gate bipolar transistor (IGBT). The IGBT combines the advantages of a power MOSFET and a bipolar junction transistor.

Junction Field-Effect Transistors

A *junction field-effect transistor (JFET)* is a simple FET with a PN junction in which output current is controlled by an input voltage. The two types of JFETs include the N-channel and P-channel. A JFET, like all FETs, contains a gate (G), drain (D), and source (S). The gate is a control element, while the drain and source provide the same function as the emitter and collector on a bipolar junction transistor. **See Figure 22-18.**

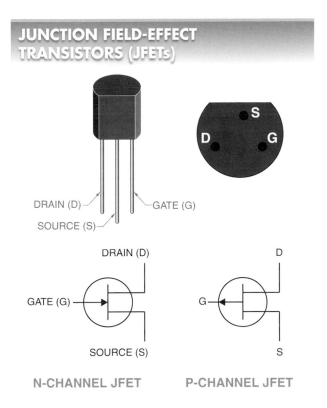

JUNCTION FIELD-EFFECT TRANSISTORS (JFETs)

N-CHANNEL JFET

P-CHANNEL JFET

Figure 22-18. The arrow in a schematic symbol for an N-channel JFET points inward. The arrow points outward for a P-channel JFET.

JFETs may be used as switches and gates. JFETs are even found in voltage regulators and current limiters. JFETs are fully compatible with other semiconductor devices, such as standard bipolar transistors, silicon-controlled rectifiers, triacs, and ICs.

JFET Operation. A JFET is a unipolar device, which differs in operation from a bipolar junction transistor. The output current of a JFET is controlled by the voltage on the gate. The gate voltage creates an electrical field, or depletion region, within the device. **See Figure 22-19.** The JFET is considered a voltage-driven device rather than a current-driven device like the bipolar junction transistor.

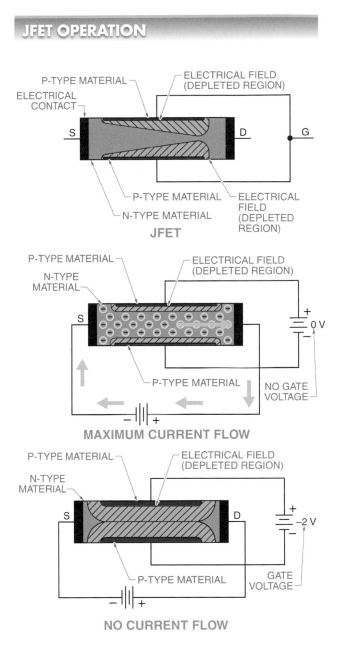

JFET OPERATION

JFET

MAXIMUM CURRENT FLOW

NO CURRENT FLOW

Figure 22-19. The output current of a JFET is controlled by an electrical field created by the input voltage.

The source and drain of the JFET are connected to a common N-type material. This common material constitutes the channel of the JFET. If a DC potential is connected between the source and the drain, current should flow in the external circuit and through the channel. At zero gate voltage, channel height is maximum and channel resistance is minimum, resulting in current flow. With a slight positive voltage, the channel height opens further, allowing maximum current flow.

The two sections of P-type material constitute the gate. Each section has its associated electrical field (depletion region). If the gate is made negative with respect to the channel, the diodes (formed by the P-type and N-type materials) become reverse biased and the depletion regions increase. At a large enough gate-to-source voltage (V_{GS}), the channel is effectively "pinched off" because the depletion regions touch. The depletion regions merge at a particular V_{GS} of somewhere between 1 V and 8 V.

The size of the depletion region is controlled by the gate-to-source voltage (V_{GS}). When V_{GS} increases, the depletion region increases. When V_{GS} decreases, the depletion region decreases.

This gate-to-source voltage controls the drain current and must always provide a reverse-bias voltage. This is quite different from a bipolar transistor. In a bipolar transistor, the junction must be forward biased so that the junction impedance is extremely low and there is current flow in the junction. In a JFET, the junction must be reverse biased so that the junction impedance is high and there is little current flow in the junction. The major advantage of this is good control using voltage rather than current. The power consumption of the JFET (in standby operation) is thousands of times less than that of a bipolar transistor controlling the same function.

JFET Output Characteristic Curves. A JFET characteristic curve shows the operating characteristics of a JFET. **See Figure 22-20.** The drain source voltage (V_{DS}) is plotted along the horizontal axis. The drain current (I_D) is plotted along the vertical axis on the characteristic curve for a typical N-channel JFET. By using one curve, the detailed information produced may be more readily obtained.

The ohmic region is the first portion of the characteristic curve of a JFET where the drain current rises rapidly. In the ohmic region, the drain current is controlled by the drain-source voltage and the resistance of the channel. The knee of the curve is the pinch-off voltage. On the flattened portion of the graph, the JFET is at the pinch-off region. In this region, the drain current is controlled primarily by the width of the channel. When the drain current begins another sharp increase, avalanche (breakdown) begins. At this point, a large amount of current begins to pass through the channel. If this current is not limited by an external resistance or load, the JFET may be damaged.

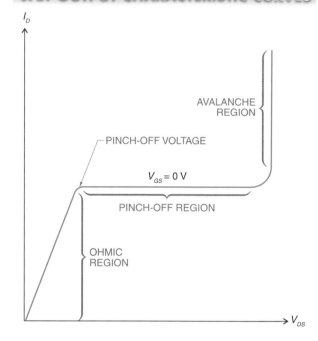

Figure 22-20. A JFET characteristic curve shows the operating characteristics of a JFET.

JFET Circuit Configurations. JFETs can be connected into three basic circuit configurations: common source, common gate, and common drain. Common-gate and common-drain circuits are constructed in a similar common element arrangement. Furthermore, the only difference between an N-channel and P-channel circuit configuration is the direction of electron flow in the external circuit and the polarity of the bias voltages. **See Figure 22-21.**

JFET Applications. JFETs are used extensively in circuits where low power and high impedance are factors. JFETs need very little power to produce a large output at the load.

JFETs are used in many amplifier configurations because they can amplify a fairly wide range of frequencies and have high input impedance. High input impedance is a definite advantage in the transfer of power.

FET multimeters use JFETs to provide high input impedances with good sensitivity. A JFET is capable of producing an input of 11 MΩ or more. In a simple FET multimeter circuit, the JFET is one "leg" in the DC bridge. Source resistor RS provides negative feedback for high linearity in the response. The circuit requires good regulation, which is provided by a zener diode. **See Figure 22-22.**

JFET CIRCUIT CONFIGURATIONS

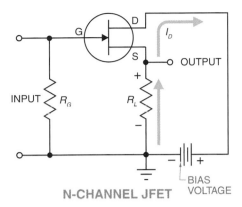

N-CHANNEL JFET

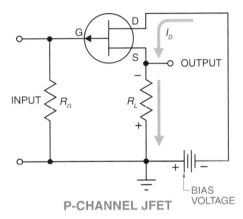

P-CHANNEL JFET

Figure 22-21. In a common-source JFET circuit, the input is across the gate and source terminals and the output is across the drain and source terminals.

Metal-Oxide Semiconductor Field-Effect Transistors

A *metal-oxide semiconductor field-effect transistor (MOSFET)* is a three-terminal or four-terminal electronic switching device with metal-oxide or polysilicon insulating material that can be used for amplification. MOSFETs have the same terminal designations (gate, drain, and source) as JFETs.

The gate voltage of a MOSFET that controls the drain current is different from that of a JFET. The main difference in operation between a JFET and MOSFET is that voltage is applied between the gate and the P and N regions of the MOSFET structure. An electric field is generated, penetrates through the oxide layer, and creates an inversion layer, or channel, at the semiconductor-insulator layers. Varying the voltage between the gate and the P and N layers controls the conductivity of this layer, allowing current to flow between the drain and source. MOSFETs can operate in either an enhancement mode or a depletion mode.

Tech Fact

MOSFETs and JFETs can be thought of as variable-resistance devices that are designed to control high-power devices and circuits. Caution must be taken when using or testing these FETS because the high circuit power can cause them to explode and cause a fire. Because they are basically resistance devices, they are checked using a DMM set for the resistance and diode check settings. After all power is turned off, the device can be tested. The manufacturer specifications list the expected resistance between the pins for the different types of devices.

DIGITAL MULTIMETERS WITH JFETs

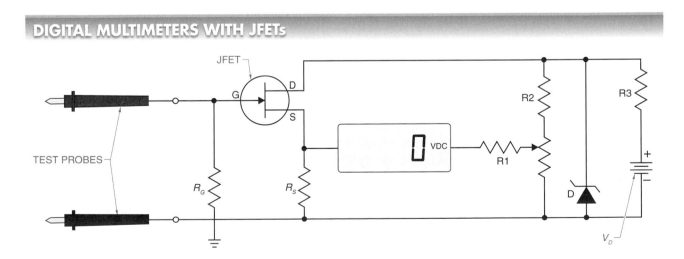

Figure 22-22. A digital multimeter can provide high input impedance with good sensitivity by using a JFET in its front-end circuit.

MOSFET Schematic Symbols and Lead Identification. MOSFETs are available as N-channel and P-channel devices. With an N-channel MOSFET, the arrow on the substrate points toward the channel. With a P-channel MOSFET, the arrow on the substrate points away from the channel. **See Figure 22-23.**

MOSFETs

DRAIN (D)

SUBSTRATE (B)

GATE (G)

SOURCE (S)

N-Channel

D

G

B

S

P-Channel

S

D   G

B

Bottom View

FOUR-TERMINAL MOSFET

D

G

S

N-Channel

D

G

S

P-Channel

THREE-TERMINAL MOSFET

International Rectifier

PACKAGING

Figure 22-23. MOSFETs are available in both N-channel and P-channel constructions.

Enhancement MOSFETs. The main body of an enhancement MOSFET is composed of a highly resistive P-type material. Two low-resistance N-type regions are diffused in the P-type region, forming the source and drain. When the source and drain are completely diffused, the surface of the MOSFET is covered with a layer of insulating material. The insulating layer of the MOSFET is silicon dioxide (SiO_2).

Holes are cut into the silicon dioxide insulating material, allowing contact with the N-type regions and thereby connecting the source and drain leads. The MOSFET construction is complete when a polysilicon contact (gate) is placed over the insulating material in a position to cover the channel from source to drain. **See Figure 22-24.**

IBM

Many electronic devices such as computers and TVs use amplifiers for signal processing.

Applying the principle of capacitance to the enhancement MOSFET, the gate can be used to produce a conductive channel from the source to the drain. The positive charge of a positive voltage placed on the gate induces a negative charge on the P-type material.

With increasing positive voltage, the holes in the P-type material are repelled until the region between the source and drain becomes an N-channel. Once the N-channel is forward biased between the source and gate, current begins to flow.

ENHANCEMENT MOSFET CAPACITIVE PROPERTIES

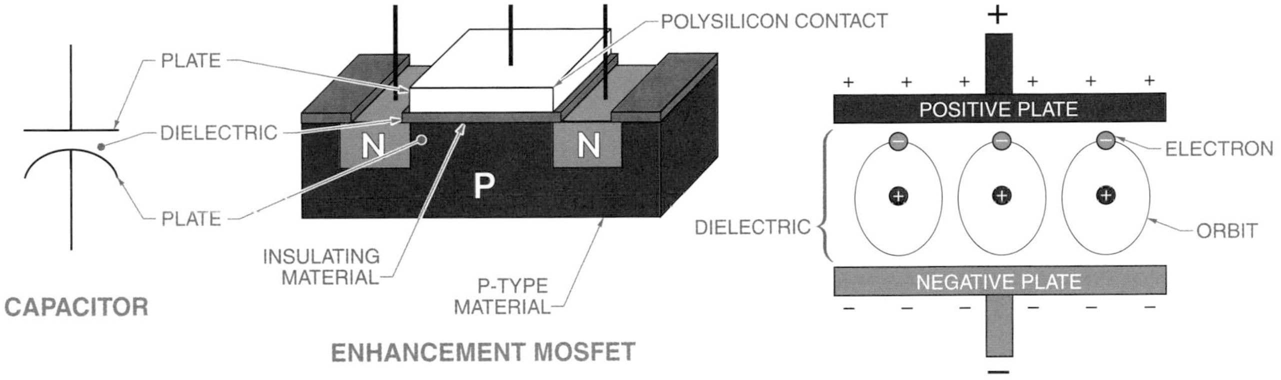

POLYSILICON CONTACT

PLATE

DIELECTRIC

PLATE

N   N

P

INSULATING
MATERIAL

P-TYPE
MATERIAL

CAPACITOR

ENHANCEMENT MOSFET

+

+ + + + + +
POSITIVE PLATE

ELECTRON

DIELECTRIC

ORBIT

NEGATIVE PLATE
− − − − − −

−

Figure 22-24. A positive voltage applied to the upper plate of a capacitor causes the dielectric to become distorted, with the electrons moving in the direction of the positive plate.

Since electrons have been added to form the N-channel, this MOSFET has enhanced current flow resulting from the application of a positive gate voltage. With a more positive gate voltage, the channel becomes wider. Therefore, current flows from source to drain due to the decreased channel resistance.

When the gate has zero voltage or a negative voltage, no enhancement effect is possible and the MOSFET does not conduct. Since the enhancement MOSFET does not conduct at zero gate voltage, it is often called a normally off MOSFET. In the schematic symbol, a broken line between terminals indicates the channel. The broken line signifies that the MOSFET is normally off. **See Figure 22-25.**

Depletion-Enhancement MOSFETs. A depletion-enhancement MOSFET is constructed similarly to the enhancement MOSFET. The main difference is the addition of a physical conducting channel between the source and drain. The presence of the channel allows current to flow from source to drain, even without a gate voltage. **See Figure 22-26.**

The depletion-enhancement MOSFET has the same capacitive effect as the enhancement MOSFET. However, when a negative voltage is applied to the gate, holes from the P-type material are attracted into the N-channel. The holes neutralize the free electrons. The result is that the N-channel is depleted, or reduced, in the number of carriers. The depletion of carriers increases channel resistance and reduces current. *Depletion mode* is the operation of a MOSFET with a negative gate voltage.

With a positive gate voltage, the N-channel MOSFET can operate in the enhancement mode. *Enhancement mode i*s the operation of a MOSFET with a positive gate voltage. The positive gate voltage widens the N-channel, causing an increase in channel current. Since the depletion-enhancement MOSFET conducts a significant current even when V_{GS} is zero, it is often called a normally on MOSFET.

Dual-Gate MOSFETs. A MOSFET can be constructed with two gates such as in a dual-gate MOSFET arrangement. Current through the MOSFET can be cut off by either gate. A MOSFET also operates on the capacitive effect. **See Figure 22-27.**

The dual-gate arrangement allows the MOSFET to be used in a variety of circuits. For example, in a gain control circuit, the audio signal is applied to Gate 1 and the gain control voltage is applied to Gate 2. The gain control voltage can then be used to control the output from Gate 2.

Power MOSFETs. Power MOSFETs exhibit the properties of small-signal MOSFETs but are designed to handle higher currents. Power MOSFETs were designed primarily for switching applications. Power MOSFETs can switch faster than bipolar transistors.

NORMALLY OFF ENHANCEMENT MOSFETs

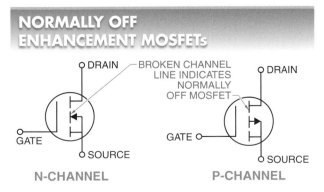

DRAIN   BROKEN CHANNEL
LINE INDICATES
NORMALLY
OFF MOSFET

DRAIN

GATE

SOURCE

GATE

SOURCE

N-CHANNEL

P-CHANNEL

Figure 22-25. The broken line in the schematic symbol indicates that the MOSFET is normally off.

DEPLETION-ENHANCEMENT MOSFETs

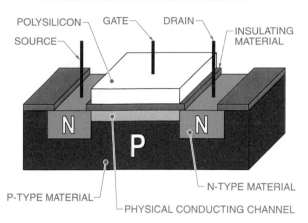

DEPLETION-ENHANCEMENT MOSFET

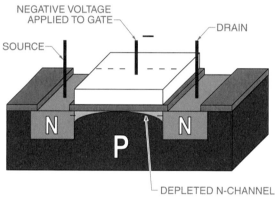

DEPLETION MODE

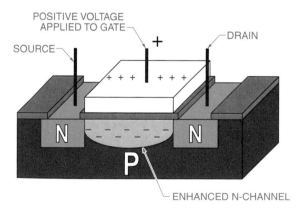

ENHANCEMENT MODE

Figure 22-26. A depletion-enhancement MOSFET has the addition of a physical conducting channel between the source and the drain. In the depletion mode, a negative voltage is applied to the gate; in the enhancement mode, a positive voltage is applied to the gate.

N-CHANNEL ENHANCEMENT MOSFETs

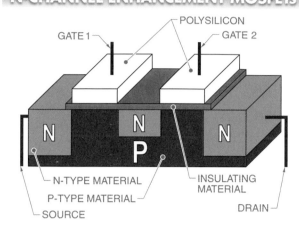

Figure 22-27. Current through a dual-gate MOSFET can be cut off by either gate.

Unlike the small-signal MOSFET, the power MOSFET is fabricated with a vertical rather than lateral structure. MOSFETs are made using the double diffused metal-oxide substrate (DMOS) process and use polysilicon gates. The gate of this device is isolated from the source by a layer of insulating silicon dioxide. When voltage is applied between the gate and source terminals, an electric field is set up within the MOSFET. This field alters the resistance between the drain and source terminals.

The DMOS power MOSFET contains an inherent PN-junction diode. Its equivalent circuit can be considered as a diode in parallel with the source-to-drain channel, as shown in the schematic symbol. **See Figure 22-28.**

DMOS ENHANCEMENT-MODE, N-CHANNEL POWER MOSFET

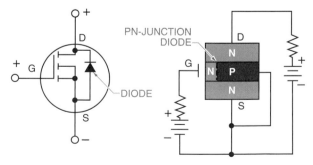

Figure 22-28. The DMOS power MOSFET contains an inherent PN-junction diode (its equivalent circuit can be considered as a diode in parallel with the source-to-drain channel).

Power MOSFETs as Switches. There are a variety of solid state switch technologies available to perform switching functions. Each switch technology, however, has strong and weak points. The ideal switch would have zero resistance in the ON state, infinite resistance in the OFF state, switch instantaneously, and would require minimum input power to make it switch. The primary characteristics that are most desirable in a solid state switch are fast switching speed, simple drive requirements, and low conduction loss. For low voltage applications, power MOSFETs offer extremely low ON resistance and approach characteristics of the desired ideal switch.

Power MOSFETs have a wide range of specifications for high-frequency switching power supplies at frequencies above 100 kHz. The advantages of using power MOSFETs over power bipolar transistors include the following:

- faster switching
- lower switching losses
- wider safe operating area (SOA)
- simple drive circuitry
- ability to be paralleled easily (the forward voltage drop increases with increasing temperature, ensuring an even distribution of current among all components)

Insulated Gate Bipolar Transistors

An *insulated gate bipolar transistor (IGBT)* is a three-terminal switching device that combines a FET for control with a bipolar transistor for switching. **See Figure 22-29.**

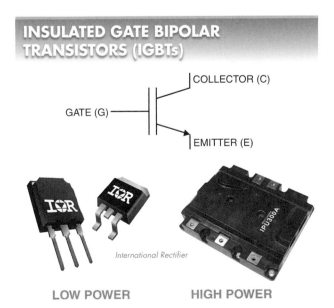

Figure 22-29. An insulated gate bipolar transistor (IGBT) is a three-terminal switching device that combines a FET with a bipolar transistor.

The main difference in construction between the power MOSFET and IGBT is the addition of an injection layer in the IGBT. Due to the presence of the injection layer, holes are injected into the highly resistive N-layer and a carrier overflow is created. This increase in conductivity of the N-layer allows the reduction of the ON-state voltage of the IGBT. **See Figure 22-30.**

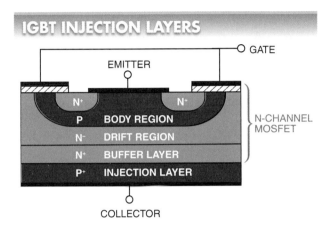

Figure 22-30. The main difference between a power MOSFET and an IGBT is the addition of an injection layer in the IGBT.

The silicon IGBT has become known as the power switch of high-voltage (greater than 500 V) and high-power (greater than 500 W) applications. The IGBT is a combination of the bipolar transistor and the MOSFET. It has the output switching and conduction characteristics of a bipolar transistor but is voltage-controlled like a MOSFET. This means it has the advantage of the high-current handling capability of a bipolar transistor with the ease of control of a MOSFET.

The IGBT is a power semiconductor device, noted for high efficiency and fast switching. The decision of whether to use an IGBT or MOSFET depends on the application. Cost, size, speed, and environmental requirements should all be considered when selecting an IGBT.

IGBT Operations

IGBTs are fast switching devices. IGBT operations consist of blocking, ON/OFF state, and latch-up operations. The safe operating area (SOA) of an IGBT protects against inductive shutoff. The two IGBT configurations include the punch-through (PT) and non-punch-through (NPT) configurations.

Blocking Operations. The ON/OFF state of an IGBT is determined by the gate voltage. If the voltage applied to the gate contact, with respect to the emitter, is less than the threshold voltage, then no MOSFET inversion layer is created and the device is turned off. When this is the case, any applied forward voltage should fall across the reverse-biased junction. The only current flow should be a small leakage current.

The forward breakdown voltage is, therefore, determined by the breakdown voltage of the junction. This is important for power devices with large voltages and currents. The breakdown voltage of the junction is dependent on the doping. A lower doping ratio results in a wider depletion region and a lower maximum electric field in the depletion region, which is why the drift region (N⁻) is doped much lighter than the body region.

The buffer layer (N⁺) is present to prevent the depletion region from extending into the bipolar collector. The benefit of this buffer layer is that it allows the thickness of the drift region to be reduced, thus reducing ON-state losses.

ON-State Operations. The ON-state of the IGBT is achieved by increasing the gate voltage so that it is greater than the threshold voltage. That increase in voltage results in an inversion layer forming under the gate, which provides a channel linking the source to the drift region of the IGBT. Electrons are then injected from the source into the drift region. At the same time, holes are injected into the drift region.

This injection causes conductivity of the drift region, where both the electron and hole densities are higher than the original N⁻ doping. This conductivity gives the IGBT its low ON-state voltage. This is possible because of the reduced resistance of the drift region. Some of the injected holes should recombine in the drift region. Others may cross the region by drift and diffusion and reach the junction of the body region where they are collected.

OFF-State Operations. Either the gate must be shorted to the emitter or a negative bias must be applied to the gate. When the gate voltage falls below the threshold voltage, the inversion layer cannot be maintained, and the supply of electrons into the drift region is blocked. At this point, the shutoff process begins. The shutoff process cannot be completed as quickly as desired due to the high concentration of minority carriers injected into the drift region during forward conduction. The collector current rapidly decreases due to the termination of the electron current through the channel. Then the collector current is reduced as the minority carriers recombine.

Latch-Up Operations. During ON-state operation, paths for current to flow in an IGBT allow holes to be injected into the drift region from the collector (P⁺). Parts of the holes disappear by recombination with electrons from the MOSFET channel. Other parts of the holes are attracted to the vicinity of the injection layer by the negative charge of electrons. These holes cross the body region and develop a voltage drop in the resistance of the body.

Once in a latch-up condition, the MOSFET gate has no control over the collector current. The only way to shut off the IGBT is to shut off the current, just as for a conventional SCR. *Note:* If latch-up is not terminated quickly, the IGBT may be destroyed by the excessive power dissipation.

Trains use IGBTs for power electronics because of their high efficiency and fast switching.

IGBT Safe Operating Area. The safe operating area (SOA) is the current-voltage limit in which a power switching device like an IGBT can be operated without being destroyed. The area is defined by the maximum collector-emitter voltage (V_{CE}) and collector current (I_C) the IGBT operation must control to protect the IGBT from damage. The types of SOAs for IGBTs are the forward-biased safe operating area (FBSOA), reverse-biased safe operating area (RBSOA), and short-circuit safe operating area (SCSOA). The two primary conditions that could affect the SOA of an IGBT are operation during a short circuit and inductive shutoff.

Protection must be in place when switching inductive loads. This can be done by the use of regular diodes, zener diodes, or resistors. The method used depends on the application. IGBTs often need this type of protection from inductive loads to prevent inductive shutoff.

Tech Fact

IGBTs that are used in variable frequency drives as the motor switching devices will sometimes fail. Failure can be reduced by using line and load reactors, limiting the distance between drive and motor, allowing enough open space for drive cooling, increasing motor acceleration/deceleration times, lowering the PWM frequency setting, ensuring proper grounding, and not loading the drive more than 80% of its rating.

PT and NPT Configurations. The two types of structures used for IGBT construction are the punch-through (PT) structure and the non-punch-through (NPT) structure. An IGBT is called a PT IGBT when there is a buffer layer (N⁺) between the injection layer (P⁺) and the drift region (N⁻). Otherwise, it is called an NPT IGBT. The buffer layer improves shutoff speed by reducing the minority-carrier injection quantity and raising the recombination rate during the switching transition. The PT IGBT has similar characteristics as the NPT IGBT for switching speed and forward voltage drop. Currently, most commercialized IGBTs are PT IGBTs.

22-3 CHECKPOINT

1. In the symbol for a JFET, if the arrow on the gate is pointing out of the device, is the JFET an N-channel or P-channel type?

2. If an output load is connected to an N-Channel JFET, the load would be connected to the source gate and which polarity of the power supply?

3. What type of MOSFETs are designed to switch higher current loads?

4. Are IGBTs available to switch high voltages and power such as a 480 V, 450 W load?

22-4 TROUBLESHOOTING INSULATED GATE BIPOLAR TRANSISTORS

To troubleshoot an IGBT, the gate-to-emitter resistance and the collector-to-emitter resistance should be measured. The leakage resistance measurements are used to determine whether the resistance is too low and the device is defective.

Gate-to-Emitter Resistance

The leakage resistance is measured between the gate (G) and emitter (E) with a jumper between the collector (C) and emitter (E). **See Figure 22-31.** When a digital multimeter (DMM) is used, the internal battery voltage should be verified to be less than 20 V. A high voltage can damage the IGBT. The resistance reading for an IGBT in good condition ranges from several megohms (MΩ) to infinity. The IGBT is considered defective if the resistance between the gate and emitter is very low.

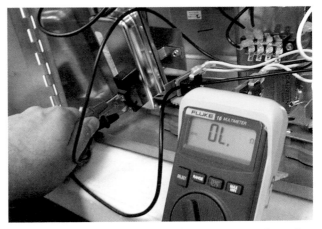

DMMs are used to check the solid-state components in modern HVAC systems.

TESTING IGBT GATE-TO-EMITTER RESISTANCE

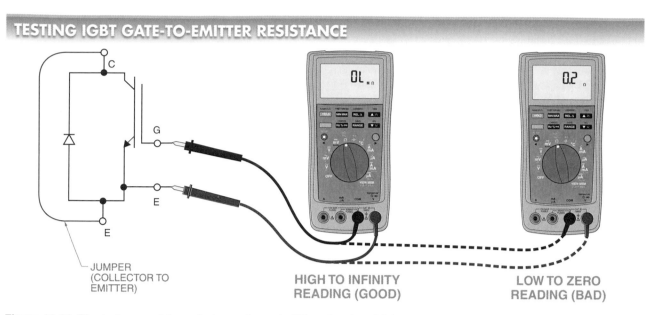

Figure 22-31. The leakage resistance between the gate (G) and emitter (E) is measured with the collector (C) and emitter (E) shorted to each other.

Collector-to-Emitter Resistance

The leakage resistance between the collector (C) and emitter (E) is measured with a jumper between the gate (G) and emitter (E). **See Figure 22-32.** The collector should be connected to the positive (red) test lead and the emitter to the negative (black) test lead. The DMM resistance reading should range from several megohms to infinity. The IGBT is considered defective if the resistance between the collector and emitter is very low. There are many advantages of IGBTs. The main advantages of an IGBT are the following:

- low ON-state voltage drop
- possibility of small chip size
- low driving power
- simple drive circuit
- wide SOA

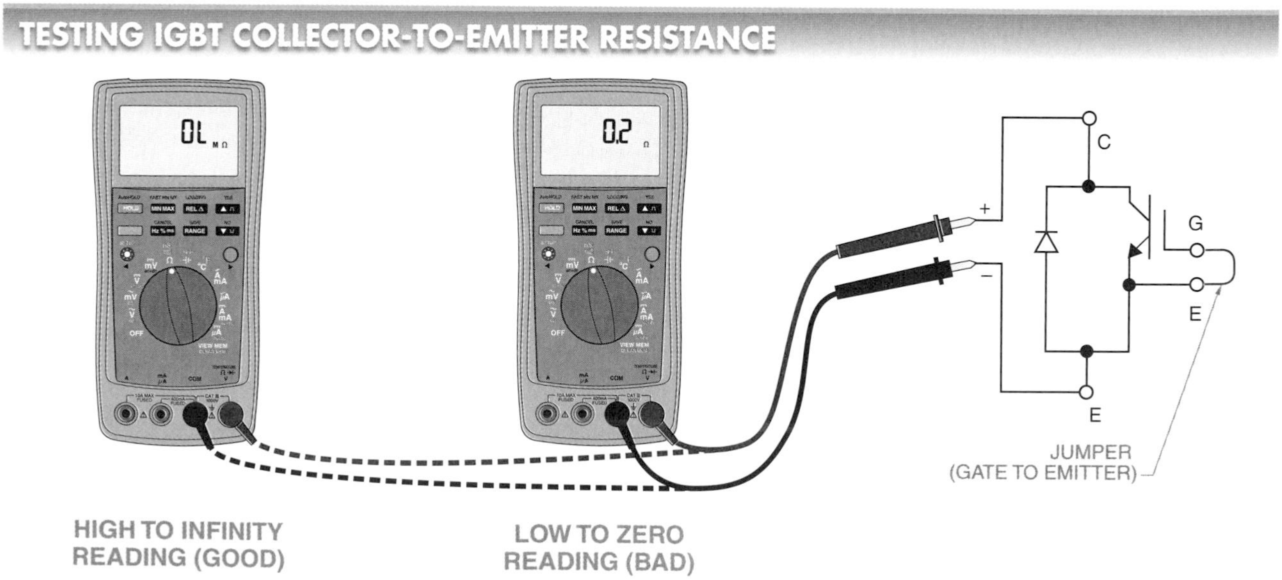

TESTING IGBT COLLECTOR-TO-EMITTER RESISTANCE

HIGH TO INFINITY READING (GOOD)

LOW TO ZERO READING (BAD)

JUMPER (GATE TO EMITTER)

Figure 22-32. The resistance between the collector (C) and emitter (E) is measured with a jumper between the gate (G) and emitter (E).

Chapter 22—Semiconductor Amplification and Switching   479

IGBTs also include disadvantages. One disadvantage of an IGBT is that the switching speed is inferior to that of a power MOSFET. The tailing of the collector current due to the minority carrier causes the shutoff speed to be slow. However, the switching speed of an IGBT is superior to that of a BJT. Another disadvantage of the IGBT is the possibility of latch-up due to the internal PNPN thyristor structure. Problems with IGBTs and other solid-state switches can be reduced by keeping their heat sinks clean for proper airflow movement.

22-4 CHECKPOINT

1. When measuring the gate-to-emitter resistance of an IGBT, is a high or low resistance expected if the device is good?
2. When measuring the collector-to-emitter resistance of an IGBT, is a high or low resistance expected if the device is good?

Additional Resources

Review and Resources

Access Chapter 22 Review and Resources for *Electrical Motor Controls for Integrated Systems* by scanning the above QR code with your mobile device.

Applying Your Knowledge

Refer to the *Electrical Motor Controls for Integrated Systems* Learner Resources for interactive Applying Your Knowledge questions.

Workbook and Applications Manual

Refer to Chapter 22 in the *Electrical Motor Controls for Integrated Systems Workbook* and the *Applications Manual* for additional exercises.

ENERGY EFFICIENCY PRACTICES

Energy-Efficient Transistors

Transistors made from silicon have been used for decades in almost every electronic device. The use of transistors drastically reduced the space required by vacuum tubes and greatly reduced the required operating power. Although transistors consume little power, they still use power to operate. As electronic devices become smaller and require more circuits, power becomes an issue for every component that is used.

New transistors made from germanium or other materials can operate at faster speeds, use less power, produce less heat, and require less space than silicon transistors. Transistors that require less space fit well into the trend of smaller and lighter electronic devices like computers and plate-screen TVs. The use of less energy increases battery life and reduces operating cost. Smaller, lower power transistors will continue to provide new possibilities in the electronics field for years to come.

Objectives

23-1
- Describe solid-state switches.
- Define and describe transistors.
- Explain how to test transistors.

23-2
- Define and describe silicon-controlled rectifiers (SCRs).
- Describe how an SCR operates.
- Explain how an SCR is used to control DC motor base speed.
- Explain how to test an SCR.

23-3
- Define and describe triacs.
- Describe how a triac operates.
- Explain how to test a triac.

23-4
- Define diac.
- Describe how a diac operates.
- Explain how to test a diac.

23-5
- Define unijunction transistor (UJT).
- Describe how a UJT operates.

Sections

23-1 Solid-State Switches

23-2 Silicon-Controlled Rectifiers

23-3 Triacs

23-4 Diacs

23-5 Unijunction Transistors

Review and Resources
atplearningresources.com/Quicklinks
Access Code: 362245

Chapter

Semiconductor Power Switching Devices

23

Mechanical switches have been used to switch AC and DC loads and circuits since switches were first used to control electrical circuits and loads. Mechanical switches have the advantages of being able to switch either AC or DC, having only two operating conditions (open or closed), and being easy to understand and troubleshoot. Mechanical switches have the disadvantages of having a much shorter operating life than solid-state switches and producing arcing at the contacts, which can be dangerous in some applications.

Solid-state switches such as silicon-controlled rectifiers (SCRs) and triacs can replace mechanical switches. Solid-state switches have a much longer operating life, can control the amount of voltage/current between being totally open or closed, and do not produce arcing since there are no contacts. Solid-state switches have the disadvantages of being able to switch only AC or DC, being harder to understand, and requiring more knowledge of the circuit and components when a technician troubleshoots them. It is important to understand solid-state switches since they are being used more often to replace mechanical switches in circuits and switching applications.

23-1 SOLID-STATE SWITCHES

Solid-state switches are electronic devices that have no moving parts (contacts). Solid-state switches can be used in most motor control applications. Advantages of solid-state switches include fast switching, no moving parts, long life, and the ability to be interfaced with electronic circuits (PLCs and PCs). However, solid-state switches must be properly selected and applied to prevent potential problems. Solid-state switches include transistors, silicon-controlled rectifiers (SCRs), triacs, diacs, and unijunction transistors (UJTs). **See Figure 23-1.**

Triacs, diacs, and unijunction transistors (UJTs), along with SCRs, are often found in the same circuitry. Triacs and SCRs are control devices. Diacs and UJTs form the triggering circuits for triacs and SCRs. Triacs, diacs, UJTs, and SCRs operate only as switches and may be used in a variety of switching applications.

Tech Fact

Solid-state switches have a switch life in the billions of cycles while mechanical switches have a switch life of about 200,000 cycles, making solid-state switches the standard for most high-volume switching applications.

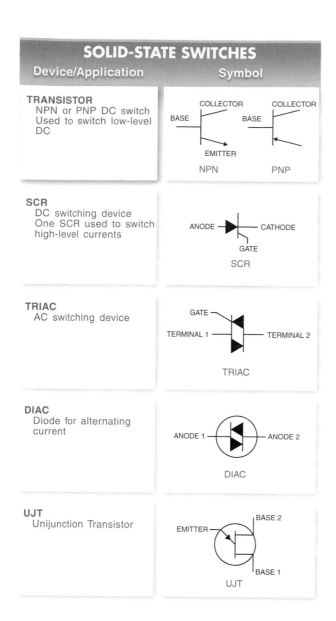

SOLID-STATE SWITCHES

Device/Application	Symbol
TRANSISTOR NPN or PNP DC switch Used to switch low-level DC	COLLECTOR / COLLECTOR, BASE / BASE, EMITTER — NPN / PNP
SCR DC switching device One SCR used to switch high-level currents	ANODE — CATHODE, GATE — SCR
TRIAC AC switching device	GATE, TERMINAL 1 — TERMINAL 2 — TRIAC
DIAC Diode for alternating current	ANODE 1 — ANODE 2 — DIAC
UJT Unijunction Transistor	EMITTER, BASE 2, BASE 1 — UJT

Figure 23-1. Solid-state switches include transistors, silicon-controlled rectifiers (SCRs), triacs, diacs, and unijunction transistors (UJTs).

Transistors

A *transistor* is a three-terminal device that controls current through the device depending on the amount of voltage applied to the base. Transistors may be NPN or PNP transistors. Transistors can be switched on and off quickly. Transistors have a very high resistance when open and a very low resistance when closed. Transistors are used to switch low-level DC only. When transistors are used as switches, a diode can be mounted across the transistor to prevent damage from high-voltage spikes (transients).

Transistors as DC Switches

Transistors were mainly developed to replace mechanical switches. Transistors have no moving parts and can switch on and off quickly. Mechanical switches have two conditions: open and closed or ON and OFF. Mechanical switches have a very high resistance when open and a very low resistance when closed.

A transistor can be made to operate like a switch. For example, a transistor can be used to turn on or off a pilot light. **See Figure 23-2.** In this circuit, the resistance between the collector (C) and the emitter (E) is determined by the current flow between the base (B) and emitter (E). When no current flows between B and E, the collector-to-emitter resistance is high, like that of an open switch. The pilot light does not glow because there is no current flow.

TRANSISTORS AS DC SWITCHES

SUPPLY VOLTAGE V_{CC}

NO BASE/EMITTER CURRENT FLOW

PILOT LIGHT OFF

NO COLLECTOR/EMITTER CURRENT FLOW

SUPPLY VOLTAGE V_{CC}

PILOT LIGHT ON

BASE/EMITTER CURRENT PRESENT I_B

COLLECTOR/EMITTER CURRENT PRESENT

Figure 23-2. A transistor can be made to operate like a switch.

If a small current flows between B and E, the collector-to-emitter resistance is reduced to a very low value, like that of a closed switch. The pilot light is switched on. A transistor switched on is normally operating in the saturation region. The *saturation region* is the maximum current that can flow in a transistor circuit. At saturation, the collector resistance is considered zero and the current is limited only by the resistance of the load.

When the circuit reaches saturation, the resistance of the pilot light is the only current-limiting device in the circuit. When the transistor is switched off, it is operating in the cutoff region. The *cutoff region* is the point at which the transistor is turned off and no current flows. At cutoff, all the voltage is across the open switch (transistor) and the collector-emitter voltage is equal to the supply voltage V_{CC}.

Transistor Applications

Transistors are used for switching because of their reliability and speed. In certain situations, transistors are also integrated with other solid-state components to form more complex devices. In each application, however, the fundamental operating principle of the transistor remains the same.

Seven-Segment Displays. By switching various combinations of transistors on or off, different numbers can be created on a seven-segment display. **See Figure 23-3.** For example, if all transistors (A through G) are switched on, an "8" should appear on the display. If all transistors, except E and D, are switched on, a "9" should appear. There is typically circuitry in addition to the seven-segment transistor devices to help decode the proper signals for the display. When all circuitry is present, it is called a seven-segment decoder/driver display or readout device.

SEVEN-SEGMENT DISPLAYS

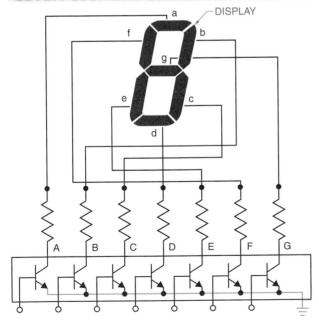

Figure 23-3. When various combinations of transistors are switched on and off, different numbers appear on a seven-segment display.

Testing Transistors. A transistor becomes defective from excessive current or temperature. A transistor normally fails due to an open or shorted junction. The two junctions of a transistor may be tested with a DMM set to measure resistance. **See Figure 23-4.** To test an NPN transistor for an open or shorted junction, the following procedure is applied:

1. Connect a DMM to the emitter and base of the transistor. Measure the resistance.

2. Reverse the DMM leads and measure the resistance. The emitter/base junction is good when the resistance is high in one direction and low in the opposite direction. *Note:* The ratio of high to low resistance should be greater than 100:1. Typical resistance values are 1 kΩ with the positive lead of the DMM on the base and 100 kΩ with the positive lead of the DMM on the emitter. The junction is shorted when both readings are low. The junction is open when both readings are high.

3. Connect the DMM to the collector and base of the transistor. Measure the resistance.

4. Reverse the DMM leads and measure the resistance. The collector/base junction is good when the resistance is high in one direction and low in the opposite direction. *Note:* The ratio of high to low resistance should be greater than 100:1. Typical resistance values are 1 kΩ with the positive lead of the DMM on the base and 100 kΩ with the positive lead of the DMM on the collector.

5. Connect the DMM to the collector and emitter of the transistor. Measure the resistance.

6. Reverse the DMM leads and measure the resistance. The collector/emitter junction is good when the resistance reading is high in both directions.

Tech Fact

Transistors were developed at Bell Labs (now AT&T) in 1947 to control electronic amplification and switch circuits. In the 1950s and 1960s, U.S. companies used transistors mostly for military applications. The Sony Electronic company was successful in using mass-produced transistors in small transistor radios. Eventually, inexpensive, mass-produced transistors eliminated the need for vacuum tubes in televisions and most electronic devices.

The same test used for an NPN transistor can be used for testing a PNP transistor. The difference is that the DMM test leads must be reversed to obtain the same results.

TESTING TRANSISTORS

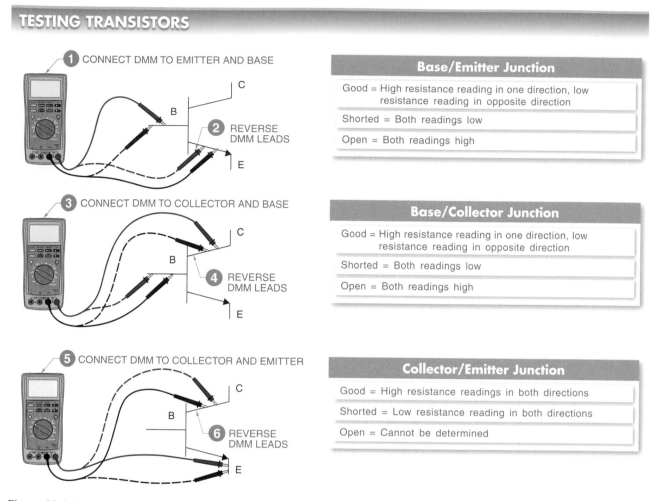

① CONNECT DMM TO EMITTER AND BASE

② REVERSE DMM LEADS

Base/Emitter Junction
Good = High resistance reading in one direction, low resistance reading in opposite direction
Shorted = Both readings low
Open = Both readings high

③ CONNECT DMM TO COLLECTOR AND BASE

④ REVERSE DMM LEADS

Base/Collector Junction
Good = High resistance reading in one direction, low resistance reading in opposite direction
Shorted = Both readings low
Open = Both readings high

⑤ CONNECT DMM TO COLLECTOR AND EMITTER

⑥ REVERSE DMM LEADS

Collector/Emitter Junction
Good = High resistance readings in both directions
Shorted = Low resistance reading in both directions
Open = Cannot be determined

Figure 23-4. A transistor normally fails due to an open or shorted junction.

23-1 CHECKPOINT

1. Can a transistor be used to switch AC?
2. What two terminals (base, emitter, or collector) make up the switching part of a transistor?
3. In Figure 23-3, which type of transistors are used, NPN or PNP?
4. When troubleshooting a transistor with a DMM set to measure resistance, what are the two possible failure conditions that can be determined?

23-2 SILICON-CONTROLLED RECTIFIERS

A *silicon-controlled rectifier (SCR)* is a four-layer (PNPN) semiconductor device that uses three electrodes for normal operation. **See Figure 23-5.** The three electrodes are the anode, cathode, and gate.

The anode and cathode of an SCR are similar to the anode and cathode of an ordinary diode. The gate serves as the control point for an SCR. SCRs are also known as thyristors.

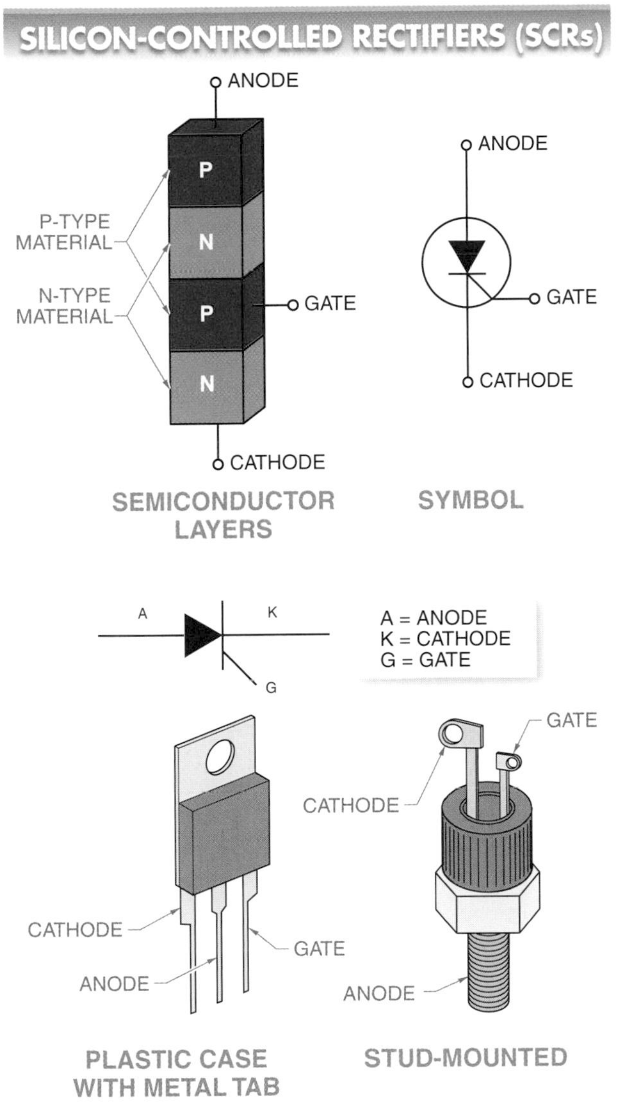

SILICON-CONTROLLED RECTIFIERS (SCRs)

ANODE

P — P-TYPE MATERIAL
N
P — N-TYPE MATERIAL
N

GATE

CATHODE

SEMICONDUCTOR LAYERS

ANODE

GATE

CATHODE

SYMBOL

A = ANODE
K = CATHODE
G = GATE

A ── K
G

GATE
CATHODE
ANODE

PLASTIC CASE WITH METAL TAB

GATE
CATHODE
ANODE

STUD-MOUNTED

Figure 23-5. A silicon-controlled rectifier (SCR) has three electrodes called the anode, cathode, and gate.

An SCR differs from an ordinary diode in that it does not pass significant current, even when forward biased, unless the anode voltage equals or exceeds the forward breakover voltage. However, when forward breakover voltage is reached, the SCR switches on and becomes highly conductive. The gate current is used to reduce the level of breakover voltage necessary for the SCR to conduct or fire.

Low-current SCRs can operate with an anode current of less than 1 A. High-current SCRs can handle load currents in the hundreds of amperes. The size of an SCR increases with an increase in current rating. **See Figure 23-6.**

SCR PACKAGES

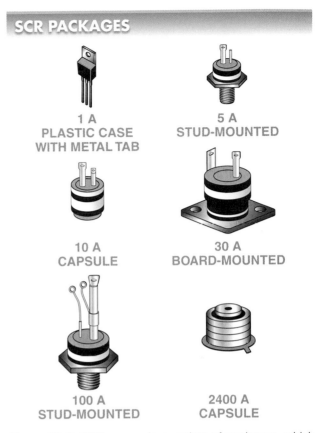

1 A
PLASTIC CASE WITH METAL TAB

5 A
STUD-MOUNTED

10 A
CAPSULE

30 A
BOARD-MOUNTED

100 A
STUD-MOUNTED

2400 A
CAPSULE

Figure 23-6. SCRs come in a variety of packages, which increase with an increase in current rating.

SCRs have voltage ratings of up to 2500 V and current ratings of up to 3000 A. SCRs are used in power switching, phase control, battery charger, and inverter circuits. In industrial applications, they are applied to produce variable DC voltages for motors (from a few to several thousand horsepower) from AC line voltage. They can also be used in some electric vehicles. SCRs may be used alone in a circuit to provide one-way current control or wired in reverse-parallel circuits to control AC line current in both directions. **See Figure 23-7.**

SCR CIRCUITS

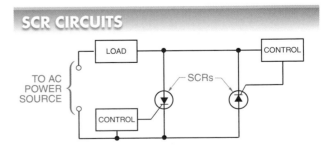

LOAD
CONTROL
TO AC POWER SOURCE
SCRs
CONTROL

Figure 23-7. SCRs may be used alone in a circuit to provide one-way current control or wired in reverse-parallel circuits to control AC line current in both directions.

Characteristic Curves

A voltage-current characteristic curve shows the operating characteristics of an SCR when its gate is not connected. **See Figure 23-8.** When an SCR is reverse biased, it operates similar to a regular semiconductor diode.

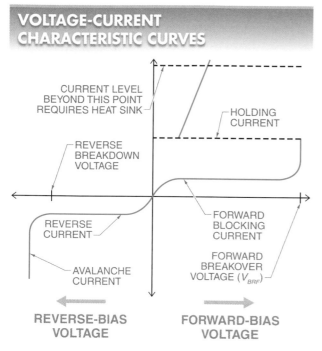

VOLTAGE-CURRENT CHARACTERISTIC CURVES

CURRENT LEVEL BEYOND THIS POINT REQUIRES HEAT SINK

HOLDING CURRENT

REVERSE BREAKDOWN VOLTAGE

REVERSE CURRENT

AVALANCHE CURRENT

FORWARD BLOCKING CURRENT

FORWARD BREAKOVER VOLTAGE (V_{BRF})

REVERSE-BIAS VOLTAGE

FORWARD-BIAS VOLTAGE

Figure 23-8. The voltage-current characteristic curve of an SCR shows how the SCR operates.

When an SCR is forward biased, there is a small amount of forward leakage current called forward blocking current. This current stays relatively constant until the forward breakover voltage is reached. At that point, the current increases rapidly. The region is often called the forward avalanche region. In the forward avalanche region, the resistance of the SCR is very small. The SCR operates similar to a closed switch, and the current is limited only by the external load resistance.

An SCR is either ON or OFF. When an applied voltage is above the forward breakover voltage (V_{BRF}), the SCR fires or turns on. The SCR remains on as long as the current stays above the holding current. When voltage across the SCR drops to a value too low to maintain the holding current, it returns to its OFF state.

Gate Control of Forward Breakover Voltage. When the gate is forward biased and current begins to flow in the gate-cathode junction, the value of forward breakover

voltage can be reduced. Increasing values of forward bias can be used to reduce the amount of forward breakover voltage necessary to get the SCR to conduct.

Once the SCR has been turned on by the gate current, the gate current loses control of the SCR forward current. Even if the gate current is completely removed, the SCR remains on until the anode voltage has been removed. The SCR also remains on until the anode voltage has been significantly reduced to a level where the current is not large enough to maintain the proper level of holding current.

Triggering Methods

An SCR normally can be triggered into conduction by applying a pulse of control current to the gate. Once turned on, an SCR remains on as long as there is a minimum level of holding current flowing through the load circuit. Once the current drops below the holding current value, the SCR turns off. **See Figure 23-9.** The correct method of turning on an SCR is to apply a proper signal to the gate of the SCR (gate turn-on).

Phase Control

The ability of an SCR to turn on at different points in the conducting cycle can be useful for varying the amount of power delivered to a load. This type of variable control is called phase control. *Phase control* is the control of the time relationship between two events when dealing with voltage and current. In this case, it is the time relationship between the trigger pulse and the point in the conducting cycle when the pulse occurs. With phase control, the speed of a motor, the brightness of a lamp, and the output of an electric-resistance heating unit can be controlled.

The Lincoln Electric Company
SCRs are used to control the power in gas tungsten arc welding machines and other medium- to high-voltage AC power control applications.

SCR TRIGGERING METHODS

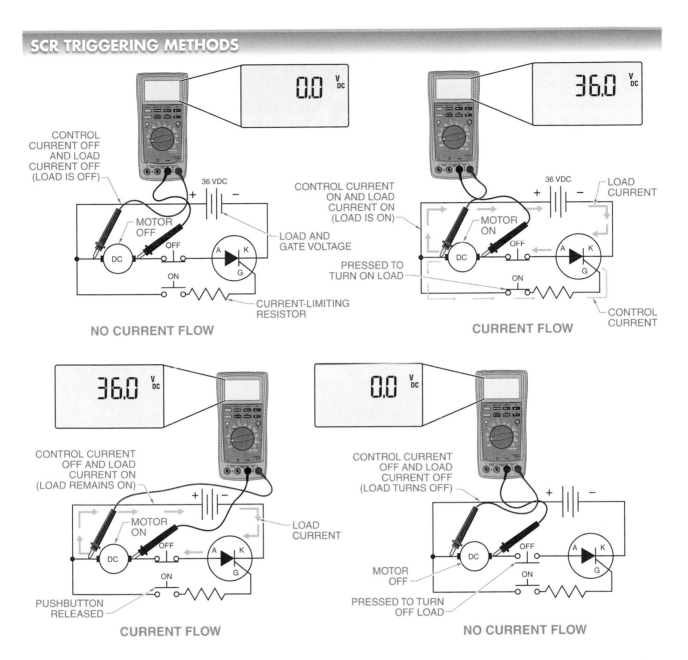

Figure 23-9. An SCR is turned on by applying a pulse of control current to the gate. Once turned on, an SCR remains on as long as there is a minimum level of holding current flowing through the load circuit.

The most basic control circuit using this principle is a half-wave phase control circuit. A *half-wave phase control circuit* is an SCR that has the ability to turn on at different points of the conducting cycle of a half-wave rectifier. **See Figure 23-10.** In this circuit, the input voltage is a standard 60 Hz line voltage.

When a negative half cycle is applied to the circuit, the SCR is reverse biased and should not conduct.

In addition, diode D1 is reverse biased and no gate current flows. Diode D2, however, is forward biased and allows capacitor C to charge to the polarity shown. No current flows through R1 since D2 is in parallel with R1. The forward-biased resistance of D2 is so much lower than R1 that almost all current passes through D2 instead of through R1. Since no current is flowing through the load, the oscilloscope should not indicate a voltage drop.

HALF-WAVE PHASE CONTROL CIRCUITS

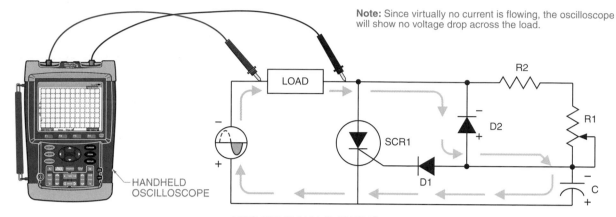

Note: Since virtually no current is flowing, the oscilloscope will show no voltage drop across the load.

HANDHELD OSCILLOSCOPE

NEGATIVE HALF CYCLE

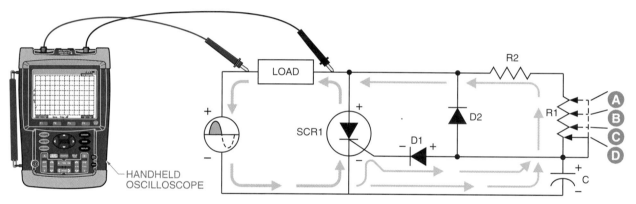

HANDHELD OSCILLOSCOPE

Note: The oscilloscope shows voltage applied to the load through the SCR for different RC time constants.

Note: C had opposite polarity on negative half cycle. C must totally discharge through R1 and R2 before it can recharge. Once C is discharged, D1 may begin conducting.

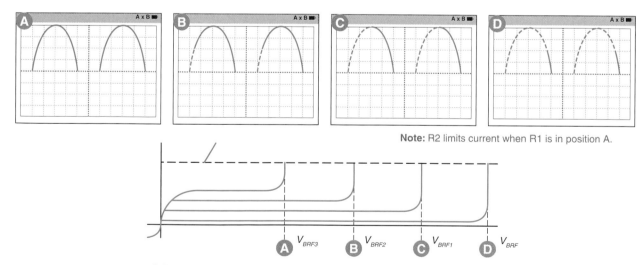

Note: R2 limits current when R1 is in position A.

SCR Characteristic Curves for Different RC Time Constants
POSITIVE HALF CYCLE

Figure 23-10. On the negative half cycle, SCR1 is reverse biased and does not conduct. On the positive half cycle of the AC input, the SCR is forward biased such that gate current will flow through diode D1 and variable resistor R1, causing the SCR to conduct.

Note: The oscilloscope on the negative half cycle should show only a slight movement of the trace. This is because the SCR is reverse biased and the current drawn through diode D2, when charging capacitor C, is small.

When the positive half cycle of the voltage source is applied, the circuit operation changes. When the positive half cycle is applied to the circuit, the SCR becomes forward biased in such a way that it conducts forward current if a gate current of sufficient strength is present. The important factor is the amount of gate current. If the gate current is too small, the SCR does not fire. The amount of gate current in this circuit is controlled by variable resistor R1 and capacitor C. Resistor R1 and capacitor C form an RC network that determines the time of charge and discharge of the capacitor.

For capacitor C to charge on the positive half cycle, it must first discharge the polarities it received on the negative half cycle. Once discharged, it may recharge with the opposite polarity. The charging and discharging rates are determined by resistor R1. The discharge and charge time of the RC network determine the time sufficient current is allowed to flow through the gate circuit to fire the SCR.

If variable resistor R1 is in position A or has no resistance, the capacitor will discharge its reverse polarity almost immediately. Current would then be allowed to pass through D1, which in turn fires the SCR. The result is that the entire positive half cycle is allowed to pass through the SCR to the load as shown by the oscilloscope display pattern correlated to position A.

If variable resistor R1 is moved to position B, where more resistance is placed in the RC network, the time of discharge for C is increased. The result is a delay in time for the gate current to reach its necessary value. Also, part of the positive half cycle is blocked as shown by the oscilloscope display pattern correlated to position B.

As more resistance is added in positions C and D, the time delay is increased. The result is that the positive half cycle is increasingly blocked while a decreasing amount of current is delivered to the load. Thus, the SCR can then deliver a varied output to the load on the positive half cycle. *Note:* To control the output in both the positive and negative half cycles, two SCRs must be used.

Controlling DC Motor Base Speed. DC motor control is one of the most suitable industrial applications of SCRs. In DC motor control applications, an SCR can be used to control the speed of a DC motor below the base speed by changing the amount of current that flows through the armature circuit.

Speed below the base speed is controlled by changing the armature voltage. *Base speed* is the speed (in rpm) at which a DC motor runs with full-line voltage applied to the armature and field. Base speed is listed on the motor nameplate.

The speed of a DC motor is controlled by varying the applied voltage across the armature and/or field. When armature voltage is controlled, the motor delivers a constant torque characteristic. When field voltage is controlled, the motor delivers a constant horsepower characteristic. **See Figure 23-11.**

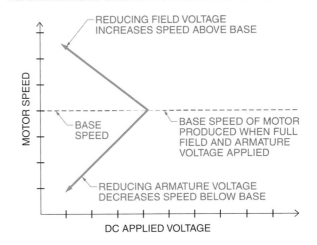

Figure 23-11. The speed of a DC motor is controlled by varying the applied voltage across the armature and/or field.

When an SCR is used to control the speed of a DC motor, the speed can be controlled from 0 rpm to the base speed. The SCR is controlled by the setting of the gate trigger circuit, which varies the on time of the SCR per cycle. This varies the amount of average current flow to the armature. **See Figure 23-12.**

The voltage applied to the SCR is AC voltage because the SCR rectifies (as well as controls) AC voltage. A rectifier circuit is required for the field circuit because the field circuit must be supplied with DC. If speed control above the base is required, the rectifier circuit in the field can also be changed to an SCR control.

Stud-Mounted SCRs. Stud-mounted SCRs are installed using the same procedure used for mounting stud-mounted diodes. Manufacturer specifications should be followed for SCR installation. Typically, stud-mounted SCRs are fastened to heat sinks to ensure good heat flow. **See Figure 23-13.**

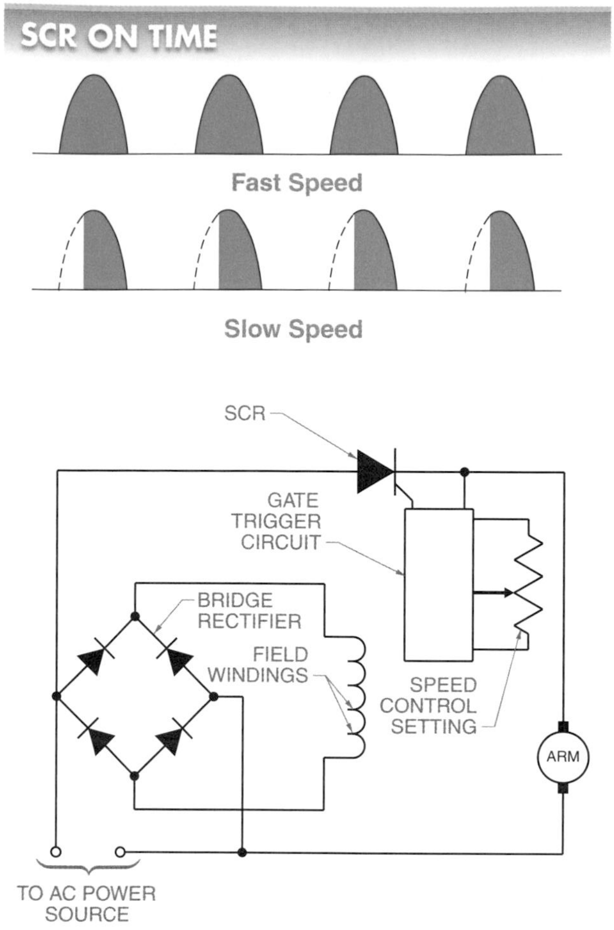

SCR ON TIME

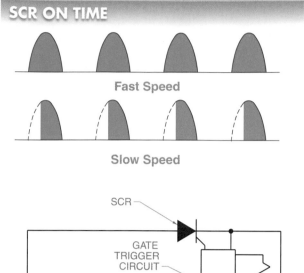

Fast Speed

Slow Speed

SCR

GATE TRIGGER CIRCUIT

BRIDGE RECTIFIER

FIELD WINDINGS

SPEED CONTROL SETTING

ARM

TO AC POWER SOURCE

Figure 23-12. An SCR is used to control the speed of a DC motor.

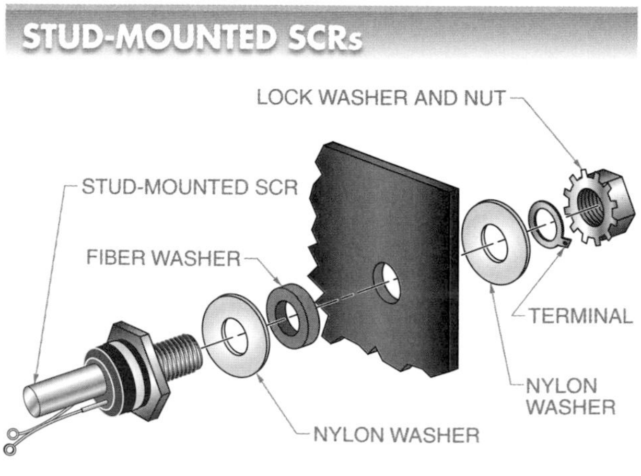

STUD-MOUNTED SCRs

LOCK WASHER AND NUT

STUD-MOUNTED SCR

FIBER WASHER

TERMINAL

NYLON WASHER

NYLON WASHER

Figure 23-13. SCRs, like regular silicon diodes, can be stud-mounted.

Testing SCRs

An oscilloscope is needed to properly test an SCR under operating conditions. A rough test using a test circuit can be made using a DMM. **See Figure 23-14.** To test an SCR using a DMM, the following procedure is applied:

1. Set the DMM on the Ω scale.
2. Connect the negative lead of the DMM to the cathode.
3. Connect the positive lead of the DMM to the anode. The DMM should read infinity.
4. Short-circuit the gate to the anode using a jumper wire. The DMM should read almost 0 Ω. Remove the jumper wire. The low resistance reading should remain.
5. Reverse the DMM leads so that the positive lead is on the cathode and the negative lead is on the anode. The DMM should read almost infinity.
6. Short-circuit the gate to the anode using a jumper wire. The resistance on the DMM should remain high.

TESTING SCRs

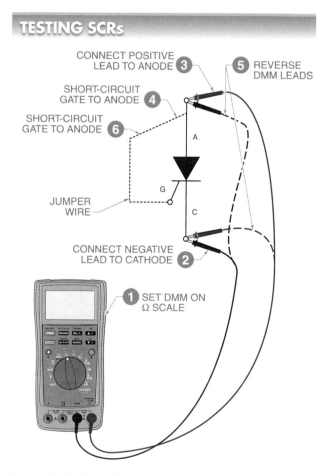

CONNECT POSITIVE LEAD TO ANODE **3**

5 REVERSE DMM LEADS

SHORT-CIRCUIT GATE TO ANODE **4**

SHORT-CIRCUIT GATE TO ANODE **6**

A

JUMPER WIRE

G

C

CONNECT NEGATIVE LEAD TO CATHODE **2**

1 SET DMM ON Ω SCALE

Figure 23-14. A rough test using a test circuit can be made on an SCR using a DMM.

23-2 CHECKPOINT

1. When switching DC loads, what is the advantage of using an SCR over a transistor?
2. Can only one SCR be used to switch an AC load?
3. Once a voltage applied to the gate of an SCR turns on the SCR and load, must the gate voltage remain on to keep the SCR and load on?

4. When an SCR is used to vary the voltage to a load, is the voltage/current on the gate increased or decreased to increase the applied voltage to the load?

23-3 TRIACS

A *triac* is a three-electrode, bidirectional AC switch that allows electrons to flow in either direction. The word "triac" is derived from the phrase "triode for alternating current." Triacs are the equivalent of two SCRs connected in a reverse-parallel arrangement with gates also connected to each other.

A triac is triggered into conduction in both directions by a gate signal in a manner similar to that of an SCR. Triacs were developed to provide a means for producing improved controls for AC power. Triacs are available in a variety of packaging arrangements. They can handle a wide range of current and voltage. Triacs generally have relatively low-current capabilities compared to SCRs. Triacs are usually limited to less than 50 A and cannot replace SCRs in high-current applications.

Triacs are considered versatile because of their ability to operate with positive or negative voltages across their terminals. Since SCRs have a disadvantage of conducting current in only one direction, controlling low power in an AC circuit is better served with the use of a triac.

Triac Construction

Although triacs and SCRs look similar, their schematic symbols are different. The terminals of a triac are the gate, terminal 1 (T1), and terminal 2 (T2). **See Figure 23-15.** There is no designation of anode and cathode. Current may flow in either direction through the main switch terminals, T1 and T2. Terminal 1 is the reference terminal for all voltages. Terminal 2 is the case or metal-mounting tab to which a heat sink can be attached.

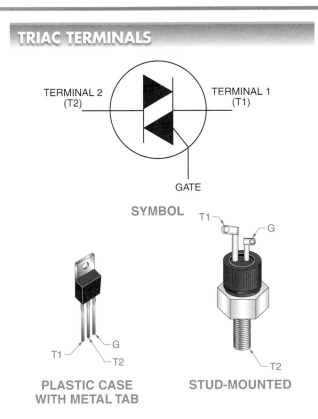

TRIAC TERMINALS

TERMINAL 2 (T2)   TERMINAL 1 (T1)

GATE

SYMBOL

T1   G

T1   G   T2   T2

PLASTIC CASE WITH METAL TAB   **STUD-MOUNTED**

Figure 23-15. Triac terminals include a gate, terminal 1 (T1), and terminal 2 (T2).

Triac Operation

Triacs block current in either direction between T1 and T2. A triac can be triggered into conduction in either direction by a momentary positive or negative pulse supplied to the gate. If the appropriate signal is applied to the triac gate, it conducts electricity.

The triac remains off until the gate is triggered at point A. **See Figure 23-16.** At point A, the trigger circuit pulses the gate and the triac is turned on, allowing current to flow. At point B, the forward current is reduced to zero and the triac is turned off. The trigger circuit can be designed to produce a pulse that varies at any point in the positive or negative half cycle. Therefore, the average current supplied to the load can vary.

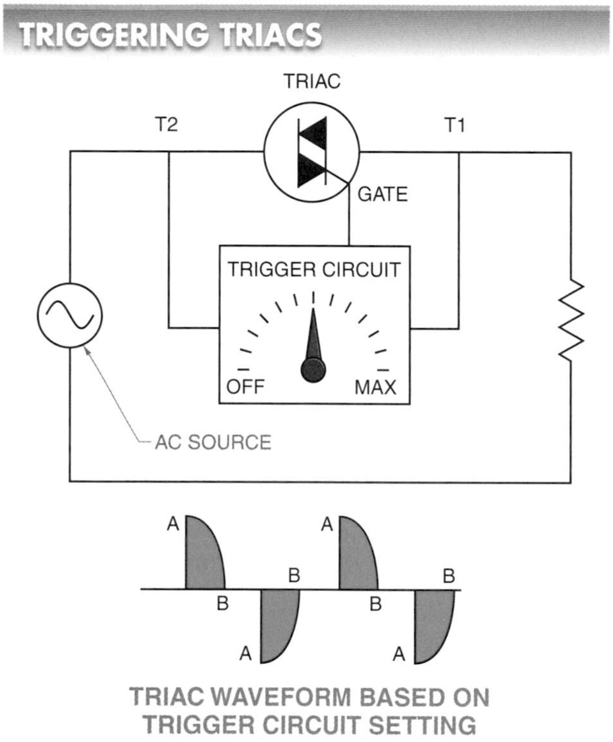

TRIGGERING TRIACS

Figure 23-16. A triac remains off until its gate is triggered.

One advantage of the triac is that virtually no power is wasted by being converted to heat. Heat is generated when current is impeded, not when current is switched off. The triac is either fully ON or fully OFF. It never partially limits current. Another important feature of the triac is the absence of a reverse breakdown condition of high voltages and high current, such as those found in diodes and SCRs. If the voltage across the triac goes too high, the triac is turned on. When turned on, the triac can conduct a reasonably high current.

Triac Characteristic Curves

The characteristics of a triac are based on T1 as the voltage reference point. The polarities shown for voltage and current are the polarities of T2 with respect to T1. The polarities shown for the gate are also with respect to T1. **See Figure 23-17.** Again, the triac may be triggered into conduction in either direction by a gate current (I_G) of either polarity.

TRIAC CHARACTERISTIC CURVE

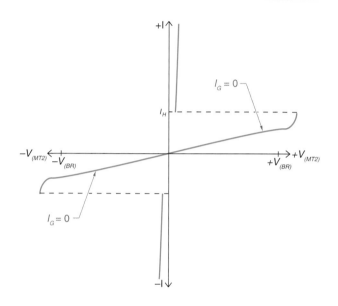

Figure 23-17. A triac characteristic curve shows the characteristics of a triac when triggered into conduction.

Triac Applications

Triacs are often used instead of mechanical switches because of their versatility. Also, where amperage is low, triacs are more economical than back-to-back SCRs.

Single-Phase Motor Starters. Often, a capacitor-start or split-phase motor must operate where arcing of a mechanical cut-out start switch is undesirable or even dangerous. In such cases, the mechanical cut-out start switch can be replaced by a triac. **See Figure 23-18.** A triac is able to operate in such dangerous environments because it does not create an arc. The gate and cut-out signal is given to the triac through a current transformer. As the motor speeds up, the current is reduced in the current transformer and the transformer no longer triggers the triac. With the triac turned off, the start windings are removed from the circuit.

SINGLE-PHASE MOTOR STARTERS

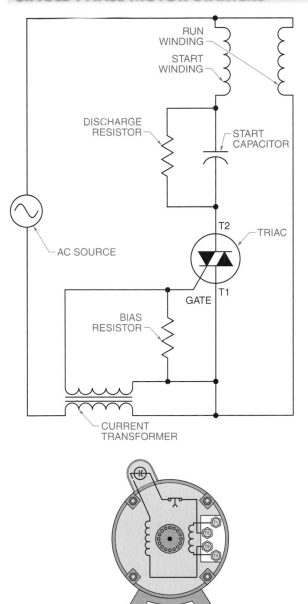

Figure 23-18. A mechanical cut-out start switch may be replaced by a triac.

Testing Triacs

Triacs should be tested under operating conditions using an oscilloscope. A DMM may be used to make a rough test with the triac out of the circuit. **See Figure 23-19.** To test a triac using a DMM, the following procedure is applied:

1. Set the DMM on the Ω scale.
2. Connect the negative lead to main terminal 1.

3. Connect the positive lead to main terminal 2. The DMM should read infinity.
4. Short-circuit the gate to main terminal 2 using a jumper wire. The DMM should read almost 0 Ω. The zero reading should remain when the lead is removed.
5. Reverse the DMM leads so that the positive lead is on main terminal 1 and the negative lead is on main terminal 2. The DMM should read infinity.
6. Short-circuit the gate of the triac to main terminal 2 using a jumper wire. The DMM should read almost 0 Ω. The zero reading should remain after the lead is removed.

TESTING TRIACS

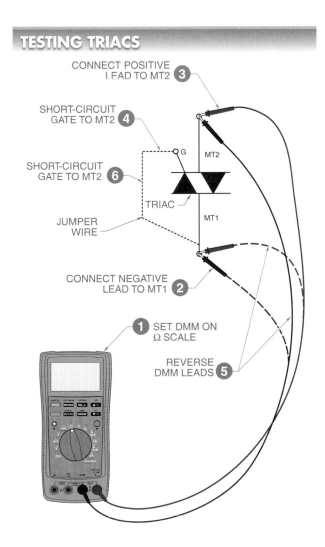

Figure 23-19. A DMM may be used to make a rough test of a triac that is out of the circuit.

23-3 CHECKPOINT

1. Can triacs be used to switch as much voltage and current as SCRs can switch?

2. Once a voltage to the gate of a triac turns on the triac and load, must the gate voltage remain on to keep the triac and load on?

23-4 DIACS

A *diac* is a three-layer, two-terminal bidirectional device that is typically used as a triggering device to control the gate current of a triac. **See Figure 23-20.** A diac is a special diode that can be triggered into conduction in either direction.

A diac has negative resistance because it does not conduct current until the voltage across it reaches breakover voltage. When a positive or negative voltage reaches the breakover voltage, the diac rapidly switches from a high-resistance state to a low-resistance state. **See Figure 23-21.** Since the diac is a bidirectional device, it is ideal for controlling a triac, which is also bidirectional.

DIACS

ANODE 1 —⬡— ANODE 2

SYMBOL

PACKAGE

P-TYPE MATERIAL

○— P | N | P —○

N-TYPE MATERIAL

SEMICOMDUCTOR LAYERS

Figure 23-20. A diac is a three-layer, two-terminal bidirectional device.

Diac Operation

Electrically, a diac operates in a manner similar to two zener diodes that are connected in series in opposite directions. The diac is used primarily as a triggering device. This operation is accomplished by the use of the negative resistance characteristic of the diac (current decreases with an increase of applied voltage).

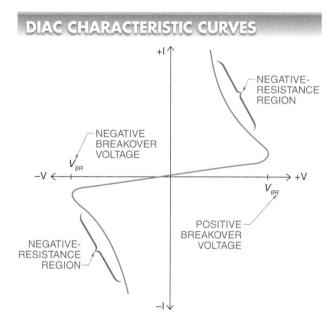

DIAC CHARACTERISTIC CURVES

+I

NEGATIVE-RESISTANCE REGION

NEGATIVE BREAKOVER VOLTAGE

V_{BR}

−V ← → +V

V_{BR}

NEGATIVE-RESISTANCE REGION

POSITIVE BREAKOVER VOLTAGE

−I

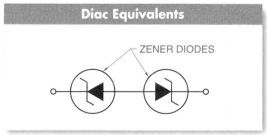

Diac Equivalents

ZENER DIODES

Figure 23-21. When a positive or negative voltage reaches the breakover voltage, a diac rapidly switches from a high-resistance state to a low-resistance state.

Diac Applications

The gate-control circuits of triacs can be improved by adding a breakover device in the gate lead, such as a diac. Using a diac in the gate-triggering circuit offers an important advantage over simple gate-control circuits. The advantage is that the diac delivers a pulse of gate current rather than a sinusoidal gate current. This results in a better-controlled firing sequence. Thus, diacs are used almost exclusively as triggering devices.

Universal Motor Speed Controllers. A diac/triac combination can be used to control the power to a universal motor. **See Figure 23-22.** In the circuit, capacitor C1 charges up to the firing voltage of the diac in either direction. Once fired, the diac applies a voltage to the gate of the triac. The triac conducts and supplies power to the motor.

UNIVERSAL MOTOR SPEED CONTROLLERS

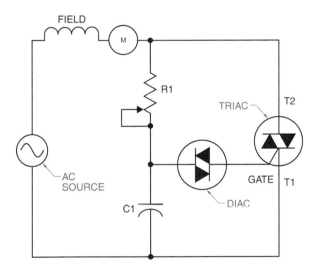

Figure 23-22. In this circuit, capacitor C1 charges up to the firing voltage of the diac in either direction. Once fired, the diac applies voltage to the gate of the triac, which conducts and applies power to the motor.

Note: The triac will conduct in either direction. Since the universal motor is basically a series DC motor, current flowing in either direction will cause rotation in only one direction. The speed may be changed by varying the resistance of potentiometer R1, which in turn varies the RC time constant.

Testing Diacs

A DMM may be used to test a diac for a short circuit. **See Figure 23-23.** To test a diac for a short circuit, the following procedure is applied:
1. Set the DMM on the Ω scale.
2. Connect the DMM leads to the leads of the diac and record the resistance reading.
3. Reverse the DMM leads and record the resistance reading.

TESTING DIACS USING A DMM

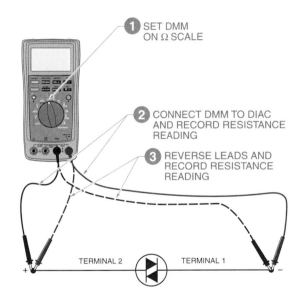

Figure 23-23. A DMM may be used to test a diac for a short circuit.

Both resistance readings should show high resistance because the diac is essentially two zener diodes connected in series. Testing a diac in this manner only shows that the component is shorted.

If a diac is suspected of being open, it should be tested using an oscilloscope. **See Figure 23-24.** To test a diac using an oscilloscope, the following procedure is applied:
1. Set up the test circuit.
2. Apply power to the circuit.
3. Adjust the oscilloscope.
4. Analyze traces.

TESTING DIACS USING AN OSCILLOSCOPE

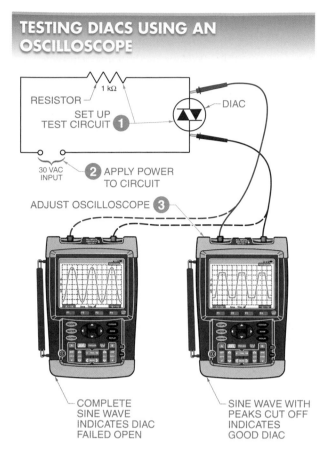

RESISTOR
1 kΩ
DIAC
1 SET UP TEST CIRCUIT

30 VAC INPUT   **2** APPLY POWER TO CIRCUIT

ADJUST OSCILLOSCOPE **3**

COMPLETE SINE WAVE INDICATES DIAC FAILED OPEN

SINE WAVE WITH PEAKS CUT OFF INDICATES GOOD DIAC

Figure 23-24. A diac should be tested using an oscilloscope if it is suspected of being open.

A diac-triac combination can be used to control power for universal motors that are used in household washing machines.

23-4 CHECKPOINT

1. Does a diac allow current flow in either direction before it reaches the breakdown voltage?
2. When troubleshooting a diac using an ohmmeter, should a good diac have a low or high resistance in both directions?

23-5 UNIJUNCTION TRANSISTORS

A *unijunction transistor (UJT)* is a three-electrode device that contains one PN junction consisting of a bar of N-type material with a region of P-type material doped within the N-type material. **See Figure 23-25.** The N-type material functions as the base and has two leads, base 1 (B1) and base 2 (B2). The lead extending from the P-type material is the emitter (E).

In the schematic symbol for a UJT, an arrowhead represents the emitter (E). Although the leads are usually not labeled, they can be easily identified because the arrowhead always points to B1. The case of a UJT may include a tab to identify the leads. **See Figure 23-26.**

A UJT is used primarily as a triggering device because it generates a pulse used to fire SCRs and triacs. Outputs from photocells, thermistors, and other transducers can be used to trigger UJTs, which in turn fire SCRs and triacs. UJTs are also used in oscillators, timers, and voltage-current sensing applications.

UNIJUNCTION TRANSISTORS (UJTs)

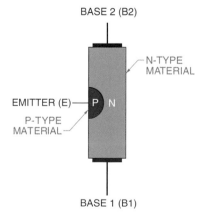

BASE 2 (B2)

N-TYPE MATERIAL

EMITTER (E) — P   N

P-TYPE MATERIAL

BASE 1 (B1)

Figure 23-25. A unijunction transistor (UJT) consists of a bar of N-type material with a region of P-type material doped within the N-type material.

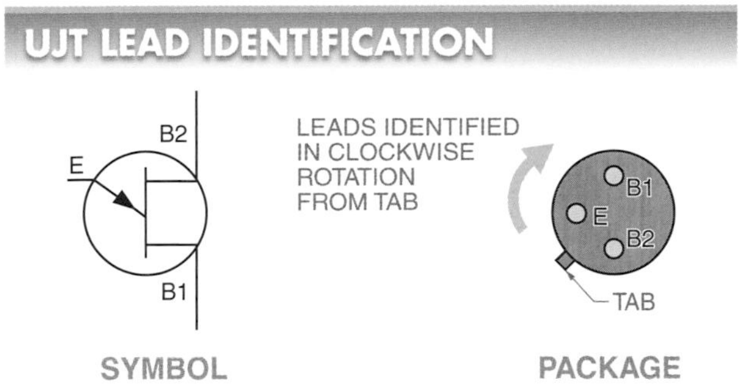

UJT LEAD IDENTIFICATION

LEADS IDENTIFIED IN CLOCKWISE ROTATION FROM TAB

SYMBOL   PACKAGE

Figure 23-26. In the schematic symbol for a UJT, an arrowhead represents the emitter (E) and always points to base 1 (B1).

UJT Biasing

In normal operation, B1 is negative and a positive voltage is applied to B2. The internal resistance between B1 and B2 is divided at E, with approximately 60% of the resistance between E and B1. The remaining 40% of the resistance is between E and B2. The net result is an internal voltage split. This split provides a positive voltage at the N-type material of the emitter junction, creating a reverse-biased emitter junction. As long as the emitter voltage remains less than the internal voltage, the emitter junction will remain reverse biased, even at a very high voltage. However, if the emitter voltage rises above this internal value, a dramatic change will take place.

When the emitter voltage is greater than the internal value, the junction becomes forward biased. Also, the resistance between E and B1 drops rapidly to a very low value. A UJT characteristic curve shows the dramatic change in voltage due to this change in resistance. **See Figure 23-27.**

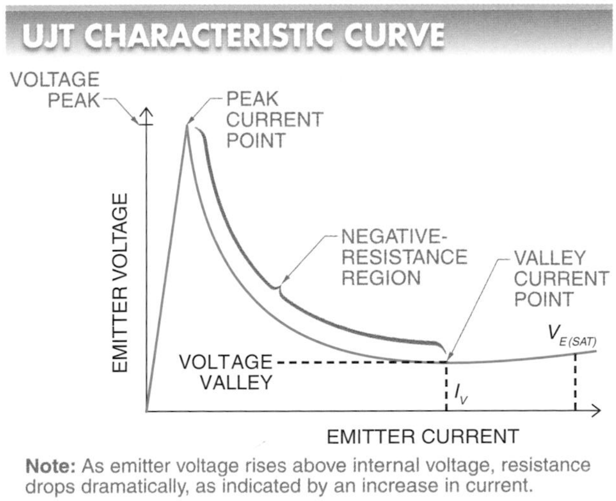

UJT CHARACTERISTIC CURVE

Note: As emitter voltage rises above internal voltage, resistance drops dramatically, as indicated by an increase in current.

Figure 23-27. A UJT characteristic curve shows the dramatic change in voltage due to a rapid change in resistance. The negative-resistance region is ideal for triggering.

Theory of Operation

As long as the E-B1 junction is reverse biased and no current flows into the emitter, the current flow in the N-type material should be minimal. This is due to the small amount of doping that creates a high resistance. When the E-B1 junction is forward biased, the junction turns on, causing carriers to be injected into the base region. These carriers create an excess of holes. Their presence in the N-type material increases conductivity, which lowers the resistance of the region. Once started, current flows easily between B1 and E. Therefore, the conductivity of this region is controlled by the flow of emitter current.

Triggering Circuits

A UJT is typically used as a triggering circuit for a triac or similar device. When switch S1 is closed, the voltage-divider action of the UJT produces a voltage between B1 and the N-type material of the emitter junction. At this same instant, the emitter voltage is zero since it is tied to capacitor C1. The emitter junction at that point is reverse biased and no current flows through the junction. As capacitor C1 begins to charge through resistor R1, the voltage across capacitor C1 should begin to increase.

For an emitter to be forward biased, it must be more positive than the base (+0.6 V for silicon or +0.2 V for germanium). Assuming a silicon crystal is used in the UJT, the junction becomes forward biased when the control voltage reaches 0.6 V beyond the junction voltage. With the junction forward biased, the internal resistance of the E-B1 region drops dramatically. This causes capacitor C1 to discharge its energy through base load resistor R3. **See Figure 23-28.** Once the capacitor has discharged enough to reduce the forward bias on the junction, the resistance of the junction returns to normal. The cycle of capacitor charging and discharging then repeats.

Each time the emitter becomes forward biased, the total resistance between B1 and B2 drops, permitting an increase in current through the UJT. As a result, a positive pulse (V_{B1}) appears at B1 and a negative pulse (V_{B2}) appears at B2 at the time the capacitor discharges.

Note: The repetition rate, or frequency, of the discharge voltage is determined by the values of resistor R3 and capacitor C1. Increasing either one makes the device run more slowly. The pulses that appear across bases B1 and B2 are very useful in triggering SCRs and triacs.

UJT TRIGGERING CIRCUITS— FORWARD BIASED

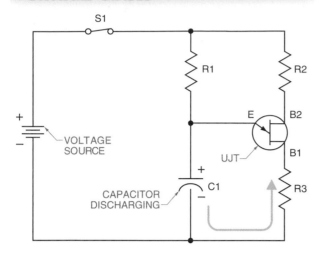

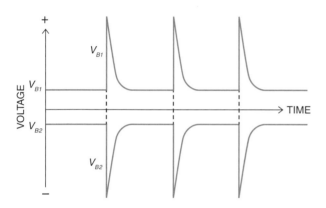

Figure 23-28. With the emitter junction forward biased, the internal resistance of the E-B1 region drops dramatically and causes capacitor C1 to discharge its energy through base load resistor R3.

UJT Applications

The UJT is used in switching and timing applications. A UJT often reduces the number of components needed to perform a given function. The number of components are often less than half of what is required when using bipolar transistors.

Emergency Flashers. A UJT can serve as a triggering circuit for an emergency flasher. As a triggering circuit, UJT Q1 provides base bias to drive transistors Q2 and Q3 through resistors R2 and R3. Transistors Q2 and Q3 are used to light an incandescent lamp load. **See Figure 23-29.**

EMERGENCY FLASHERS

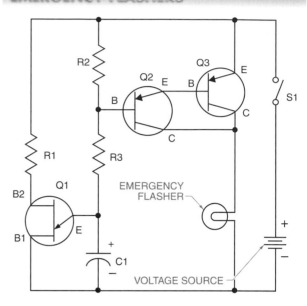

Figure 23-29. A UJT can serve as a triggering circuit for an emergency flasher.

The circuit repetition rate (frequency) is determined by the characteristics of the UJT, supply voltage, and emitter RC time constant of Q1. To change the flashing rate, the value of capacitor C1 must be changed. As capacitor C1 increases in value, the flashing rate decreases. As capacitor C1 decreases in value, the flashing rate increases.

Tech Fact

A UJT can be considered as a diode connected to a voltage divider network. UJTs have the ability to be used as relaxation oscillators. An oscillator is a circuit that produces a repetitive electronic signal, such as a sine wave, without AC input signals. Relaxation oscillators are characterized internally by short, sharp pulses of waveforms that can potentially trigger gates.

23-5 CHECKPOINT

1. Can a UJT be used to trigger both SCRs and triacs?
2. What part of a UJT acts like the gate of an SCR or triac by controlling the flow of current?

Additional Resources

Review and Resources

Access Chapter 23 Review and Resources for *Electrical Motor Controls for Integrated Systems* by scanning the above QR code with your mobile device.

Applying Your Knowledge

Refer to the *Electrical Motor Controls for Integrated Systems* Learner Resources for interactive Applying Your Knowledge questions.

Workbook and Applications Manual

Refer to Chapter 23 in the *Electrical Motor Controls for Integrated Systems Workbook* and the *Applications Manual* for additional exercises.

ENERGY EFFICIENCY PRACTICES

Controlling Heat

All solid-state power switching devices produce heat, which is caused by voltage drop across the solid-state switch and current flow through the device. For this reason, solid-state switching devices use heat sinks to transfer the heat away from the switching device and out into the surrounding air. The higher the required switched current, the larger the required heat sinks and the greater the produced heat by the device.

The heat that radiates off the heat sinks can cause problems with electronic components if it is not completely removed from the air surrounding the heat sink and components. To solve this problem, the heat is usually removed from the area by using a circulating fan or an enclosure that has a built-in air conditioner. Additional energy is required to operate the fan or air conditioner that removes the heat produced by the solid-state switches. The larger the system, the greater the amount of energy required to remove heat.

The amount of energy required to remove the heat can be lowered by reducing the size of or eliminating the requirement for fans and air-conditioned cabinets. Energy usage can be lowered through the following steps:

Carlo Gavazzi Inc.

- Ensure proper mounting and spacing. Since heat rises naturally upwards (by convection), it is important to mount the heat sink to provide the most air up and around the heat sink. The solid-state switch should be mounted low in the cabinet, away from the rising heat.
- Ensure proper enclosure ventilation. Use open vents at the bottom of the enclosure to allow cooler air in and open vents at the top of the enclosure to allow hotter air out. This will aid in the natural convection cooling of the heat sinks.
- Increase enclosure size. If there is enough space, a larger enclosure can be used to better dissipate the heat and reduce or eliminate the need for fans or air conditioning units.
- Ensure proper enclosure location. It is best to mount the enclosure in an open air environment that allows air to flow completely around the enclosure. Even a small space behind the enclosure (if it is not mounted to wall) can help increase the effects of cooling.

Sections

Objectives

24-1

- Describe photoelectric devices.
- Define and describe PIN photo-diodes.
- Define and describe phototransistors.
- Define and describe light-activated silicon-controlled rectifiers (LASCRs).
- Define and describe phototriacs.
- Define and describe light-emitting diodes (LEDs).
- Define and describe optocouplers.

24-2

- Define photoelectric sensor.
- Describe AC photoelectric sensors.
- Describe DC photoelectric sensors.

24-3

- Define and describe fiber optics.

24-4

- Describe how photoelectric sensors are used.
- Define photoelectric scanning and list the most common scanning techniques.
- Explain the difference between modulated and unmodulated light.
- Explain response time and sensitivity adjustment.
- Explain the difference between dark-operated and light-operated photoelectric controls.

24-5

- Explain how to properly mount photoelectric sensors.

Review and Resources
atplearningresources.com/Quicklinks
Access Code: 362245

Chapter

Photoelectric Semiconductors, Fiber Optics, and Light-Based Applications

24

Electricity has been used to provide light, heat, rotational motion, linear motion, sound, and many other outputs for over 150 years. Since the earliest usages of electricity, mechanically activated switches such as pushbuttons, limit switches, and liquid level switches have been used to control when an electrical load is ON or OFF. Mechanically activated switches require some type of physical contact to open or close the contacts. For example, a manually operated switch requires a person to move the switch, a limit switch requires something to move the switch's lever, level switches require a liquid to move the switch's float, etc.

As electrical control circuits were used in more and more applications, the need developed for switches that required no physical contact with the switch to detect an object. Those applications included the detection of hot objects, very large or heavy objects, very small or light objects, and objects that must be detected at a distance without being touched. Photoelectric switches were developed that could detect objects at a far distance without touching them by using a beam of light. Proximity switches were developed that could detect closer objects without touching them by using a magnetic or capacitive field. Since photoelectric and proximity switches are commonly used today, it is important to understand their theory of operation, common usages, advantages and disadvantages, and installation and troubleshooting procedures.

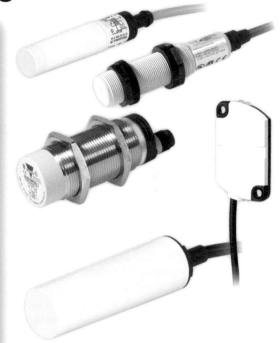

24-1 PHOTOELECTRIC DEVICES

Photoelectric devices typically contain solid-state output switches. A solid-state switch has no moving parts (contacts). A solid-state switch uses a triac, SCR, NPN (current sink) transistor, or PNP (current source) transistor output to perform the switching function. **See Figure 24-1.** The triac output is used for switching AC loads. The SCR output is used for switching high-power DC loads. The current sink and current source transistor outputs are used for switching low-power DC loads.

Solid-state switches include normally open (NO), normally closed (NC), or combination switching outputs.

Tech Fact

Common photoelectric or proximity switch cylindrical housings are available in 5 mm, 8 mm, 12 mm, 18 mm, 25 mm, and 30 mm sizes (though 18 mm is the most common size). The millimeter (mm) rating refers to the diameter of the sensor's heat-sensing head. The larger the diameter of the sensing head, the farther the sensing distance.

OUTPUT SWITCHING DEVICES

Device		Use
TRIAC		Switch AC loads
SCR		Switch high-power DC loads
NPN TRANSISTOR	PNP TRANSISTOR	Switch low-power DC loads

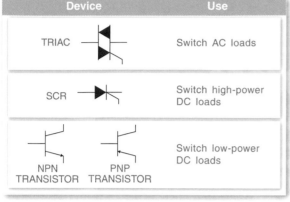

PROXIMITY SWITCH USING A SOLID-STATE SWITCHING DEVICE

OUTPUT SWITCHING DEVICE

SENSING FIELD

TRIGGER CIRCUIT

COIL OSCILLATOR

SOLID-STATE SWITCHES

Figure 24-1. A solid-state switch uses a triac, SCR, NPN (current sink) transistor, or PNP (current source) transistor output to perform the switching function.

Solid-state devices are usually connected (interfaced) through the use of electrical devices. They can also be interfaced through light-driven systems. Once light rays have passed through the optical fiber, they must be detected and converted back into electrical signals. The detection and conversion is accomplished with light-activated devices such as PIN photodiodes, phototransistors, light-activated SCRs, and phototriacs. The light source must be properly matched to the light-activated device to operate effectively. The source must also be of sufficient intensity to drive the light-activated device.

PIN Photodiodes

A *PIN photodiode* is a diode with a large intrinsic region sandwiched between P-type and N-type regions. PIN stands for P-type material, insulator, and N-type material. **See Figure 24-2.** The operation of a PIN photodiode is based on the principle that light radiation, when exposed to a PN junction, momentarily disturbs the structure of the PN junction. The disturbance is due to a hole created when a high-energy photon strikes the

PN junction and causes an electron to be ejected from the junction. Thus, light creates electron-hole pairs that act as current carriers. PIN photodiodes are used in gas detectors, spectrometers, and gas analyzers.

PIN PHOTODIODES

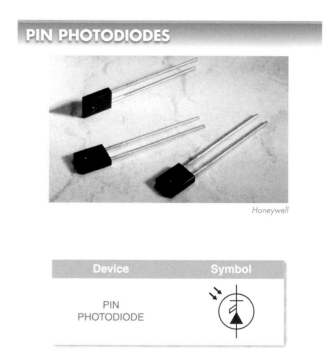

Honeywell

Device	Symbol
PIN PHOTODIODE	

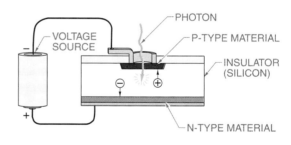

PHOTON

VOLTAGE SOURCE

P-TYPE MATERIAL

INSULATOR (SILICON)

N-TYPE MATERIAL

Figure 24-2. A PIN photodiode is a diode with a large intrinsic region sandwiched between P-type and N-type regions.

Phototransistors

A *phototransistor* is a device that combines the effect of a photodiode and the switching capability of a transistor. The symbol for a phototransistor is similar to the symbol for a standard transistor except with the addition of two arrows that represent a light source pointing into the transistor. **See Figure 24-3.** A phototransistor, when connected in a circuit, is placed in series with the bias voltage so that it is forward biased.

PHOTOTRANSISTORS

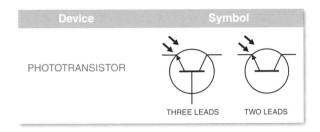

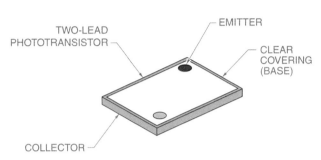

Figure 24-3. A phototransistor is a device that combines the effect of a photodiode and the switching capability of a transistor.

In a two-lead phototransistor, the base lead is replaced by a clear covering that allows light to fall on the base region. Light falling on the base region causes current to flow between the emitter and collector. The collector-base junction is enlarged and works as a reverse-biased photodiode controlling the phototransistor. The phototransistor conducts more or less current, depending on the light intensity. If the light intensity increases, resistance decreases and more emitter-to-base current is created. Although the base current is relatively small, its amplifying capability is used to control the large emitter-to-collector current. The collector current depends on the light intensity and the DC current gain of the phototransistor. In darkness, the phototransistor is switched off with the remaining leakage current (collector dark current).

Light-Activated SCRs

A *light-activated SCR (LASCR)* is an SCR that is activated by light. The symbol of an LASCR is identical to the symbol of a regular SCR with one difference. The only difference is that arrows are added in the LASCR symbol to indicate a light-sensitive device. **See Figure 24-4.**

LIGHT-ACTIVATED SCRs (LASCRs)

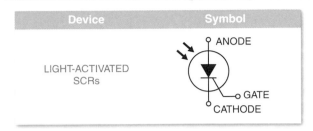

Figure 24-4. A light-activated SCR (LASCR) is an SCR that is activated by light.

Similar to a photodiode, a very low level of current is in an LASCR. Even the largest LASCRs are limited to a maximum of a few amps. When larger current requirements are necessary, an LASCR can be used as a trigger circuit for a standard high-power switching SCR. The primary advantage of an LASCR over an SCR is its ability to provide isolation. Because the LASCR is triggered by light, it provides complete isolation between the input signal and the output load current.

Phototriacs

A *phototriac* is a triac that is activated by light. **See Figure 24-5.** The gate of a phototriac is light sensitive. It triggers the triac at a specified light intensity. In darkness, the triac is not triggered. The remaining leakage current is referred to as peak blocking current. A phototriac is bilateral and designed to switch AC signals.

PHOTOTRIACS

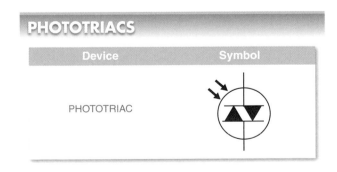

Figure 24-5. A phototriac is a triac that is activated by light.

Light-Emitting Diodes

A *light-emitting diode (LED)* is a semiconductor diode that produces light when current flows through it. As electrons move across the depletion region, they give up extra kinetic energy. The extra energy is converted to light. An electron must acquire additional energy to get through the depletion region. This additional energy comes from the positive field of the anode. If the positive field is not strong, the electron will not get through the depletion region and no light will be emitted. **See Figure 24-6.**

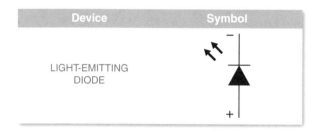

LIGHT-EMITTING DIODES (LEDs)

Device	Symbol
LIGHT-EMITTING DIODE	

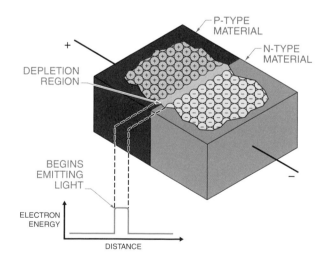

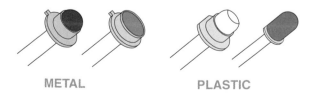

METAL   PLASTIC

Figure 24-6. A light-emitting diode (LED) is a semiconductor diode that produces light when current flows through it.

For a standard silicon diode, a minimum of 0.6 V must be present before the diode conducts. For a germanium diode, 0.3 V must be present before the diode conducts. Most LED manufacturers make a larger depletion region that requires 1.5 V for the electrons to get across.

LED Construction. LED manufacturers normally use a combination of gallium and arsenic with silicon or germanium to construct LEDs. By adding other impurities to the base semiconductor and adjusting them, the manufacturers can produce different wavelengths of light. LEDs are capable of producing infrared light. *Infrared light* is light that is not visible to the human eye. LEDs may emit a visible red or green light. Colored plastic lenses are available if different colors are desired. As with standard semiconductor diodes, there is a method for determining which end of an LED is the anode and which end is the cathode. The cathode lead is identified by the flat side of the device or a notch that may have been cut into the ridge.

A colored plastic lens focuses the light produced at the junction of the LED. Without the lens, the small amount of light produced at the junction is diffused and becomes virtually unusable as a light source. The size and shape of the LED package determines how it is positioned for proper viewing.

The schematic symbol for an LED is exactly like that of a photodiode, but the arrows point away from the diode. The LED is forward biased and a current-limiting resistor is normally present to protect the LED from excessive current.

Laser Diodes. A *laser diode* is a diode similar to an LED but with an optical cavity that is required for lasing production (emitting coherent light). The optical cavity is formed by coating opposite sides of a chip to create two highly reflective surfaces. **See Figure 24-7.**

Optocouplers

An *optocoupler* is an electrically isolated device that consists of an IRED as the input stage and an NPN phototransistor as the output stage. An optocoupler is normally constructed as a dual inline package. **See Figure 24-8.**

An optocoupler uses a glass dielectric sandwich to separate input from output. The coupling medium between the LED and sensor is the transmitting glass. This provides one-way transfer of electrical signals from the LED to the photodetector (phototransistor) without an electrical connection between the circuitry containing the devices.

LASER DIODES

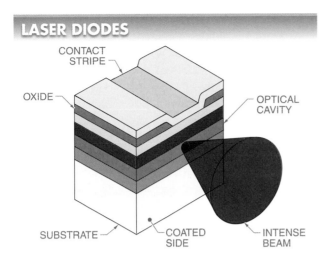

Figure 24-7. A laser diode is similar to an LED but with an optical cavity that is required for lasing production (emitting coherent light).

OPTOCOUPLERS

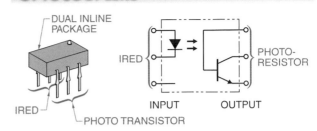

Figure 24-8. An optocoupler consists of an IRED as the input stage and an NPN phototransistor as the output stage.

Input and output devices are always spectrally matched by their wavelengths for maximum transfer characteristics. The signal cannot go back in the opposite direction because the emitters and detectors cannot reverse their operating functions.

24-1 CHECKPOINT

1. What solid-state device is used to switch high-power DC loads?
2. What distinguishes the symbol for a phototransistor from the symbol for a standard transistor?
3. If an LASCR is required to switch a higher power load, how is the LASCR used?
4. What is the difference between the symbol for a photodiode and the symbol for an LED?

24-2 PHOTOELECTRIC SENSORS

A *photoelectric sensor (photoelectric switch)* is a solid-state sensor that can detect the presence of an object without touching the object. A photoelectric sensor detects the presence of an object by means of a beam of light. Photoelectric sensors use solid-state outputs to control the flow of electric current. The solid-state output of a photoelectric sensor may be a thyristor, NPN transistor, or PNP transistor. The thyristor output is used for switching AC circuits. The NPN and PNP transistor outputs are used for switching DC circuits. The output selected depends on specific application needs. Considerations that affect the solid-state output include the following:

- Voltage type to be switched—The voltage may be AC or DC.
- Amount of current to be switched—Most proximity sensors can only switch a maximum of a few hundred milliamperes. An interface is needed if higher current switching is required. The solid-state relay is the most common interface used with photoelectric and proximity sensors.

- Electrical requirements of the device to which the output of the proximity sensor is to be connected—Compatibility with a controller such as a programmable controller may require that a certain type of solid-state output be used as the input to a specific controller.
- Required polarity of the switched DC output—NPN outputs deliver a negative output and PNP outputs deliver a positive output.
- Electrical characteristics such as load current, operating current, and minimum holding current—Solid-state outputs are never completely open or closed.

Tech Fact

When troubleshooting or testing a transistor and the output type is not identified on the device, the first step can be to test it as if it is an NPN transistor. With an NPN transistor output, the transistor outputs a negative reading. The negative (black) lead of a DMM is then connected to the transistor output to verify whether the transistor is operating by outputting a voltage when the sensor is activated.

AC Photoelectric Sensors

AC photoelectric sensors switch AC circuits. An AC photoelectric sensor is connected in series with the load that it controls. The sensor is connected between line 1 (L1) and the load to be controlled. **See Figure 24-9.** Because AC photoelectric sensors are connected in series with the load, special precautions must be taken. The three main factors considered when connecting AC photoelectric sensors include load current, operating (residual) current, and minimum holding current.

Load Current. *Load current* is the amount of current drawn by a load when energized. Because a solid-state sensor is wired in series with the load, the current drawn by the load must pass through the solid-state sensor. For example, if a load draws 5 A, the sensor must be able to safely switch 5 A. Five amperes burns out most solid-state proximity sensors because they are normally rated for a maximum of less than 0.5 A. An electromechanical or solid-state relay must be used as an interface to control the load if a solid-state sensor must switch a load above its rated maximum current. A solid-state relay is the preferred choice for an interface with a sensor. **See Figure 24-10.**

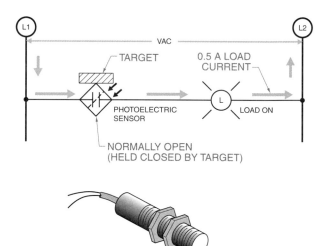

PHOTOELECTRIC SENSOR LOAD CURRENT

Figure 24-10. Load current is the amount of current drawn by a load when energized and flows through AC photoelectric sensors.

Operating Current. *Operating current (residual or leakage current)* is the amount of current a sensor draws from the power lines to develop a field that can detect a target. When a sensor is in the OFF condition (target not detected), a small amount of current passes through both the sensor and the load. This operating current is required for the solid-state detection circuitry housed within the sensor. Operating currents are normally in the range of 1.5 mA to 7 mA for most sensors. **See Figure 24-11.**

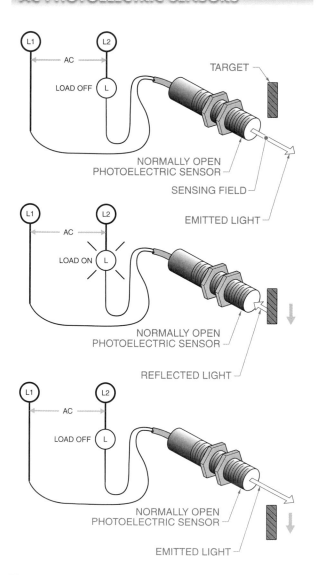

AC PHOTOELECTRIC SENSORS

Figure 24-9. AC photoelectric sensors are connected in series with the loads they control.

OPERATING CURRENT

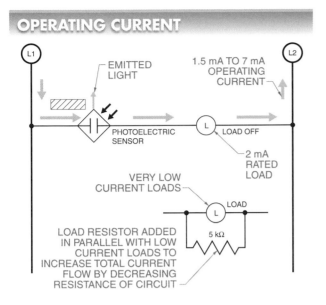

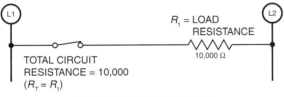

EQUIVALENT CIRCUIT
WITHOUT LOAD RESISTOR

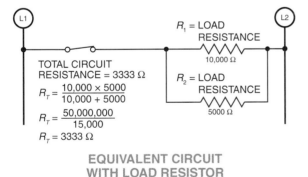

EQUIVALENT CIRCUIT
WITH LOAD RESISTOR

Figure 24-11. Operating current is the amount of current a sensor draws from the power lines to develop a field that can detect a target.

The small operating current normally does not have a negative effect on low-impedance loads or circuits such as mechanical relays, solenoids, and magnetic motor starters. However, the operating current may be enough to activate high-impedance loads such as programmable controllers, electronic timers, and other solid-state devices. In this case, the load is activated regardless of

whether a target is present or not. This problem may be corrected by placing a load resistor in parallel with low current loads. The resistance value should be selected to ensure that the effective load impedance (load plus resistor) is reduced to a level that prevents false triggering due to the operating current. It must also ensure that the minimum current required to operate the load is provided. This resistance value is normally in the range of 4.5 kΩ to 7.5 kΩ. A general rule is to use a 5 kΩ, 5 W resistor for most conditions.

Minimum Holding Current. *Minimum holding current* is the minimum amount of current required to keep a sensor operating. When the sensor has been triggered and is in the ON condition (target detected), the current drawn by the load must be sufficient to keep the sensor operating. Minimum holding currents range from 3 mA to 20 mA for most solid-state sensors. The amount of current a load draws must be correct for the proper operation of a sensor. Excessive current (operating current) burns up the sensor. Low current (minimum holding current) prevents proper operation of the sensor. **See Figure 24-12.**

MINIMUM HOLDING CURRENT

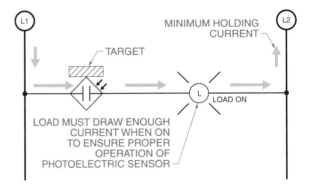

Figure 24-12. Minimum holding current is the minimum amount of current required to keep a sensor operating.

Series/Parallel Connections. All AC two-wire photoelectric sensors may be connected in series or parallel to provide both AND and OR control logic. When connected in series (AND logic), all sensors must be activated to energize the load. When connected in parallel (OR logic), any one sensor that is activated energizes the load. **See Figure 24-13.**

PHOTOELECTRIC SENSOR CONNECTIONS

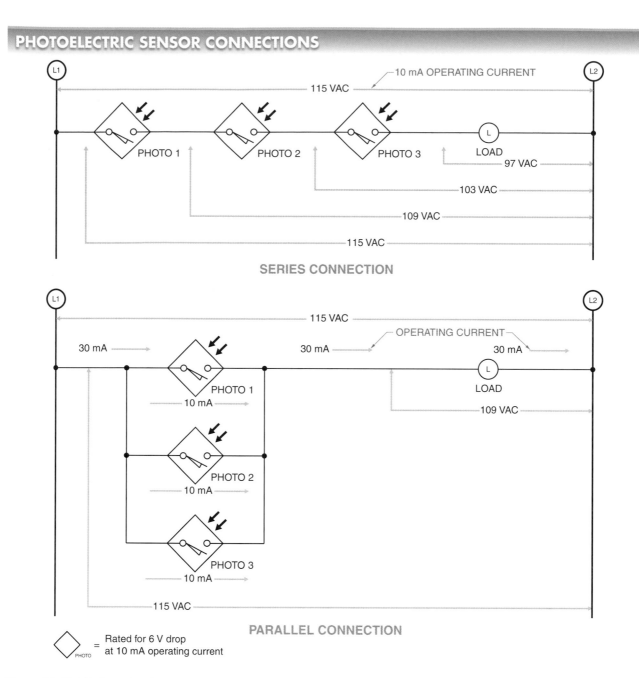

Figure 24-13. AC photoelectric sensors may be connected in series or parallel.

As a general rule, a maximum of three sensors may be connected in series to provide AND logic. Factors that limit the number of AC two-wire sensors that may be wired in series to provide AND logic include the following:

• AC supply voltage—Generally, the higher the supply voltage is, the higher the number of sensors that may be wired in series.

• Voltage drop across the sensor—Voltage drop varies for different sensors. The lower the voltage drop, the higher the number of sensors that may be connected in series.

• Minimum operating load voltage—This voltage varies depending on the load that is controlled. For every proximity sensor added in series with the load, less supply voltage is available across the load.

As a general rule, a maximum of three sensors may be connected in parallel to provide OR logic. Factors that limit the number of AC two-wire sensors that may be wired in parallel to provide OR logic include the following:

- Photoelectric and proximity switch operating current—The total operating current flowing through a load is equal to the sum of each sensor's operating current. The total operating current must be less than the minimum current required to energize the load.
- Amount of current that a load draws when energized—The total amount of current a load draws must be less than the maximum current rating of the lowest rated sensor. For example, if three sensors rated at 125 mA, 250 mA, and 275 mA are connected in parallel, the maximum rating of the load cannot exceed 125 mA.

DC Photoelectric Sensors

Photoelectric sensors that switch DC circuits normally use transistors as the switching element. The sensors use NPN transistors or PNP transistors. For most applications, the exact transistor used does not matter, as long as the switch is properly connected into the circuit. However, NPN transistor sensors are far more common than PNP transistor sensors.

NPN Transistor Switching. When using an NPN transistor, the load is connected between the positive terminal of the supply voltage and the output terminal (collector) of the sensor. When the sensor detects a target, current flows through the transistor and the load is energized. **See Figure 24-14.**

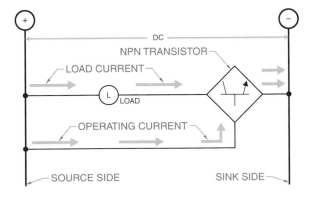

NPN TRANSISTOR SWITCHING

Figure 24-14. When an NPN (current sink) transistor is used, the load is connected between the positive terminal of the supply voltage and the output terminal of the sensor.

Output devices that use an NPN transistor as the switching element are current sink devices. The negative terminal of a DC system is the sink due to conventional current flowing into it. A current sinking switch "sinks" the current from the load. *Conventional current flow* is the movement of electrons from positive to negative. When working with transistors and other solid-state devices, some manufacturers use conventional current flow. However, electron current flow is used throughout this textbook. Regardless of the method, both engineers and technicians must follow schematics and observe polarity so that devices are properly installed.

PNP Transistor Switching. When using a PNP transistor, the load is connected between the negative terminal of the supply voltage and the output terminal (collector) of the sensor. When the sensor (current source) detects a target, current flows through the transistor and the load is energized. **See Figure 24-15.**

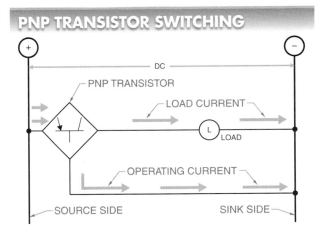

PNP TRANSISTOR SWITCHING

Figure 24-15. When a PNP (current source) transistor is used, the load is connected between the negative terminal of the supply voltage and the output terminal of the sensor.

24-2 CHECKPOINT

1. Can a transistor output be used in a photoelectric switch to switch an AC load?
2. Can a triac output be used in a photoelectric switch to switch an DC load?
3. Does the operating current of a two-wire photoelectric switch affect low impedance or high impedance loads more?
4. As a general rule, how many two-wire photoelectric sensors can be connected in series or parallel?

24-3 FIBER OPTICS

Fiber optics is a technology that uses a thin, flexible glass or plastic optical fiber to transmit light. Fiber optics is most commonly used as a transmission link. As a transmission link, it connects two electronic circuits consisting of a transmitter and a receiver. **See Figure 24-16.**

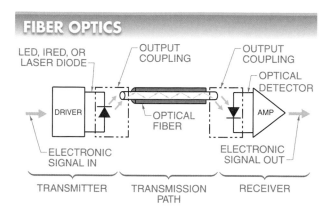

Figure 24-16. Fiber optics connect two electronic circuits consisting of a transmitter and a receiver.

Communication is the transmission of information from one point to another by means of electromagnetic waves. Fiber-optic technology has been developed to use light-wave communication to transmit data. This communication is carried out by transmitting modulating light through strands of glass (or plastic in certain applications) over various distances.

The light-emitting and receiving components in fiber optics are located remotely at the housing of the control, while only the optical fiber cables are exposed to the severe environmental or hazardous areas. The optical fiber cables are not adversely affected by electrical noise because they carry signals in the form of light.

The central part of the transmitter is its source. The source consists of a light-emitting diode (LED), infrared-emitting diode (IRED), or laser diode that changes electrical signals into light signals. The receiver normally contains a photodiode that converts light back into electrical signals. The receiver output circuit also amplifies the signal and produces the desired results, such as voice transmission or video signals. Advantages of optical fiber cables include large bandwidth, low cost, low power consumption, low loss (attenuation), electromagnetic interference (EMI) immunity, small size, light weight, and security (the inability to be tapped by unauthorized users).

Optical Fibers

Optical fibers consist of a core, cladding, and protective jacket. **See Figure 24-17.** The *core* is the actual path for light in an optical fiber cable. The core is normally made of glass but may occasionally be constructed of plastic. *Cladding* is the first layer of protection for the glass or plastic core of an optical fiber cable. A glass or plastic cladding layer is bonded to the core. The cladding is enclosed in a jacket for additional protection.

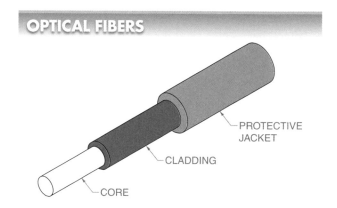

Figure 24-17. Optical fibers consist of a core, cladding, and protective jacket.

Fiber Connectors

The ideal interconnection of one fiber to another is an interconnection that has two fibers that are optically and physically identical. These two fibers are held together by a connector or splice that squarely aligns them on their center axes. The joining of the fibers is so nearly perfect that the interface between them has no influence on light propagation. A perfect connection is limited by variations in fibers and the high tolerances required in the connector or splice. These two factors also affect cost and ease of use.

Fiber-Connecting Hardware

Splices and fiber interconnections are often more of a negative factor than poor quality materials because of alignment problems that can arise, such as fiber gap, lateral offset, and angular misalignment. The elimination of alignment problems can be accomplished through proper installation of fiber splices, connectors, and couplers. **See Figure 24-18.**

OPTICAL FIBER CABLE CONNECTION

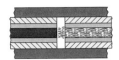
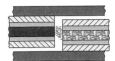
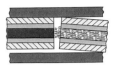

PROPER
CONNECTION

FIBER
GAP

LATERAL
OFFSET

ANGULAR
MISALIGNMENT

Figure 24-18. Improper connection of optical fiber cables can result in poor or no transmission.

24-3 CHECKPOINT

1. What two materials can be used for core of optical fiber?

2. Why are signals sent over an optical fiber cable not affected by electrical noise such as external magnetic fields produced by motors and other coils?

3. What are three poor splice problems that can occur and cause reduced data transfer in fiber optics?

24-4 PHOTOELECTRIC SENSOR APPLICATIONS

Photoelectric sensors can detect most materials and have a longer sensing distance than ultrasonic and proximity sensors. Depending on the model, photoelectric sensors can detect objects from several millimeters to over 100′ away. The maximum sensing distance of any sensor is determined by the size, shape, color, and character of the surface of the object to be detected. Many sensors include an adjustable sensing distance, making it possible to exclude detection of the background of the object. Photoelectric sensors are used in applications in which the object to be detected may be excessively light, heavy, hot, or untouchable. **See Figure 24-19.**

Photoelectric sensors are used in a variety of applications for the detection of almost any object. Along a production line, photoelectric sensors are used for counting, positioning, and sorting objects, as well as providing safety. In security systems, they are used to detect the presence or removal of an object. Photoelectric sensors are also used to detect the presence of vehicles at tollgates, parking areas, and truck loading bays. In most applications, photoelectric sensors are used as inputs into timers, relays, counters, programmable controllers, and motor control circuits.

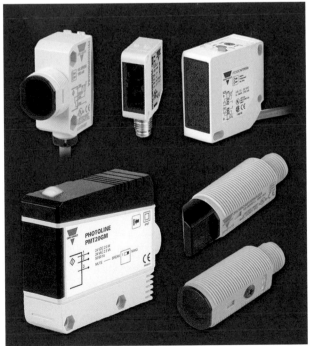

Carlo Gavazzi, Inc.

Various types of photoelectric sensors are used in the manufacturing industry to detect objects along a production line.

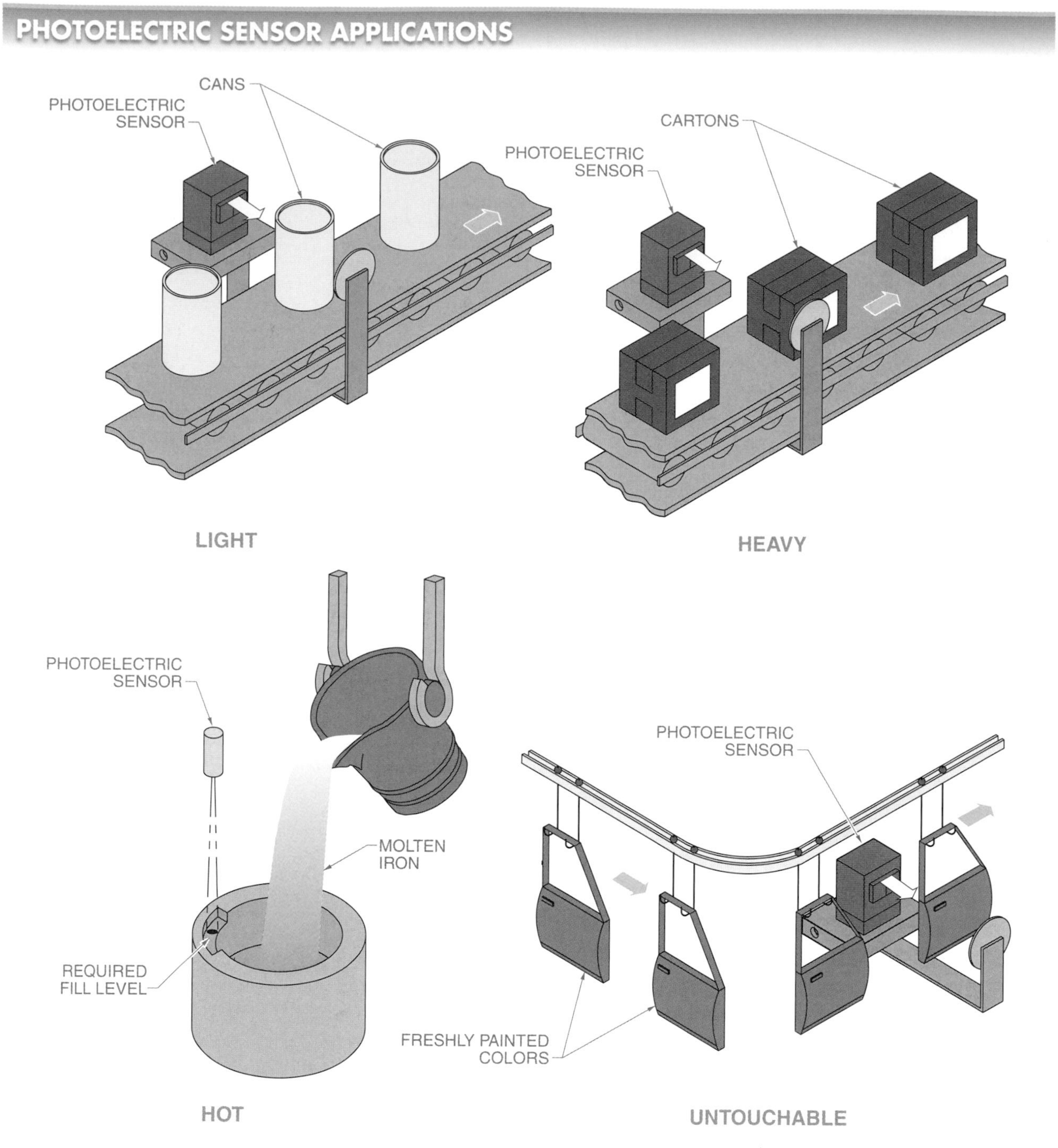

PHOTOELECTRIC SENSOR APPLICATIONS

Figure 24-19. Photoelectric sensors are used to detect objects without touching the objects.

Scanning Techniques

Photoelectric sensors consist of two separate major components: a light source (phototransmitter) and a photosensor (photoreceiver). The light source emits a beam of light and the photosensor detects the beam of light.

The light source and photosensor may be housed in the same enclosure or in separate enclosures. *Scanning* is the process of using a light source and photosensor together to measure a change in light intensity when a target is present in, or absent from, the transmitted light beam.

When the photosensor detects the target, it sends a signal to the control circuit. The control circuit processes the signal and activates a solid-state output switch (thyristor or transistor). The output switch energizes or de-energizes a solenoid, relay, magnetic motor starter, or other load.

The phototransmitter and photoreceiver may be set up for several different scanning techniques. The best technique depends on the particular application. The scanning method used for an application depends on the environment of the scanning area. In many applications, several methods of scanning work, but normally one method is the best. **See Figure 24-20.**

Factors that determine the best scanning technique include the following:

- Scanning distance—An application may require the target be a few millimeters or several feet away.
- Size of target—The target may be as small as a needle or as large as a truck.

- Reflectance level of target—All targets reflect the transmitted light. However, light-colored targets reflect more of the transmitted light than dark-colored targets.
- Target positioning—Targets may enter the detection area in the same position or in different positions.
- Differences in color and reflective properties between the background and the target—The transmitted light beam is reflected by the background as well as the target.
- Changes in the ambient light intensity—The photosensor may be affected by the amount of natural (or artificial) light at the detection area.
- Condition of the surrounding air—The transmitted light beam is affected by the quality (amount of impurities) of the air that the transmitted light beam must travel through. Impurities reduce the range of the transmitted light beam.

SCANNING METHODS

Methods	Features		
	Configuration	Advantages	Disadvantages
Direct	Transmitter on one side sends signal to receiver on other side; object to be detected passes between the transmitter and receiver	Reliable performance in contaminated areas; long range scanning; most well-defined, effective beam of all scanning techniques	Wiring and alignment required for both transmitter and receiver; high installation cost
Retroreflective	Transmitter and receiver are housed in one package and are placed on same side of object to be detected; signal from transmitter is reflected to receiver by retroreflector	Ease of installation in that wiring on only one side is required; alignment need not be exact; more tolerant to vibration	Sensitive to contamination since light source must travel to retroreflector and back; hard to detect transparent or translucent materials; not good for small part detection
Polarized	Transmitter and receiver housed in one package and placed on side of object to be detected; special lens is used to filter light beam to project it in one plane only	Only depolarized light from transmitter is detected, ignoring other unwanted light sources	Detection distance and plane of detection limited
Specular	Transmitter sends signal to receiver by reflecting signal of object to be detected; transmitter and receiver are not housed in same package and receiver must be positioned precisely to receive reflected light	Good for detecting shiny versus dull surfaces; depth of field can be changed by changing transmitter/receiver angle	Wiring required for both transmitter and receiver; proper alignment is important
Diffuse	Transmitter and receiver are housed in one package; object being detected reflects signal back to detector; no retroreflector is used	Ease of installation in that wiring on only one side is required since detected object returns signal; exact alignment is not critical; best scanning technique for transparent or translucent materials	Limited range since object is used to reflect transmitted light; performance changes from one type of object to be detected to another
Convergent Beam	Transmitter and receiver are housed in one package and are placed on side of object to be detected; light beam is focused to a fixed point in front of controller	Detection point is fixed so that objects before or beyond focal point are not detected	Detection point is very small, not allowing for that much variation in distance that may be caused by such factors as vibration

Figure 24-20. The scanning method used for an application depends on the environment of the scanning area.

The most common scanning techniques used with photoelectric switches are the direct, retroreflective, polarized, specular, diffuse, and convergent beam scanning methods.

Direct Scan. *Direct scan (transmitted beam, thru-beam, opposed scan)* is a method of scanning in which the transmitter and receiver are placed opposite each other so that the light beam from the transmitter shines directly at the receiver. The target must pass directly between the transmitter and receiver. The target size should be at least 50% of the diameter of the receiver lens to block enough light for detection. **See Figure 24-21.** For very small targets, a special converging lens or aperture may be used.

DIRECT SCAN

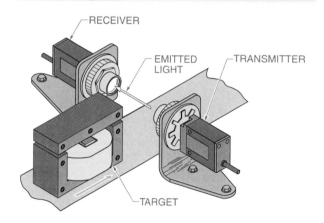

Figure 24-21. Direct scan is a method of scanning in which the target is detected as it passes between the transmitter and receiver.

The direct scan method should generally be the first choice for scanning targets that block most of the light beam. Because the light beam travels in only one direction, direct scan provides long-range sensing and works well in areas with heavy dust, dirt, mist, etc. Direct scan may be used at distances of over 100′.

Retroreflective Scan. *Retroreflective scan (retro scan)* is a method of scanning in which the transmitter and receiver are housed in the same enclosure and the transmitted light beam is reflected back to the receiver from a reflector. The light beam is directed at a reflector that returns the light beam to the receiver when no target is present. When a target blocks the light beam, the output switch is activated. **See Figure 24-22.**

RETROREFLECTIVE SCAN

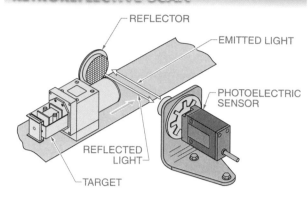

Figure 24-22. Retroreflective scan is a method of scanning in which the target is detected as it passes between the photoelectric sensor and reflector.

Alignment is not critical with retroreflective scan. Reflector misalignment up to 15° does not normally affect operation. This makes it a good choice for high-vibration applications. Retroreflective scan is used in applications in which sensing is possible from only one side at distances up to about 40′. Retroreflective scan does not work well in applications that require the detection of translucent or transparent materials.

Polarized Scan. *Polarized scan* is a method of scanning in which the receiver responds only to the depolarized reflected light from corner cube reflectors or polarized sensitive reflective tape. The light source (emitter) and photoreceiver in a polarized scanner are located on the same side of the object to be detected. A special lens filters the beam of light from the emitter so that it is projected in one plane only. **See Figure 24-23.** The receiver ignores the light reflected from most varieties of shrink-wrap materials, shiny luggage, aluminum cans, or common reflective objects. Thus, the receiver picks up the reflection from the reflector but cannot pick up the reflection from most shiny targets.

Specular Scan. *Specular scan* is a method of scanning in which the transmitter and receiver are placed at equal angles from a highly reflective surface. The angle at which light strikes the reflective surface equals the angle at which it reflects from the surface. This is similar to billiards in which a ball leaves the cushion at an angle equal to the angle at which it struck the cushion. **See Figure 24-24.**

POLARIZED SCAN

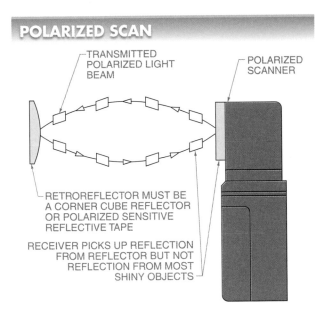

Figure 24-23. Polarized scan uses a special lens that filters the beam of light from the emitter so it is projected in one plane only.

SPECULAR SCAN

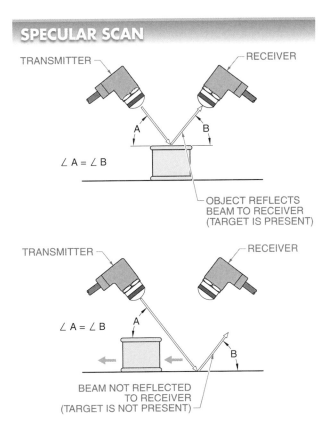

Figure 24-24. Specular scan is a method of scanning in which the transmitter and receiver are placed at equal angles from a highly reflective surface.

Specular scan distinguishes between shiny and nonshiny (matte) surfaces. For example, specular scan may detect a break when a newspaper is being printed. The newspaper may be moved over a stainless steel plate in which a photoelectric sensor is positioned. When a break occurs, the photoelectric sensor detects the break and stops the press.

Diffuse Scan. *Diffuse scan (proximity scan)* is a method of scanning in which the transmitter and receiver are housed in the same enclosure and a small percentage of the transmitted light beam is reflected back to the receiver from the target. In diffuse scan, the transmitter and the receiver are placed in the same enclosure so the receiver picks up some of the diffused (scattered) light. The target detected may be large or small. **See Figure 24-25.**

DIFFUSE SCAN

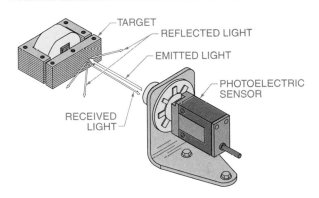

Figure 24-25. Diffuse scan is a method of scanning in which the target is detected when some of the emitted light is reflected and received.

Diffuse scan is used in color mark detection to detect the amount of light that is reflected from a printed surface. Color marks (registration or index marks) are used for registering a specific location on a product. For example, registration marks are used in packaging applications to determine the cutoff point and to identify the point for adding printed material.

The color of a registration mark is selected to provide enough contrast so that the diffuse scanner can detect the difference between the registration mark and the background material. Black marks against a white background provide the best contrast. However, to provide a better selection of sensors, manufacturers offer transmitters with infrared, visible red, green, or white light sources. By using different colors of transmitted light, many different color registration marks may be used with different color backgrounds.

Convergent Beam Scan. *Convergent beam scan* is a method of scanning that simultaneously focuses and converges a light beam to a fixed focal point in front of a photoreceiver. **See Figure 24-26.** Glass optical fiber cables are used with photoelectric sensors to conduct the transmitted light into and out of the sensing area. The optical fiber cables are used as light pipes.

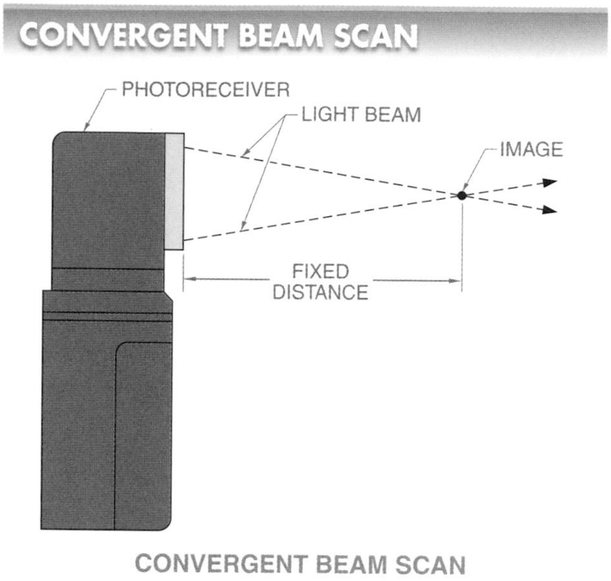

CONVERGENT BEAM SCAN

CONVERGENT BEAM SCAN

Figure 24-26. Convergent beam scan is a method of scanning that simultaneously focuses and converges a light beam to a fixed focal point in front of a photoreceiver.

Glass optical fiber cables can withstand much higher temperatures than plastic optical fiber (POF) cables. Glass cables can typically withstand temperatures up to 500°F. Plastic cables are usually limited to about 158°F. Plastic cable is more suitable than glass cable for applications that require severe bending over short distances because plastic is less susceptible to breakage than glass. Although plastic cable is less expensive than glass cable, glass performs better than plastic, offering better transmission quality at higher speeds over longer distances.

The control beam is transmitted through an optical fiber cable and returned to the receiver through a separate cable either combined in the same cable assembly (known as a bifurcated fiber bundle) or within a separate cable assembly. Retroreflective scan and diffuse scan use a bifurcated cable and direct scan uses two separate cables (emitter and receiver). **See Figure 24-27.** Scan distances commonly vary from 0.4″ to 54″, depending on the scanning technique. An optical lens accessory that attaches to certain cable ends significantly increases scan distances.

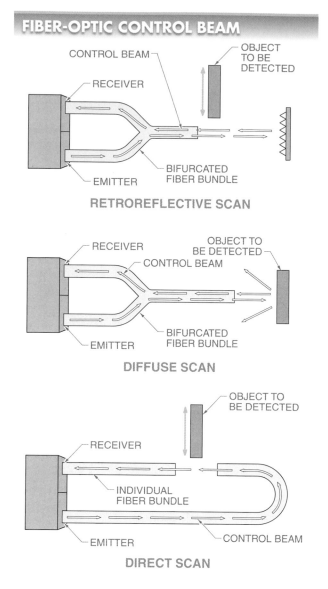

FIBER-OPTIC CONTROL BEAM

RETROREFLECTIVE SCAN

DIFFUSE SCAN

DIRECT SCAN

Figure 24-27. Fiber optics use transparent fibers of glass or plastic to conduct and guide light energy.

Fiber-optic controllers are available in different sizes and configurations. **See Figure 24-28.** Combining the optical fiber cables with photoelectric sensors enables their use in limited mounting spaces and for small parts detection and detection in applications having high temperature, high vibration, or high electrical noise levels.

Convergent beam scanning is used to detect products that are inches away from another reflective surface. It is a good choice for edge-guiding or positioning clear or translucent materials. The well-defined beam makes convergent beam scanning a good choice for sensing the position of opaque materials.

FIBER-OPTIC CONTROLLERS

Honeywell

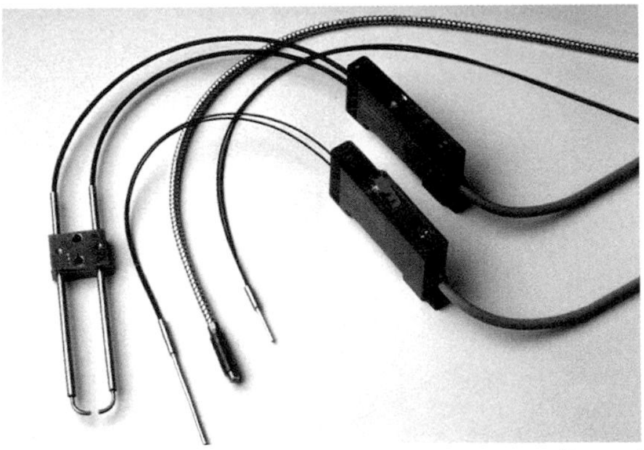

Figure 24-28. Fiber-optic controllers are available in different sizes and configurations.

MODULATED AND UNMODULATED LIGHT

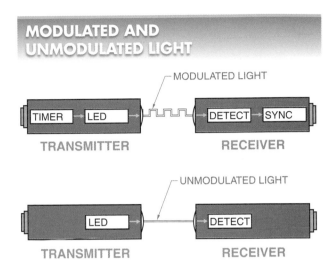

Figure 24-29. Modern photoelectric sensors use modulated or unmodulated light.

The optical system of a convergent beam scanner can only sense light reflected back from an object in its focal point. The scanner is blind a short distance before and beyond the focal point. Operation is possible when highly reflective backgrounds are present. Convergent beam scanning is used for detecting the presence or absence of small objects while ignoring nearby background surfaces.

Parts on a conveyor can be sensed from above while the conveyor belt is ignored. Parts may also be sensed from the side without the detection of guides or rails directly in back of the object. Convergent beam scanning can detect the presence of fine wire, resistor leads, needles, bottle caps, pencils, stack height of materials, and fill level of clear liquids. It is also capable of sensing bar code marks against a contrasting background.

Modulated and Unmodulated Light

Although some older photoelectric transmitters used white light (unmodulated), modern photoelectric light sources produce infrared light, which can be either modulated or unmodulated. Most photoelectric sensors today use modulated infrared light. **See Figure 24-29.**

In a modulated light source, the light source is turned on and off at a very high frequency, normally several kilohertz (kHz). The control responds to this modulated frequency rather than just the intensity of the light. Because the receiver circuitry is tuned to the phototransmitter modulating frequency, the control does not respond to ambient light. This feature also helps to reject other forms of light (noise). A modulated light source should always be considered first when a photoelectric sensors is used.

In an unmodulated light source, the light beam is constantly ON and is not turned on and off. Unmodulated light is considered when the scanning range is very short and when dirt, dust, and bright ambient light conditions are not a problem. Unmodulated light sources are also used for high-speed counting because the beam is continually transmitting and responds quickly. Most manufacturers offer both types of photoelectric sensors.

Tech Fact

Although most sensors use a modulated light that is unaffected by natural or artificial light, high-speed sensors often use an unmodulated light because they have a higher switching rate. Since natural or artificial light can cause detection problems, it is best to shield the sensing head from outside light by placing a light blocking cover over the sensing head area.

Response Time

Response time is the number of pulses (objects) per second a controller can detect. The response time of a photoelectric control must be considered when the object to be detected moves past the beam at a very high speed or when the object to be detected is not much bigger than the effective beam of the controller. This information is listed in the specification sheet of the photoelectric control. For example, a photoelectric control may have an activating frequency of 10 pulses per second. This means the photoelectric control, on average, can detect an object passing by it every $\frac{1}{10}$ (0.1) sec. **See Figure 24-30.**

RESPONSE TIME

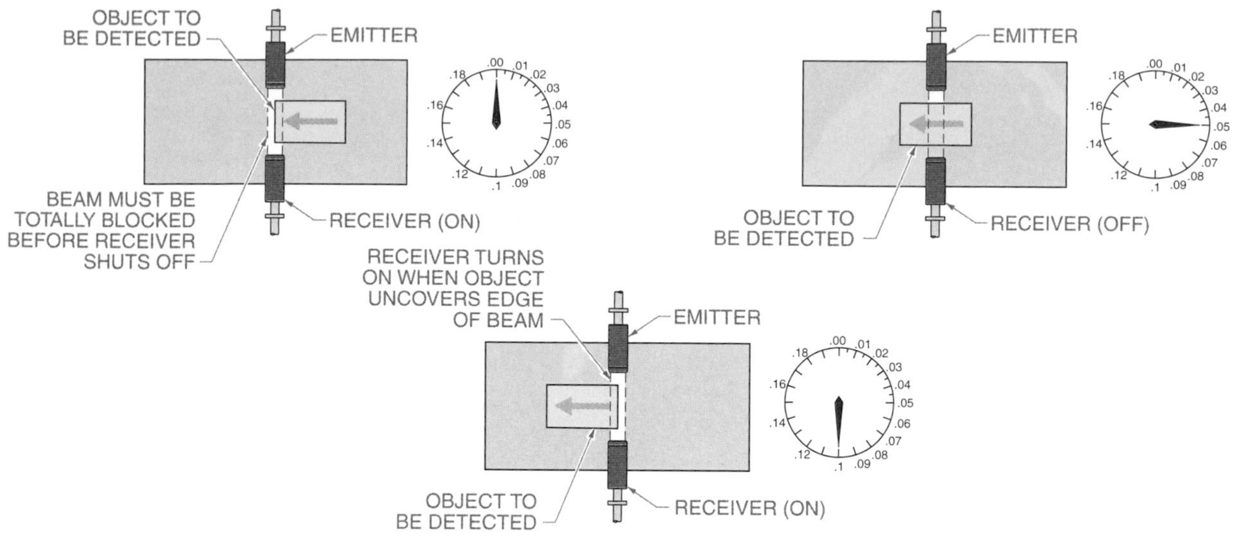

Figure 24-30. The response time of a photoelectric control is the number of pulses per second the sensor can detect.

The beam must be totally blocked before the receiver shuts off. The receiver turns on when the object uncovers an edge of the beam. This has the effect of shortening the size of the object to be detected as seen by the photoelectric control.

The length of time that an object breaks the beam is found by applying the following formula:

$$t = \frac{w - D}{s}$$

where

t = time object takes to break beam (in sec)

w = width of object moving through beam (in in.)

D = effective beam diameter (in in.)

s = speed of object (in in./sec)

Example: Calculating Object Beam Break Time

What is the length of time it takes an object that is 2¼″ wide to pass a ¼″ diameter beam when the object is moving 2″ per second?

$$t = \frac{w - D}{s}$$

$$t = \frac{2.25 - 0.25}{2}$$

$$t = \frac{2}{2}$$

$$t = \textbf{1 sec}$$

A photoelectric control rated for 10 pulses per second may be used because the object takes 1 sec to pass the photoelectric sensor. If the speed of the object is increased to 10″ per second, the length of time it takes the object to move past the photoelectric sensor is 0.2 sec ([2.25 − 0.25] ÷ 10 = 0.2 sec). A photoelectric control rated for 10 pulses per second may be used because it can detect an object passing by it every ¹⁄₁₀ (0.1) sec and the object takes 0.2 sec to pass the photoelectric sensor.

Sensitivity Adjustment

Many photoelectric and proximity sensors have a sensitivity adjustment screw that determines the operating point or the intensity of light that triggers the output. **See Figure 24-31.** This adjustment allows the sensitivity to be set between a minimum and maximum range. The adjustment is made after the unit has been installed and the minimum and maximum settings for the application are experimentally determined. The sensitivity adjustment is normally set halfway between the minimum and maximum points.

A low sensitivity setting may be desirable, especially when there is bright ambient light or electrical noise interference, or when detecting translucent objects. Reducing the sensitivity may prevent false triggering of the control in these conditions. Some photoelectric sensors include a two-color LED. When the LED is red, the photoelectric sensor is operating at its maximum range. For optimum performance, the photoelectric sensor should be operated when the LED is green.

SENSITIVITY ADJUSTMENT

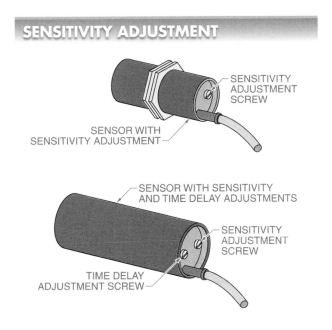

Figure 24-31. Many photoelectric sensors have adjustment screws for setting sensitivity of operation and may also have a screw to adjust time delay.

Dark-Operated/Light-Operated

Basic photoelectric controls are designed to be dark-operated. A *dark-operated photoelectric control* is a photoelectric control that energizes the output switch when a target is present (breaking the beam). Some photoelectric controls include an optional feature that allows the control to be set in a light-operated mode. A *light-operated photoelectric control* is a photoelectric control that energizes the output switch when the target is missing (removed from the beam). **See Figure 24-32.**

Dark-operated (DO) or light-operated (LO) mode is usually set using a selector switch located on the photoelectric control. The selector switch is typically marked with a light-operated/dark-operated (LO/DO) position or an invert (INV) position. For example, in a security system, the dark-operated mode is used to activate the switch contacts that sound an alarm when a person walks into the beam. The light-operated mode is used to activate the switch contacts that sound an alarm when a person removes an object (toolbox, painting, etc.) from the light beam.

PHOTOELECTRIC CONTROL

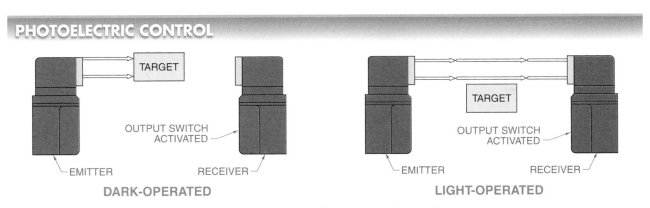

Figure 24-32. A photoelectric control can have a dark-operated or light-operated mode.

24-4 CHECKPOINT

1. What type of scanning method works best when other reflective surfaces are in the same area but must not affect the sensor detecting the actual target reflector?

2. What type of scanning method provides the longest detection distance because the light travels in only one direction and is considered the best first choice when selecting a detection method?

3. What type of scanning method works best when detecting a color mark used to register the location of a product or point on a product?

4. Does a modulated or unmodulated light source work best in a high ambient light area, such as when a machine is used under a skylight?

5. When detecting an object, when is the known response time of a photoelectric sensor important?

6. When an attempt is made to detect an object once it breaks the light beam, should a dark-operated or light-operated mode be used?

24-5 PHOTOELECTRIC CONTROL APPLICATIONS

Photoelectric sensors can be used in a variety of control applications for the detection of almost any object. For example, photoelectric sensors are used along a production line for counting, positioning, sorting, and safety. They are used in security systems to detect the presence or removal of an object. Photoelectric sensors are also used to detect the presence of vehicles at tollgates, parking areas, and truck loading bays. In most applications, photoelectric sensors are used as inputs into timers, relays, counters, programmable controllers, and motor control circuits.

Height and Distance Monitoring

Photoelectric sensors may be used to monitor a truck loading bay for clearance and distance. **See Figure 24-33.** In this application, any truck 14'-0" or larger must be unloaded at another bay. (The dimensions vary depending on particular needs.) In the control circuit, photoelectric sensor 1 (PHOTO 1) turns on an alarm in line 2 if the truck is too high. Photoelectric sensor 2 (PHOTO 2) starts a recycle timer (TR) that flashes a yellow light on and off at a distance of 2' from the dock. Photoelectric sensor 3 (PHOTO 3) starts a recycle timer that turns on a red light at a distance of 6" from the dock.

The best scanning method for this application is direct scan with modulated controls. The photoelectric sensor is connected for a dark-operated controller, which allows for operation only when the truck blocks the beam.

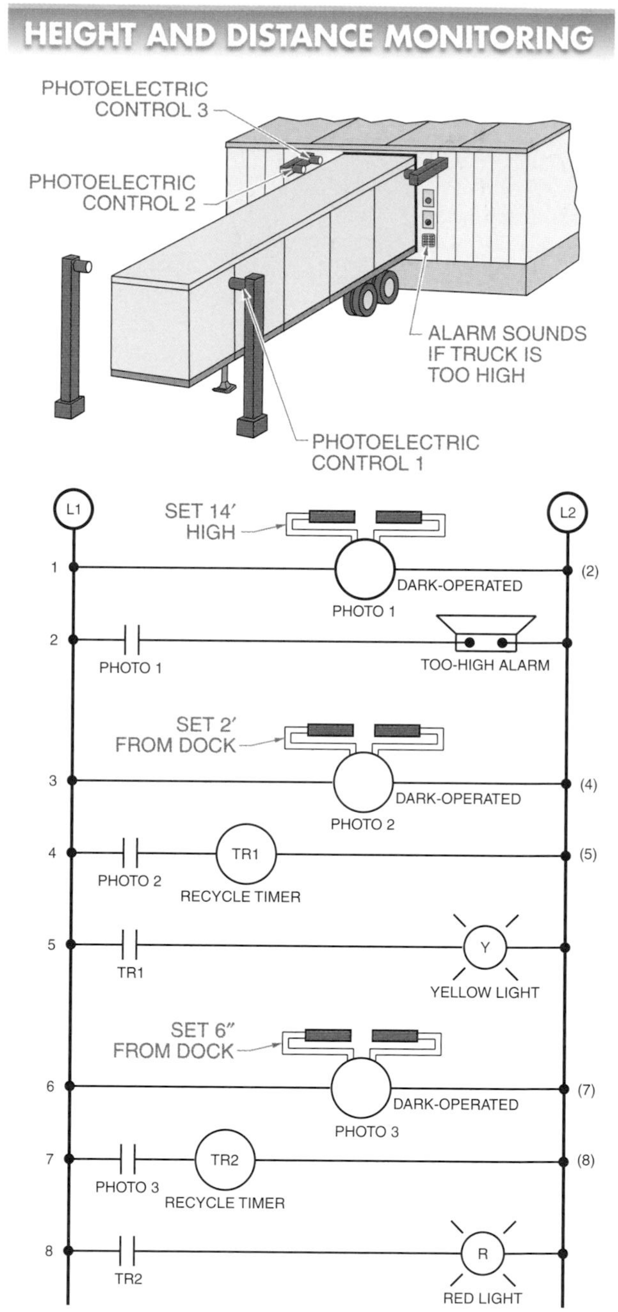

Product Monitoring

Photoelectric sensors may be used to detect a backup of a product on a conveyor line. **See Figure 24-34.** In this application, three photoelectric sensors are used to turn on a warning light and turn off the conveyor motor if required.

Photoelectric sensor 1 (PHOTO 1) turns on a warning light, which indicates that product is at the end of the conveyor line. At this time, an operator may remove the product or wait until more products are on the line. This allows the operator to be time-efficient. If the product backs up to photoelectric sensor 2 (PHOTO 2), a recycle timer is activated, which flashes the warning light. At this time, the operator should unload the conveyor. If the conveyor is not unloaded and product backs up to photoelectric sensor 3 (PHOTO 3), an on-delay timer is activated, which stops the conveyor motor after a few seconds, thus preventing a problem.

Figure 24-33. Photoelectric sensors may be used to monitor a truck loading bay for clearance and distance.

All upstream conveyors and machines must also be turned off to prevent a jam. In this application, retroreflective scan is used for ease of installation and because of vibration of the conveyor line.

PRODUCT MONITORING

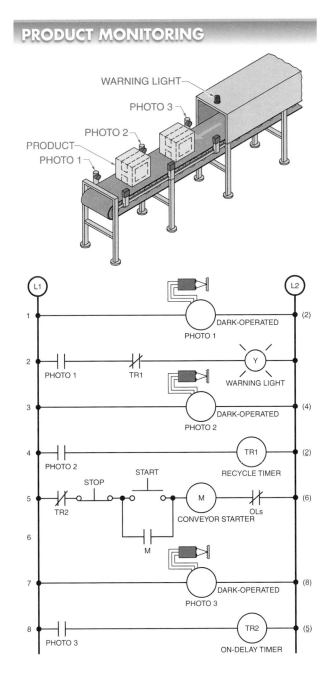

Figure 24-34. Photoelectric sensors may be used to detect a backup of a product on a conveyor line.

Mounting Photoelectric Sensors

A photoelectric sensor transmits a light beam. The light beam detects the presence (or absence) of an object. Only part of the light beam is effective when detecting the object. The *effective light beam* is the area of light that travels directly from the transmitter to the receiver. The object is not detected if the object does not completely block the effective light beam.

The receiver is positioned to receive as much light as possible from the transmitter when photoelectric sensors are mounted. When more light is available at the receiver, greater operating distances are allowed and more power is available for the system to see through dirt and debris in the air and on the transmitter and receiver lenses. The transmitter is mounted on the clean side of the detection zone because light scattered by debris on the transmitter lens affects the system more than light scattered by debris on the receiver lens. **See Figure 24-35.**

MOUNTING PHOTOELECTRIC SENSORS

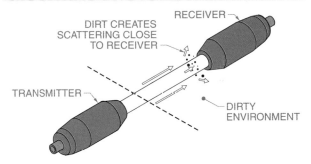

CORRECT SENSOR MOUNTING

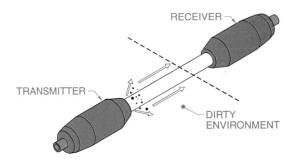

INCORRECT SENSOR MOUNTING

Figure 24-35. A transmitter is mounted on the clean side of the detection zone because light scattered by dirt on the transmitter lens affects the system more than light scattered by dirt on the receiver lens.

24-5 CHECKPOINT

1. In the truck loading bay application shown in Figure 24-33, is it best to use a modulated or unmodulated light source?

2. In the conveyor line application shown in Figure 24-34, what type of scanning method is used to detect the product?

Additional Resources

Review and Resources

Access Chapter 24 Review and Resources for *Electrical Motor Controls for Integrated Systems* by scanning the above QR code with your mobile device.

Applying Your Knowledge

Refer to the *Electrical Motor Controls for Integrated Systems* Learner Resources for interactive Applying Your Knowledge questions.

Workbook and Applications Manual

Refer to Chapter 24 in the *Electrical Motor Controls for Integrated Systems Workbook* and the *Applications Manual* for additional exercises.

ENERGY EFFICIENCY PRACTICES

Photovoltaics

Photovoltaics is solar energy technology that uses the unique properties of semiconductors to convert solar radiation into electricity. Photovoltaics is an environmentally friendly technology that produces energy with no noise or pollution, while conserving nonrenewable energy sources. Photovoltaic (PV) systems use silicon wafers that are sensitive to light and produce a small direct current when exposed to light. When PV cells are combined into large systems (modules), they produce a significant amount of electricity.

Electricity supplied by PV systems displaces electricity supplied from other technologies. Finite supplies of fossil fuels and a demand for renewable, clean energy are making PV systems an increasingly attractive alternative energy source. In addition to the potential financial savings, many PV system owners place a high importance on producing clean "green" energy.

PV systems are flexible and can be adapted to many different applications. The modular nature of PV system components makes them easy to expand for increased capacity. PV systems are extremely reliable and have long operating life with minimal maintenance. PV systems can be used in almost any application where electricity is required and can support DC loads, AC loads, or both.

Objectives

25-1
- Define relay and solid-state relay (SSR).
- Describe the circuits that make up a solid-state circuit.
- Explain how SSRs are used to control different circuits.
- Describe the different solid-state switching methods.
- Explain why temperature must be considered when using an SSR.
- Describe relay current and voltage problems.

25-2
- Compare electromechanical relays (EMRs) and solid-state relays (SSRs).

25-3
- Explain how to troubleshoot a solid-state relay (SSR) that fails to turn off a load.
- Explain how to troubleshoot an SSR that fails to turn on a load.
- Explain how to troubleshoot erratic relay operation.

25-4
- Define and describe solid-state motor starters.
- Explain how to wire the power and control circuits of a solid-state motor starter and how it is protected from overcurrents and overloads.
- Describe solid-state motor starting.
- Describe how a solid-state motor starter is programmed.
- Describe the different motor stopping modes.
- Compare the different reduced-voltage starting methods.

Review and Resources
atplearningresources.com/Quicklinks
Access Code: 362245

Chapter
Solid-State Relays
and Starters

Electromechanical relays (EMRs) have been used in control circuits since electricity was used to operate solenoids, lamps, small motors, and heating elements. Typically, EMR contacts were limited to approximately 15 A. As current requirements increased, EMR contacts increased in size and became known as lighting contactors and heating contactors. EMRs are still only designed to handle currents of approximately 15 A or less. Contactors are basically just relays with high-power contacts. When motor overload protection is added to a contactor, it is called a motor starter. If a motor starter includes mechanical contacts operated by a coil, it is called a magnetic motor starter.

As the current switching capabilities of solid-state switching devices improved, solid-state switching began to replace mechanical switching. Solid-state switches that are designed to perform the same control functions as mechanical relays are called solid-state relays (SSRs). When motor overload protection is added to a solid-state switch it is called a solid-state starter, solid-state motor starter, or soft starter. If a solid-state starter also includes speed control, it is called a motor drive.

25-1 SOLID-STATE RELAYS

A *relay* is a device that controls one electrical circuit by opening and closing another circuit. A small voltage applied to a relay results in a larger voltage being switched. A *solid-state relay (SSR)* is a switching device that has no contacts and switches entirely by electronic means. An SSR uses a silicon-controlled rectifier (SCR), triac, or transistor output instead of mechanical contacts to switch the controlled power. The output is optically coupled to a light-emitting diode (LED) light source inside the relay. The relay is turned on when the LED is energized, usually with low-voltage DC power. **See Figure 25-1.**

The industrial control market has moved to solid-state electronics. Due to declining cost, high reliability,

and immense capability, solid-state devices are replacing many devices that operate on mechanical and electrical principles. The selection of a solid-state relay is based on the electrical, mechanical, and cost characteristics of each device and the required application.

SSR Circuits

An SSR circuit consists of an input circuit, a control circuit, and an output (load-switching) circuit. These circuits may be used in any combination to provide many different solid-state switching applications. **See Figure 25-2.**

SOLID-STATE RELAYS (SSRs)

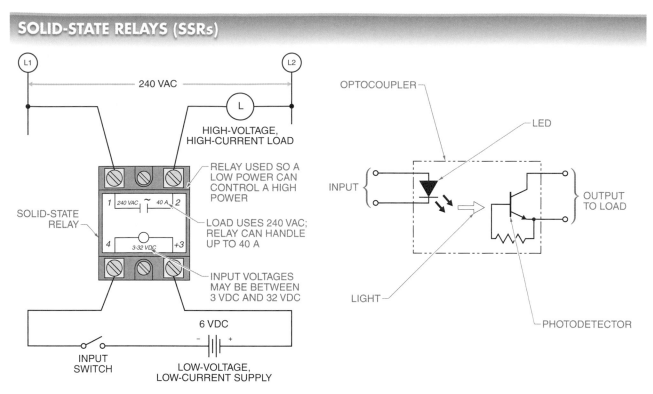

Figure 25-1. A solid-state relay (SSR) is an electronic switching device that has no moving parts.

SSR CIRCUITS

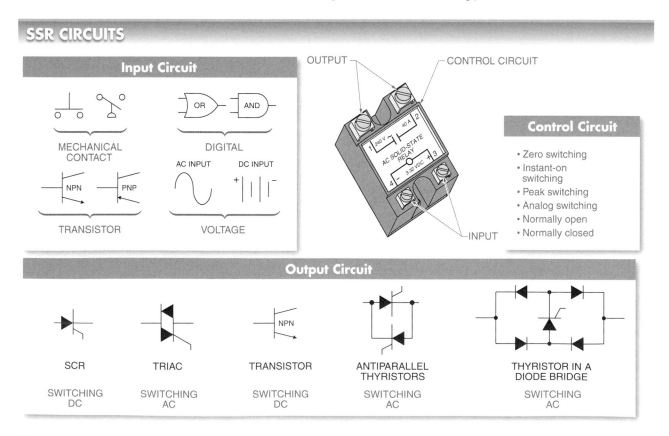

Figure 25-2. An SSR circuit consists of an input circuit, a control circuit, and an output (load-switching) circuit.

Input Circuits. An *input circuit* is the part of an SSR to which the control component is connected. The input circuit performs the same function as the coil of an EMR. The input circuit is activated by applying a voltage to the input of the relay that is higher than the specified pickup voltage of the relay. The input circuit is deactivated when a voltage less than the specified minimum dropout voltage of the relay is applied. Some SSRs have a fixed input voltage rating, such as 12 VDC. Most SSRs have an input voltage range, such as 3 VDC to 32 VDC. The voltage range allows a single SSR to be used with most electronic circuits.

Carlo Gavazzi Inc.

Solid-state relays are used to control circuits electronically, or without the use of contacts.

The input voltage of an SSR may be controlled (switched) through mechanical contacts, transistors, digital gates, etc. Most SSRs may be switched directly by low-power devices, which include integrated circuits, without adding external buffers or current-limiting devices. Variable-input devices, such as thermistors, may also be used to switch the input voltage of an SSR.

Control Circuits. A *control circuit* is the part of an SSR that determines when the output component is energized or de-energized. The control circuit functions as the interface between the input and output circuits. In an SSR, the interface is accomplished by electronic circuitry inside the SSR. In an EMR, the interface is accomplished by a magnetic coil that closes a set of mechanical contacts.

When the control circuit receives the input voltage, the circuit is switched or not switched depending on whether the relay is a zero switching, instant-on, peak switching, or analog switching relay. Each relay is designed to turn on the load-switching circuit at a predetermined voltage point. For example, a zero switching relay allows the load to be turned on only after the voltage across the load is at or near zero. The zero switching function provides a number of benefits, such as the elimination of high inrush currents on the load.

Output (Load-Switching) Circuits. The output (load-switching) circuit of an SSR is the load switched by the SSR. The output circuit performs the same function as the mechanical contacts of an electromechanical relay. However, unlike the multiple output contacts of EMRs, SSRs normally have only one output contact.

Most SSRs use a thyristor as the output switching component. Thyristors change from the OFF state (contacts open) to the ON state (contacts closed) very quickly when their gate switches on. This fast switching action allows for high-speed switching of loads. The output switching device used depends on the type of load to be controlled. Different outputs are required when switching DC circuits than are required when switching AC circuits. Common output switching devices used in SSRs include the following:

- SCRs are used to switch high-current DC loads.
- Triacs are used to switch low-current AC loads.
- Transistors are used to switch low-current DC loads.
- Antiparallel thyristors are used to switch high-current AC loads. They are able to dissipate more heat than a triac.
- Thyristors in diode bridges are used to switch low-current AC loads.

SSR Circuit Capabilities

An SSR can be used to control most of the same circuits that an EMR is used to control. Because an SSR differs from an EMR in function, the control circuit for an SSR differs from that of an EMR. This difference is how the relay is connected into the circuit. An SSR performs the same circuit functions as an EMR but with a slightly different control circuit.

Two-Wire Control. An SSR may be used to control a load using a momentary control such as a pushbutton. **See Figure 25-3.** In this circuit, a pushbutton signals the SSR, which turns on the load. The pushbutton must be held down to keep the load turned on. The load is turned off when the pushbutton is released. This circuit is identical in operation to the standard two-wire control circuit used with EMRs, magnetic motor starters, and contactors. For this reason, the pushbutton could be changed to any manual, mechanical, or automatic control device for simple ON/OFF operation. The same circuit may be used for liquid level control if the pushbutton is replaced with a float switch.

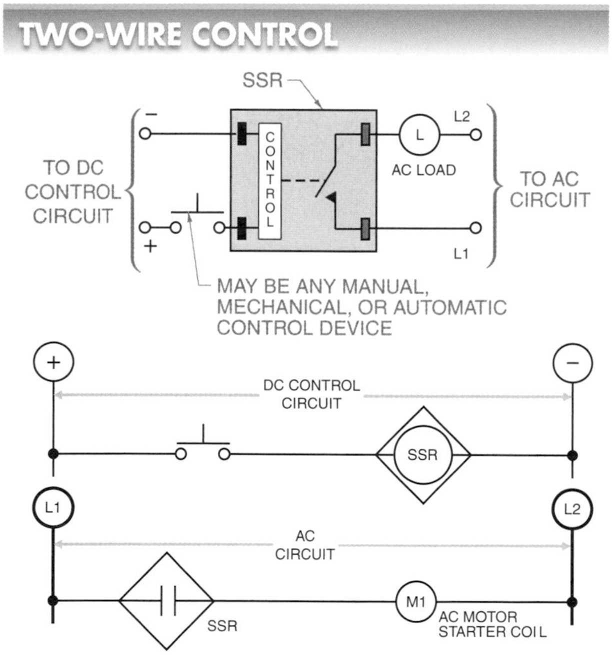

There must be a flow of a definite minimum current to turn on the SCR. This is accomplished when the start pushbutton is pressed. Once the gate of the SCR has voltage applied, the SCR is latched in the ON condition and allows the DC control voltage to pass through even after the start pushbutton is released. Resistor R1 is used as a current-limiting resistor for the gate and is determined by gate current and supply voltage.

Tech Fact

When a motor drive is programmed, the control circuit must be programmed for two-wire or three-wire operation. The term "two-wire" means that a switch can perform two functions. The term "three-wire" means that a switch can perform only one function. All three-wire switches require a second switch to control the load.

TWO-WIRE CONTROL

Figure 25-3. An SSR may be used to control a load using a momentary control such as a pushbutton.

THREE-WIRE MEMORY CONTROL

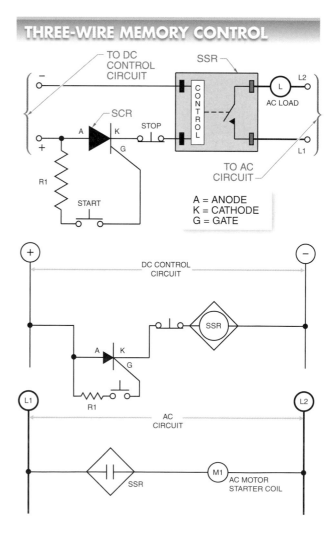

A = ANODE
K = CATHODE
G = GATE

Figure 25-4. An SSR may be used with an SCR to latch a load in the ON condition.

Three-Wire Memory Control. An SSR may be used with an SCR to latch a load in the ON condition. **See Figure 25-4.** This circuit is identical in operation to the standard three-wire memory control circuit. An SCR is used to add memory after the start pushbutton is pressed. An SCR acts as a current-operated OFF-to-ON switch. The SCR does not allow the DC control current to pass through until a current is applied to its gate.

Equivalent NC Contacts. An SSR may be used to simulate an equivalent normally closed (NC) contact condition. An NC contact must be electrically made because most SSRs have the equivalent of a normally open (NO) contact. This is accomplished by allowing the DC control voltage to be connected to the SSR through a current-limiting resistor (R). **See Figure 25-5.** The load is held in the ON condition because the control voltage is present on the SSR. The selector switch is moved to turn off the load. This allows the DC control voltage to take the path of least resistance and electrically remove the control voltage from the relay. This also turns off the load until the pushbutton is released.

Transistor Control. SSRs are also capable of being controlled by electronic control signals from logic gates and transistors. **See Figure 25-6.** In this circuit, the SSR is controlled through an NPN transistor that receives its signal from IC logic gates, etc. Two resistors (R1 and R2) are used as current-limiting resistors.

EQUIVALENT NC CONTACTS

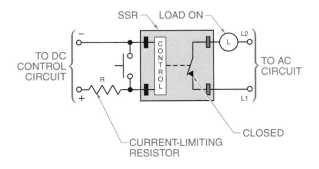

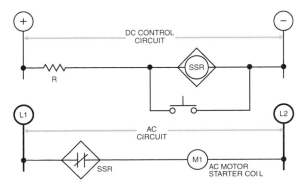

Figure 25-5. An SSR with a current-limiting resistor may be used to simulate an equivalent normally closed (NC) contact condition.

TRANSISTOR CONTROL

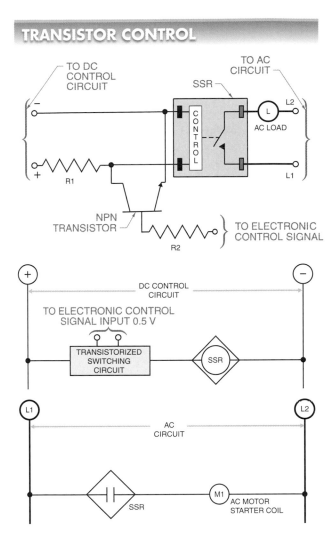

Figure 25-6. SSRs may be controlled by electronic control signals from logic gates and transistors.

Series and Parallel Control. SSRs can be connected in series or parallel to create multicontacts that are controlled by one input device. Multicontact SSRs may also be used. Three SSR control inputs may be connected in parallel so that when the switch is closed, all three are actuated. **See Figure 25-7.** This controls the 3ϕ circuit.

In this application, the DC control voltage across each SSR is equal to the DC supply voltage because they are connected in parallel. When a multicontact SSR is used, there is only one input that controls all output switches.

SSRs can be connected in series to control a 3ϕ circuit. **See Figure 25-8.** The DC supply voltage is divided across the three SSRs when the switch is closed. For this reason, the DC supply voltage must be at least three times the minimum operating voltage of each relay.

PARALLEL-CONNECTED AND MULTICONTACT SSRs

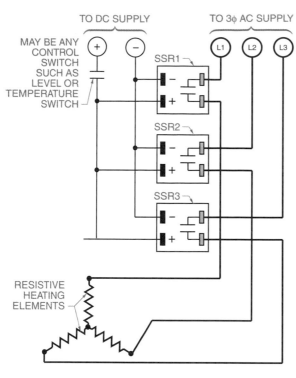

PARALLEL-CONNECTED SSRs

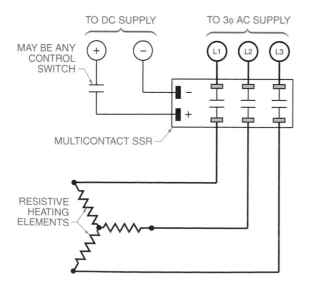

MULTICONTACT SSRs

Figure 25-7. Three SSRs may be connected in parallel to control a 3φ circuit or a multicontact SSR may be used.

SERIES-CONNECTED SSRs

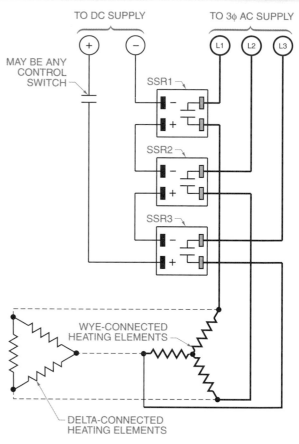

SERIES-CONNECTED SSRs

Figure 25-8. Three SSRs may be connected in series to control a 3φ circuit. When SSRs are connected in series, the DC supply must be three times the minimum operating voltage of each relay.

SSR Switching Methods

The SSR used in an application depends on the load to be controlled. The different SSRs are designed to properly control certain loads. The four basic SSRs are the zero switching (ZS) relay, instant-on (IO) relay, peak switching (PS) relay, and analog switching (AS) relay.

Zero Switching. A *zero switching relay* is an SSR that turns on the load when the control voltage is applied and the voltage at the load crosses zero (or within a few volts of zero). The relay turns off the load when the control voltage is removed and the current in the load crosses zero. **See Figure 25-9.**

ZERO SWITCHING SSRs

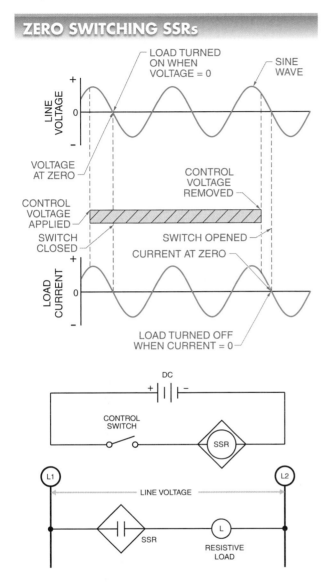

Figure 25-9. A zero switching relay turns on the load when the control voltage is applied and the voltage at the load crosses zero.

The zero switching relay is the most widely used relay. Zero switching relays are designed to control resistive loads. Zero switching relays control the temperature of heating elements, soldering irons, extruders for forming plastic, incubators, and ovens. Zero switching relays control the switching of incandescent lamps, tungsten lamps, flashing lamps, and programmable controller interfacing.

Instant-On Switching. An *instant-on switching relay* is an SSR that turns on the load immediately when the control voltage is present. This allows the load to be turned on at any point on the AC sine wave. The relay

turns off when the control voltage is removed and the current in the load crosses zero. Instant-on switching is exactly like electromechanical switching because both switching methods turn on the load at any point on the AC sine wave. **See Figure 25-10.**

INSTANT-ON SWITCHING SSRs

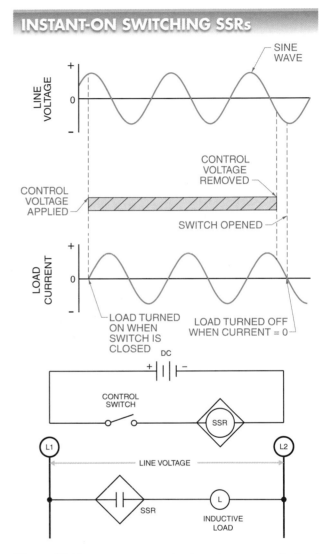

Figure 25-10. An instant-on switching relay turns on the load immediately when the control voltage is present.

Instant-on relays are designed to control inductive loads. In inductive loads, voltage and current are not in phase and the loads turn on at a point other than the zero voltage point that is preferred. Instant-on relays control the switching of contactors, magnetic valves and starters, valve position controls, magnetic brakes, small motors (used for position control), 1ϕ motors, small 3ϕ motors, lighting systems, programmable controller interfaces, and phase controls.

Peak Switching. A *peak switching relay* is an SSR that turns on the load when the control voltage is present and the voltage at the load is at its peak. The relay turns off when the control voltage is removed and the current in the load crosses zero. Peak switching is preferred when the voltage and current are about 90° out of phase because switching at peak voltage is switching at close to zero current. **See Figure 25-11.**

Analog Switching. An *analog switching relay* is an SSR that has an infinite number of possible output voltages within the rated range of the relay. An analog switching relay has a built-in synchronizing circuit that controls the amount of output voltage as a function of the input voltage. This allows for a ramp-up function of the load. In a ramp-up function, the voltage at the load starts at a low level and is increased over a period of time. The relay turns off when the control voltage is removed and the current in the load crosses zero. **See Figure 25-12.**

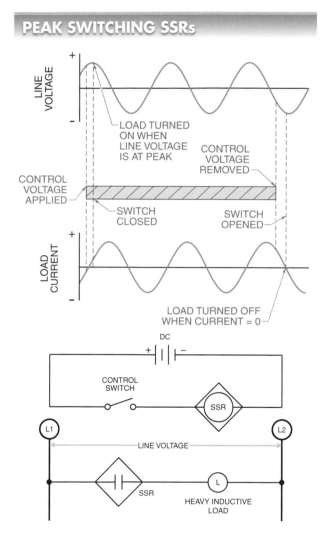

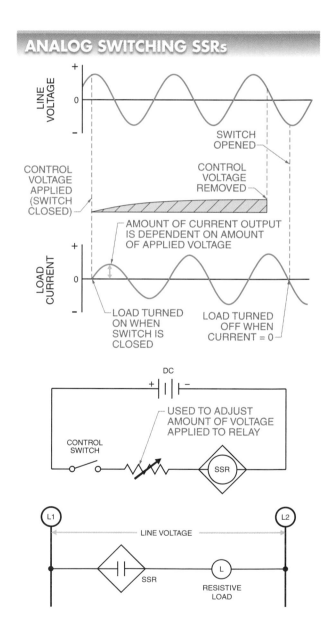

Figure 25-11. A peak switching relay turns on the load when the control voltage is present and the voltage at the load is at its peak.

Figure 25-12. An analog switching relay has an infinite number of possible output voltages within the rated range of the relay.

Peak switching relays control transformers and other heavy inductive loads and limit the current in the first half period of the AC sine wave. Peak switching relays control the switching of transformers, large motors, DC loads, high inductive lamps, magnetic valves, and small DC motors.

A typical analog switching relay has an input control voltage of 0 VDC to 5 VDC. These low and high limits correspond respectively to no switching and full switching on the output load. For any voltage between 0 VDC and 5 VDC, the output is a percentage of the available output voltage. However, the output is normally nonlinear when compared to the input. Also, the manufacturer's data must be checked.

Analog switching relays are designed for closed-loop applications. One closed-loop application is a temperature control with feedback from a temperature sensor to the controller. In a closed-loop system, the amount of output is directly proportional to the amount of input signal. For example, if there is a small temperature difference between the actual temperature and the set temperature, the load (heating element) is given low power. However, if there is a large temperature difference between the actual temperature and the set temperature, the load (heating element) is given high power. This relay may also be used for starting high-power incandescent lamps to reduce the inrush current.

SSR Temperature Considerations

Temperature rise is the largest problem in applications that use an SSR. As temperature increases, the failure rate of SSRs increases. As temperature increases, the number of operations of an SSR decreases. The higher the heat in an SSR, the more problems occur. **See Figure 25-13.**

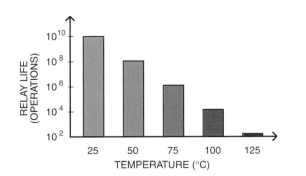

Figure 25-13. As temperature increases, the number of operations of an SSR decreases.

The failure rate of most SSRs doubles for every 10°C temperature rise above an ambient temperature of 40°C. An ambient temperature of 40°C is considered standard by most manufacturers.

Solid-state relay manufacturers specify the maximum relay temperature permitted. The relay must be properly cooled to ensure that the temperature does not exceed the specified maximum safe value. Proper cooling is accomplished by installing the SSR to the correct heat sink. A heat sink is chosen based on the maximum amount of load current controlled.

Heat Sinks. The performance of an SSR is affected by ambient temperature. The ambient temperature of a relay is a combination of the temperature of the relay location and the type of enclosure used. The temperature inside an enclosure may be much higher than the ambient temperature of an enclosure that allows good air flow.

The temperature inside an enclosure increases if the enclosure is located next to a heat source or in the sun. The electronic circuit and SSR also produce heat. Forced-air cooling is required in some applications.

Selecting Heat Sinks. A low resistance to heat flow is required to remove the heat produced by an SSR. The opposition to heat flow is thermal resistance. *Thermal resistance (R_{TH})* is the ability of a device to impede the flow of heat. Thermal resistance is a function of the surface area of a heat sink and the conduction coefficient of the heat sink material. Thermal resistance (R_{TH}) is expressed in degrees Celsius per watt (°C/W).

Heat sink manufacturers list the thermal resistance of heat sinks. The lower the thermal resistance value, the more easily the heat sink dissipates heat. The larger the thermal resistance value, the less effectively the heat sink dissipates heat. The thermal resistance value of a heat sink is used with an SSR load current/ambient temperature chart to determine the size of the heat sink required. **See Figure 25-14.**

THERMAL RESISTANCE

Heat Sink Selections		
Type	H × W × L (mm)	R$_{TH}$ (°C/W)
01	15 × 79 × 100	2.5
02	15 × 100 × 100	2.0
03	25 × 97 × 100	1.5
04	37 × 120 × 100	0.9
05	40 × 60 × 150	0.5
06	40 × 200 × 150	0.4

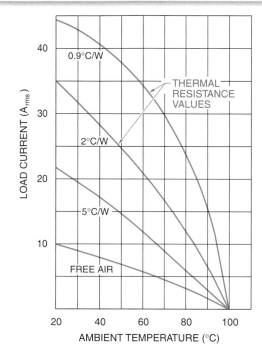

40 A RELAY LOAD CURRENT/AMBIENT
TEMPERATURES

Figure 25-14. The thermal resistance value of a heat sink is used with an SSR load current/ambient temperature chart to determine the size of the heat sink required.

A relay can control a large amount of current when a heat sink with a low thermal resistance number is used. A relay can control the least amount of current when no heat sink (free-air mounting) is used. Heat conduction through a relay and into a heat sink can be maximized with the following recommendations:
• Use heat sinks made of a material that has a high thermal conductivity. Silver has the highest thermal conductivity rating. Copper has the highest practical thermal conductivity rating. Aluminum has a good thermal conductivity rating and is the most cost-effective and widely used heat sink.

• Keep the thermal path as short as possible.
• Use the largest cross-sectional surface area in the smallest space.
• Always use thermal grease or pads between the relay housing and the heat sink to eliminate air gaps and aid in thermal conductivity.

Mounting Heat Sinks. A heat sink must be correctly mounted to ensure proper heat transfer. A heat sink can be properly mounted using the following recommendations:

• Choose a smooth mounting surface. The surfaces between a heat sink and a solid-state device should be as flat and smooth as possible. Ensure that the mounting bolts and screws are securely tightened.
• Locate heat-producing devices so that the temperature is spread over a large area. This helps prevent higher temperature areas.
• Use heat sinks with fins to achieve as large a surface area as possible.
• Ensure that the heat from one heat sink does not add to the heat from another heat sink.
• Always use thermal grease between the heat sink and the solid-state device to ensure maximum heat transfer.

Relay Current Problems

The overcurrent passing through an SSR must be kept below the maximum load current rating of the relay. An overload protection fuse is used to prevent overcurrents from damaging an SSR. An overload protection fuse opens the circuit when the current is increased to a higher value than the nominal load current. The fuse should be an instantaneous fuse used for the protection of semiconductors. **See Figure 25-15.**

Relay Voltage Problems

Most AC power lines contain voltage spikes superimposed on the voltage sine wave. Voltage spikes are produced by the switching of motors, solenoids, transformers, motor starters, contactors, and other inductive loads. Large spikes are also produced when lightning strikes the power distribution system.

The output element of a relay can exceed its breakdown voltage and turn on for part of a half period if overvoltage protection is not provided. This short turn-on can cause problems in the circuit.

RELAY OVERCURRENT PROTECTION

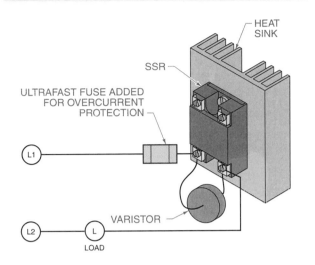

Figure 25-15. An overload protection fuse opens the circuit when the current is increased to a higher value than the nominal load current.

Varistors are added to the relay output terminals to prevent an overvoltage problem. A varistor should be rated 10% higher than the line voltage of the output circuit. The varistor bypasses the transient current. **See Figure 25-16.**

RELAY OVERVOLTAGE PROTECTION

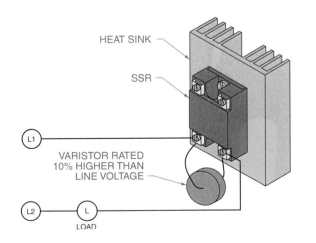

Figure 25-16. Varistors are added to relay output terminals to prevent an overvoltage problem.

Voltage Drop. In all series circuits, the total circuit voltage is dropped across the circuit components. The higher the resistance of any component, the higher the voltage drop. The lower the resistance of any component, the lower the voltage drop. Thus, an open switch that has a meter connected across it shows a very high voltage drop because the meter and open switch have very high resistance when compared to the load. Conversely, a closed switch that has a meter connected across it shows a very low voltage drop because the meter is closed and closed switches have very low resistance when compared to the load.

Carlo Gavazzi Inc.
A solid-state contactor with an integrated fuse holder is used for short-circuit protection for both the solid-state semiconductor and the load cables.

A voltage drop in the switching component is unavoidable in an SSR. The voltage drop produces heat. The larger the current passing through the relay, the greater the amount of heat produced. The generated heat affects relay operation and can destroy the relay if it is not removed. **See Figure 25-17.**

VOLTAGE DROP HEAT

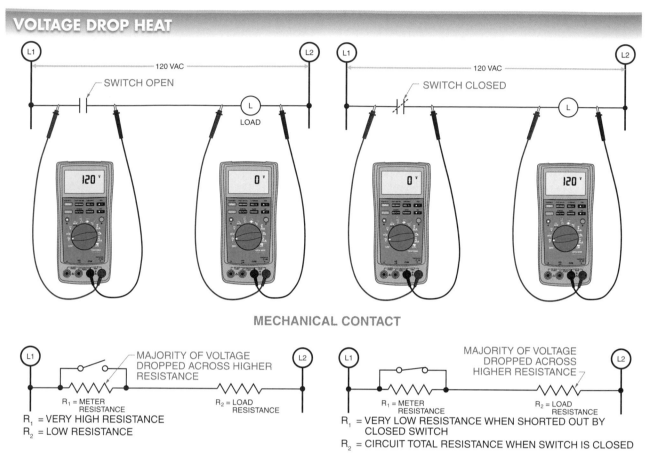

MECHANICAL CONTACT

MECHANICAL CONTACT EQUIVALENT CIRCUIT

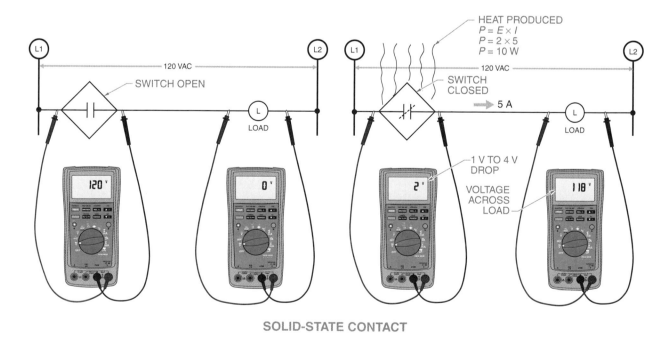

SOLID-STATE CONTACT

Figure 25-17. The voltage drop in the switching component of an SSR produces heat and can destroy the relay if it is not removed.

The voltage drop in an SSR is usually 1 V to 1.6 V, depending on the load current. For small loads (less than 1 A), the heat produced is safely dissipated through the relay's case. High-current loads require a heat sink to dissipate the extra heat. **See Figure 25-18.**

For example, if the load current in a circuit is 1 A and the SSR switching device has a 2 V drop, the power generated is 2 W. The 2 W of power generates heat that can be dissipated through the relay's case.

However, if the load current in a circuit is 20 A and the SSR switching device has a 2 V drop, the power generated in the device is 40 W. The 40 W of power generates heat that requires a heat sink to safely dissipate the heat.

SSR VOLTAGE DROP		
Load Current (in A)	Voltage Drop (in V)	Power at Switch (in W)
1	2	2
2	2	4
5	2	10
10	2	20
20	2	40
50	2	100

Figure 25-18. For small loads (less than 1 A), the heat produced in an SSR is safely dissipated through the relay's case.

25-1 CHECKPOINT

1. What two types of thyristor are used in an SSR to switch DC output loads?
2. What is a zero-switching relay?
3. What is a peak switching relay?
4. What type of SSR is designed to control resistive loads?
5. How can a heat problem be reduced when using an SSR?
6. What device can be added to the relay output terminals to reduce any problems transients may cause?

25-2 ELECTROMECHANICAL AND SOLID-STATE RELAY COMPARISON

EMRs and SSRs are designed to provide a common switching function. An EMR provides switching through the use of electromagnetic devices and sets of contacts. An SSR depends on electronic devices such as SCRs and triacs to switch without contacts. In addition, the physical features and operating characteristics of EMRs and SSRs are different. **See Figure 25-19.**

An equivalent terminology chart is used as an aid in the comparison of EMRs and SSRs. Because the basic operating principles and physical structures of the devices are so different, it is difficult to find a direct comparison of the two. Differences arise almost immediately both in the terminology used to describe the devices and in their overall ability to perform certain functions. **See Figure 25-20.**

Advantages and Limitations

EMRs and SSRs are used in many applications. The relay used depends on the electrical requirements, cost requirements, and life expectancy of the application. Although SSRs have replaced EMRs in many applications, EMRs are still very common. EMRs offer many advantages that make them cost-effective. However, they have limitations that restrict their use in some applications.

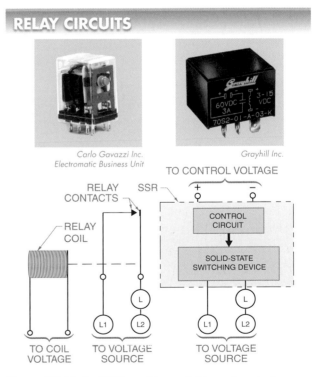

Figure 25-19. An EMR provides switching using electromagnetic devices. An SSR depends on SCRs and triacs to switch without contacts.

EMR/SSR EQUIVALENT TERMINOLOGY CHART

EMRs		SSRs	
Term	**Definition**	**Term**	**Definition**
Coil Voltage	Minimum voltage necessary to energize or operate relay; also referred to as pickup voltage	Control Voltage	Minimum voltage required to gate or activate control circuit of SSR; generally, a maximum value is also specified
Coil Current	Amount of current necessary to energize or operate relay	Control Current	Minimum current required to turn on solid-state control circuit; generally, a maximum value is also specified
Holding Current	Minimum current required to keep a relay energized	Control Current	
Dropout Voltage	Maximum voltage at which the relay is no longer energized	Control Voltage	
Pull-In Time	Amount of time required to operate (open or close) relay contacts after coil voltage is applied	Turn-On Time	Elapsed time between application of control voltage and application of voltage to load circuit
Dropout Time	Amount of time required for the relay contacts to return to their normal de-energized position after coil voltage is removed	Turn-Off Time	Elapsed time between removal of control voltage and removal of voltage from load circuit
Contact Voltage Rating	Maximum voltage rating that contacts of relay are capable of safely switching	Load Voltage	Maximum output voltage-handling capability of an SSR
Contact Current Rating	Maximum current rating that contacts of relay are capable of safely switching	Load Current	Maximum output current-handling capability of an SSR
Surge Current	Maximum peak current which contacts on a relay can withstand for short periods of time without damage	Surge Current	Maximum peak current which an SSR can withstand for short periods of time without damage
Contact Voltage Drop	Voltage drop across relay contacts when relay is operating (usually low)	Switch-On Voltage Drop	Voltage drop across SSR when operating
Insulation Resistance	Amount of resistance measured across relay contacts in open position	Switch-Off Resistance	Amount of resistance measured across an SSR when turned off
No equivalent or comparison		OFF-State Leakage Current	Amount of leakage current through SSR when turned off but still connected to load voltage
No equivalent or comparison		Zero Current Turn-Off	Turn-off at zero crossing of load current that flows through an SSR; a thyristor turns off only when current falls below minimum holding current; if input control is removed when current is a higher value, turn-off is delayed until next zero current crossing
No equivalent or comparison		Zero Voltage Turn-On	Initial turn-on occurs at a point near zero crossing of AC line voltage; if input control is applied when line voltage is at a higher value, initial turn-on is delayed until next zero crossing

Figure 25-20. An equivalent terminology chart is used as an aid in the comparison of EMRs and SSRs.

EMR advantages include the following:

- normally have multipole, multithrow contact arrangements
- contacts can switch AC or DC
- low initial cost
- very low contact voltage drop, thus no heat sink is required
- very resistant to voltage transients
- no OFF-state leakage current through open contacts

EMR limitations include the following:

- contacts wear, thus having a limited life
- short contact life when used for rapid switching applications or high-current loads
- generate electromagnetic noise and interference on the power lines
- poor performance when switching high inrush currents

SSRs provide many advantages such as small size, fast switching, long life, and the ability to handle complex switching requirements. SSRs have some limitations that restrict their use in some applications.

SSR advantages include the following:

- very long life when properly applied
- no contact to wear
- no contact arcing to generate electromagnetic interference
- resistant to shock and vibration because they have no moving parts
- logic compatible with programmable controllers, digital circuits, and computers
- very fast switching capability
- different switching modes (zero switching, instant-on, etc.)

SSR limitations include the following:

- normally only one contact available per relay
- heat sink required due to voltage drop across switch
- can switch only AC or DC
- OFF-state leakage current when switch is open
- normally limited to switching only a narrow frequency range such as 40 Hz to 70 Hz

Input Signals

The application of voltage to the input coil of an electromagnetic device creates an electromagnet that is capable of pulling in an armature with a set of contacts attached to control a load circuit. It takes more voltage and current to pull in the coil than to hold it in due to the initial air gap between the magnetic coil and the armature. The specifications used to describe the energizing and de-energizing process of an electromagnetic device are coil voltage, coil current, holding current, and drop-out voltage.

An SSR has no coil or contacts and requires only minimum values of voltage and current to turn it on and off. The two specifications needed to describe the input signal for an SSR are control voltage and control current.

The electronic nature of an SSR and its input circuit allows easy compatibility with digitally controlled logic circuits. Many SSRs are available with minimum control voltages of 3 V and control currents as low as 1 mA, which makes them ideal for a variety of current state-of-the-art logic circuits.

Response Time

One of the significant advantages of an SSR over an EMR is its response time (ability to turn on and turn off). An EMR may be able to respond hundreds of times per minute, but an SSR is capable of switching thousands of times per minute with no chattering or bounce.

DC switching time for an SSR is in the microsecond range. AC switching time for an SSR, with the use of zero-voltage turn-on, is less than 9 ms. The reason for this advantage is that the SSR may be turned on and turned off electronically much more rapidly than a relay may be electromagnetically pulled in and dropped out.

The higher speeds of SSRs have become increasingly more important as industry demands higher productivity from processing equipment. The more rapidly the equipment can process or cycle its output, the greater its productivity.

Voltage and Current Ratings

EMRs and SSRs have certain limitations that determine how much voltage and current each device can safely handle. The values vary from device to device and from manufacturer to manufacturer. Data sheets are used to determine whether a given device can safely switch a given load. The advantages of SSRs are that they have a capacity for arcless switching, have no moving parts to wear out, and are totally enclosed, thus allowing them to be operated in potentially explosive environments without special enclosures. The advantage of EMRs is that the contacts can be replaced if the device receives an excessive surge current. In an SSR, the complete device must be replaced if there is damage.

Voltage Drop

When a set of contacts on an EMR closes, the contact resistance is normally low unless the contacts are pitted or corroded. However, because an SSR is constructed of semiconductor materials, it opens and closes a circuit by increasing or decreasing its ability to conduct. Even at full conduction, an SSR presents some residual resistance, which can create a voltage drop of up to approximately 1.5 V in the load circuit. This voltage drop is usually considered insignificant because it is small in relation to the load voltage and in most cases presents no problems. This unique feature may have to be taken into consideration when load voltages are small. A method of removing the heat produced at the switching device must be used when load currents are high.

Insulation and Leakage

The air gap between a set of open contacts provides an almost infinite resistance through which no current flows. Due to their unique construction, SSRs provide a very high but measurable resistance when turned off. SSRs have a switched-off resistance not found on EMRs.

It is possible for small amounts of current (OFF-state leakage) to pass through an SSR because some conductance is still possible even though the SSR is turned off. OFF-state leakage current is not found on EMRs.

OFF-state leakage current is the amount of current that leaks through an SSR when the switch is turned off, normally about 2 mA to 10 mA. The rating of OFF-state leakage current in an SSR is usually determined at 200 VDC across the output and should not usually exceed more than 200 mA at this voltage.

25-2 CHECKPOINT

1. Does an SSR or EMR have the advantage of usually including multiple contacts?
2. Does an SSR or EMR have the advantage of a much longer operating life when properly used in an application?
3. Does an SSR or EMR have the advantage of being able to switch both AC and DC loads using the same device?
4. Does an SSR or EMR have the advantage of being able to switch a wider range of frequencies?

25-3 TROUBLESHOOTING SOLID-STATE RELAYS

Troubleshooting an SSR is accomplished by either the exact replacement method or the circuit analysis method. The *exact replacement method* is a method of SSR replacement in which a bad relay is replaced with a relay of the same type and size. The exact replacement method involves making a quick check of the relay's input and output voltages. The relay is assumed to be the problem and is replaced when there is only an input voltage being switched.

The *circuit analysis method* is a method of SSR replacement in which a logical sequence is used to determine the reason for a failure. Steps are taken to prevent the problem from recurring once the reason for a failure is known. The circuit analysis method of troubleshooting is based on three improper relay operations:
• The relay fails to turn off the load.
• The relay fails to turn on the load.
• The relay operates erratically.

Relay Fails to Turn Off Load

A relay may not turn off the load to which it is connected when the relay fails. This condition occurs either when the load is drawing more current than the relay can withstand, the heat sink for the relay is too small, or transient voltages are causing breakover at the relay output. A *transient voltage* is a temporary, unwanted voltage in an electrical circuit. Overcurrent permanently shorts the relay's switching device if the load draws more current than the rating of the relay. High temperature causes thermal runaway of the relay's switching device if the heat sink does not remove the heat.

If the power lines are likely to have transients (usually from inductive loads connected on the same line), the relay should be replaced with one of a higher voltage rating and/or a transient suppression device should be added to the circuit. **See Figure 25-21.**

To troubleshoot an SSR that fails to turn off a load, the following procedure is applied:

1. Disconnect the input leads from the SSR. See Step 3 if the relay load turns off. The relay is the problem if the load remains on and the relay is normally open.

2. Measure the voltage of the circuit that the relay is controlling. The line voltage should not be higher than the rated voltage of the relay. Replace the relay with a relay that has a higher voltage rating if the line voltage is higher than the relay's rating. Check to ensure that the relay is rated for the type of line voltage (AC or DC) that is being used.

3. Measure the current drawn by the load. The current draw must not exceed the relay's rating. For most applications, the current draw should not be more than 75% of the relay's maximum rating.

4. Reconnect the input leads and measure the input voltage to the relay at the time when the control circuit should turn the relay off. The control circuit is the problem and needs to be checked if the control voltage is present. The relay is the problem if the control voltage is removed and the load remains on. Before changing the relay, ensure that the control voltage is not higher than the relay's rated limit when the control circuit delivers the supply voltage. Ensure that the control voltage is not higher than the relay's rated drop-out voltage when the control circuit removes the supply voltage. This condition may occur in some control circuits using solid-state switching.

Relay Fails to Turn On Load

A relay may fail to turn on the load to which it is connected when the relay fails. This condition occurs when the relay's switching device receives a very high voltage spike or the relay's input is connected to a higher-than-rated voltage. A high voltage spike blows open the relay's switching device, preventing the load from turning on. Excessive voltage on the relay's input side destroys the relay's electronic circuit.

Carlo Gavazzi Inc.

Three-phase relays may be used for direct and delta switching of motor loads and direct switching of motor loads with shunting provided by an electromechanical contactor.

RELAY FAILS TO TURN OFF LOAD

CONTROL CIRCUIT

1 DISCONNECT INPUT LEADS FROM RELAY

SSR

120 °

L1

L2

0 °

L

LOAD

MEASURE VOLTAGE OF POWER CIRCUIT 2

4 RECONNECT INPUT LEADS AND MEASURE VOLTAGE WHEN CONTROL CIRCUIT SHOULD TURN RELAY OFF

1.5 A

3 MEASURE CURRENT DRAWN BY LOAD

Figure 25-21. A relay may not turn off the load to which it is connected when the relay fails.

If the power lines are likely to have high voltage spikes, the relay should be replaced with one that has a higher voltage and current rating and/or a transient suppression device should be added to the circuit. **See Figure 25-22.** To troubleshoot an SSR that fails to turn on a load, the following procedure is applied:

1. Measure the input voltage when the relay should be turned on. Troubleshoot the circuit ahead of the relay's input if the voltage is less than the relay's rated pickup voltage. The circuit ahead of the relay is the problem if the voltage is greater than the relay's rated pickup voltage. The higher voltage may have destroyed the relay. The relay may be a secondary problem caused by the primary problem of excessive applied voltage. Correct the high-voltage problem before replacing the relay. The relay or output circuit is the problem if the input voltage is within the pickup limits of the relay.

2. Measure the voltage at the output of the relay. The relay is probably the problem if the relay is not switching the voltage. See Step 3. The problem is in the output circuit if the relay is switching the voltage. Check for an open circuit in the load.

3. Insert a DMM set to measure current in series with the input leads of the relay. Measure the current when the relay should turn on. The relay input is open if no current is flowing. Replace the relay. The relay is bad if the current flow is within the relay's rating. Replace the relay. The control circuit is the problem if current is flowing but is less than that required to operate the relay.

Tech Fact

Before replacing an SSR, it should be verified that the relay type best matches the load type being switched. After replacing the SSR, current is measured to verify that it is at least 80% less than the relay's rating.

RELAY FAILS TO TURN ON LOAD

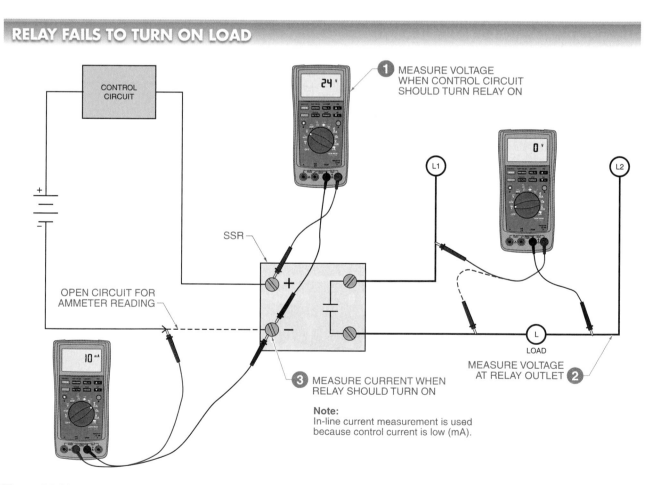

Figure 25-22. A relay may fail to turn on the load to which it is connected when the relay fails.

Erratic Relay Operation

Erratic relay operation is the proper operation of a relay at times and the improper operation of the relay at other times. Erratic relay operation is caused by mechanical problems (loose connections), electrical problems (incorrect voltage), or environmental problems (high temperature). **See Figure 25-23.** To troubleshoot erratic relay operation, the following procedure is applied:

1. Check all wiring and connections for proper wiring and tightness. Loose connections cause many erratic problems. No sign of burning should be present at any terminal. Burning at a terminal usually indicates a loose connection.

2. Ensure that the input control wires are not next to the output line or load wires. The noise carried on the output side may cause unwanted input signals. These can cause a false turn-on through an induced magnetic field in the output conductors, which induces a signal into the input conductors.

3. The relay may be half-waving if the load is a chattering AC motor or solenoid. *Half-waving* is a phenomenon that occurs when a relay fails to turn off because the current and voltage in the circuit reach zero at different times. Half-waving is caused by the phase shift inherent in inductive loads. The phase shift makes it difficult for some solid-state relays to turn off. Connecting an RC or another snubber circuit across the output load should allow the relay to turn off. An *RC circuit* is a circuit in which resistance (R) and capacitance (C) are used to help filter the power in a circuit.

Tech Fact

SSRs can fail in short-circuit condition on the output power side. This creates an unsafe situation because, if the circuit is shorted, unexpected current is flowing in the circuit and loads are either turned on or powered, which could cause injury. Input voltage is measured to verify that there is none and output voltage is measured to verify that the relay is OFF on the power side with no output voltage to the loads.

TROUBLESHOOTING ERRATIC RELAY OPERATION

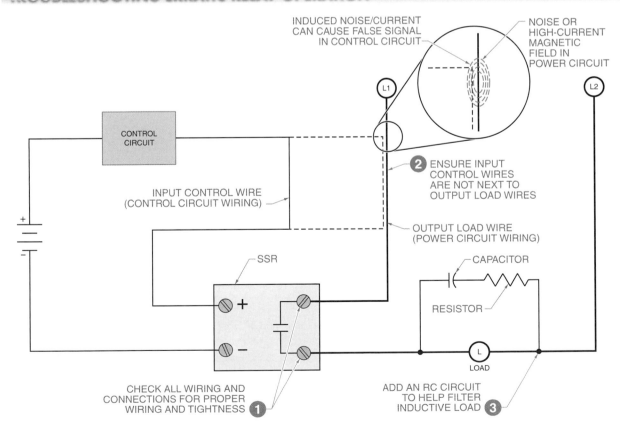

Figure 25-23. Erratic relay operation is the proper operation of a relay at times and the improper operation of the relay at other times.

25-3 CHECKPOINT

1. What three conditions can prevent an SSR from turning off?

2. Why is it important to separate the input control wires of an SSR from its output line or load wires?

25-4 SOLID-STATE MOTOR STARTERS

A *solid-state motor starter* is an electronically operated switch (contactor) that uses solid-state components to eliminate mechanical contacts and includes motor overload protection. Solid-state motor starters are connected into a circuit after the disconnect/overcurrent protection device and before the motor. Solid-state motor starters include motor overload protection and are controlled by the same switches (pushbuttons, pressure switches, etc.) as electromechanical starters. **See Figure 25-24.** Solid-state motor starters consist of solid-state components such as SCRs or triacs that allow current flow when they are conducting and stop current flow when they are not conducting.

Solid-State Motor Starter Sections

Solid-state motor starters have terminals for connecting the incoming supply power (L1/R, L2/S, L3/T) and terminals for connecting a motor (T1/U, T2/V, T3/W). Solid-state motor starters also include a terminal strip for connecting external inputs (pushbuttons, proximity switches, etc.). Solid-state motor starters also include a dual inline package (DIP) switchboard for programming starter functions (starting mode/time, stopping mode, etc.), and potentiometers for adjusting motor full-load current (in amps) and trip class. Solid-state motor starters may also include LEDs to provide visual indication of circuit conditions. **See Figure 25-25.**

SOLID-STATE MOTOR STARTERS

MOTOR CONTROL ENCLOSURE

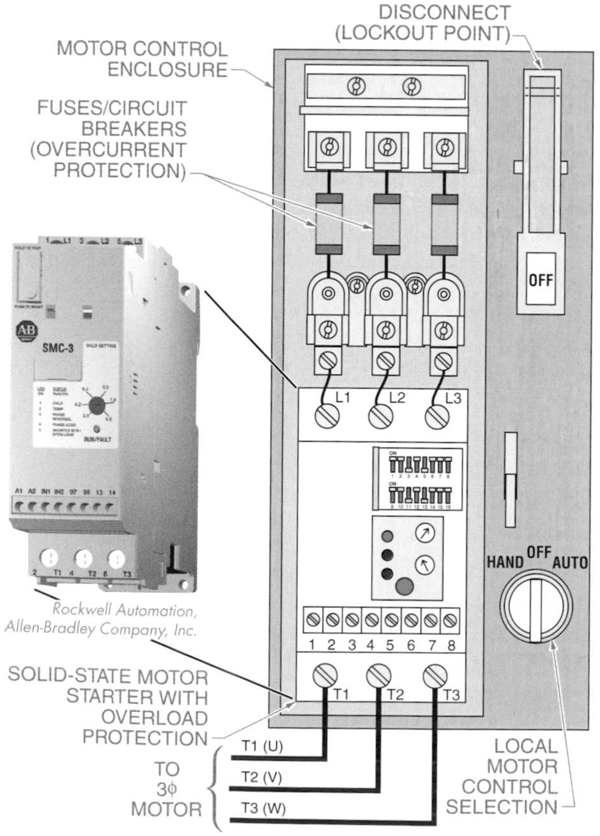

DISCONNECT (LOCKOUT POINT)

FUSES/CIRCUIT BREAKERS (OVERCURRENT PROTECTION)

OFF

Rockwell Automation, Allen-Bradley Company, Inc.

SOLID-STATE MOTOR STARTER WITH OVERLOAD PROTECTION

HAND OFF AUTO

TO 3ϕ MOTOR
- T1 (U)
- T2 (V)
- T3 (W)

LOCAL MOTOR CONTROL SELECTION

Figure 25-24. Solid-state motor starters eliminate electromechanical components by using solid-state components to turn a motor on and off.

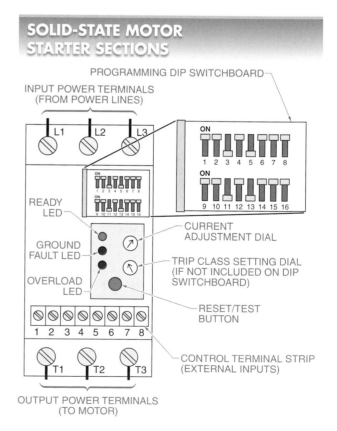

SOLID-STATE MOTOR STARTER SECTIONS

PROGRAMMING DIP SWITCHBOARD

INPUT POWER TERMINALS (FROM POWER LINES)

L1 L2 L3

ON
1 2 3 4 5 6 7 8

ON
9 10 11 12 13 14 15 16

READY LED

GROUND FAULT LED

OVERLOAD LED

CURRENT ADJUSTMENT DIAL

TRIP CLASS SETTING DIAL (IF NOT INCLUDED ON DIP SWITCHBOARD)

RESET/TEST BUTTON

CONTROL TERMINAL STRIP (EXTERNAL INPUTS)

OUTPUT POWER TERMINALS (TO MOTOR)

Figure 25-25. Solid-state motor starters have a control terminal strip, input and output power terminals, dials for current and trip class adjustment, and programming DIP switches.

Wiring Power and Control Circuits

The power circuit of a solid-state motor starter is wired by bringing power from the fuses/circuit breaker into the starter. The incoming power must be at the same voltage level for which the motor is rated or wired. **See Figure 25-26.** The control circuit is wired to the control terminal strip located on the starter. The control circuit voltage is less than the power circuit voltage (typically 12 VDC, 12 VAC, 24 VDC, or 24 VAC). The control terminal strip includes a connection for external control voltage (when required), connections for external control switches (pushbuttons, temperature switches, etc.), and connections for output contacts (alarms, indicating lamps, etc.) that can be used for controlling external loads.

Setting Overload Protection

Motors must be protected from overcurrents and overloads. Fuses and circuit breakers (normally located in the motor disconnect) are used to protect a motor from overcurrents (short circuits and high operating currents). Overloads located in the motor starter protect the motor from overload current caused when the load on the motor is greater than the design torque rating of the motor. Overloads can be thermal overloads (heaters) or solid-state overloads.

Solid-state overloads use a current transformer (CT) to monitor each power line. Solid-state overloads are set by selecting a current limit based on full-load current ratings listed on the motor nameplate and the trip class setting (class 10, 15, 20, or 30). The current limit is set by adjusting the current adjustment dial located on the starter. The *trip class setting* is the length of time it takes for an overload relay to trip and remove power from the motor. The lower the trip class setting, the faster the trip time of the solid-state overload. The higher the trip class setting, the slower the trip time of the solid-state overload.

The trip class setting is based on the motor application (type of load placed on the motor). The trip class setting may be adjusted using a trip class setting dial located near the current adjustment dial or by using DIP switches. **See Figure 25-27.** *Cold trip* is the trip point from the time the motor starts until the first time the overloads trip (motor operating below nameplate rated current). *Hot trip* is the trip point after the overloads have tripped and have been reset (motor operating near or over nameplate rated current).

Tech Fact

Magnetic motor starter overloads usually have a class 20 trip rating. Solid-state starters or drives have a class 10 trip rating. A slower trip time is usually programmed by selecting a class number or, if a number is unavailable, a "fast/minimum" or "slow/maximum" rating.

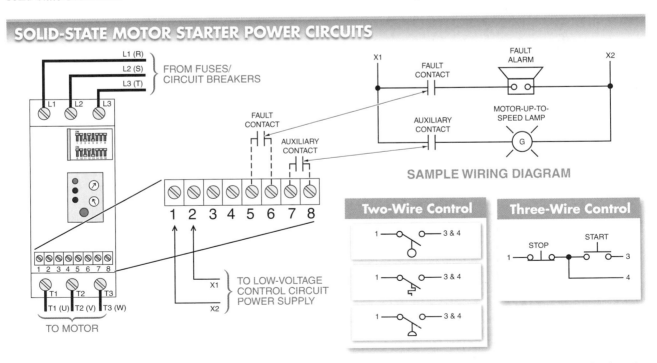

SOLID-STATE MOTOR STARTER POWER CIRCUITS

Figure 25-26. The power circuit of a solid-state motor starter is wired by bringing power from the fuses/circuit breaker into the starter. The control circuit is wired to the control terminal strip located on the starter.

SOLID-STATE OVERLOAD SETTINGS

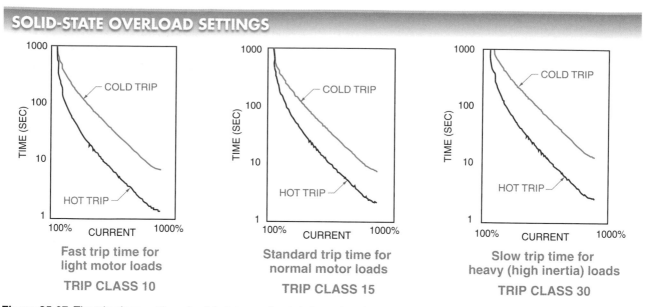

Fast trip time for
light motor loads
TRIP CLASS 10

Standard trip time for
normal motor loads
TRIP CLASS 15

Slow trip time for
heavy (high inertia) loads
TRIP CLASS 30

Figure 25-27. The trip class setting of solid-state overloads is based on the motor application (type of load placed on the motor).

Solid-State Starting

A solid-state reduced-voltage starter ramps up motor voltage as the motor accelerates, instead of applying full voltage instantaneously like across-the-line starters do. Solid-state starters reduce inrush current compared to the high inrush current across-the-line starters produce. Solid-state starters also minimize starting torque, which can damage some loads connected to the motor, and smooth motor acceleration. **See Figure 25-28.**

Solid-state starting provides a smooth, stepless acceleration in applications that require it, such as starting conveyors, compressors, pumps, and a wide range of other industrial applications because of its unique switching capability. The advantage of silicon-controlled rectifiers is that they are small in size, are rugged, and have no contacts. Unlimited lifetime operation can be expected when silicon-controlled rectifiers operate within their specifications.

SOLID-STATE STARTING

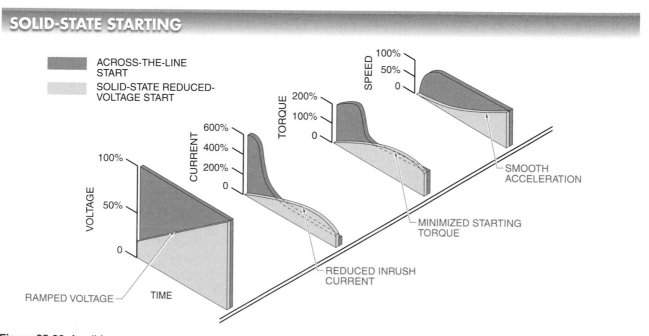

Figure 25-28. A solid-state starter ramps up voltage, reduces inrush current, minimizes starting torque, and smooths acceleration.

Electronic Control Circuitry. A solid-state controller determines to what degree the SCRs should be triggered on to control the voltage, current, and torque applied to a motor. A solid-state controller also includes current-limiting fuses and current transformers for protection of the unit. The current-limiting fuses are used to protect the SCRs from excess current. The current transformers are used to feed information back to the controller. Heat sinks and thermostat switches are also used to protect the SCRs from high temperatures.

The controller also provides the sequential logic necessary for interfacing other control functions of the starter, such as line loss detection during acceleration. The controller is turned off if any voltage is lost or too low on any one line. This may happen if one line opens or a fuse blows.

Solid-State Starting Circuits. A typical solid-state starting circuit consists of both start and run contactors connected in the circuit. The start contactor contacts C1 are in series with the SCRs and the run contactor contacts C2 are in parallel with the SCRs. **See Figure 25-29.**

SOLID-STATE STARTING CIRCUITS

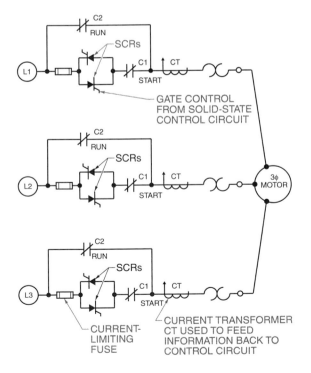

Figure 25-29. An SCR circuit with reverse-parallel wiring of SCRs provides maximum control of an AC load.

The start contacts C1 close and the acceleration of the motor is controlled by triggering on the SCRs when the starter is energized. The SCRs control the motor until it approaches full speed, at which time the run contacts C2 close, connecting the motor directly across the power line. At this point, the SCRs are turned off, and the motor runs with full power applied to the motor terminals.

Motor Starting Modes

Electromechanical and solid-state motor starters can be used to start a motor. When an electromechanical motor starter is used, the motor is connected to the full supply voltage. When a motor is connected to full supply voltage, the motor has the highest possible current draw, the highest possible torque applied to the load, and the shortest acceleration time. This operating condition may be acceptable for some loads. However, many loads cannot be started with high starting torque because they control light loads (small parts, etc.) or delicate loads (paper rolls, etc.). High starting current can also damage the power distribution system and trip breakers or blow fuses. Solid-state motor starters can be programmed for different starting modes to help reduce problems caused by full-voltage starting. Motor starting modes include soft start, soft start with start boost, and current-limit start. **See Figure 25-30.**

Soft Start. Soft start is the most common solid-state starting method. When a starter is set for soft start, the motor is gradually accelerated over a programmable time period, normally 0 sec to 30 sec. Common start time periods include 2 sec, 4 sec, 6 sec, 8 sec, or 16 sec. The starting torque is adjustable to a percent of the motor's locked rotor torque. Common starting torque settings include 15%, 25%, 50%, or 60%. A soft start helps cushion the stress applied to loads connected to the motor.

Soft starting is achieved by increasing the motor voltage in accordance with the setting of the ramp-up control. A potentiometer is used to set the ramp-up time (normally 1 sec to 20 sec). Soft stopping is achieved by decreasing the motor voltage in accordance with the setting of the ramp-down control. A second potentiometer is used to set the ramp-down time (normally 1 sec to 20 sec). A third potentiometer is used to adjust the starting level of motor voltage to a value at which the motor starts to rotate immediately when soft starting is applied. **See Figure 25-31.**

PROGRAMMED START MODES

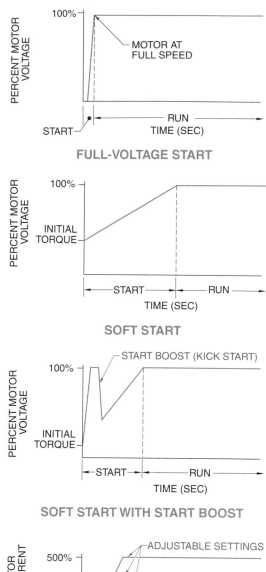

FULL-VOLTAGE START

SOFT START

SOFT START WITH START BOOST

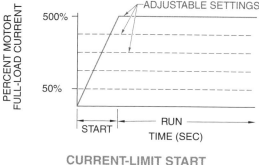

CURRENT-LIMIT START

Figure 25-30. Solid-state motor starters can be programmed for different starting modes to help reduce problems caused by full-voltage starting.

SOFT STARTERS

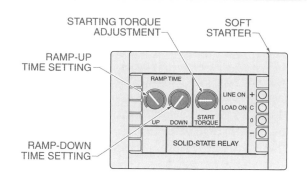

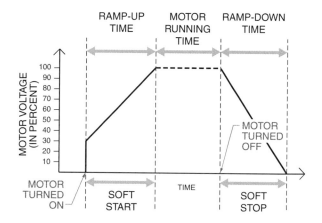

Figure 25-31. A soft starter is a device that provides a gradual voltage increase (ramp up) during motor starting and a gradual voltage decrease (ramp down) during motor stopping.

Like any solid-state switch, a soft starter produces heat that must be dissipated for proper operation. The heat dissipation requires large heat sinks when high-current loads (motors) are controlled. For this reason, a contactor is often added in parallel with a soft starter. The soft starter is used to control the motor when the motor is starting or stopping. The contactor is used to short out the soft starter when the motor is running. This allows for soft starting and soft stopping without the need for large heat sinks during motor running. The soft starter includes an output signal that is used to control the time when the contactor is ON or OFF. **See Figure 25-32.**

Soft Start with Start Boost. When a solid-state motor starter is set for soft start with start boost, the motor is given a current pulse during starting to provide additional starting torque for loads that are difficult to start. The boost time is usually adjustable from 0 sec to 2 sec. The start boost is normally applied when there is a problem starting a motor using only a soft start.

SOFT STARTER CONTACTOR CIRCUITS

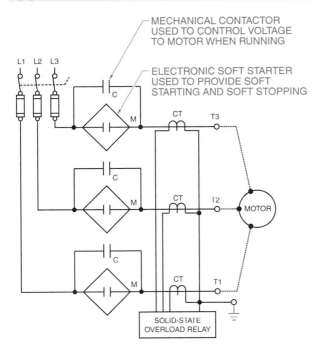

MECHANICAL CONTACTOR USED TO CONTROL VOLTAGE TO MOTOR WHEN RUNNING

ELECTRONIC SOFT STARTER USED TO PROVIDE SOFT STARTING AND SOFT STOPPING

Figure 25-32. A contactor is used with a soft starter to control the voltage to the motor when the motor is running.

Programming Solid-State Motor Starters

A solid-state motor starter must be programmed for proper operation before any power is applied to the starter. A solid-state motor starter is programmed by setting each DIP switch to a predetermined position based on motor and application requirements. The number of DIP switch parameters can range from a few parameters (4 to 6) to numerous parameters. The higher the number of parameters that are available, the greater the number of applications for which the starter can be used. **See Figure 25-33.**

DIP switch parameters include motor starting mode (start time, soft start, start boost, etc.) and the operation of auxiliary contacts (when they are open or closed). Each DIP switch setting must be understood and checked before any power is applied to the starter because some settings can be critical to protecting workers, the motor, and the system.

For example, the overload reset function can be placed in a manual or automatic mode. In the manual mode, the reset button on the starter must be pressed before the motor can be restarted manually (by external pushbuttons, etc.). However, in the automatic reset mode, the starter automatically restarts the motor after a short time period if the external control switch

(pressure switch, etc.) is still closed. This can cause a safety hazard if the person working on or around the system does not know the motor may automatically restart. For this reason, it is important to always refer to the manufacturer literature regarding the setting and meaning of each DIP switch position.

Motor Stopping Modes

Electromechanical and solid-state motor starters can be used to stop a motor when power is removed. When an electromechanical starter is used, the motor coasts to a stop at a rate determined by the load connected to the motor. Solid-state motor starters can be programmed for different stopping modes. This allows greater application flexibility and protection of the motor/load. Motor stopping modes include soft stop, pump control, and brake stop. **See Figure 25-34.**

Soft Stop. Soft stop is the most common solid-state motor starter stopping method. Soft stops allow for an extended controlled stop. In a soft stop, the deceleration time is controlled by the starter, not the load. The soft stop mode is designed for friction loads that tend to stop suddenly when voltage is removed from the motor.

Pump Control. The pump control mode is used to reduce surges that occur when centrifugal pumps are started and stopped. The pump control mode produces smooth acceleration and deceleration of motors and pumps. Common motor and pump starting times range from a few seconds to 30 sec. Common motor and pump stopping times range from a few seconds to 120 sec, depending on the size of the motor and pump.

Brake Stop. Some applications require a fast motor stop. The brake stop mode provides motor braking for a faster stop than a coast stop or soft stop. The amount of braking (and thus braking time) is programmed based on the application requirements. When using the brake stop mode, the longest time is set first and adjusted downward as needed.

Starting Method Comparison

Several starting methods are available when an industrial application calls for the use of reduced-voltage starting. The amount of reduced current, the amount of reduced torque, and the cost of each starting method must be considered when the appropriate starting method is selected. The selection is not simply a matter of selecting the starting method that reduces the current the most. If the starting torque is reduced too much, the motor will not start and the motor overloads will trip.

PROGRAMMING SOLID-STATE MOTOR STARTERS

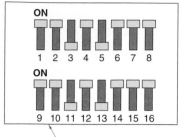

PROGRAMMING DIP SWITCH SELECTION BOARD

Position Number	Description
1	Start time
2	Start time
3	Start mode (current-limit or soft start)
4	Current-limit start setting (when selected) or soft start initial torque setting (when selected)
5	Current-limit start setting (when selected) or soft start initial torque setting (when selected)
6	Soft stop
7	Soft stop
8	Not used
9	Start boost (kick start)
10	Start boost (kick start)
11	Overload class section
12	Overload class section
13	Overload reset
14	Auxiliary relay #1 (normal or up-to-speed)
15	Optional auxiliary relay #2 (normal or up-to-speed)
16	Phase rotation check

Start Time

DIP Switch 1	DIP Switch 2	Time (Seconds)
OFF	OFF	2
ON	OFF	5
OFF	ON	10
ON	ON	15

Start Mode

DIP Switch 3	Setting
OFF	Current limit
ON	Soft start

Soft Start Initial Torque Setting

DIP Switch 4	DIP Switch 5	Initial Torque % LRT
OFF	OFF	15
ON	OFF	25
OFF	ON	35
ON	ON	65

Current-Limit Start Setting

DIP Switch 4	DIP Switch 5	Current Limit % FLA
OFF	OFF	150
ON	OFF	250
OFF	ON	350
ON	ON	450

Soft Stop

DIP Switch 6	DIP Switch 7	Setting
OFF	OFF	Coast-to-reset
ON	OFF	100% of start time
OFF	ON	200% of start time
ON	ON	300% of start time

Start Boost (Kick Start)

DIP Switch 9	DIP Switch 10	Time (Seconds)
OFF	OFF	OFF
ON	OFF	0.5
OFF	ON	1.0
ON	ON	1.5

Overload Class Selection

DIP Switch 11	DIP Switch 12	Trip Class
OFF	OFF	OFF
ON	OFF	10
OFF	ON	15
ON	ON	20

Overload Reset

DIP Switch 13	Reset
OFF	Manual
ON	Automatic

Auxiliary Relay #1

DIP Switch 14	Setting
OFF	Normal
ON	Up-to-speed

Optional Auxiliary Relay #2

DIP Switch 15	Setting
OFF	Normal
ON	Up-to-speed

Phase Rotation Check

DIP Switch 16	Setting
OFF	Enabled
ON	Disabled

Figure 25-33. A solid-state motor starter is programmed by setting each DIP switch to a predetermined position based on motor and application requirements.

PROGRAMMED STOP MODES

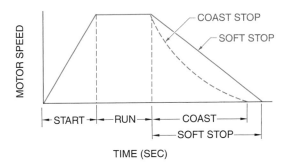

SOFT STOP

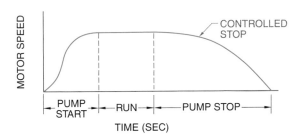

PUMP CONTROL STOP (AND START)

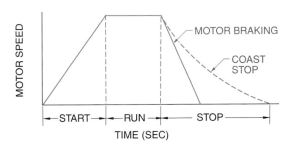

BRAKE STOP

Figure 25-34. Solid-state motor starters can be programmed for different stopping modes to allow greater application flexibility and protection of the motor/load.

A general comparison can be made of the percentage of reduced current for each type of starting method compared to across-the-line starting. **See Figure 25-35.** The amount of reduced current is adjustable when using solid-state or autotransformer starting. Autotransformer starting uses taps so the amount of reduced current is somewhat adjustable. Solid-state starting is adjustable throughout its range. Some primary resistor starters are adjustable, while others are not. Part-winding and wye-delta starting are not adjustable.

PERCENTAGE OF FULL-VOLTAGE LINES STARTING CURRENT

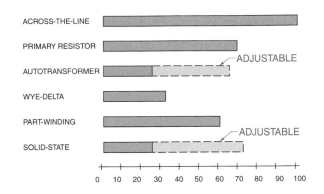

Figure 25-35. The different methods of reduced-voltage starting produce different percentages of full-voltage current.

A general comparison can also be made of the amount of reduced torque for each type of starting method compared to across-the-line starting. **See Figure 25-36.** The amount of reduced torque is adjustable when using the autotransformer starting method. The autotransformer starting method has taps, so the amount of reduced torque is somewhat adjustable. The motor overloads trip if the load requires more torque than the motor can deliver. The torque requirements of the load must be taken into consideration when a starting method is selected.

STARTING METHOD COMPARISON

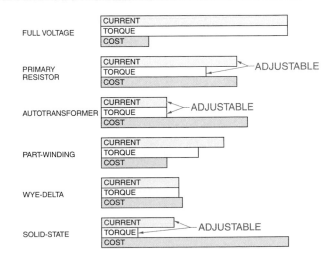

Figure 25-36. The different methods of reduced-voltage starting produce different amounts of reduced torque.

A general comparison can also be made of the costs for each type of starting method compared to across-the-line starting. Although reducing the amount of starting current or starting torque in comparison to the load requirements is the primary consideration for selecting a starting method, installation cost should also be considered. The installation costs for solid-state motor starters tend to be the highest.

The primary resistor starting method is used when it is necessary to restrict inrush current to predetermined increments. Primary resistors can be built to meet almost any current inrush limitation. Primary resistors also provide smooth acceleration and can be used where it is necessary to control starting torque. Primary resistor starting may be used with any motor.

The autotransformer starting method provides the highest possible starting torque per ampere of line current and is the most effective means of motor starting for applications where the inrush current must be reduced with a minimum sacrifice of starting torque.

Three taps are provided on the transformers, making it field adjustable. Its cost must be considered since the autotransformer is the most expensive type of transformer. Autotransformer starting can be used with any motor.

The part-winding starting method is simple in construction and economical in cost. Part-winding starting provides a simple method of accelerating fans, blowers, and other loads involving low starting torque. The part-winding starting method requires a nine-lead wye motor. Its cost is low because no external resistors or transformers are required.

The wye-delta starting method is particularly suitable for applications involving long accelerating times or frequent starts. Wye-delta starting is commonly used for high-inertia loads such as centrifugal air conditioning units, although it can be used in applications where low starting torque is necessary or where low starting current and low starting torque are permissible. The wye-delta starting method requires a special six-lead motor.

25-4 CHECKPOINT

1. Will a class 10 trip allow more or less time for an overload condition to exist before removing power from the motor?

2. Why is a contactor added in parallel with solid-state motor starters?

Additional Resources

Review and Resources

Access Chapter 25 Review and Resources for *Electrical Motor Controls for Integrated Systems* by scanning the above QR code with your mobile device.

Applying Your Knowledge

Refer to the *Electrical Motor Controls for Integrated Systems* Learner Resources for interactive Applying Your Knowledge questions.

Workbook and Applications Manual

Refer to Chapter 25 in the *Electrical Motor Controls for Integrated Systems Workbook* and the *Applications Manual* for additional exercises.

ENERGY EFFICIENCY PRACTICES

Solid-State Relays and Starters

Both mechanical relay contacts and solid-state relay (SSR) switches can be used to turn incandescent lamps and heating elements on and off. If either mechanical relay contacts or an SSR switch turns the lamp or heating element completely on or off, there are no energy savings between the two. However, if an analog SSR is used, energy usage can be reduced by reducing the voltage to the load.

When voltage is lowered to the level required to power an incandescent lamp, the lamp produces less light and uses less energy. When voltage is lowered to the level required to power a heating element, the heating element produces less heat and uses less energy. In many lighting and heating applications, there are times where full light or full heat is not required. For example, an analog SSR can be used to provide full light output during times when a room is occupied but reduced light output (25%, 50%, 75%, etc.) when the room is not occupied. Room occupancy can be determined through the use of a motion sensor or time-based sensor programmed to allow a light to turn on or off at a predetermined time of day.

Likewise, there are times when full heat may not be required. For example, when electric heat is used to heat the water in hot water tank (heater), the amount of time required to bring the water to the desired temperature depends upon the heating element output (in watts) and the amount of water in the tank to heat. An analog SSR can be used to apply full power during peak usage times when recovery time needs to be short due to high usage, such as in the morning hours when there would typically be a high demand for hot water, and reduce the wattage during nonpeak times, such as late night hours when there would be minimal demand for hot water.

Objectives

26-1
- Define and describe motor drives.
- Compare magnetic motor starter and motor drive control circuits.

26-2
- Describe the parameter groups available in motor drives.

26-3
- Describe DC motor drives and list the methods of braking a DC motor.

26-4
- Describe AC motor drives and list their main sections.
- Describe the converter, DC bus, and inverter of an AC motor drive.
- Describe the relationship between voltage and frequency in an AC motor.
- Explain how to control motor speed and torque.
- Describe motor frequencies and pulse width modulation (PWM).
- List and describe different AC motor stopping methods.
- Describe electronic and programmed overloads.
- Explain how to test an AC motor drive, motor, and load.

26-5
- Explain how to measure voltage and current in a motor drive circuit.
- Explain how to troubleshoot reversing motor circuits.

Sections

Review and Resources
atplearningresources.com/Quicklinks
Access Code: 362245

Chapter
Motor Drives

Improvements are continuously being made in the performance, operating speed, and applications of circuit controls. Functions once performed mechanically are now performed electronically. Just as computers have replaced typewriters and calculators have replaced slide rules, electronic motor drives are used to replace mechanical magnetic motor starters. Both magnetic motor starters and motor drives can turn a motor on and off and provide overload protection from high operating currents. However, a motor drive can also provide speed control through its electronic circuits, a controlled acceleration and deceleration time, different types of stopping methods (braking, coasting, etc.), and many more motor operating preset conditions.

Most motor drives also provide and display operating conditions such as the motor's voltage, current, frequency, and power draw during operation in addition to the drive's operating temperature and any recorded system faults. Another advantage of a drive is that when motor reversing is required as part of the control circuit, a second magnetic motor starter is required but already provided with a signal motor starter. Understanding how both magnetic motor starters and motor drives are connected, operate, and control motor functions is important for any application using an electric motor.

26-1 MOTOR DRIVES

A *motor drive* is an electronic unit designed to control the speed of a motor using solid-state components. Motor drives may be AC or DC drives. Both AC and DC drives are used to control motor direction, speed, and acceleration/deceleration time. AC drives control 3ϕ AC motors and DC drives control DC motors.

AC drives are the most common drives because they can easily vary the speed of a 3ϕ motor. Three-phase motors require little or no maintenance compared to 1ϕ AC motors, which contain centrifugal switches and capacitors, and DC motors, which contain brushes and commutators. Another reason AC drives have become common is because smaller drives (less than 3 HP) can be powered by 1ϕ, 115 V or 230 V and control a 3ϕ motor.

These drives are ideal for residential and light commercial HVAC systems and other motor control applications that require speed control and/or energy-saving advantages of 3ϕ motors.

Drives and Magnetic Motor Starters

Magnetic motor starters are used to turn motors on and off and provide overload protection. A magnetic motor starter also provides additional control contacts (NO and NC auxiliary contacts) in addition to power contacts (L1/T1, L2/T2, L3/T3) used to switch the motor on and off. Magnetic motor starters work well in applications that require basic ON/OFF motor control.

In addition to turning motors on and off and providing overload protection, motor drives also provide speed control, timed acceleration and deceleration, motor-starting boost, fault monitoring, programmable set speeds, different stopping methods, and many other motor control functions. The amount of motor control a drive provides depends on the drive used. Basic drives provide a minimum of ON/OFF, speed control, and overload protection. In addition, most drives provide display information (voltage, current, frequency, and drive temperature), motor and circuit fault information (under- or overvoltages, under- or overcurrents, ground faults, and phase losses), and advanced programming features. Advanced programming features include setting multiple preset speeds, limiting the motor maximum and minimum speeds, and providing motor braking.

The cost of using a magnetic motor starter or a motor drive is about the same for any horsepower (HP) rating. The cost of a motor drive for any given horsepower increases with drives that offer advanced features, such as the ability to connect to and communicate with a computer (for programming or for printing parameter settings and faults), provide multiple programmable output contacts, include analog inputs (0 V to 10 V, 4 mA to 20 mA, variable resistance), and include advanced parameter settings. Some drives may include built-in PLC/PLR functions such as internal timers and counters. An AC motor can be controlled with either a magnetic motor starter or a motor drive. The required level of control depends on the application where the motor is used. **See Figure 26-1.**

MAGNETIC MOTOR STARTERS AND MOTOR DRIVES

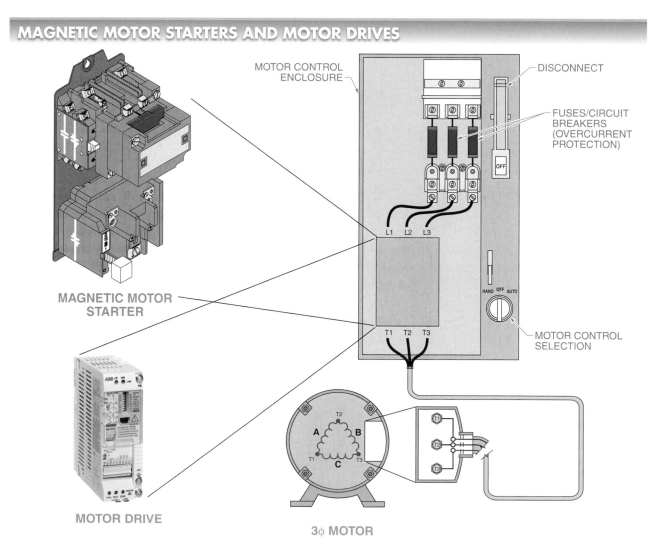

Figure 26-1. A magnetic motor starter or a motor drive can be used to control a motor.

Magnetic Motor Starter Control Circuits

When using a magnetic motor starter to control a motor, the starter is wired following a standard line diagram. For example, a pump motor can be controlled by a magnetic motor starter that uses a control circuit. The control circuit includes a three-position selector switch (HAND/OFF/AUTO) and a liquid level switch to control the motor starter. **See Figure 26-2.** When the selector switch is placed in the HAND position, the motor is ON. When the selector switch is placed in the AUTO position, the motor is ON only when the liquid level switch contacts are closed (liquid at or above switch level). In the OFF position, the motor is OFF regardless of the position of the liquid level switch contacts.

The magnetic motor starter can turn the motor on and off, but it cannot set the acceleration and deceleration times for the motor. Also, the motor starter cannot be used to set the motor speed, provide circuit condition readouts, or display circuit or motor faults. However, a magnetic motor starter does provide overload protection. Some models may provide phase-loss detection that turns the motor off if one of the three phases is lost.

Tech Fact

Motor drive control circuits are 24 VDC, while modern magnetic motor starter control circuits are 24 VAC. When troubleshooting motor control circuits, it is important to measure the voltage to determine the type (AC or DC) and the level so that the proper PPE can be used.

MAGNETIC MOTOR STARTER CONTROL CIRCUITS

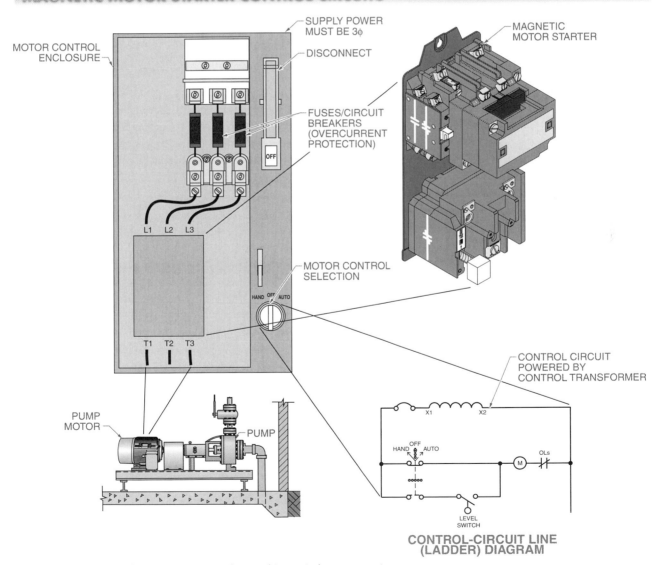

Figure 26-2. A magnetic motor starter can be used to control a pump motor.

Motor Drive Control Circuits

When using a motor drive to control a motor, the drive is wired following the control-circuit wiring diagram provided by the drive manufacturer. For example, a pump motor can be controlled by a motor drive that uses a control circuit that includes a three-position selector switch (HAND/OFF/AUTO) and a liquid level switch to control a motor starter. **See Figure 26-3.** When the selector switch is placed in the HAND position, the motor is ON. When the selector switch is placed in the AUTO position, the motor is ON only when the liquid level switch contacts are closed (liquid at or above the level switch). In the OFF position, the motor is OFF regardless of the position of the liquid level switch contacts.

In a magnetic motor starter, the selector and level switches are connected to the motor starter coil. In the motor drive controller, the selector switches and level switches are connected to the input terminals. Another difference is that the control circuit for the magnetic motor starter is normally powered by a step-down control transformer added between the power circuit and the control circuit. The control circuit for the motor drive does not require a control transformer because the motor drive provides the control circuit power at the terminals where the control devices are connected. The control voltage of a motor drive is normally 12 VDC or 24 VDC, which provides additional safety because of the low voltage. The motor drive also provides overload protection in addition to other motor control features.

MOTOR DRIVE CONTROL CIRCUITS

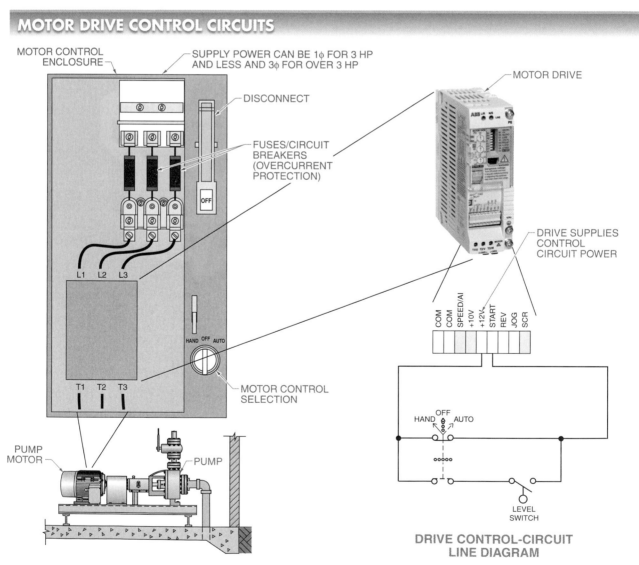

Figure 26-3. A motor drive can be used to control a pump motor.

One of the major advantages of using a drive instead of a motor starter is that the drive offers motor speed control. Speed control can be added to a motor control circuit through several different methods. For example, a potentiometer can be used to control the speed of a motor. A *potentiometer* is a variable-resistance electric device that divides voltage proportionally between two circuits. A potentiometer (with resistance of 5 kΩ to 10 kΩ) can be connected directly to the motor drive control terminals as specified by the manufacturer. **See Figure 26-4.**

In addition to using a potentiometer for speed control, most motor drives allow a voltage or current input to control the speed of a motor. For example, a 0 VDC to 10 VDC supply voltage can be connected to the motor drive control input terminals. At 5 VDC, the motor runs at 50% speed. At 7.5 VDC, the motor runs at 75% speed. Also, a 4 mA to 20 mA DC supply current can be connected to the motor drive control input terminals. At 8 mA, the motor runs at 50% speed. At 12 mA, the motor runs at 75% speed. In industrial systems, standard control-circuit power is supplied at 0 V to 10 V and 4 mA to 20 mA.

MOTOR SPEED CONTROL

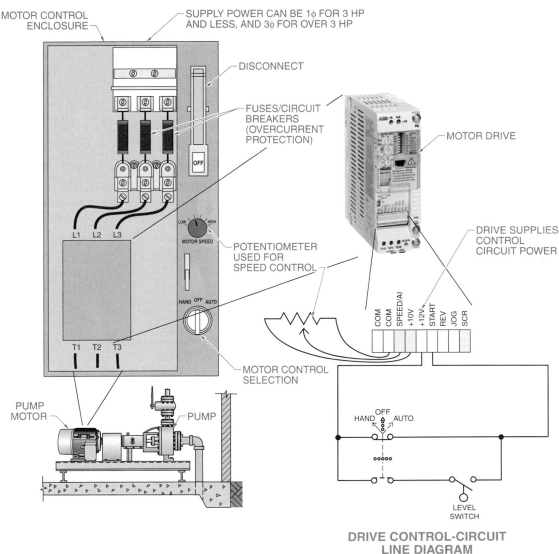

Figure 26-4. A potentiometer can be added to the control circuit to control the speed of a motor.

Motor control and drive functions are set by placing switches in certain positions or by programming through a keypad. The simplest drives include switches for setting the most common motor/drive system requirements. On advanced drives, a keypad may be used to enter the motor/drive system requirements. The keypad is normally located on the drive and may be permanent or removable. Removable keypads reduce cost because one keypad can be used to program many drives. **See Figure 26-5.**

Tech Fact

Drives that set several switches in one of two different positions instead of using keypad programming to set drive operating parameters require an understanding of the abbreviations used on each switch position. A load-type switch may be labeled "P&F," which stands for pump or fan load, or "CT," which stands for constant torque load. A PWM setting may be silent run ON or OFF. The OFF setting is a lower operating frequency (5 kHz, etc.) and the ON setting is a higher operating frequency (16 kHz, etc.). The manufacturer's operating manual that describes each setting must be used when any switch positions are changed.

MOTOR CONTROL AND DRIVE FUNCTIONS

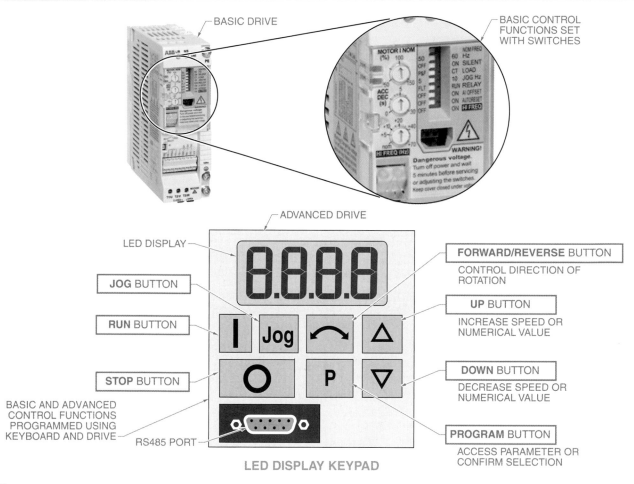

Figure 26-5. Motor control and drive functions are set by using switches on the drive or by a keypad.

26-1 CHECKPOINT

1. If only 1φ power is available, can a motor drive be used to control a fractional horsepower 3φ motor?

2. Is a step-down control transformer required to provide a lower control voltage when a motor drive is used?

3. What is the typical control voltage level and type (AC or DC) available on most motor drives?

4. What are the two methods used to set motor control and drive functions in a motor drive?

26-2 PROGRAMMING MOTOR DRIVES

A motor drive is programmed to control a motor by setting various parameters. For example, a motor drive can control motor acceleration time and deceleration/stopping method and time. All motor drives contain a number of parameters to control a motor. Basic motor drives can contain approximately 30 parameters, while advanced motor drives can contain over 100 parameters.

Regardless of manufacturer, electric motor drives share many common parameters. Although the parameter names may vary, the functions performed are the same. For example, acceleration time and ramp-up time are two different parameter names for the same function. Numbers assigned to a parameter and its location in a menu structure also vary based on manufacturer.

Electric motor drives are shipped with factory settings (defaults) for most parameters. Default parameters are normally the most conservative and frequently used parameter values that create the least amount of risk to equipment and personnel.

Default values are identified in the instruction manual of an electric motor drive and function properly for most drive applications. However, some parameters must be programmed for proper and safe motor drive system operation. For example, parameters for motor nameplate data are not factory set and must be programmed into the motor drive.

Parameter Menu Formatting

Parameter menu formatting varies based on manufacturer. Some manufacturers list parameters in numerical order and other manufacturers arrange parameters by file and group based on function. Some assign a number to a parameter in addition to a file and group designation. For example, a drive manufacturer may arrange groups of parameters into files with a letter such as d, P, A, and F. The letter designates the parameter group as display (d), basic programming (P), and advanced programming (A). In addition, the letter F designates the group of fault codes (F). **See Figure 26-6.**

Some drive parameters are designated as display parameters and others as editing parameters. *Display parameters* are parameters that allow drive or motor operating conditions, such as applied voltage, current draw, and internal drive temperature, to be viewed but not changed. *Editing parameters* are parameters that can be programmed or adjusted to set a drive for a specific application and motor.

MOTOR DRIVE PARAMETER GROUPS

Menu	Description
d	**Display Group (View Only)** Consists of commonly viewed drive operating conditions
P	**Basic Program Group** Consists of most commonly used programmable functions
A	**Advanced Program Group** Consists of remaining programmable functions
F	**Fault Designator** Consists of list of codes for specific fault conditions; displayed only when fault is present

Figure 26-6. Motor drive parameter groups may consist of display, basic programming, advanced programming, and fault codes.

Display Parameters. Display parameters are used to give a visual display of operating conditions, which can be used when installing, testing, operating, and troubleshooting a circuit using a motor drive. **See Figure 26-7.** Display parameters normally include the following:

- drive output frequency
- motor current draw
- drive output voltage
- drive DC bus voltage
- drive internal temperature
- motor fault
- drive elapsed operating time
- motor operating speed

DISPLAY PARAMETERS

Display Parameters

No.	Parameter	Min/Max	Display/Options			
d001	[Output Freq]	0.0/[Maximum Freq]	0.1 Hz			
d002	[Commanded Freq]	0.0/[Maximum Freq]	0.1 Hz			
d003	[Output Current]	0.00/(Drive Amps x 2)	0.01 Amps			
d004	[Output Voltage]	0/Drive Rated Volts	1 VAC			
d005	[DC Bus Voltage]	Based on Drive Rating	1 VDC			
d006	[Drive Status]	0/1 (1 = Condition True)	Bit 3 Decelerating	Bit 2 Accelerating	Bit 1 Forward	Bit 0 Running
d007- d009	[Fault x Code]	F2/F122	F1			

Figure 26-7. Display parameters are used to give a visual display of operating conditions.

Basic Programming Parameters. Although most motor drives include numerous parameters to customize a drive to a motor and application, normally most parameters do not need to be reprogrammed from the factory default settings. To simplify programming, some manufacturers group the most commonly programmed parameters together to make programming easier. **See Figure 26-8.** Basic programming parameters normally require technical information listed on a motor nameplate. Basic programming parameters also include basic circuit operating conditions, such as motor acceleration and deceleration time. Basic programming parameters normally include the following:

- motor nameplate voltage
- motor nameplate current
- motor nameplate frequency
- desired motor acceleration time
- desired motor deceleration time
- motor stopping mode (ramp, coast, and brake)
- circuit control (drive keypad and external pushbutton operation)

Advanced Programming Parameters. Although basic programs normally allow for good motor drive system operation, there are applications that require special operating conditions. Advanced programming parameters allow for customizing a motor drive application. **See Figure 26-9.** Advanced programming parameters normally include the following:

- drive output contacts operation (when a motor is at set speed and during motor acceleration)
- multiple acceleration and deceleration times (different from the basic programmed times)

- preset motor speeds
- type and time of braking applied to a motor
- preventing a motor from operating in reverse direction
- jog time and speed
- number of times the motor drive attempts to restart a motor after a fault
- analog inputs (0 VDC to 10 VDC and 4 mA to 20 mA)

Saftronics Inc.

Motor drives use solid-state components to control the speed of a motor.

BASIC PROGRAMMING PARAMETERS

Basic Programming Parameters

◯ = Stop drive before changing this parameter.

No.	Parameter	Min/Max	Display/Options	Default
P031 ◯	[Motor NP Volts] Set to the motor nameplate rated volts.	20/Drive Rated Volts	1 VAC	Based on Drive Rating
P032 ◯	[Motor NP Hertz] Set to the motor nameplate rated frequency.	10/240 Hz	1 Hz	60 Hz
P033	[Motor OL Current] Set to the maximum allowable motor current.	0.0/(Drive Rated Amps x 2)	0.1 Amps	Based on Drive Rating
P034	[Minimum Freq] Sets the lowest frequency the drive will output continuously.	0.0/240.0 Hz	0.1 Hz	0.0 Hz
P035 ◯	[Maximum Freq] Sets the highest frequency the drive will output.	0/240 Hz	1 Hz	60 Hz
P036 ◯	[Start Source] Sets the control scheme used to start the drive.	0/5	0 = "Keypad" [1] 3 = "2-W Lvl Sens" 1 = "3-Wire" 4 = "2-W Hi Speed" 2 = "2-Wire" 5 = "Comm Port"	0

[1] When active, the Reverse key is also active unless disabled by A095 [Reverse Disable].

Figure 26-8. Basic programming parameters consist of the most commonly used programming parameters.

ADVANCED PROGRAMMING PARAMETERS

Advanced Programming Parameters

No.	Parameter	Min/Max	Display/Options		Default
A051 A052 ◯	[Digital In1 Sel] I/O Terminal 05 [Digital In1 Sel] I/O Terminal 05	0/26	0 = "Not Used" 1 = "Acc 2 & Dec 2" 2 = "Jog" 3 = "Aux Fault" 4 = "Reset Freq" 5 = "Local" 6 = "Comm Port" 7 = "Clear Fault"	8 = "RampStop,CF" 9 = "CoastStop,CF" 10 = "DCInjStop,CF" 11 = "Jog Forward" 12 = "Jog Reverse" 13 = "10V in Ctrl" 14 = "20mA In Ctrl" 26 = "Anlg Invert"	4
A055	[Relay Out Sel]	0/21	0 = "Ready/Fault" 1 = "At Frequency" 2 = "MotorRunning" 3 = "Reverse" 4 = "Motor Overld" 5 = "Ramp Reg"	6 = "Above Freq" 7 = "Above Cur" 8 = "Above DCVolt" 9 = "Retries Exst" 10 = "Above Anlg V" 20 = "ParamControl" 21 = "NonRec Fault"	0
A056	[Relay Out Level]	0.0/9999	0.1		0.0
A067	[Accel Time 2]	0.0/600.0 Secs	0.1 Secs		20.0 Secs
A068	[Decel Time 2]	0.1/600.0 Secs	0.1 Secs		20.0 Secs
A069	[Internal Freq]	0.0/240.0 Hz	0.1 Hz		60.0 Hz
A070 A071 A072 A073	[Preset Freq 0][1] [Preset Freq 1] [Preset Freq 2] [Preset Freq 3]	0.0/240.0 Secs	0.1 Hz		0.0 Hz 5.0 Hz 10.0 Hz 20.0 Hz

[1] To activate [Preset Freq 0] set P038 [Speed Reference] to option 4.

Input State of Digital in 1 (I/O Terminal 05)	Input State of Digital in 2 (I/O Terminal 06)	Frequency Source	Accel/Decel Parameter Used[2]
0	0	[Preset Freq 0]	[Accel Time 1]/[Decel Time 1]
1	0	[Preset Freq 1]	[Accel Time 1]/[Decel Time 1]
0	1	[Preset Freq 2]	[Accel Time 2]/[Decel Time 2]
1	1	[Preset Freq 3]	[Accel Time 2]/[Decel Time 2]

[2] When a Digital Input is set to "Accel 2 & Decel 2," and the input is active, that input overrides the settings in this table.

No.	Parameter	Min/Max	Display/Options	Default
A078	[Jog Frequency]	0.0/[Maximum Freq]	0.1 Hz	10.0 Hz
A079	[Jog Accel/Decel]	0.1/600.0 Secs	0.1 Secs	10.0 Secs

Figure 26-9. Advanced programming parameters consist of parameters for customizing motor drive applications that require special operating conditions.

Fault Codes. When drives are used in a system, troubleshooting can be more difficult and time consuming than when magnetic motor starters are used. To quickly identify a system fault, motor drives can detect and display common circuit faults. **See Figure 26-10.** A motor drive that monitors and automatically turns a motor off if there is a problem helps maintain a safe system for equipment and operators. A drive that displays a fault helps a technician determine the problem when troubleshooting or testing a system. Common motor drive faults include the following:

- power loss
- overvoltage or undervoltage condition
- motor overload condition
- high drive operating temperature condition
- ground fault condition
- loss of phase
- phase-to-ground short
- drive overload

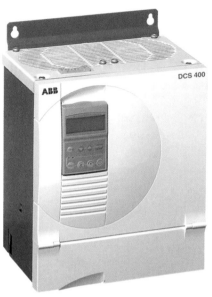

ABB Inc., Drives and Power Electronics
DC motor drives can use fieldbus adapters to communicate their braking conditions to a PLC or building automation system.

FAULT CODES

Fault Codes

To clear fault, press the Stop key, cycle power or set A100 [Fault Clear] to 1 or 2.

No.	Fault	Description
F2	Auxiliary Input[1]	Check remote wiring.
F3	Power Loss	Monitor the incoming AC line for low voltage or line power interruption.
F4	UnderVoltage[1]	Monitor the incoming AC line for low voltage or line power interruption.
F5	OverVoltage[1]	Monitor the AC line for high line voltage or transient conditions. Bus overvoltage can also be caused by motor regeneration. Extend the decel time or install dynamic brake option.
F6	Motor Stalled[1]	Increase [Accel Time x] or reduce load so drive output current does not exceed the current set by parameter A089 [Current Limit].
F7	Motor Overload[1]	An excessive motor load exists. Reduce load so drive output current does not exceed the current set by parameter P033 [Motor OL Current].
F8	Heatsink OvrTmp[1]	Check for blocked or dirty heat sink fins. Verify that ambient temperature has not exceeded 40°C (104°F) for IP 30/NEMA 1/UL Type 1 installations or 50°C (122°F) for Open type installations. Check fan.
F12	HW OverCurrent[1]	Check programming. Check for excess load, improper DC boost setting, DC brake volts set too high or other causes of excess current.
F13	Ground Fault	Check the motor and external wiring to the drive output terminals for a grounded condition.

[1] When active, the Reverse key is also active unless disabled by A095 [Reverse Disable].

Figure 26-10. Fault codes are used to detect and display common circuit faults.

26-2 CHECKPOINT

1. What do manufacturers call the preset factory motor drive parameter settings?
2. Can all drive parameters be reprogrammed (changed)?
3. Where is the overload protection current setting information that is entered into basic parameters of the motor drive found?
4. If a drive automatically turns off a motor because of a problem, is there a way to determine what the problem may be before any tests are run or measurements taken?

26-3 DC MOTOR DRIVES

DC motors are normally used in applications that require precise motor speed control. However, since AC motor drives offer comparable speed control in most applications where DC motors were once used, AC motor drives are used to reduce the high maintenance cost of the DC motors. DC motor drives are normally used where a DC motor control circuit is in use and requires an upgrade from a mechanical motor control.

The speed of a DC motor is proportional to the applied DC voltage at the motor. The DC voltage applied to a motor can be varied by tapping different battery connections (on older portable equipment), using a variable DC power supply, using a variable resistor (tapped or adjustable potentiometer), or using electronic controls (such as SCRs). DC motor drives use electronic controls to vary the amount of DC power applied to a motor. The amount of applied voltage is proportional to motor speed and the amount of applied current is proportional to motor torque. **See Figure 26-11.**

Motor speed is controlled by controlling the voltage to a DC motor. The higher the applied voltage, the faster a DC motor rotates. DC motor drives normally control the voltage applied to a motor over the range of 0 VDC to the maximum nameplate voltage rating of the motor. If a DC motor drive can deliver more voltage than the rating of the motor, the drive should be set to limit the output voltage to prevent damage to the motor.

Motor torque is controlled by controlling the amount of current in the armature of a DC motor. Motor torque is proportional to the current in the armature. DC motor drives are designed to control the amount of voltage and current applied to the armature of DC motors to produce desired torque and prevent motor damage. The ideal operating condition is to deliver current to a motor to produce enough torque to operate the load without overloading the motor, motor drive, or electrical distribution system.

Powering DC Motor Drives

DC motor drives deliver a controlled DC output to DC motors. DC motor drives are available for connection to DC or AC power supplies. AC is used to power most DC motor drives because AC is readily available at most locations, except where portable equipment is used. AC can be easily rectified (changed from AC to DC) by a DC motor drive. **See Figure 26-12.**

For optimal motor performance, rectified DC should be as smooth as possible (containing little variation in amplitude of waveform). When AC power is from a 1ϕ source (115/230 VAC), a filter circuit (capacitors) should be used to smooth the DC waveform. When a 3ϕ power supply is used, a smooth DC waveform can be obtained by using six diodes (two per phase).

Dynamic and Regenerative Braking

When a motor is turned off, the load on the motor shaft normally determines the amount of time it takes the motor to coast to a complete stop. The greater the load on a motor, the longer the deceleration time of the motor. In some applications, the length of time a motor takes to decelerate from full speed to a complete stop is not an issue. In other applications, a motor must be stopped in a shorter length of time than coasting allows.

The two major methods of braking a DC motor faster than coasting are dynamic braking and regenerative braking. DC motor drives include dynamic and/or regenerative braking. Both types of braking take advantage of the fact that DC motors become DC generators when disconnected from the power supply. In such a case, the rotating armature and stationary field act as a DC generator and produce DC voltage. The energy produced from the DC motor acts as a DC generator and is called counter electromotive force (CEMF).

DC MOTOR DRIVE VOLTAGE AND CURRENT

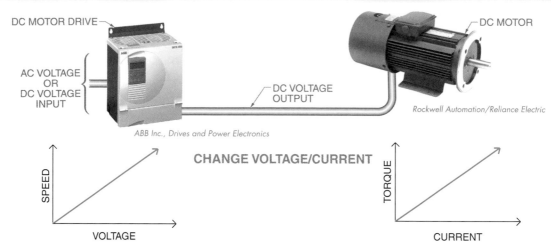

DC MOTOR DRIVE

AC VOLTAGE
OR
DC VOLTAGE
INPUT

ABB Inc., Drives and Power Electronics

DC VOLTAGE
OUTPUT

DC MOTOR

Rockwell Automation/Reliance Electric

CHANGE VOLTAGE/CURRENT

SPEED

VOLTAGE

TORQUE

CURRENT

Figure 26-11. DC motor drives control the amount of DC voltage and current applied to a motor.

POWERING DC MOTOR DRIVES

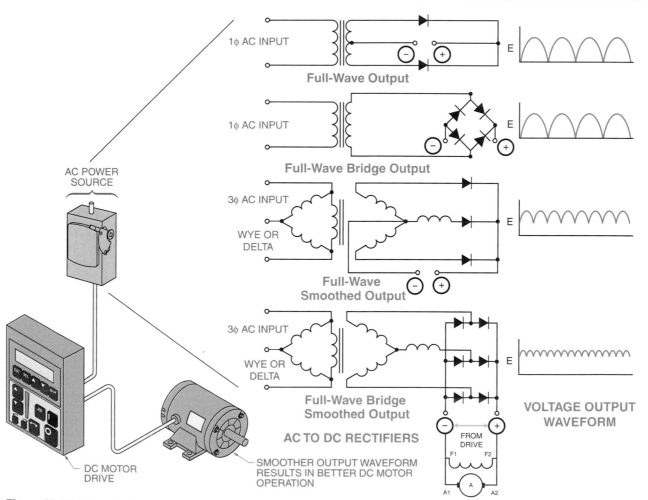

AC POWER
SOURCE

1ϕ AC INPUT

Full-Wave Output

1ϕ AC INPUT

Full-Wave Bridge Output

3ϕ AC INPUT

WYE OR
DELTA

Full-Wave
Smoothed Output

3ϕ AC INPUT

WYE OR
DELTA

Full-Wave Bridge
Smoothed Output

AC TO DC RECTIFIERS

E

E

E

E

VOLTAGE OUTPUT
WAVEFORM

FROM
DRIVE

F1   F2

A

A1   A2

DC MOTOR
DRIVE

SMOOTHER OUTPUT WAVEFORM
RESULTS IN BETTER DC MOTOR
OPERATION

Figure 26-12. DC motor drives are normally powered by an AC power supply.

Dynamic braking is a method of motor braking in which a DC motor is reconnected to act as a generator immediately after it is turned off. A DC motor drive can provide dynamic braking by connecting a resistor across the motor armature after the motor input power is removed. The resistor, referred to as a braking resistor, dissipates the rotating energy from the motor at the resistor. The smaller the resistance, the less braking time and greater amount of heat produced at the resistor. Dynamic braking cannot provide a controlled stop unless several different resistors are used. **See Figure 26-13.**

However, when a motor lowers a load, the motor force (shaft torque) may need to be in the down or up direction. For very light loads, the motor may need to produce a downward force to lower the load. For most loads, gravity attempts to pull the load down faster than the speed of the motor. To prevent this and provide a smooth deceleration, the motor must provide an upward torque that is in the opposite direction the motor shaft is rotating (lowering the load). Reverse torque prevents the load from falling faster than the set motor speed. DC regenerative control drives can detect this condition and automatically adjust the torque of a motor to maintain a controlled speed regardless of the load. **See Figure 26-14.**

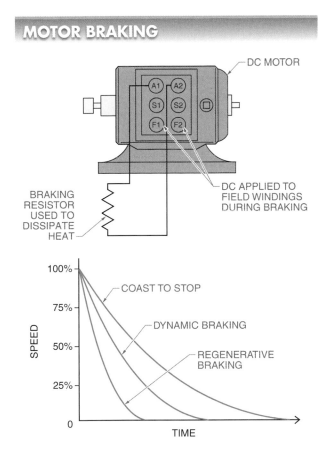

Figure 26-13. Dynamic braking uses a DC motor as a DC generator during braking.

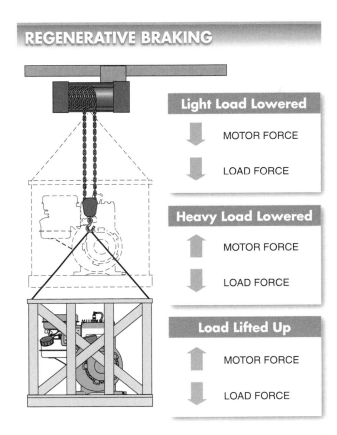

Figure 26-14. Regenerative control allows a controlled speed regardless of the force/pull of the load on a motor shaft.

Regenerative braking is a method of motor braking in which the regenerated power of a DC motor that is coming to a stop is returned to the input power supply. Regenerative braking requires a method to transfer the energy produced by a rotating motor shaft back to the main power supply. For example, when a motor lifts a load, the motor force (shaft torque) must always be in the forward (up) direction to overcome the downward force of the load.

26-3 CHECKPOINT

1. If changing the applied frequency to an AC motor changes the motor speed, how is the speed of a DC motor changed?
2. Can dynamic braking provide a controlled stop of a motor using one resistor?

26-4 AC MOTOR DRIVES

AC motor drives are referred to as variable frequency drives, adjustable frequency drives, inverter drives, vector drives, direct torque control drives, and closed loop drives. Regardless of how an AC drive is referred to, its primary function is to convert the incoming supply power to an altered voltage level and frequency that can safely control the motor connected to the drive.

AC motor drives are designed to operate 3φ AC motors regardless of whether the drive is designed for 1φ power (115 VAC or 230 VAC), 3φ power, or DC power. The speed of an AC motor is determined by the number of stator poles and the frequency of the AC power supply. AC motor drives control the speed of a motor by varying the frequency of the power applied to the motor. **See Figure 26-15.** The lower the frequency applied to a motor, the slower the motor speed. For example, an AC motor rated for 1730 rpm at 60 Hz operates at 1730 rpm at 60 Hz, 865 rpm at 30 Hz, and 432.5 rpm at 15 Hz.

AC Motor Drive Construction

The three main sections of an AC motor drive are the converter, DC bus, and inverter. The converter (rectifier) receives incoming AC voltage and changes it to DC voltage. If the AC input voltage is different from AC output voltage sent to a motor, the converter must first step up or step down the AC voltage to the proper voltage source level. For example, an electric motor drive supplied with 115 VAC that delivers 230 VAC to a motor requires a step-up transformer to increase the input voltage. A drive supplied with 230 VAC would be stepped down to deliver 115 VAC. **See Figure 26-16.**

The DC bus filters the voltage and maintains the proper DC voltage level. The DC bus may also deliver DC to the inverter for conversion back to AC. The inverter controls the speed of a motor by controlling frequency and controls motor torque by controlling the voltage sent to the motor.

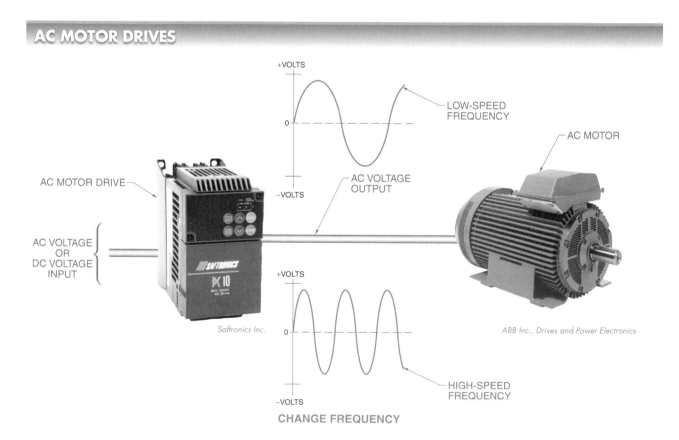

Figure 26-15. An AC motor drive controls the speed of a motor by varying the frequency of the power applied to the motor.

AC DRIVE SECTIONS

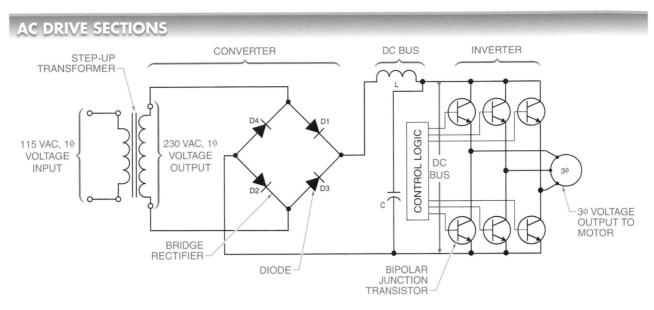

Figure 26-16. The three main sections of an AC motor drive are the converter, DC bus, and inverter.

Converters

A *converter* is an electronic device that changes AC voltage into DC voltage. Converters in electric motor drives are 1φ full-wave rectifiers, 1φ bridge rectifiers, or 3φ full-wave rectifiers. Small AC motor drives supplied with 1φ power use 1φ full-wave or bridge rectifiers. Most electric motor drives are supplied with 3φ power, which requires 3φ full-wave rectifiers. **See Figure 26-17.**

AC Motor Drive Power Converter Requirements. In order for a converter to deliver the proper DC voltage to the DC bus of an AC motor drive, the converter must be connected to the proper power supply. AC motor drives operate satisfactorily only when connected to the proper power supply. The power supply must be at the correct voltage level and frequency and must also provide enough current to operate an AC motor drive at full power. When a power supply cannot deliver enough current, the available voltage to an AC motor drive drops when the drive is required to deliver full power. Current to an AC motor drive is limited by the size of the conductors to the drive, fuse and circuit breaker sizes, and the transformer(s) delivering power to the system.

Supply voltage to an AC motor drive must be checked when additional loads or drives are installed, serviced, or added to a system. To determine whether an AC motor drive is underpowered, the voltage at the drive is measured under no-load and full-load operating conditions. **See Figure 26-18.** A voltage drop greater than 3% between no-load and full-load conditions indicates that the AC motor drive is underpowered and/or overloaded.

THREE-PHASE FULL-WAVE RECTIFIERS

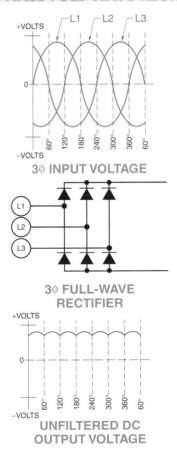

Figure 26-17. Most AC motor drives use 3φ full-wave rectifiers to convert AC voltage to DC voltage to supply the DC bus.

TROUBLESHOOTING UNDERPOWERED AC MOTOR DRIVES

Figure 26-18. AC motor drive voltage should be measured under no-load and full-load operating conditions to determine whether a drive is underpowered.

Voltage drop is found by applying the following formula:

$$V_D = V_{NL} - V_{FL}$$

where

V_D = voltage drop (in V)

V_{NL} = no-load voltage (in V)

V_{FL} = full-load voltage (in V)

Example: Calculating Voltage Drop

What is the voltage drop when an AC motor drive is measured to have 230 V with no load and 226 V under full load?

$$V_D = V_{NL} - V_{FL}$$
$$V_D = 230 - 226$$
$$V_D = \mathbf{4\ V}$$

ABB Inc., Drives and Power Electronics

Remote keypads display motor drive operating data and can be used to modify parameters.

Voltage drop percentage is found by applying the following formula:

$$V_\% = \frac{V_D}{V_{NL}} \times 100$$

where

$V_\%$ = percentage of voltage drop (in %)

V_D = voltage drop (in V)

V_{NL} = no-load voltage (in V)

100 = constant

Example: Calculating Voltage Drop Percentage

What is the voltage drop percentage when an AC motor drive has a 4 V drop with a 230 V no-load measurement?

$$V_\% = \frac{4}{230} \times 100$$

$$V_\% = 0.01739 \times 100$$

$$V_\% = \mathbf{1.739\%}$$

When the supply voltage to an AC motor drive is measured, it is recommended to check the measured voltage against the rated input voltage of the drive. Large-horsepower AC motor drives are connected to high voltage to reduce the amount of current required.

AC voltage sources vary due to fluctuations within the power distribution system. AC loads, including motor drives and motors, are designed to operate within a specified voltage range. Operating outside the specified voltage range can cause a motor drive to operate improperly and/or incur damage over time.

Electrical loads operating at low voltages are less likely to be damaged than loads operating at high voltages. Operating at a voltage less than the rated voltage causes lamps to dim, heating elements to produce less heat, computers to lose memory and/or reboot, and motors to produce less torque. Although operating at less than rated voltage is not desirable for electrical loads, it normally does not cause damage. Operating at higher than rated voltage causes lamps to fail, heating elements to burn out (open), computer circuits to become permanent damaged, and motor insulation to be destroyed.

Tech Fact

Most drives display a "low voltage" fault when the drive turns off the motor due to a low voltage condition. When troubleshooting a low-voltage fault, measurements are taken of the voltage into the drive, the voltage out of the drive (using the MIN/MAX mode during a complete cycle of the motor to determine whether there is a power feed or drive output problem), and the drive's DC bus voltage to eliminate a drive problem.

AC loads are rated for proper operation at a voltage that is ±10% of the device's rated voltage. Because higher voltages are more damaging, some devices with higher voltage ratings have a +5% to –10% voltage rating to protect them from the high voltage side.

DC Buses

DC buses filter and maintain the proper voltage level. DC buses (links) include DC filter components and are supplied with DC voltage by the converter. The capacitors and inductors in the DC bus filter and maintain the proper voltage level. DC bus voltage is typically about 1.4 times the AC supply voltage to an AC motor drive.

Circuit Protection. A bridge rectifier receives incoming AC supply power and converts the AC voltage to fixed DC voltage. The fixed DC voltage powers the DC bus of the AC motor drive. To prevent damage to the diodes in the converter and to the AC motor drive electronic circuits, protection against transient voltages must be included in the drive.

A transient voltage is a high-energy, high-voltage, short-duration spike in an electrical system. All electrical systems experience some type of transient voltage. Lightning strikes and utility switching cause high-energy-level transient voltages. High-energy-level transients seldom occur but are quite damaging to equipment if allowed to travel through a power distribution system and into electrical equipment. Low-energy-level transients are transient voltages commonly caused when motors and equipment are switched on and off. Low-energy-level transients occur often but do not cause immediate equipment damage. Low-energy-level transients cause malfunctions such as processing errors and damage to equipment over time.

The electronic circuits of an AC motor drive require protection against transient voltages. Protection methods include proper motor drive wiring, grounding, shielding for power lines, and surge suppressors. A *surge suppressor* is an electrical device that provides protection from transient voltages by limiting the level of voltage allowed downstream from the surge suppressor. Surge suppressors are installed at service entrance panels, distribution panels feeding motor drives, and/or the incoming power lines to a drive. Normally, a surge suppressor consists of metal-oxide varistors (MOVs) connected to the converter of a motor drive. **See Figure 26-19.**

METAL-OXIDE VARISTOR (MOV) SURGE SUPPRESSION

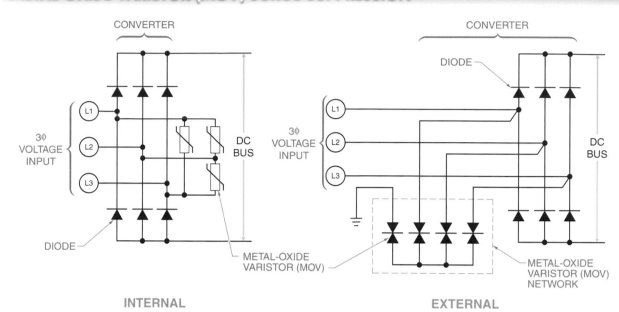

Figure 26-19. Metal-oxide varistors (MOVs) are added to the converter of an AC motor drive to reduce the amount of transient voltage entering the motor drive.

MOVs are designed for surge suppression of damaging transient voltages. When high-voltage transients enter an AC motor drive, MOVs change electrical state from high resistance (open switch) to low resistance (closed switch). In a low-resistance state, MOVs absorb and/or divert transient voltage spikes. MOVs limit the level of transient voltages so voltages do not exceed the maximum voltage rating of the rectifier diodes.

Capacitors. A *capacitor* is an electrical device designed to store a voltage charge by means of an electrostatic field. *Capacitance (C)* is the ability to store energy in the form of an electrical charge. Capacitors in a DC bus are charged from rectified DC voltage produced by the converter. Capacitors oppose a change in voltage and, when DC bus voltage starts to drop, they discharge a voltage back into the system to stop the drop in voltage. The main function of capacitors in a DC bus is to maintain proper voltage levels when voltage fluctuates. **See Figure 26-20.**

Inverters

An *inverter* is an electronic device that changes DC voltage into AC voltage. Inverters in an AC motor drive are the most important part of the drive because the inverter determines the voltage level, voltage frequency, and amount of current that a motor receives. AC motor

drive manufacturers are continuously developing inverters that can control motor speed and torque with the fewest problems. The main problem for manufacturers is to find a high-current, fast-acting solid-state switch that has the least amount of power loss (voltage drop).

DC BUS FILTER CAPACITORS

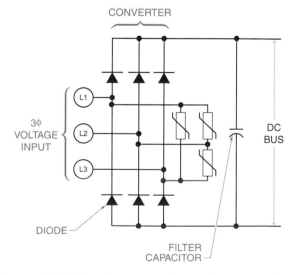

Figure 26-20. Capacitors are added to filter or smooth the DC bus voltage.

Voltage and Frequency

The voltage applied to the stator of an AC motor must be decreased by the same amount as the frequency. The motor heats excessively and damage occurs to the windings if the voltage is not reduced when frequency is reduced. The motor does not produce its rated torque if the voltage is reduced more than required. The ratio between the voltage applied to the stator and the frequency of the voltage applied to the stator must be constant. This ratio is referred to as the volts-per-hertz (V/Hz) ratio (constant volts-per-hertz characteristic). The *volts-per-hertz (V/Hz) ratio* is the relationship between voltage and frequency that exists in a motor. The motor develops rated torque if this relationship is kept constant (linear).

The volts-per-hertz ratio for an induction motor is found by dividing the rated nameplate voltage by the rated nameplate frequency. To find the volts-per-hertz ratio for an AC induction motor, the following formula is applied:

$$V/Hz = \frac{V}{Hz}$$

where

V/Hz = volts-per-hertz ratio

V = rated nameplate voltage (in V)

Hz = rated nameplate frequency (in Hz)

Example: Calculating Volts-per-Hertz Ratio

What is the volts-per-hertz ratio if a motor nameplate rates an AC motor for 230 VAC, 60 Hz operation?

$$V/Hz = \frac{V}{Hz}$$
$$V/Hz = \frac{230}{60}$$
$$V/Hz = \mathbf{3.83}$$

Above approximately 15 Hz, the amount of voltage required to keep the volts-per-hertz ratio linear is a constant value. Below 15 Hz, the voltage applied to the motor stator may be boosted to compensate for the large power loss AC motors have at low speed. The amount of voltage boost depends on the motor. **See Figure 26-21.**

A motor drive can be programmed to apply a voltage boost at low motor speeds to compensate for the power loss at low speeds. The voltage boost gives the motor additional rotor torque at very low speeds. The amount of torque boost depends on the voltage boost programmed into the motor drive. The higher the voltage boost, the greater the motor torque. **See Figure 26-22.**

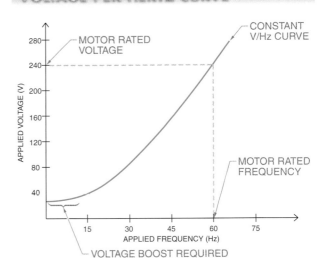

Figure 26-21. Below 15 Hz, the voltage applied to the motor stator may be boosted to compensate for the large power loss AC motors have at low speed.

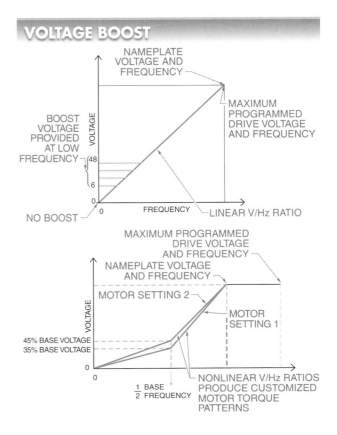

Figure 26-22. Motor drives can be programmed to apply a voltage boost at low motor speeds and to change the standard linear volts-per-hertz ratio to a nonlinear ratio.

Motor drives can also be programmed to change the standard linear volts-per-hertz ratio to a nonlinear ratio. A nonlinear ratio produces a customized motor torque pattern that is required by the load operating characteristics. For example, a motor drive can be programmed for two nonlinear ratios that can be applied to fan or pump motors. Fans and pumps are normally classified as variable torque/variable horsepower (VT/VH) loads. VT/VH loads require varying torque and horsepower at different speeds.

Controlling Motor Speed and Torque

AC motors produce work to drive a load by a rotating shaft. The amount of work produced is a function of the amount of torque produced by the motor shaft and the speed of the shaft. The primary function of all motor drives is to control the speed and torque of a motor.

To safely control a motor, an AC motor drive must monitor electrical characteristics such as motor current, motor voltage, drive temperature, and other operating conditions. All motor drives are designed to remove power when there is a problem. Some drives allow conditions and faults to be monitored and displayed. In addition to controlling motor speed and torque, an AC motor drive can include additional specialty functions that are built-in, programmed, or sent to the drive through on-board communication with a PC or PLC.

Controlling frequency (in Hz) to an AC motor controls the speed of the motor. AC motor drives control frequency applied to a motor over the range 0 Hz to several hundred hertz. AC motor drives are programmed for a minimum operating speed and a maximum operating speed to prevent damage to a motor or driven load. Damage occurs when a motor is driven faster than its rated nameplate speed. AC motors should not be driven at speeds greater than 10% of the rated nameplate speed.

Volts per hertz (V/Hz) is the relationship between voltage and frequency that exists in a motor and is expressed as a ratio. Controlling the V/Hz ratio applied to an AC motor controls motor torque. An AC motor develops rated torque when the V/Hz ratio is maintained. During acceleration (any speed between 1 Hz and 60 Hz), the motor shaft delivers constant torque because the voltage is increased at the same rate as the frequency. Once an AC motor drive reaches the point of delivering full motor voltage, increasing the frequency does not increase torque on the motor shaft because voltage cannot be increased further to maintain the V/Hz ratio. **See Figure 26-23.**

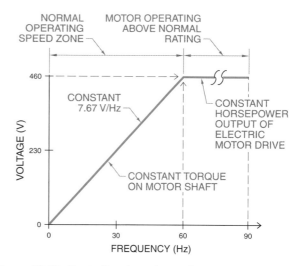

Figure 26-23. Controlling the volts-per-hertz ratio (V/Hz) applied to an AC motor controls motor torque.

Carrier Frequencies

The *carrier frequency* is the frequency that controls the number of times solid-state switches in the inverter of a motor drive with pulse width modulation turn on and off. The higher the carrier frequency, the more individual pulses there are to reproduce the fundamental frequency. The *fundamental frequency* is the frequency of the voltage used to control motor speed. The number of carrier frequency pulses per fundamental frequency is found by applying the following formula:

$$P = \frac{F_{CARR}}{F_{FUND}}$$

where
P = pulses
F_{CARR} = carrier frequency
F_{FUND} = fundamental frequency

Example: Calculating Carrier Frequency Pulses

What is the number of pulses per fundamental frequency when a carrier frequency of 1 kHz is used to produce a 60 Hz fundamental frequency?

$$P = \frac{F_{CARR}}{F_{FUND}}$$

$$P = \frac{1000}{60}$$

$$P = \mathbf{16.66}$$

The fundamental frequency is the frequency of voltage that a motor uses, but the carrier frequency actually delivers the fundamental frequency voltage to the motor. The carrier frequency of most motor drives can range from 1 kHz to approximately 16 kHz. A carrier frequency of 6 kHz used to produce a 60 Hz fundamental frequency would have 100 individual pulses per fundamental cycle. **See Figure 26-24.** The higher the carrier frequency, the closer the output sine wave is to a pure fundamental frequency sine wave.

Motor noise is a problem in motor drive applications, such as in HVAC systems where the noise can carry throughout an entire building. Increasing the frequency to a motor above the standard 60 Hz increases the noise produced by the motor. Noise is noticeable in the 1 kHz to 2 kHz range since it is within the range of human hearing and is amplified by the motor. A motor connected to an AC motor drive delivering a 60 Hz fundamental frequency with a carrier frequency of 2 kHz is approximately three times louder than the same motor connected directly to a pure 60 Hz sine wave with a magnetic motor starter.

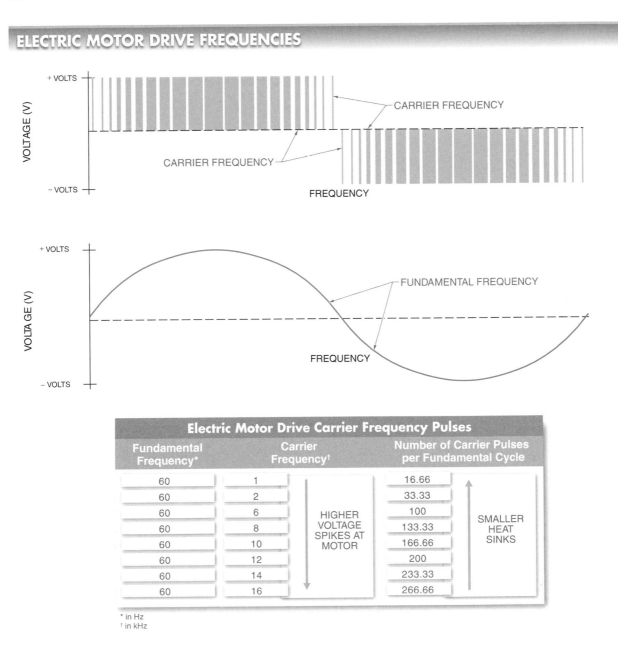

ELECTRIC MOTOR DRIVE FREQUENCIES

Electric Motor Drive Carrier Frequency Pulses		
Fundamental Frequency*	Carrier Frequency†	Number of Carrier Pulses per Fundamental Cycle
60	1	16.66
60	2	33.33
60	6	100
60	8	133.33
60	10	166.66
60	12	200
60	14	233.33
60	16	266.66

HIGHER VOLTAGE SPIKES AT MOTOR

SMALLER HEAT SINKS

* in Hz
† in kHz

Figure 26-24. Carrier frequencies of AC motor drives range from 1 kHz to 16 kHz.

Manufacturers have raised the carrier frequency beyond the range of human hearing to eliminate noise. High carrier frequencies cause greater power loss (thermal loss) in an AC motor drive due to the solid-state switches in the inverter. AC motor drives should be slightly derated or the size of heat sinks should be increased due to the increase in thermal loss. Derating a motor drive decreases the power rating of the drive. Increasing the heat sink size adds to the cost of the drive. **See Figure 26-25.**

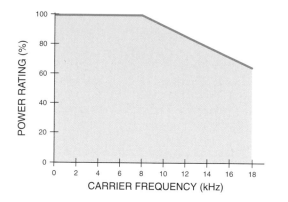

CARRIER FREQUENCY POWER DERATING CURVE

Figure 26-25. An AC motor drive may need to be derated at higher carrier frequencies to protect the insulation of a motor.

Higher carrier frequencies are better than lower carrier frequencies to a certain extent. For example, a 6 kHz to 8 kHz carrier frequency simulates a pure sine wave better than a 1 kHz to 3 kHz carrier frequency and reduces motor temperature. Since the voltage delivered to the motor better simulates a pure sine wave, less heat is produced. A slight decrease in motor temperature increases insulation life.

Pulse Width Modulation

Pulse width modulation (PWM) is a method of controlling the amount of voltage sent to a motor by converting the DC voltage into fixed values of individual DC pulses. AC motor drives must control the amount of voltage produced in order to control the speed and torque of a motor. PWM controls the amount of voltage output. The fixed-value pulses are produced by the high-speed switching of transistors,

normally insulated-gate bipolar transistors (IGBTs), on and off. By varying the width of each pulse (time ON) and/or by varying the frequency, the voltage can be increased or decreased. The greater the width of individual pulses, the higher the DC voltage output. **See Figure 26-26.** PWM of DC voltage is also used to reproduce AC sine waves.

When PWM is used with AC voltage, two IGBTs are used for each phase. One IGBT is used to produce positive pulses and another IGBT is used to produce negative pulses of a sine wave. Since AC drives are normally used to control 3ϕ motors, six IGBTs (two per phase) are used to simulate 3ϕ power. **See Figure 26-27.** Higher IGBT switching frequencies better simulate AC sine waves than lower frequencies. As simulated AC sine waves become more accurate, motors produce less heat.

Motor Stopping Methods

An AC motor drive decelerates a motor at a controlled rate by placing an electric load on the motor. The advantage in using a motor drive to apply a braking force is that maintenance is kept to a minimum because there are no parts that come in contact during braking. Applications that require a braking force to hold a load for a period of time after the motor has stopped may use the motor drive to stop the motor and a friction brake to hold the motor shaft/load. Friction brakes may also be used with motor drive braking in applications that require an emergency stop function.

AC motor drives can be programmed for different braking (stopping) methods. Common AC motor drive stopping methods include ramp stop, coast stop, DC brake stop, and soft stop (S-curve) stopping methods. **See Figure 26-28.**

PULSE WIDTH MODULATION

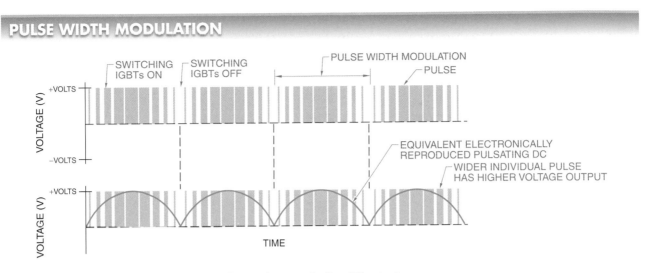

Figure 26-26. Pulse width modulation is used to produce a pulsating DC output.

IGBT-PRODUCED SINE WAVES

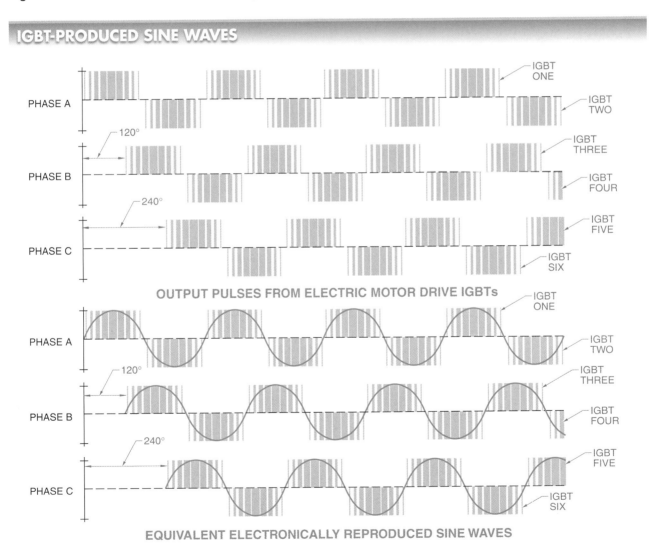

Figure 26-27. IGBTs with pulse width modulation are used to produce simulated 3ϕ power.

MOTOR DRIVE STOPPING METHODS

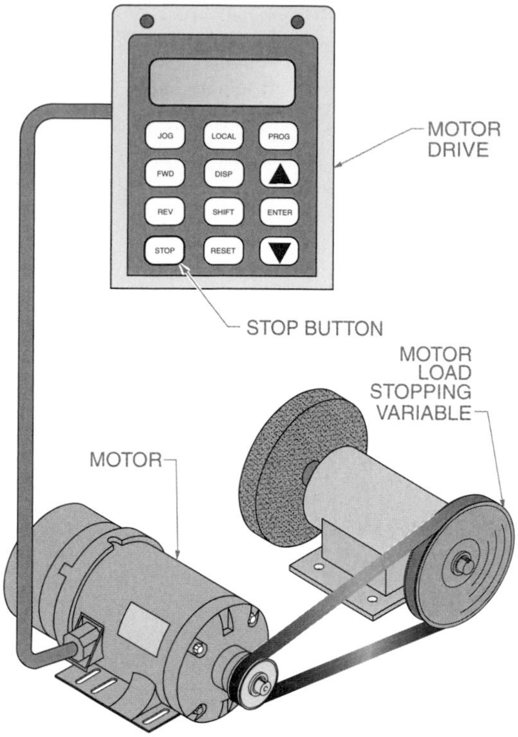

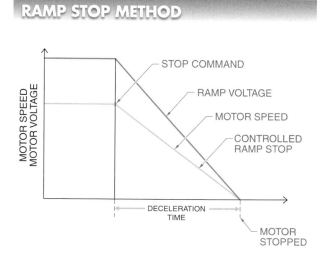

Stopping Method	Load Determines Stopping Time	Drive Determines Stopping Time
Ramp		X
Coast	X	
DC brake		X
Soft stop		X

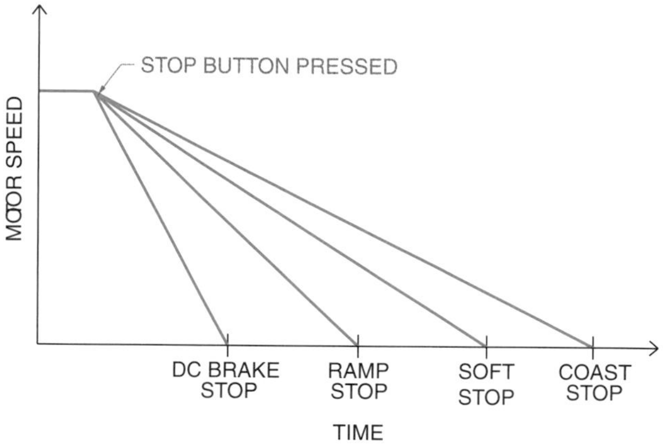

Figure 26-28. Motor drive stopping methods include ramp stop, coast stop, DC brake stop, and soft stop (S-curve) methods.

Ramp Stop. *Ramp stop* is a stopping method in which the level of voltage applied to a motor is reduced as the motor decelerates. Ramp stop is normally the factory (default) setting for controlling the stopping of a motor. When the motor drive receives a stop command, the drive maintains control of the motor speed by controlling the voltage on the motor stator. This allows for a smooth stop from any speed. The length of time the motor drive takes to stop the motor is controlled by the deceleration time parameter on the motor drive. The default setting for the deceleration is normally 10 sec, but it can be set from a few seconds (or less) to several minutes. **See Figure 26-29.**

Coast Stop. *Coast stop* is a stopping method in which the motor drive shuts off the voltage to a motor, allowing the motor to coast to a stop. When the coast stop method is used, the drive does not have any control of the motor after the stop command is entered. The length of time the motor takes to stop depends on the load connected to the motor.

RAMP STOP METHOD

Figure 26-29. A ramp stop provides a smooth, controlled deceleration by reducing the voltage applied to a motor.

DC Brake Stop. *DC brake stop (DC injection braking)* is a stopping method in which a DC voltage is applied to the stator winding of a motor after a stop command is entered. Unlike ramp stopping, the applied DC voltage is held at the level entered into the DC hold volts parameter on the motor drive. The applied DC voltage is maintained on the motor stator for the length of time entered into the DC hold time parameter on the drive. **See Figure 26-30.**

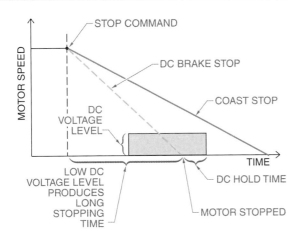

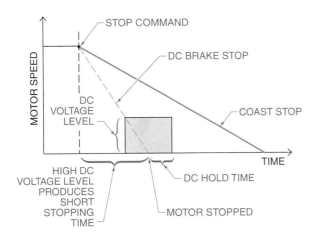

Figure 26-30. In the DC brake stop method, the DC hold level (amount of applied voltage) determines the stopping time of the motor.

The DC brake stop method provides a fast stop and can apply a braking force to hold the motor after the motor is stopped. However, the DC brake stop method cannot take the place of a friction brake in applications

that require the motor shaft to be held for a long period of time after stopping. This is because the DC hold time parameter normally has a maximum setting of approximately 15 sec. A longer time period produces excessive heat in the motor windings.

Soft Stop (S-Curve). *Soft stop (S-curve)* is a stopping method in which the programmed deceleration time is doubled and the stop function is changed from a ramp slope to an S-curve slope. The soft stop method provides a very soft stopping operation. A very soft stop may be required to prevent light loads (such as empty cans or bottles) from tipping. In such applications, the motor drive can be reprogrammed for a soft stop (S-curve) stopping method. **See Figure 26-31.**

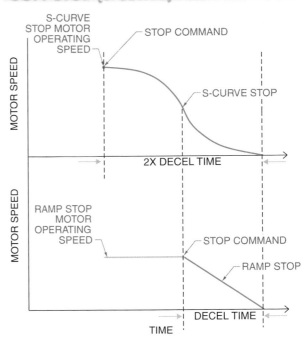

Figure 26-31. The soft stop (S-curve) method has a doubled stopping time and an S-curve slope reduction in voltage.

Electronic Overloads

New motor starters normally include an electronic overload instead of heaters. An *electronic overload* is a device that has built-in circuitry to sense changes in current and temperature. An electronic overload monitors the current in the load (motor, heating elements, etc.) directly by measuring the current in the power lines leading to the load. The electronic overload is built directly into the motor starter. **See Figure 26-32.**

ELECTRONIC OVERLOADS

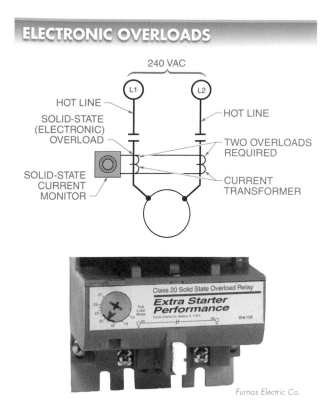

Furnas Electric Co.

Figure 26-32. An electronic overload has built-in circuitry that senses changes in current and temperature.

An electronic overload measures the strength of the magnetic field around a wire instead of converting the current into heat. The higher the current in the wire leading to a motor, the stronger the magnetic field produced. An electronic circuit is used to activate a disconnecting device that opens the starter power contacts. Electronic overloads have an adjustable range. The setting is based on the nameplate current listed on the motor.

Programmed Overloads

Magnetic motor starters protect motors from an overload by the addition of overload heaters to the starter. The motor nameplate current is used to select an overload size from a chart provided by the manufacturer of the magnetic motor starter. An electronic overload protects a motor from an overload by the setting of an adjustment dial on the starter overload block to the motor nameplate current.

When motor drives are used to control motors, motors are protected from an overload by programming the motor nameplate current into the drive. On basic motor drives, motor overload protection is set with an adjustment dial on the motor drive. On advanced motor drives, motor overload protection is programmed into the motor drive using a keypad. **See Figure 26-33.**

MOTOR DRIVE OVERLOAD PROTECTION

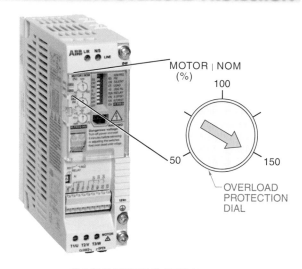

BASIC MOTOR DRIVE

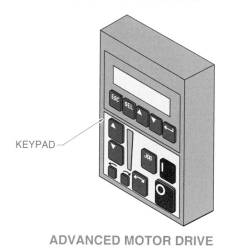

ADVANCED MOTOR DRIVE

Figure 26-33. Motor overload protection may be programmed into a motor drive by setting a dial or using a keypad.

When setting motor overload protection using the dial on a motor drive, the maximum rated continuous output current of the motor drive and the full-load current (nameplate current) of the motor is used to determine the dial setting. For example, if the maximum rated continuous output current of a motor drive is 10 A and the nameplate current of a motor is 10 A, the overload protection dial of the motor drive is set to 100%. **See Figure 26-34.**

MOTOR DRIVE OVERLOAD PROTECTION DIALS

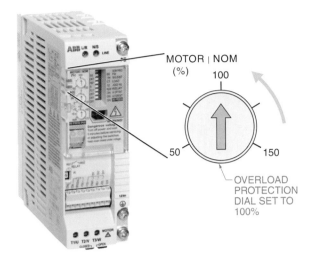

MOTOR | NOM
(%)
100
50
150

OVERLOAD
PROTECTION
DIAL SET TO
100%

Figure 26-34. The overload protection dial is set to 100% if the maximum rated continuous output current of the motor drive is equal to the full-load current (nameplate current) of the motor.

Saftronics Inc.

Advanced parameter programming on some motor drives allows customized overload protection.

Motor nameplate current is the amount of current a motor draws when the motor is operating at its nameplate power (HP or kW) rating. Normally, motors do not operate at 100% of their power rating. They also draw less current than the motor nameplate current. *Motor drive current rating* is the maximum continuous output current a motor drive can deliver for an extended period of time. A motor drive can deliver more current than its maximum continuous output current rating for short periods of time. For this reason, the motor drive overload protection dial can be set higher than 100%. However, the overload protection should generally remain under 100%. High current produces more heat, which damages motor drives and motors. Motor current percentage is found by applying the following formula:

$$I_m\% = \left(\frac{I_{mn}}{I_{mrd}} \right) \times 100$$

where

$I_m\%$ = motor current percentage

I_{mn} = motor nameplate current (in A)

I_{mrd} = maximum rated drive current (in A)

Example: Calculating Motor Current Percentage

What is the motor current percentage for a motor with a nameplate current of 8.2 A and a maximum rated motor drive current of 12 A?

$$I_m\% = \left(\frac{I_{mn}}{I_{mrd}} \right) \times 100$$

$$I_m\% = \left(\frac{8.2}{12} \right) \times 100$$

$$I_m\% = (0.683) \times 100$$

$$I_m\% = \textbf{68.3\%}$$

When programming motor overload protection, motor nameplate current is programmed into a motor drive. On advanced motor drives, motor nameplate current is entered into the motor drive as a drive parameter using a keypad. The motor drive monitors current drawn by the motor and turns off the motor if there is an overload. **See Figure 26-35.**

Motor drives with advanced parameter programming may have customized overload protection for a given application. Some motor drives can be programmed to automatically attempt to restart a motor after an overload. The number of automatic restarts permitted (normally no more than 9) and the amount of time (normally 1 sec to several minutes) between each automatic restart can be programmed.

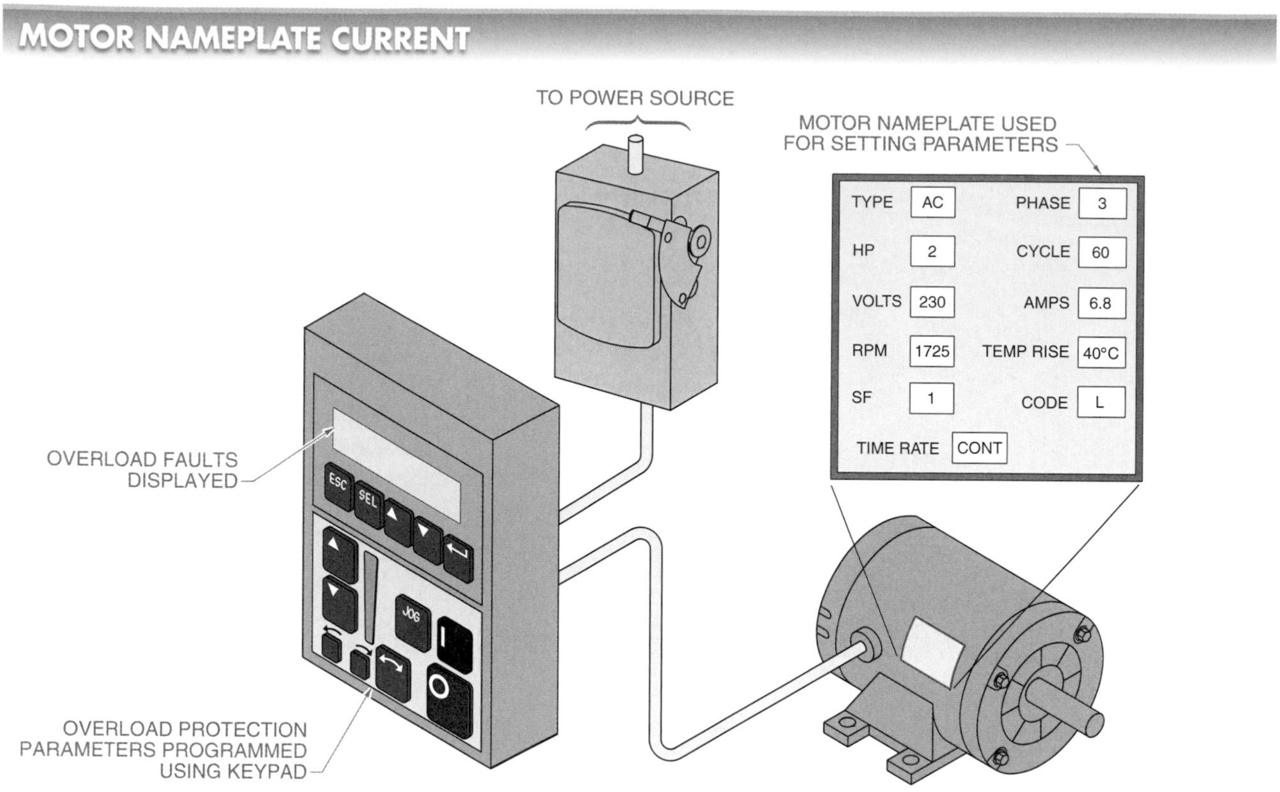

MOTOR NAMEPLATE CURRENT

Figure 26-35. Advanced motor drives normally contain a keypad that is used to enter drive parameters, such as motor nameplate current.

AC Motor Drive Load Test

An AC motor drive load test is a test used to verify that a motor drive and motor function properly together to rotate a driven load. An AC motor drive set to factory defaults and controlled by an integral keypad is tested with a motor connected. At this point, inputs and outputs are not tested. **See Figure 26-36.** To test an AC motor drive, motor, and load, the following procedure is applied:

1. If the AC motor drive is ON, push the STOP button.

2. Turn the disconnect off. Lockout/tagout the disconnect.

3. Wait for the DC bus capacitors to discharge. Do not manually discharge the capacitors. Remove the AC motor drive cover. Use a DMM to verify that the AC line voltage is not present. Use a DMM to verify that the DC bus capacitors have discharged. Do not rely on the DC bus and charge LED(s).

4. Reconnect the load conductors to their previous locations on the power terminal strip in order to maintain correct motor rotation. Incorrect motor rotation causes damage in certain applications. Reinstall the AC motor drive cover.

5. Remove the lockout/tagout from the disconnect.

6. Stand to the side of the disconnect and AC motor drive when energizing in case of a major failure. Turn on the disconnect. Do not push the START button. The AC motor drive LED display or clear text display should activate.

7. Program the appropriate parameters into the AC motor drive with motor nameplate data.

8. Program the display mode to show motor drive output frequency.

9. For the safety of personnel and equipment, a technician should monitor and control the motor drive as another technician monitors the motor and load during the test. Do not start the motor drive until a check is made to ensure that personnel are not at risk from the load.

10. Stand to the side of the disconnect and motor drive when energizing in case of a major drive failure. Push the START button. The LED display should ramp up to a low frequency. If the LED display shows 0 Hz, push the RAMP UP button until 5 Hz is displayed.

11. Increase the frequency of the motor to 60 Hz using the RAMP UP button. The motor and load should accelerate smoothly to 60 Hz. Any unusual noises or vibrations must be recorded along with the frequency at which the occurrence appeared. Unusual noises or vibrations indicate alignment problems or require the use of the skip frequency parameter to avoid unwanted mechanical resonance.

12. Remove the motor drive cover. **CAUTION:** Dangerous voltage levels exist when the motor drive cover is removed and the drive is energized. Exercise extreme caution and use the appropriate personal protective equipment.

13. Measure and record the current in each of the three load conductors using a true-rms clamp-on ammeter. True-rms clamp-on ammeters are required because the current waveform of a motor drive is not a perfect sine wave. Current readings are taken at 60 Hz because the motor nameplate current is based on 60 Hz. Current readings of the three load conductors should be equal or very close to each other. A problem with the load conductors or motor is present if the current readings of the load conductors are not equal or very close to each other.

An *overloaded motor* is a motor that has a current reading greater than 105% of the nameplate current rating. There is a problem with the motor or the load if the current readings are greater than 105% of the nameplate current rating. Reinstall the motor drive cover.

14. Decrease the speed of the motor to 0 Hz using the RAMP DOWN button. The motor and driven load must decelerate smoothly to 0 Hz. Any unusual noises or vibrations must be recorded along with the frequency at which the occurrence appeared. Unusual noises or vibrations indicate alignment problems, or can require the use of the skip frequency parameter to avoid unwanted mechanical resonance.

15. Push the STOP button.

16. If the motor drive, motor, and load performed without any problems, the motor drive, motor, and load are not the source of the problem. The motor drive inputs should be tested.

17. There is a problem if the motor drive, motor, and load did not perform correctly. The problem may be the motor drive parameters, motor, or load.

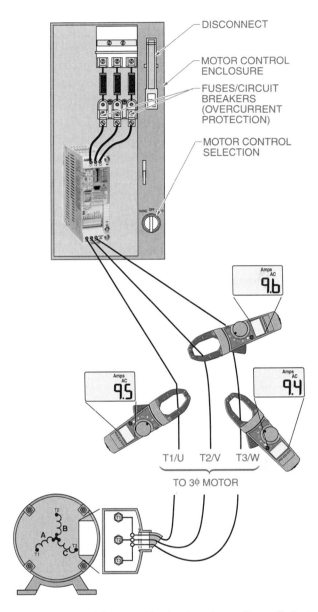

Figure 26-36. An AC motor drive load test is used to verify that a motor drive and motor function properly together to rotate a driven load.

Motor Load Applications

The difference between the various motor load classifications must be understood when a motor drive is programmed for a given application. Normally, a motor drive can be matched to an application by setting switches on the drive to constant or variable torque.

A motor drive is programmed to match an application by modifying the drive settings. Motor drive settings are modified by changing the position of dual in-line package (DIP) switches. **See Figure 26-37.** The most common settings include the following:

- NOM FREQ Hz—This setting sets the motor nominal frequency to 50 Hz or 60 Hz, depending on the power frequency.
- JOG Hz—This setting sets the jogging frequency to 5 Hz or 10 Hz.
- RELAY—This setting sets the drive state that the normally open contact of the relay output indicates (NO closed and NC open). FLT indicates that the contact is opened during a fault. RUN indicates that the contact is closed while the motor is running.
- LOAD—The load switch requires an understanding of the type of load the motor is operating. The P&F (pump and fan) setting is used for variable-torque loads, such as pumps, fans, blowers, and mixers. The CT (constant torque) setting is used for constant-torque loads, such as conveyors and load-lifting equipment.

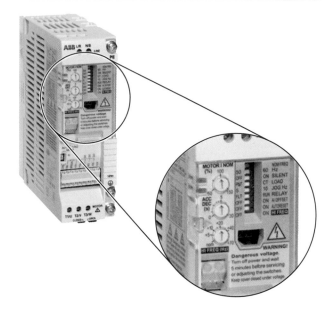

MOTOR DRIVE DIP SWITCHES

Figure 26-37. Motor drives can be matched to an application by setting DIP switches on the drive appropriately.

26-4 CHECKPOINT

1. What is the function of the converter section of a drive?
2. What is the function of the inverter section of a drive?
3. What is the V/Hz ratio of a 480 VAC, 60 Hz motor?
4. Does increasing the carrier frequency produce more or less voltage spikes (transients) at the motor?
5. Does increasing the carrier frequency require a smaller or larger heat sink?
6. If the nameplate current of a motor is listed at 23 A and the drive current rating is listed as 50 A, what setting (in %) should the motor drive's overload protection be set or programmed at?

26-5 TROUBLESHOOTING MOTOR DRIVE CIRCUITS

Troubleshooting is the systematic elimination of the various parts of a system or process to locate a malfunctioning part. When a motor circuit is not operating properly, voltage and current measurements are taken to help determine or isolate the problem. Voltage measurements are taken to establish that the voltage is present and at the correct level. Voltage measurements may help determine circuit problems such as blown fuses, improper grounding, contacts not closing, etc. However, voltage measurements alone do not indicate the true condition of a motor because the voltage may be correct at the motor

terminals but the motor may be faulty. *Note:* When voltage measurements are taken, the required protection and safety equipment should always be worn and the proper and safe procedures should always be applied. To measure voltage, the following procedure is applied:

1. Measure the voltage at the disconnect. With the power ON, measure the voltage between each power line (L1-L2, L2-L3, and L1-L3). The voltage between the power lines should be within 2% (2 V per 100 V). If the power lines are not within 2%, there is a power supply problem. **See Figure 26-38.**

2. Check the fuses or circuit breakers. Replace any blown fuses (or reset tripped circuit breakers). *Note:* Ensure that the disconnect is in the OFF position and the motor circuit is in the OFF condition before replacing fuses or resetting circuit breakers.

3. With the power ON, measure the voltage into the drive (L1-L2, L2-L3, and L1-L3). The voltage into the drive (or motor starter) should be within 2% (2 V per 100 V). If the voltage into the drive is not within 2%, there is a problem between the disconnect and the drive.

4. With power ON, measure the voltage out of the motor drive (T1-T2, T2-T3, and T1-T3). The voltage out of the drive should be within 2% (2 V per 100 V) when the motor is at full speed. If the voltage out of the motor drive is not within 2%, there is a problem with the motor drive (or control circuit). Turn the power off and apply a lockout/tagout before making any repairs.

Current measurements are taken to determine the condition of a motor. Current measurements indicate if a motor is underloaded (receiving less than nameplate rated current), fully loaded (receiving nameplate current), or overloaded (receiving higher than nameplate rated current). *Note:* When taking current measurements, the required protection and safety equipment should always be worn and proper and safe procedures should always be applied. To measure current, the following procedure is applied:

1. With power ON, measure the current in the lines leading to the motor (T1, T2, and T3). **See Figure 26-39.** The current on each line should be within 10% of each of the other lines and should be less than the motor nameplate rated current. If the current is equal to the nameplate rated current, the motor is fully loaded and may have a problem or the motor may be undersized for the application. If the current is higher than the nameplate rated current, there is a problem with the motor that must be repaired. Turn off the power and apply a lockout/tagout before making any repairs.

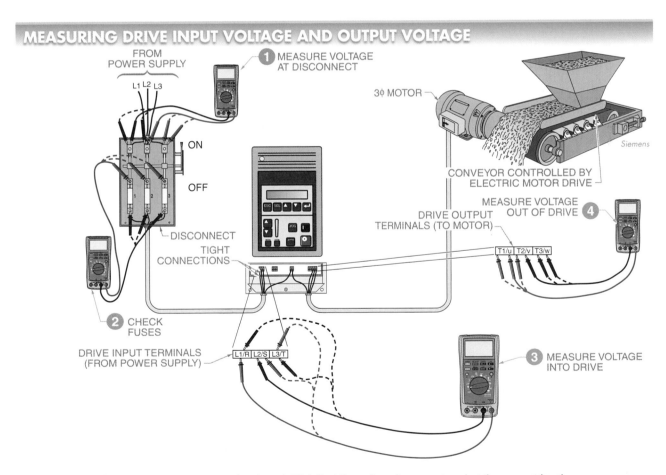

Figure 26-38. Voltage measurements are taken to establish that the voltage is present and at the correct level.

MEASURING MOTOR CURRENT

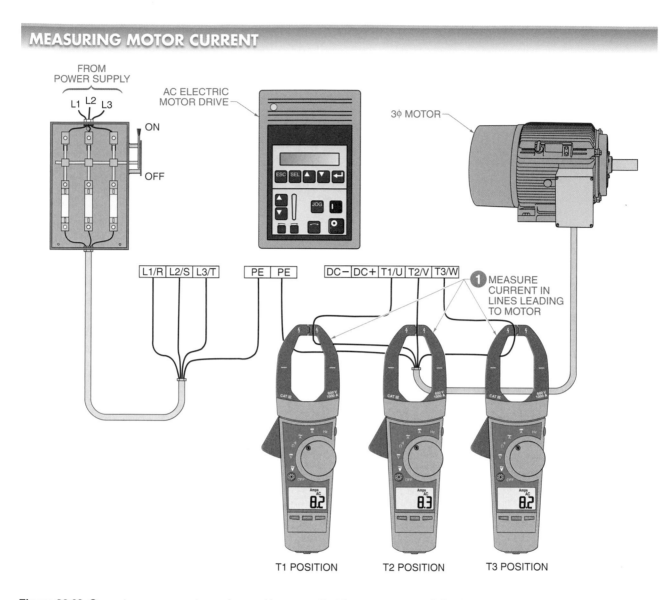

Figure 26-39. Current measurements can be used to ensure that the current on each line is within 10% of each of the other lines.

Saftronics Inc.

Motor drive and enclosure fabrication requires that all devices and components be installed in a neat and organized manner.

Troubleshooting Reversing Circuits

When a reversing motor circuit does not operate properly, the problem may be electrical or mechanical. The control circuit and power circuit are tested using a DMM to check for proper electrical operation. Troubleshooting starts inside the control cabinet when testing reversing control circuits or power circuits. When troubleshooting reversing control circuits, a line diagram is used to illustrate circuit logic and a wiring diagram is used to locate the actual test points at which a DMM is connected to the circuit. **See Figure 26-40.**

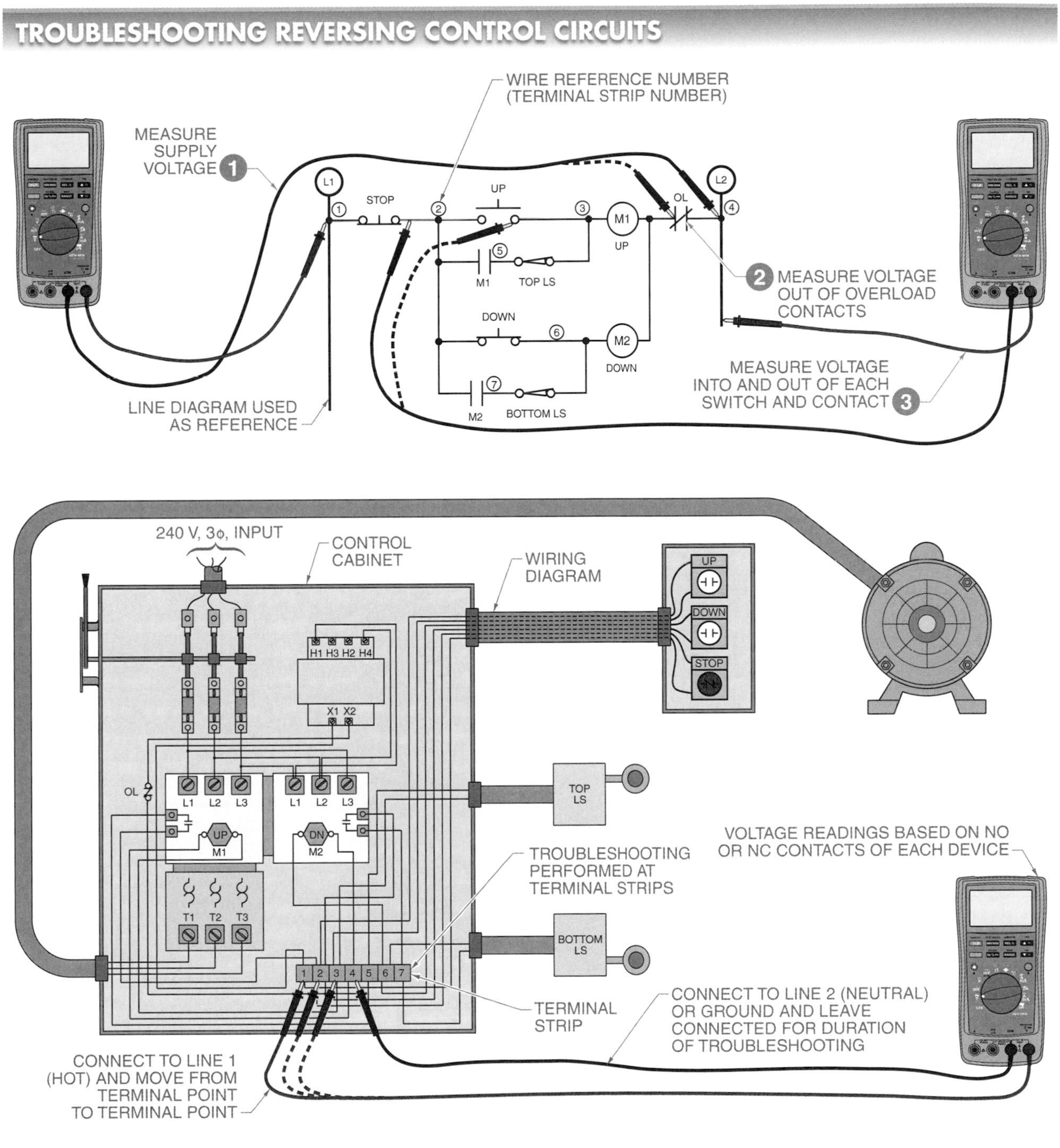

TROUBLESHOOTING REVERSING CONTROL CIRCUITS

Figure 26-40. When troubleshooting reversing control circuits, a line diagram is used to illustrate circuit logic and a wiring diagram is used to locate the actual test points at which a DMM is connected.

To troubleshoot a reversing control circuit, apply the following procedure:

1. Measure the supply voltage of the control circuit by connecting a DMM set to measure voltage between line 1 (hot conductor) and line 2 (neutral conductor).

The voltage must be within 10% of the control circuit rating. Test the power circuit if the voltage is not correct. The control circuit voltage rating is determined by the voltage rating of the loads used in the control circuit (motor starter coils, etc.).

2. Measure the voltage out of the overload contacts to ensure the contacts are closed. The contacts are tripped or are faulty if no voltage is present. Reset the overloads if tripped. Overloads are installed to protect the motor during operation. The control circuit does not operate when the overloads are tripped.

3. Measure the voltage into and out of the control switch or contacts. Normally closed switches (stop pushbuttons, etc.) should have a voltage output before they are activated. Normally open switches (start pushbuttons, memory contacts, etc.) should have a voltage output only after they are activated.

When wiring the control circuit using a motor drive, the control devices (pushbutton, etc.) are wired to the drive control terminal strip. The power at the terminal strip is usually already stepped down to less than 30 V, making the control circuit safe and simple to wire. A PLC can be used to control the motor through the motor drive, either by using the PLC output contact or by using direct PLC/drive communication through a designated port (serial port, etc.). **See Figure 26-41.**

Electric Motor Drive Wiring. An electric motor drive can be used to control various functions of a motor. The various functions of a motor normally include the following:

- starting
- stopping
- jogging
- speed control
- motor direction control
- acceleration time
- deceleration time
- overload protection
- braking force
- programmable output contacts
- voltage, current, power, and frequency metering and display
- preselection of multiple remote controls

Motor drives eliminate the need for forward and reversing starters because the motor drive can be used to select motor direction. The direction of the motor can be selected using the keypad on the motor drive or external pushbuttons connected to the drive input terminals. **See Figure 26-42.** The motor drive internal circuit and parameter settings can be used to prevent changing motor direction before the motor has come to a full stop.

When wiring the power circuit using a motor drive, the incoming power is connected to L1, L2, and L3 (sometimes marked as R, S, and T). The motor is connected to T1, T2, and T3 (sometimes marked as U, V, and W). The motor ground, drive ground, and power supply ground are all connected to form a common ground. The power circuit is simplified when using a motor drive because there is no need for a reversing part of the power circuit. All reversing functions are performed internally within the drive. **See Figure 26-43.**

Tech Fact

When wiring a 230 V to 460 V motor drive power circuit, copper wire should be used that has at least a 600 V/167°F (75°C) rating and a current rating that is at least 1.5 times higher than the drive's maximum current output rating. The drive's maximum output current rating should be at least 1.25 times the motor's nameplate rated current. If a motor has a service factor amperage (SFA) rating, it can be higher than the standard current rating on the motor and should be used since the motor can safely draw higher current without damage.

Motor Lead Length. In a typical electrical system, distance between components can affect the operation of the system. The primary limit to the distance between a magnetic motor starter and a motor is the voltage drop of the conductors. The voltage drop of conductors should not exceed 3% for many types of motor circuits.

When an AC motor drive is used to control a motor, the distance between the drive and the motor may be limited by other factors besides the voltage drop of conductors. Conductors between an AC motor drive and motor have line-to-line (phase-to-phase) capacitance and line-to-ground (phase-to-ground) capacitance. Longer conductors produce higher capacitance. The capacitance produced by conductors causes high voltage spikes in the voltage to a motor. Since voltage spikes are reflected into the system, the voltage spikes are often called reflective wave spikes. As the length of conductors increase and/or the motor drive output carrier frequency increases, voltage spikes increase. **See Figure 26-44.**

Determining the length of cable at which voltage spikes become a problem may be difficult. Normally, when lengths of cable between an AC motor drive and motor are less than 100′, problems do not occur. Small-horsepower motors and multiple motors connected to one motor drive are more susceptible to voltage spikes. Voltage spikes are problems because they stress and wear out motor insulation. When voltage spikes become a problem, motor lead length should be reduced and/or filters that suppress voltage spikes should be added. Reducing the carrier frequency also reduces reflective wave voltage spikes. Inverter-rated motors that have spike-resistant insulation also reduce the damage caused by voltage spikes.

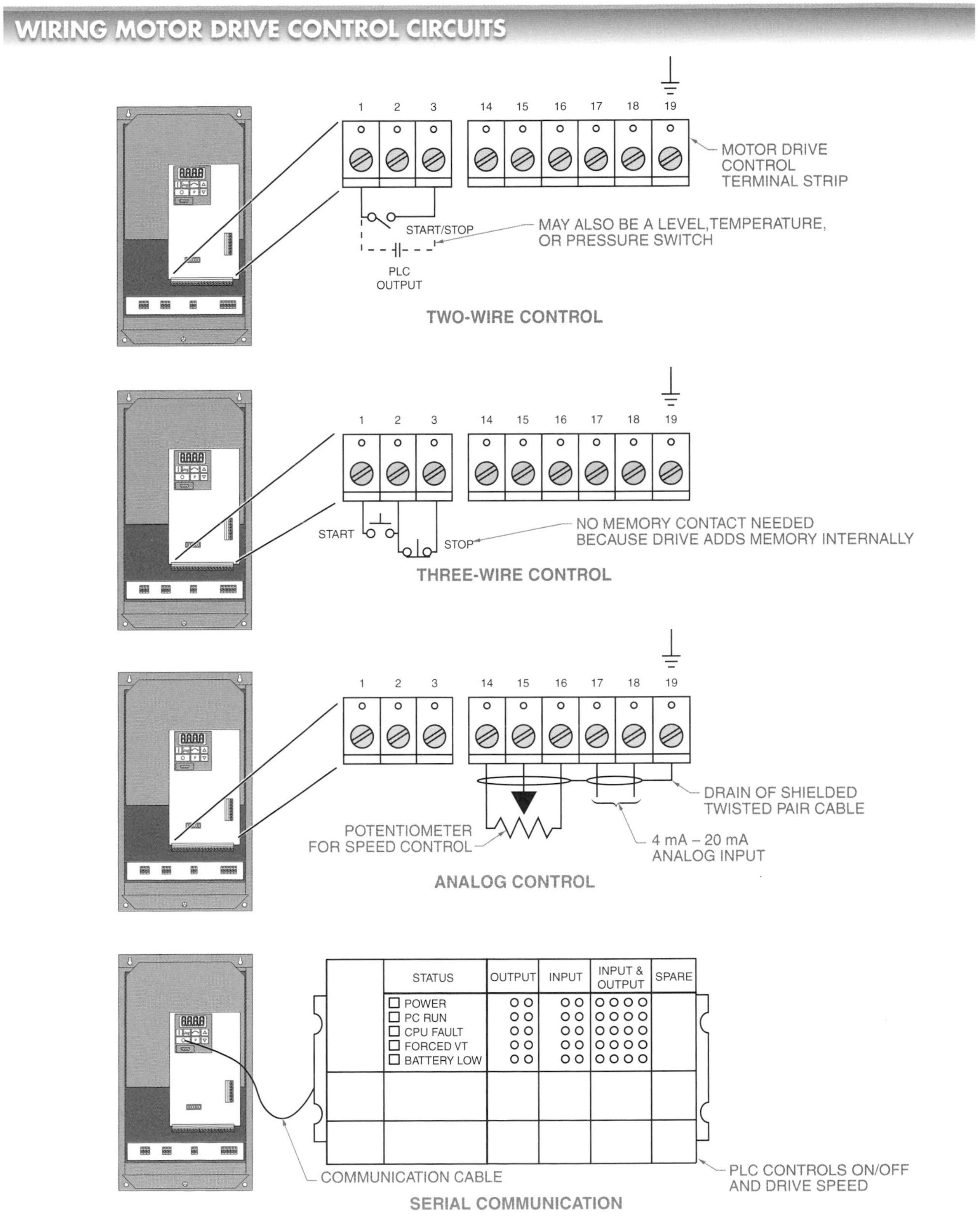
WIRING MOTOR DRIVE CONTROL CIRCUITS
1 2 3 14 15 16 17 18 19
MOTOR DRIVE CONTROL TERMINAL STRIP
START/STOP
MAY ALSO BE A LEVEL, TEMPERATURE, OR PRESSURE SWITCH
PLC OUTPUT
TWO-WIRE CONTROL
1 2 3 14 15 16 17 18 19
START
STOP
NO MEMORY CONTACT NEEDED BECAUSE DRIVE ADDS MEMORY INTERNALLY
THREE-WIRE CONTROL
1 2 3 14 15 16 17 18 19
DRAIN OF SHIELDED TWISTED PAIR CABLE
POTENTIOMETER FOR SPEED CONTROL
4 mA – 20 mA ANALOG INPUT
ANALOG CONTROL
STATUS
OUTPUT
INPUT
INPUT & OUTPUT
SPARE
POWER
PC RUN
CPU FAULT
FORCED VT
BATTERY LOW
COMMUNICATION CABLE
PLC CONTROLS ON/OFF AND DRIVE SPEED
SERIAL COMMUNICATION

Figure 26-41. A PLC can be used to control a motor through a motor drive by either using the PLC output contact or by using direct PLC/drive communication through a designated port.

MOTOR DRIVE MOTOR DIRECTION

CONTROLLED MOTOR

PROGRAMMING KEYS

HMI KEYPAD

INCREASE MOTOR SPEED

UP/INCREASE PARAMETER SETTING

DOWN/DECREASE PARAMETER SETTING

ESC   SEL

MOTOR JOG

DECREASE MOTOR SPEED

JOG

MOTOR ON

JOG BUTTONS

FORWARD/REVERSE

MOTOR OFF

MOTOR DRIVE USED TO CONTROL MOTOR DIRECTION

CONTROLLED MOTOR

Figure 26-42. A motor drive may be used to select the direction of a motor.

WIRING MOTOR DRIVE POWER CIRCUITS

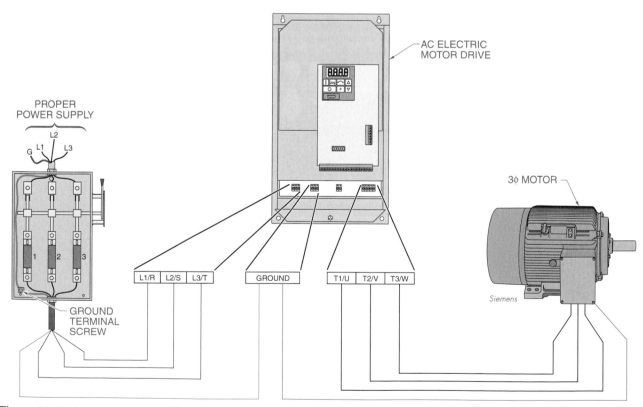

PROPER POWER SUPPLY

G L1 L2 L3

1 2 3

GROUND TERMINAL SCREW

AC ELECTRIC MOTOR DRIVE

3φ MOTOR

Siemens

| L1/R | L2/S | L3/T | | GROUND | | T1/U | T2/V | T3/W |

Figure 26-43. A motor drive simplifies a reversing power circuit because all reversing functions are performed internally within the drive.

VOLTAGE SPIKES

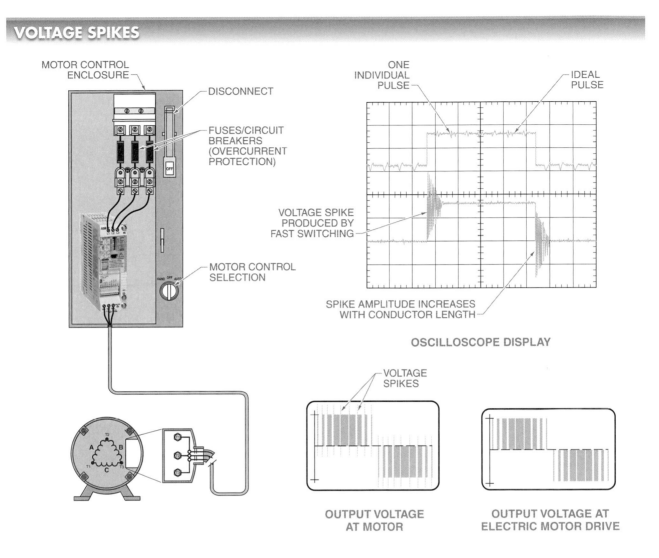

Figure 26-44. Longer lead lengths between an AC motor drive and a motor and/or higher carrier frequencies produce higher voltage spikes.

26-5 CHECKPOINT

1. When testing the voltage into a drive and at the motor, the voltage between each line should not vary more than what percentage?

2. When testing the current into a drive and at the motor, the current between each line should not vary more than what percentage?

3. How are the power lines L1, L2, and L3 sometimes marked on a motor drive?

4. How are the motor connection lines T1, T2, and T3 sometimes marked on a motor drive?

Additional Resources

Review and Resources

Access Chapter 26 Review and Resources for *Electrical Motor Controls for Integrated Systems* by scanning the above QR code with your mobile device.

Applying Your Knowledge

Refer to the *Electrical Motor Controls for Integrated Systems* Learner Resources for interactive Applying Your Knowledge questions.

Workbook and Applications Manual

Refer to Chapter 26 in the *Electrical Motor Controls for Integrated Systems Workbook* and the *Applications Manual* for additional exercises.

ENERGY EFFICIENCY PRACTICES

Motor Drive Energy Savings

Motor drives provide energy savings when used with variable-speed and variable-torque loads. Motor drives are commonly installed on pumps and fans in HVAC applications. When a motor drive is installed to control a pump or fan motor, the speed of the motor can be varied to control the flow rate of the pump or fan, allowing less energy to be used. Reducing the amount of energy used by pump and fan motors reduces the utility bills of a facility and also reduces the emissions of greenhouse gases from power plants.

Motor drives provide a number of performance benefits over standard motor control systems. The benefits of motor drives include significant energy savings when used for pump and fan applications, precise control of motor speed and torque, and elimination of high starting currents and high transient voltages resulting from large motors turning on and off because motor drives ramp up motor speed when starting and ramp down motor speed with stopping.

Motor drive manufacturers provide software programs to calculate the cost savings realized when using motor drives with pumps and fans. Information such as motor horsepower, motor efficiency, cost of electricity, cost of the motor drive, and length of time the load is operated at different speeds is entered into the program. The cost-savings program calculates the annual cost savings and the payback period. In many cases, the software is available free or as a download from the website of the motor drive manufacturer.

Objectives

27-1
- Explain the differences between discrete parts manufacturing and process manufacturing.
- Define and describe the different programmable controllers.
- Describe how programmable controllers are selected and configured.

27-2
- Describe the power supply of a programmable controller.
- Describe the input/output (I/O) sections of a programmable controller.
- Define and describe the processor section of a programmable controller.
- Describe the programming section and describe programming devices, symbols, and languages.
- Describe how to develop a typical program for a controller.
- List and describe the status and fault indicators included in programmable controllers.
- Describe the force and disable commands.
- Explain how programmable controllers are used within a network.
- Explain how programmable timers work.

27-3
- List and describe different programmable controller applications.
- Describe programmable controller circuits.

27-4
- Explain how to troubleshoot input modules and devices.
- Explain how to troubleshoot output modules and devices.

Programmable
Controllers

27

The first electrical circuits, such as basic switches that controlled lamps for general lighting, were simple circuits with few inputs and output. However, electrical circuits continued to grow in size, function, and complexity. Today, the electrical circuits in some manufacturing plants are interconnected to form systems with hundreds of inputs and outputs. These systems are interconnected to form a continuous process designed to safely and efficiently produce parts and products at an ever-increasing speed.

Just as microelectronics have allowed computers to become smaller over the years as operating speed and functions increased, programmable controllers have allowed control circuits to become smaller as system operating speed and functions increased. Programmable controllers eliminate the need for external timers, counters, and relays. Programmable controllers also allow the entire circuit to be designed and downloaded using a keyboard/mouse and monitor. Because programmable controllers have become the standard control device in most circuits that include more than a few inputs and outputs, understanding programmable controllers is important for any person working in the electrical or electronic field today.

27-1 PROGRAMMABLE CONTROLLERS

Programmable controllers were developed to solve the problem of automating larger and more complex control systems. The first programmable controllers where developed to eliminate individual control relays (CRs) and timers (TRs) by consolidating them into one unit that could be programmed and reprogrammed as required by the application. The first programmable controllers were called programmable logic controllers (PLCs) and quickly developed into control devices that could be used to control and monitor almost any industrial type control circuit.

As PLCs became more popular, less expensive, and easier to program and use, a much smaller and less expensive type of programmable controller was developed to control any smaller electrical system in residential, commercial, or industrial applications. This smaller but powerful programmable controller is called a programmable logic relay (PLR). PLRs are basically small PLCs that include the same control-type functions, but on a smaller scale. The main advantage of a PLR is that they are very easy to program and use with smaller control circuits.

Just as the improvements and increased versatility of PLCs led to the development and usage of PLRs, PLCs led to the development and usage of a higher-end control system that could tie the technologies of PLC industrial control technology with personal computer (PC) technology. This computer integrated control system is called a programmable automation controller (PAC). The main advantage of PACs is that they combine the latest technologies, programming methods, and communication networks of PCs with the latest technologies of PLCs.

Regardless of what a programmable controller is called (PLC, PLR, or PAC) or its size and complexity, they all share common platforms and technology. For example, all three programmable controller types are designed to have digital and analog controls connected to their input section and have their output section control motors, lamps, and other electrical loads. Understanding a PLC makes understanding and using PLRs and PACs easier. Each system is designed to control large or small systems, processes, and machines so they are safer, are easier to program, can monitor the circuit/component, and be interconnected together to form larger local or remote control systems. **See Figure 27-1.**

Tech Fact

When selecting an automated controller for an application, a PLR, PLC, or PAC can be selected. If a controller is fixed and the number of required inputs and outputs is not likely to change, select the simplest to program. Consider starting with a PLR. If the controller is used in an application that requires changes, may require expansion, or needs to be a part of a larger system, start with an expandable PLC with interchangeable I/O modules.

Programmable Controller Usage

Industrial electrical systems designed to produce products are commonly divided into discrete parts manufacturing and process manufacturing. In discrete parts manufacturing and process manufacturing, PLCs, PLRs, and PACs have become the standard components used to control the operation from start to finish.

Discrete Parts Manufacturing. The discrete parts manufacturing market produces durable goods such as automobiles, washers, refrigerators, and tractors. Discrete parts manufacturing is done primarily by stand-alone machines that bend, drill, punch, grind, and shear metals. All of these machines can be automated with programmable controllers. **See Figure 27-2.**

DISCRETE PARTS MANUFACTURING

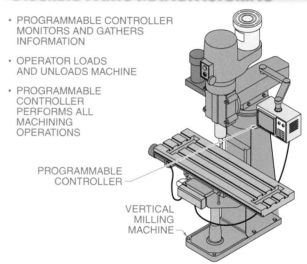

- PROGRAMMABLE CONTROLLER MONITORS AND GATHERS INFORMATION
- OPERATOR LOADS AND UNLOADS MACHINE
- PROGRAMMABLE CONTROLLER PERFORMS ALL MACHINING OPERATIONS

PROGRAMMABLE CONTROLLER

VERTICAL MILLING MACHINE

Figure 27-2. A programmable controller can be used to control all electrical functions on a machine used in discrete parts manufacturing.

PROGRAMMABLE CONTROLLER SIZE AND REQUIREMENTS

SYSTEM CONTROL SIZE

- NUMBER OF INPUTS/OUTPUTS
- UNIT EXPANDABILITY
- TYPE OF INPUT (DISCRETE, ANALOG, AND MOTION)

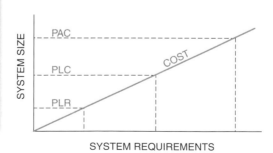

SYSTEM REQUIRMENTS

- EXPANDABLE NEEDS
- REMOTE (NETWORK) CONTROL AND MONITORING
- SYSTEM COMMUNICATION OPTIONS
- ABILITY TO INTERGRATE SYSTEM HARDWARE AND SOFTWARE

Figure 27-1. Programmable controllers include PLCs, PLRs, and PACs.

A programmable controller allows each machine to have its own unique capability using standard hardware. This allows easy modification of the controls when the functional requirements of the machine change. Modular replacement of programmable controllers reduces downtime of the machine. The use of programmable controller helps reduce startup and debug time and allows manufacturers to incorporate additional user requirements for changes in machine operations after startup.

Today, programmable controllers have become the standard for machine builders. Increased capabilities in a reduced size allow current programmable controllers to control one machine or to connect with many machines in any network configuration.

In addition to allowing each machine to have its own unique capabilities, a programmable controller can also be used to interface and control the operation of all or parts of the machines along a production line. Programmable controllers can be used to control the speed of a production line, divert production to other lines when there is a problem, make product changes, and maintain documentation such as inventory and losses. **See Figure 27-3.**

Process Manufacturing. The process manufacturing industry produces consumables such as food, gas, paint, pharmaceutical products, paper, and chemicals. Most of these processes require systems to blend, cook, dry, separate, or mix ingredients. **See Figure 27-4.**

Automation is required for opening and closing valves and controlling motors in the proper sequence and at the correct time. A programmable controller allows for easy modifications to the system if the time, temperature, or flow requirements of the products change.

Today's programmable controllers control process manufacturing activities such as the conveying, palletizing, storing, and treatment of products. They also control the alarms, interlocks, and preventive maintenance functions for the system. A programmable controller can also generate reports that are used to determine production efficiency.

Programmable controller manufacturers offer a variety of units from microunits to very large units. **See Figure 27-5.** A microcontroller or small controller is the best choice for machines and processes that have limited capability and little potential for future expansion.

Programmable controllers are used in manufacturing processes where the synchronization of various pieces of mechanical equipment is important.

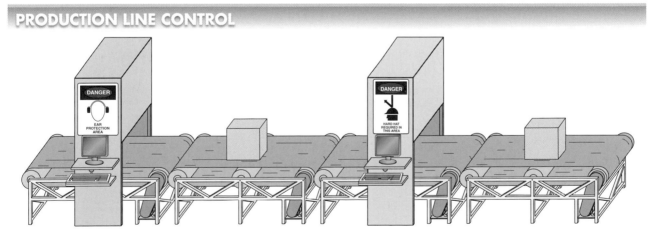

PRODUCTION LINE CONTROL

Figure 27-3. Programmable controllers can be used to control the speed of a production line, divert production to other lines when there is a problem, make product changes, and maintain documentation such as inventory and losses.

PROCESS MANUFACTURING—PAPER MILL

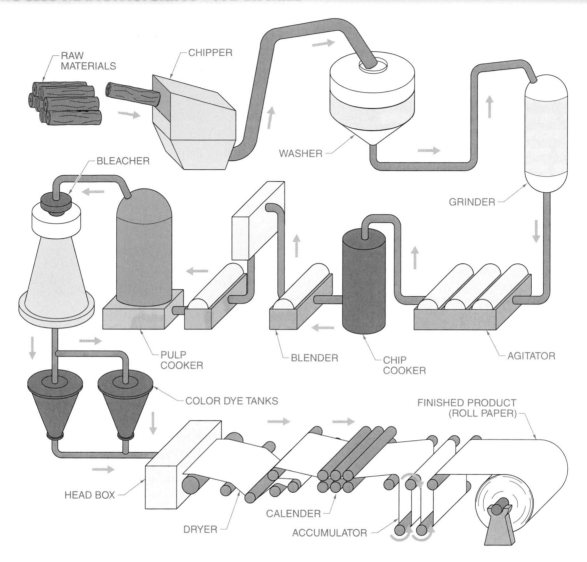

Figure 27-4. The process manufacturing industry produces consumables such as paper, food, gas, paint, pharmaceutical products, and chemicals, which require systems to blend, cook, dry, separate, or mix ingredients.

PROGRAMMABLE CONTROLLER UNITS

Omron Electronics, Inc.

Figure 27-5. Programmable controller manufacturers offer a variety of controllers for machines and processes that have limited capability and little potential for future expansion or for processes that have complex control requirements.

Types of Programmable Controllers

The type of programmable controller that is used depends upon the application, size and number of loads to be controlled, need for monitoring and reprogramming the circuit, and amount of control required (ON/OFF to anywhere in between). Budget cost and the knowledge level of personnel who will need to program the circuit/system and keep the system operating and upgraded must also be considered. The larger and more complex the system, the greater the need for more precise programming, control, flexibility, and monitoring. Types of programmable controllers include programmable logic controllers (PLCs), programmable logic relays (PLRs), and programmable automation controllers (PACs).

Programmable Logic Controllers (PLCs). A *programmable logic controller (PLC)* is a solid-state control device that is designed to be programmed and reprogrammed to automatically control industrial processes or machine circuits. Although PLCs can be used to control almost any type of electrical control circuit, they are primarily designed for the harsher environments found in industrial applications. **See Figure 27-6.** A PLC consists of a power supply, processor section (central processing unit or CPU), input section, and output section.

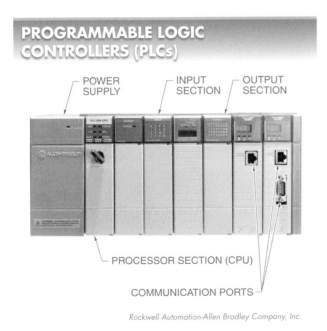

PROGRAMMABLE LOGIC
CONTROLLERS (PLCs)

POWER SUPPLY

INPUT SECTION

OUTPUT SECTION

PROCESSOR SECTION (CPU)

COMMUNICATION PORTS

Rockwell Automation-Allen Bradley Company, Inc.

Figure 27-6. A PLC is a solid-state control device that is designed to be programmed and reprogrammed to automatically control motors in industrial processes or machine circuits.

PLCs are designed to control circuits with many inputs and output with individual input and output (I/O) modules that are designed to be connected to 8, 16, 32, or 64 devices. Individual modules can be added and grouped together in a 4, 6, 8, 12, or more units within a rack. There, I/O capabilities can be further expanded by adding additional expansion modules and/or interconnecting different PLCs into a common control system, allowing applications with over 1000 I/Os to be controlled.

Smaller PLCs, also called stand-alone PLCs, include the same basic parts as larger PLCs, but they do not allow for interchanging or adding additional modules to them. They can be grouped together to for larger systems or an "expansion" module can sometimes be added, but they are designed to operate smaller circuits that have no more I/O than they have available.

Most PLCs are designed to be programmed using line (ladder) programming or function blocks. Programming is done using the manufacturers software designed for the type of PLC being used. Although programs vary based on the PLC size and complexity, they all share common programming devices (inputs, outputs, timers, counters, etc.) and operating functions (circuit monitoring, I/O forcing, etc.).

Programmable Logic Relays (PLRs). A *programmable logic relay (PLR)* is a solid-state control device that includes internal relays, timers, counters, and other control functions that can be programmed and reprogrammed to automatically control small residential, commercial, and industrial circuits. PLRs are basically small PLCs that are designed for more general control applications that require fewer inputs (usually less than 10) and fewer outputs (usually less than 6). Because PLRs are basically small PLCs, some manufacturers call them "smart relays" or list them as "nano" or "micro" PLCs. Regardless of what the manufacturers call them, these PLCs program and operate the same as a small circuit PLC. **See Figure 27-7.**

Like PLCs, PLRs are designed to be programmed using line (ladder) programming or function blocks. Although they can be programmed using only the keys and display included on the unit, programming and circuit monitoring are best accomplished by connecting the PLR to a PC because the local viewing window is small and can only display smaller sections of the larger program. However, the local display and operating keys are good for programming passwords and language into the unit (though the unit will usually offer several common languages) and for viewing/changing operating functions such as run or stop modes.

PROGRAMMABLE LOGIC RELAYS (PLRs)

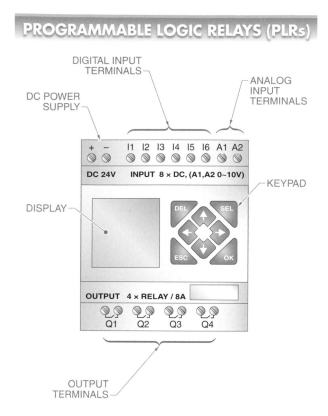

Figure 27-7. A PLR is a solid-state control device, similar to a PLC, that includes internal relays, timers, and counters.

Programmable Automation Controllers (PACs).
A *programmable automation controller (PAC)* is a combination of a PLC and a PC-based CPU control device that is designed to operate in an industrial environment. PACs can be programmed and reprogrammed using standard PLC languages (ladder or function blocks) or PC languages (structured text, sequence function charts, etc.) to automatically control a large industrial process. PACs are interconnected into a larger system that usually includes human-machine interfaces (HMIs), remote monitors, data acquisition, motion control, or vision control-type devices that are part of the total PAC operating system. Like PLCs, PACs are designed to operate in harsh environments found in industrial applications.

PACs are available in self-contained units that include the power supply, CPU, and input or output parts designed for individual control functions. They are also available in rack-mounted units that can form units to control entire systems. **See Figure 27-8.**

Like PLC units, PAC units can include basic I/O modules, such as digital or analog I/O modules and

temperature modules. In addition, PAC units can also include specialized modules, such as serial communication modules, servo/stepper motor control modules, and data acquisition modules. Process voltage (0 VDC to 10 VDC) and current (4 mA to 20 mA) analog control modules are also available for usage with control inputs like thermocouples, resistance temperature detectors (RTDs), pressure monitors, flow monitors, distance detectors, strain gauges, and other analog output devices.

Programmable Controller Selection

In general, a PLC is considered the first choice for controlling industrial electrical circuits and systems. A PLR can be considered when a system does not have a lot of loads (outputs) or control switches (inputs) and requires less internal controller timers, counters, etc. Likewise, a PAC can be considered when more advance control is required to work within a large system and PC-based network, such as a large process control system.

A good rule-of-thumb is that when the use of a PLC is being considered, a PLR can be used if the control circuit is small enough and will not be operating in a harsh industrial-type environment. A PAC can be used if the control system also requires remote monitoring, HMIs, data acquisition of process conditions, motion control, servo/stepper motor control, and/or vision control.

The development of PLCs, PLRs, and PACs began slowly in the 1960s but continues today at a faster rate. Before there was any type of computer-based control circuits or information-type systems, electrical relays were used to develop control circuit logic and typewriters were used to produce written documents. The development of PLCs, PLRs, and PACs parallels the development of PCs.

As the capabilities and operating speeds of programmable controllers and PCs increased, both controllers and PCs were developed into larger units (mainframes, etc.) and smaller units (laptops, etc.). Larger controllers added better analog I/O controls, the ability to tie individual units into a centrally controlled system, and improved circuit/application control functions such as the ability to use latch/unlatch, real-time clock control of loads, logic and math functions within the programs, etc. One of the improvements in the software programming packages for controllers was the addition of the ability to program control circuits using logic function blocks and/or ladder diagrams.

PROGRAMMABLE AUTOMATION CONTROLLERS (PACs)

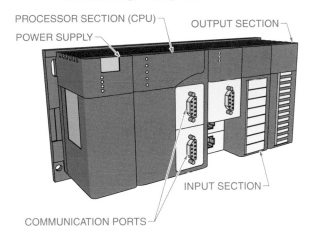

Figure 27-8. A PAC is a combination of a PLC and a PC-based CPU control device that is designed to operate in an industrial environment.

As programmable controllers were used in more control circuits, the need arose for a small/inexpensive controller that could be used for small industrial applications and small residential and commercial applications. The small/inexpensive controller that was developed was the programmable logic relay (PLR) because it was designed similar to the original controllers, which were basically relay replacements with added timers and counters. The small/inexpensive (often less than $100.00) PLRs functioned similar to how the first, much larger controllers were designed to functioned.

As industrial applications continued to grow in size, the need arose for a more computer-based control unit for monitoring requirements, locally and remotely interfacing networks, positioning and inspection, and other precise control functions. The latest PLC technology was combined with the latest PC technology to create a programmable automation controller (PAC).

PLCs, PLRs, and PACs continue to grow in their usage, function, speed, and ability to connect into different networking systems so individual units and parts can all communicate with each other. PLCs, PLRs, and PACs all share the same basic control requirements, which include the following:

- controlling circuit outputs based on circuit input conditions
- digital (OPEN/CLOSED) inputs connected to their input section
- analog (infinite position) inputs connected to their input section
- digital (ON/OFF) outputs controlled by their output section
- analog (infinite operating condition) outputs controlled by their output section
- a method or methods of programming the control circuit using PCs
- a PC to monitor the circuit, print documentation, and provide system/component troubleshooting features
- expandable and interconnected into larger systems

PACs can be used to control the fill of industrial-sized tanks.

Programmable Controller Configurations

Programmable controllers can be used as stand-alone control devices or configured into a system. A programmable controller can be configured into a system that uses a PC, a handheld programming unit, an operator interface panel, other programmable controllers, or other devices that connect into an electrical system. The amount and level of control between the components in a programmable controller depends on the system operating requirements and cost. The configuration of a programmable controller's system is determined by the type and quantity of required I/O devices, required communication between devices, programming types and requirements, and any future needs. **See Figure 27-9.**

PROGRAMMABLE CONTROLLER SYSTEM CONFIGURATIONS

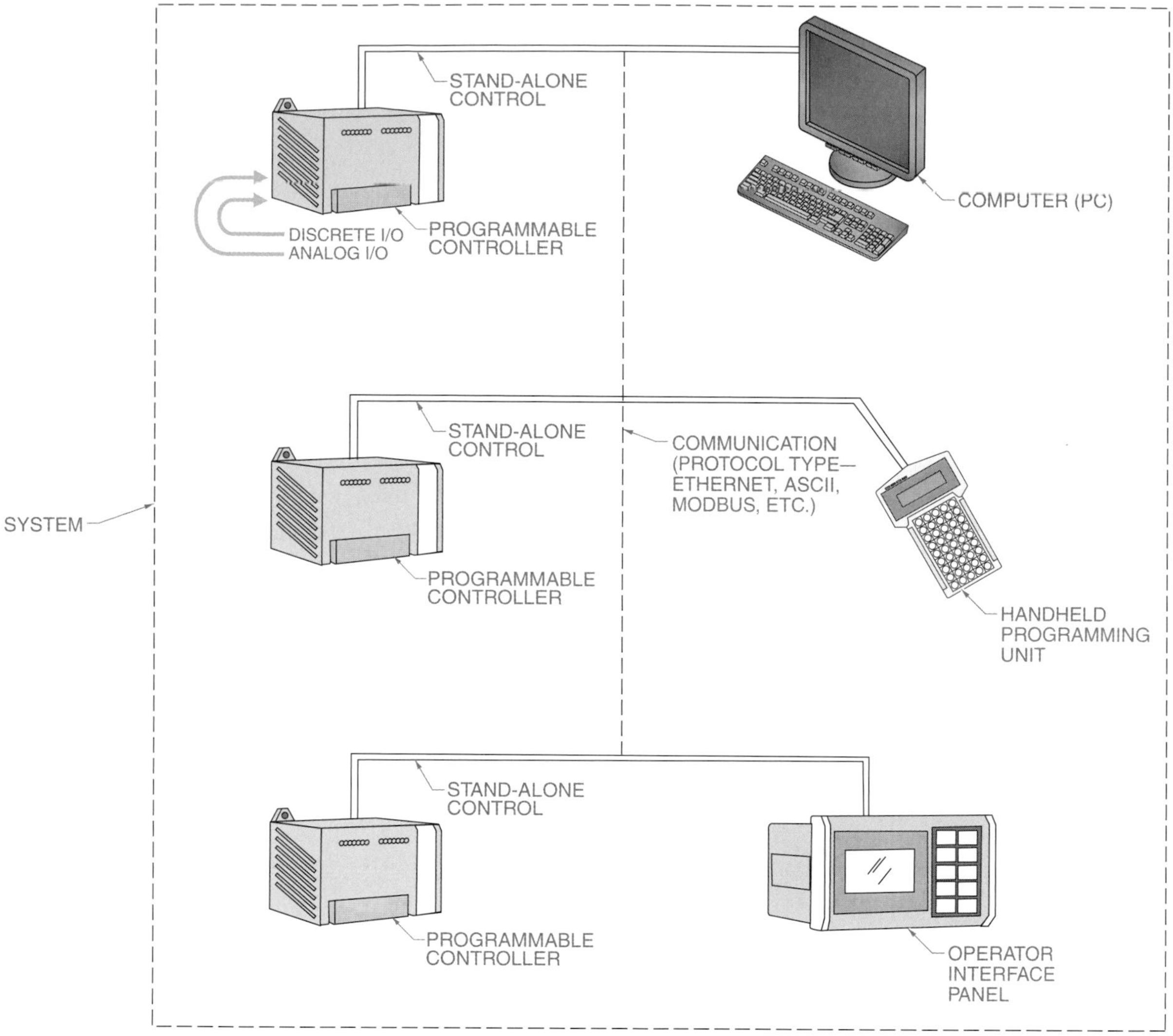

Figure 27-9. The configuration of a programmable controller system is determined by the type and quantity of required I/O devices, required communication between devices, programming types and requirements, and any future needs.

27-1 CHECKPOINT

1. A soft drink bottling plant is an example of what type of manufacturing?

2. A furniture plant is an example of what type of manufacturing?

3. Does the use of a PLR, PLC, or PAC to control a system eliminate the need for inputs and outputs?

4. Does the use of a PLR, PLC, or PAC to control a system eliminate the need for stand-alone (external) timers and counters?

27-2 PROGRAMMABLE CONTROLLER SECTIONS

All programmable controllers have four basic sections. The four basic sections of a programmable controller include the power supply, input/output (I/O) section, processor section, and programming section. **See Figure 27-10.**

The programs used for discrete parts manufacturing and process manufacturing are stored in and retrieved from memory as required. Programmable controller sections are interconnected and work together to allow the controller to accept inputs from a variety of sensors, make a logical decision as programmed, and control outputs such as motor starters, solenoids, valves, and drives.

Power Supplies

The power supply provides the necessary voltage levels that are required for the internal operations of the programmable controller. In addition, it may provide power for the I/O modules. The power supply can be a separate unit or built into the processor section. It takes the incoming voltage (normally 120 VAC or 240 VAC) and changes the voltage as required (normally 5 VDC to 32 VDC).

The power supply must provide constant output voltage free of transient voltage spikes and other electrical noise. The power supply also charges an internal battery in the programmable controller to prevent memory loss when external power is removed. The operating life of lithium batteries, which are commonly used as power supplies for programmable controllers, is from 3 years to 5 years.

Input/Output (I/O) Sections

The input/output (I/O) sections function as the eyes, ears, and hands of a programmable controller. The input section is designed to receive information from pushbuttons, temperature switches, pressure switches, photoelectric and proximity switches, and other sensors. The output section is designed to deliver the output voltage required to control loads such as alarms, lights, solenoids, and starters.

The input section receives incoming signals (normally at a high voltage level) and converts them to low-power digital signals that are sent to the processor section. The processor then registers and compares the incoming signals to the program.

The output section receives low-power digital signals from the processor and converts them into high-power signals. These high-power signals can drive industrial loads that light, move, grip, rotate, extend, release, heat, and perform other functions.

The I/O sections can either be located on the programmable controller (onboard) or be part of expansion modules. Onboard I/Os are a permanent part of the programmable controller package. Expansion modules are removable units that include inputs, outputs, or combinations of inputs and outputs.

Onboard I/Os usually include a fixed number of inputs and outputs that define the limits of the programmable controller. For example, a small programmable controller may include up to 16 inputs and eight outputs. This means that up to 16 inputs and eight outputs may be connected to the controller. Programmable controllers that use expansion modules allow the total number of inputs and/or outputs to be changed by changing or adding modules. Onboard programmable controllers are normally used for individual machines and small systems. Expansion programmable controllers are normally used for large systems or small systems that require flexible changes.

Discrete I/Os. Discrete I/Os are the most common inputs and outputs. Discrete I/Os use bits that represent a separate and distinct signal each, such as ON/OFF, open/closed, or energized/de-energized. The processor reads this as the presence or absence of power.

Examples of discrete inputs are pushbuttons, selector switches, joysticks, relay contacts, starter contacts, temperature switches, pressure switches, level switches, flow switches, limit switches, photoelectric switches, and proximity switches. Discrete outputs include lights, relays, solenoids, starters, alarms, valves, heating elements, and motors.

Data I/Os. In many applications, more complex information is required than the simple discrete I/O is capable of producing. For example, measuring temperature may be required as an input into the programmable controller and numerical data may be required as an output. Data I/Os are inputs and outputs that produce or receive a variable signal. They may be analog, which allows for monitoring and controlling analog voltages and currents, or they may be digital, such as binary-coded decimal (BCD) inputs and outputs.

PROGRAMMABLE CONTROLLER SECTIONS

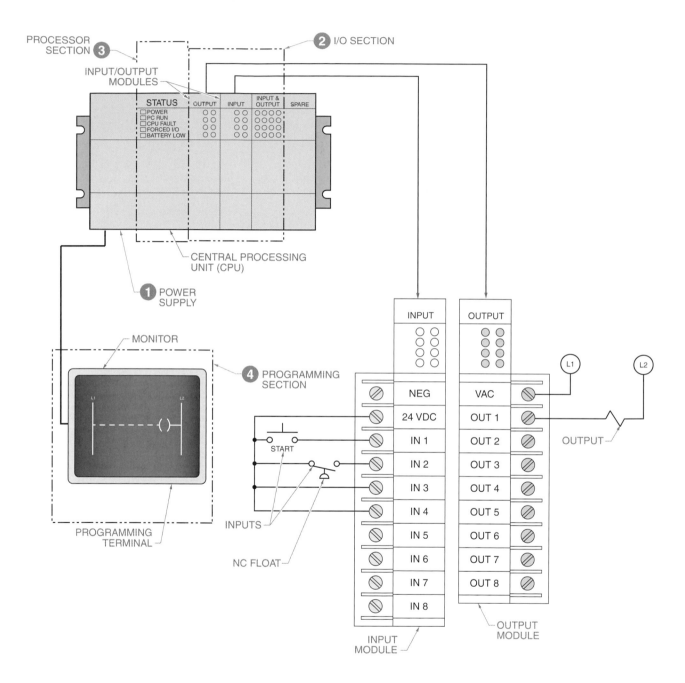

Figure 27-10. The four basic parts of a controller include the power supply, I/O section, processor section, and programming section.

When an analog signal such as voltage or current is input into an analog input card, the signal is converted from analog to digital by an analog-to-digital (A/D) converter. The converted value, which is proportional to the analog signal, is sent to the processor section. After the processor has processed the information according to the program, the processor outputs the information to a digital-to-analog (D/A) converter. The converted signal can provide an analog voltage or current output that can be used or displayed on an instrument in a variety of processes and applications.

Examples of data inputs are potentiometers, rheostats, encoders, bar code readers, and temperature, level, pressure, humidity, and wind speed transducers. Examples of data outputs are analog meter displays, digital meter displays, stepper motor signals, variable voltage outputs, and variable current outputs.

I/O Capacity. The size of a programmable controller is based on the I/O capacity. Common I/O capacities of different sizes of programmable controllers include the following:

- mini/micro (32 or fewer I/Os, but may have up to 64)
- small (64 to 128 I/Os, but may have up to 256)
- medium (256 to 512 I/Os, but may have up to 1023)
- large (1024 to 2048 I/Os, but may have many thousands more on very large units)

The I/Os may be directly connected to the programmable controller or may be in a remote location. I/Os in a location remote from the processor section can be hardwired to the programmable controller, multiplexed over a pair of wires, or sent by a fiber-optic cable. In any case, the remote I/O is still under the control of the central processor section. Common programmable controllers may have 16, 32, 64, 128, or 256 remote I/Os.

Fiber-optic communication modules route signals from the inputs to the processor section and then to the outputs. Fiber-optic communication modules are unaffected by noise interference and are commonly used for process applications in the food and petrochemical industries. They are also used in hazardous locations.

Solid-State Input Controls. Programmable controllers can have many types of inputs, including pushbuttons, level switches, temperature controls, and photoelectric controls. Inputs such as pushbuttons and temperature controls are normally easy to input. However, more complex solid-state control inputs such as proximity and photoelectric switch inputs require special consideration because of their function.

Solid-state proximity and photoelectric controls are used in many automated systems. **See Figure 27-11.** These controls normally have a solid-state output that is ideal for input into programmable controllers. Photoelectric controls can be input into programmable controllers for detection, inspection, monitoring, counting, and documentation. Available outputs include two- and three-wire types with thyristor and transistor outputs that can be connected individually or in series/parallel combinations.

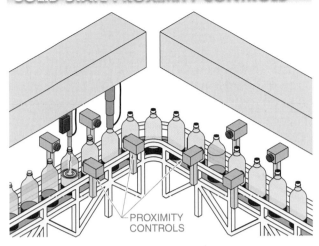

SOLID-STATE PROXIMITY CONTROLS

PROXIMITY CONTROLS

Figure 27-11. Solid-state proximity controls normally have a solid-state output that is ideal for input into programmable controllers.

Two-Wire Thyristor Outputs. Two-wire thyristor outputs are available in a supply voltage range of 20 VAC to 270 VAC at about 180 mA to 500 mA range in either NO or NC versions. Two-wire thyristor outputs have only two wires and are wired in series with the load, which is similar to a mechanical switch. **See Figure 27-12.**

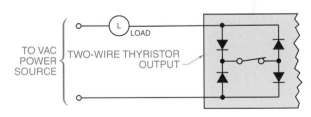

TWO-WIRE THYRISTOR OUTPUTS

LOAD

TO VAC POWER SOURCE

TWO-WIRE THYRISTOR OUTPUT

Figure 27-12. Two-wire thyristor outputs have only two wires and are wired in series with the load, which is similar to a mechanical switch.

The power to two-wire thyristor output sensors is received through the load when the load is not being operated. As with any thyristor output device, some consideration must be given to off-state leakage current and minimum load current. Unlike a mechanical switch, the proximity sensor consumes current in the inactivated mode. The current is small enough that most industrial loads are unaffected. However, this leakage current may be enough to activate the load on some high-impedance loads and controllers.

The leakage current problem can be corrected by placing a load resistor across the input device. **See Figure 27-13.** The resistor value should be chosen to ensure that minimum load current is exceeded and the effective load impedance is reduced. This prevents off-state leakage current turn-on. This resistance value is normally in the range of 4.5 kΩ to 7.5 kΩ. A general rule is to use a 5 kΩ, 5 W resistor for most applications.

LOAD RESISTORS

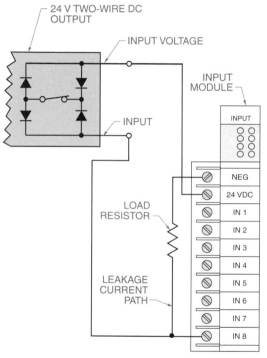

Figure 27-13. A load resistor may be required when connecting a sensor to the controller to prevent leakage current from being input into the controller.

Electrical Noise Suppression. *Electrical noise* is unwanted signals that are present on a power line. Electrical noise enters through input devices, output devices, and power supply lines. Unwanted noise pickup may be reduced by placing the programmable controller away from noise-generating equipment such as motors, motor starters, welders, and drives.

Noise suppression should be included in every programmable controller installation because it is impossible to eliminate noise in an industrial environment. Certain sensitive input devices (analog and digital) require a shielded cable to reduce electrical noise.

A shielded cable uses an outer conductive jacket (shield) to surround the two inner signal-carrying conductors. The shield blocks electromagnetic interference. The shield must be properly grounded to be effective. Proper grounding includes grounding the shield at only one point. A shield grounded at two points tends to conduct current between the two grounds. Low-voltage DC signals must not be routed near high-voltage (120 V) AC signals. If the signals must cross, it is important to ensure that the cables cross at 90° to minimize interference. **See Figure 27-14.**

SHIELDED CABLE

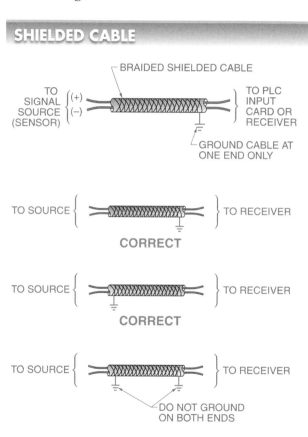

Figure 27-14. A shielded cable uses an outer conductive jacket (shield) that blocks magnetic interference from the two inner signal-carrying conductors.

A high-voltage spike is produced when inductive loads such as motors, solenoids, and coils are turned off. These spikes may cause problems in a programmable controller. High-voltage spikes should be suppressed to prevent problems. A snubber circuit is used to suppress a voltage spike. Typical snubber circuits use an RC (resistor/capacitor), MOV (metal-oxide varistor), or diode, depending on the load. **See Figure 27-15.**

SNUBBER CIRCUITS

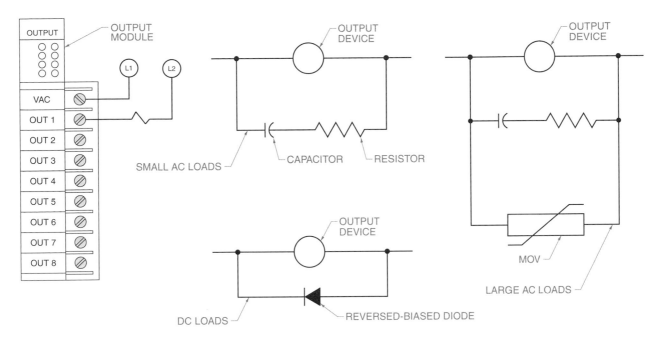

Figure 27-15. Snubber circuits are used to suppress voltage spikes in controllers.

Three-Wire Transistor Outputs. Three-wire transistor outputs are available in a supply range of 10 VDC to 40 VDC at about 200 mA. These sensors are easily interfaced with other electronic circuitry and programmable controllers. Output sensor types consist of either an open collector NPN or PNP transistor. Both normally open (NO) or normally closed (NC) versions are available.

These output devices receive their power to operate through two of the leads (positive and negative, respectively) from the power source. The third lead is used to switch power to the load either using the same source of power as the proximity switch or an independent source of power. **See Figure 27-16.** When an independent source is used, one lead of that source is common with one lead of the source used to power the output device. The voltage level must be within the specifications of the output device used when using an independent power source for the load.

Processor Sections

The *processor section* is the section of a programmable controller that organizes all control activity by receiving inputs, performing logical decisions according to the

program, and controlling the outputs. The processor section is the brain of the programmable controller and can be referred to as the central processing unit (CPU). **See Figure 27-17.**

Large automated systems use multiple programmable controllers.

THREE-WIRE TRANSISTOR OUTPUT SENSORS

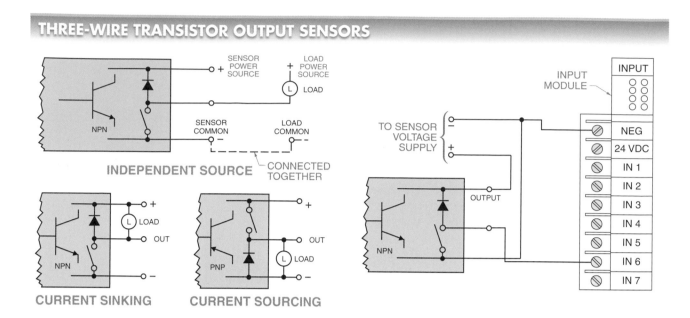

Figure 27-16. Three-wire transistor output sensors use either NPN or PNP transistors to control the load.

PROCESSOR SECTIONS

Rockwell Automation, Allen-Bradley Company, Inc.

Figure 27-17. The processor section organizes all control activity by receiving inputs, performing logical decisions, and controlling the outputs.

The processor section evaluates all input signals and levels. This data is compared to the memory in the programmable controller, which contains the logic of how the inputs are interconnected in the circuit. The interconnections are programmed into the processor by the programming section. The processor section controls the outputs based on the input conditions and the program. The processor continuously examines the status of the inputs and outputs and updates them according to the program. **See Figure 27-18.**

Scan is the process of evaluating the I/O status, executing the program, and updating the system. *Scan time* is the time it takes a programmable controller to make a sweep of the program. Scan time is normally given as the time per 1 kilobyte (kB) of memory and is normally listed in milliseconds (ms). Scanning is a continuous and sequential process of checking the status of inputs, evaluating the logic, and updating the outputs.

The processor section of a programmable controller has different modes. The different modes allow the controller to be taken on-line (system running) or off-line (system on standby). Processor modes include the program, run, and test modes. **See Figure 27-19.** The program mode is used for developing the logic of the control circuit. In the program mode, the circuit is monitored and the program is edited, changed, saved, and transferred.

The run mode is used to execute the program. In the run mode, the circuit may be monitored and the inputs and outputs forced. Program changes cannot normally be made in the run mode. The test mode is used to check the program without energizing output circuits or devices. In the test mode, the circuit is monitored and inputs and outputs are forced in the program (without actually energizing the load connected to the output).

PROCESSOR MONITORING

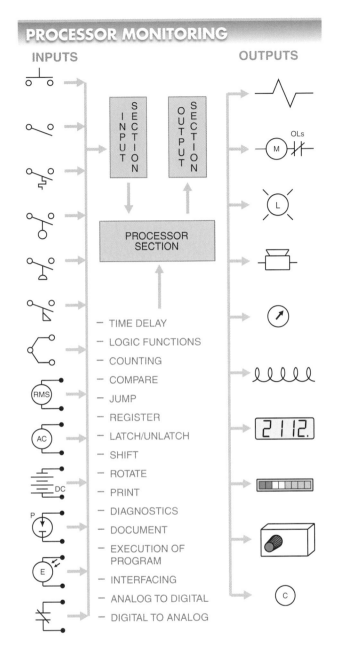

Figure 27-18. The processor continuously examines the status of the inputs and outputs and updates them according to the program.

WARNING: A programmable controller is switched from the program mode to the run mode by placing the controller in the run mode. The machine or process is started when the controller is placed in the run mode. Extreme care must be taken to ensure that no damage to personnel or equipment occurs when switching the controller to the run mode. Only qualified personnel should change processor modes. Key-operated switches should always be used in any dangerous application.

PROCESSOR MODES

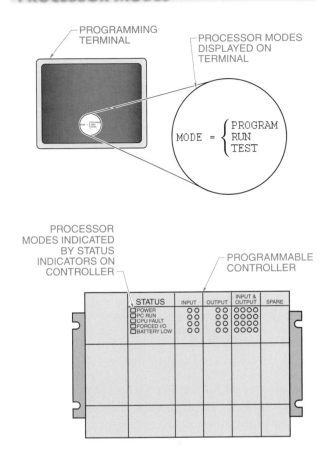

Figure 27-19. The processor section of a controller has different modes that allow the controller to be taken on-line (system running) or off-line (system on standby).

Programming Sections

The *programming section* is the section of a programmable controller that allows input into the controller through a keyboard. The processor must be given exact, step-by-step directions. This means things such as load, set, reset, clear, enter in, move, and start timing must be communicated to the processor. The programming of a controller involves the programming device that allows access to the processor and the programming language that allows the operator to communicate with the processor section.

Programming Devices. Programming devices vary in size, capability, and function. Programming devices are available as simple, small text display units or complex color displays with monitoring and graphic capabilities. **See Figure 27-20.**

PROGRAMMING DEVICES

Siemens Corporation

Figure 27-20. Programming devices are available as simple, small text display units or complex color displays with monitoring and graphics capabilities.

A programming device may be connected permanently to the controller or connected only while the program is being entered. Once a program is entered, the programming device is no longer needed, except to make changes in the program or for monitoring functions. Some programmable controllers are designed to use an existing personal computer (PC) for programming. *Off-line programming* is the use of a PC to program a programmable controller that is not in the run mode. This permits the computer to be used for other purposes when not being used with the programmable controller.

Programming Symbols. Controller programs are designed using controller software. Controller software uses different types of symbols, letters, and numbers to designate each component. Components such as inputs, outputs, relays, timers, and counters each have their own symbol and address (assigned values and numbers). When designing and programming a circuit using a programmable controller, component symbols are selected and placed on the screen as the circuit is developed. The symbols are commonly selected from the tool palette that is displayed on the computer screen. **See Figure 27-21.**

Standard symbols used in programming a circuit include NO inputs, NC inputs, and standard outputs. Inputs are pushbuttons, limit switches, pressure

switches, and other devices that are used to send information to a circuit. Standard outputs include lamps, solenoids, motor starters, alarms, and other devices that are used to perform work or give an indication of circuit operation. All controller programs also include expanded (special) components that can be programmed into a circuit. Special components include timers, counters, logic functions, and common control functions (latch, unlatch, etc.). Although exact symbols differ slightly with each manufacturer, most have common symbols, shapes, and designations.

Programming Languages. The first programmable controllers used line diagrams as a language to input information. Line diagrams are still commonly used as a language for programmable controllers throughout the world. Other languages used for programmable controllers include Boolean, Functional Blocks, and English Statement. Line diagrams and Boolean are basic programmable controller languages. Functional Blocks and English Statement are higher-level languages required to execute more powerful operations such as data manipulation, diagnostics, and report generation.

A line diagram is drawn as a series of rungs. Each rung contains one or more inputs and the output (or outputs) controlled by the inputs. The rung relates to the machine or process controls. The programming instructions communicate the desired logic to the processor.

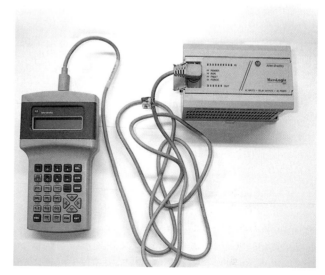

A program can be entered into a programmable controller through a handheld programming device.

PROGRAMMING SYMBOLS

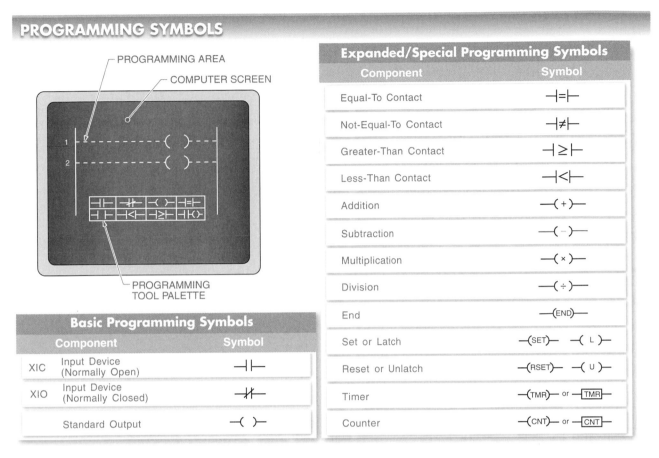

Figure 27-21. When designing and programming a circuit using a controller, component symbols are selected and placed on the screen as the circuit is developed.

Basic logic functions are used to enter the logical operation of the circuit into the processor section. **See Figure 27-22.** The program is entered into the programmable controller through the keyboard.

Programming a controller follows a logical process. Inputs and outputs are entered into the controller in the same manner as if connecting them through hardwiring. The difference in programming is that although a circuit is the same, each manufacturer has a different method of entering that circuit. There are more similarities than differences from manufacturer to manufacturer.

Programmable Controller Line Diagrams. Except for a few differences, programmable controller line diagrams are similar to standard hardwired line diagrams. Programmable controller line diagrams have two vertical power lines that represent L1 and L2, except no voltage potential exists between the two lines. Horizontal lines represent the current paths between the vertical power lines. These horizontal lines are referred to as rungs. Each rung may have several input and output devices. **See Figure 27-23.**

Input devices are either NO or NC devices. The NO devices are referred to as "examined if closed" (XIC) contacts. The NC devices are referred to as "examined if open" (XIO) contacts. A *controller scan* is one execution cycle of a line diagram. A typical controller scan starts in the upper-left corner of the line diagram and scans from left to right and top to bottom. During each scan, the NO contacts are examined for being closed (XIC) and the NC contacts are examined for being opened (XIO).

Tech Fact

Since PLC programs use only generic symbols and not defined symbols, it is best to write in the on-screen meaning of the PLC function when programming it. It is also best to add a written description to each line, such as "Pressing START button turns ON pump motor 1 and pressing STOP button or an overpressure turns OFF pump motor 1".

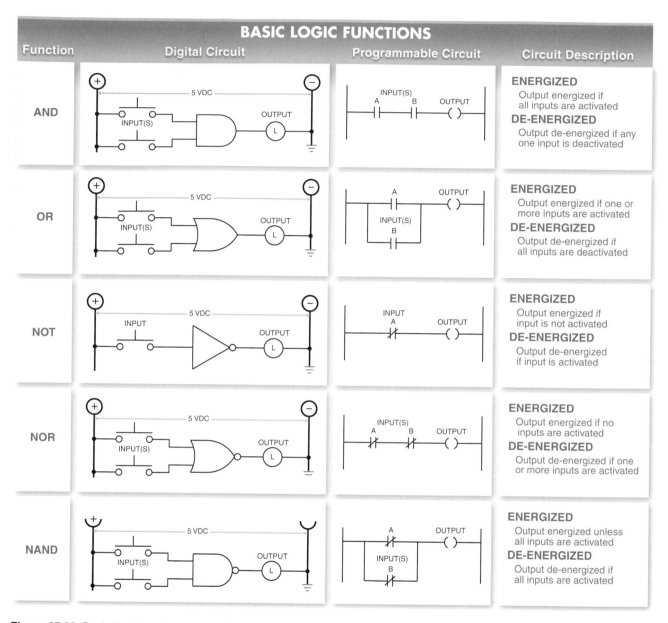

Figure 27-22. Basic logic functions are used to enter the logical operation of the circuit into the processor section.

Programming Rules. When programming a controller circuit, basic rules must be followed if the circuit is to be accepted by the software before it is downloaded. **See Figure 27-24.** Basic controller circuit programming rules include the following:

Rule 1: Inputs (NO, NC, and special) are placed on the left side of the circuit between the left rung and the output. Outputs are placed on the right side of the circuit.

Rule 2: Only one output can be placed on a rung. This means that outputs can be placed in parallel but never in series.

Rule 3: Inputs can be placed in series, parallel, or in series/parallel combinations.

Rule 4: Inputs can be programmed at multiple locations in the circuit. The same input can be programmed as NO and/or NC at multiple locations.

Rule 5: Standard outputs cannot be programmed at multiple locations in the circuit. There is a special output called an "or-output" that allows an output to be placed in more than one location, but only if the or-out special function is identified when programming the output.

PROGRAMMABLE CONTROLLER LINE DIAGRAMS

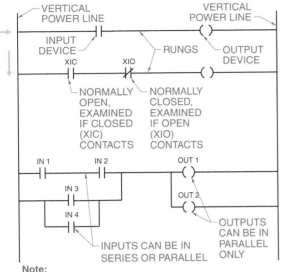

Note:
PLC scan is from left to right and top to bottom.
Note:
Programmable controller line diagrams do not use dots to represent connections.

Figure 27-23. Programmable controller diagrams follow the same basic rules as standard hardwired line diagrams.

Developing Typical Programs. Several steps must be taken before a program can be entered into a programmable controller. The first step is to develop the logic that is required of the circuit into a line diagram. **See Figure 27-25.** In this circuit, pressing any one of the three start pushbuttons energizes the motor starter. Once the motor starter is energized, the start pushbutton may be released. The motor starter remains energized because the M1 contact closes and provides a parallel path for current flow around the start pushbuttons. Pressing one of the stop pushbuttons stops the flow of current through the motor starter and de-energizes it.

The line diagram shows the logic of the circuit but not the actual location of each component. A wiring diagram shows the location of the components in an electrical circuit. **See Figure 27-26.** The wiring diagram of the three start/stop pushbutton stations shows the location of each pushbutton.

The phantom line around each start/stop pushbutton station indicates that the two pushbuttons are located in the same enclosure. Each pushbutton in the wiring diagram is connected in the exact manner as in the line diagram. Any additions (or changes) to this hardwired control circuit require that the circuit be rewired.

PROGRAMMING RULES

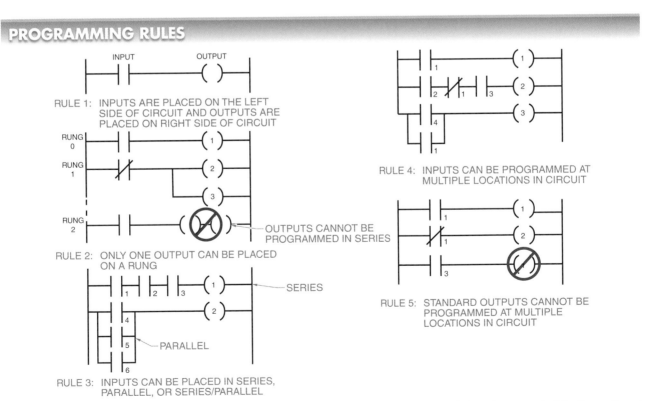

Figure 27-24. Basic rules must be followed when programming a controller if the circuit is to be accepted by the software before downloading.

DEVELOPING CIRCUIT LOGIC

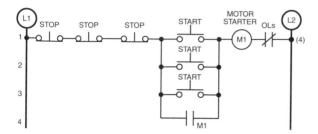

Figure 27-25. Circuit logic must be developed into a line diagram to enter the circuit into a programmable controller.

WIRING DIAGRAMS

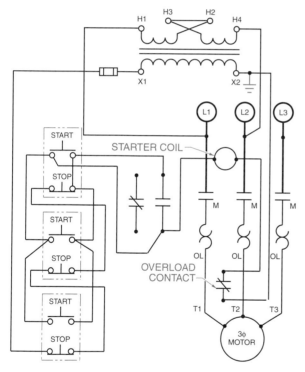

Figure 27-26. A wiring diagram shows the location of the components in an electrical circuit.

The second step is to take the line diagram and convert it into a programming diagram. A *programming diagram* is a line diagram that better matches the language of the programmable controller. Like a standard (hardwired) line diagram, a programming diagram shows the flow of current through the control circuit. The programming diagram does not use

distinct symbols for each I/O. Instead, there are two basic symbols for inputs and one basic symbol for outputs. One of the input symbols represents NO inputs and the other represents NC inputs. **See Figure 27-27.**

PROGRAMMING DIAGRAMS

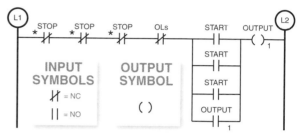

*Note:
1. Hardwired normally open pushbutton is programmed as normally closed.
2. Hardwired normally closed pushbutton is programmed as normally open.

Figure 27-27. A programming diagram is a line diagram that better matches the language of the controller.

In this circuit, pressing any one of the three start pushbuttons energizes the motor starter. Once the motor starter is energized, the start pushbutton may be released and remain energized. This is because the contacts of output 1 close and provide a parallel path for current flow around the start pushbuttons. The motor starter de-energizes when the current flow to output 1 is de-energized by the stop pushbutton being pressed.

The controller wiring diagram is very different from a hardwired wiring diagram. **See Figure 27-28.** In a controller wiring diagram, each input is wired to a designated input terminal and each output is wired to a designated output terminal. The way the inputs and outputs are connected does not determine the logic of the circuit's operation. The logic is controlled by the way the circuit is programmed into the programmable controller. Any changes to the circuit are made by changing the program, not the wiring of the inputs and outputs.

The third step is to enter the desired logic of the circuit into the programmable controller. Every manufacturer has a slightly different set of steps and functions to enter the program into the programmable controller. The program is entered in the program mode and then saved for future use.

CONTROLLER WIRING DIAGRAM

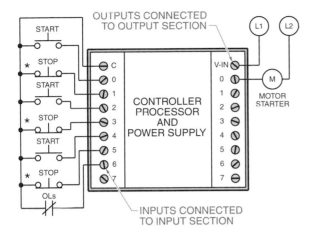

OUTPUTS CONNECTED TO OUTPUT SECTION

START

STOP

START

STOP

START

STOP

OLs

CONTROLLER PROCESSOR AND POWER SUPPLY

L1 L2

V-IN

0 M
1 MOTOR STARTER
2
3
4
5
6
7

INPUTS CONNECTED TO INPUT SECTION

*Note:
1. Hardwired normally open pushbutton is programmed as normally closed.
2. Hardwired normally closed pushbutton is programmed as normally open.

Figure 27-28. In a controller wiring diagram, each input is wired to a designated input terminal and each output is wired to a designated output terminal.

Storing and Documentation. Once a program has been developed it may be necessary to store the program outside of the programmable controller or document the program by printing it out. **See Figure 27-29.** This allows for a means of storing and retrieving control programs, which makes for fast changes in a process or operation. A program is commonly stored in a file on a hard drive, CD-ROM, or USB flash drive. For example, one file may have the program for filling 8 oz bottles and a second file may have the program for filling 16 oz bottles.

When a change from one size bottle to another is required, the programmable controller is loaded with the correct file to start the line with all the proper control settings. Even if the programmable controller is not likely to ever have its program changed, the program should be stored on a CD-ROM, flash drive, or using a secure cloud-based backup. This ensures the safety of the program in the event of a problem.

Once a program has been entered into the programmable controller, a copy of the program and other circuit documentation can be made by connecting the controller to a printer. The printout can be used as a hard copy of the program for documentation and future reference.

STORING AND DOCUMENTATION

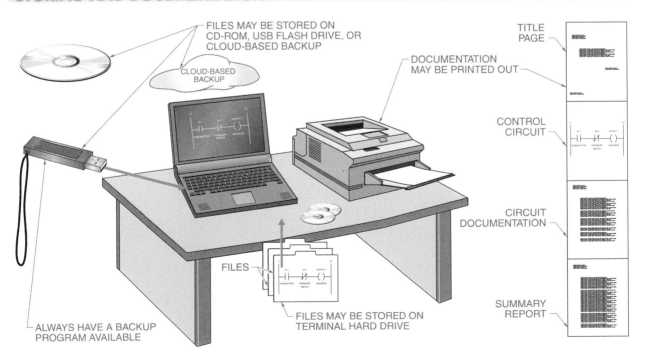

Figure 27-29. A program is commonly stored in a file on a hard drive, CD-ROM, or USB memory card.

Controller Status and Fault Indicators. All programmable controllers include LED indicator lamps that show the condition of the components in the controller. Indicator lamps can show the condition of the inputs and outputs or the operating condition of the controller. **See Figure 27-30.**

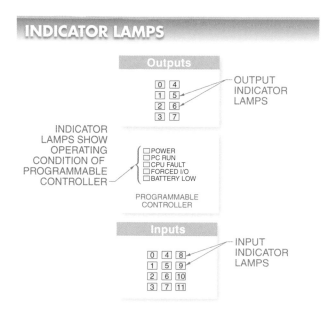

Figure 27-30. Programmable controller indicator lamps show the operating condition of the inputs, outputs, and the controller.

An input/output (I/O) indicator lamp shows the status of the I/O devices. The input indicator lamps are energized when an electrical signal is received at an input terminal. This occurs when an input contact is closed or a signal is present. The output indicator lamps are energized when an output device is energized. Each input and output on the controller has its own indicator lamp.

Most programmable controllers have several status indicator lamps that show the operating condition of the controller. The status indicator lamps commonly include power, PC run, CPU fault, forced I/O, and battery low. **See Figure 27-31.**

The power lamp indicates when power is applied to the programmable controller and when the processor is energized. The power lamp should normally be energized.

The PC run lamp indicates when the processor is in the run mode. The PC run lamp may be energized at all times in run mode or may flash on and off as the programmable controller is running and processing I/O data.

The CPU fault lamp indicates that there is an error in the programmable controller system. Programmable controllers are normally designed to de-energize all outputs when the CPU fault lamp is energized. On most controllers, an error message that indicates the error is also displayed.

The forced I/O lamp indicates when the programmable controller is in the forced operating mode. In the forced operating mode, the inputs and/or outputs are being forced on or off through the software. Extreme caution must be used when inputs and outputs are forced, and any time the forced I/O lamp is energized.

Programmable controllers can be used in amusement parks to control ride functions.

The battery low lamp indicates a low battery charge problem or that it is time to replace the battery. The battery is used to maintain processor memory during a power failure. The battery should be replaced as recommended by the manufacturer or every five years.

Force and Disable. The force command opens or closes an input device or turns on or off an output device. The force command is designed for use when troubleshooting the system. When an I/O device is forced, the circuit can be checked using software. **See Figure 27-32.**

An input device may be forced to test the circuit operation. An input device may also be forced when service is required on a defective input device. The defective input device may be forced on until the device is serviced if the input device is not critical to production. The force command is removed after the device is fixed.

STATUS INDICATOR CONDITIONS

Status Indicator	Problem	Possible Cause	Corrective Action
■ POWER ■ PC RUN ☐ CPU FAULT ☐ FORCED I/O ☐ BATTERY LOW	Normal situation		None
☐ POWER ☐ PC RUN ☐ CPU FAULT ☐ FORCED I/O ☐ BATTERY LOW	No or low system power	Blown fuse, tripped CB, or open circuit	Test line voltage at power supply; (line voltage must be within 10% of the controller's rated voltage); check for proper power supply jumper connections when voltage is correct; replace the power supply module when module has power coming into it but is not delivering correct power
■ POWER ☐ PC RUN ☐ CPU FAULT ☐ FORCED I/O ☐ BATTERY LOW	Programmable controller not in run mode	Improper mode selection; faulty memory module, memory loss, or memory error caused by a high voltage surge, short circuit, or improper grounding	Check line voltage and use an ohmmeter to check system ground
■ POWER ☐ PC RUN ■ CPU FAULT ☐ FORCED I/O ☐ BATTERY LOW	Fault in controller	Faulty memory module, memory loss, or memory error caused by a high-voltage surge, short circuit, or improper grounding	Turn off power and restart system; remove power and replace the memory module when fault indicator is still ON
■ POWER ☐ PC RUN ☐ CPU FAULT ☐ FORCED I/O ▨ BATTERY LOW	Fault in controller due to inadequate or no power	Loss of memory when power is OFF and battery charge was inadequate to maintain memory	Replace battery and reload program
■ POWER ■ PC RUN ☐ CPU FAULT ■ FORCED I/O ☐ BATTERY LOW	System does not operate as programmed	Input or output device(s) in forced condition	Monitor program and determine forced input and output device(s); disable forced input or output device(s) and test system
■ POWER ■ PC RUN ☐ CPU FAULT ☐ FORCED I/O ☐ BATTERY LOW	System does not operate	Defective input device, input/output device, output module, or program	Monitor program and check condition of status indicators on the input and output modules; reload program when there is a program error

Figure 27-31. Status indicator lamps show the operating condition of the controller.

An output device turns on regardless of the logic of the programmed circuit when the force on command is used. The output device remains on until the force off command is used. Care must be taken when using the force command because it overrides all safety features designed for the program.

The disable command prevents an output device from operating. The disable command is the opposite of the force command. The disable command is used to prevent one or all of the output devices from operating. It is important to ensure that all force and disable commands are removed before a system is returned to normal operation.

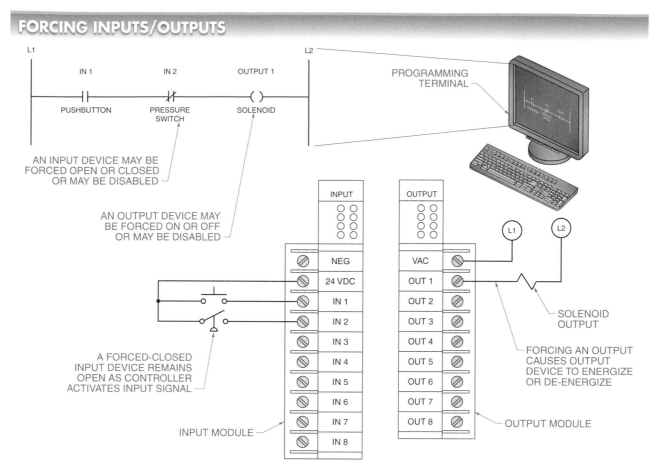

FORCING INPUTS/OUTPUTS

AN INPUT DEVICE MAY BE FORCED OPEN OR CLOSED OR MAY BE DISABLED

AN OUTPUT DEVICE MAY BE FORCED ON OR OFF OR MAY BE DISABLED

A FORCED-CLOSED INPUT DEVICE REMAINS OPEN AS CONTROLLER ACTIVATES INPUT SIGNAL

INPUT MODULE

PROGRAMMING TERMINAL

SOLENOID OUTPUT

FORCING AN OUTPUT CAUSES OUTPUT DEVICE TO ENERGIZE OR DE-ENERGIZE

OUTPUT MODULE

Figure 27-32. When an I/O device is forced, the circuit can be checked using software.

Controller Communication Networks. Programmable controllers may be used as stand-alone control devices that control their outputs (solenoids, motor starters, etc.) as their inputs (limit switches, etc.) send signals to the controller. Controllers may also be connected through communication ports to other devices such as HMIs, PCs, and other controllers. **See Figure 27-33.**

In a typical network system, the programmable controllers become part of a large control system. Field input devices (limit switches, photoelectric switches, etc.) and output devices (solenoids, motor starters, etc.) are connected to the controllers. The controller is then connected to the local area network (LAN) system. Smart I/O devices can be directly connected to the LAN system.

The LAN system is a collection of data and power lines that are used to communicate information among individual devices and to supply power to individual devices connected to the system. Depending on the size of the process, there can be any number of

LAN systems. LAN systems can be connected to information networks through other programmable controllers. The information network system is also a collection of data and power lines that are used to communicate information among individual devices and to supply power to individual devices connected to the system. The information network devices are used to control the LAN devices and allow the flow of information between the central computing system and each local I/O device.

Each network system monitors and controls variables such as time, temperature, speed, weight, voltage, current, power, flow rate, level, volume, density, color, brightness, and pressure. These variables can be controlled, measured, displayed, and recorded. This system allows for a closed loop operation. A *closed loop operation* is an operation that has feedback from the output to the input. Monitoring the outputs and sending information to the inputs controls the system so that the inputs are automatically adjusted to meet the needs of the outputs.

NETWORKING CONTROLLERS

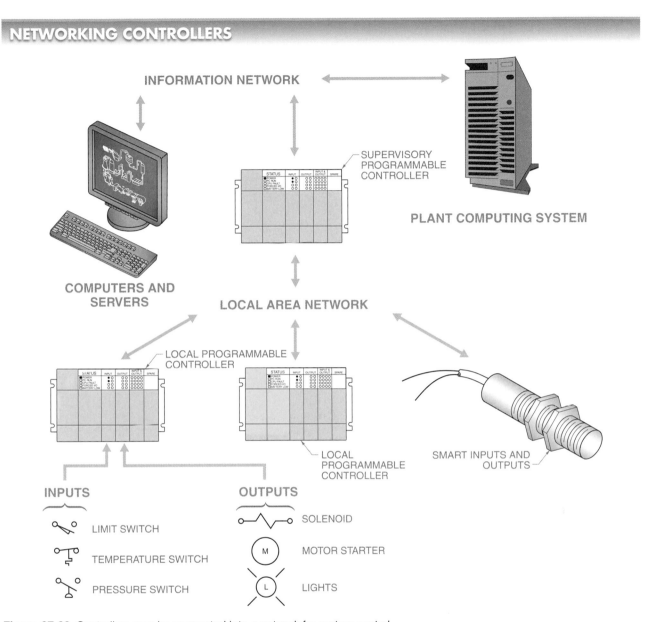

INFORMATION NETWORK

COMPUTERS AND SERVERS

SUPERVISORY PROGRAMMABLE CONTROLLER

PLANT COMPUTING SYSTEM

LOCAL AREA NETWORK

LOCAL PROGRAMMABLE CONTROLLER

LOCAL PROGRAMMABLE CONTROLLER

SMART INPUTS AND OUTPUTS

INPUTS

LIMIT SWITCH

TEMPERATURE SWITCH

PRESSURE SWITCH

OUTPUTS

SOLENOID

M   MOTOR STARTER

L   LIGHTS

Figure 27-33. Controllers may be connected into a network for system control.

A *fieldbus* is an open, serial, two-way communication network that connects with high-level information devices. These information devices include process control valves and transmitters, flowmeters, and other complex field devices that are typically used with control equipment in process control applications.

Fieldbus networks are gradually being used to replace the more common analog networks that are based on the 4 mA to 20 mA standard for analog devices. Fieldbus provides greater accuracy and repeatability in process applications, as well as adding bidirectional communication between the field devices and the programmable controller.

PROFIBUS is a process bus network capable of communicating information between a master controller (or host) and a smart slave process field device, as well as from one host to another. The master controllers control the bus network and determine the data communication on the bus. The slave devices are typically I/O devices, control valves, drives, or transmitters. Slave devices are also called passive stations. PROFIBUS networks require a separate controller or computer to implement control strategies.

MODBUS is a messaging structure with master-slave communication between smart devices and is independent of the physical interconnecting method. MODBUS works equally well using RS-232, RS-485, optical fiber, radio, and cellular communication. The message contains the address of the slave, the command, the data, and a checksum.

DeviceNet is a network with a controller connected to field devices over a two-wire network in a trunk line configuration with either single drops off the trunk or branched drops through multiport interfaces at the device locations. DeviceNet is typically used for communication between controlled devices such as motors and stop/start stations. A DeviceNet byte-wide network can support 64 nodes and a maximum of 2048 field I/O devices.

ControlNet is a real-time device network providing for high-speed transport of time-critical I/O data, messaging, peer-to-peer communications, and the uploading and downloading of programs. Transmission speed is 5 Mbits/sec. The protocol is a producer/consumer communications model that uses no addresses.

Programming Programmable Timers. All programmable controllers include internal programmable timers (timing functions). The programmable timers are programmed, not wired, into the control circuit. When programming a programmable timer, the type of timer (on-delay or off-delay), a preset time value (15 sec, 2 min, etc.), and a time base (second, hundredth of a second, etc.) are selected. Preset time is the length of time for which a timer is set. Programmable timers must be identified and set the same as a stand-alone timer. **See Figure 27-34.**

The method of timer identification in the circuit and program varies with different programmable controller manufacturers. The timer instructions include a timer reference number and a preset timer value. The timer reference number specifies the timer number as used in the control circuit and program. The preset timer value is the required time delay value for the application in which the timer is used. Once programmed, the on-screen timer display normally shows the timer type, number, time base, preset time setting, and accumulated time. Accumulated time is the length of time that has passed since the timer started.

Programmable timers use status bits to control the timer. A status bit is an indicator used to show an ON or OFF state. The status bit can be changed from high to low, ON to OFF, or true to false. Status bits include enable and done bits. An *enable bit* is a controlling contact or input

that sends power to the timer (or counter) and starts the timing of the timer (or the counting of the counter). A *done bit* is the controlled contact (or output) that changes from open or closed (OFF to ON) after the preset time (or number of counts) has been reached.

PROGRAMMING PROGRAMMABLE TIMERS

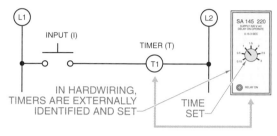

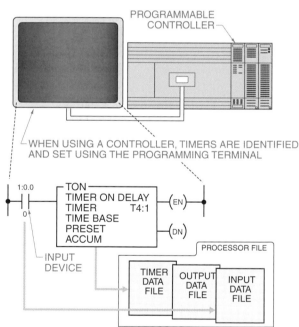

Figure 27-34. Programmable timers must be identified and set the same as stand-alone timers.

Tech Fact

Stand-alone timers generally have only one function. Programmable timers have clearly marked function settings. Timers on PLCs are the programmable type and usually include different timing functions such as ON Delay, OFF Delay, and so on. Timer operation graphs provided in the OEM manual illustrate each timer function using step-by-step procedures and illustrations.

The enable bit is ON (true) any time there is an input signal applied to the timer. The enable bit is OFF (false) or reset at all other times. The timer contacts are activated and the timer no longer times when the accumulated value reaches the preset value. The done bit is turned on when the timer contacts are activated. The done bit is OFF when the timer contacts are not activated.

In an on-delay timer, the done bit is ON (true) when the accumulated value equals the preset value. The done bit is OFF (false) when the enable bit is OFF. In an off-delay timer, the done bit is ON (true) when the enable bit is ON. The done bit is OFF (false) only after the enable bit is OFF and the preset time delay passes. **See Figure 27-35.**

A PC can be used to program a programmable controller using application software.

PROGRAMMABLE TIMERS

INPUT   TIMER BITS
(EN)
(DN)

INPUT PRESENT
(SWITCH CLOSED)

INPUT REMOVED
(SWITCH OPEN)

ENABLE BIT ON   ENABLE BIT OFF

DONE BIT ON   DONE BIT OFF

T = PREPROGRAMMED TIME PERIOD

ON-DELAY

INPUT   TIMER BITS
(EN)
(DN)

INPUT PRESENT
(SWITCH CLOSED)

INPUT REMOVED
(SWITCH OPEN)

ENABLE BIT ON   ENABLE BIT OFF

DONE BIT ON   DONE BIT OFF

T = PREPROGRAMMED TIME PERIOD

OFF-DELAY

Figure 27-35. In an on-delay programmable timer, the done bit is ON (true) when the accumulated value equals the preset value. In an off-delay timer, the done bit is ON (true) when the enable bit is ON.

27-2 CHECKPOINT

1. What are the four basic sections of a PLC?
2. A potentiometer is an example of what type of PLC input?
3. A selector switch is an example of what type of PLC input?
4. When a PLC uses an NPN transistor input or output, is the NPN switching also called current sink or current source?
5. Is a programmed XIC input an NO or NC input?
6. In a PC-controlled system, can an output device like a solenoid be turned on using the programming device (programming keypad)?

27-3 PROGRAMMABLE CONTROLLER APPLICATIONS

Programmable controllers are useful in increasing production and improving overall plant efficiency. Programmable controllers can control individual machines and link the machines together into a system. The flexibility provided by a programmable controller has allowed its use in many applications for manufacturing and process control.

Process control has gone through many changes. In the past, process control was mostly accomplished through manual control. Flow, temperature, level, pressure, and other control functions were monitored and controlled at each stage by production workers.

Today, an entire process can be automatically monitored and controlled with few or no workers involved through the use of programmable controllers. Process applications in which PLCs are used include the following:

- grain operations that involve storage, handling, and bagging
- syrup refining that involves product storage tanks, pumping, filtration, clarification, evaporators, and all fluid distribution systems
- fats and oils processing that involves filtration units, cookers, separators, and all charging and discharging functions
- dairy plant operations that involve all-process control from raw milk delivered to finished dairy products
- oil and gas production and refining from the well pumps in the fields to finished product delivered to the customer
- bakery applications from raw material to finished product
- beer and wine processing, including the required quality control and documentation procedures

Controller Timer Applications

An example of a controller timer application is a stoplight circuit. A stoplight circuit is a circuit that uses timers to sequence when the red, yellow, and green lamps turn on and off. A stoplight timing circuit can use stand-alone timers or programmable timers to control the timing sequence.

The circuit is drawn as a standard line diagram when stand-alone timers are used. Four on-delay timers are used for a basic stoplight timing sequence. **See Figure 27-36.** The stoplight circuit shows the basic operation of sequencing three lamps, and the time values have been reduced to seconds. Once the circuit start

pushbutton is pressed and released, the lamps turn on and off in the following sequence:

1. The red lamp turns on for 30 sec.
2. The red lamp turns off and the yellow lamp turns on for 5 sec.
3. The yellow lamp turns off and the green lamp turns on for 40 sec.
4. The green lamp turns off and the yellow lamp turns on for 5 sec.
5. The yellow lamp turns off and the red lamp turns on for 30 sec.
6. Timer 4 resets the circuit and the sequence starts over.

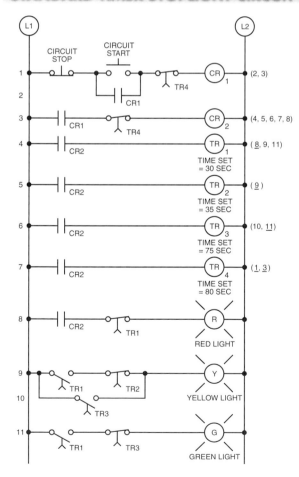

Figure 27-36. Four on-delay stand-alone timers are used for a basic stoplight timing sequence application.

The circuit is drawn as a controller line diagram when a programmable timer is used. The controller line diagram is similar to the standard line diagram, but the difference is that a standard line diagram is drawn by hand or by using a computer program. The programmable timer line diagram is automatically drawn on the computer screen as the circuit is programmed. **See Figure 27-37.**

PROGRAMMABLE TIMER STOPLIGHT CIRCUIT

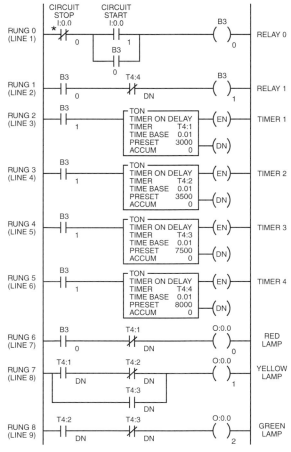

*Note:
1. Hardwired normally open pushbutton is programmed as normally closed.
2. Hardwired normally closed pushbutton is programmed as normally open.

Figure 27-37. A programmable timer line diagram is automatically drawn on the computer screen as the circuit is programmed.

In this diagram, each line of the control circuit is referred to as a rung. Starting with the number 0, each rung is identified by a number. Input, output, relay, and timer identification varies by manufacturer. The following is a common manufacturer identification method:

- The inputs are addressed as I:0.0-0 (stop button) and I:0.0-1 (start button).
- The outputs are addressed as 0:0.0-0 (red lamp), 0:0.0-1 (yellow lamp), and 0:0.0-2 (green lamp).
- The relays are addressed as B3-0 (first relay), and B3-1 (second relay).
- The timers are addressed as T4:1 (timer 1), T4:2 (timer 2), T4:3 (timer 3), and T4:4 (timer 4).

Input and Output Address Identification. Every time an I/O device on a programmable controller is programmed, the device must be assigned an address (instruction). The address links external inputs and outputs to data files and processor files within the controller. Although each manufacturer assigns their own addresses to inputs and outputs, there are more similarities than differences among most manufacturers. For example, a typical manufacturer addressing numbering system uses number/letter assignments. **See Figure 27-38.** The numbering/letter assignments are made as follows:

I = Input (pushbutton, limit switch, etc.)

O = Output (solenoid, lamp, motor starter, etc.)

T = Timer (internal PLC timer)

C = Counter

: = Slot number (physical slot number [1, 2, 3, etc.] of the I/O module)

For example, I:0/1 identifies an input in slot number 0 at terminal 1 and 0:0/4 identifies an output in slot number 0 at terminal 4.

Welding

In discrete parts manufacturing, welding is often a major part of the system. Programmable controllers may be used to control and automate industrial welding processes. **See Figure 27-39.** In this application, the programmable controller can control the length of the weld and the power required to produce the correct weld. The controller is programmed to allow the weld to occur only if all inputs and conditions are correct. These inputs and conditions include the following:

- presence and correct position of all the parts
- the correct weld cycle speed and power setting
- the correct rate of speed on the line for the given application
- proper functioning of all interlocks and safety features

INPUT AND OUTPUT ADDRESS IDENTIFICATION

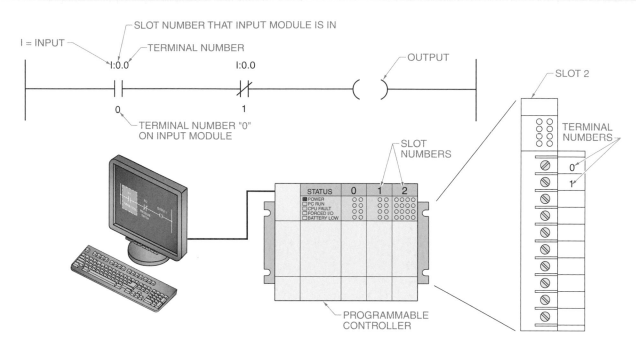

Figure 27-38. Numbers and letters are used to assign addresses to inputs, outputs, timers, and other internal and external components.

WELDING

FANUC Robotics North America

Figure 27-39. Programmable controllers can be used to control and automate industrial welding processes.

In addition, the controller can be used to determine whether parts are running low and can be set to automatically turn the line on and off as required. Documentation of production efficiency can be generated for quality control and inventory requirements.

A programmable controller may be used to control and interlock many welders. Welders at one station may require more power than is available if all the welders are ON simultaneously. In this case, a large power draw can cause poor-quality welds. A requirement for a system using many welders is to limit the amount of power being consumed at any one time. This is accomplished by time-sharing the power feed to each welder.

A programmable controller may be programmed for a maximum power draw. The controller can determine whether power is available when a welder requires power. The weld takes place if the correct power level is available. If not, the controller remembers the request and permits the welder to proceed with the weld cycle when power is available. The controller can also be programmed to determine which welder has priority.

Machine Control

Controls must be synchronized when machines are linked together to form an automated system. **See Figure 27-40.** In this application, each machine may be controlled by a programmable controller, with another programmable controller synchronizing the operation. This situation is likely if the machines have been purchased from different manufacturers. In this case, each machine may include a controller to control all the functions on that machine only. If the machines are purchased from one manufacturer or designed in-plant, it is possible to use one large controller to control each machine and synchronize the process.

Industrial Robot Control

Programmable controllers are ideal devices for controlling any industrial robot. **See Figure 27-41.** A programmable controller can be used to control all operations such as rotate, grip, withdraw, extend, and lift. A controller is recommended because most robots operate in an industrial environment.

Fluid Power Control

Fluid power cylinders are normally chosen when a linear movement is required in an automated application. Pneumatic cylinders are common because they are easy to install and most plants have access to compressed air. Pneumatics work well for most robot grippers, drives, and positioning cylinders, as well as machine loading and unloading and tool-working applications. Hydraulic cylinders are used when a manufacturing process requires high forces. Hydraulic systems of several thousand psi are often used to punch, bend, form, and move components.

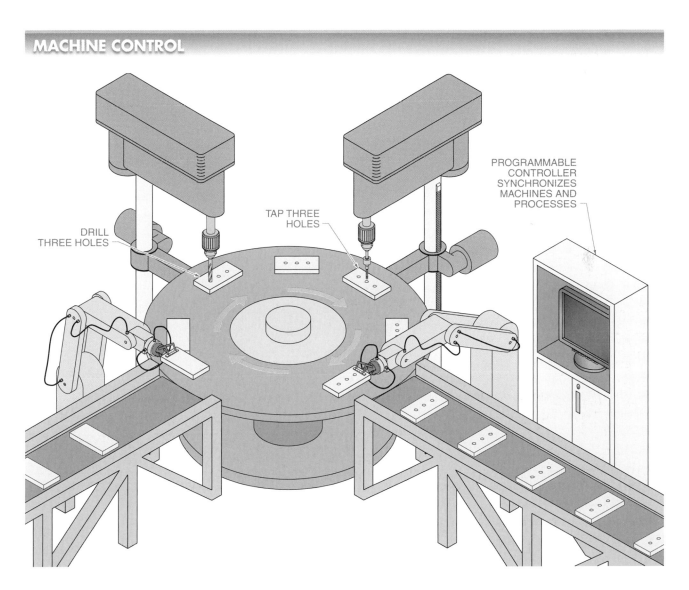

MACHINE CONTROL

Figure 27-40. Controllers are used to control and synchronize individual machine operations with other machines.

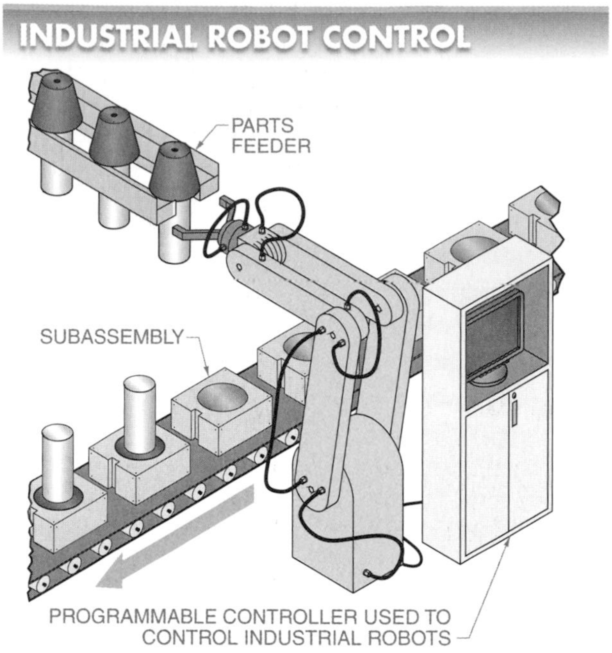

INDUSTRIAL ROBOT CONTROL

PARTS FEEDER

SUBASSEMBLY

PROGRAMMABLE CONTROLLER USED TO CONTROL INDUSTRIAL ROBOTS

Figure 27-41. Controllers can be used to control the operations of an industrial robot.

Programmable controllers may be used to control linear and rotary actuators in an industrial fluid power circuit. **See Figure 27-42.** In this system, the output module of the programmable controller is connected to control the four solenoids. Solenoid A moves the cylinder in, solenoid B moves the cylinder out, solenoid C rotates the rotary actuator in the forward direction, and solenoid D rotates the rotary actuator in the reverse direction. The controller is used to control the energizing or de-energizing of the solenoids. Solenoids control the directional control valves, which control the actuators.

Tech Fact

When a PLC is used to control a fluid power circuit, the PLC outputs control solenoids that switch valve positions. Solenoids can have a high inrush current when energized and produce high voltage spikes (transients) when de-energized. Both the inrush current and the transients can destroy the PLC output, shorten its operating life, or cause false signals within the system. Using a surge suppressor protects against transients, and using a higher current-rated output contact protects against inrush current.

FLUID POWER CONTROL

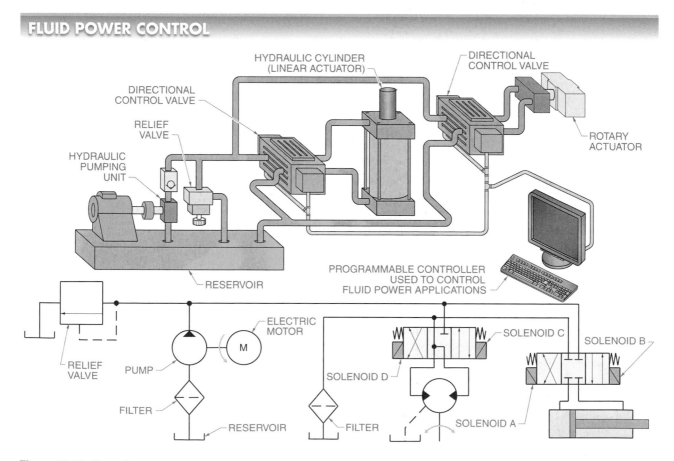

Figure 27-42. Controllers can be used to control linear and rotary actuators in an industrial fluid power circuit.

Industrial Drive Control

Motors are often directly connected to the power lines and operate at a set speed. As systems become more automated, variable motor speed is required. Adjustable speed controllers are available to control the speed of AC and DC motors. These controllers are normally manually set for the desired speed, but many allow for automatic control of the set speed. A PLC may be used to control AC drives. **See Figure 27-43.** The drives can accept frequency and direction commands in a BCD format that the PLC can provide with a BCD output module.

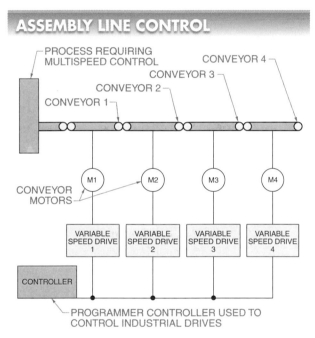

ASSEMBLY LINE CONTROL

Figure 27-43. Controllers can be used to control and synchronize the speed of conveyors on an assembly line.

Pulp and Paper Industries. Pulp and paper industries use programmable controllers to control each production process and diagnose problems in the system. **See Figure 27-44.** Pulp and paper production processes can involve equipment that covers a large area. The control of pulp and paper production processes is ideal for a programmable controller because most control logic includes start/stop, time delay, count sequential, and interlock functions.

A programmable controller allows for the required inputs and outputs, which, when multiplexed, can transmit multiple signals over a single pair of wires. The basic operation of a paper mill is to receive raw material such as logs, pulpwood, or chips and then process, size, store, and deliver the material. This operation includes a large

conveyor system with diverter gates, overtravel switches, speed control, and interlocking. A break in any part of the system can shut down the entire system. Attempts to find a fault can be time-consuming since the system covers a large area. To solve this problem, a programmable controller with fault diagnostics can be used to analyze the system and give an alarm and printout of where the problem exists with suggested solutions.

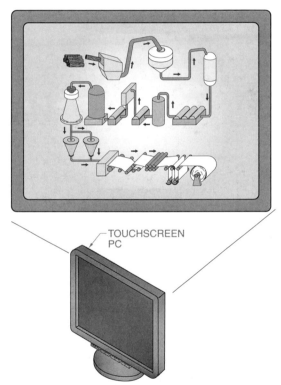

PAPER MILL PROCESS CONTROL

Fluke Corporation

Figure 27-44. Controllers in a paper mill control each process and diagnose problems in the system.

Batch Process Control Systems. Batch processing blends sequential, step-by-step functions with continuous closed-loop control. Because systems are made up of many parts, batch process control is essentially system control. Individual programmable controllers can be used to control each part and step of the process, with additional controllers and computers supervising the total operation.

In a batch process control system, an operator interface is used for instrumentation or other monitoring functions. An operator interface is added as part of the system. This interface may be in the form of an instrumentation and process control station, an HMI, or any other type of interface. To aid in interfacing and monitoring a programmable-based system, a serial port is used for monitoring and programming a system using a computer. Thus, the individual solenoids, motor starters, and heating elements at each process step are directly controlled by the local programmable controllers while the host computer supervises all of the controllers. **See Figure 27-45.**

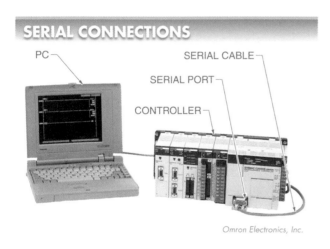

Omron Electronics, Inc.

Figure 27-45. A serial port is used for monitoring and programming a system using a computer.

Programmable Controller Circuits

Control circuits that do not use programmable controllers for control functions have been used for over 100 years. However, these control circuits do not allow for much flexibility or change. Modern electrical circuits are usually designed with change in mind. Changes may include the way the circuit operates or additional safety features. With a controller, changes in an electrical circuit can easily be made by changing the program.

A basic forward/reversing circuit is an example of a circuit that may require changes. In a basic forward/reversing circuit, very little circuit logic is required. **See Figure 27-46.** In this circuit, the forward pushbutton operates the forward starter coil, and the reverse pushbutton operates the reverse starter coil. This circuit operates satisfactorily if no operator error occurs.

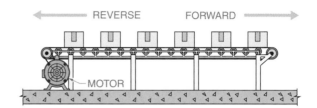

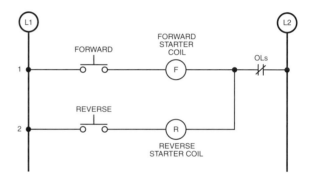

Figure 27-46. In a basic forward/reversing circuit, very little circuit logic is required.

However, if an operator presses both pushbuttons at the same time, both starter coils energize. This causes a short circuit in the power circuit. Interlocking is added to solve this problem. Interlocking prevents the operator from energizing both starter coils at the same time. **See Figure 27-47.**

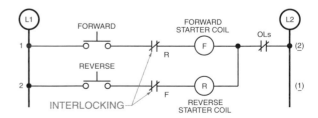

Figure 27-47. Interlocking is added to a circuit to prevent the operator from energizing both starter coils at the same time.

In a hardwired circuit, auxiliary contacts must be added and wired to interlock the circuit. A programmable controller allows interlocking of the circuit with a simple change in the program and no additional components. **See Figure 27-48.**

PROGRAMMABLE CONTROLLER INTERLOCKING

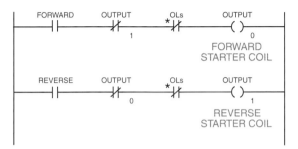

*Note:
1. Hardwired normally open pushbutton is programmed as normally closed.
2. Hardwired normally closed pushbutton is programmed as normally open.

Figure 27-48. A controller allows interlocking of the circuit with a simple change in the program and no additional components.

Another change that might be required is the addition of memory to the circuit. In a hardwired circuit, auxiliary contacts are required. In a controller circuit, the program is changed. **See Figure 27-49.**

The timing circuit for a stoplight can use stand-alone timers or programmable timers to control the timing sequence of the light.

CIRCUIT MEMORY

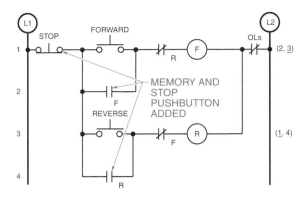

HARDWIRED CIRCUIT

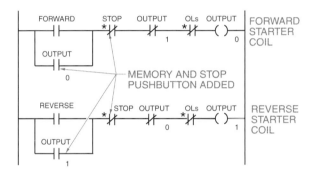

PROGRAMMABLE CONTROLLER CIRCUIT

*Note:
1. Hardwired normally open pushbutton is programmed as normally closed.
2. Hardwired normally closed pushbutton is programmed as normally open.

Figure 27-49. The addition of memory to a circuit requires the addition of components and wiring if the circuit is hardwired or changing the program if the circuit is controlled by a programmable controller.

In a controller circuit, many circuit changes can be programmed. However, additional inputs and outputs must be wired to the controller. For example, if a light is required to indicate the direction of motor (or product) travel, the light must be physically wired to the controller. This requirement is one of the similarities between hardwired and programmable controller circuits. **See Figure 27-50.**

WIRING ADDITIONAL OUTPUTS

HARDWIRED CIRCUIT

PROGRAMMABLE CONTROLLER CIRCUIT

Figure 27-50. In a programmable controller circuit, additional inputs and outputs must be wired to the controller.

Controller Wiring. The use of a programmable controller for motor control allows for greater flexibility and circuit monitoring of a motor and control circuit. A programmable controller can monitor and control all motor control functions, but it cannot directly monitor and display motor parameters such as voltage, current, frequency, and power.

When a programmable controller is used to control a circuit, the controller is connected into the control circuit. The power circuit does not change. What does change is that the control circuit inputs (pushbuttons, limit switches, and overload contacts) are wired to the controller input module and the control circuit outputs (motor starter coils and indicator lamps) are wired to the controller output module. **See Figure 27-51.**

The circuit operation (logic) is programmed using the controller software and the circuit is downloaded to the controller. The controller software can monitor and display the condition (ON or OFF) of the circuit inputs and outputs. If changes to the control circuit are required, they can be reprogrammed and downloaded without changing the circuit wiring.

When a programmable controller is used to program inputs, the actual input type (normally open or normally closed) and the way the input is programmed must be considered. This is because when using a programmable controller, an input can be wired normally open and programmed either normally open or normally closed. Likewise, an input can be wired normally closed and programmed either normally closed or normally open.

Programmable controllers are commonly used in large industrial facilities to control electric motors, valves, pneumatic or hydraulic cylinders, magnetic relays, and other devices.

CONTROLLER WIRING

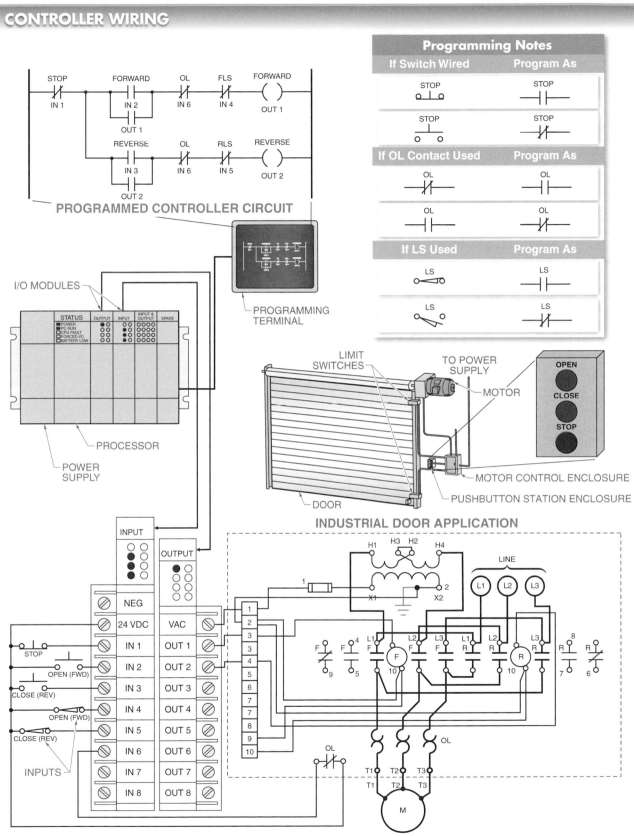

Figure 27-51. When using a controller to control a circuit, the control circuit inputs are wired to the controller input module and the control circuit outputs are wired to the programmable controller output module.

27-3 CHECKPOINT

1. In the circuit in Figure 27-36, how many inputs must be connected to the PLC input section?
2. In the circuit in Figure 27-36, would the circuit start pushbutton be programmed as an XIO or XIC contact?

3. In the circuit in Figure 27-36, how many outputs must be connected to the PLC output section?
4. In the circuit in Figure 27-50, what are three ways to stop (turn off reversing starter) the lowering of the door?

27-4 TROUBLESHOOTING PROGRAMMABLE CONTROLLERS

Troubleshooting programmable controllers normally involves finding a problem in the hardware or software. Most hardware problems are found in the I/O sections of the controller and usually can be found using standard DMMs. Software problems require knowledge of the specific program and type of manufacturer equipment used.

Troubleshooting Input Modules

Signals and information are sent to a programmable controller using input devices such as pushbuttons, limit switches, level switches, and pressure switches. Input devices are connected to the input module of the controller. Terminal screws at the back of the input module allow the connection of the input devices. The controller does not receive the proper information if the input module is not operating correctly. **See Figure 27-52.** To troubleshoot the input module of a programmable controller, the following procedure is applied:

1. Measure the supply voltage at the input module to ensure that there is power supplied to the input device(s). Test the main power supply of the controller when there is no power.
2. Measure the voltage from the control switch. Connect the DMM directly to the same terminal screw to which the input device is connected. The DMM should read the supply voltage when the control switch is closed. The DMM should read the full supply voltage when the control device uses mechanical contacts. The DMM should read nearly the full supply voltage when the control device is solid-state. Full supply voltage is not read because 0.5 V to 6 V is dropped across the solid-state control device. The DMM should read zero or little voltage when the control switch is open.

3. Monitor the status indicators on the input module. The status indicators should illuminate when the DMM indicates the presence of supply voltage.
4. Monitor the input device symbol on the programming terminal monitor. The symbol should be highlighted when the DMM indicates the presence of supply voltage. Replace the control device if the control device does not deliver the proper voltage. Replace the input module if the control device delivers the correct voltage but the status indicator does not illuminate.

CAUTION: A vibrating voltage tester (Wiggy®) must never be used to measure voltage levels on a programmable controller. Vibrating voltage testers contain a solenoid. When the test leads of a vibrating voltage tester are removed, the collapsing field of the solenoid can damage the solid-state components of programmable controller I/O modules.

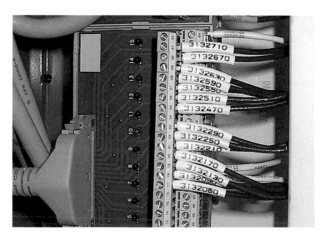

Troubleshooting and replacing old devices is more efficient when conductors are clearly labeled.

TROUBLESHOOTING INPUT MODULES

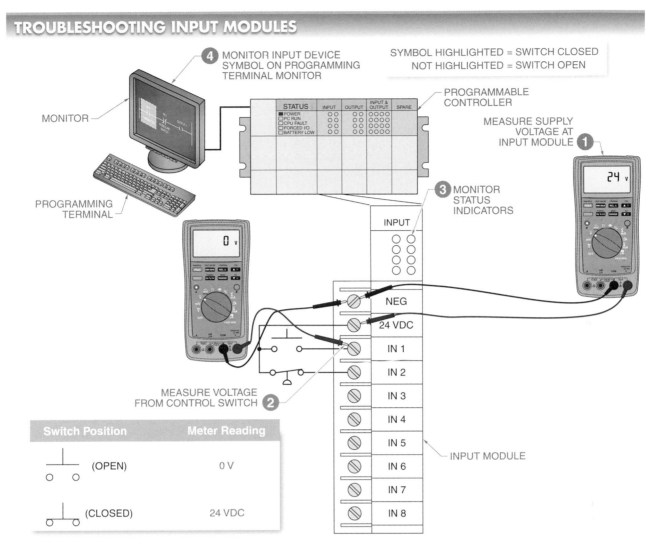

4 MONITOR INPUT DEVICE SYMBOL ON PROGRAMMING TERMINAL MONITOR

SYMBOL HIGHLIGHTED = SWITCH CLOSED
NOT HIGHLIGHTED = SWITCH OPEN

MONITOR

PROGRAMMING TERMINAL

PROGRAMMABLE CONTROLLER

MEASURE SUPPLY VOLTAGE AT INPUT MODULE **1**

STATUS	INPUT	OUTPUT	INPUT & OUTPUT	SPARE
POWER				
PC RUN				
CPU FAULT				
FORCED I/O				
BATTERY LOW				

24 v

3 MONITOR STATUS INDICATORS

INPUT

0 v

NEG

24 VDC

IN 1

IN 2

MEASURE VOLTAGE FROM CONTROL SWITCH **2**

IN 3

IN 4

IN 5

INPUT MODULE

IN 6

IN 7

IN 8

Switch Position	Meter Reading
(OPEN)	0 V
(CLOSED)	24 VDC

Figure 27-52. A controller does not receive the proper information if the input module is not operating correctly.

Troubleshooting Input Devices

Input devices such as pushbuttons, limit switches, pressure switches, and temperature switches are connected to the input module(s) of a programmable controller. Input devices send information and data concerning circuit and process conditions to the controller. The processor receives the information from the input devices and executes the program. All input devices must operate correctly for the circuit to operate properly. **See Figure 27-53.** To troubleshoot an input device of a programmable controller, the following procedure is applied:

1. Place the controller in the test or program mode. This step prevents the output devices from turning on. Output devices are turned on when the controller is placed in the run mode.

2. Monitor the input devices using the input status indicators (located on each input module), the programming terminal monitor, or the data file. A *data file* is a group of data values, such as inputs, timers, counters, and outputs, that are displayed as a group and whose status may be monitored.

3. Manually operate each input, starting with the first input. Never reach into a machine when manually operating an input. Always use a wooden stick or other nonconductive device.

The input status indicator located on the input module should illuminate and the input symbol should be highlighted in the control circuit on the monitor screen when an NO input device is closed. The bit status on the programming terminal monitor screen should be set to "1," indicating a high voltage or the presence of voltage.

TROUBLESHOOTING INPUT DEVICES

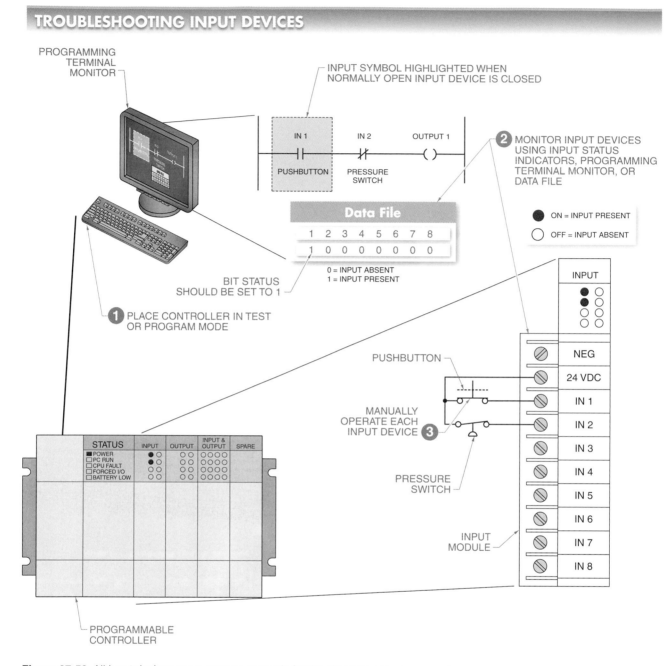

Figure 27-53. All input devices must operate correctly for the circuit to operate properly.

The input status indicator located on the input module should turn off and the input symbol should no longer be highlighted in the control circuit on the monitor screen when an NC input device is open. The bit status on the programming terminal monitor screen should be set to "0," indicating a low voltage or the absence of voltage.

The next input device should be selected and tested when the status indicator and associated bit status match. Each input device should be tested until all inputs have been tested. The input device and output device should be tested when the status indicator and associated bit status do not match.

Tech Fact

During capacitor discharging, a residual charge can cause an arc when working on a PLC module, which can cause damage to the modules electronic circuits and/or an electrical shock. Hold modules by the edges to avoid contact and use a voltmeter to ensure that no voltage is present before working on any PLC module.

Troubleshooting Output Modules

A programmable controller turns on and off the output devices (loads) in the circuit according to the program. The output devices are connected to the output module of the programmable controller. No work is produced in the circuit when the output module or the output devices are operating incorrectly. When an output device does not operate, the problem may lie in the output module, output device, or controller. **See Figure 27-54.** To troubleshoot the output module of a programmable controller, the following procedure is applied:

1. Measure the supply voltage at the output module to ensure that there is power supplied to the output devices. Test the main power supply of the controller when there is no power.

2. Measure the voltage delivered from the output module. Connect the DMM directly to the same terminal screw to which the output device is connected. The DMM should read the supply voltage when the program energizes the output device. The DMM should read full supply voltage when the output module uses mechanical contacts. The DMM should read almost full supply voltage when the output module uses a solid-state switch. Full voltage is not read because 0.5 V to 6 V is dropped across the solid-state switch. The DMM should read zero or little voltage when the program de-energizes the output device.

3. Monitor the status indicators on the output module. The status indicators should be energized when the DMM indicates the presence of supply voltage.

4. Monitor the output device symbol on the programming terminal monitor. The output device symbol should be highlighted when the DMM indicates the presence of supply voltage. Replace the output module when the output module does not deliver the proper voltage. Troubleshoot the output device when the output module does deliver the correct voltage but the output device does not operate.

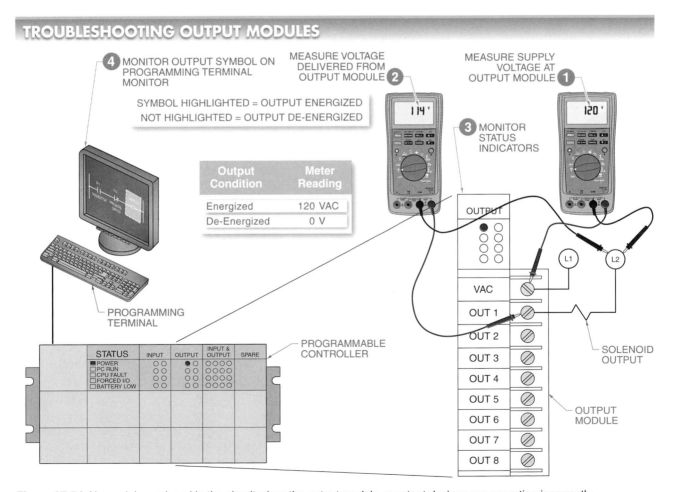

Figure 27-54. No work is produced in the circuit when the output module or output devices are operating incorrectly.

Troubleshooting Output Devices

Output devices such as motor starters, solenoids, contactors, and lights are connected to the output modules of a controller. An output device performs the work required for the application. The processor energizes and de-energizes the output devices according to the program. All output devices must operate correctly for the circuit to operate properly. **See Figure 27-55.** To troubleshoot an output device of a programmable controller, the following procedure is applied:

1. Place the controller in the test or program mode. Placing the controller in the test or program mode prevents the output devices from turning on. Output devices turn on when the controller is placed in the run mode.

2. Monitor the output devices using the output status indicators (located on each output module), the programming terminal monitor, or the data file.

3. Activate the input that controls the first output device. Check the program displayed on the monitor screen to determine which input activates which output device. Never reach into a machine to activate an input.

The next output device should be selected and tested when the status indicator and associated bit status match. Each output device should be tested until all output devices have been tested. The input device and output device should be tested when the status indicator and associated bit status do not match.

TROUBLESHOOTING OUTPUT DEVICES

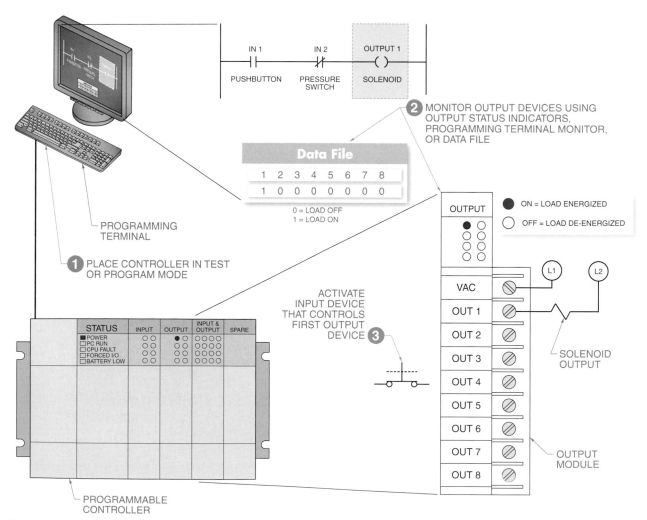

Figure 27-55. All output devices such as motor starters, solenoids, and contactors must operate correctly for the circuit to operate properly.

27-4 CHECKPOINT

1. In a displayed input data file, what condition (open or closed) does a "1" displayed in a limit switch input box mean?

2. In a displayed output data file, what condition (ON or OFF) does a "1" displayed in a solenoid output box mean?

Additional Resources

Review and Resources

Access Chapter 27 Review and Resources for *Electrical Motor Controls for Integrated Systems* by scanning the above QR code with your mobile device.

Applying Your Knowledge

Refer to the *Electrical Motor Controls for Integrated Systems* Learner Resources for interactive Applying Your Knowledge questions.

Workbook and Applications Manual

Refer to Chapter 27 in the *Electrical Motor Controls for Integrated Systems Workbook* and the *Applications Manual* for additional exercises.

ENERGY EFFICIENCY PRACTICES

PLC Advantages

PLCs offer many cost- and time-saving advantages over other methods of industrial process control. A major advantage of PLCs is that they can be programmed and reprogrammed as processes change. This reduces the control-circuit wiring requirements and the rewiring requirements when a circuit must be modified. In addition, PLCs use solid-state components to control a variety of devices. Solid-state components are very reliable, since they contain no moving parts.

PLCs include programmable memory, software timers/counters, software control relays, diagnostic indicators, and modular I/O interfaces. Programmable memory simplifies changes and permits flexibility when programming a PLC. Software timers/counters and control relays eliminate the costs associated with the hardware, wiring, and space normally required for these physical components. Diagnostic indicators reduce troubleshooting time. The modular architecture allows for easy installation and flexibility since subassemblies can be removed easily for replacement or repair. This reduces hardware costs, minimizes process downtime, and allows for future process expansion or modification.

Sections

Objectives

28-1
- Define and describe centralized power distribution.
- Define distribution substation and describe its three main functions.
- Define and describe switchgear and switchboards.
- Define and describe panelboards.
- Describe motor control centers.
- Define and describe busways.
- Describe grounding in a power distribution system.

28-2
- Describe transformer ratings and overloading.
- Describe different transformer connections.
- Describe transformer installation and conductor identification.
- Describe harmonic distortion and nonlinear loads.

28-3
- Explain how phase problems affect power quality.
- Describe how AC and DC motor voltage variations affect power quality.
- Describe how voltage surges and voltage unbalance affect power quality.
- Describe how current unbalance affects power quality.
- Describe how power factor affects power quality.

28-4
- Define electrical grid and smart grid and explain how electrical grids are updated.
- Describe alternatives to aging centralized power distribution systems.
- Describe how power grids are metered and monitored.

Review and Resources
atplearningresources.com/Quicklinks
Access Code: 362245

Chapter
Power Distribution and Smart Grid Systems

The amount of electricity used has grown over the last 100 years, and new electrical/electronic devices are continuously being developed. Electricity is used in almost every residential, commercial, industrial, and transportation application. Before electricity can be used, it must be generated, distributed, and delivered safely, reliably, and economically to the points of use. It is important to understand how a power distribution system operates, problems that may occur along the system, how measurements are taken, the actions needed to increase reliability and reduce downtime, and new methods of power distribution.

Although devices that use electricity continue to increase in number, complexity, and use, the means of generating and distributing electrical power has not changed much over time. However, as the amount of available electrical power reaches its maximum and costs increase, the way of producing and distributing power is changing. New power sources include wind farms and photovoltaic power plants. Upgrading older systems and adding new electronic monitoring along the distribution system help integrate new sources of power into the system, reduce power losses, and reduce power outages.

28-1 ELECTRICAL POWER DISTRIBUTION

Today, most electrical power is distributed through a network of transmission lines (conductors), substations (transformers), and generating equipment from relatively large, centralized power-generating stations directly to the customer. These large, centralized power-generating stations are located near abundant energy sources such as coal, oil, and natural gas. Large, centralized power-generating stations are also located near natural resources. For example, nuclear power-generating stations are located near large sources of water to easily cool the reactor.

The electrical power created in a large, centralized power-generating station travels through many stages before it is used by loads. The organization that produces and/or distributes electricity to customers is normally referred to as a utility. However, the organizational structure of utilities has been changing as they divide their capabilities into separate business models such as electricity producers and electricity distributors.

Centralized Power Transmission and Distribution

Centralized power distribution is the process of delivering electrical power that is generated at a large centralized location to customers. The power transmission and distribution system, from the power generation source to the customer's loads, must be in good working order and properly maintained. **See Figure 28-1.**

CENTRALIZED POWER TRANSMISSION AND DISTRIBUTION

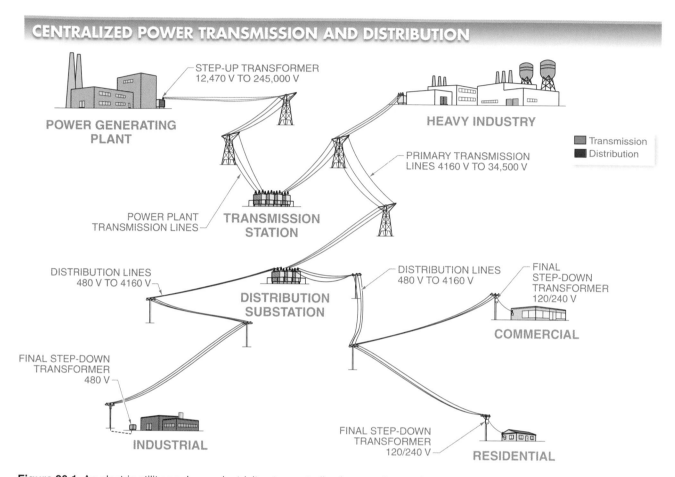

Figure 28-1. An electric utility produces electricity at a centralized generating station and transmits and distributes it to industrial, commercial, and residential customers through power lines, substations, and transformers.

Power transmission is the process of delivering electrical power from a power-generating plant to a substation. *Power distribution* is the process of delivering electrical power from a substation to the customer's service-entrance equipment. Power control, protection, transformation, transmission, distribution, and regulation must occur when electrical power is delivered to the customer. A centralized power transmission and distribution system commonly includes the following parts:

- Step-up transformers—The generated voltage is stepped up to transmission voltage level. The transmission voltage level is normally between 12.47 kV and 245 kV.

- Generating station transmission lines—The 12.47 kV to 245 kV generating station transmission lines deliver power to the transmission substations.

- Transmission substations—The voltage is transformed to a lower primary (feeder) voltage. The primary voltage level is normally between 4.16 kV and 34.5 kV.

- Primary transmission lines—The 4.16 kV to 34.5 kV primary transmission lines deliver power to the distribution substations and heavy industry.

- Distribution substations—The voltage is transformed down to utilization voltages. Utilization voltage levels range from 480 V to 4.16 kV.

- Distribution lines—Power is carried from the distribution substation along the street or rear lot lines to the final step-down transformer.

- Final step-down transformers—Voltage is transformed to the required voltage, such as 480 V or 120/240 V. The final step-down transformers may be installed on poles, on grade-level pads, or in underground vaults. The secondary of the final step-down transformer is connected to service-entrance cables that deliver power to service-entrance equipment. The number and size of transformers used to step down (reduce) the voltage before it is used by the customer's power distribution system depends on the customer's power requirements.

Transmission Lines and Towers. A *transmission line* is a conductor that carries large amounts of electrical power at high voltages over long distances. **See Figure 28-2.** Aerial transmission lines must be spaced far enough apart and elevated in order to be safe. The transmission voltage level varies depending on the required transmission distance and amount of power carried. The longer the distance or higher the transmitted power, the higher the transmitted voltage.

TRANSMISSION LINES

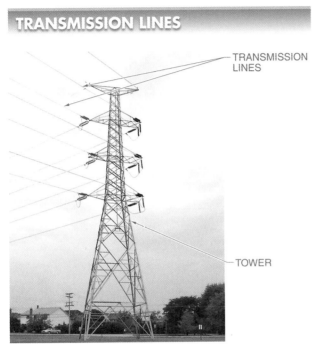

TRANSMISSION LINES

TOWER

Figure 28-2. Transmission lines safely carry large amounts of electrical power at high voltages over long distances.

Annual electricity transmission and distribution losses may result in losses as much as 6% to 7% of the generated electrical power, mostly attributed to heat. Transmission line voltages can vary from a few kilovolts to hundreds of kilovolts. Transmission-line conductor sizes are based on the amount of current they can safely carry without overheating. For a given power level, the amount of current varies inversely with the amount of voltage. **See Figure 28-3.** Increasing the transmitted voltage reduces the power losses between the utility and the customer. Power loss can be reduced up to 75% when transmitted voltage is doubled. When power is transmitted at high voltages, the required size and weight of the conductors is reduced. Therefore, higher transmitted voltages allow for reduced conductor size, allow more power to be transmitted, and result in lower material cost.

POWER, VOLTAGE, AND CURRENT RELATIONSHIP

Power*	Voltage†	Current‡
10,000	20,000	0.5
10,000	10,000	1
10,000	5000	2
10,000	2500	4
10,000	1250	8
10,000	480	20.83
10,000	240	41.66
10,000	120	83.33

* in W
† in V
‡ in A

VOLTAGE VARIES INVERSELY WITH CURRENT

Figure 28-3. For a given power level, the amount of current varies inversely with the amount of voltage.

Distribution Substations

A *distribution substation* is an outdoor facility located close to the point of electrical service use and is used for changing voltage levels, providing a central place for system switching, monitoring protection, and redistributing power. Distribution substations take high transmitted voltages and reduce the voltage for appropriate distribution levels. Distribution substations normally operate at lower voltages than transmission substations. Distribution substation output voltages are normally between 12 kV and 13.8 kV. **See Figure 28-4.**

Distribution substations serve as a source of voltage transformation and control along the distribution system. Distribution substations include the following functions:

- receiving voltage generated and adjusting it to a level appropriate for further transmission or customer use

- providing a switching point where different connections may be made

- providing a safe point in the distribution grid for disconnecting the power in the event of a problem

- providing a convenient place to take measurements and check the operation of the distribution system

Tech Fact

The highest current at which a power line can be operated is called the transmission line real-time rating. Transmission line real-time ratings are based on wind speed, ambient temperature, and solar radiation. This rating determines the maximum current that can be transmitted without violating safety codes, decreasing network reliability, or damaging equipment. When live current exceeds this rating, the conductors heat, causing the line to sag and increase in resistance.

DISTRIBUTION SUBSTATIONS

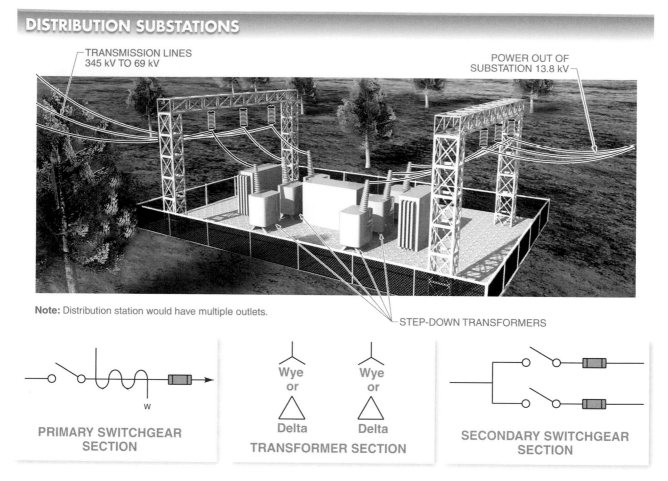

Figure 28-4. Distribution substations provide a convenient place along the distribution system for maintenance, checks, and fine adjustments.

Distribution substations have three main sections: primary switchgear, transformer, and secondary switchgear. Depending on the function of the substation, the primary or secondary switchgear may be the high-voltage or low-voltage section. In step-up substations, the primary switchgear is the low-voltage section and the secondary switchgear is the high-voltage section. In step-down substations, the primary switchgear is the high-voltage section and the secondary switchgear is the low-voltage section. The sections of a distribution substation also normally include circuit breakers and interrupter switches.

Power Distribution to Point of Use

Electrical power is delivered to residential, commercial, and industrial buildings through distribution lines. **See Figure 28-5.** These distribution lines are terminated at commercial and industrial buildings through switchboards. Switchboards are high-power electrical equip-

ment that switch and divide circuits in the building distribution system. Switchboards may contain a panel or an assembly of panels containing electrical switches, meters, and overcurrent protective devices (OCPDs). Once power is delivered to a building, switchboards further distribute the power to where it is required within the building. The power for residential customers is generally terminated and distributed through a panelboard.

The electrical service provided by a utility may be overhead or lateral. *Overhead service* is an electrical service in which service-entrance conductors are run through the air from the utility pole to the building. *Service lateral* is an electrical service in which service-entrance conductors are run underground from the utility system to the service point. A *service point* is the point of connection between the facilities of the utility and the premises wiring. Typically, the switchboard is the last point on the power distribution system from the utility company and the beginning of most building power distribution systems.

POWER DISTRIBUTION TO POINTS OF USE

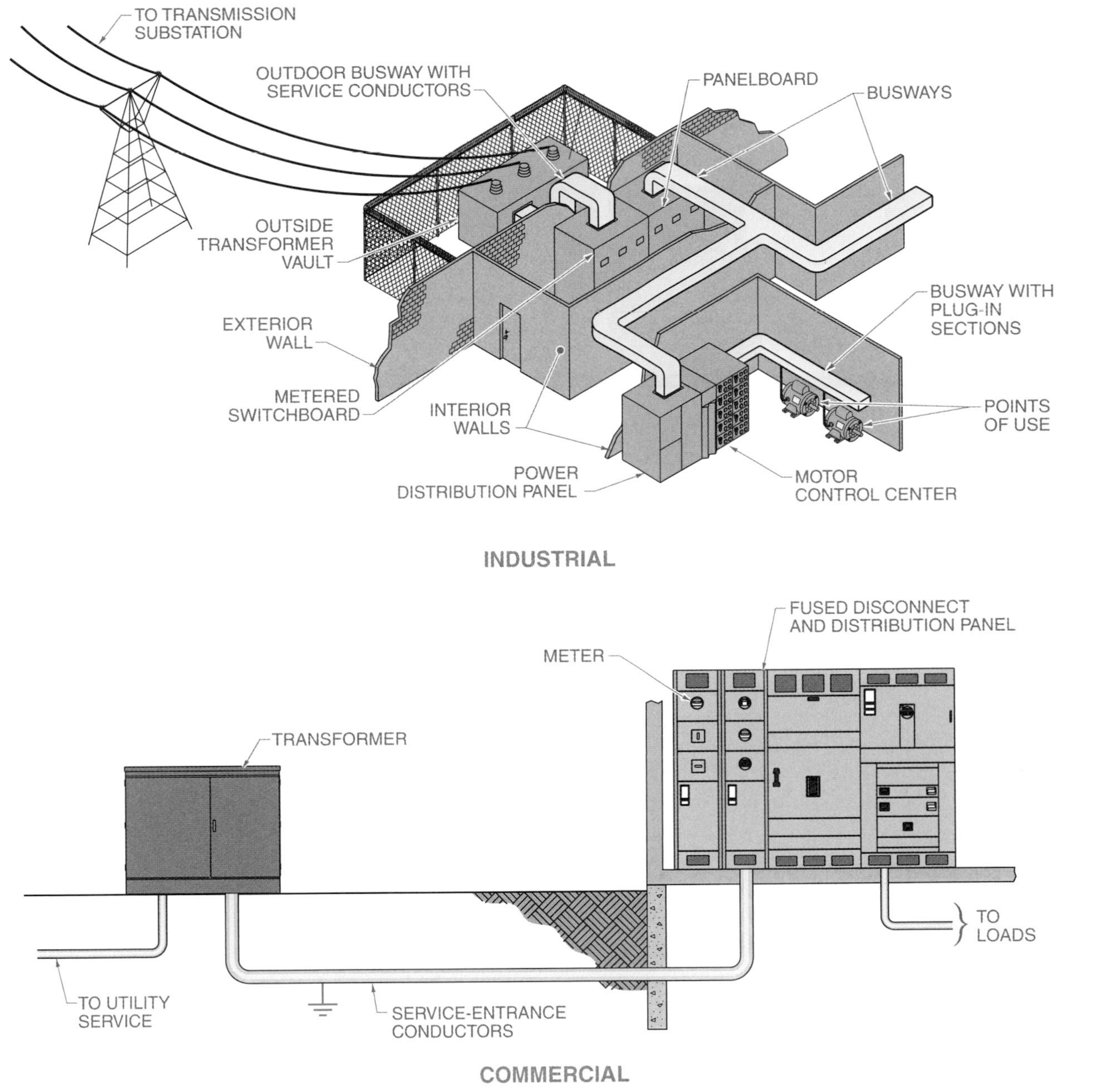

TO TRANSMISSION
SUBSTATION

OUTDOOR BUSWAY WITH
SERVICE CONDUCTORS

PANELBOARD

BUSWAYS

OUTSIDE
TRANSFORMER
VAULT

BUSWAY WITH
PLUG-IN
SECTIONS

EXTERIOR
WALL

METERED
SWITCHBOARD

INTERIOR
WALLS

POINTS
OF USE

POWER
DISTRIBUTION PANEL

MOTOR
CONTROL CENTER

INDUSTRIAL

FUSED DISCONNECT
AND DISTRIBUTION PANEL

METER

TRANSFORMER

TO
LOADS

TO UTILITY
SERVICE

SERVICE-ENTRANCE
CONDUCTORS

COMMERCIAL

Figure 28-5. Power is fed through circuit breakers in the panelboard and routed through cables or busways to power distribution panels to the points of use.

Switchboards

Electrical power is delivered to industrial, commercial, and residential buildings through a distribution and transmission system. Once the power is delivered to a building, it is up to the building electrician to further distribute the power to where it is required within the building.

A *switchboard* is a large floor-mounted panel or assembly of panels in which electrical switches,

OCPDs, buses, and instruments are mounted. **See Figure 28-6.** *Switchgear* is any high-powered electrical device that switches or interrupts devices or circuits in a building distribution system. The terms switchgear and switchboards are often used interchangeably or when referring to both switchgear and switchboard equipment. A switchboard is typically the link between the power delivered to a building

SWITCHBOARDS

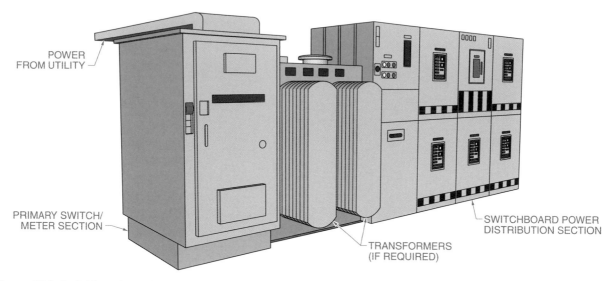

Figure 28-6. Switchboards are large floor-mounted panels or assemblies of panels in which electrical switches, OCPDs, buses, and instruments are mounted.

or property and the start of the building's power distribution system. The switchboard may be the last point on the power distribution system for the utility and the beginning of the distribution system for the building electrician. Switchboards are rated by the manufacturer for a maximum voltage and current output. For example, a switchboard may have a 600 V rating and a bus rating up to 5000 A.

In addition to dividing the incoming power, a switchboard may contain all the equipment needed for controlling, monitoring, protecting, and recording the functions of the substation. Switchboards are designed for use in service-entrance and distribution.

A service-entrance switchboard has space and mounting provisions for metering equipment (as required by the local power company), overcurrent protection, and a means of disconnect for the service conductors. Provisions for grounding the service neutral conductor when a ground is needed are also provided. A distribution switchboard contains the protective devices and feeder circuits required to distribute the power throughout a building. A distribution switchboard may contain either circuit breakers or fused switches.

A distribution switchboard has the space and mounting provisions required by the local power company. Building power distribution systems are used to deliver the required type (DC, 1ɸ, or 3ɸ) and level (120 V, 230 V, 460 V, etc.) of power to the loads connected to the system. Metering equipment for the power used by the building's tenants is

also installed at this location. To meter the incoming power, the switchboard must have a watt-hour meter to measure power usage. Metering equipment is always located on the incoming line side of the disconnect. The metering equipment compartment cover is sealed to prevent power from being tapped ahead of the power company metering equipment. **See Figure 28-7.**

DISTRIBUTION SWITCHBOARDS

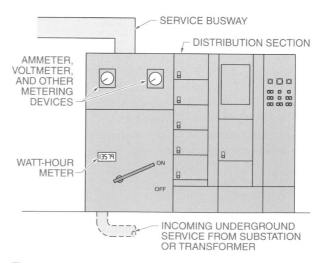

Figure 28-7. A distribution switchboard contains the protective devices and feeder circuits required to distribute power throughout a building.

Other meters and indicator lights, such as voltmeters, ammeters, and wattmeters, may also be built into the meter equipment compartment. In most cases, these components are optional, depending on the application and the plant requirements. A voltmeter is used to indicate the various incoming and outgoing voltages to the maintenance personnel. An ammeter is used to indicate the various current levels throughout the system. A wattmeter is used to indicate the power used throughout the system. These instruments can be indicating types, recording types, or a combination of both types. Recording instruments are used to track various values during a period of time.

In addition to measuring the voltage, current, and power of a system, a switchboard also controls the power. Control is achieved through the use of switches, fuses, circuit breakers, and overcurrent and overvoltage relays that can disconnect the power. These devices protect the distribution system in the event of fault.

Switchboards that have more than six switches or circuit breakers must include a main switch to protect or disconnect all circuits. Switchboards with more than one but not more than six switches or circuit breakers do not require a main switch. In a switchboard with more than six switches or breakers, the service-entrance section of the switchboard may have feeder circuits added to the rated capacity of the main. A switchboard with a main section can easily contain more than one distribution system. The system requirement depends on the number of feeder circuits entering the building. More information is available in NEC® Article 240.

In addition to distributing the power throughout a building, the distribution section of a switchboard may contain provisions for motor starters and other control devices. **See Figure 28-8.** The addition of starters and controls to the switchboard allows for motors to be connected to the switchboard. This combination can be used when the motors to be controlled are located near the switchboard. The combination also allows for high-current loads such as motors to be connected to the source of power without further power distribution.

Panelboards and Branch Circuits

A *panelboard* is a wall-mounted distribution cabinet containing a group of overcurrent and short-circuit protective devices for lighting, appliance, or power distribution branch circuits. Wall-mounted panelboards are discernible from switchboards because switchboards are normally freestanding. **See Figure 28-9.**

DISTRIBUTION SECTIONS

General Electric Company

Figure 28-8. The distribution section of a switchboard may contain provisions for motor starters and other devices.

PANELBOARDS

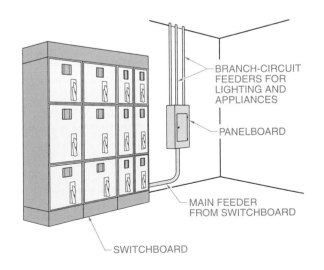

BRANCH-CIRCUIT FEEDERS FOR LIGHTING AND APPLIANCES

PANELBOARD

MAIN FEEDER FROM SWITCHBOARD

SWITCHBOARD

Figure 28-9. A panelboard is a wall-mounted distribution cabinet containing overcurrent and short-circuit protection devices.

A panelboard is normally supplied with power by a switchboard and further divides the power distribution system into smaller parts. Panelboards are the part of the distribution system that provides the last centrally located protection for the final power run to the load and its control circuitry. Panelboards are classified according to their use in the distribution system.

A panelboard provides the required circuit control and overcurrent protection for all circuits and power-consuming loads connected to the distribution system. **See Figure 28-10.** Panelboards are located throughout a plant or building, providing the necessary protection for the branch circuits that feed the loads.

A *branch circuit* is the portion of the distribution system between the final OCPD and the outlet or load connected to it. The basic requirements for panelboards and OCPDs are given in the NEC® Articles 240 and 408 and must be met for individual applications. In addition, local, city, and country regulations must also be met.

OCPDs used for protecting branch circuits include fuses or circuit breakers. OCPDs must provide for proper overload and short-circuit protection. The size (in amperes) of the OCPD is based on the rating of the panelboard and load. OCPDs must protect the load and be within the rating of the panelboard. If the OCPD exceeds the capacity of the busbars in the panelboard, the panelboard is undersized for the load(s) that are to be connected.

Panelboard Installation. Panelboards are the main place where most indoor electrical circuits begin. The panelboard must have a sufficient current and voltage rating and must be properly installed and grounded. An electrical shock or fire is possible if the panelboard is not properly installed and grounded. In addition, the system can be overloaded and a fire can result if the panelboard is not properly sized. All panelboard installations should follow NEC® and NFPA requirements. For example, NEC® Articles 408 and 409 cover the installation of switchboards and panelboards. **See Figure 28-11.** These requirements protect persons working around panelboards, help prevent fires, and reduce the chance of an electrical shock.

Tech Fact

A neutral ground connection is made by connecting the neutral bus in a main panelboard to the ground bus with a main bonding jumper. The ground bus is connected to the earth grounding system. Additional neutral-to-ground connections should not be made in any other subpanels, receptacles, or equipment. If a neutral-to-ground connection is made, a parallel path for the normal return neutral current is created. This will produce an unsafe condition because current will flow through all the grounds and can cause electrical shocks.

PANELBOARD CIRCUITS AND OVERCURRENT PROTECTION

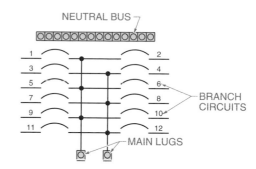

1ϕ, THREE-WIRE WITH CIRCUIT BREAKERS

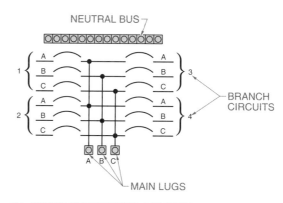

3ϕ, FOUR-WIRE WITH CIRCUIT BREAKERS

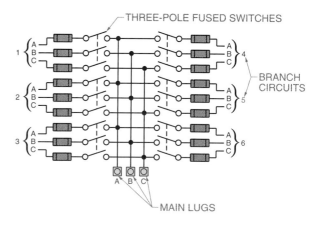

3ϕ POWER DISTRIBUTION

Figure 28-10. A panelboard provides the required circuit control and overcurrent protection for all circuits and power-consuming loads connected to the distribution system.

PANELBOARDS—NEC® ARTICLE 408

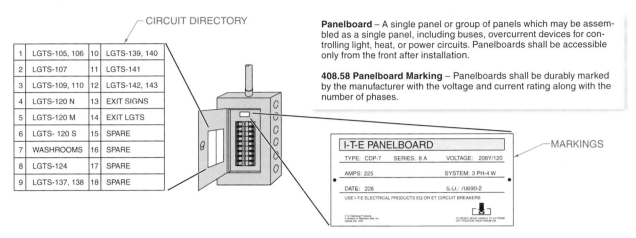

CIRCUIT DIRECTORY

1	LGTS-105, 106	10	LGTS-139, 140
2	LGTS-107	11	LGTS-141
3	LGTS-109, 110	12	LGTS-142, 143
4	LGTS-120 N	13	EXIT SIGNS
5	LGTS-120 M	14	EXIT LGTS
6	LGTS- 120 S	15	SPARE
7	WASHROOMS	16	SPARE
8	LGTS-124	17	SPARE
9	LGTS-137, 138	18	SPARE

Panelboard – A single panel or group of panels which may be assembled as a single panel, including buses, overcurrent devices for controlling light, heat, or power circuits. Panelboards shall be accessible only from the front after installation.

408.58 Panelboard Marking – Panelboards shall be durably marked by the manufacturer with the voltage and current rating along with the number of phases.

MARKINGS

I-T-E PANELBOARD

TYPE: CDP-7   SERIES: 8 A   VOLTAGE: 208Y/120

AMPS: 225   SYSTEM: 3 PH-4 W

DATE: 226   S.O.: 70690-2

USE I-T-E ELECTRICAL PRODUCTS EQ OR ET CIRCUIT BREAKERS

I-T-E Electrical Products
A division of Siemens-Allis, Inc.
Atlanta GA, USA

TO RESET, MOVE HANDLE TO EXTREME
OFF POSITION THEN THROW ON

408.30 The panelboard rating shall not be less than the minimum feeder capacity per Article 220. They shall be marked by the manufacturer with the voltage, current rating, number of phases, and manufacturer's name or trademark. A circuit directory shall be provided on the face or inside the door.

408.36 Ex 1. Individual protection is not required for a panelboard used as service equipment with multiple disconnecting means according to 230.71.

408.36 Ex 2. Each panelboard shall be protected on the supply side by no more than two main CBs or two sets of fuses with a combined rating that does not exceed the panelboard's rating. The panelboard shall not contain more than 42 overcurrent devices.

408.36 Ex 3. For existing installations, individual protection is not required if the lighting and appliance branch-circuit panelboard is in an individual residential occupancy.

408.36 (A) Panelboards with snap switches rated 30 A or less shall have overcurrent protection not in excess of 200 A.

408.37 Panelboards in damp or wet locations shall comply with 312.2 (A).

408.38 Panelboards shall be mounted in enclosures designed for the purpose and shall be dead–front.

408.39 Except as permitted for services, panelboard fuses shall be installed on the load side of switches.

408.40 Metal panelboard cabinets and frames shall be grounded per Article 250.

Figure 28-11. NEC® Article 408 covers the installation of panelboards.

Motor Control Centers

In a power distribution system, many different kinds of loads are connected to the system. The loads vary considerably from application to application, as does their degree of control. For example, a lamp may be connected to a system that requires only a switch for control, along with proper protection. However, other loads such as motors may require complicated and lengthy control and protection circuits. As a control circuit becomes more complicated, the more difficult it is to wire into the system.

The most common loads requiring simple and complex control are electric motors. Simplifying and consolidating motor control circuits is required because an electric motor is an integral part for almost all production and industrial applications. To do this, the incoming power, control circuitry, required overload protection, required overcurrent protection, and any transformation of power are combined into one convenient motor control center. **See Figure 28-12.**

MOTOR CONTROL CENTERS

UE Systems, Inc.

Figure 28-12. A motor control center provides a central location for the wiring, control, and troubleshooting of motor control circuits.

A motor control center combines individual control units into standard modular structures. Power for a motor control center is normally supplied from a panelboard or switchboard. A motor control center is different from a switchboard that contains motor panels because the motor control center is a modular structure designed specifically for plug-in control units and motor controls.

A motor control center receives the incoming power and delivers it to the control circuit and motor loads. The motor control center provides space for the control and load wiring in addition to providing required control components. The control inputs into the motor control center are the control devices such as pushbuttons, liquid level switches, limit switches, and other devices that provide a signal. The output of the motor control center is the wire connecting the motors. All other control devices are located in the motor control center. Control devices include relays, control transformers, motor starters, overload protective devices, overcurrent protective devices, timers, and counters.

One advantage of a motor control center is that it provides one convenient place for installing and troubleshooting control circuits. This is especially useful in applications that require individual control circuits to be related to the control circuits. An example is an assembly line in which one machine feeds the next machine.

A second advantage of a motor control center is that individual units can be easily removed, replaced, added to, and interlocked at one central location. Manufacturers of motor control centers produce factory-preassembled units to meet all the standard motor functions, such as start/stop, reversing, reduced-voltage starting, and speed control. This leaves only the connection of the control devices, such as pushbuttons, limit switches, level switches, and pressure switches, and the motor control center.

Common preassembled motor control center panels along with their schematic diagrams are available from the manufacturer. **See Figure 28-13.** The only wiring required from the electrician is the connections to control inputs, terminal blocks, and the motor. The motor is connected to T1, T2, and T3. The control inputs are connected to the terminal blocks marked 1, 2, and 3. If a two-wire control, such as a liquid level switch, is connected to the circuit, it is connected to terminals 1 and 3 only.

Also provided on preassembled motor control center panel are predrilled holes to allow for easy additions to the circuit. These holes match the manufacturer's standard devices, and most manufacturers provide templates

for easy layout and circuit design. Preassembled motor control center panels should not be removed or installed unless the disconnect for that unit is in the OFF position. *Note:* It is important to always verify that power is OFF by using a DMM set to measure voltage before removing or installing any electrical equipment.

MOTOR CONTROL CENTER SCHEMATIC DIAGRAMS

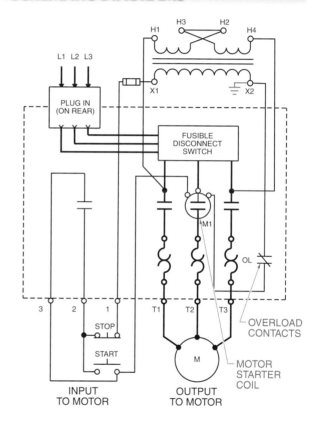

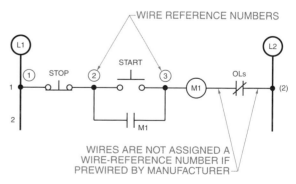

Figure 28-13. Common preassembled motor control center panels are available from the factory, along with corresponding schematic diagrams.

Power Distribution System One-Line Diagrams

A one-line diagram for a power distribution system is an electrical drawing that uses single lines and graphic symbols to illustrate the current path, voltage values, circuit disconnects, fuses, circuit breakers, transformers, and panelboards. One-line diagrams use the most basic symbols because the intent of the drawing is to illustrate as clearly as possible the flow of current throughout the building distribution system and where each component or device connects into the system. One-line diagrams are also used when designing large commercial and industrial installations to show the path of electrical power throughout a building.

One-line diagrams are also used when troubleshooting distribution system problems such as loss of power, low voltage, blown fuses, tripped circuit breakers, and poor power quality. They are also used to determine power shut-off points, future expansion capacity, and where emergency back-up generators or secondary power systems are connected into the system. A one-line diagram is helpful when troubleshooting a power system and can show the entire distribution system or specific parts of a system. **See Figure 28-14.**

ONE-LINE DIAGRAMS

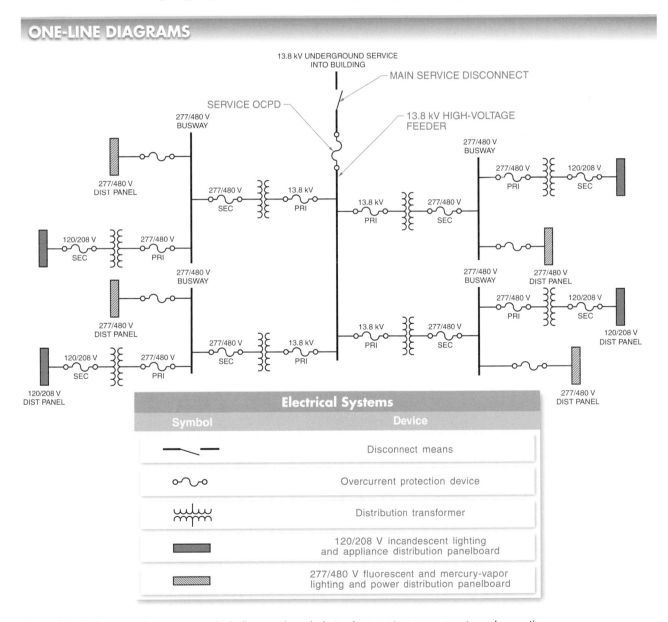

Electrical Systems	
Symbol	**Device**
╱	Disconnect means
o⌒⌒o	Overcurrent protection device
⌇⌇⌇	Distribution transformer
▬	120/208 V incandescent lighting and appliance distribution panelboard
▨	277/480 V fluorescent and mercury-vapor lighting and power distribution panelboard

Figure 28-14. One-line diagrams use single lines and symbols to show system components and operation.

For example, a one-line diagram may show a 13.8 kV feed into a building and the transformers used for the distribution of specific voltages. High voltages are used for the distribution of large amounts of power using small conductor sizes. The high voltage is then stepped down to low voltage levels and delivered to distribution panels. The distribution panels route power to individual loads such as industrial equipment, motors, lamps, and computers.

Feeders and Busways

The electrical distribution system of a building must transport electrical power from the source of power to the loads. In large buildings, distribution may be over large areas with many different electrical requirements throughout the building. **See Figure 28-15.** In many cases where it is common to shift production machinery, the distribution system must be changed from time to time. A *busway* is a metal-enclosed distribution system of busbars available in prefabricated sections. Prefabricated fittings, tees, elbows, and crosses simplify the connection and reconnection of the distribution system. When sections are bolted together, electrical power is available at many locations and throughout the system.

A busway does not have exposed conductors. This is because the power in a plant distribution system is at a high level. To offer protection from the high voltage, the conductors of a busway are supported with insulating blocks and covered with an enclosure to prevent accidental contact. A typical busway distribution system provides for fast connection and disconnection of machinery. Busways enable manufacturing plants to be retooled or re-engineered without major changes in the distribution system.

The most common length of busways is 10′. Shorter lengths are used as needed. Prefabricated elbows, tees, and crosses make it possible for the electrical power to run up, down, and around corners and to be tapped off from the distribution system. This allows the distribution system to have maximum flexibility with simple and easy connections as work is performed on installations.

The two basic types of busways are feeder and plug-in busways. **See Figure 28-16.** Feeder busways deliver the power from the source to a load-consuming device. Plug-in busways serve the same function as feeder busways, but they also allow load-consuming devices to be conveniently added along the bus structure. A plug-in power panel is used on a plug-in busway system.

ELECTRICAL DISTRIBUTION SYSTEMS

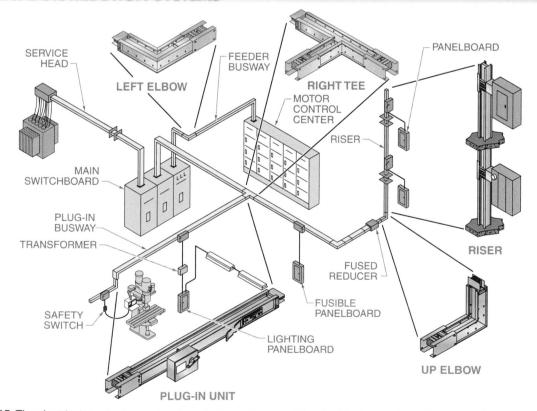

Figure 28-15. The electrical distribution system in a plant must transport the electrical power from the source of supply to the loads.

FEEDER AND PLUG-IN BUSWAYS

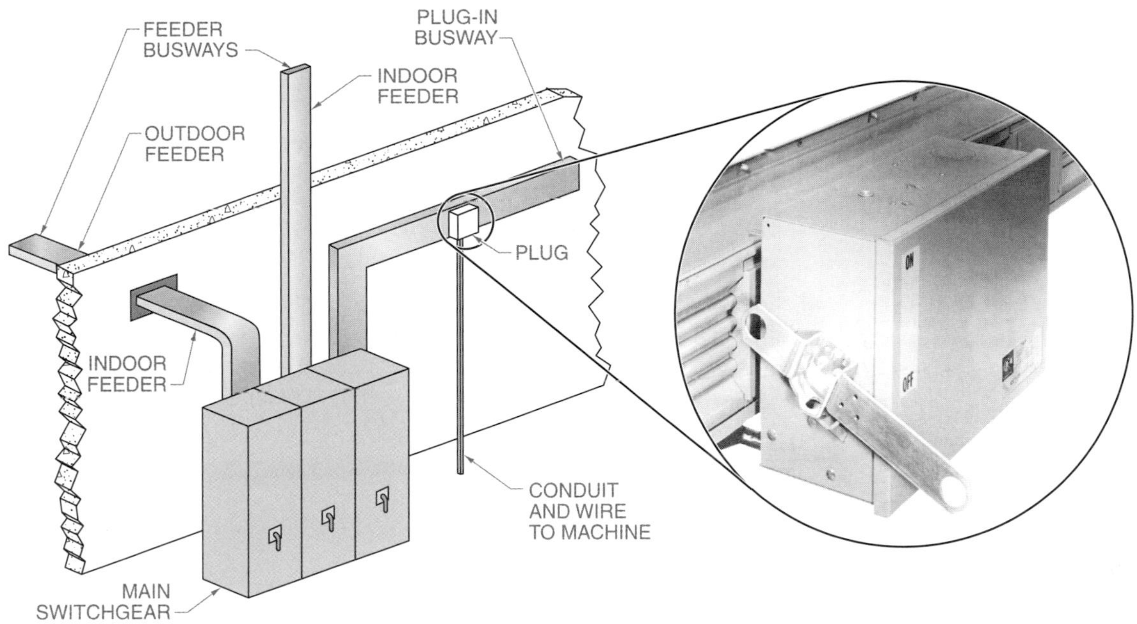

Figure 28-16. The two basic types of busways are feeder and plug-in busways.

The three general types of plug-in power panels used with busways are fusible switches, circuit breakers, and specialty plugs such as duplex receptacles with circuit breakers and twist-lock receptacles. The conduit and wire are run to a machine or load from the fusible switches and circuit breaker plug-in panels. Generally, power cords may be used only for portable equipment.

The loads connected to the power distribution system are often portable or unknown at the time of installation. For this reason, the power distribution system must often terminate in such a manner as to provide for a quick connection of a load in the future. To accomplish this, an electrician installs receptacles throughout the building or plant to serve the loads as required. With these receptacles, different loads can be connected easily.

Because the distribution system wiring and protective devices determine the size of the load that can be connected to it, a method is required for distinguishing the rating in voltage and current of each termination. This is especially true in industrial applications that require a variety of different currents, voltages, and phases.

The National Electrical Manufacturers Association (NEMA) has established a set of standard plug and receptacle configurations that clearly indicate the type of termination. **See Appendix.** The standard configurations enable the identification of the voltage and current rating of any receptacle or plug simply by looking at the configuration. Plug-in bus disconnect switches must not be removed or installed unless the disconnect switch is in the OFF position.

Grounding

Equipment grounding is required throughout an entire power distribution system. All non-current-carrying metal parts including conduit, raceways, transformer cases, and switchgear enclosures must be connected to ground. The objective of grounding is to limit the voltage of all metal parts to ground and establish an effective ground-fault current path. **See Figure 28-17.**

Grounding is accomplished by connecting the non-current-carrying metal to a ground with an approved grounding conductor and fitting. A ground bus is a network that ties solidly to grounding electrodes. A *grounding electrode* is a conducting object through which a direct connection to earth is established.

The ground bus must be connected to the grounding electrodes in several spots. The size of the ground bus is determined by the amount of current that flows through the grounding system and the length of time the current flows.

TYPES OF GROUNDING SYSTEMS

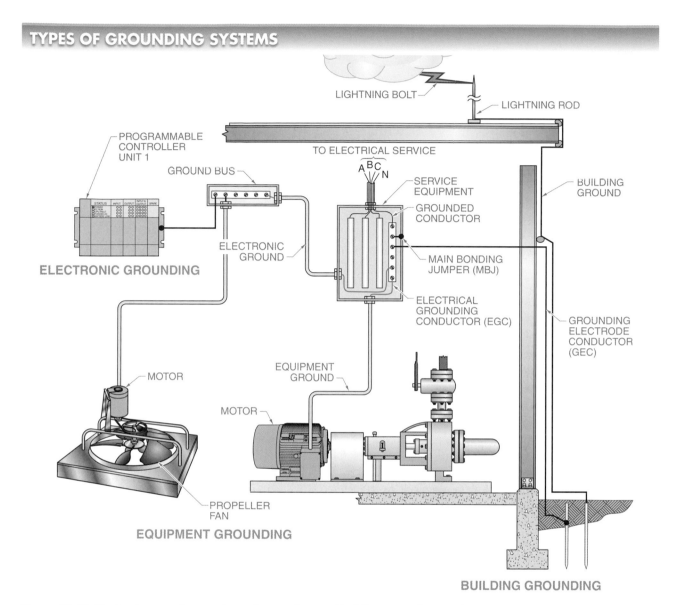

Figure 28-17. Electronic grounding, equipment grounding, and building grounding are the three types of grounding needed to create a safe work environment for individuals.

In addition to grounding all non-current-carrying metal, lightning arresters may be needed. A *lightning arrester* is a device that protects transformers and other electrical equipment from voltage surges caused by lightning. A lightning arrester provides a path over which the surge can pass to ground before it has a chance to damage electrical equipment.

Tech Tip

Never assume that a metal enclosure is properly grounded. An electrical shock can be caused by ungrounded enclosures when the metal handle is touched to turn off a circuit at a disconnect switch.

28-1 CHECKPOINT

1. How does increasing the transmitted voltage at a higher level reduce conductor size and allow more transmitted power?
2. What is the purpose of a distribution substation?
3. What is the purpose of a switchboard?
4. What type of basic diagram is used to show the path of the power distribution system through transformers, disconnects, and OCPDs?

28-2 DISTRIBUTION TRANSFORMERS

Distribution transformers are step-down transformers that reduce high transmitted voltage to usable residential, commercial, and industrial voltage levels. Although power is efficiently transmitted at high voltages, the high voltages are not safe for use in practical applications. Distribution transformers are normally rated from 1.5 kVA to 500 kVA and deliver a secondary voltage of 115 V, 120 V, 208 V, 230 V, 240 V, 460 V, or 480 V.

Transformer Ratings

Transformers are designed to transform power at one voltage level to power at another voltage level. In an ideal transformer, there is no loss or gain. Energy is simply transferred from the primary circuit to the secondary circuit. For example, if the secondary of a transformer requires 500 W of power to operate the loads connected to it, the primary must deliver 500 W. However, all transformers have some power loss, normally 0.5% to 8%.

Standard practice with transformers is to use voltage and current ratings, not wattage ratings. It is also standard practice to rate a transformer for its output capabilities since it is the output of the transformer that the loads are connected to. Thus, transformers are rated by their volt-ampere (VA) or kilovolt-ampere (kVA) output. Small transformers can be rated in either VA or kVA. Large transformers are rated in kVA. For example, a 50 VA transformer may be rated as a 50 VA transformer or a 0.05 kVA transformer. A 5000 VA transformer is rated as a 5 kVA transformer.

Transformer Overloading

All transformers have a power rating in VA or kVA. The power rating of a transformer indicates the amount of power the transformer can safely deliver. However, like most electrical devices, this rating is not an absolute value. For example, a 100 VA rated transformer will not be permanently damaged or destroyed if it is required to deliver 110 VA for a short period of time.

The heat produced by the power damages or destroys the transformer. The heat damages or destroys the transformer by breaking down the insulation, causing short circuits within the windings of the transformer. For this reason, temperature is the limiting factor in loading transformers. The more power the transformer must deliver, the higher the temperature produced at the transformer.

Distribution transformers are used to reduce high transmitted voltage to lower voltage levels that can be used for residential, commercial, and industrial applications.

Transformers are used to deliver power to a set number of loads. For example, a transformer can be used to deliver power to a school. The power delivered by the transformer changes as loads in the school are switched on and off. At certain times, such as during the night, the power output required from the transformer may be low. At other times, such as during school hours, the power output required from the transformer may be high. *Peak load* is the maximum output required of a transformer. **See Figure 28-18.**

TRANSFORMER LOAD CYCLES

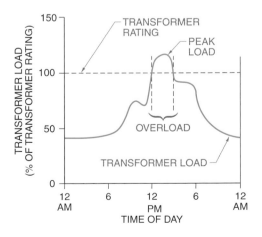

Figure 28-18. The power output required from a transformer varies based on the time of day.

A transformer is overloaded when it is required to deliver more power than its rating. A transformer is not damaged when overloaded for a short time period. This is because the heat storage capacity of a transformer ensures a relatively slow increase in internal temperature.

Transformer manufacturers list the length of time a transformer may be safely overloaded at a given peak level. For example, a transformer that is overloaded three times its rated current has a permissible overload time of about six minutes. **See Figure 28-19.**

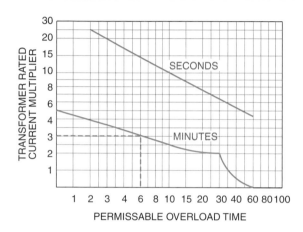

TRANSFORMER OVERLOADING

Figure 28-19. Transformer overloading occurs when a transformer is required to deliver more power than its rating allows.

Determining 3ϕ Transformer Current Draw

The current draw of a 3ϕ transformer is calculated similarly to the current draw of a 1ϕ transformer. The only difference is that the constant (1.732) for 3ϕ power is included in the formula. To calculate current draw of a 3ϕ transformer when kVA capacity and voltage are known, the following formula is applied:

$$I = \frac{(kVA_{CAP} \times 1000)}{(E \times 1.732)}$$

where

I = current (in A)

kVA_{CAP} = transformer capacity (in kVA)

1000 = constant

E = voltage (in V)

$1.732 = \sqrt{3}$ (constant)

Example: Calculating 3ϕ Transformer Current Draw

What is the current draw of a 3ϕ, 45 kVA transformer at 480 V?

$$I = \frac{(kVA_{CAP} \times 1000)}{(E \times 1.732)}$$

$$I = \frac{(45 \times 1000)}{(480 \times 1.732)}$$

$$I = \frac{45,000}{831.36}$$

$$I = \mathbf{54.13}$$

Three-Phase Transformer Connections

Three 1ϕ transformers may be connected to develop 3ϕ voltage. The three transformers may be connected in a wye or delta configuration. A *wye configuration* is a transformer connection that has one end of each transformer coil connected together. The remaining end of each coil is connected to the incoming power lines (primary side) or used to supply power to the load (secondary side). A *delta configuration* is a transformer connection that has each transformer coil connected end-to-end to form a closed loop. Each connecting point is connected to the incoming power lines or used to supply power to the load. The voltage output and type available for the load is determined by whether the transformer is connected in a wye or delta configuration. **See Figure 28-20.**

Three-Phase, Delta-to-Delta Connections. Three transformers may be connected in a delta-to-delta connection. A delta-to-delta transformer connection is used to supply 3ϕ voltage on the secondary. In a delta-to-delta connection, each transformer is connected end-to-end. **See Figure 28-21.**

The advantage of a delta-to-delta connection is that if one transformer is disabled, the other two may be used in an open-delta connection for emergency power. The rating of the open-delta bank is 57.7% of the original bank of three transformers, but 3ϕ power will be available until repairs are made.

One of the delta transformers is center-tapped to supply both 3ϕ voltage and 1ϕ voltage. Single-phase voltage at 120/240 V is available when the transformer is center-tapped. However, because only one transformer is tapped, it carries all of the 1ϕ, 120/240 V load and ⅓ of the 3ϕ, 240 V load. The other two transformers each carry ⅓ of the 3ϕ, 240 V load. For this reason, this connection should be used in applications that require a large amount of 3ϕ power and a small amount of 1ϕ power.

TRANSFORMER CONFIGURATIONS

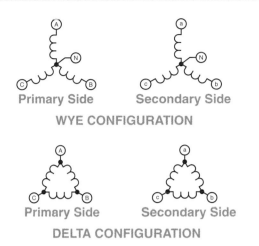

Figure 28-20. Transformers may be connected in a wye or delta configuration.

Three-Phase, Wye-to-Wye Connections. Three transformers may be connected in a wye-to-wye connection. A wye-to-wye transformer connection is used to supply both 1ϕ and 3ϕ voltage. In a wye-to-wye transformer connection, the ends of each transformer are connected together. **See Figure 28-22.**

The advantage of a wye-connected secondary is that the 1ϕ power draw may be divided equally over the three transformers. Each transformer carries ⅓ of the 1ϕ and 3ϕ power when the loads are divided equally. A disadvantage of a wye-to-wye connection is that interference with telephone circuits may result.

Tech Fact

Although most power transformers are the aboveground type and are designed to be mounted on pads or poles, underground transformers are available and are used in large facilities to eliminate overhead power lines.

THREE-PHASE, DELTA-TO-DELTA CONNECTIONS

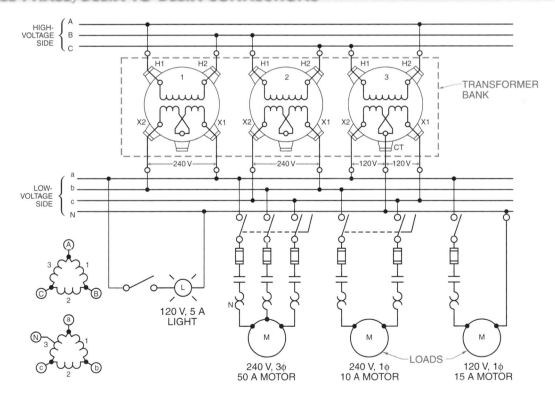

Figure 28-21. Three transformers may be connected in a delta-to-delta configuration in which each transformer is connected end-to-end.

THREE-PHASE, WYE-TO-WYE CONNECTIONS

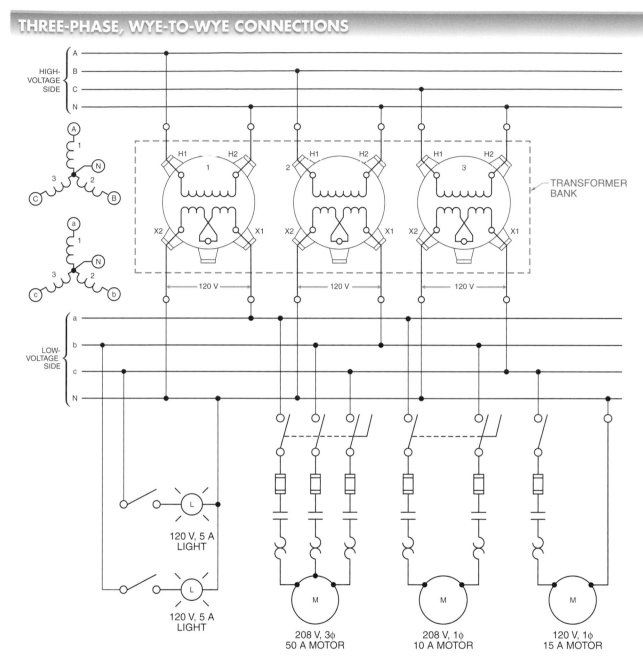

Figure 28-22. Three transformers may be connected in a wye-to-wye configuration in which the ends of each transformer are connected together.

Delta-to-Wye or Wye-to-Delta Connections. Transformers may also be connected in a delta-to-wye or wye-to-delta connection. The connection used depends on the incoming supply voltage, the requirements of the loads, and the practice of the local power company. A delta-to-wye transformer connection delivers the same voltage output as the wye-to-wye transformer connection, thought it differs from a wye-to-wye connection in that the primary is supplied from a delta system. A wye-to-delta transformer connection delivers the same voltage output as the delta-to-delta transformer connection, though it differs from a delta-to-delta connection in that the primary is supplied from a wye system.

Single-Phase Transformer Parallel Connections

Additional power is required when the capacity of a transformer is insufficient for the power requirements of the load(s). Additional power may be obtained by changing the overloaded transformer to a larger size (higher kVA rating) or adding a second transformer in parallel with the overloaded transformer. The best and most efficient method is to replace the overloaded transformer with a larger one. However, in some applications, it is easier to add a second transformer in parallel. These include systems where extra power is needed only temporarily or a larger transformer is not available.

Single-phase transformers may be connected in parallel as long as certain conditions are met. In order to connect 1ϕ transformers in parallel, the following conditions must be met:

• Primary and secondary voltage ratings are identical.

• Frequencies are the same.

• Tap settings are identical.

• Impedance of either transformer is within ±7% (93% to 107%) of the other. **See Figure 28-23.**

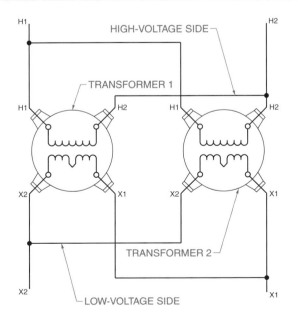

SINGLE-PHASE TRANSFORMER PARALLEL CONNECTIONS

Figure 28-23. As long as certain conditions are met, 1ϕ transformers may be connected in parallel.

The total power rating of two compatible 1ϕ transformers connected in parallel is equal to the sum of the individual power ratings. To calculate the total power rating of two 1ϕ transformers connected in parallel, the following formula is applied:

$$kVA_T = kVA_1 + kVA_2$$

where

kVA_T = total rating of transformer combination (in kVA)

kVA_1 = rating of transformer 1 (in kVA)

kVA_2 = rating of transformer 2 (in kVA)

Example: Calculating Total Power Rating— Parallel-Connected 1ϕ Transformers

What is the total output rating of two compatible 1ϕ, 5 kVA transformers connected in parallel?

$$kVA_T = kVA_1 + kVA_2$$
$$kVA_T = 5 + 5$$
$$kVA_T = \textbf{10 kVA}$$

Three-Phase Transformer Parallel Connections

Similar to 1ϕ transformers, 3ϕ transformers may also be connected in parallel. In order to connect 3ϕ transformers in parallel, the following conditions must be met:

• Primary and secondary voltage ratings are identical.

• Frequencies are the same.

• Tap settings are identical.

• Impedance of either transformer is within ±7% (93% to 107%) of the other.

• Angular displacement of the transformer banks is the same. For example, both banks must have a 0°, 30°, or 180° angular displacement. Standard angular displacements are 0° for wye-to-wye or delta-to-delta connected banks and 30° for wye-to-delta or delta-to-wye connected banks. **See Figure 28-24.**

Calculating the total power rating of two compatible 3ϕ transformers connected in parallel is similar to calculating the power rating of two compatible 1ϕ transformers connected in parallel. The total power rating of two compatible 3ϕ transformers connected in parallel equals the sum of the individual power ratings (kVA).

THREE-PHASE TRANSFORMER PARALLEL CONNECTIONS

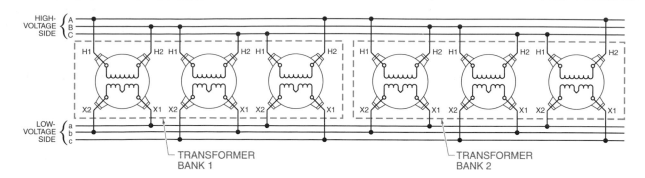

Figure 28-24. Like 1φ transformers, 3φ transformers may also be connected in parallel as long as certain conditions are met.

Transformer Load Balancing

The loads connected to a transformer should be connected so that the transformer is as electrically balanced as possible. Electrical balance occurs when loads on a transformer are placed so that each coil of the transformer carries the same amount of current. **See Figure 28-25.**

Tech Fact

Alternating current produced at generating plants is transformed to a higher voltage to allow efficient transmission of electrical power between power stations and end users. Changes that an electric utility makes to power delivery can affect the operation of in-plant transformers. For example, a new area substation can boost the delivered voltage. New factories may increase the local load and decrease the voltage available.

TRANSFORMER LOAD BALANCING

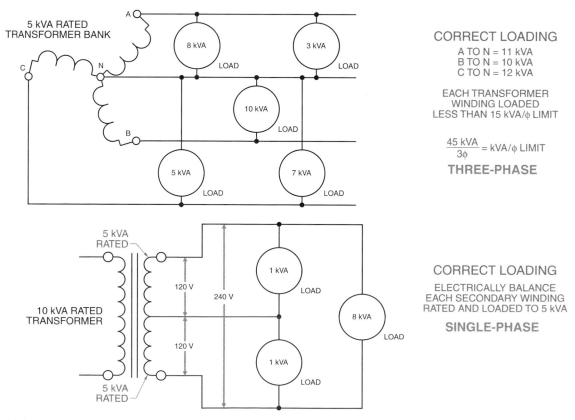

Figure 28-25. The loads connected to a transformer should be connected so that the transformer is as electrically balanced as possible.

Transformer Installation

Extreme care must be taken when work is performed around transformers because of the high voltage present. The proper protective equipment must be used and all plant safety procedures must be followed. All transformer installations should follow National Electrical Code (NEC®) and manufacturer requirements. For example, NEC® Article 450.21 covers the installation of indoor dry-type transformers based on their kVA and voltage ratings. **See Figure 28-26.** These requirements protect persons working around the transformer, help prevent fires, and reduce the chance of an electrical shock.

Conductor Identification and Color Coding

Conductors (wires) are covered with an insulating material that is available in different colors. The advantage of using different colors on conductors is that the function of each conductor can be easily determined. Some colors have a definite meaning. For example, the color green always indicates a conductor used for grounding. Other colors may have more than one meaning depending on the circuit. For example, a red conductor may be used to indicate a hot wire in a 230 V circuit or switched wire in a 115 V circuit. **See Figure 28-27.** Conductor color coding makes balancing loads among the different phases easier and helps during troubleshooting.

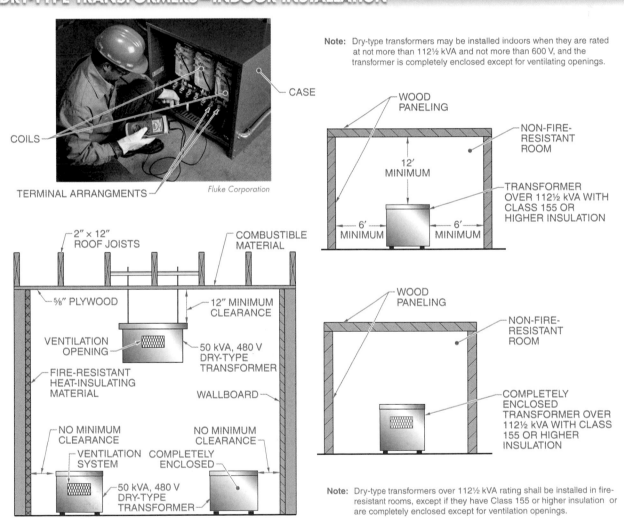

DRY-TYPE TRANSFORMERS—INDOOR INSTALLATION

Figure 28-26. All transformer installations should follow NEC® and manufacturer requirements.

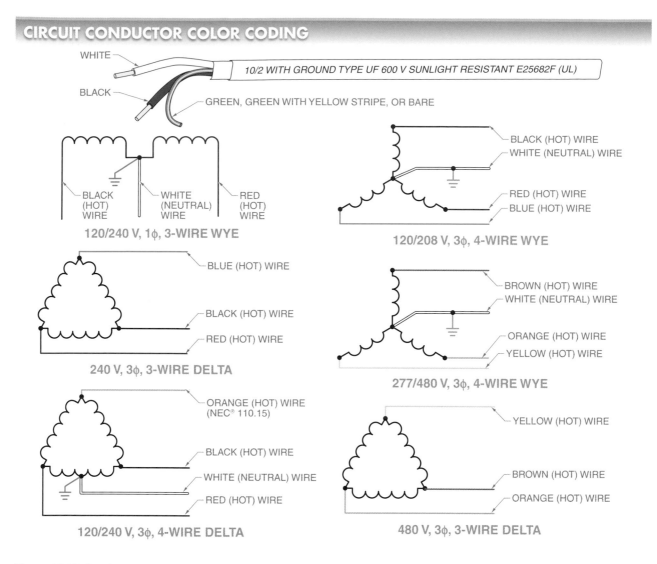

Figure 28-27. Conductors are covered with an insulating material that is available in different colors to enable easy identification of the function of each conductor.

Harmonic Distortion and Nonlinear Loads

Harmonic distortion is related to wave shape. Nonlinear loads create waves with frequencies that are multiples of the basic system frequency. For example, the second harmonic frequency of a 60 Hz sine wave is 120 Hz, the third harmonic frequency is 180 Hz, and so on. These higher-frequency harmonic components superimpose on the fundamental frequency, distorting the waveform. **See Figure 28-28.** The higher-frequency components are called harmonics. The change of a wave caused by these harmonics is called harmonic distortion.

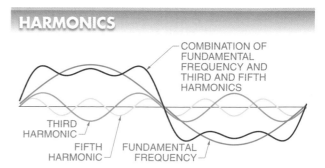

Figure 28-28. When harmonics combine with the fundamental frequency, the resulting distorted waveform creates a power quality problem.

Wave quality can be affected by the presence of a wide range of disturbances throughout the transmission and distribution network. One of the most important factors causing these disturbances is the connection of nonlinear loads such as variable-frequency drives and electronically switched power supplies. **See Figure 28-29.** Harmonics cause extra heat in motors and transformers. They can also sometimes create audible noise.

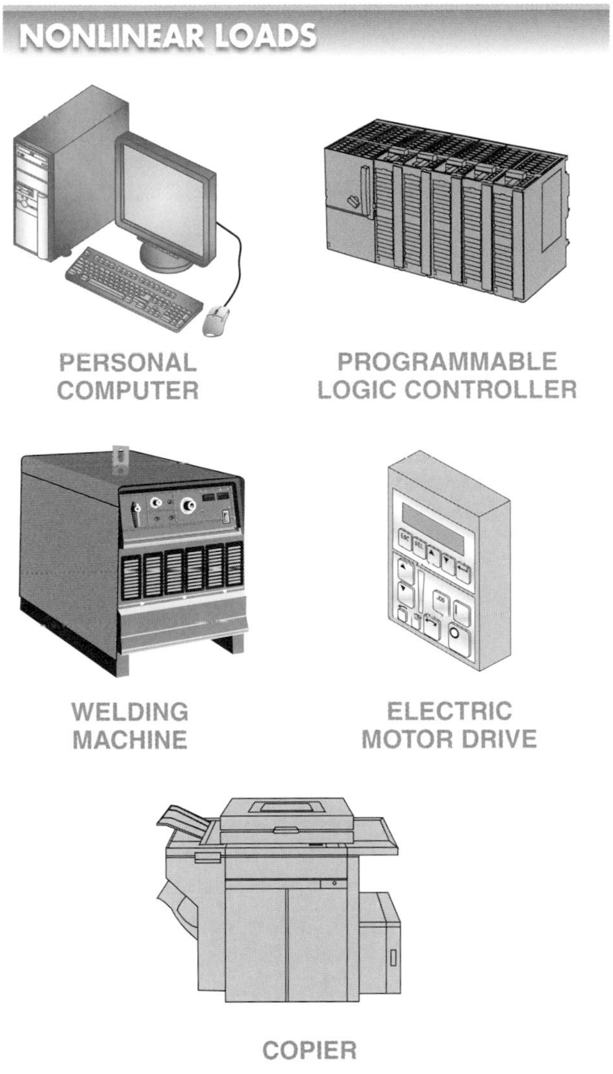

NONLINEAR LOADS

PERSONAL COMPUTER

PROGRAMMABLE LOGIC CONTROLLER

WELDING MACHINE

ELECTRIC MOTOR DRIVE

COPIER

Figure 28-29. In nonlinear loads, current is not a pure proportional sine wave because current is drawn in short pulses.

Harmonic Filters

To reduce harmonic distortion in power distribution systems and circuits, a harmonic filter can be connected to the system. **See Figure 28-30.** A *harmonic*

filter is a device used to reduce harmonic frequencies and harmonic distortion in a power distribution system. Three-phase harmonic filters are installed between the transformer and the distribution panel. Harmonic filters should be installed as close as possible to nonlinear loads such as large motor drives. In most systems, a large system harmonic filter is installed in the main service panel.

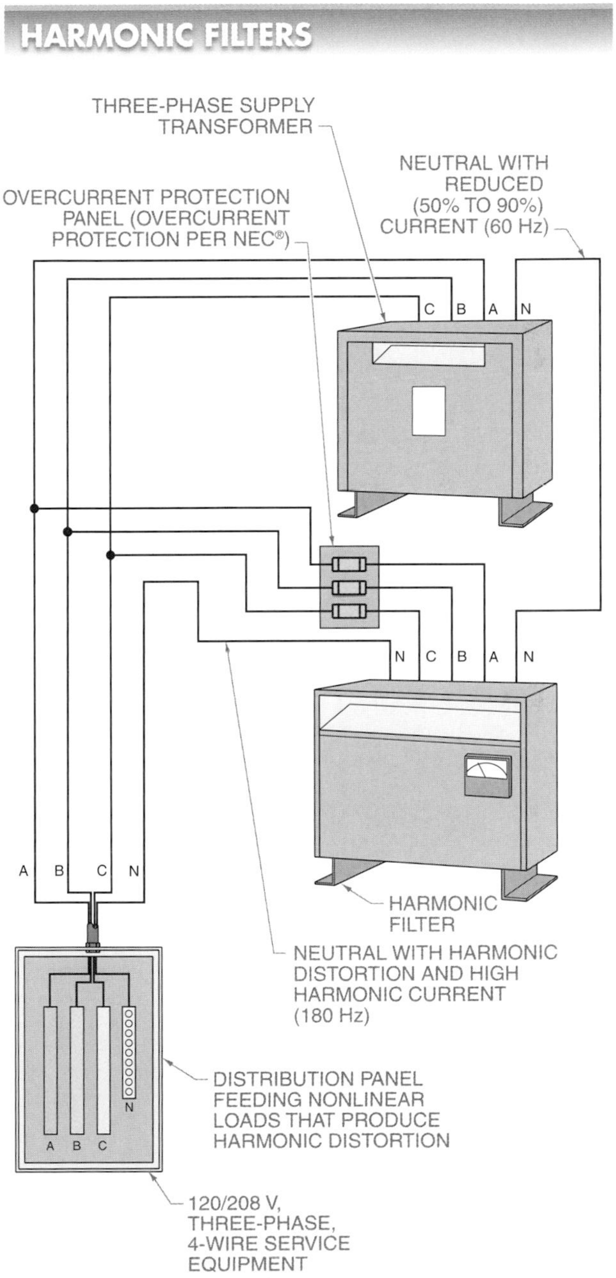

HARMONIC FILTERS

THREE-PHASE SUPPLY TRANSFORMER

OVERCURRENT PROTECTION PANEL (OVERCURRENT PROTECTION PER NEC®)

NEUTRAL WITH REDUCED (50% TO 90%) CURRENT (60 Hz)

HARMONIC FILTER

NEUTRAL WITH HARMONIC DISTORTION AND HIGH HARMONIC CURRENT (180 Hz)

DISTRIBUTION PANEL FEEDING NONLINEAR LOADS THAT PRODUCE HARMONIC DISTORTION

120/208 V, THREE-PHASE, 4-WIRE SERVICE EQUIPMENT

Figure 28-30. Three-phase harmonic filters are installed between the transformer and distribution panel to reduce harmonic frequencies and total harmonic distortion.

28-2 CHECKPOINT

1. What type of transformer connection ties the ends of each of the three 1φ transformers together?

2. What type of transformer connection ties the end of one of the 1φ transformers to the beginning of the next and so on until the three transformers are connected in parallel?

3. If the phase-to-phase voltage is 208 V and the phase-to-neutral voltage is 120 V, what type of 3φ transformer bank is the voltage being supplied from?

4. If the phase-to-phase voltage is 230 V and the phase-to-neutral voltage is 115 V, what type of 3φ transformer bank is the voltage being supplied from?

28-3 POWER QUALITY PROBLEMS

Correcting power quality problems requires an understanding of all power distribution system components and how a problem in one component can cause problems in other components. Distribution system components include transformers, distribution lines, switchboards, panelboards, disconnects, circuit breakers, fuses, and receptacles to deliver, control, and protect the system. Electrical power distribution systems must deliver quality power to loads if the loads are to operate properly for their rated lifetime and performance.

Quality power is power delivered to a load that is within the load specified voltage, is capable of delivering enough current under any operating condition, and includes minimal, not damaging, changes. Poor quality power is power delivered to a load that includes excessive or damaging changes such as voltage drops, voltage unbalance, voltage fluctuations, current unbalance, transients, and harmonic distortion. **See Figure 28-31.**

Single Phasing

Single phasing is the operation of a motor that is designed to operate on three phases but is only operating on two phases because one phase is lost. Single phasing occurs when one of the three lines leading to a 3φ motor does not deliver voltage to the motor. Single phasing is the maximum condition of voltage unbalance in a distribution system.

Single phasing occurs when one phase opens on either the primary or secondary power distribution system. A phase opens when one fuse blows, when there is a mechanical failure within the switching equipment, or

when lightning takes out one of the lines. Single phasing can go undetected in some systems because a 3φ motor running on two phases can still run in low torque applications. A motor that is single phasing will draw all of its current from two lines.

Voltage measurements taken at a motor do not normally indicate a single-phasing condition. The open winding in the motor generates voltage almost equal to the phase voltage that is lost. In this case, the open winding acts as the secondary of a transformer, while the two windings connected to power act as the primary.

Fluke Corporation
All components of an electrical system must be properly maintained in order to deliver quality power.

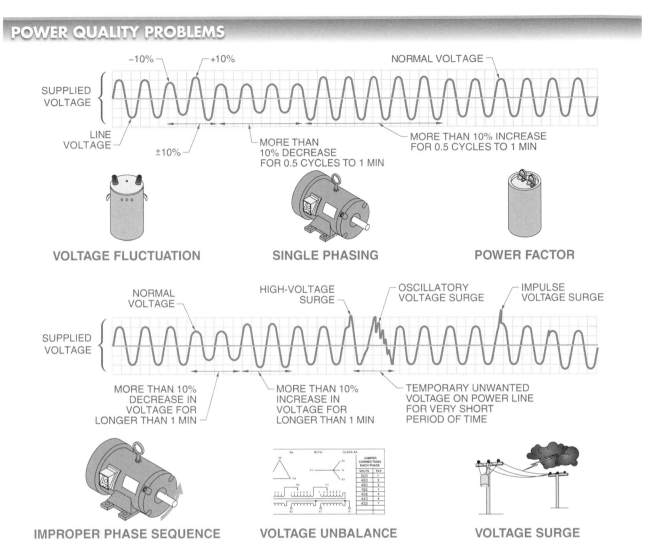

POWER QUALITY PROBLEMS

Figure 28-31. Power quality problems can damage electrical equipment and lead to unsafe operating conditions.

Single phasing is reduced through the use of properly sized dual-element fuses and heaters. An electronic phase-loss monitor is used to detect phase loss in motor circuits and other types of circuits in which a single-phasing condition cannot be allowed to exist for even a short period of time. The monitor activates a set of contacts to drop out the starter coil when a phase loss is detected.

The severe blackening of one delta winding or two wye windings of the three 3φ windings will be observable when a motor has failed due to single phasing. The coil or coils that experienced the voltage loss will be in the best condition. The damage occurs in the other coils because of overcurrent. **See Figure 28-32.**

Single phasing is distinguished from voltage unbalance by the severity of the damage. Voltage unbalance causes less blackening (but over more windings, normally)

than single phasing and little or no distortion. Single phasing causes burns and distortion to one winding.

Improper Phase Sequence

Improper phase sequence is the changing of the sequence of any two phases (phase reversal) in a 3φ motor circuit. Improper phase sequence reverses the motor rotation. Reversing motor rotation can damage driven machinery or injure personnel. Phase reversal can occur when modifications are made to a power distribution system or when maintenance is performed on electrical conductors or switching equipment. The NEC® requires phase reversal protection on all personnel transportation equipment such as moving walkways, escalators, and ski lifts. **See Figure 28-33.**

SINGLE-PHASE MOTOR DAMAGE

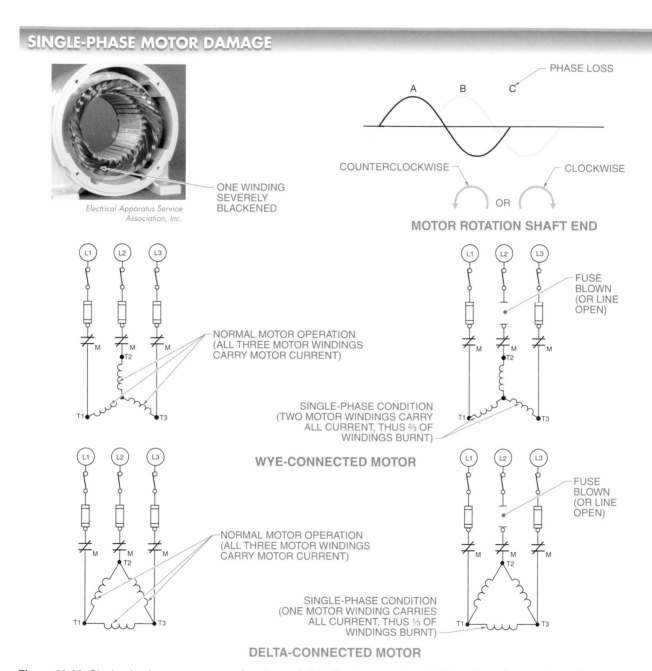

Electrical Apparatus Service Association, Inc.

ONE WINDING SEVERELY BLACKENED

PHASE LOSS

A B C

COUNTERCLOCKWISE OR CLOCKWISE

MOTOR ROTATION SHAFT END

L1 L2 L3

NORMAL MOTOR OPERATION (ALL THREE MOTOR WINDINGS CARRY MOTOR CURRENT)

M M M
T2
T1 T3

L1 L2 L3

FUSE BLOWN (OR LINE OPEN)

M M M
T2
T1 T3

SINGLE-PHASE CONDITION (TWO MOTOR WINDINGS CARRY ALL CURRENT, THUS ⅔ OF WINDINGS BURNT)

WYE-CONNECTED MOTOR

L1 L2 L3

NORMAL MOTOR OPERATION (ALL THREE MOTOR WINDINGS CARRY MOTOR CURRENT)

M M M
T2
T1 T3

L1 L2 L3

FUSE BLOWN (OR LINE OPEN)

M M M
T2
T1 T3

SINGLE-PHASE CONDITION (ONE MOTOR WINDING CARRIES ALL CURRENT, THUS ⅓ OF WINDINGS BURNT)

DELTA-CONNECTED MOTOR

Figure 28-32. Single-phasing causes severe burning and distortion to one or two windings depending on the configuration.

IMPROPER PHASE SEQUENCE

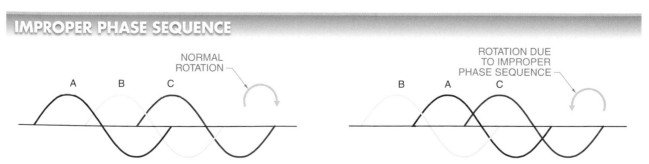

NORMAL ROTATION

A B C

ROTATION DUE TO IMPROPER PHASE SEQUENCE

B A C

Figure 28-33. Improper phase sequence is the changing of the sequence of any two phases (phase reversal) in a 3φ motor circuit.

Phase Unbalance

Phase unbalance is the unbalance that occurs when power lines are out of phase. Phase unbalance of a 3ϕ power system occurs when 1ϕ loads are applied, which causes one or two of the lines to carry more or less of the load. An electrician balances the load of a 3ϕ power system during the installation process. A power quality meter can be used to check phase unbalance on power lines. An unbalance begins to occur when additional 1ϕ loads are added to the system. This unbalance causes the 3ϕ lines to move out of phase so the lines are no longer 120 electrical degrees apart. **See Figure 28-34.**

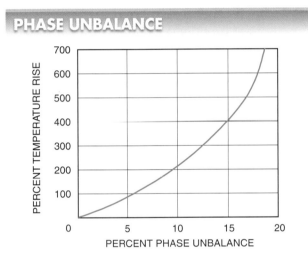

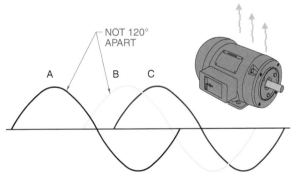

Figure 28-34. Phase unbalance is the unbalance that occurs when power lines are out of phase.

Phase unbalance causes 3ϕ motors to run at temperatures higher than their listed ratings. The greater the phase unbalance, the greater the temperature rise. High temperatures cause insulation breakdown and other related problems. A 3ϕ motor operating in an unbalanced circuit cannot deliver its rated horsepower. For example, a phase unbalance of 3% causes a motor to work at 90% of its rated power. This requires the motor to be derated. **See Figure 28-35.**

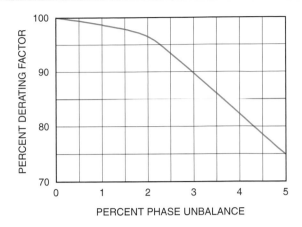

Figure 28-35. A motor operating on a circuit that has phase unbalance must be derated.

AC Voltage Variations

Motors are rated for operation at specific voltages. Motor performance is affected when the supply voltage varies from a motor's rated voltage. A motor operates satisfactorily with a voltage variation of ±10% from the voltage rating listed on the motor nameplate. **See Figure 28-36.**

VOLTAGE VARIATION CHARACTERISTICS		
Performance Characteristics	10% Above Rated Voltage	10% Below Rated Voltage
Starting current	+10% to +12%	−10% to −12%
Full-load current	−7%	+11%
Motor torque	+20% to +25%	−20% to −25%
Motor efficiency	Little change	Little change
Speed	+1%	−1.5%
Temperature rise	+3%C to −4%C	+6%C to +7%C

Figure 28-36. A motor operates satisfactorily with a voltage variation of ±10% from the voltage rating listed on the motor nameplate.

AC Frequency Variations

Motors are rated for operation at specific frequencies. Motor performance is affected when the frequency varies from a motor's rated frequency. A motor operates satisfactorily with a frequency variation of ±5% from the frequency rating listed on the motor nameplate. **See Figure 28-37.**

FREQUENCY VARIATION CHARACTERISTICS

Performance Characteristics	5% Above Rated Frequency	5% Below Rated Frequency
Starting current	−5% to −6%	+5% to +6%
Full-load current	−1%	+1%
Motor torque	−10%	+11%
Motor efficiency	Slight increase	Slight increase
Speed	+5%	−5%
Temperature rise	Slight decrease	Slight decrease

Figure 28-37. A motor operates satisfactorily with a frequency variation of ±5% from the frequency rating listed on the motor nameplate.

DC Voltage Variations

DC motors should be operated on pure DC power. *Pure DC power* is power obtained from a battery or DC generator. DC power is also obtained from rectified AC power. Most industrial DC motors obtain power from a rectified AC power supply. DC power obtained from a rectified AC power supply varies from almost pure DC power to half-wave DC power.

DC motor operation is affected by a change in voltage. The change may be intentional, as in a speed-control application, or it may be caused by variations in the power supply. The power supply voltage normally should not vary by more than 10% of a motor's rated voltage. Motor speed, current, torque, and temperature are affected when the DC voltage varies from the motor rating. **See Figure 28-38.**

Voltage Surges

A *voltage surge* is a higher-than-normal voltage that temporarily exists on one or more power lines. Lightning is a major cause of large voltage surges. A lightning surge on a power line comes from a direct lightning hit or induced voltage. The lightning energy moves in both directions on the power lines, much like a rapidly moving wave.

A traveling surge of lightning energy causes a large voltage rise in a short period of time. The large voltage is impressed on the first few turns of the motor windings, destroying the insulation and burning out the motor. An electrician will be able to observe the burning and opening of the first few turns of the windings that occur when a motor has failed due to a voltage surge. The rest of the windings will appear normal, with little or no damage. **See Figure 28-39.**

Lightning arresters with the proper voltage rating and connection to an excellent ground ensure maximum voltage surge protection. Surge protectors are also available. Surge protectors are placed on equipment or throughout the distribution system.

Voltage surges can also occur due to the normal switching of high-power circuits. Voltage surges that occur due to the switching of high-power circuits are of lesser magnitude than lightning strikes and normally do not cause motor problems. A surge protector should be used on computer equipment circuits to protect sensitive electronic components.

Tech Fact

Never assume that phase A, phase B, and phase C are the same throughout a structural distribution system. Instead, a phase sequence test instrument is used to identify which lines are powered and which power lines are phase A, phase B, and phase C.

DC MOTOR PERFORMANCE CHARACTERISTICS

Performance Characteristics	Voltage 10% Below Rated Voltage		Voltage 10% Above Rated Voltage	
	Shunt	Compound	Shunt	Compound
Starting current	−15%	−15%	+15%	+15%
Full-load current	−5%	−6%	+5%	+6%
Motor torque	+12%	+12%	−8%	−8%
Motor efficiency	Decreases	Decreases	Increases	Increases
Speed	Increases	Increases	Decreases	Decreases
Temperature rise	Increases	Increases	Decreases	Decreases

Figure 28-38. Motor speed, current, torque, and temperature are affected if the DC voltage varies from the motor rating.

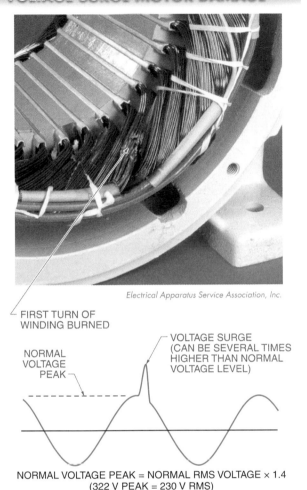

VOLTAGE SURGE MOTOR DAMAGE

Electrical Apparatus Service Association, Inc.

FIRST TURN OF
WINDING BURNED

VOLTAGE SURGE
(CAN BE SEVERAL TIMES
HIGHER THAN NORMAL
VOLTAGE LEVEL)

NORMAL
VOLTAGE
PEAK

NORMAL VOLTAGE PEAK = NORMAL RMS VOLTAGE × 1.4
(322 V PEAK = 230 V RMS)

Figure 28-39. A voltage surge causes burning and opening of
the first few turns of the windings.

Voltage Unbalance

Voltage unbalance, also known as voltage imbalance, is
the unbalance that occurs when voltages at the terminals
of an electric motor or other 3ϕ load are not equal. Voltage unbalance causes motor windings to overheat, resulting in thermal deterioration of the windings. When a 3ϕ
motor fails due to voltage unbalance, one or two of the
stator windings become blackened. **See Figure 28-40.**

The problem with voltage unbalance within a power
distribution system is that a small amount of voltage
unbalance can cause a high current unbalance in loads
such as electric motors. In general, voltage unbalance
should not be more than 1%. Whenever there is a 2% or
greater voltage unbalance, corrective action should be
taken. This may include repositioning loads to balance

the current draw on the three power lines if the problem
is within the building. The problem is within the building if the unbalance deteriorates when loads are ON and
improves when loads are OFF. If the unbalance is at the
main power entrance at all times, the problem is most
likely with the utility system and the utility company
should be notified.

VOLTAGE UNBALANCE

LOW
VOLTAGE

A   B   C

UNBALANCED
VOLTAGE

ONE OR TWO
BLACKENED
WINDINGS

*Electrical Apparatus
Service Association, Inc.*

Figure 28-40. Voltage unbalance within a power distribution
system can cause high current unbalance in loads such as
electric motors.

*Understanding how a power distribution system works is important
when troubleshooting problems and maintaining operation.*

Voltage unbalance can be determined through the use of a test instrument or meter. First, the meter is set to measure AC voltage for an AC circuit or DC voltage for a DC circuit. Next, the meter is tested on a known energized source to verify that the meter is in proper working condition before measurements are taken. Then, to find voltage unbalance, the following procedure is applied:

1. Measure the voltage between each incoming power line. The readings are taken from L1 to L2, L1 to L3, and L2 to L3. **See Figure 28-41.**

2. Add the voltages.

3. Find the voltage average by dividing the sum of the voltages by 3.

4. Find the voltage deviation by subtracting the voltage average from the voltage with the largest deviation.

5. Find voltage unbalance by applying the following formula:

$$V_u = \frac{V_d}{V_a} \times 100$$

where

V_u = voltage unbalance (in %)
V_d = voltage deviation (in V)
V_a = voltage average (in V)
100 = percentage

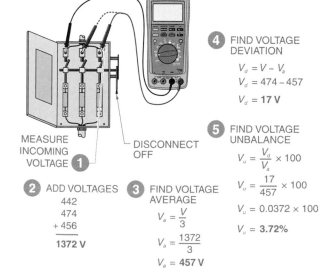

Figure 28-41. Voltage unbalance is the unbalance that occurs when the voltages at different motor terminals are not equal.

Note: The proper PPE for the location and type of test must always be worn. It must be verified that the test instrument (DMM or voltage tester) has the minimum CAT III rating for the measuring location.

Example: Calculating Voltage Unbalance

Calculate the voltage unbalance of a feeder system with the following voltage readings: L1 to L2 = 442 V, L1 to L3 = 474 V, and L2 to L3 = 456 V.

1. Measure the voltage between each incoming power line. Incoming voltages are 442 V, 474 V, and 456 V.

2. Add the voltages.

442 V + 474 V + 456 V = 1372 V

3. Find the voltage average.

$$V_a = \frac{V}{3}$$

$$V_a = \frac{1372}{3}$$

$$V_a = 457 \text{ V}$$

4. Find the voltage deviation.

$$V_d = V - V_a$$
$$V_d = 474 - 457$$
$$V_d = 17 \text{ V}$$

5. Find voltage unbalance.

$$V_u = \frac{V_d}{V_a} \times 100$$

$$V_u = \frac{17}{457} \times 100$$

$$V_u = 0.0372 \times 100$$

$$V_u = \mathbf{3.72\%}$$

An electrician will be able to observe the blackening of one delta stator winding or two wye stator windings that occurs when a motor has failed due to voltage unbalance. The winding with the largest voltage unbalance will be the darkest.

Current Unbalance

Current unbalance, also known as current imbalance, is the unbalance that occurs when the currents on each of the three power lines of a 3ϕ power supply are not equal. Current unbalance from overloading one or two of the 3ϕ power lines can cause voltage unbalances. This can cause voltage unbalances on all loads connected within a building. A 2% voltage unbalance can cause an 8% or higher current unbalance.

Current unbalances should not exceed 10%. Any time current unbalance exceeds 10%, the system should be tested for voltage unbalance. Likewise, any time a voltage unbalance is more than 1%, the system should be tested for a current unbalance. Current unbalance is determined in the same manner as voltage unbalance, except that current measurements are used. **See Figure 28-42.**

Current unbalance can be determined through the use of a test instrument or meter. First, the meter is set to measure AC current for an AC circuit or DC current for a DC circuit. Next, the meter is tested on a known energized source to verify that the meter is in proper working condition before measurements are taken. Then, to find the percentage of current unbalance in a circuit, the following procedure is applied:

1. Measure current on each of the incoming power lines.

2. Add all current values together.

3. Calculate the current average by taking the sum of current measurements and dividing by the number of measurements taken.

4. Calculate the largest current deviation by subtracting the lowest current measurement from the current average.

5. Calculate the current unbalance by dividing the largest current deviation by the current average and multiplying by 100.

Note: The proper PPE for the location and type of test must always be worn. It must be verified that the test instrument (clamp-on ammeter or DMM with current clamp-on attachment) has the minimum CAT III rating for the measuring location.

DETERMINING CURRENT UNBALANCE

POWER QUALITY CABINET CONTAINING TRANSFORMERS, REACTORS, POWER FACTOR CORRECTION CAPACITORS, AND FILTERS

ASI Robicon

DISCONNECT

Siemens

L1/R
L2/S
L3/T

MEASURE INCOMING CURRENT ON EACH POWER LINE ①

② ADD CURRENT VALUES

58
53
+ 57
—————
I=168 A

③ FIND CURRENT AVERAGE

$I_a = \dfrac{I}{3}$

$I_a = \dfrac{168}{3}$

$I_a = 56 \text{ A}$

④ FIND LARGEST CURRENT DEVIATION

$I_d = I - I_a$

$I_d = 56 - 53$

$I_d = 3 \text{ A}$

⑤ FIND CURRENT UNBALANCE

$I_u = \dfrac{I_d}{I_a} \times 100$

$I_u = \dfrac{3}{56} \times 100$

$I_u = 0.0535 \times 100$

$I_u = 5.35\%$

Note: Perform test measurement on known energized source to verify meter working condition.

Figure 28-42. Current unbalance is determined in the same manner as voltage unbalance, except that current measurements are used.

Example: Calculating Current Unbalance

Calculate the current unbalance of a feeder system with the following current readings: L1 = 58 A, L2 = 53 A, and L3 = 57 A.

1. Measure the current for each incoming power line. Incoming currents are 58 A, 53 A, and 57 A.

2. Add the currents.

   $$58 \text{ A} + 53 \text{ A} + 57 \text{ A} = 168 \text{ A}$$

3. Find the current average.

   $$I_a = \frac{I}{3}$$
   $$I_a = \frac{168}{3}$$
   $$I_a = 56 \text{ A}$$

4. Find the current deviation.

   $$I_d = I - I_a$$
   $$I_d = 56 - 53$$
   $$I_d = 3 \text{ A}$$

5. Find current unbalance.

   $$I_u = \frac{I_d}{I_a \times 100}$$
   $$I_u = \frac{3}{56 \times 100}$$
   $$I_u = 0.0535 \times 100$$
   $$I_u = \mathbf{5.35\%}$$

Power Factor

A utility and/or customer can increase the available power by correcting the system's power factor. Poor power factor is caused by motor loads that cause current to lag voltage. Poor power factor can be improved by adding power factor-correcting capacitor banks in the utility's distribution system and/or in the customer's facility. Utilities add capacitor banks to their distribution system to reduce the amount of apparent power and increase the amount of true power. Large power-consuming commercial and industrial plants add capacitor banks to improve their power factor and reduce the power factor penalty imposed by utilities. **See Figure 28-43.**

POWER FACTOR CORRECTION

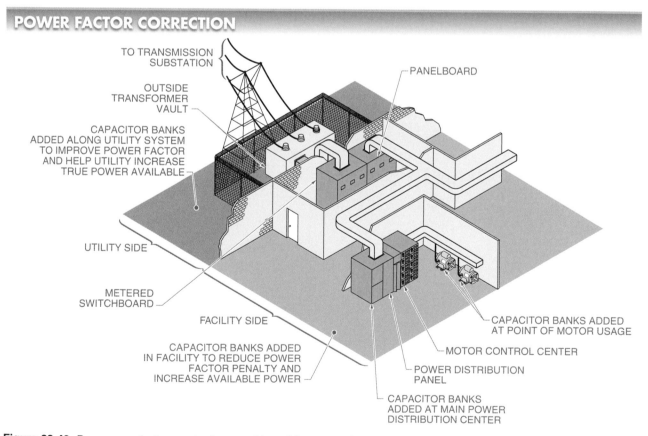

Figure 28-43. Poor power factor can be improved by adding power-factor-correcting capacitor banks in the utility distribution system and/or customer facility.

Theoretically, capacitors should improve the power factor to 1 (100%). In practical use however, capacitors are used to correct the power factor to approximately 95%. The addition of excessive capacitance into a circuit causes the voltage to leak current and cause poor power factor because power factor is less than 1 (100%) any time voltage and current are out of phase. Improving power factor reduces the electric bill and increases system capacity. **See Figure 28-44.**

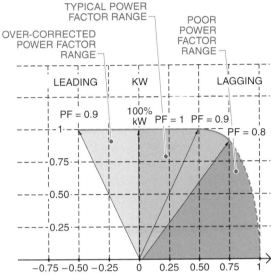

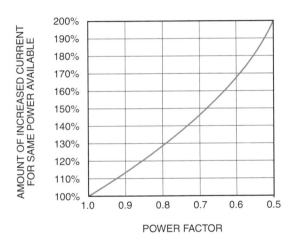

Transmission lines carry electrical power from distribution substations to step-down transformers.

Figure 28-44. Improving power factor reduces the electric bill when a power factor penalty is imposed and increases the system capacity.

28-3 CHECKPOINT

1. If one of the three coil windings inside a motor is heavily blackened, what is the most likely cause of the problem?

2. What type of system protection is required on all personnel transportation equipment?

3. Voltage unbalance should not exceed what percent?

4. Current unbalance should not exceed what percent?

28-4 SMART GRIDS

An *electrical grid* is a network that delivers electrical power from the power plants where it is generated to customers. **See Figure 28-45.** An electrical grid includes wires, substations, transformers, and switches. A *smart grid* is an electrical grid that uses computer-based remote control and automation to deliver electrical power from where it is generated to customers. In order to improve the delivery of electrical power, the continual developments in smart grid technology can be used to make a power distribution system more intelligent, efficient, and secure. A smart grid normally includes the following upgrades:

- two-way communication between devices and locations from generation to consumption
- more efficient transmission of power between generation and consumption
- capability of the system to meet peak demand capacity by automatically controlling when some power-consuming devices may be temporarily removed from the grid
- the ability to automatically redirect power sources as needed
- constant monitoring of the individual and total systems for consumption, which allows the utility and customer instant access to consumption information at their locations
- increases to the total system efficiency and security to ensure the delivery of clean, safe, quality power
- the ability to automate problem recognition for issues that occur in the power grid and to correct these issues to reduce or eliminate outages

Upgrading the Electrical Grid

Upgrades to the electrical grid allow each critical component, such as substations and transformers, to have two-way communication. In this case, each device or location would be capable of calling for assistance if it were experiencing problems. It could also be called back to see the extent of the problem. For example, if a critical component such as a transformer was overloaded or overheating, a message would go out to a monitoring station. When this information is received, the power could be rerouted until the transformer is inspected or replaced. The monitoring station could also remotely call up the transformer and, with proper diagnostic software, determine its current condition and efficiency.

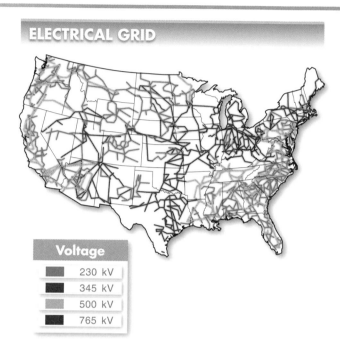

Figure 28-45. An electrical grid is a network that delivers electrical power from where it is generated to its point of use.

Two-way communication also allows one or more power sources on the generation side of the electrical grid to be accessed remotely and routed to a location needed by the customer, depending on the type and amount of power needed. The power may come from a centralized power plant, wind farm, or photovoltaic (PV) array.

Traditional centralized power distribution systems are aging and cannot keep up with the high demand being placed on the systems. Distributed power generation, interactive distributed generation, and microgrids are used to provide alternatives to aging centralized power distribution systems and relieve some of the stress of high demand.

Distributed Power Generation

Distributed power generation is the use of small-scale power generation technologies located close to the loads that are being served. Distributed power generation systems can include PV arrays, wind turbines, biodiesel generators, and other relatively small-scale power systems. **See Figure 28-46.** A distributed power generation system may serve as the only source of power for residential, commercial, industrial, or remotely accessible areas. A distributed power generation system may be a stand-alone system or combined with a centralized power distribution system.

DISTRIBUTED POWER GENERATION

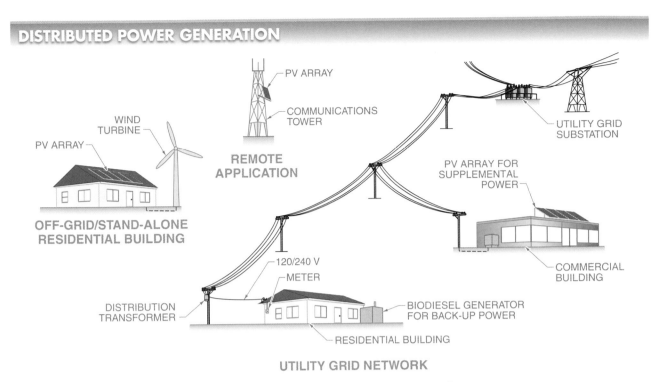

Figure 28-46. Distributed power generation produces electricity close to where it is used.

Interactive Distributed Generation

Interactive distributed generation systems are also small power generating systems. These systems must be interactive to connect to the utility's distribution grid for two-way power. **See Figure 28-47.** Interactive distributed generation of electrical power is increasingly common as a supplement to traditional centralized power generation. This increases the diversity and security of the electrical energy supply and benefits both customers and electric utilities.

For customers, these systems can provide power to on-site loads and back up their stand-alone systems in the event of a utility power outage. For utilities, the additional power sources supplying capacity during peak loads increases the utility's capacity to serve customers without the need to build new power plants.

For example, back-up generators are generators that are placed in a fixed location and connected to the power distribution system through a manual or automatic transfer switch. When utility power is out, the back-up generator provides power without the need to unplug devices from receptacles and plug them into the generator. Backup generators normally supply power only to the loads that must have power during a power outage.

A transfer switch detects when utility power has been removed, disconnects the utility distribution system

from the loads, and connects the generator to the loads. Transfer switches and generator control circuits can be manually operated or completely automatic, starting the generator and automatically controlling the power as needed. **See Figure 28-48.**

Distributed power generation systems such as wind turbines can be used as stand-alone systems or connect to a centralized power distribution system.

INTERACTIVE DISTRIBUTED GENERATION

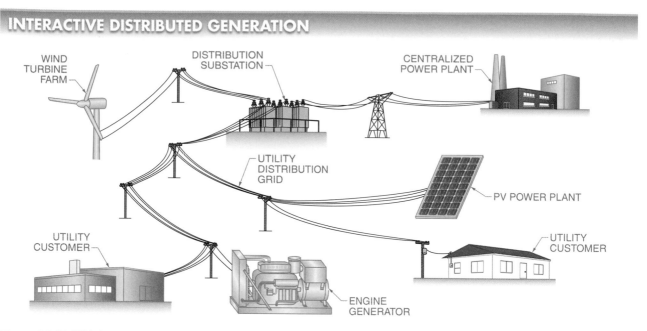

Figure 28-47. With interactive distributed generation, utility customers are served by both the centralized power plant and the power exported from interconnected distribution generators.

AUTOMATIC TRANSFER SWITCH CONNECTIONS

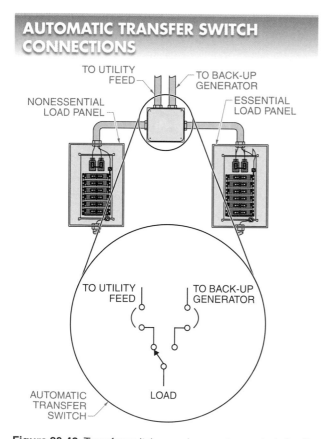

Figure 28-48. Transfer switches and generator control circuits can be manually operated, or they can be totally automatic, starting the generator and automatically controlling the power as needed.

Microgrids

Microgrids are small-scale versions of a centralized power distribution system. Like the centralized power distribution system, microgrids generate, distribute, and regulate electrical power to the customer. Smart microgrids can also be a reliable way to integrate renewable resources on a local level. A microgrid can also function as a stand-alone system. **See Figure 28-49.** Most microgrids are used for industrial, commercial, and institutional applications. These systems use diesel and natural gas-powered cogeneration power plants for most of their electrical power.

MICROGRIDS

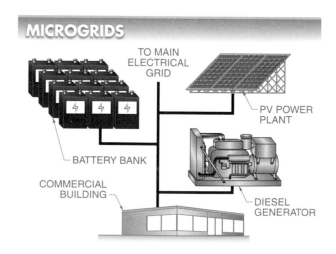

Figure 28-49. Microgrids can use renewable energy or a diesel generator to function as stand-alone systems.

Microgrids may also include the ability to use renewable energy sources such as wind and solar energy, which can be isolated from the larger grid in case the primary power source is unavailable. This ensures the operation of critical infrastructure during an emergency or power outage. When a microgrid is disconnected from the main distribution network, it is said to be operating in "island mode" during faults in the main grid. Microgrid installations can be found in a variety of locations such as rural electrification systems for remote communities to theme parks and small cities.

DC Microgrids. Traditionally, microgrids are based on AC current. However, the increased use of DC current in personal devices, business equipment, and industrial processes has caused manufacturers to rethink how these devices receive their primary power. In the case of solar power, the output is already DC, which is converted to AC and then back to DC after distribution. This conversion involves more equipment and results in substantial energy loss.

One method to minimize energy conversion losses that is being pursued is the use of a DC microgrid to directly distribute primary power to a device or throughout a home, business, or industrial complex. The net result will be to minimize energy conversion losses. Further research and development of DC microgrid technology will lead to more cost-effective uses of electrical energy.

Advanced Metering Infrastructure (AMI)

Advanced metering infrastructure (AMI) consists of new sensors, communication networks, and data management systems that are used to modernize the electrical grid and provide new capabilities to utilities and customers. An AMI system is composed of sensors, smart meters, meter data management (MDM) systems, wide-area communication, and home area networks (HANs).

Smart meters equipped with power quality monitoring capabilities enable more rapid detection, diagnosis, and resolution of power quality problems. AMI provides for self-healing by helping outage management systems detect and locate failures quickly and accurately. Power generation and storage options at the customer level can be monitored and controlled. AMI data provides the information needed to greatly improve asset management and operations.

Before AMI, utilities typically learned of outages when customers reported problems. Often repair crews were dispatched only to find problems were on the customer's side of the meter. An AMI solution provides automatic notification of outages. For example, smart meters alert the utility when a customer loses power. This helps pinpoint outages quickly. An AMI system also accurately records and sends messages on behalf of each meter and sensor when power is restored so utilities can see whether outages have been restored.

AMI implementation also reduces the need for everyday meter readings since readings are taken automatically and regularly. Meter technicians used to be the first defense against energy theft. New systems fill this void with built-in analysis tools to detect possible theft or tampering. By analyzing real-time data and comparing it with historical trends from the same meter or similar customers, MDM systems can identify patterns likely to suggest theft or tampering. The system then automatically generates work reports for field managers to investigate. MDM systems can be linked with supervisory control and data acquisition (SCADA) management systems.

Phasor Measurements

As more power is delivered through the current electrical grid, renewable energy sources such as solar energy and wind are raising concerns on how to maintain electrical grid balance and stability. Balance can only be maintained and power quality increased through the precise measurement of voltage and current waveforms at multiple points on the grid. Phasor measurement is a technology that can help maintain stability in the power grid.

A *phasor measurement unit (PMU)* is a device that measures electrical waveforms on the electrical grid. The unit of measure is the phasor. PMUs are also referred to as synchrophasors. PMUs provide information to operators and planners to measure the state of the electrical grid and manage power quality.

PMUs are precise grid measurement devices, taking measurements as often as 30 times per second. SCADA data is only taken around 4 times per second. PMU measurements are time stamped to a common time reference provided through GPS tracking satellites. Coordinated Universal Time (UTC) time stamps allow PMU data from different utilities to be synchronized and combined together to give a comprehensive view of all utilities effected. PMUs provide more precise information for analysis.

The use of this type of measuring system creates an accurate and reliable source for locating line faults. For example, with time of signal travel, the signal sent down the power line can be used to pinpoint faults to within a few feet. A similar technique is used to determine faults in a fiber optic cable by sending light through the cable and determining the time of the return signal from the problem.

Real-Time Monitoring and Sensing

Real-time monitoring and sensing is an essential component of a smart grid. The power industry has only recently begun to move to real-time monitoring systems to provide up-to-date information using two-way communication. At the utility level advancements are occurring more rapidly on the transmission side as compared to the distribution side.

The next-generation networked sensors have measuring and processing capabilities that help locate a faulted line and identify parts of the grid that might be susceptible to outages before they occur. For example, next-generation networked sensors can highlight power fluctuations that might be the result of a tree limbs contacting a power line.

Advantages of real-time monitoring and sensors include the following:

- outage detection and notification
- reduced energy costs
- power quality throughout the grid
- meter tampering and energy theft
- deterrence of organized attacks on the grid
- more targeted and efficient maintenance programs
- vegetation intrusion on power lines
- reduced power-delivery energy losses
- power factor correction
- minimization of environmental impact

Another move the power industry has embraced is employing digital electronics in metering. At the customer level electronic metering is still in development. One of the concerns of introducing new devices is interoperability and the idea that certain technologies may be discarded if different technologies became standard. It is certain, however, that in the future no meters will be based on electromechanical technology. Instead, meters will be digital and will communicate with both the customer and utility. This technology will also use the Internet. **See Figure 28-50.**

Harvesting Energy for Grid Sensors

For sensors on the smart grid, the power may not be available across the entire power distribution system. This is true on many transmission towers. Batteries are often used to offset the lack of power. However, batteries eventually need to be recharged. To meet this challenge, a new power method called power harvesting is being introduced as part of the emerging smart grid technology. *Power harvesting* is the process of obtaining power from the surrounding environment. **See Figure 28-51.** The primary sources of power harvesting are solar energy, vibration, magnetic fields, thermal (heat) energy, and radio frequency (RF) waves.

New power sources are essential for the widespread use of sensors throughout the smart grid. Many challenges face these new technologies. For example, a few of these challenges include the energy storage of solar-based sensors for use at night, the ability to charge batteries in subzero temperatures, and the useful life of rechargeable batteries.

Digital Protective Devices

Microprocessor-based digital relays provide metering data, status information, and fault location, in addition to protection functions. The data is accessed through relay communication ports, local displays, or other human-machine interfaces (HMIs). Many individuals within a utility organization use the data. For example, operators may need to know the fault location for a particular electrical disturbance. Fault type and fault location, unavailable until the creation of fault-locating digital relays, are now required by most operating and dispatch centers to guide system restoration.

Distribution substations are located in close proximity to where the electrical service is provided.

REAL-TIME MONITORING

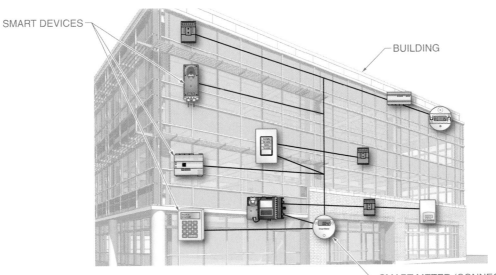

Figure 28-50. Smart meters allow for remote control by owners and utilities.

POWER HARVESTING

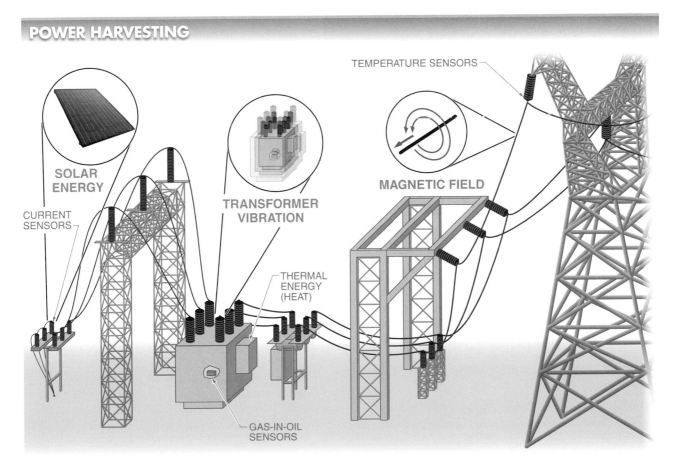

Figure 28-51. Power harvesting is the process of obtaining power from the surrounding environment and using it to power sensors.

Digital Relays. Digital relays use microprocessors to measure power (currents and voltage) and process this information through their internal logic to control the operation of a circuit breaker. A digital relay is also referred to as a microprocessor-based relay. A digital relay consists of sensors that monitor voltage and current and input this information into a data acquisition module that converts the signal from analog to digital output. **See Figure 28-52.** The digital information is then analyzed by the data-processing module to determine what outputs are needed. These outputs can be used to activate the digital relay or send feedback information to the operator on how the relay is performing. Most digital relays have automated self-test functions that verify the correct operation of the relay. When a self-test detects a problem, a message indicating some type of failure is transmitted.

point on the power waveform to generate a low-current pulse in the power line. Then the pulse closer analyzes the pulse to determine whether the contacts should reclose or remain open. Pulse closing technology is a breakthrough in overhead distribution system protection up to 27 kV. Pulse closing is a superior alternative to conventional reclosing. It significantly reduces stress on system components and improves power quality by reducing the voltage sags experienced by customers upstream of a fault.

Pulse closers are completely self-contained. **See Figure 28-53.** Embedded voltage and current sensors provide 3ϕ monitoring of line current and 3ϕ monitoring of line voltage on both sides of each interrupter. The units provide a complete set of protection and control functions, with comprehensive diagnostics.

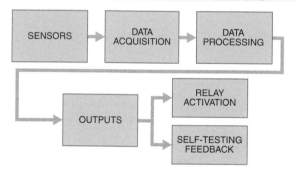

Figure 28-52. Digital relays process information received from sensors and determine an appropriate output to control a circuit breaker.

Digital relays allow the utility to change protection settings and logic based on the time of day and day of the week. Protection requirements may change as the system load changes. Original settings are usually quite sensitive. Other settings may be less sensitive and tolerate higher loading. Other settings may be engaged when the digital relay triggers. When conditions return to normal for a period of time, the digital relay returns to its original settings.

Pulse Closing Technology. *Pulse closing technology* is a unique means for verifying that the line is clear of faults before power interrupters are reclosed. Pulse closing technology works by rapidly closing and reopening the power interrupter contacts at a certain

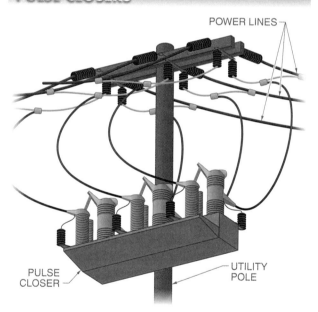

Figure 28-53. Pulse closers verify that a power line is clear of faults before reclosing power interrupters.

The communication module can provide a short-distance, secure Wi-Fi communication link to a nearby personal computer. A secure Wi-Fi connection can be used with a laptop computer to quickly and easily access and transfer detailed logs, geographical data, and other operational information over the wireless link. A global positioning system (GPS) chip set in the communication module provides 1 ms accurate time-stamping of events to speed event analysis. It also provides location data.

With RF communications, system operation checks, diagnostics, downloads, and even uploads of new software can be accomplished. This communication system eliminates costly field visits to the equipment after it has been installed.

A memory module, installed in the base, backs up configuration data and site-specific information such as the device identifier, sensor calibration data, and operation counter reading. This system also allows passwords and security clearances to be established.

28-4 CHECKPOINT

1. What device detects when utility power is lost, disconnects the utility lines, and reconnects the loads to a backup power supply?
2. What measurement device measures the electrical waveforms on the distribution system to provide operating information to the operators?
3. What are some examples of power harvesting used to operate smart devices without using batteries?
4. Why is pulse closing technology used?

Additional Resources

Review and Resources

Access Chapter 28 Review and Resources for *Electrical Motor Controls for Integrated Systems* by scanning the above QR code with your mobile device.

Applying Your Knowledge

Refer to the *Electrical Motor Controls for Integrated Systems* Learner Resources for interactive Applying Your Knowledge questions.

Workbook and Applications Manual

Refer to Chapter 28 in the *Electrical Motor Controls for Integrated Systems Workbook* and the *Applications Manual* for additional exercises.

ENERGY EFFICIENCY PRACTICES

Power Monitoring

Producing, transmitting, and using electrical energy incurs cost. A part of this cost at every stage is the cost of wasted energy that is consumed and not used to produce any useful work such as light or rotary motion. In the past, wasted energy was considered part of a system and its cost was built into the price. Today, the trend is to charge inefficient energy users more and efficient energy users less. This is possible because the usage of energy can be monitored and recorded at every step from production to its final usage in residential, commercial, and industrial locations.

Through the use of monitoring equipment along the system that monitors and transmits real-time operating data from every point, utilities can maximize the produced energy and minimize wasted energy. Monitoring and controlling the system allows utilities to produce less energy and reduce the cost to customers. For example, utilities that are allowed to switch on loads at nonpeak times, such as late night or early morning, can charge customers less than if the loads are turned on during peak usage times.

Likewise, large commercial and industrial customers that become part of the energy-saving programs can save energy and reduce cost. For example, as poor power factor is improved and high peak demands are reduced, a lower rate is applied to the electric bill. Poor power factor can be improved by using some synchronous motors instead of all induction motors and power factor correction capacitors. High peak demands can be reduced by reducing HVAC systems and other nonproduction loads during high production times.

Objectives

29-1
- Define preventive maintenance (PM) and describe the types of maintenance work.
- Define work order and describe the different types of work.
- Define and describe inventory control and logbook.

29-2
- Define and describe alignment and misalignment.
- Describe the common causes of misalignment.
- Explain how to adjust and align a machine.

29-3
- Define bearing and describe the different types of bearings.
- Describe how to remove bearings.
- Describe how to install bearings.

29-4
- Define and describe flexible belt drives.
- Describe pulley misalignment.
- Explain how to adjust belt tension.

29-5
- Explain how to test failing insulation.
- Explain how to perform an insulation spot test.
- Explain how to perform a dielectric absorption test.
- Explain how to perform an insulation step voltage test.

Review and Resources
atplearningresources.com/Quicklinks
Access Code: 362245

Chapter
Preventive
Maintenance Systems

Electrical equipment, circuits, and systems are designed to last for their rated life expectancy, provided they are properly installed and maintained. However, environmental conditions such as dirt, high and low temperatures, and moisture can contribute to a shorter operating life. Operational conditions such as misalignment, loose or tight belts, poor coupling, etc. can also contribute to a shorter operating life. In order to determine potential problems before they cause damage or shorten operating life, steps should be taken to prevent or minimize existing or potential problems. Understanding how to identify, document, and correct or prevent possible problems should be a part of any maintenance program.

29-1 PREVENTIVE MAINTENANCE PRINCIPLES

Preventive maintenance (PM) is maintenance performed to keep machines, assembly lines, production operations, and plant operations running with little or no downtime. Preventive maintenance is a combination of unscheduled and scheduled work required to maintain equipment in peak operating condition. Preventive maintenance increases efficiency, reduces cost, and minimizes health and safety problems. Preventive maintenance can also be used to document compliance with environmental, health, and safety regulations. In addition, data pertaining to maintenance costs, parts, time, and breakdowns can be used to make purchase decisions.

Types of Maintenance Work

Maintenance work is commonly divided into the general categories of facilities maintenance and industrial

maintenance. *Facilities maintenance* is maintenance performed on systems and equipment in hotels, schools, office buildings, and hospitals. Facilities maintenance includes work performed on heating, ventilating, and air conditioning (HVAC) equipment, fire protection systems, and security systems. *Industrial maintenance* is maintenance performed on production systems and equipment in industrial settings. Industrial maintenance includes work performed in food processing plants, parts manufacturing plants, foundries, timber and pulp mills, mines, and other industrial settings.

The primary difference between facilities maintenance and industrial maintenance is downtime cost. In facilities maintenance, downtime cost is normally less than in industrial maintenance. For example, if an HVAC system in an office building fails, workers may become uncomfortable but normally continue to work until the system is repaired.

Downtime cost is much higher in an industrial setting because production is affected. Production is either stopped or substandard products are produced when equipment malfunctions or fails in an industrial setting. A malfunction or failure in an industrial setting must be corrected immediately to reduce downtime cost.

Some settings may require a combination of facilities maintenance and industrial maintenance. For example, the health and safety concerns of patients in a health care facility require a combination of facilities maintenance and industrial maintenance.

Work Orders

Preventive maintenance requires a consistent, accurate flow of information. Preventive maintenance may use a computerized maintenance management system (CMMS) or a paper-based system. Both systems can be purchased as a package or developed in-house. The preventive maintenance system selected depends on the operating budget, plant size, personnel considerations, and the objective of the system. Large facilities require a CMMS. Small facilities can be managed from paper-based systems or a basic CMMS.

A *work order* is a document that details work required to complete specific maintenance tasks. Work orders are also used to organize, schedule, and monitor work tasks. Work orders commonly include the time, date, name of equipment, area where the equipment is located, work description, approximate time to complete the work, and safety requirements. Some work orders also list the steps for completing the task. Work orders can be generated using paper forms or a computerized maintenance management system (CMMS). **See Figure 29-1.**

Work Priority. *Work priority* is the order in which work is done based on its importance. The most important work is done first, followed by less important work. Work priority is indicated on the work order. Work priority methods vary from plant to plant. A three-level work priority system is commonly used. The first priority is work relating to safety, downtime, and production efficiency. The second priority is periodic maintenance. The third priority is long-term projects. Some plants identify work priority by completion time as emergency, within two weeks, or timely. **See Figure 29-2.**

The size of a plant and the number of available personnel dictate work priority procedures. For example, some plants list preventive maintenance tasks as the highest priority, with only designated workers responding to emergency calls. Regardless of work priority procedures, a maintenance technician must always be alert for signs of potential maintenance problems. For example, a change in the sound of a motor may indicate potential failure. The smell of hot electrical insulation requires immediate investigation. A glance at gauges can identify a problem.

WORK ORDERS

PRIORITY ☐ IMMEDIATE ☐ 24 HOURS ☐ TIMELY	MAINTENANCE REQUEST FORM		WO #: ______ REQUESTED BY: ______
DATE:	TIME: am pm	AREA:	EQUIPMENT:
ASSIGNED TO:		WORK DESCRIPTION:	
SAFETY ☐ TAGOUT REQUIRED ☐ LOCKOUT REQUIRED ☐ HOTWORK PERMIT REQUIRED ☐ CONFINED SPACE ENTRY PERMIT REQUIRED			
START DATE ______			
START TIME ______			
COMPLETION DATE ______			
COMPLETION TIME ______			
TECHNICIAN INITIALS ______		☐ DOWNTIME: __ DAYS __ HOURS __ MONTHS ☐ PARTS ORDERED - REQUISITION #:	

Figure 29-1. A work order generated using paper forms or a computerized maintenance management system includes information required to complete specific maintenance tasks.

Unscheduled Maintenance. *Unscheduled maintenance* is unplanned service performed by a maintenance technician that includes emergency work and breakdown maintenance. *Emergency work* is work performed to correct an unexpected malfunction on equipment that has received some scheduled maintenance. Emergency work orders are issued to repair damaged equipment immediately. **See Figure 29-3.** Keeping a log of emergency work provides information that can improve maintenance procedures or equipment design by identifying common equipment problems.

Breakdown maintenance is service on failed equipment that has not received scheduled maintenance such as cleaning or lubrication. For example, light bulbs are serviced on breakdown maintenance because it is less costly to replace a bad bulb than to predict a bulb failure by testing. Breakdown maintenance is the least sophisticated maintenance work and normally is used only on equipment that is inexpensive and noncritical to plant operations. However, if applied to the wrong equipment, breakdown maintenance can be the most expensive maintenance work. For example, centrifugal pump bearings should be maintained and replaced according to manufacturer recommendations. Excessively worn bearings allowed to operate until failure can result in costly shaft assembly replacement and/or pump damage.

WORK PRIORITIES

MAINTENANCE REQUEST

Please type or print legibly.

Date: 11/11

Originator: Chris Williams   Location: Packing Area   Phone: 1556

Action Needed: ____ Emergency _____ Within 2 weeks ✓ Timely _____ Complete By _____

If cannot complete by _____ contact originator.

Description of work: Fluorescent lights flickering above packing area. Some lights are out completely.

Figure 29-2. Work priority on a work order dictates which maintenance task must be completed first.

EMERGENCY WORK ORDERS

EMERGENCY WORK ORDER   No. 821463

MACHINE Box Former   Date 3/12

Time Called 2:10   (AM) PM   Mechanical ☒

Time Arrived 2:15   (AM) PM   Employee ☐

Time Finished 2:22   (AM) PM   Other ☐

Machine Downtime 12 min

Production Time Lost 12 min

Number of Employees Affected by Shutdown _____

Reason for Stoppage Vacuum grip failed due to bad solenoid. Loose connection. Grip also out of place.

Maintenance Technician Jean Smith

Production Supervisor Gerald Brown

Figure 29-3. Emergency work orders are issued to repair damaged or malfunctioning equipment immediately.

Scheduled Maintenance. *Scheduled maintenance* is work that is planned and scheduled for completion. Scheduled maintenance is performed to minimize emergency work and ensure reliable and efficient equipment operation to maintain required quality standards. Scheduled maintenance work includes periodic maintenance, corrective work, and project work.

Periodic maintenance includes tasks completed at specific time intervals to prevent breakdowns and production inefficiency. Periodic maintenance is scheduled based on time intervals, such as daily, weekly, monthly, or quarterly, or it is based on hours of equipment operation. **See Figure 29-4.** Periodic maintenance tasks commonly include the following:

- inspection of equipment for conditions, such as unusual noise, leaks, or excessive heat, that indicate potential problems
- lubrication of equipment at scheduled intervals
- adjustments and equipment parts replacement to maintain proper operating condition
- checks of the electrical, hydraulic, and mechanical systems of operating equipment

Periodic work orders specific to one piece of equipment or several pieces of equipment are scheduled at specific intervals throughout the year. If CMMS software is used, a master schedule can be created with work orders projected automatically by day, week, month, or year. **See Figure 29-5.**

Corrective work is repair of a known problem before a breakdown occurs. Corrective work is requested, discovered during periodic inspections, or discovered while performing other maintenance tasks. Data that includes work completed, supplies used, cause of problem, costs, and time for completion is recorded when corrective work is completed.

Project work is work on long-term projects that require advanced planning and more time than typical maintenance tasks. Project work commonly includes rebuilding or modifying equipment, renovating structures, or installing new equipment.

Vibration analysis is a type of periodic maintenance used to analyze the condition of equipment.

PERIODIC MAINTENANCE

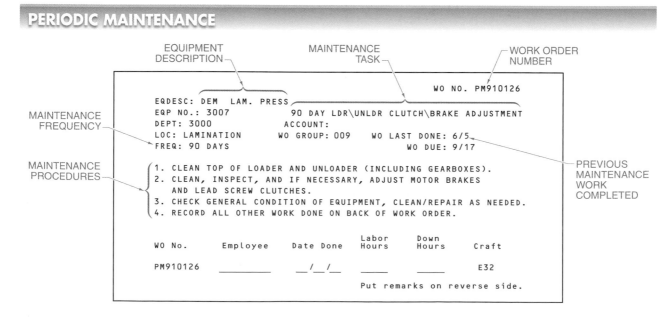

Figure 29-4. Periodic maintenance consists of tasks scheduled at specific time intervals to maintain equipment.

COMPUTERIZED MAINTENANCE MANAGEMENT SYSTEMS (CMMSs)

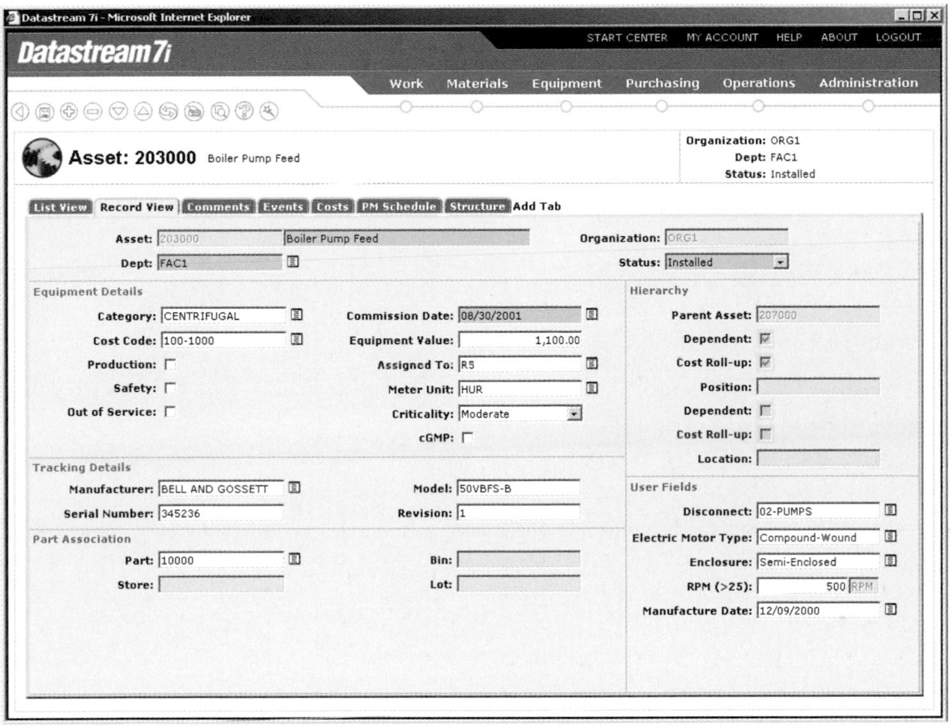

Datastream Systems, Inc.

Figure 29-5. CMMS software can be used to develop a master schedule with work orders projected automatically by day, week, month, or year.

Inventory Control

Inventory control is the organization and management of commonly used parts, vendors and suppliers, and purchasing records in a preventive maintenance system. **See Figure 29-6.** In some systems, replacement parts are scanned with a barcode reader and recorded in a computer. In an inventory control system, a part is assigned a number that is printed in computer code on a stick-on label. A barcode reader is passed over the label when a part is removed. The inventory control computer records that one part has been removed and subtracts one part from the total number to keep an up-to-date record of parts available. The computer can also be used to issue a purchase order for the number of parts required when supplies are close to depletion.

Logbooks

A *logbook* is a book or electronic file that documents all work performed during a shift and lists information needed to complete work by maintenance technicians on other shifts. **See Figure 29-7.** Maintenance technicians should begin each day by reviewing the logbooks from previous shifts. Based on information in the log and the quantity and type of work orders issued, the maintenance technician makes a list of all work to be completed. The work is then prioritized as required. Emergency or high-priority work requests are handled as they occur during a shift.

Maintenance technicians are responsible for a wide range of integrated industrial systems and are expected to perform multiple maintenance tasks. Competent technicians must understand the operating principles of various systems. Fundamental maintenance procedures and practices must be understood and followed in a systematic manner. Preventive maintenance consists of proper alignment of machine and motor shafts, anchoring of motors and machinery, maintaining bearings, maintaining flexible belt drives, and performing preventive maintenance tests.

INVENTORY CONTROL

Hand Held Products

Figure 29-6. Inventory control requires the organization and management of parts commonly used for maintenance tasks.

LOGBOOKS

MAINTENANCE LOG

DATE: _3/5_   NAME: _Pat Williams_   SHIFT: _4 PM to 12 PM_

TIME	TASK	COMMENTS
1. 4-6	PM of oven #3, conveyors #5, 6, 7, & 8, pan washer.	No unusual conditions.
2. 6-7	Assisted Jean with mixer #6 motor replacement.	Motor overheated when mixer locked up.
3. 7-7:20	Dinner	
4. 7:20-9:30	Repair chain guard on conveyor #2 as per work order 35-556.	Used portable welder. Reorder 6010 welding rod.
5. 9:30-12	Repaired pie filler with Jean & Chris.	Retimed conveyor and pie trays with pie indexer. Worn parts. Ordered new conveyor chain, On-Off switch.

Graveyard Shift — Check timing of indexer.
Parts will arrive about 3 AM.
Filler keeps going out of time.
Check side mounted limit switch.

Figure 29-7. Logbooks are used by maintenance technicians to record tasks completed during a shift.

29-1 CHECKPOINT

1. Why is industrial maintenance generally more costly than facilities maintenance, even if the repair time and cost of repair components is the same?

2. What are two examples of unscheduled maintenance that might need to be performed?

29-2 ALIGNMENT

Alignment is the condition where the centerlines of two machine shafts are placed within specified tolerances. The objective of proper alignment is to connect two shafts under operating conditions so that all forces that cause damaging vibration between the two shafts and their bearings are minimized.

Misalignment is the condition where the centerlines of two machine shafts are not aligned within specified tolerances. Misalignment may be offset or angular. *Offset misalignment* is a condition where two shafts are parallel but are not on the same axis. *Angular misalignment* is a condition where one shaft is at an angle to the other shaft. Shaft misalignment is normally a combination of offset and angular misalignment. **See Figure 29-8.**

Properly aligned rotating shafts reduce vibration and add many years of service to equipment seals and bearings. As a rule of thumb, misalignment of a coupling by 0.0004″ can shorten its life by 50%. A *coupling* is a device that connects the ends of rotating shafts. Couplings require accurate alignment of the mating shafts. Couplings are classified as rigid or flexible and are the most common and least expensive method of connecting two shafts.

A *rigid coupling* is a device that joins two precisely aligned shafts within a common frame. A *flexible coupling* is a coupling within a resilient center, such as rubber or oil, that flexes under temporary torque or misalignment due to thermal expansion. Flexible couplings can allow enough vibration to cause excessive wear to seals and bearings. Where flexible couplings are used, shaft alignment should be as accurate as it would be if solid couplings were used. **See Figure 29-9.**

COUPLINGS

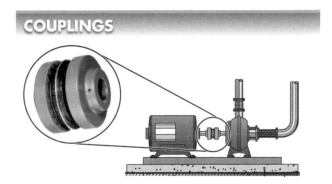

Figure 29-9. Couplings allow drive and driven equipment connection and provide protection against misalignment, vibration, and shock.

MISALIGNMENT

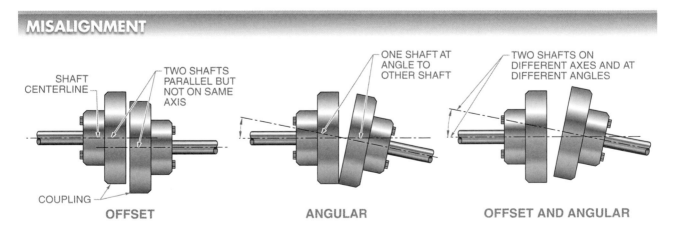

Figure 29-8. Misalignment can be offset or angular, though it is normally a combination of the two.

Machinery to be aligned that is connected to an electrical power supply must be locked out first. Before working on the equipment, the electrical functions should be challenged by testing the start switch. *Challenging* is the process of pressing the start switch of a machine to determine whether the machine starts when it is not supposed to start. Upon completion of the lockout challenge, all switches should be placed in the OFF position. All equipment energy sources must be controlled through lockout, tagout, and blockout procedures. **See Figure 29-10.**

Perfectly aligned motors connected to driven machines lose less than 1% of their transmitted torque and produces less noise and vibration than improperly aligned motors. Each additional degree of misalignment increases torque loss by 1% or greater, increases noise, and increases vibration.

Alignment of two machine shafts is accomplished by properly anchoring the machines. *Anchoring* is a means of fastening a machine securely to a base or foundation. Machine anchoring must consider piping and plumbing and the condition of anchoring components such as bolts and washers. Firm but adjustable anchoring of machines on a base plate is accomplished using proper mechanical fasteners, such as bolts, screws, and nuts. Improper anchoring includes bolt-bound machines, excessive bolt body or length, and improper washers that create a dowel effect. **See Figure 29-11.**

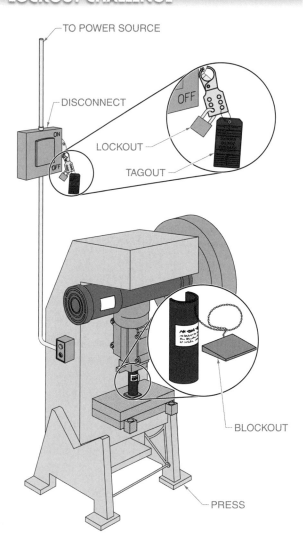

LOCKOUT CHALLENGE

TO POWER SOURCE

DISCONNECT

OFF

LOCKOUT

TAGOUT

ON

OFF

BLOCKOUT

PRESS

Figure 29-10. Equipment energy sources are controlled through lockout, tagout, and blockout procedures.

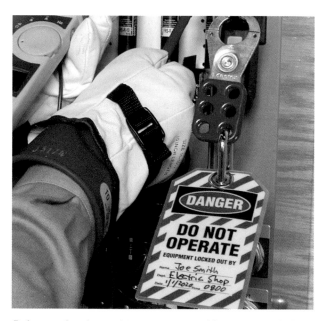

DANGER
DO NOT OPERATE
EQUIPMENT LOCKED OUT BY
Name Joe Smith
Dept. Electric Shop
Date 1/1/2020 Time 0800

Before performing maintenance on machinery, proper lockout/ tagout procedures must be used when disconnecting the power to the machinery.

Vibration from misaligned shafts has a direct effect on the operating costs of a plant. Misaligned shafts require more power, create premature seal damage, and cause excessive force on bearings. This leads to early bearing, seal, or coupling failure. Misalignment is generally caused by improper machine foundations, pipe strain, soft foot, or thermal expansion.

MACHINE ANCHORING

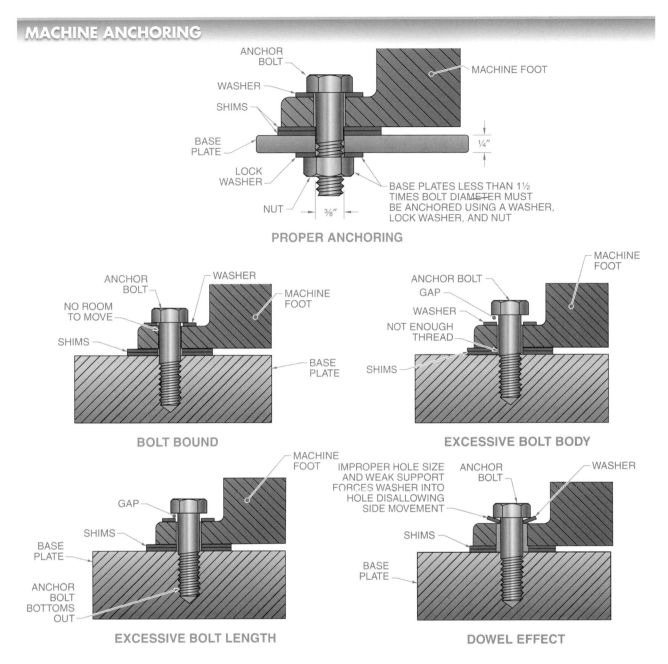

Figure 29-11. Firm but adjustable anchoring of machines on base plates is accomplished using proper mechanical fasteners, such as bolts, screws, and nuts.

Machine Foundations

Alignment of all equipment begins with the foundation and base plate to which the equipment is anchored. A *foundation* is an underlying base or support. A *base plate* is a rigid steel support for firmly anchoring and aligning two or more rotating devices. Foundations must be level and strong enough to provide support. The feet of machines must be checked for cracks, breaks, rust,

corrosion, or paint. Machines should be bolted to a base plate, not anchored to concrete. The contacting surfaces between machines and the base plate must be smooth, flat, and free of paint, rust, and foreign material. An adjustable base plate simplifies the installation, maintenance, and replacement of equipment. An *adjustable base plate* is a mounting base that allows equipment to be easily moved over a short distance. **See Figure 29-12.**

MACHINE FOUNDATIONS

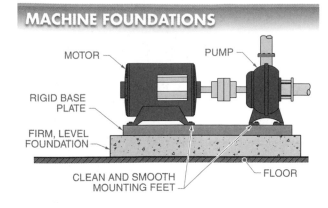

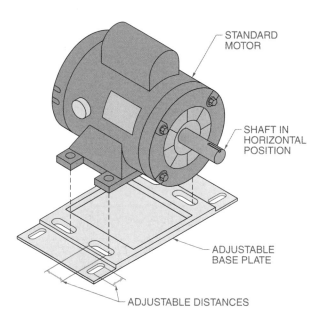

Figure 29-12. A clean, firm, and level machine base plate and foundation are required for proper alignment to ensure minimal flexing between machines.

Pipe Strain

Pipe and conduit connections can produce enough force to affect machine alignment if improperly installed. Thermal expansion created by the temperature of liquids and reaction forces from piped products can produce enough force to affect machine alignment. To ensure that any transmission of outside force does not affect the proper alignment of machines, machines should be initially aligned unattached from any piping if possible. Therefore, all plumbing must be properly aligned and have its own permanent support even when unattached. In some cases, flexible plumbing connections are necessary to separate stresses and vibrations between pump/motor and product lines. **See Figure 29-13.**

PIPING SUPPORTS

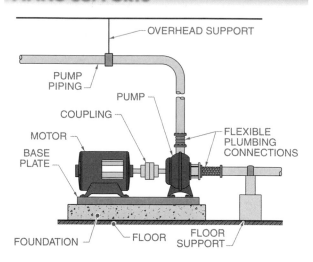

Figure 29-13. Pump piping must be independently supported to prevent angular forces from working against bearing and alignment tolerances.

Soft Foot

Soft foot is a condition that occurs when one or more machine feet do not make complete contact with a base plate. Distorted frames create internal misalignment due to soft foot, which is a major reason for bearing failure. The internal misalignment and distortion loads the bearings and deflects the shaft. Soft foot also creates difficulty in shaft alignment. Before aligning any machine that has soft foot, the machine must be shimmed for equal and parallel support on all feet.

As a soft-foot bolt is tightened, the shaft of the machine is deflected, which loads the bearings. The deflection can cause enough vibration and pressure to damage the bearings, seals, and shaft. A shaft that rotates at 1800 rpm in a machine with a soft-foot condition deflects 30 times each second (108,000 times per hour). Soft-foot tolerance must be within 0.002″ of shaft movement. Soft foot may be parallel, angular, springing, or induced. **See Figure 29-14.**

Thermal Expansion

Thermal expansion is a dimensional change in a substance due to a change in temperature. For proper alignment, two shafts must be on the same horizontal and vertical plane under operating conditions. However, there can be a significant change in physical dimensions when there is a temperature change from resting conditions to operating conditions and thermal expansion moves the load shaft.

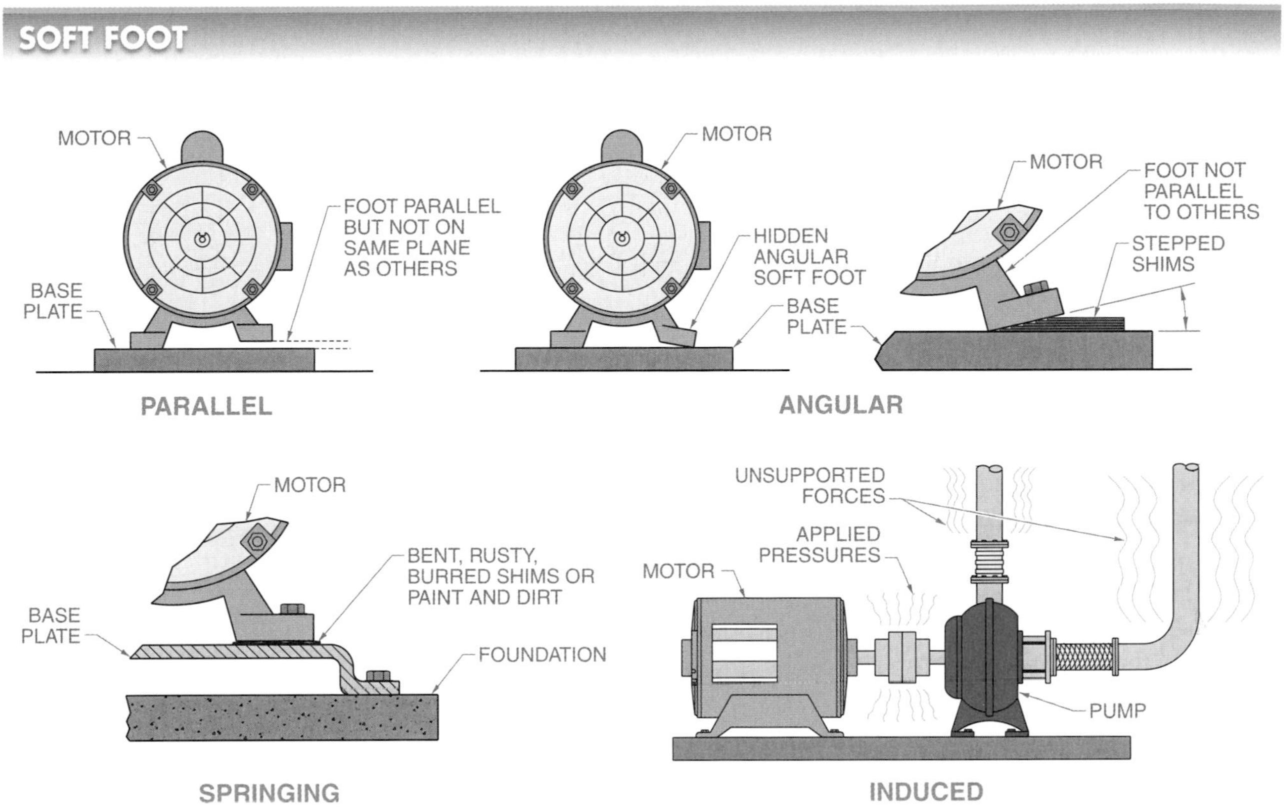

SOFT FOOT

PARALLEL

MOTOR
FOOT PARALLEL BUT NOT ON SAME PLANE AS OTHERS
BASE PLATE

ANGULAR

MOTOR
HIDDEN ANGULAR SOFT FOOT
BASE PLATE

MOTOR
FOOT NOT PARALLEL TO OTHERS
STEPPED SHIMS

SPRINGING

MOTOR
BENT, RUSTY, BURRED SHIMS OR PAINT AND DIRT
BASE PLATE
FOUNDATION

INDUCED

UNSUPPORTED FORCES
APPLIED PRESSURES
MOTOR
PUMP

Figure 29-14. Soft foot is a condition that occurs when one or more machine feet do not make complete contact with a base plate.

Temperature changes may be caused by the temperature of product being pumped, excessive room or ambient temperature, or heat from a loaded motor. Pump manufacturers often provide the amount of thermal expansion for a given temperature change.

Machine Adjustments

Shims are used as spacers between machine feet and the base plate. The feet of a machine must be firmly anchored to the base plate without creating excessive force or movement between mating shafts. It is rare for any machine to have all of its feet in contact with the base plate and also be within tolerance. Shims and spacers are used to adjust the height of a machine. Shim stock is manufactured in thicknesses ranging from 0.0005″ to 0.125″ and can be purchased as a sheet or roll in precut shapes. Spacers are used for filling spaces 0.25″ or greater.

One machine is normally chosen as a stationary machine (SM) and another machine is chosen as a machine to be shimmed (MTBS). It is important to choose the appropriate machine to shim and to use the proper tools

and components for accurate, fast, and damage-free alignment. The proper use of shim stock and spacers is required for machine adjustments. **See Figure 29-15.**

STATIONARY MACHINE SHIMS

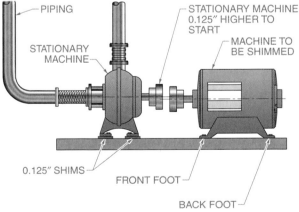

PIPING
STATIONARY MACHINE
STATIONARY MACHINE 0.125″ HIGHER TO START
MACHINE TO BE SHIMMED
0.125″ SHIMS
FRONT FOOT
BACK FOOT

Figure 29-15. A machine may be chosen as the machine to be shimmed (MTBS) because it is easier to move than the stationary machine (SM).

Normally the heaviest machine or the machine attached to plumbing is the SM. The motor is normally the MTBS. Regardless of which machine is shimmed, the SM must initially be higher than the MTBS to allow for proper vertical alignment. It is common practice to install the SM using 0.125″ shims under each foot. This practice requires raising the MTBS. However, it prevents any vertical movement requirements of the SM.

Alignment Methods

Five methods are available to align machinery, each having its own degree of accuracy. The five methods include straightedge, rim-and-face, reverse dial, electronic reverse dial, and laser rim-and-face methods. **See Figure 29-16.**

All alignment techniques require that a specific order of adjustments be made. The specific order of adjustments for shaft alignment is angular in the vertical plane (up and down angle), parallel in the vertical plane (up and down offset), angular in the horizontal plane (side to side angle), and parallel in the horizontal plane (side to side offset).

Once angular in the vertical plane and parallel in the vertical plane have been corrected, they generally are not lost when angular in the horizontal plane and parallel in the horizontal plane are in the process of being corrected. This step-by-step process is used regardless of the alignment method. Each corrective move should be double-checked for accuracy.

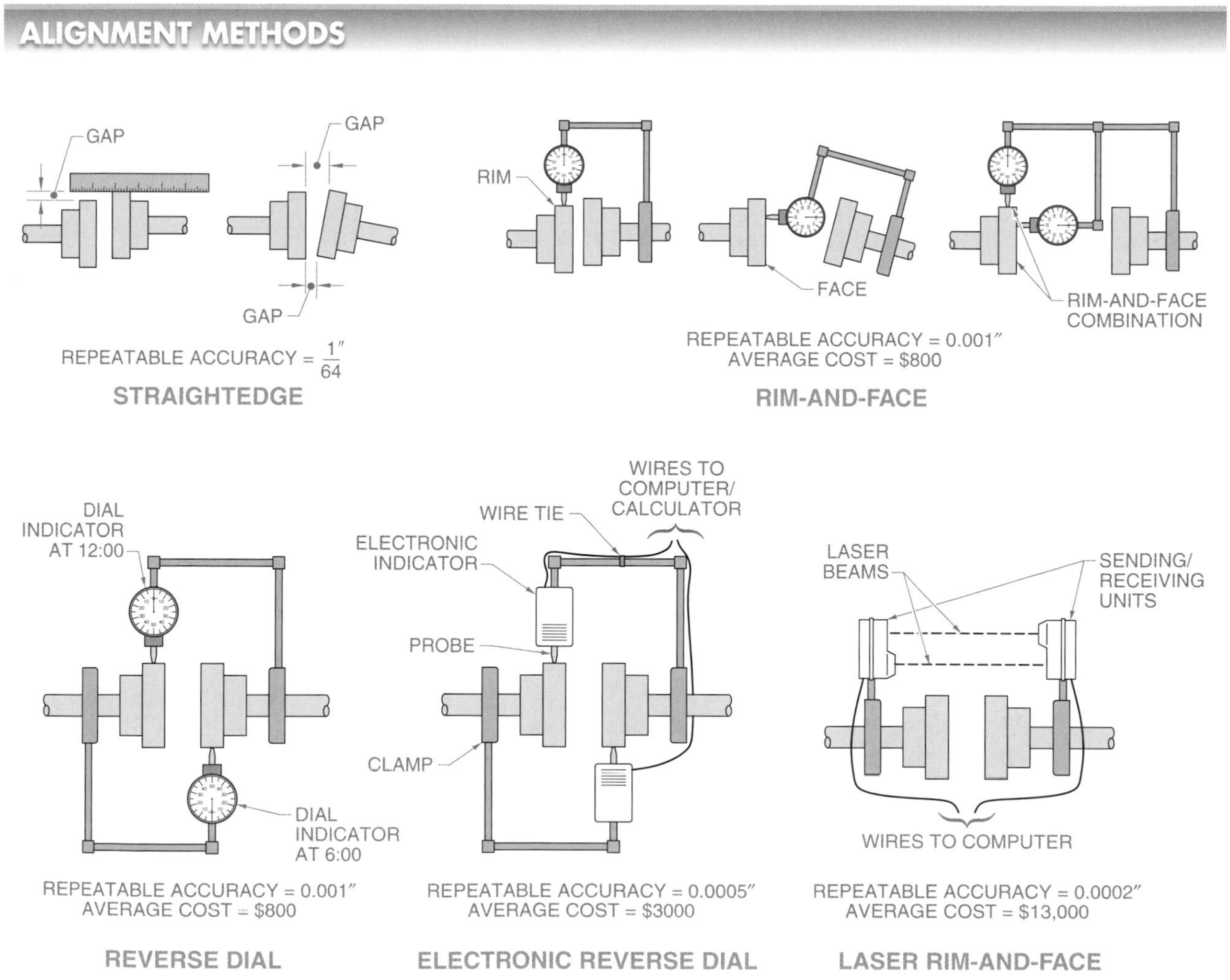

ALIGNMENT METHODS

REPEATABLE ACCURACY = $\frac{1″}{64}$

STRAIGHTEDGE

REPEATABLE ACCURACY = 0.001″
AVERAGE COST = $800

RIM-AND-FACE

REPEATABLE ACCURACY = 0.001″
AVERAGE COST = $800

REVERSE DIAL

REPEATABLE ACCURACY = 0.0005″
AVERAGE COST = $3000

ELECTRONIC REVERSE DIAL

REPEATABLE ACCURACY = 0.0002″
AVERAGE COST = $13,000

LASER RIM-AND-FACE

Figure 29-16. The five methods of aligning machinery are the straightedge, rim-and-face, reverse dial, electronic reverse dial, and laser rim-and-face methods.

The choice of alignment method is based on cost, accuracy required, ease of use, and time required to perform the alignment. The accuracy of any alignment is based on the individual doing the alignment and the alignment method used. For example, straightedge measurements are normally made without the knowledge of coupling irregularities and require the feel of thickness gauge measurements. Therefore, the accuracy of straightedge alignment is approximately $\frac{1}{64}''$.

Dial indicators and electronic measuring devices (except laser) measure in the thousandths of an inch, which allow for an accuracy of alignment within $0.001''$. Laser alignment methods are generally the most precise and quickest, with an accuracy of $0.0002''$. It is important to always follow manufacturer recommendations.

Motors that are not mounted properly are more likely to fail from mechanical problems. To ensure a long life span, a motor should be mounted so that it is kept as clean as possible to reduce the chance of damaging material reaching a motor. Standard motors are designed to be mounted with the shaft horizontal. The horizontal position is the best operating position for motor bearings. A specifically designed motor can be used with vertical mounting. Motors designed to operate vertically are more expensive and require more preventive maintenance.

29-2 CHECKPOINT

1. What are the two types of misalignment that require correction?
2. Which type of alignment method provides the highest alignment accuracy?

29-3 BEARINGS

A *bearing* is a component used to reduce friction and maintain clearance between stationary and moving parts in a motor or machine. Motor bearings are mounted in the end bell at each end of a motor. Bearings guide and position moving parts to reduce friction, vibration, and temperature. The length of time a machine retains proper operating efficiency and accuracy depends on proper bearing selection, installation, handling, and maintenance procedures. Bearings are available with many features and incorporate the same basic parts.

Bearings are designed to support radial, axial, and radial and axial loads. A *radial load* is a load in which the applied force is perpendicular to the axis of rotation. For example, a rotating shaft resting horizontally on, or being supported by, a bearing surface at each end has a radial load due to the weight of the shaft itself. An *axial load* is a load in which the applied force is parallel to the axis of rotation. For example, a rotating vertical shaft has an axial load due to the weight of the shaft itself.

A radial and axial load occurs when a combination of radial loads and axial loads is present. For example, the shaft of a fan blade is supported horizontally (radial load) and is pulled or pushed (axial load) by the fan blade. Bearings are classified as rolling-contact (antifriction) or plain bearings.

Rolling-Contact Bearings

A *rolling-contact (antifriction) bearing* is a bearing that contains rolling elements that provide a low-friction support surface for rotating or sliding surfaces. Rolling-contact bearings include ball, roller, and needle bearings. A *ball bearing* is an antifriction bearing that permits free motion between a moving part and a fixed part by means of balls confined between inner and outer rings. A *roller bearing* is an antifriction bearing that has parallel or tapered steel rollers confined between inner and outer rings. A *needle bearing* is an antifriction roller-type bearing with long rollers of small diameter. **See Figure 29-17.**

Tech Fact

Although metal is the most common bearing material, bearings are available in other materials for specific applications. Ceramic bearings are used in food-processing environments because they can withstand the extreme heat and cold required and are resistant to the corrosive processing and cleaning solutions often used. Also, standard bearing lubricants cannot be used in food-processing environments. Ceramic bearings that require no lubrication can be used or a lubricant that is FDA-approved for food processing can be used.

ROLLING-CONTACT (ANTIFRICTION) BEARINGS

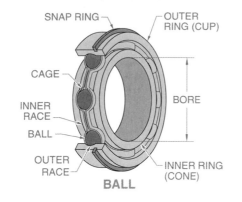

BALL

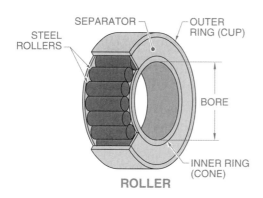

ROLLER

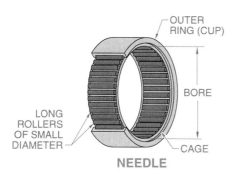

NEEDLE

Figure 29-17. Rolling-contact (antifriction) bearings include ball, roller, and needle bearings.

Plain Bearings

A *plain bearing* is a bearing in which the shaft turns and is lubricated by a sleeve. Plain bearings are used in areas of heavy loads where space is limited. Plain bearings are quieter, less costly, and, if kept lubricated,

have little metal fatigue compared to other bearings. **See Figure 29-18.** Plain bearings may support radial and axial (thrust) loads. In addition, plain bearings can conform to the part in contact with the bearing because of the sliding rather than rolling action. This allows the plain bearing to yield to any abnormal operating condition rather than distort or damage the shaft or journal. A *journal* is the part of a shaft, such as an axle or spindle, that moves in a plain bearing. The sliding motion of a shaft or journal, whether it is rotating or reciprocating, is generally against a softer, lower-friction bearing material.

Bearing Selection

No universal bearing exists that can be used for every function and application required in industry. In many cases, a review of the machine function and its bearing requirements may indicate whether proper bearings are being used. When a replacement bearing is chosen by comparing it to a removed bearing instead of from an equipment manual or parts book, certain factors other than dimensions must be observed. Factors to be considered include the exact replacement part number, the type and position of any seal, the direction of force and positioning of a required high shoulder, and whether a retaining ring is required.

Equipment exposed to outdoor elements will require more frequent maintenance to prolong its life.

PLAIN BEARINGS

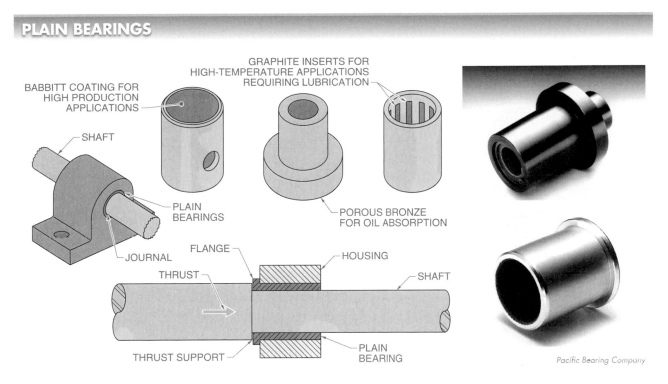

Figure 29-18. Plain bearings provide sliding contact between mating surfaces.

Provisions made for thermal expansion within a machine are generally published by the manufacturer and are listed as space tolerances for between housing, bearing components, and shaft. Greater space tolerances are allowed for plain bearings than for rolling-contact bearings because plain bearings are more susceptible to damage from higher temperatures.

Bearing Removal

A firm, solid contact must be made for bearing removal. Bearings should always be removed from a shaft with even pressure against the ring that was press fit. Bearings are removed from shafts using bearing pullers, gear pullers, arbor presses, or manual impact. These methods enable easy bearing removal and reduce the damage to the bearing. **See Figure 29-19.**

Extreme caution must be taken to prevent damage to any bearing part. Most damage during removal goes unnoticed. The use of a hammer and chisel to pry a bearing off of its shaft normally results in damage and contamination. Any bearing that was difficult to remove should be discarded because it probably was damaged during removal.

Proper tools and maintenance procedures are required when removing bearings. Many bearing failures are due to contaminants that have worked their way into or around a bearing before it has been placed in operation. Workbenches, tools, clothing, wiping cloths, and hands must be clean and free from dust, dirt, and other contaminants.

When a bearing has been removed and taken apart for maintenance, the parts should be cleaned and inspected. Cleaning is accomplished by dipping or washing the housing, shaft, bearing, spacers, and other parts in a clean, nonflammable cleaning solvent. All traces of dirt, grease, oil, rust, or any other foreign matter must be removed. Caution should be taken when using part-cleaning solutions. Seals, O-rings, and other soft materials may deteriorate due to incompatibility.

All parts should be wiped with a clean towel soaked in lightweight oil and then wrapped or covered to protect them from dust and dirt. All parts must be inspected for nicks, burrs, or corrosion on shaft seats, shoulders, or faces. All bearing components should be inspected for indication of abnormality or obvious defects, such as cracks and breaks. Any nicks, corrosion, rust, and scuffs on shaft or housing surfaces should be removed. Any worn spacers, shafts, bearings, or housings must be replaced.

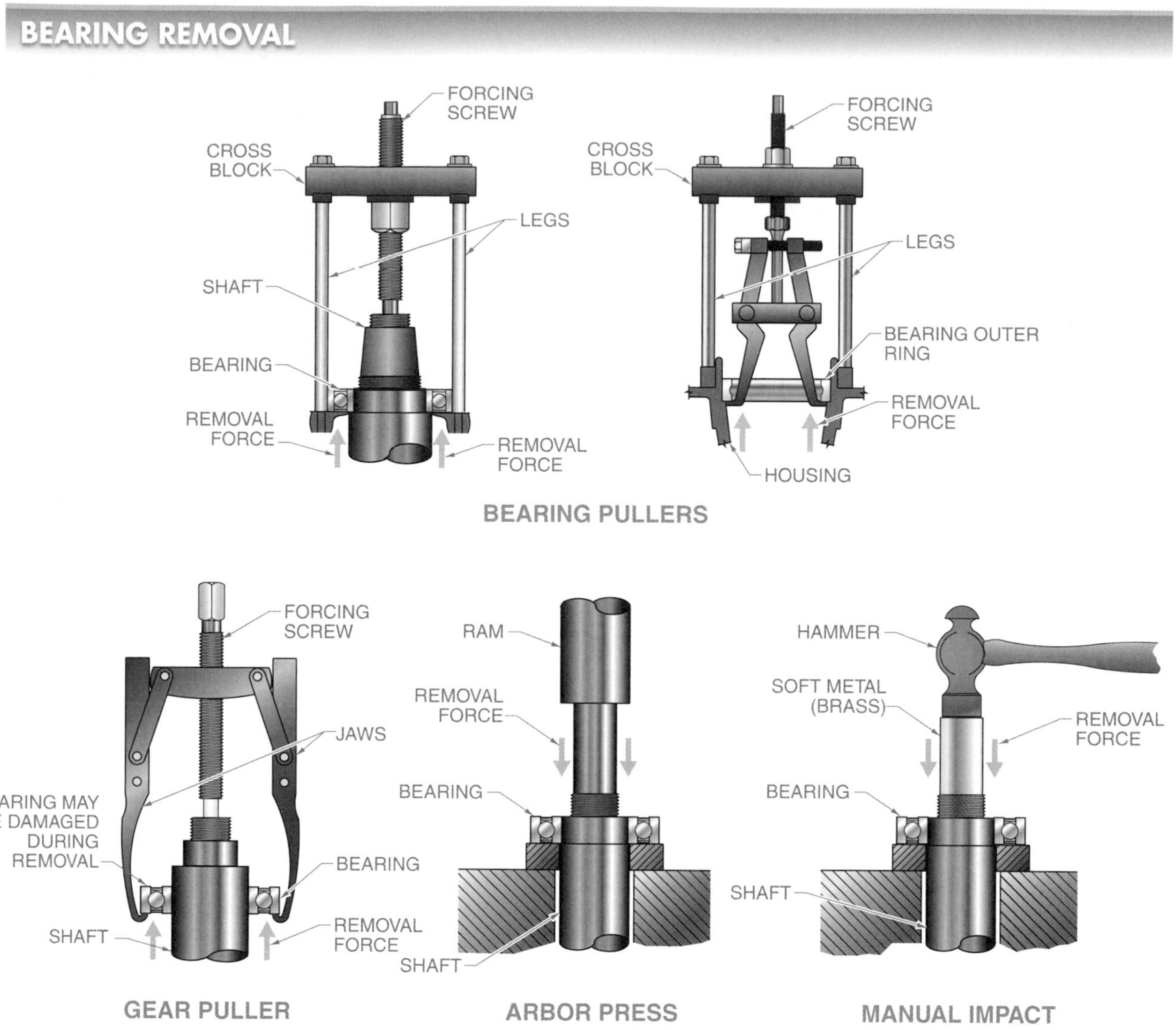

BEARING REMOVAL

Figure 29-19. Bearings are removed from shafts using bearing pullers, gear pullers, arbor presses, or manual impact.

Bearing Installation

Bearings must be installed properly to ensure proper life and service. Precautions that must be taken when bearings are mounted include knowing the function of the bearing; keeping all bearings wrapped or in the original sealed container until needed; treating reusable bearings as new; maintaining clean tools, hands, and work surfaces; and working in a clean environment. More bearings fail because of poor mounting practices than from malfunction during their useful life.

Bearing mounting procedures affect the performance, durability, and reliability of a motor. Precautions should be taken to allow a bearing to perform without excessive temperature rise, noise from misalignment or vibration, and shaft movement.

Bearings are mounted with one rotating ring installed as a press fit over the shaft and the other rotating ring installed as a push fit in the bearing housing. *Press fit* is a bearing installation where the bore of the inner rotating ring is smaller than the diameter of the shaft and considerable force must be used to press the bearing onto the shaft. *Push fit* is a bearing installation where the diameter of the outer fixed ring is smaller than the diameter of the bearing housing and the ring can be pushed in by hand.

During a press-fit installation, force must be applied uniformly on the face or ring that is to be press fit. This can be accomplished by using a piece of tubing, a steel plate, and a hammer or by using an arbor press. **See Figure 29-20.** Wood should not be used because of the possibility of contaminating a bearing with wood splinters or fibers. A push fit allows the outer ring to be slid into the bearing housing by hand.

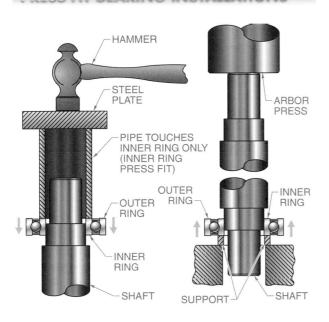

PRESS FIT BEARING INSTALLATIONS

HAMMER

STEEL PLATE

PIPE TOUCHES INNER RING ONLY (INNER RING PRESS FIT)

ARBOR PRESS

OUTER RING

OUTER RING

INNER RING

INNER RING

INNER RING

SHAFT

SUPPORT

SHAFT

Figure 29-20. During a press-fit installation, uniform force must be applied on the face that is to be press fit using a piece of tubing, a steel plate, and a hammer or using an arbor press.

Bearings that are designed for thrust loads must be installed in the correct direction to prevent the load from separating the bearing components. These bearings have a face and back side for ease in identifying the thrust direction. The back side receives the thrust and is marked with the bearing number, tolerance, manufacturer, and, in some cases, the word "thrust."

After a bearing is installed on a motor, the bearing must be lubricated if required, and tested. Many smaller bearings are factory sealed and should not be lubricated. Motors are normally lubricated at the factory to provide long operation under normal service conditions without relubrication. Excessive and frequent lubrication can damage a motor. The appropriate time period between lubrication depends on the motor service conditions, ambient temperature, and the environment. The lubrication instructions provided with a motor should always be followed. These instructions are normally listed on the nameplate or terminal box cover. If lubrication instructions are not available, plain bearings and rolling-contact bearings should be relubricated according to a set schedule.

Many bearings are fitted with a shield that helps to contain the grease inside the bearing. Shielded bearings require regular lubrication. Devices used for lubricating bearings include grease fittings, pressure cups, oil cups, and oil wicks. **See Figure 29-21.**

Improper lubrication is a major cause of bearing failure. Improper lubrication includes underlubrication, overlubrication, lubricant contamination, and mixing lubricants. Underlubrication causes immediate wear to a bearing. If the bearing is underlubricated, the metal surfaces touch, causing rapid failure.

Contamination causes many bearing failures. Dirt or other impurities can enter a bearing during operation, lubrication, and/or assembly. Dirt is abrasive to bearing components and causes premature failure of the bearing. Bearing tolerances are such that a solid particle of a few thousandths of an inch (0.001″ to 0.003″) lodged between the housing and the outer ring can distort raceways enough to reduce critical clearances.

Machine Run-In

A machine run-in check should be made after bearing assembly is complete. A run-in check starts with a hand check of the torque of the machine shaft. For safety reasons, the power must be locked out when manually rotating a machine shaft. Unusually high torque normally indicates a problem with a tight fit, misalignment, or improper assembly of machine parts. Machine power is restored and noise levels recorded. High noise levels may indicate excessive loading or damaged bearings. The problem must be corrected before continuing.

Final checks are accomplished by measuring machine temperatures. High initial temperatures are common because bearings are packed with grease, which can produce excessive friction when the motor is first started. Run-in temperatures should decrease to within recommended ranges. Any machine with temperatures that continue to run high should be corrected before proceeding. Continued high temperatures are normally a sign of tight fit, misalignment, or improper assembly.

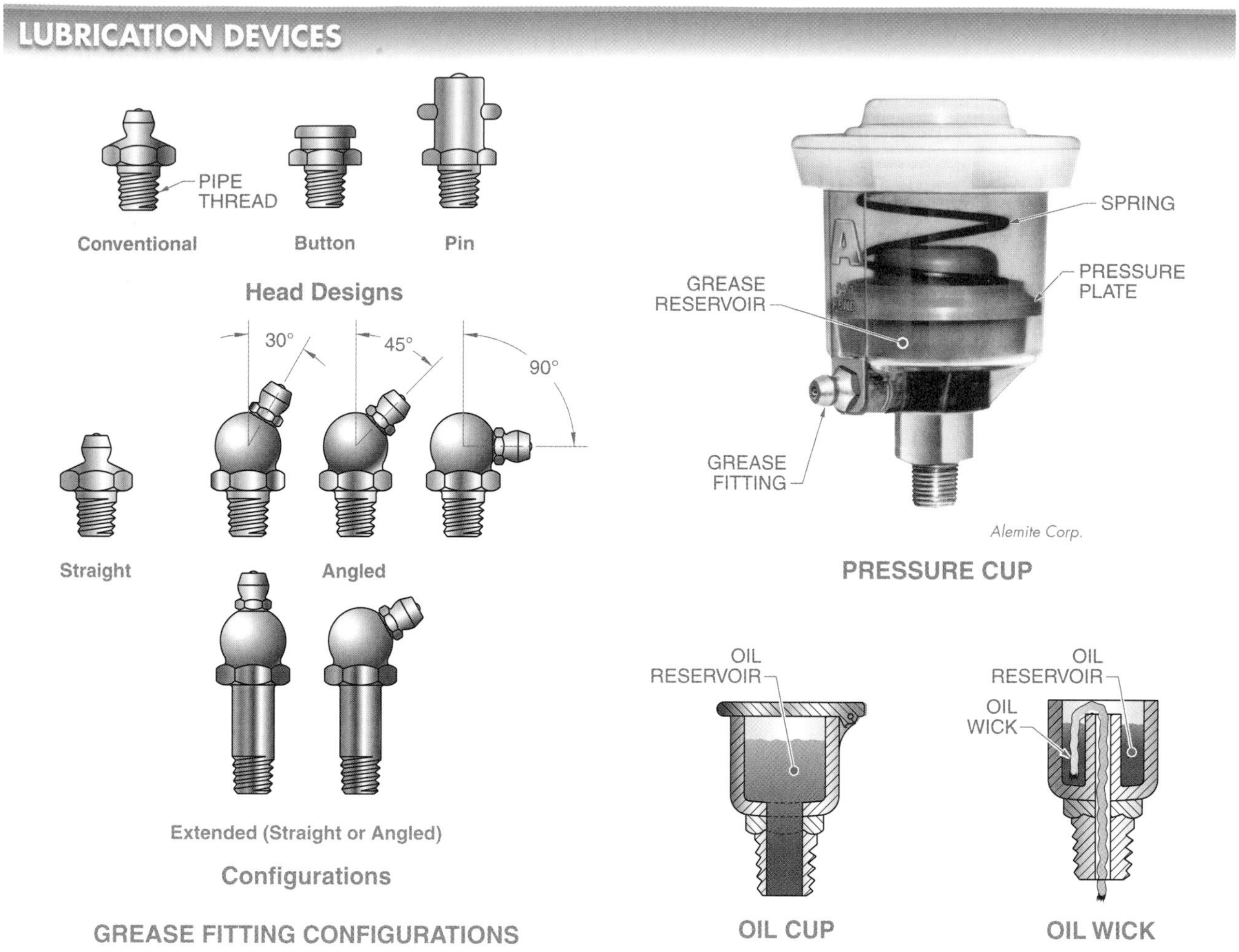

LUBRICATION DEVICES

Head Designs

Straight   Angled

Extended (Straight or Angled)

Configurations

GREASE FITTING CONFIGURATIONS

Alemite Corp.

PRESSURE CUP

OIL CUP   **OIL WICK**

Figure 29-21. Devices used for lubricating bearings include grease fittings, pressure cups, oil cups, and oil wicks.

29-3 CHECKPOINT

1. Does a rolling-contact bearing or a plain bearing allow for a greater space tolerance?

2. Is it recommended that motor bearings be over-lubricated to ensure there is enough lubricant for proper operation?

29-4 FLEXIBLE BELT DRIVES

A *flexible belt drive* is a system in which resilient flexible belts are used to drive one or more shafts. Flexible belt drives are one of the most common drive systems used in industry. Belts are attached to a motor shaft and a load shaft. As the motor shaft rotates, friction between the belt and a pulley provides the torque needed to turn the other shaft. Flexible belt drives are relatively inexpensive, quiet, and easy to maintain. Flexible belt drives also provide a wide range of speed and torque. The material used for a belt is normally selected based on the application. Belts normally have tensile members that run the length of the belt to provide tensile strength for the belt. Belts commonly used in industry include flat belts, V-belts, and timing belts. **See Figure 29-22.**

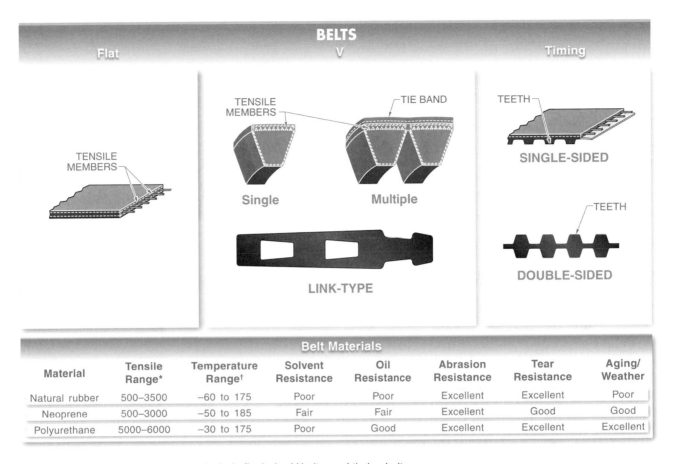

BELTS

| Flat | V | Timing |

TENSILE MEMBERS

TENSILE MEMBERS

TIE BAND

TEETH

Single

Multiple

SINGLE-SIDED

TEETH

LINK-TYPE

DOUBLE-SIDED

Belt Materials

Material	Tensile Range*	Temperature Range†	Solvent Resistance	Oil Resistance	Abrasion Resistance	Tear Resistance	Aging/ Weather
Natural rubber	500–3500	−60 to 175	Poor	Poor	Excellent	Excellent	Poor
Neoprene	500–3000	−50 to 185	Fair	Fair	Excellent	Good	Good
Polyurethane	5000–6000	−30 to 175	Poor	Good	Excellent	Excellent	Excellent

Figure 29-22. Belts used in industry include flat belts, V-belts, and timing belts.

Belts should be protected from the surrounding environment. A protective cover should be designed to keep objects and foreign substances such as grease, oil, and dirt from contacting the belt or pulleys. Foreign material on a belt causes glazing. *Glazing* is a slick, polished surface caused by dirt and other debris being rubbed on the surface of a belt. A glazed belt has reduced friction with the pulleys, resulting in belt slippage and a loss of power transmission. Glazed belts should be replaced and the pulleys inspected for possible damage.

A *V-belt* is a continuous-power transmission belt with a trapezoidal cross section. V-belts are made of molded fabric and rubber for body and bending action. V-belts also contain fiber or steel cord reinforcement called tension members as their major pulling-strength material. V-belts are resilient and able to absorb many shocks because of their construction. V-belts are generally classified as standard or high capacity. Standard V-belts are designated as A, B, C, D, or E. High-capacity V-belts are designated as 3V, 5V, or 8V. The letter or number designation also indicates the cross-sectional dimension and thickness of the belt. **See Figure 29-23.**

V-belts run in a pulley (sheave) with a V-shaped groove. V-belts transmit power through the wedging action of the tapered sides of the belt in the pulley groove. The wedging action results in an increased coefficient of friction. V-belts do not normally contact the bottom of the pulley. A pulley or belt should be replaced if it has worn enough that the belt touches the bottom of the pulley. The belt will become shiny and will slip and burn if allowed to bottom out. More than one belt may be used if additional power transmission is required. However, each belt must be the same type and size.

V-belt replacement, whether for preventive maintenance or equipment breakdown, starts with proper identification and sizing of the belt being replaced. A technician can prevent many premature belt failures by selecting the proper belt, belt size, and installation procedure. The technician must be sure to follow all manufacturer replacement specifications when installing a new belt.

V-BELT CLASSIFICATIONS

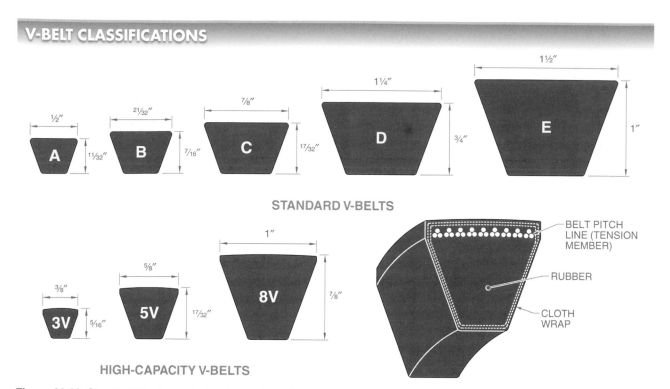

Figure 29-23. Standard V-belts are designated as A, B, C, D, or E, and high-capacity V-belts are designated as 3V, 5V, or 8V.

Pulleys

A *sheave* is a grooved wheel used to hold a V-belt. A pulley consists of one or more sheaves and a frame or block to hold the sheaves. V-belt pulleys may be fixed-bore or tapered-bore pulleys. Pulleys are used to change the speed of a driven load relative to the motor speed. A pulley of one size is placed on the drive shaft and a pulley of another size is placed on the driven shaft. A V-belt connects the two pulleys to transfer the torque. V-belts that are not aligned properly are destroyed prematurely due to excessive side wear, broken or stretched tension members, or rolling over in the pulley. Pulley misalignment may be offset, nonparallel, or angular. **See Figure 29-24.**

Offset misalignment is a condition where two shafts are parallel but the pulleys are not on the same axis. Off-set misalignment may be corrected using a straightedge along the pulley faces. The straightedge may be a solid object, such as a ruler, square, or metal object, or it may be a string. Offset misalignment must be within 1/10″ per foot of drive center distance.

Nonparallel misalignment is misalignment where two pulleys or shafts are not parallel. Nonparallel misalignment is also corrected using a string or straightedge. The device connected to the pulley that touches the straightedge at one point is rotated to bring it parallel with the other pulleys so that the two pulleys touch the straightedge at four points.

Angular misalignment is a condition where two shafts are parallel but at different angles with the horizontal plane. Angular misalignment is corrected using a level placed on top of the pulley, parallel with the pulley shaft. Angular misalignment must not exceed 1/2″.

Proper alignment and tension of V-belts produces long and trouble-free belt operation. Excessive tension produces excessive strain on belts, bearings, and shafts, which causes premature wear. Too little tension causes belt slippage. Belt slippage causes excessive heat and premature belt and pulley wear. The best tension for a V-belt is the lowest tension at which the belt does not slip under peak loads. Belt dressing should not be used as a remedy for belt slippage due to pulley wear. Two methods used for tensioning a belt include the visual adjustment and belt deflection methods. **See Figure 29-25.**

The *visual adjustment method* is a belt tension method in which the tension is adjusted by observing the slight sag at the slack side of the belt. The visual adjustment method is used to roughly adjust belt tension. The belt is placed on the pulleys without forcing the belt over the pulley flange. Tension is applied to the belt by increasing the drive center distance until the belt is snug. The machine is run for approximately 5 min to seat the belt. Slight sag is normal. A squeal or slip indicates that the belt is too loose.

PULLEYS

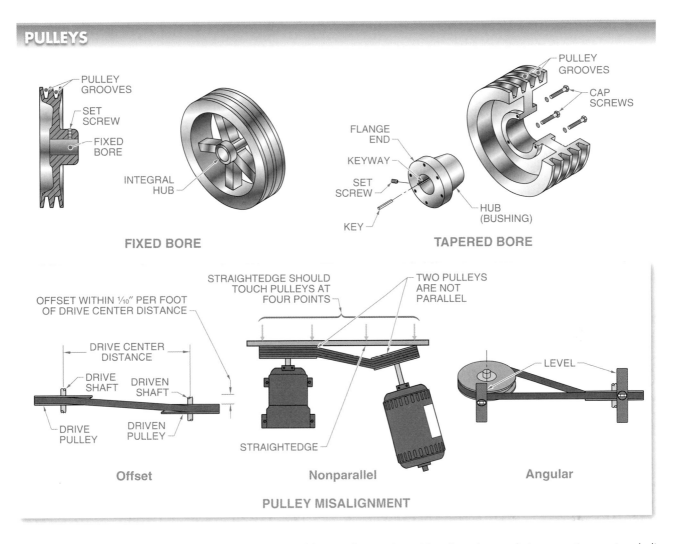

FIXED BORE

TAPERED BORE

PULLEY MISALIGNMENT

Offset

Nonparallel

Angular

Figure 29-24. V-belt pulleys may be fixed-bore or tapered-bore pulleys and must be aligned properly to prevent premature belt wear or failure.

V-BELT TENSIONING METHODS

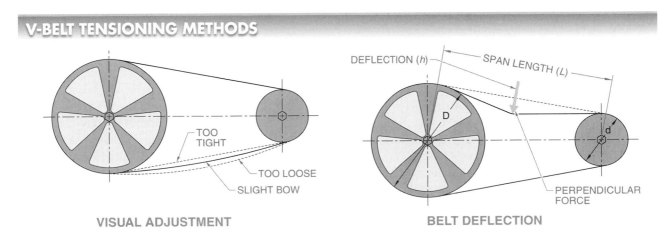

VISUAL ADJUSTMENT

BELT DEFLECTION

Figure 29-25. Flexible belt drives must be checked for correct belt tension.

The *belt deflection method* is a belt tension method in which the tension is adjusted by measuring the deflection of the belt. The belt is installed and tension is applied by increasing the drive center distance. The span length (L) between the pulley centers and the deflection height (h) is measured. Proper deflection height is $\frac{1}{64}$″ per inch of span length. A perpendicular force at the midpoint of the span length is applied to measure for proper tension. Proper deflection height is found by applying the following formula:

$$h = L \times \frac{1}{64}″$$

where

h = deflection height (in in.)

L = span length (in in.)

$\frac{1}{64}″$ = constant ($0.0156″$)

Example: Calculating Deflection Height

What is the proper belt deflection of an assembly using a 10″ pulley and a 5″ pulley having a span length of 36″?

$$h = L \times \frac{1}{64}″$$
$$h = 36 \times 0.0156$$
$$h = \mathbf{0.562″}$$

New V-belts seat rapidly during the first few hours of operation. New belts should be checked and retensioned following the first 24 hr and 72 hr of operation. During retensioning, the deflection height from the normal position is determined by the use of a straightedge or string stretched across the pulley tops. Belt tension tools are also available. Belt tension tools allow for faster, simpler, and more accurate tension checks.

29-4 CHECKPOINT

1. Are standard V-belt sizes designated by a number or letter?

2. When testing belt tension, is $\frac{1}{8}″$, $\frac{1}{32}″$, or $\frac{1}{64}″$ considered a proper deflection height?

29-5 PREVENTIVE MAINTENANCE TESTS

A good preventive maintenance program can detect and eliminate problems before they create downtime. A main part of an electrical system preventive maintenance program is the testing of equipment and conductor insulation. The condition of conductor insulation is a good general indicator of the condition of the equipment and electrical system.

Failing insulation must be corrected so that a system does not fail at an inopportune time. In general, as any system is operated over a long period, conductor insulation quality deteriorates at a predictable rate. By taking resistance measurements over time, conductor insulation failure (or expected life) can be predicted.

The two basic test instruments used to measure insulation resistance include the hipot (high potential or high voltage) tester and megohmmeters. **See Figure 29-26.** A *hipot tester* is a test instrument that measures insulation resistance by measuring leakage current. Hipot testers apply a high test voltage between two different conductors or

between a conductor and ground and measure the leakage current. An excessive amount of leakage current indicates a low resistance or breakdown in insulation. Tests using hipot testers normally involve applying a test voltage that is several times higher than the specified operating voltage of the cable or device being tested.

A megohmmeter performs the same basic test as a hipot tester in that the meter also applies a high test voltage to the circuit or component being tested, but it displays the measurement as resistance in ohms (normally MΩ). Megohmmeters are manufactured in a variety of styles. Most megohmmeters have a function switch or selector switch to choose the appropriate test voltage. The megohmmeter display can be an analog display or a digital display. The power source of a megohmmeter can be a hand crank, battery power, or 120 VAC. Some models have dual power sources such as a hand crank and 120 VAC. The three basic types of insulation tests include insulation spot test, dielectric absorption test, and insulation step voltage test.

INSULATION TEST INSTRUMENTS

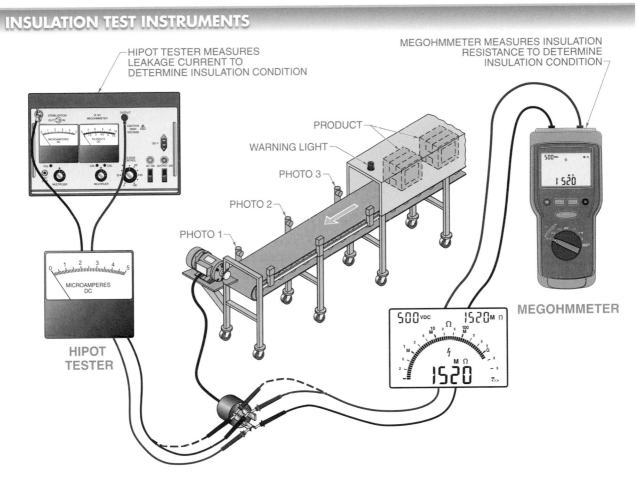

HIPOT TESTER MEASURES
LEAKAGE CURRENT TO
DETERMINE INSULATION CONDITION

MEGOHMMETER MEASURES INSULATION
RESISTANCE TO DETERMINE
INSULATION CONDITION

PRODUCT

WARNING LIGHT

PHOTO 3

PHOTO 2

PHOTO 1

MEGOHMMETER

HIPOT
TESTER

Figure 29-26. Two test instruments normally used to measure insulation resistance are hipot (high potential or high voltage) testers and megohmmeters.

Insulation Spot Tests

An *insulation spot test* is a test that checks motor insulation over the life of the motor. An insulation spot test is taken when the motor is placed in service and every six months thereafter. The test should also be taken after a motor is serviced. **See Figure 29-27.** To perform an insulation spot test, the following procedure is applied:

1. Connect a megohmmeter to measure the resistance of each winding lead to ground. Record the readings after 60 sec. Service the motor if a reading does not meet the minimum acceptable resistance. If all readings are above the minimum acceptable resistance, record the lowest meter reading on an insulation spot test graph. The lowest reading is used because a motor is only as good as its weakest point.

2. Discharge the motor windings.

3. Repeat steps 1 and 2 every six months.

The results of the test are then interpreted to determine the condition of the insulation. Point A represents the motor insulation condition when the motor was placed in service. Point B represents effects of aging and contamination on the motor insulation. Point C represents motor insulation failure. Point D represents motor insulation condition after being rewound.

Tech Fact

Motor insulation typically fails because of contamination, normal thermal aging, and overheating due to overloaded operating conditions, vibration, and overvoltage spikes. To help prevent insulation failure, use the correct enclosure for the application, keep the motor and cooling vents clean, use properly sized overload protection to prevent overheating before it occurs, and reduce overvoltage spikes (usually caused by VFDs, lightning, and switching). Use snubbers and line reactors before the VFD and load reactors before the motor.

INSULATION SPOT TESTS

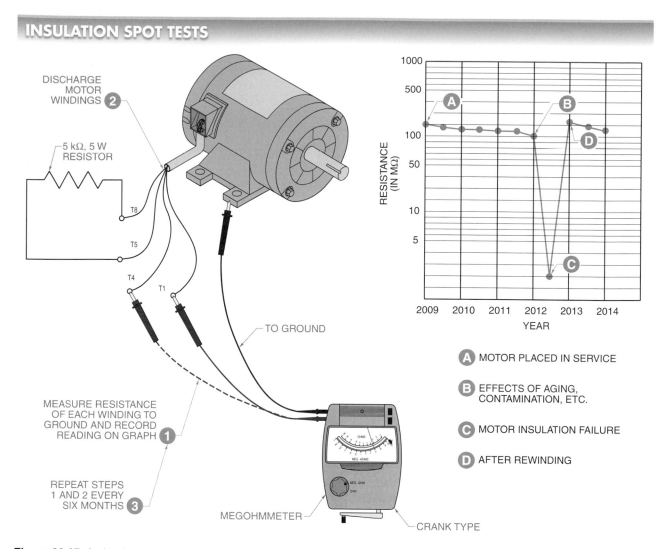

Figure 29-27. An insulation spot test checks motor insulation over the life of the motor.

Dielectric Absorption Tests

A *dielectric absorption test* is a test that checks the absorption characteristics of moist or contaminated insulation. The test is performed over a period of 10 min. **See Figure 29-28.** To perform a dielectric absorption test, the following procedure is applied:

1. Connect a megohmmeter to measure the resistance of each winding lead to ground. Service the motor if a reading does not meet the minimum acceptable resistance. If all readings are above the minimum acceptable resistance, record the lowest meter reading on a dielectric absorption test graph. Record the readings every 10 sec for the first minute and every minute thereafter for 10 min.

2. Discharge the motor windings.

The results of the test are then interpreted to determine the condition of the insulation. The slope of the curve shows the condition of the insulation. Good insulation (Curve A) shows a continual increase in resistance. Moist or cracked insulation (Curve B) shows a relatively constant resistance.

A polarization index is obtained by dividing the value of the 10 min reading by the value of the 1 min reading. The polarization index is an indication of the condition of the insulation. A low polarization index indicates excessive moisture of contamination. **See Figure 29-29.**

For example, if the 1 min reading of Class B insulation is 80 MΩ and the 10 min reading is 90 MΩ, the polarization index is 1.125 (90 ÷ 80 = 1.125). In this example, the insulation contains excessive moisture or contamination.

DIELECTRIC ABSORPTION TESTS

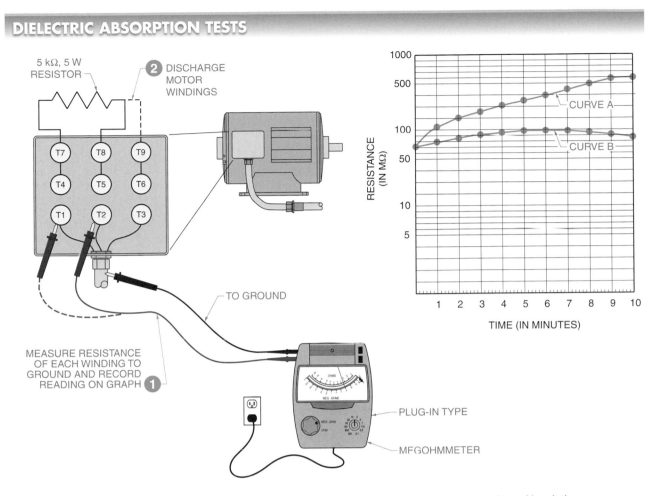

Figure 29-28. A dielectric absorption test checks the absorption characteristics of moist or contaminated insulation.

MINIMUM ACCEPTABLE POLARIZATION INDEX VALUES

Insulation	Value
Class A	1.5
Class B	2.0
Class F	2.0

Figure 29-29. The polarization index is an indication of the condition of insulation. A low polarization index indicates excessive moisture or contamination.

Insulation Step Voltage Tests

An *insulation step voltage test* is a test that creates electrical stress on internal insulation cracks to reveal aging or damage not found during other motor insulation tests. An insulation step voltage test is performed only after an insulation spot test. **See Figure 29-30.**

To perform an insulation step voltage test, apply the following procedure:

1. Set the megohmmeter to 500 V and connect it to measure the resistance of each winding lead to ground. Take each resistance reading after 60 sec. Record the lowest reading.

2. Place the meter leads on the winding that has the lowest reading.

3. Set the megohmmeter on increments of 500 V, starting at 1000 V and ending at 5000 V. Record each reading after 60 sec.

4. Discharge the motor windings.

The results of the test are then interpreted to determine the condition of the insulation. The resistance of good insulation that is thoroughly dry (Curve A) remains approximately the same at different voltage levels. The resistance of deteriorated insulation (Curve B) decreases substantially at different voltage levels.

INSULATION STEP VOLTAGE TESTS

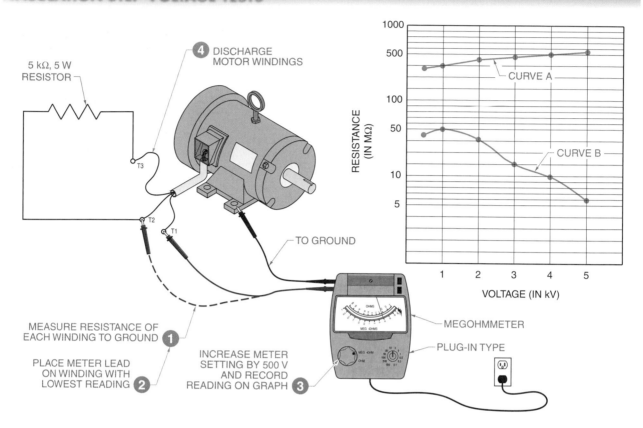

5 kΩ, 5 W
RESISTOR

④ DISCHARGE
MOTOR WINDINGS

T3

T2

T1

MEASURE RESISTANCE OF
EACH WINDING TO GROUND ❶

PLACE METER LEAD
ON WINDING WITH
LOWEST READING ❷

INCREASE METER
SETTING BY 500 V
AND RECORD
READING ON GRAPH ❸

TO GROUND

MEGOHMMETER

PLUG-IN TYPE

CURVE A

CURVE B

RESISTANCE
(IN MΩ)

VOLTAGE (IN kV)

Figure 29-30. An insulation step voltage test is a test that creates electrical stress on internal insulation cracks to reveal aging or damage not found during other motor insulation tests.

29-5 CHECKPOINT

1. Which test is used to measure and record the insulation resistance of a motor over time, such as every six months?

2. In a dielectric absorption test, should the measured resistance decrease each minute of testing if the insulation is good?

Additional Resources

Review and Resources

Access Chapter 29 Review and Resources for *Electrical Motor Controls for Integrated Systems* by scanning the above QR code with your mobile device.

Applying Your Knowledge

Refer to the *Electrical Motor Controls for Integrated Systems* Learner Resources for interactive Applying Your Knowledge questions.

Workbook and Applications Manual

Refer to Chapter 29 in the *Electrical Motor Controls for Integrated Systems Workbook* and the *Applications Manual* for additional exercises.

ENERGY EFFICIENCY PRACTICES

Electrical Equipment Heating Effects

All electrical equipment produces heat. Heat is produced in the loads (lamps, motors, heating elements, etc.) as they convert electrical energy into some other form of energy (light, sound, rotary motion, etc.). Heat is also produced in the conductors (wires), connections, conduit, control devices, and all equipment from the point of generation to usage within the system. A properly designed and installed system allows for the expected heat and provides enough space, insulation, and cooling to prevent problems. However, loose connections, undersized conductors, overloaded circuits, and faulty components and equipment can cause overheating.

Overheating can cause injury, fires, and is a waste of energy. Even small amounts of overheating that do not cause injury or fire are a waste of energy and must be corrected. Maintenance must be performed when electrical equipment is operating at higher-than-normal temperatures to avoid equipment failure and fire. Most electrical equipment and conductors have a maximum temperature rating. Temperature rises of 50°F (28°C) or more must be investigated. Temperature rises of 100°F (56°C) or more require immediate action (immediate shutdown of the system and repair of the fault).

A thermal imaging camera/meter measures heat energy by measuring the infrared (IR) energy emitted by a material and displays the image and temperatures as colors and numerical values. Thermal imaging cameras can detect problems, potential problems, and wasted heat energy not visible to the naked eye. Thermal imaging cameras should be used as a part of the preventive maintenance and energy-auditing program.

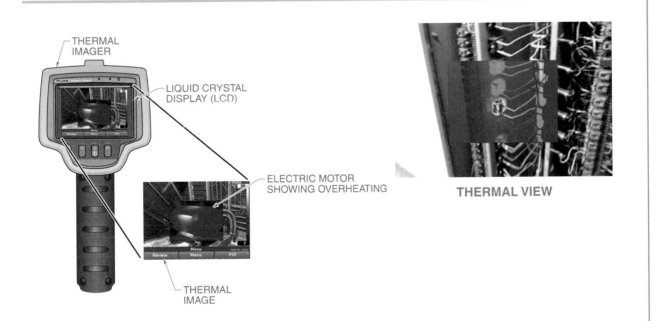

THERMAL IMAGER

LIQUID CRYSTAL DISPLAY (LCD)

ELECTRIC MOTOR SHOWING OVERHEATING

THERMAL IMAGE

THERMAL VIEW

Objectives

30-1
- Define predictive maintenance (PDM) and list the different types.
- Define and describe common PDM procedures.

30-2
- Describe the different resources available to maintenance technicians.

Review and Resources
atplearningresources.com/Quicklinks
Access Code: 362245

708

Chapter 30

Predictive Maintenance

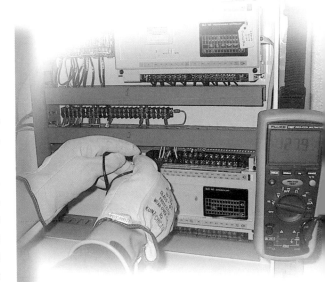

The understanding of electrical theory, components, and circuits allows for the proper design, installation, and troubleshooting of electrical equipment, circuits, and systems. Equipment, circuits, and systems should operate safely and satisfactorily for many years when they are properly manufactured, designed, and installed. If a problem occurs and shuts down the circuit or system, proper troubleshooting skills will help a technician determine the problem and what corrective action to take to solve the problem. However, circuits and systems that are not operating can cause downtime, product loss, unsafe conditions, and additional costs to fix. It is better and less expensive to prevent a problem than to wait until a problem occurs. In order to prevent a problem, potential problems must first be predicted so corrective action can be taken before the problem occurs.

30-1 PREDICTIVE MAINTENANCE MONITORING

Predictive maintenance (PDM) is the monitoring of wear conditions and equipment characteristics against a predetermined tolerance to predict possible malfunctions or failures. Equipment operation data is gathered and analyzed to show trends in performance and component characteristics. Corrective repairs are made as required.

Predictive maintenance requires a substantial investment in training and equipment and is most commonly used for expensive or critical equipment. Data collected through equipment monitoring is analyzed to check whether values are within acceptable tolerances. **See Figure 30-1.** If values are within tolerances, data is continually collected and analyzed on a regular basis. Maintenance procedures are performed if values are outside acceptable tolerances.

The equipment is closely monitored after maintenance procedures are performed. If a problem recurs, the equipment application and design are analyzed, with changes made as required.

In predictive maintenance, the monitoring of equipment can be random, scheduled, or continuous. *Random monitoring* is the unscheduled monitoring of equipment as required. *Scheduled monitoring* is the monitoring of equipment at specific time intervals. *Continuous monitoring* is the monitoring of equipment at all times. Common predictive maintenance procedures include visual and auditory inspection, vibration analysis, lubricating oil analysis, thermography, ultrasonic analysis, and electrical analysis.

PREDICTIVE MAINTENANCE FLOW CHART

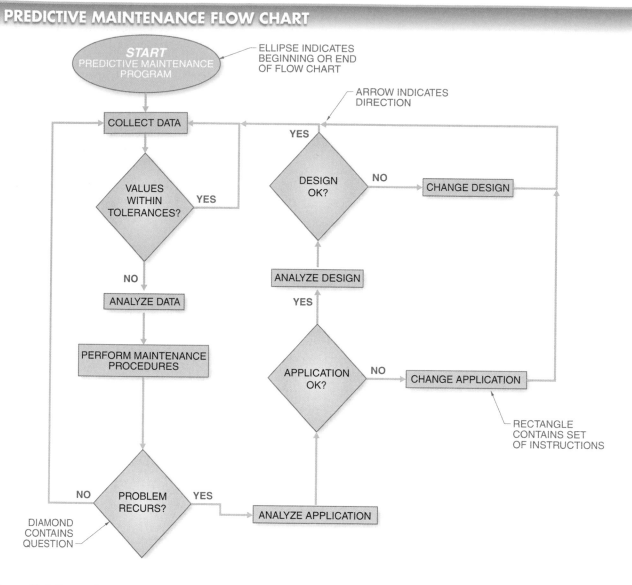

Figure 30-1. Data collected through predictive maintenance procedures is analyzed to check whether values are within acceptable tolerances.

Visual and Auditory Inspection

Visual and auditory inspection is the analysis of the appearance and sound of operating equipment. **See Figure 30-2.** Visual and auditory inspection is the simplest predictive maintenance procedure performed in a plant and requires no tools or equipment. Visual and auditory inspection is most effective when a potential problem is obvious to a trained maintenance technician. Extraordinary operating characteristics are noted and the equipment is scheduled for the required maintenance.

Vibration Analysis

Vibration analysis is the monitoring of equipment vibration characteristics to analyze the equipment condition. **See Figure 30-3.** Equipment failure is frequently caused by worn parts. Worn parts produce increased vibration or noise. An estimate of vibration problems shows that 50% to 60% of damaging machinery vibrations are a result of shaft misalignment, 30% to 40% of damaging vibrations are the result of equipment unbalance, and 20% are the result of resonance. *Resonance* is the magnification of vibration and its noise by 20% or more.

Figure 30-2. Maintenance technicians routinely check the appearance and sound of operating equipment using visual and auditory inspection.

VIBRATION ANALYSIS

UE Systems, Inc.

Figure 30-3. Vibration analysis measures noise or vibration produced by worn parts in equipment.

Lubricating Oil Analysis

Lubricating oil analysis is a predictive maintenance technique that detects and analyzes the presence of acids, dirt, fuel, and wear particles in lubricating oil to predict equipment failure. Lubricating oil analysis is performed on a scheduled basis. An oil sample is taken from a machine to determine the condition of the lubricant and moving parts. Oil samples are commonly sent to a company that specializes in lubricating oil analysis. **See Figure 30-4.**

Wear Particle Analysis. *Wear particle analysis* is the study of wear particles present in lubricating oil. While lubricating oil analysis focuses on the condition of lubricating oil, wear particle analysis focuses on the size, frequency, shape, and composition of particles produced from worn parts. Equipment condition is assessed by monitoring the wear particles in lubricating oil.

Test instruments can be used to collect data during equipment monitoring.

Thermography

Thermography is the use of temperature-indicating devices to detect temperature changes in operating equipment. Inadequate or excessive temperatures can predict potential problems. Temperature is normally measured in degrees Fahrenheit (°F) or degrees Celsius (°C). Temperature-indicating devices commonly used in industrial maintenance are portable or stationary. **See Figure 30-5.**

LUBRICATING OIL ANALYSIS

PREDICT/DLI

Central:
9555 Rockside Road, Suite 350
Cleveland, OH 44125-6231

West Coast:
253 Winslow Way West
Bainbridge Island, WA 98110

LUBRICANT CONDITION REPORT®

MERCURY MACHINE PROCESSES, INC.
3482 Lynwood Avenue
Cincinnati, OH 45208

CRITICAL

SAMPLE ID: USE-DRYER #4-GB
SAMPLE DESC: C1 AGITATOR
COMPONENT: GEARCASE SPEED REDUCER
MANUFACTURER: FALK
LUBRICANT: MOBIL MOBILGEAR 627
RESERVOIR CAP: 65 Gallons [246.04 Liters]
LUBE TIME: 1508 Hours MACHINE TIME: 52898 Hours

SAMPLE DATE: 07/15
REC'D DATE: 07/16
REPORT DATE: 07/16
1st SAMPLE: 04/15
PREV SAMPLE: 06/15 [C]
NO. SAMPLES: 25
PROGRAM: P1

RECOMMENDATIONS:
Iron and copper levels continue to increase and suggest severe corrosion and pitting of the gear teeth, bearings, and any bushings. The physical properties indicate that this lubricant is degraded and can no longer support this agitator. The anti-wear additive package is depleted and it appears that this unit has entered a CRITICAL lubricant condition. Also, the viscosity is higher than expected. This further indicates a degraded lubricant. Consider inspecting this unit for damage and pitting of the gear teeth and bearing components.

PHYSICAL PROPERTIES:

Sample Date	12/10	01/08	06/15	07/15	Reference
Viscosity (cSt)	10.2	13.4	39.8	14.7	11.0
% Water(IR)	2.10	2.50	< 0.10	5.00	n/a
% Oxidation	32.84	37.31	100.00	100.00	n/a
% Nitration	2.99	5.97	67.16	59.70	n/a
% Sulfate	2.99	2.99	17.91	29.85	n/a
% Solvents	< 0.01	0.12	< 0.01	< 0.01	n/a
Refrigerant	Negative	Negative	Negative	Negative	n/a
% Insolubles	< 0.1	< 0.1	5.66	< 0.1	n/a
TAN/TBN					

TRACE ELEMENTS: (ppm)

Equipment Wear	Iron	68	122	799	1773	0
	Chromium	3	5	10	11	0
	Molybdenum	0	0	0	0	0
	Aluminum	2	3	0	0	0
	Copper	78	98	500	1110	0
	Lead	3	5	1	11	1
	Tin	32	43	30	60	0
	Silver	0	0	0	0	0
	Nickel	3	3	0	0	1
	Vanadium	0	0	2	0	2
	Titanium	0	0	0	0	0
Contaminants	Silicon	45	79	0	0	0
	Sodium	85	126	0	0	1
	Potassium	2	0	0	0	1
Additives	Boron	23	22	0	0	28
	Magnesium	22	21	0	0	0
	Calcium	29	28	0	0	31
	Barium	7	5	0	0	11
	Phosphorus	242	123	5	1	391
	Zinc	33	32	0	0	11
	Cadmium	0	0	0	0	0

9999999-15
010003

JKM-00387177
90-95026-042

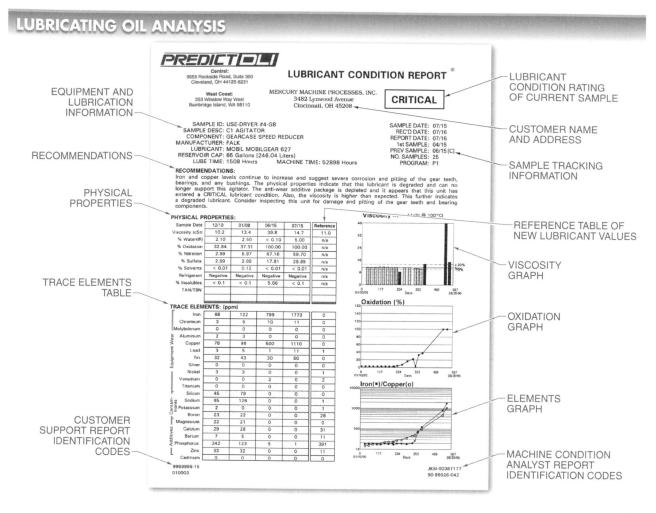

Predict/DLI

Figure 30-4. Lubricating oil analysis is used to predict potential equipment malfunction or failure.

INFRARED THERMOMETER

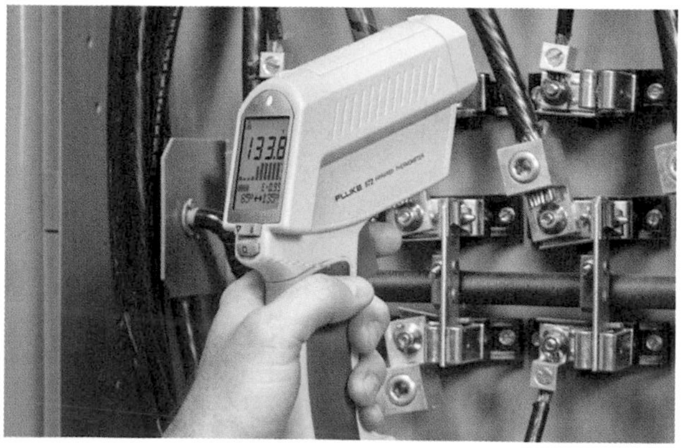

Fluke Corporation

Figure 30-5. An infrared thermometer is a temperature-indicating device that senses infrared radiation emissions to measure temperature without touching an object.

A *thermal imager* is a device that detects heat patterns in the infrared-wavelength spectrum without making direct contact with equipment. The thermal imager displays an electronically processed image of a target with different colors representing the various temperatures of the heat patterns. **See Figure 30-6.**

For example, a quick inspection of an electric motor may reveal that a bearing is significantly warmer than the motor casing. This suggests the possibility of a problem with lubrication or alignment. An alignment problem may also be indicated when one side of the coupling is warmer than the opposite side. **See Figure 30-7.**

Ultrasonic Analysis

Ultrasonic analysis is an analysis that uses high vibration frequencies to create an image or reading. Ultrasonic analysis is similar to vibration analysis. With amplification and proper sound isolation and analysis techniques, ultrasonic analysis can be used for flow measurement of liquid and gas and for leak detection. A transducer senses the vibration created by flowing liquid or gas that produces a signature without intrusion in the line or vessel. The signature is then compared with data collected on similar equipment or data collected at different times. **See Figure 30-8.**

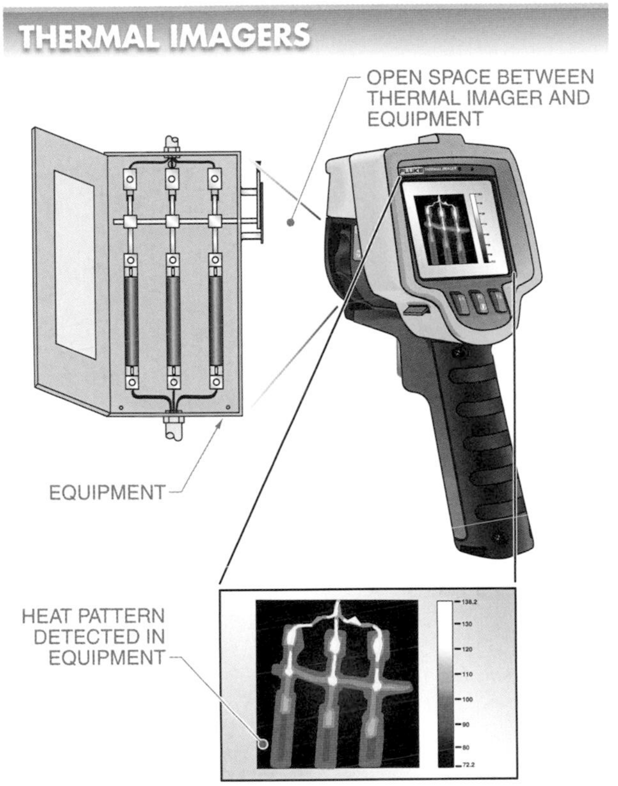

THERMAL IMAGERS

OPEN SPACE BETWEEN THERMAL IMAGER AND EQUIPMENT

EQUIPMENT

HEAT PATTERN DETECTED IN EQUIPMENT

Figure 30-6. A thermal imager is a device that detects heat patterns in the infrared-wavelength spectrum without making direct contact with the equipment.

TROUBLESHOOTING MOTOR BEARINGS

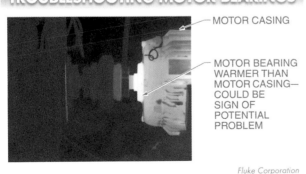

MOTOR CASING

MOTOR BEARING WARMER THAN MOTOR CASING— COULD BE SIGN OF POTENTIAL PROBLEM

Fluke Corporation

Figure 30-7. A motor bearing that is significantly warmer than the motor casing suggests the possibility of a problem with lubrication or alignment.

ULTRASONIC LEAK DETECTORS

Amprobe, Advanced Test Products

Figure 30-8. Ultrasonic leak detectors are used to measure high vibration frequencies when testing for leaks.

Internal flaws or cracks in metal can be located using ultrasonic testers that consist of a transmitting and receiving unit. Ultrasonic waves from the transmitter are passed through the metal being tested. Internal flaws or cracks distort the sound waves as they pass through the metal. The distortions are interpreted by the receiver, which generates a picture or readout of the flaw.

Electrical Analysis

Electrical analysis is an analysis that uses electrical monitoring equipment and/or test instruments to evaluate the quality of electrical power delivered to equipment and the performance of electrical equipment. Electrical monitoring equipment can be permanently installed or it can be portable to allow analysis of equipment condition on location. For example, the condition of a specific electric motor can be checked using an insulation spot test, which uses a megohmmeter to measure the resistance of motor winding insulation. Over time, the effects of heat, aging, and/or contamination can cause insulation breakdown. The values obtained with a test at scheduled time increments are compared with benchmark acceptable readings to predict useful life and motor insulation failure.

30-1 CHECKPOINT

1. What are the three types of equipment monitoring?
2. What type of inspection can be done without using any test equipment?
3. What type of inspection can be done using test equipment?

30-2 MAINTENANCE TECHNICIAN RESOURCES

Maintenance involves the use of information and communication technology, which includes operator manuals, technical service bulletins, troubleshooting reports, electronic monitoring systems, computers, telephones, information from machine operators, and advice from other technicians. Most manufacturers supply maintenance and troubleshooting recommendations and symptom diagnostic guides with their equipment.

Manufacturer information is traditionally found in the operations and maintenance (operator) manual. **See Figure 30-9.** Equipment sales personnel can be excellent sources of maintenance assistance. Equipment sales personnel may also have contact with manufacturer representatives or company engineers who have useful suggestions. Trade journals and magazines are often excellent sources of maintenance information when specific maintenance suggestions cannot be found. Magazine advertisements often contain manufacturer contact numbers and may include manufacturer contact cards that allow a maintenance technician to obtain information from a variety of manufacturers. Often, specific maintenance procedures are discussed in articles or case studies from industry.

Codes and standards may be used by state and local authorities when dictating preventive maintenance requirements. The National Electrical Manufacturers Association (NEMA) produces recommendations for establishing a preventive maintenance system for industrial equipment in NEMA ICS 1.3, *Preventive Maintenance of Industrial Control and Systems Equipment.* The National Fire Protection Association (NFPA) publishes electrical maintenance standards in *NFPA 70B: Recommended Practice for Electrical Equipment Maintenance.* Preventive maintenance requirements are also given in the ASME International Boiler and Pressure Vessel Code, Section VI, *Recommended Rules for the Care and Operation of Heating Boilers,* and Section VII, *Recommended Rules for the Care of Power Boilers.*

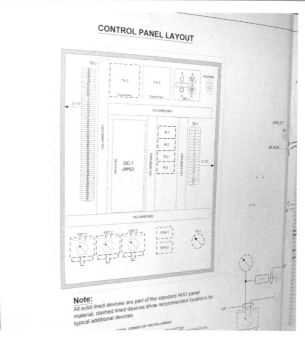

Note:
All solid lined devices are part of the standard AHU panel material, dashed lined devices show recommended locations for typical additional devices.

Electrical diagrams must be updated every time there is a change to the system. This includes the addition of new equipment or the removal of old equipment.

The American Society of Heating, Refrigeration, and Air-Conditioning Engineers (ASHRAE) standards cover topics such as refrigeration, indoor air quality, ventilation standards, building operation and maintenance, and energy efficiency. Some regional and local building and mechanical codes also specify maintenance requirements.

Tech Fact

Electronic versions of service manuals for new equipment can usually be found on the OEM's website. Since hard copy manuals may get lost and electronic files may not be available after a certain time, it is best to download all manuals so that they are always available.

OPERATION AND MAINTENANCE MANUALS

6.17 DRIVE MOTOR LUBRICATION

Induction squirrel cage motors have antifriction ball or roller bearings front and rear. At extended intervals they require lubrication.

The periods between greasings of the motor bearings can vary, primarily with the severity of the service conditions under which the motor operates. As a general rule, the following applies:

Frequency of Lubrication-Normal Environments

Motor Size	Lubrication Interval
25-40 HP	3 Months (or 1000 hr)

NOTE: For severe duty - Dusty locations
                    - High ambient
                       temperatures

Reduce time intervals in preceding table to ½ the listed value.

Lubrication Procedure

> ⚠ **CAUTION**
>
> **Grease should be added when the motor is stopped and power disconnected.**

When greasing, stop motor and remove inlet and outlet plugs. Inlet grease gun fittings and spring-loaded outlets are arranged at each end on the motor housing. Use a hand lever grease gun. Determine the quantity of grease delivered with each stroke of the lever. Add grease in the following quantity:

Motor Frame Size	Lubrication Amount	
	in^3	oz
256-286	1.0	0.8
324-326	1.5	1.2

Do not expect grease to appear at the outlet. If it does, discontinue greasing immediately.

> ⚠ **CAUTION**
>
> **Overgreasing is a major cause of bearing and motor failure. Ensure dirt and contaminants are not introduced when adding grease.**

Run motor for about ten minutes before replacing outlet plug. Certain TEFC motors have a spring relief outlet fitting on the fan end. If the outlet plug is not accessible at surface of hood, it is the spring relief type and need not be removed when greasing.

A major cause of motor bearing failure is overgreasing. The quantity of grease added should be carefully controlled. Small motors must be greased with a lesser amount of grease than large motors.

Recommended Motor Greases
(or equivalents)

Chevron SRI . . . . Standard Oil
                              of California
Premium RB . . . . Texaco
Unirex N2 . . . . Exxon
Dolium R . . . . Shell
Rykon Premium . . . . American Oil

> ⚠ **CAUTION**
>
> **Never mix greases. Mixing greases can cause motor failure.**

Figure 30-9. Preventive maintenance task recommendations are located in the operation and maintenance manuals for each piece of equipment.

Operator manuals may contain troubleshooting information printed in chart form. The chart lists symptoms, possible causes, and suggestions for repairing a problem. **See Figure 30-10.** For example, a symptom of incorrect motor rotation may be caused by the wiring. The solution is to interchange two of the motor conductors.

Troubleshooting information may also be presented as a flow chart. **See Figure 30-11.** The flow chart is read by answering yes or no to questions about the problem at the beginning of the chart and then following an arrow to the next appropriate question. For example, if the answer to the first question is no, then the appropriate action is to turn the power off and check to ensure that the power is OFF. If the answer to the question is yes, the appropriate action is to check to ensure the power is OFF. The arrows are followed and the questions are answered by replying yes or no and following the respective paths that lead to the problem. The flow chart condenses lengthy word descriptions into a chart for easy problem solving.

ELECTRIC MOTOR DRIVE TROUBLESHOOTING CHARTS

Electric Motor Drive Troubleshooting Matrix			
Faults			
Symptom/Fault Code	Problem	Cause	Solution
Electric motor drive overvoltage fault	Electric motor drive overvoltage	Deceleration time is too short	Increase deceleration time
		High input voltage (voltage swell)	*See incoming power troubleshooting matrix*
		Load is overhauling motor	Add dynamic braking resistor and/or increase deceleration time
Component Failures			
Electric motor drive does not turn on. Blown fuse or tripped breaker	Defective converter section (rectifier semiconductor)	High input voltage (voltage swell)	Replace converter section semiconductor or replace electric motor drive *See also incoming power matrix*
		Electric motor drive cooling fan is defective	Replace converter section semiconductor and cooling fan or replace electric motor drive

Figure 30-10. Troubleshooting information may be printed in chart form.

MOTOR TROUBLESHOOTING FLOW CHART

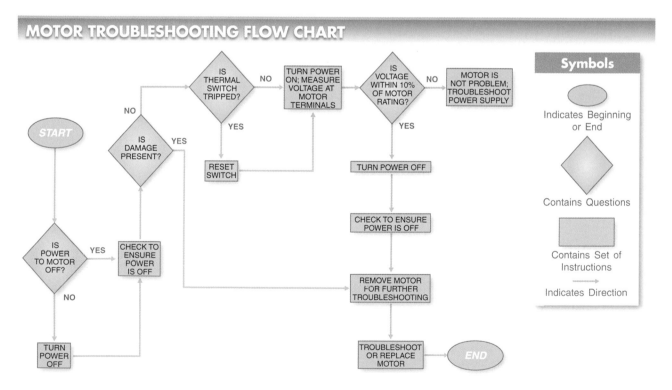

Figure 30-11. Troubleshooting information is often presented as a flow chart.

Technical Service Bulletins and Troubleshooting Reports

A *troubleshooting report* is a record of a specific problem that occurs with a particular piece of equipment. **See Figure 30-12.** Maintenance technicians can evaluate troubleshooting reports to improve their own troubleshooting abilities. Troubleshooting reports that are incorporated into a plant preventive maintenance system become part of the equipment history for each machine. If the cause of the problem is discovered, modifications to the machine or adjustments to the machine preventive maintenance system are made.

A troubleshooting report is filled out for each breakdown or equipment problem immediately after the problem is solved. The information about the problem is filed manually or entered into a computer for future reference. The next time a particular machine requires troubleshooting, the technician can access the machine troubleshooting report to learn whether the symptoms of the current problem have previously occurred. Troubleshooting reports can result in a tremendous saving of time. Each troubleshooting report should include standard information such as the individuals who worked on the problem, department, equipment identification number, problems, symptoms, causes, repair procedures, and preventive maintenance actions.

Out-of-Plant Services

Out-of-plant services, such as outside advice, are often necessary when decisions must be made or work must be completed that requires expertise not found in a plant maintenance crew. In addition, the installation of new equipment or major renovations may require companies with specialized tools and expertise. Outside companies can install new equipment or make major renovations more efficiently than a plant maintenance crew. In addition, outside companies provide warranties for their work should problems occur during the equipment break-in period.

TROUBLESHOOTING REPORTS

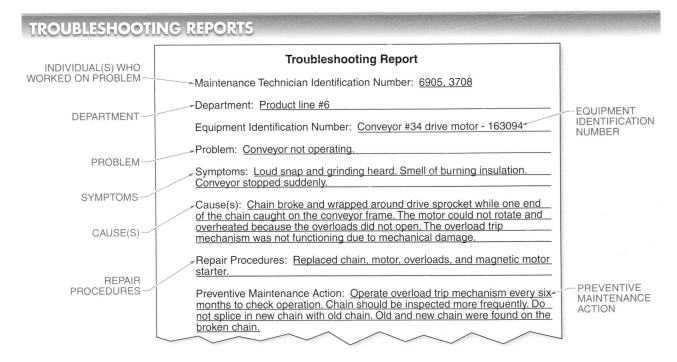

Figure 30-12. A troubleshooting report is a record of a specific problem that occurs in a particular piece of equipment.

30-2 CHECKPOINT

1. How does a flow chart direct a person to information that contains instructions on what to try or do?

2. What are the acronyms of some of the organizations that produce recommendations, standards, codes, and requirements for preventive maintenance?

Additional Resources

Review and Resources

Access Chapter 30 Review and Resources for *Electrical Motor Controls for Integrated Systems* by scanning the above QR code with your mobile device.

Applying Your Knowledge

Refer to the *Electrical Motor Controls for Integrated Systems* Learner Resources for interactive Applying Your Knowledge questions.

Workbook and Applications Manual

Refer to Chapter 30 in the *Electrical Motor Controls for Integrated Systems Workbook* and the *Applications Manual* for additional exercises.

ENERGY EFFICIENCY PRACTICES

Total Quality Maintenance

Total quality maintenance (TQM) is maintenance that incorporates information from a predictive maintenance program into the design and purchase specifications of equipment to reduce overall maintenance costs. In a TQM plant, maintenance technicians collaborate with engineering and purchasing personnel. Equipment operators perform routine equipment maintenance as a standard task. Maintenance technicians perform more complex maintenance tasks. The goal of TQM is zero breakdowns and zero tolerance from established plant quality standards.

Data from a facility's preventive maintenance program and plant survey is used to develop the master equipment list. Periodic work orders are developed and scheduled. The predictive maintenance work is completed and documented. Equipment history is developed from predictive maintenance work and corrective, emergency, and project work completed. Equipment design, modification, and purchase are based on equipment history and input from affected personnel. Any equipment change data is added to the master equipment list.

MotorMaster+ is a software program provided by the U.S. Department of Energy that allows comparison of various motors and motor characteristics and can be used when designing, modifying, and replacing equipment. Motors can be selected based on size, speed, enclosure type, and required features. The software provides economic comparisons of new motors, new motors versus existing motors, or replacement motors versus rewound motors. The software also allows maintenance technicians to enter nameplate and test data from each motor in a plant and run a batch analysis of the best upgrade opportunities based on new and existing efficiencies, annual operating hours, energy rates, percent full load, and utility rebate criteria.

Of the millions of motors of more than 1 HP in service in U.S. plants today, if energy-efficient motors replaced 20% as they failed, billions of dollars of energy would be saved, and the release of millions of tons of air pollution would be prevented each year.

Appendix

METRIC PREFIXES

Multiples and Submultiples	Prefixes	Symbols	Meaning
$1,000,000,000,000 = 10^{12}$	tera	T	trillion
$1,000,000,000 = 10^{9}$	giga	G	billion
$1,000,000 = 10^{6}$	mega	M	million
$1000 = 10^{3}$	kilo	k	thousand
$100 = 10^{2}$	hecto	h	hundred
$10 = 10^{1}$	deka	d	ten
Unit $1 = 10^{0}$			
$0.1 = 10^{-1}$	deci	d	tenth
$0.01 = 10^{-2}$	centi	c	hundredth
$0.001 = 10^{-3}$	milli	m	thousandth
$0.000001 = 10^{-6}$	micro	μ	millionth
$0.000000001 = 10^{-9}$	nano	n	billionth
$0.000000000001 = 10^{-12}$	pico	p	trillionth

METRIC CONVERSIONS

Initial Units	Final Units											
	giga	mega	kilo	hecto	deka	base unit	deci	centi	milli	micro	nano	pico
giga		3R	6R	7R	8R	9R	10R	11R	12R	15R	18R	21R
mega	3L		3R	4R	5R	6R	7R	8R	9R	12R	15R	18R
kilo	6	3L		1R	2R	3R	4R	5R	6R	9R	12R	15R
hecto	7L	4L	1L		1R	2R	3R	4R	5R	8R	11R	14R
deka	8L	5L	2L	1L		1R	2R	3R	4R	7R	10R	13R
base unit	9L	6L	3L	2L	1L		1R	2R	3R	6R	9R	12R
deci	10L	7L	4L	3L	2L	1L		1R	2R	5R	8R	11R
centi	11L	8L	5L	4L	3L	2L	1L		1R	4R	7R	10R
milli	12L	9L	6L	5L	4L	3L	2L	1L		3R	6R	9R
micro	15L	12L	9L	8L	7L	6L	5L	4L	3L		3R	6R
nano	18L	15L	12L	11L	10L	9L	8L	7L	6L	3L		3R
pico	21L	18L	15L	14L	13L	12L	11L	10L	9L	6L	3L	

COMMON PREFIXES

Symbol	Prefix	Equivalent
G	giga	1,000,000,000
M	mega	1,000,000
k	kilo	1000
base unit	—	1
m	milli	0.001
μ	micro	0.000001
n	nano	0.000000001
p	pico	0.000000000001
Z	impedance	ohms — Ω

THREE-PHASE VOLTAGE VALUES

For 208 V × 1.732, use 360
For 230 V × 1.732, use 398
For 240 V × 1.732, use 416
For 440 V × 1.732, use 762
For 460 V × 1.732, use 797
For 480 V × 1.732, use 831
For 2400 V × 1.732, use 4157
For 4160 V × 1.732, use 7205

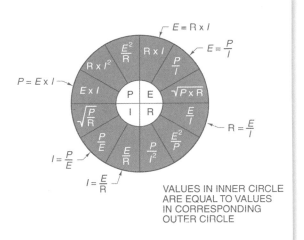

VALUES IN INNER CIRCLE
ARE EQUAL TO VALUES
IN CORRESPONDING
OUTER CIRCLE

OHM'S LAW AND POWER FORMULA

POWER FORMULA ABBREVIATIONS AND SYMBOLS

P = Watts	V = Volts
I = Amps	VA = Volt Amps
A = Amps	φ = Phase
R = Ohms	√ = Square Root
E = Volts	

POWER FORMULAS — 1φ, 3φ

Phase	To Find	Use Formula	Example		
			Given	Find	Solution
1φ	I	$I = \dfrac{VA}{V}$	32,000 VA, 240 V	I	$I = \dfrac{VA}{V}$ $I = \dfrac{32,000 \text{ VA}}{240 \text{ V}}$ $I = \textbf{133 A}$
1φ	VA	$VA = I \times V$	100 A, 240 V	VA	$VA = I \times V$ $VA = 100 \text{ A} \times 240 \text{ V}$ $VA = \textbf{24,000 VA}$
1φ	V	$V = \dfrac{VA}{I}$	42,000 VA, 350 A	V	$V = \dfrac{VA}{I}$ $V = \dfrac{42,000 \text{ VA}}{350 \text{ A}}$ $V = \textbf{120 V}$
3φ	I	$I = \dfrac{VA}{V \times \sqrt{3}}$	72,000 VA, 208 V	I	$I = \dfrac{VA}{V \times \sqrt{3}}$ $I = \dfrac{72,000 \text{ VA}}{360 \text{ V}}$ $I = \textbf{200 A}$
3φ	VA	$VA = I \times V \times \sqrt{3}$	2 A, 240 V	VA	$VA = I \times V \times \sqrt{3}$ $VA = 2 \times 416$ $VA = \textbf{832 VA}$

CAPACITORS

Connected in Series		Connected in Parallel	Connected in Series/Parallel
Two Capacitors	Three or More Capacitors		
$C_T = \dfrac{C_1 \times C_2}{C_1 + C_2}$ where C_T = total capacitance (in μF) C_1 = capacitance of capacitor 1 (in μF) C_2 = capacitance of capacitor 2 (in μF)	$\dfrac{1}{C_T} = \dfrac{1}{C_1} + \dfrac{1}{C_2} + \ldots$	$C_T = C_1 + C_2 + \ldots$	1. Calculate the capacitance of the parallel branch. 2. Calculate the capacitance of the series combination. $C_T = \dfrac{C_1 \times C_2}{C_1 + C_2}$

TEMPERATURE CONVERSIONS

Convert °C to °F	Convert °F to °C
$°F = (1.8 \times °C) + 32$	$°C = \dfrac{°F - 32}{1.8}$

UNITS OF POWER

Power	W	ft lb/s	HP	kW
Watt	1	0.7376	0.341×10^{-3}	0.001
Foot-pound/sec	1.356	1	0.818×10^{-3}	1.356×10^{-3}
Horsepower	745.7	550	1	0.7457
Kilowatt	1000	736.6	1.341	1

BRANCH CIRCUIT VOLTAGE DROP

$$\%V_D = \frac{V_{NL} - V_{FL}}{V_{FL}} \times 100$$

where
$\%V_D$ = percent voltage drop (in volts)
V_{NL} = no-load voltage drop (in volts)
V_{FL} = full-load voltage drop (in volts)
100 = constant

STANDARD SIZES OF FUSES AND CBs

NEC® 240.6(A) lists standard ampere ratings of fuses and fixed-trip CBs as follows:
15, 20, 25, 30, 35, 40, 45,
50, 60, 70, 80, 90, 100, 110,
125, 150, 175, 200, 225,
250, 300, 350, 400, 450,
500, 600, 700, 800,
1000, 1200, 1600,
2000, 2500, 3000, 4000, 5000, 6000

VOLTAGE CONVERSIONS

To Convert	To	Multiply By
rms	Average	0.9
rms	Peak	1.414
Average	rms	1.111
Average	Peak	1.567
Peak	rms	0.707
Peak	Average	0.637
Peak	Peak-to-peak	2

TYPICAL MOTOR EFFICIENCIES

HP	Standard Motor (%)	Energy-Efficient Motor (%)	HP	Standard Motor (%)	Energy-Efficient Motor (%)
1	76.5	84.0	30	88.1	93.1
1.5	78.5	85.5	40	89.3	93.6
2	79.9	86.5	50	90.4	93.7
3	80.8	88.5	75	90.8	95.0
5	83.1	88.6	100	91.6	95.4
7.5	83.8	90.2	125	91.8	95.8
10	85.0	90.3	150	92.3	96.0
15	86.5	91.7	200	93.3	96.1
20	87.5	92.4	250	93.6	96.2
25	88.0	93.0	300	93.8	96.5

ELECTRICAL/ELECTRONIC ABBREVIATIONS/ACRONYMS

Abbr/ Acronym	Meaning	Abbr/ Acronym	Meaning	Abbr/ Acronym	Meaning
A	Ammeter; Ampere; Anode; Armature	FU	Fuse	PNP	Positive-Negative-Positive
AC	Alternating Current	FWD	Forward	POS	Positive
AC/DC	Alternating Current; Direct Current	G	Gate; Giga; Green; Conductance	POT.	Potentiometer
A/D	Analog to Digital	GEN	Generator	P-P	Peak-to-Peak
AF	Audio Frequency	GRD	Ground	PRI	Primary Switch
AFC	Automatic Frequency Control	GY	Gray	PS	Pressure Switch
Ag	Silver	H	Henry; High Side of Transformer; Magnetic Flux	PSI	Pounds Per Square Inch
ALM	Alarm			PUT	Pull-Up Torque
AM	Ammeter; Amplitude Modulation	HF	High Frequency	Q	Transistor
AM/FM	Amplitude Modulation; Frequency Modulation	HP	Horsepower	R	Radius; Red; Resistance; Reverse
		Hz	Hertz	RAM	Random-Access Memory
ARM.	Armature	I	Current	RC	Resistance-Capacitance
Au	Gold	IC	Integrated Circuit	RCL	Resistance-Inductance-Capacitance
AU	Automatic	INT	Intermediate; Interrupt	REC	Rectifier
AVC	Automatic Volume Control	INTLK	Interlock	RES	Resistor
AWG	American Wire Gauge	IOL	Instantaneous Overload	REV	Reverse
BAT.	Battery (electric)	IR	Infrared	RF	Radio Frequency
BCD	Binary Coded Decimal	ITB	Inverse Time Breaker	RH	Rheostat
BJT	Bipolar Junction Transistor	ITCB	Instantaneous Trip Circuit Breaker	rms	Root Mean Square
BK	Black	JB	Junction Box	ROM	Read-Only Memory
BL	Blue	JFET	Junction Field-Effect Transistor	rpm	Revolutions Per Minute
BR	Brake Relay; Brown	K	Kilo; Cathode	RPS	Revolutions Per Second
C	Celsius; Capacitiance; Capacitor	L	Line; Load; Coil; Inductance	S	Series; Slow; South; Switch
CAP.	Capacitor	LB-FT	Pounds Per Foot	SCR	Silicon Controlled Rectifier
CB	Circuit Breaker; Citizen's Band	LB-IN.	Pounds Per Inch	SEC	Secondary
CC	Common-Collector Configuration	LC	Inductance-Capacitance	SF	Service Factor
CCW	Counterclockwise	LCD	Liquid Crystal Display	1 PH; 1φ	Single-Phase
CE	Common-Emitter Configuration	LCR	Inductance-Capacitance-Resistance	SOC	Socket
CEMF	Counter Electromotive Force	LED	Light Emitting Diode	SOL	Solenoid
CKT	Circuit	LRC	Locked Rotor Current	SP	Single-Pole
CONT	Continuous; Control	LS	Limit Switch	SPDT	Single-Pole, Double-Throw
CPS	Cycles Per Second	LT	Lamp	SPST	Single-Pole, Single-Throw
CPU	Central Processing Unit	M	Motor; Motor Starter; Motor Starter Contacts	SS	Selector Switch
CR	Control Relay			SSW	Safety Switch
CRM	Control Relay Master	MAX.	Maximum	SW	Switch
CT	Current Transformer	MB	Magnetic Brake	T	Tera; Terminal; Torque; Transformer
CW	Clockwise	MCS	Motor Circuit Switch	TB	Terminal Board
D	Diameter; Diode; Down	MEM	Memory	3 PH; 3φ	Three-Phase
D/A	Digital to Analog	MED	Medium	TD	Time Delay
DB	Dynamic Braking Contactor; Relay	MIN	Minimum	TDF	Time Delay Fuse
DC	Direct Current	MN	Manual	TEMP	Temperature
DIO	Diode	MOS	Metal-Oxide Semiconductor	THS	Thermostat Switch
DISC.	Disconnect Switch	MOSFET	Metal-Oxide Semiconductor Field-Effect Transistor	TR	Time Delay Relay
DMM	Digital Multimeter			TTL	Transistor-Transistor Logic
DP	Double-Pole	MTR	Motor	U	Up
DPDT	Double-Pole, Double-Throw	N; NEG	North; Negative	UCL	Unclamp
DPST	Double-Pole, Single-Throw	NC	Normally Closed	UHF	Ultrahigh Frequency
DS	Drum Switch	NEUT	Neutral	UJT	Unijunction Transistor
DT	Double-Throw	NO	Normally Open	UV	Ultraviolet; Undervoltage
DVM	Digital Voltmeter	NPN	Negative-Positive-Negative	V	Violet; Volt
EMF	Electromotive Force	NTDF	Nontime-Delay Fuse	VA	Volt Amp
F	Fahrenheit; Fast; Field; Forward; Fuse	O	Orange	VAC	Volts Alternating Current
FET	Field-Effect Transistor	OCPD	Overcurrent Protection Device	VDC	Volts Direct Current
FF	Flip-Flop	OHM	Ohmmeter	VHF	Very High Frequency
FLC	Full-Load Current	OL	Overload Relay	VLF	Very Low Frequency
FLS	Flow Switch	OZ/IN.	Ounces Per Inch	VOM	Volt-Ohm-Milliammeter
FLT	Full-Load Torque	P	Peak; Positive; Power; Power Consumed	W	Watt; White
FM	Fequency Modulation	PB	Pushbutton	w/	With
FREQ	Frequency	PCB	Printed Circuit Board	X	Low Side of Transformer
FS	Float Switch	PH; φ	Phase	Y	Yellow
FTS	Foot Switch	PLS	Plugging Switch	Z	Impedance

AC MOTOR CHARACTERISTICS

Motor Type 1φ	Typical Voltage	Starting Ability (Torque)	Size (HP)	Speed Range (rpm)	Cost*	Typical Uses
Shaded-pole	115 V, 230 V	Very low 50% to 100% of full load	Fractional 1/2 HP to 1/3 HP	Fixed 900, 1200, 1800, 3600	Very low 75% to 85%	Light-duty applications such as small fans, hair dryers, blowers, and computers
Split-phase	115 V, 230 V	Low 75% to 200% of full load	Fractional 1/3 HP or less	Fixed 900, 1200, 1800, 3600	Low 85% to 95%	Low-torque applications such as pumps, blowers, fans, and machine tools
Capacitor-start	115 V, 230 V	High 200% to 350% of full load	Fractional to 3 HP	Fixed 900, 1200, 1800	Low 90% to 110%	Hard-to-start loads such as refrigerators, air compressors, and power tools
Capacitor-run	115 V, 230 V	Very low 50% to 100% of full load	Fractional to 5 HP	Fixed 900, 1200, 1800	Low 90% to 110%	Applications that require a high running torque such as pumps and conveyors
Capacitor-start-and-run	115 V, 230 V	Very high 350% to 450% of full load	Fractional to 10 HP	Fixed 900, 1200, 1800	Low 100% to 115%	Applications that require both a high starting and running torque such as loaded conveyors
3φ Induction	230 V, 460 V	Low 100% to 175% of full load	Fractional to over 500 HP	Fixed 900, 1200, 3600	Low 100%	Most industrial applications
Wound rotor	230 V, 460 V	High 200% to 300% of full load	1/2 HP to 200 HP	Varies by changing resistance in rotor	Very high 250% to 350%	Applications that require high torque at different speeds such as cranes and elevators
Synchronous	230 V, 460 V	Very low 40% to 100% of full load	Fractional to 250 HP	Exact constant speed	High 200% to 250%	Applications that require very slow speeds and correct power factors

* based on standard 3φ induction motor

DC AND UNIVERSAL MOTOR CHARACTERISTICS

Motor Type	Typical Voltage	Starting Ability (Torque)	Size (HP)	Speed Range (rpm)	Cost*	Typical Uses
DC Series	12 V, 90 V, 120 V, 180 V	Very high 400% to 450% of full load	Fractional to 100 HP	Varies 0 to full speed	High 175% to 225%	Applications that require very high torque such as hoists and bridges
Shunt	12 V, 90 V, 120 V, 180 V	Low 125% to 250% of full load	Fractional to 100 HP	Fixed or adjustable below full speed	High 175% to 225%	Applications that require better speed control than a series motor such as woodworking machines
Compound	12 V, 90 V, 120 V, 180 V	High 300% to 400% of full load	Fractional to 100 HP	Fixed or adjustable	High 175% to 225%	Applications that require high torque and speed control such as printing presses, conveyors, and hoists
Permanent-magnet	12 V, 24 V, 36 V, 120 V	Low 100% to 200% of full load	Fractional	Varies from 0 to full speed	High 150% to 200%	Applications that require small DC-operated equipment such as automobile power windows, seats, and sun roofs
Stepping	5 V, 12 V, 24 V	Very low** 0.5 to 5000 oz/in.	Size rating is given as holding torque and number of steps	Rated in number of steps per sec (maximum)	Varies based on number of steps and rated torque	Applications that require low torque and precise control such as indexing tables and printers
AC/DC Universal	115 VAC, 230 VAC, 12 VDC, 24 VDC, 36 VDC, 120 VDC	High 300% to 400% of full load	Fractional	Varies 0 to full speed	High 175% to 225%	Most portable tools such as drills, routers, mixers, and vacuum cleaners

* based on standard 3φ induction motor

** torque is rated as holding torque

OVERCURRENT PROTECTIVE DEVICES

Motor Type	Code Letter	FLC (%)			
		Motor Size	TDF	NTDF	ITB	ITCB
AC*	—	—	175	300	150	700
AC*	A	—	150	150	150	700
AC*	B–E	—	175	250	200	700
AC*	F–V	—	175	300	250	700
DC	—	1/8 to 50 HP	150	150	150	150
DC	—	Over 50 HP	150	150	150	175

* full-voltage and resistor starting

FULL-LOAD CURRENTS—DC MOTORS

Motor Rating (HP)	Current (A)	
	120 V	240 V
¼	3.1	1.6
⅓	4.1	2.0
½	5.4	2.7
¾	7.6	3.8
1	9.5	4.7
1½	13.2	6.6
2	17	8.5
3	25	12.2
5	40	20
7½	48	29
10	76	38

FULL-LOAD CURRENTS—1φ, AC MOTORS

Motor Rating (HP)	Current (A)	
	115 V	230 V
⅙	4.4	2.2
¼	5.8	2.9
⅓	7.2	3.6
½	9.8	4.9
¾	13.8	6.9
1	16	8
1½	20	10
2	24	12
3	34	17
5	56	28
7½	80	40

FULL-LOAD CURRENTS—3φ, AC INDUCTION MOTORS

Motor Rating (HP)	Current (A)			
	208 V	230 V	460 V	575 V
¼	1.11	0.96	0.48	0.38
⅓	1.34	1.18	0.59	0.47
½	2.2	2.0	1.0	0.8
¾	3.1	2.8	1.4	1.1
1	4.0	3.6	1.8	1.4
1½	5.7	5.2	2.6	2.1
2	7.5	6.8	3.4	2.7
3	10.6	9.6	4.8	3.9
5	16.7	15.2	7.6	6.1
7½	24.0	22.0	11.0	9.0
10	31.0	28.0	14.0	11.0
15	46.0	42.0	21.0	17.0
20	59	54	27	22
25	75	68	34	27
30	88	80	40	32
40	114	104	52	41
50	143	130	65	52
60	169	154	77	62
75	211	192	96	77
100	273	248	124	99
125	343	312	156	125
150	396	360	180	144
200	—	480	240	192
250	—	602	301	242
300	—	—	362	288
350	—	—	413	337
400	—	—	477	382
500	—	—	590	472

MOTOR FRAME DIMENSIONS

Frame No.	Shaft U	Shaft V	Key W	Key T	Key L	A	B	D	E	F	BA
48	½	1½*	flat	3/64	—	5⅝*	3½*	3	2⅛	1⅜	2½
56	⅝	1⅞*	3/16	3/16	1⅜	6½*	4¼*	3½	2 7/16	1½	2¾
143T	⅞	2	3/16	3/16	1⅜	7	6	3½	2¾	2	2¼
145T	⅞	2	3/16	3/16	1⅜	7	6	3½	2¾	2½	2¼
182	⅞	2	3/16	3/16	1⅜	9	6½	4½	3¾	2¼	2¾
182T	1⅛	2½	¼	¼	1¾	9	6½	4½	3¾	2¼	2¾
184	⅞	2	3/16	3/16	1⅜	9	7½	4½	3¾	2¾	2¾
184T	1⅛	2½	¼	¼	1¾	9	7½	4½	3¾	2¾	2¾
203	¾	2	3/16	3/16	1⅜	10	7½	5	4	2¾	3⅛
204	¾	2	3/16	3/16	1⅜	10	8½	5	4	3¼	3⅛
213	1⅛	2¾	¼	¼	2	10½	7½	5¼	4¼	2¾	3½
213T	1⅜	3⅛	5/16	5/16	2⅜	10½	7½	5¼	4¼	2¾	3½
215	1⅛	2¾	¼	¼	2	10½	9	5¼	4¼	3½	3½
215T	1⅜	3⅛	5/16	5/16	2⅜	10½	9	5¼	4¼	3½	3½
224	1	2¾	¼	¼	2	11	8¾	5½	4½	3⅜	3½
225	1	2¾	¼	¼	2	11	9½	5½	4½	3¾	3½
254	1⅛	3⅛	¼	¼	2⅜	12½	10¾	6¼	5	4⅛	4¼
254U	1⅜	3½	5/16	5/16	2¾	12½	10¾	6¼	5	4⅛	4¼
254T	1⅝	3¾	⅜	⅜	2⅞	12½	10¾	6¼	5	4⅛	4¼
256U	1⅜	3½	5/16	5/16	2¾	12½	12½	6¼	5	5	4¼
256T	1⅝	3¾	⅜	⅜	2⅞	12½	12½	6¼	5	5	4¼
284	1¼	3½	¼	¼	2¾	14	12½	7	5½	4¾	4¾
284U	1⅝	4⅝	⅜	⅜	3¾	14	12½	7	5½	4¾	4¾
284T	1⅞	4⅜	½	½	3¼	14	12½	7	5½	4¾	4¾
284TS	1⅝	3	⅜	⅜	1⅞	14	12½	7	5½	4¾	4¾
286U	1⅝	4⅝	⅜	⅜	3¾	14	14	7	5½	5½	4¾
286T	1⅞	4⅜	½	½	3¼	14	14	7	5½	5½	4¾
286TS	1⅝	3	⅜	⅜	1⅞	14	14	7	5½	5½	4¾
324	1⅝	4⅝	⅜	⅜	3¾	16	14	8	6¼	5¼	5¼
324U	1⅞	5⅜	½	½	4¼	16	14	8	6¼	5¼	5¼
324S	1⅝	3	⅜	⅜	1⅞	16	14	8	6¼	5¼	5¼
324T	2⅛	5	½	½	3⅞	16	14	8	6¼	5¼	5¼
324TS	1⅞	3½	½	½	2	16	14	8	6¼	5¼	5¼
326	1⅝	4⅝	⅜	⅜	3¾	16	15½	8	6¼	6	5¼
326U	1⅞	5⅜	½	½	4¼	16	15½	8	6¼	6	5¼
326S	1⅝	3	⅜	⅜	1⅞	16	15½	8	6¼	6	5¼
326T	2⅛	5	½	½	3⅞	16	15½	8	6¼	6	5¼
326TS	1⅞	3½	½	½	2	16	15½	6	6¼	6	5¼
364	1⅞	5⅜	½	½	4¼	18	15¼	9	7	5⅝	5⅞
364S	1⅝	3	⅜	⅜	1⅞	18	15¼	9	7	5⅝	5⅞
364U	2⅛	6⅛	½	½	5	18	15¼	9	7	5⅝	5⅞

* not NEMA standard dimensions

MOTOR FRAME LETTERS

Letter	Designation
G	Gasoline pump motor
K	Sump pump motor
M and N	Oil burner motor
S	Standard short shaft for direct connection
T	Standard dimensions established
U	Previously used as frame designation for which standard dimensions are established
Y	Special mounting dimensions required from manufacturer
Z	Standard mounting dimensions except shaft extension

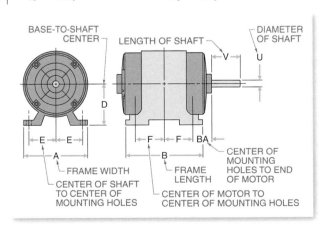

3φ, 230 V MOTORS AND CIRCUITS — 240 V SYSTEM

| Size of motor | | Motor overload protection — Low-peak or Fusetron® | | Switch 115% minimum or HP rated or fuse holder size | Minimum size of starter | Controller termination temperature rating | | | | Minimum size of copper wire and trade conduit | |
| | | | | | | 60°C | | 75°C | | | |
HP	Amp	Motor less than 40°C or greater than 1.15 SF (Max fuse 125%)	All other motors (Max fuse 115%)			TW	THW	TW	THW	Wire size (AWG or kcmil)	Conduit (inches)
½	2	2½	2¼	30	00	•	•	•	•	14	½
¾	2.8	3½	3²/₁₀	30	00	•	•	•	•	14	½
1	3.6	4½	4	30	00	•	•	•	•	14	½
1½	5.2	6¼	5⁶/₁₀	30	00	•	•	•	•	14	½
2	6.8	8	7½	30	0	•	•	•	•	14	½
3	9.6	12	10	30	0	•	•	•	•	14	½
5	15.2	17½	17½	30	1	•	•	•	•	14	½
7½	22	25	25	30	1					10	½
10	28	35	30	60	2	•	•	•		8	¾
									•	10	½
15	42	50	45	60	2	•	•	•		6	1
									•	6	¾
20	54	60	60	100	3	•	•	•	•	4	1
25	68	80	75	100	3	•	•			3	1¼
								•		3	1
									•	4	1
30	80	100	90	100	3	•	•	•		1	1¼
									•	3	1¼
40	104	125	110	200	4	•	•	•		2/0	1½
									•	1	1¼
50	130	150	150	200	4	•	•	•		3/0	2
									•	2/0	1½
75	192	225	200	400	5	•	•	•		300	2½
									•	250	2½
100	248	300	250	400	5	•	•	•		500	3
									•	350	2½
150	360	450	400	600	6	•	•	•		300-2/φ*	2-2½*
									•	4/0-2/φ*	2-2*

* two sets of multiple conductors and two runs of conduit required

3φ, 460 V MOTORS AND CIRCUITS — 480 V SYSTEM

Size of motor — HP	Amp	Motor overload protection — Low-peak or Fusetron® — Motor less than 40°C or greater than 1.15 SF (Max fuse 125%)	All other motors (Max fuse 115%)	Switch 115% minimum or HP rated or fuze holder size	Minimum size of starter	60°C TW	60°C THW	75°C TW	75°C THW	Wire size (AWG or kcmil)	Conduit (inches)
$\frac{1}{2}$	1	$1\frac{1}{4}$	$1\frac{1}{8}$	30	00	•	•	•	•	14	$\frac{1}{2}$
$\frac{3}{4}$	1.4	$1\frac{6}{10}$	$1\frac{6}{10}$	30	00	•	•	•	•	14	$\frac{1}{2}$
1	1.8	$2\frac{1}{4}$	2	30	00	•	•	•	•	14	$\frac{1}{2}$
$1\frac{1}{2}$	2.6	$3\frac{2}{10}$	$2\frac{6}{10}$	30	00	•	•	•	•	14	$\frac{1}{2}$
2	3.4	4	$3\frac{1}{2}$	30	00	•	•	•	•	14	$\frac{1}{2}$
3	4.8	$5\frac{6}{10}$	5	30	0	•	•	•	•	14	$\frac{1}{2}$
5	7.6	9	8	30	0	•	•	•	•	14	$\frac{1}{2}$
$7\frac{1}{2}$	11	12	12	30	1	•	•	•	•	14	$\frac{1}{2}$
10	14	$17\frac{1}{2}$	15	30	1	•	•	•	•	14	$\frac{1}{2}$
15	21	25	20	30	2	•	•	•	•	10	$\frac{1}{2}$
20	27	30	30	60	2	•	•	•		8	$\frac{3}{4}$
									•	10	$\frac{1}{2}$
25	34	40	35	60	2	•	•	•		6	1
									•	9	$\frac{3}{4}$
30	40	50	45	60	3	•	•	•		6	1
									•	9	$\frac{3}{4}$
40	52	60	60	100	3	•	•	•		4	1
									•	6	1
50	65	80	70	100	3	•	•	•		3	$1\frac{1}{4}$
									•	4	1
60	77	90	80	100	4	•	•	•		1	$1\frac{1}{4}$
									•	3	$1\frac{1}{4}$
75	96	110	110	200	4	•	•	•		1/0	$1\frac{1}{2}$
									•	1	$1\frac{1}{4}$
100	124	150	125	200	4	•	•	•		3/0	2
									•	2/0	$1\frac{1}{2}$
125	156	175	175	200	5	•	•	•		4/0	2
									•	3/0	2
150	180	225	200	400	5	•	•	•		300	$2\frac{1}{2}$
									•	4/0	2
200	240	300	250	400	5	•	•	•		500	3
									•	350	$2\frac{1}{2}$
250	302	350	325	400	6	•	•	•		4/0-2/φ*	2-2*
									•	3/0-2/φ*	2-2*
300	361	450	400	600	6	•	•	•		300-2/φ*	2-1$\frac{1}{2}$*
									•	4/0-2/φ*	2-2*

* two sets of multiple conductors and two runs of conduit required

NEMA ENCLOSURE CLASSIFICATION

Type	Use	Service Conditions	Tests	Comments	Type
1	Indoor	No unusual	Rod entry, rust resistance		
3	Outdoor	Windblown dust, rain, sleet, and ice on enclosure	Rain, external icing, dust, and rust resistance	Does not provide protection against internal condensation or internal icing	1
3R	Outdoor	Falling rain and ice on enclosure	Rod entry, rain, external icing, and rust resistance	Does not provide protection against dust, internal condensation, or internal icing	
4	Indoor/outdoor	Windblown dust and rain, splashing water, hose-directed water, and ice on enclosure	Hosedown, external icing, and rust resistance	Does not provide protection against internal condensation or internal icing	4
4X	Indoor/outdoor	Corrosion, windblown dust and rain, splashing water, hose-directed water, and ice on enclosure	Hosedown, external icing, and corrosion resistance	Does not provide protection against internal condensation or internal icing	
6	Indoor/outdoor	Occasional temporary submersion at a limited depth			4X
6P	Indoor/outdoor	Prolonged submersion at a limited depth			
7	Indoor locations classified as Class I, Groups A, B, C, or D, as defined in the NEC®	Withstand and contain an internal explosion of specified gases, contain an explosion of specified gases, contain an explosion sufficiently so an explosive gas-air mixture in the atmosphere is not ignited	Explosion, hydrostatic, and temperature	Enclosed heat-generating devices shall not cause external surfaces to reach temperatures capable of igniting explosive gas-air mixtures in the atmosphere	7
9	Indoor locations classified as Class II, Groups E or G, as defined in the NEC*	Dust	Dust penetration, temperature, and gasket aging	Enclosed heat-generating devices shall not cause external surfaces to reach temperatures capable of igniting explosive gas-air mixtures in the atmosphere	9
12	Indoor	Dust, falling dirt, and dripping noncorrosive liquids	Drip, dust, and rust resistance	Does not provide protection against internal condensation	12
13	Indoor	Dust, spraying water, oil, and noncorrosive coolant	Oil explosion and rust resistance	Does not provide protection against internal condensation	

IEC ENCLOSURE CLASSIFICATION

IEC Publication 529 describes standard degrees of protection that enclosures of a product must provide when properly installed. The degree of protection is indicated by two letters, IP, and two numerals. International Standard IEC 529 contains descriptions and associated test requirements to define the degree of protection that each numeral specifies. The following table indicates the general degrees of protection. For complete test requirements refer to IEC 529.

First Numeral*†	Second Numeral*†
Protection of persons against access to hazardous parts and protection against penetration of solid foreign objects.	Protection against liquids‡ under test conditions specified in IEC 529.
0 Not protected	0 Not protected
1 Protection against objects greater than 50 mm in diameter (hands)	1 Protection against vertically falling drops of water (condensation)
2 Protection against objects greater than 12.5 mm in diameter (fingers)	2 Protection against falling water with enclosure tilted 15°
3 Protection against objects greater than 2.5 mm in diameter (tools, wires)	3 Protection against spraying of falling water with enclosure tilted 60°
4 Protection against objects greater than 1.0 mm in diameter (tools, small wires)	4 Protection against splashing water
5 Protection against dust (dust may enter during test but must not interfere with equipment operation or impair safety)	5 Protection against low-pressure water jets
6 Dusttight (no dust observable inside enclosure at end of test)	6 Protection against powerful water jets
	7 Protection against temporary submersion
	8 Protection against continuous submersion

Example: IP41 describes an enclosure that is designed to protect against the entry of tools or objects greater than 1 mm in diameter, and to protect against vertically dripping water under specified test conditions.

* All first and second numerals up to and including numeral 6 imply compliance with the requirements of all preceding numerals in their respective series. Second numerals 7 and 8 do not imply suitability for exposure to water jets unless dual coded; e.g., IP_5/IP_7
† The IEC permits use of certain supplementary letters with the characteristic numerals. If such letters are used, refer to IEC 529 for an explanation.
‡ The IEC test requirements for degrees of protection against liquid ingress refer only to water

INDUSTRIAL ELECTRICAL SYMBOLS . . .

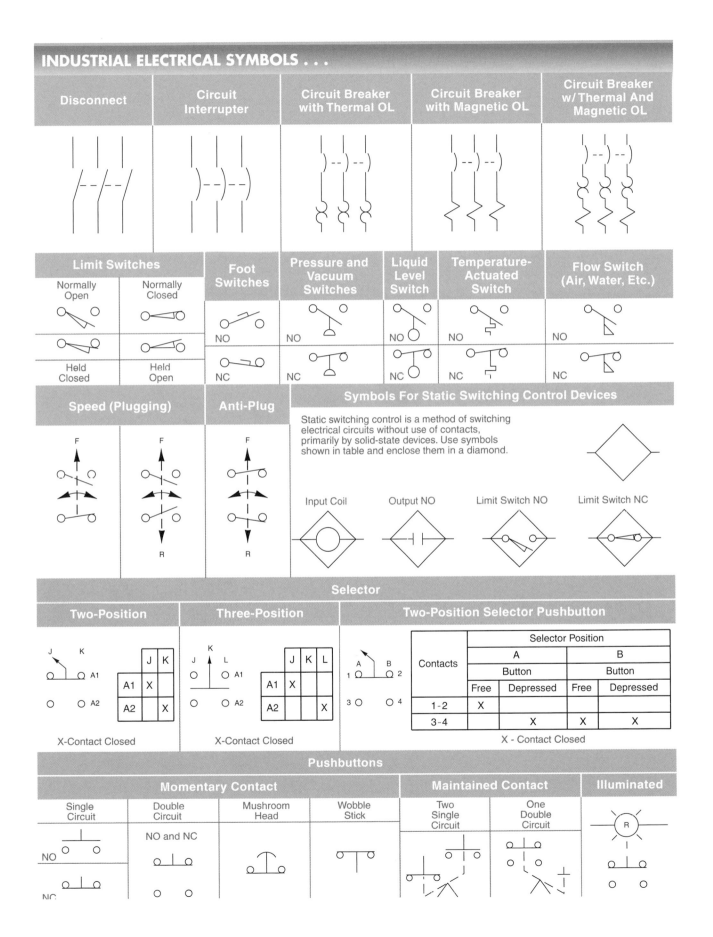

| Disconnect | Circuit Interrupter | Circuit Breaker with Thermal OL | Circuit Breaker with Magnetic OL | Circuit Breaker w/ Thermal And Magnetic OL |

| Limit Switches | | Foot Switches | Pressure and Vacuum Switches | Liquid Level Switch | Temperature-Actuated Switch | Flow Switch (Air, Water, Etc.) |

Normally Open — Normally Closed

Held Closed — Held Open

NO / NC

| Speed (Plugging) | Anti-Plug | Symbols For Static Switching Control Devices |

Static switching control is a method of switching electrical circuits without use of contacts, primarily by solid-state devices. Use symbols shown in table and enclose them in a diamond.

Input Coil   Output NO   Limit Switch NO   Limit Switch NC

Selector

| Two-Position | Three-Position | Two-Position Selector Pushbutton |

Two-Position:

	J	K
A1	X	
A2		X

X-Contact Closed

Three-Position:

	J	K	L
A1	X		
A2			X

X-Contact Closed

Two-Position Selector Pushbutton:

Contacts	Selector Position			
	A		B	
	Button		Button	
	Free	Depressed	Free	Depressed
1-2	X			
3-4		X	X	X

X - Contact Closed

Pushbuttons

| Momentary Contact | | | | Maintained Contact | | Illuminated |

| Single Circuit | Double Circuit | Mushroom Head | Wobble Stick | Two Single Circuit | One Double Circuit | |

NO and NC

NO

NC

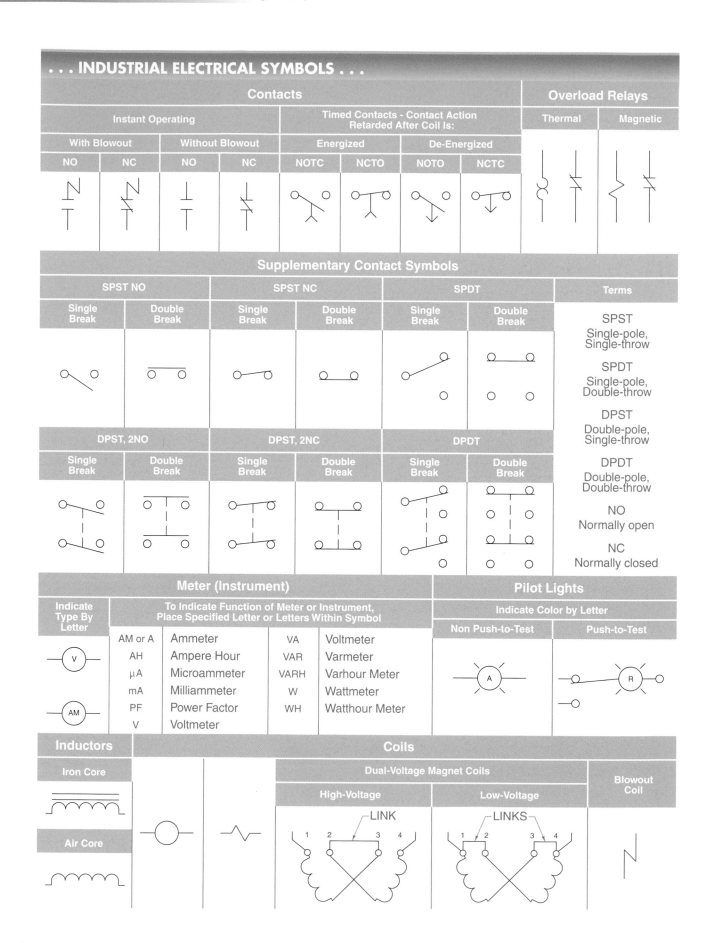

. . . INDUSTRIAL ELECTRICAL SYMBOLS . . .

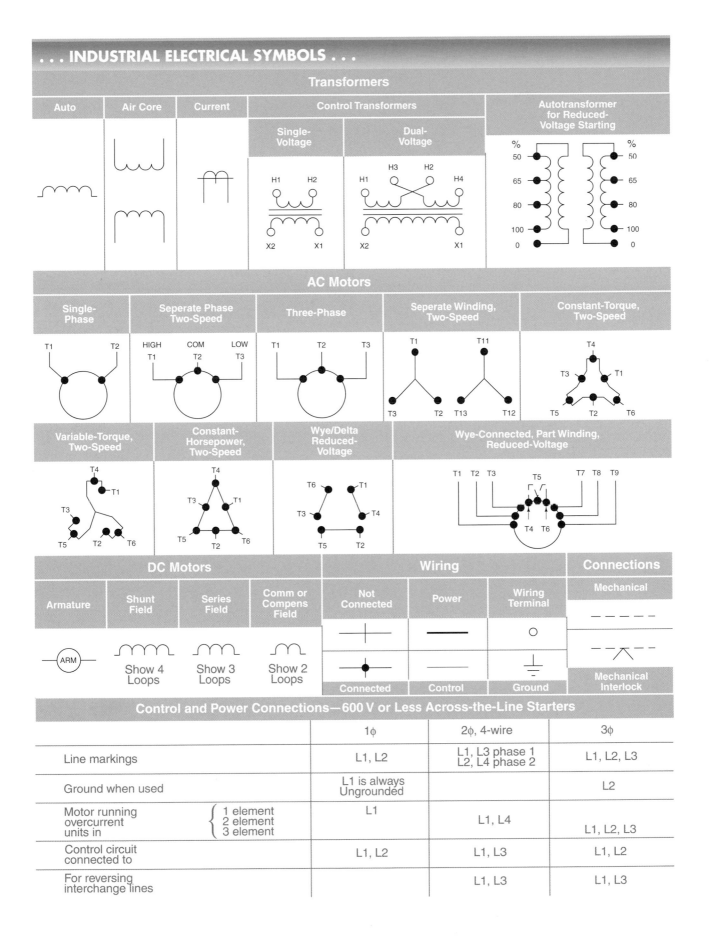

. . . INDUSTRIAL ELECTRICAL SYMBOLS . . .

Transformers

Auto	Air Core	Current	Control Transformers		Autotransformer for Reduced-Voltage Starting
			Single-Voltage	**Dual-Voltage**	
			H1 H2 / X2 X1	H1 H3 H2 H4 / X2 X1	% 50 65 80 100 0 / % 50 65 80 100 0

AC Motors

Single-Phase	Seperate Phase Two-Speed	Three-Phase	Seperate Winding, Two-Speed	Constant-Torque, Two-Speed
T1 T2	HIGH COM LOW / T1 T2 T3	T1 T2 T3	T1 T11 / T3 T2 T13 T12	T4 / T3 T1 / T5 T2 T6

Variable-Torque, Two-Speed	Constant-Horsepower, Two-Speed	Wye/Delta Reduced-Voltage	Wye-Connected, Part Winding, Reduced-Voltage
T4 / T1 / T3 / T5 T2 T6	T4 / T3 T1 / T5 T2 T6	T6 T1 / T3 T4 / T5 T2	T1 T2 T3 T5 T7 T8 T9 / T4 T6

DC Motors / Wiring / Connections

DC Motors				Wiring			Connections
Armature	Shunt Field	Series Field	Comm or Compens Field	Not Connected	Power	Wiring Terminal	**Mechanical**
ARM	Show 4 Loops	Show 3 Loops	Show 2 Loops	╪	▬▬▬	○	─ ─ ─ ─
				Connected	**Control**	**Ground**	**Mechanical Interlock**

Control and Power Connections—600 V or Less Across-the-Line Starters

		1φ	2φ, 4-wire	3φ
Line markings		L1, L2	L1, L3 phase 1 L2, L4 phase 2	L1, L2, L3
Ground when used		L1 is always Ungrounded		L2
Motor running overcurrent units in	1 element 2 element 3 element	L1	L1, L4	L1, L2, L3
Control circuit connected to		L1, L2	L1, L3	L1, L2
For reversing interchange lines			L1, L3	L1, L3

. . . INDUSTRIAL ELECTRICAL SYMBOLS

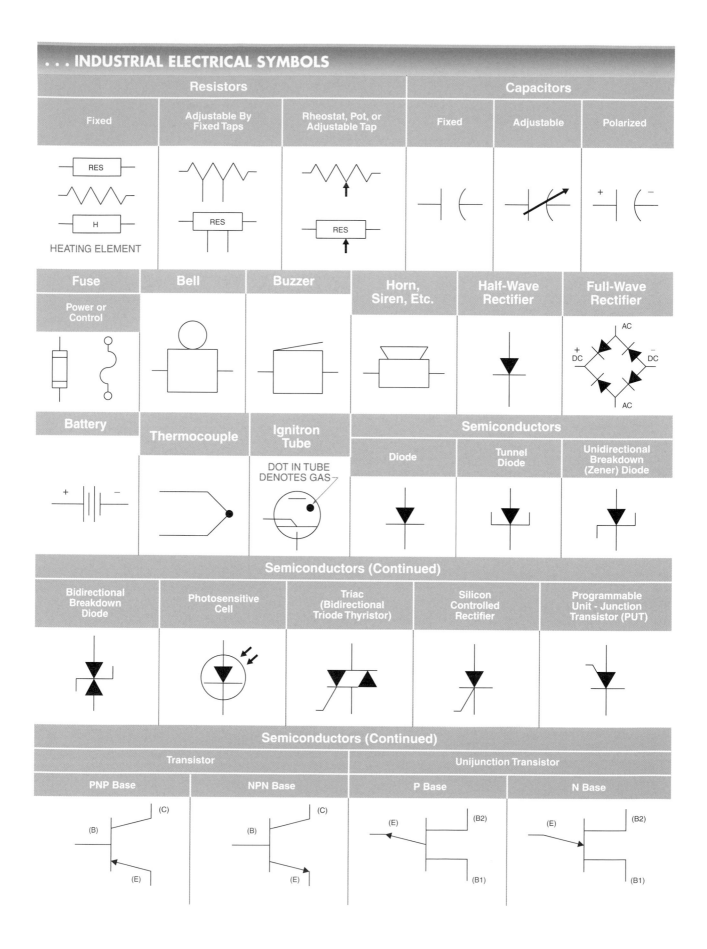

Resistors			Capacitors		
Fixed	Adjustable By Fixed Taps	Rheostat, Pot, or Adjustable Tap	Fixed	Adjustable	Polarized

HEATING ELEMENT

Fuse	Bell	Buzzer	Horn, Siren, Etc.	Half-Wave Rectifier	Full-Wave Rectifier
Power or Control					

Battery	Thermocouple	Ignitron Tube	Semiconductors		
		DOT IN TUBE DENOTES GAS	Diode	Tunnel Diode	Unidirectional Breakdown (Zener) Diode

Semiconductors (Continued)

Bidirectional Breakdown Diode	Photosensitive Cell	Triac (Bidirectional Triode Thyristor)	Silicon Controlled Rectifier	Programmable Unit - Junction Transistor (PUT)

Semiconductors (Continued)

Transistor		Unijunction Transistor	
PNP Base	NPN Base	P Base	N Base

NON-LOCKING WIRING DEVICES

2-Pole, 3-Wire

Wiring Diagram	NEMA ANSI	Receptacle Configuration	Rating
	5-15 C73.11		15 A 125 V
	5-20 C73.12		20 A 125 V
	5-30 C73.45		30 A 125 V
	5-50 C73.46		50 A 125 V
	6-15 C73.20		15 A 250 V
	6-20 C73.51		20 A 250 V
	6-30 C73.52		30 A 250 V
	6-50 C73.53		50 A 250 V
	7-15 C73.28		15 A 277 V
	7-20 C73.63		20 A 277 V
	7-30 C73.64		30 A 277 V
	7-50 C73.65		50 A 277 V

4-Pole, 4-Wire

Wiring Diagram	NEMA ANSI	Receptacle Configuration	Rating
	18-15 C73.15		15 A 3φY 120/208 V
	18-20 C73.26		20 A 3φY 120/208 V
	18-30 C73.47		30 A 3φY 120/208 V
	18-50 C73.48		50 A 3φY 120/208 V
	18-60 C73.27		60 A 3φY 120/208 V

3-Pole, 3-Wire

Wiring Diagram	NEMA ANSI	Receptacle Configuration	Rating
	10-20 C73.23		20 A 125/250 V
	10-30 C73.24		30 A 125/250 V
	10-50 C73.25		50 A 125/250 V
	11-15 C73.54		15 A 3φ 250 V
	11-20 C73.55		20 A 3φ 250 V
	11-30 C73.56		30 A 3φ 250 V
	11-50 C73.57		50 A 3φ 250 V

3-Pole, 4-Wire

Wiring Diagram	NEMA ANSI	Receptacle Configuration	Rating
	14-15 C73.49		15 A 125/250 V
	14-20 C73.50		20 A 125/250 V
	14-30 C73.16		30 A 125/250 V
	14-50 C73.17		50 A 125/250 V
	14-60 C73.18		60 A 125/250 V
	15-15 C73.58		15 A 3φ 250 V
	15-20 C73.59		20 A 3φ 250 V
	15-30 C73.60		30 A 3φ 250 V
	15-50 C73.61		50 A 3φ 250 V
	15-60 C73.62		60 A 3φ 250 V

DC MOTORS AND CIRCUITS

Size of Motor		Motor overload protection dual-element fuse		Switch 115% minimum or HP rated or fuse holder size	Minimum size of starter	Controller termination temperature rating				Minimum size of copper wire and trade conduit	
		Motor less than 40°C or greater than 1.15 SF (Max fuse 125%)	All other motors (Max fuse 115%)			60° C		75° C		Wire size (AWG or kcmil)	Conduit (inches)
HP	AMP					TW	THW	TW	THW		
90 V											
¼	4.0	5	4½	30	0	•	•	•	•	14	½
⅓	5.2	6¼	5⁶⁄₁₀	30	0	•	•	•	•	14	½
½	6.8	8	7½	30	0	•	•	•	•	14	½
¾	9.6	12	10	30	0	•	•	•	•	14	½
1	12.2	15	12	30	0	•	•	•	•	14	½
120 V											
¼	3.1	3½	3½	30	0	•	•	•	•	14	½
⅓	4.1	5	4½	30	0	•	•	•	•	14	½
½	5.4	6¼	6	30	0	•	•	•	•	14	½
¾	7.6	9	8	30	0	•	•	•	•	14	½
1	9.5	10	10	30	0	•	•	•	•	14	½
1½	13.2	15	15	30	1	•	•	•	•	14	½
2	17	20	17½	30	1	•	•	•	•	12	½
5	40	50	45	60	2	•	•	•		6	¾
									•	8	¾
10	76	90	80	100	3	•	•	•	•	2	1
										3	1
180 V											
¼	2	2½	2¼	30	0	•	•	•	•	14	½
⅓	2.6	3²⁄₁₀	2⁸⁄₁₀	30	0	•	•	•	•	14	½
½	3.4	4	3½	30	0	•	•	•	•	14	½
¾	4.8	6	5	30	0	•	•	•	•	14	½
1	6.1	7½	7	30	0	•	•	•	•	14	½
1½	8.3	10	9	30	1	•	•	•	•	14	½
2	10.8	12	12	30	1	•	•	•	•	14	½
3	16	20	17½	30	1	•	•	•	•	12	½
5	27			60	1	•		•		8	½
							•			8	¾
									•	10	½

1ϕ MOTORS AND CIRCUITS

Size of Motor		Motor overload protection dual-element fuse		Switch 115% minimum or HP rated or fuse holder size	Minimum size of starter	Controller termination temperature rating				Minimum size of copper wire and trade conduit	
		Motor less than 40°C or greater than 1.15 SF (Max fuse 125%)	All other motors (Max fuse 115%)			60° C		75° C		Wire size (AWG or kcmil)	Conduit (inches)
HP	AMP					TW	THW	TW	THW		
115 V (120 V system)											
1/6	4.4	5	5	30	00	•	•	•	•	14	1/2
1/4	5.8	7	6 1/4	30	00	•	•	•	•	14	1/2
1/3	7.2	9	8	30	00	•	•	•	•	14	1/2
1/2	9.8	12	10	30	00	•	•	•	•	14	1/2
3/4	13.8	15	15	30	00	•	•	•	•	14	1/2
1	16	20	17 1/2	30	00	•	•	•	•	14	1/2
1 1/2	20	25	20	30	01	•	•	•	•	12	1/2
2	24	30	25	30	01	•	•	•	•	10	1/2
230 V (240 V system)											
1/6	2.2	2 1/2	2 1/2	30	00	•	•	•	•	14	1/2
1/4	2.9	3 1/2	3 2/10	30	00	•	•	•	•	14	1/2
1/3	3.6	4 1/2	4	30	00	•	•	•	•	14	1/2
1/2	4.9	5 6/10	5 6/10	30	00	•	•	•	•	14	1/2
3/4	6.9	8	7 1/2	30	00	•	•	•	•	14	1/2
1	8	10	9	30	00	•	•	•	•	14	1/2
1 1/2	10	12	10	30	0	•	•	•	•	14	1/2
2	12	15	12	30	0	•	•	•	•	14	1/2
3	17	20	17 1/2	30	1	•	•	•	•	12	1/2
5	28	35		60	2	•		•		8	3/4
							•			8	1/2
									•	10	1/2
7 1/2	40	50	45	60	2	•	•	•		6	3/4
									•	8	3/4
10	50	60	60	60	3	•	•	•		4	1
									•	6	3/4

IEC—REFERENCE CHART

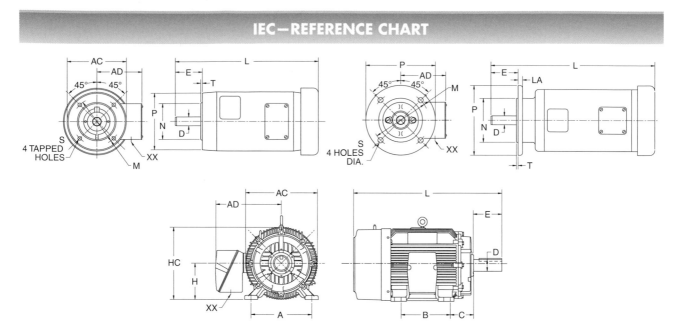

IEC Frame	Type	Foot Mounting*				Shaft*			B5 Flange*					B14 Face*					General*				
		A	B	C	H	D	E	LA	M	N	P	S	T	M	N	P	S	T	L	AC	AD	HC	XX
63	300	3.397	3.150	1.570	2.480	0.433	0.906	0.313	4.528	3.740	5.512	0.354	0.118	2.953	2.362	3.540	M5	0.098	†	4.690	4‡ 4.567‡	4.760 5.375‡	0.500 0.880‡
71	300 400	4.409	3.543	1.770	2.800	0.551	1.181	0.313	5.118	4.331	6.299	0.393	0.138	3.347	2.756	4.130	M6	0.098	†	4.690 5.960‡	4	5.140 5.880‡	0.690 0.844‡
80	400 500	4.921	3.937	1.969	3.150	0.748	1.575	0.500	6.496	5.118	7.874	0.430	0.138	3.937	3.150	4.724	M6	0.118	†	5.690 6.614‡	4.510 5.120	6 6.380‡	0.880 0.844‡
90	S L	5.511	3.937 4.921	2.205	3.543	0.945	1.969	0.500	6.496	5.118	7.874	0.472	0.138	4.530	3.740	5.512	M8	0.118	†	6.614 5.687‡	5.120 4.250‡	6.810 6.531‡	0.880 0.844‡
100	S L	6.300	4.409 5.512	2.480	3.937	1.102	2.362	0.562	8.465	7.087	9.840	0.560	0.160	5.108	4.331	6.299	M8	0.138	†	7.875	5.875 6.060‡	7.906 9.440‡	1.062
112	S M	7.480	4.488 5.512	2.760	4.409	1.102	2.362	0.562	8.465	7.087	9.840	0.560	0.160	5.108	4.331	6.299	M8	0.138	†	7.875	5.875	8.437	1.062
132	S M	8.504	5.512 7.008	3.504	5.197	1.496	3.150	0.562	10.433	9.055	11.811	0.560	0.160	6.496	5.118	7.874	M8	0.138	†	9.562	7.375	10.062	1.062
160	M L	10	8.268 10	4.252	6.299	1.654	4.331	0.787	11.811	9.842	13.780	0.748	0.200	8.465	7.087	9.840	M12	0.160	†	12.940	9.510	12.940	1.375
180	M L	10.984	9.488 10.984	4.764	7.087	1.890	4.331		11.811	9.842	13.780	0.748	0.200						†	15.560	13.120	14.640	2.008
200	L M	12.520	10.512 12.008	5.236	7.874	2.165	4.331		13.780	11.811	15.748	0.748							†	17.375	14.125	16.375	2.500
225	S M	14.016	11.260 12.244	5.866	8.858	2.362	5.512		15.748	13.780	17.716	0.748							†	19.488	15.079	19.016	2.500
250	S M	15.984	12.244 13.740	6.614	9.843	2.756	5.512												†	20.472	17.992	20.197	2.500
280	S M	17.992	14.488 16.496	7.485	11.025	3.150	6.693												†	24.252	19.567	22.874	2.500
315	S M	20	16 18	8.500	12.400	3.346	6.693												†	29.900	26.880	28.840	4
355	S L	24	19.690 24.800	10	13.980	3.346	6.693												†	29.900	26.880	28.320	4

* In in.
† Contact manufacturer for "L" dimensions
‡ DC motor

COUPLING SELECTIONS

Coupling Number	Rated Torque (lb-in)	Maximum Shock Torque (lb-in)
10-101-A	16	45
10-102-A	36	100
10-103-A	80	220
10-104-A	132	360
10-105-A	176	480
10-106-A	240	660
10-107-A	325	900
10-108-A	525	1450
10-109-A	875	2450
10-110-A	1250	3500
10-111-A	1800	5040
10-112-A	2200	6160

V-BELTS

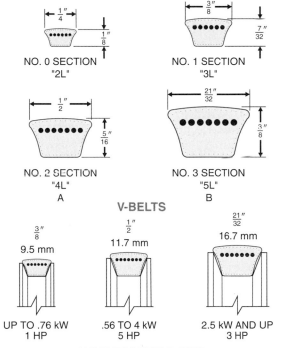

NO. 0 SECTION "2L"

NO. 1 SECTION "3L"

NO. 2 SECTION "4L"
A

NO. 3 SECTION "5L"
B

V-BELTS

9.5 mm

11.7 mm

16.7 mm

UP TO .76 kW
1 HP

.56 TO 4 kW
5 HP

2.5 kW AND UP
3 HP

V-BELTS/MOTOR SIZE

TYPICAL MOTOR POWER FACTORS

HP	Speed (rpm)	Power Factor at		
		½ Load	¾ Load	Full Load
0–5	1800	0.72	0.82	0.84
5.01–20	1800	0.74	0.84	0.86
20.1–100	1800	0.79	0.86	0.89
100.1–300	1800	0.81	0.88	0.91

COMMON SERVICE FACTORS

Equipment	Service Factors
Blowers	
Centrifugal	1.00
Vane	1.25
Compressors	
Centrifugal	1.25
Vane	1.50
Conveyors	
Uniformly loaded or fed	1.50
Heavy-duty	2.00
Elevators	
Bucket	2.00
Freight	2.25
Extruders	
Plastic	2.00
Metal	2.50
Fans	
Light-duty	1.00
Centrifugal	1.50
Machine tools	
Bending roll	2.00
Punch press	2.25
Tapping machine	3.00
Mixers	
Concrete	2.00
Drum	2.25
Paper mills	
De-barking machines	3.00
Beater and pulper	2.00
Bleacher	1.00
Dryers	2.00
Log haul	2.00
Printing presses	1.50
Pumps	
Centrifugal—general	1.00
Centrifugal—sewage	2.00
Reciprocating	2.00
Rotary	1.50
Textile	
Batchers	1.50
Dryers	1.50
Looms	1.75
Spinners	1.50
Woodworking machines	1.00

Glossary

A

abbreviation: A letter or combination of letters that represents a word.

AC sine wave: A symmetrical waveform that contains 360 electrical degrees.

actuator: The part of a limit switch that transfers the mechanical force of the moving part to the electrical contacts.

AC voltage: Voltage that reverses its direction of flow at regular intervals.

adjustable base plate: A mounting base that allows equipment to be easily moved over a short distance.

alignment: The condition where the centerlines of two machine shafts are placed within specified tolerances.

alternating current (AC): Current that reverses its direction of flow at regular intervals.

alternating current (AC) motor: A motor that uses alternating current to produce rotation.

alternation: A half of a cycle.

ambient temperature: The temperature of the air surrounding a motor.

ampere: The number of electrons passing a given point in one second.

amplification: The process of taking a small signal and making it larger.

analog display: An electromechanical device that indicates a value by the position of a pointer on a scale.

analog multimeter: A meter that can measure two or more electrical properties and displays the measured properties along calibrated scales using a pointer.

analog switching relay: An SSR that has an infinite number of possible output voltages within the rated range of the relay.

anchoring: A means of fastening a machine securely to a base or foundation.

AND gate: A device with an output that is high only when both of its inputs are high.

angular misalignment: 1. A condition where one shaft is at an angle to the other shaft. **2.** A condition where two shafts are parallel but at different angles with the horizontal plane.

arc blast: An explosion that occurs when the air surrounding electrical equipment becomes ionized and conductive.

arc chute: A device that confines, divides, and extinguishes arcs drawn between contacts opened under load.

arc flash: An extremely high-temperature discharge produced by an electrical fault in the air.

arc flash boundary: The distance from exposed energized conductors or circuit parts where bare skin would receive the onset of a second-degree burn.

arcing: The discharge of an electric current across a gap, such as when an electric switch is opened.

arc-rated hood: An eye and face protection device that covers the entire head and is used for protection from arc blast and arc flash.

arc suppressor: A device that dissipates the energy present across opening contacts.

armature: 1. The movable part of a solenoid. **2.** The movable coil of wire in a generator that rotates through the magnetic field. **3.** The rotating part of a DC motor.

asymmetrical recycle timer: A timer that has independent adjustments for the on and off time periods.

atom: The smallest particle that an element can be reduced to and still keep the properties of that element.

axial load: A load in which the applied force is parallel to the axis of rotation.

B

ball bearing: An antifriction bearing that permits free motion between a moving part and a fixed part by means of balls confined between inner and outer rings.

bar graph: A graph composed of segments that function as an analog pointer.

base plate: A rigid steel support for firmly anchoring and aligning two or more rotating devices.

base speed: The speed (in rpm) at which a DC motor runs with full-line voltage applied to the armature and field.

bearing: A component used to reduce friction and maintain clearance between stationary and moving parts in a motor or machine.

bellows: A cylindrical device with several deep folds that expand or contract when pressure is applied.

belt deflection method: A belt tension method in which the tension is adjusted by measuring the deflection of the belt.

bench oscilloscope: A test instrument that displays the shape of a voltage waveform and is used mostly for bench testing electrical and electronic circuits.

bimetallic overload relay: An overload relay that resets automatically.

bipolar device: A device in which both holes and electrons are used as internal carriers for maintaining current flow.

block diagram: A diagram that shows the relationship between individual sections, or blocks, of a circuit or system.

branch circuit: The portion of the distribution system between the final OCPD and the outlet or load connected to it.

branch circuit identifier: A two-piece test instrument that includes a transmitter that is plugged into a receptacle and a receiver that provides an audible indication when located near the circuit to which the transmitter is connected.

break: The number of separate places on a contact that open or close an electrical circuit.

breakdown maintenance: Service on failed equipment that has not received scheduled maintenance such as cleaning or lubrication.

breakdown torque (BDT): The maximum torque a motor can provide without an abrupt reduction in motor speed.

brownout: A reduction of the voltage level by a power company to conserve power during times of peak usage or excessive loading of the power distribution system.

brush: The sliding contact that rides against the commutator segments or slip rings and is used to connect the armature to the external circuit.

busway: A metal-enclosed distribution system of busbars available in prefabricated sections.

C

capacitance (C): The ability of a component or circuit to store energy in the form of an electrical charge.

capacitive circuit: A circuit in which current leads voltage (voltage lags current).

capacitive filter: A circuit consisting of a capacitor and resistor connected in parallel.

capacitive level switch: A level switch that detects the dielectric variation when the product is in contact (proximity) with the probe and when the product is not in contact with the probe.

capacitive proximity sensor: A sensor that detects either conductive or nonconductive substances.

capacitive reactance (X_C): The opposition to current flow by a capacitor.

capacitor: An electric device that stores electrical energy by means of an electrostatic field.

capacitor motor: A 1ϕ AC motor that includes a capacitor in addition to the running and starting windings.

carrier frequency: The frequency that controls the number of times solid-state switches in the inverter of a motor drive with pulse width modulation turn on and off.

caution signal word: A word used to indicate a potentially hazardous situation which, if not avoided, may result in minor or moderate injury.

centralized power distribution: The process of delivering electrical power that is generated at a large centralized location to customers.

challenging: The process of pressing the start switch of a machine to determine whether the machine starts when it is not supposed to start.

circuit analysis method: A method of SSR replacement in which a logical sequence is used to determine the reason for a failure.

circuit breaker: An overcurrent protection device with a mechanism that may manually or automatically open the circuit when an overload condition or short circuit occurs.

cladding: The first layer of protection for the glass or plastic core of an optical fiber cable.

clamp-on ammeter: A meter that measures the current in a circuit by measuring the strength of the magnetic field around a single conductor.

closed loop operation: An operation that has feedback from the output to the input.

coast stop: A stopping method in which the motor drive shuts off the voltage to a motor, allowing the motor to coast to a stop.

cold trip: The trip point from the time the motor starts until the first time the overloads trip (motor operating below nameplate rated current).

communication: The transmission of information from one point to another by means of electromagnetic waves.

commutator: A ring made of segments that are insulated from one another.

complementary metal-oxide semiconductor (CMOS) ICs: A group of ICs that employ MOS transistors.

compound-wound generator: A generator that includes series and shunt field windings.

conductive probe level switch: A level switch that uses liquid to complete the electrical path between two conductive probes.

conductor: A material that has very little resistance to current flow and permits electrons to move through it easily.

confined space: A space large enough and so configured that an employee can physically enter and perform assigned work, that has limited or restricted means for entry and exit, and is not designed for continuous employee occupancy.

constant horsepower/variable torque (CH/VT) load: A load that requires high torque at low speeds and low torque at high speeds.

constant torque/variable horsepower (CT/VH) load: A load in which the torque requirement remains constant.

contact block: The part of the pushbutton that is activated when the operator is pressed.

contact-controlled timer: A timer that does not require the control switch to be connected in line with the timer coil.

contact life: The number of times the contacts of a relay switch the load controlled by the relay before malfunctioning.

contactor: A control device that uses a small control current to energize or de-energize the load connected to it.

contact protection circuit: A circuit that protects contacts by providing a nondestructive path for generated voltage as a switch is opened.

continuity tester: A test instrument that tests for a complete path for current to flow.

continuous monitoring: The monitoring of equipment at all times.

control circuit: The part of an SSR that determines when the output component is energized or de-energized.

controller scan: One execution cycle of a line diagram.

ControlNet: A real-time device network providing for high-speed transport of time-critical I/O data, messaging, peer-to-peer communications, and the uploading and downloading of programs.

control switch: A switch that controls the flow of current in a circuit.

control transformer: A transformer that is used to step down the voltage to the control circuit of a system or machine.

conventional current flow: The movement of electrons from positive to negative.

convergent beam scan: A method of scanning that simultaneously focuses and converges a light beam to a fixed focal point in front of a photoreceiver.

converter: An electronic device that changes AC voltage into DC voltage.

core: The actual path for light in an optical fiber cable.

corrective work: Repair of a known problem before a breakdown occurs.

counter: A counting device that accounts for the total number of inputs entering into it and can provide an output (mechanical or solid-state contacts) at predetermined counts in addition to displaying the counted value.

coupling: A device that connects the ends of rotating shafts.

current (I): The amount of electrons flowing through an electrical circuit.

current unbalance: Also known as current imbalance, is the unbalance that occurs when the currents on each of the three power lines of a 3ϕ power supply are not equal.

cutoff region: The point at which the transistor is turned off and no current flows.

cycle: One complete positive and negative alternation of a wave form.

D

danger signal word: A word used to indicate an imminently hazardous situation which, if not avoided, results in death or serious injury.

dark-operated photoelectric control: A photoelectric control that energizes the output switch when a target is present (breaking the beam).

dashpot timer: A timer that provides time delay by controlling how rapidly air or liquid is allowed to pass into or out of a container through an orifice (opening) that is either fixed or variable in diameter.

data file: A group of data values, such as inputs, timers, counters, and outputs, that are displayed as a group and whose status may be monitored.

DC brake stop (DC injection braking): A stopping method in which a DC voltage is applied to the stator winding of a motor after a stop command is entered.

DC compound motor: A DC motor with the field connected in both series and parallel (shunt) with the armature.

DC permanent-magnet motor: A motor that uses magnets, not a winding, for the field poles.

DC power supply: A device that converts alternating current (AC) to regulated direct current (DC) for use in electrical circuits.

DC series motor: A DC motor that has the series field coils connected in series with the armature.

DC shunt motor: A DC motor that has the field connected in parallel (shunt) with the armature.

DC voltage: Voltage that flows in one direction only.

deadband (differential): The amount of pressure that must be removed before the switch contacts reset for another cycle after the setpoint has been reached and the switch has been actuated.

dead short: A short circuit that opens the circuit as soon as the circuit is energized or when the section of the circuit containing the short is energized.

decibel (dB): A unit of measure used to express the relative intensity of sound.

delta configuration: A transformer connection that has each transformer coil connected end-to-end to form a closed loop.

delta (Δ) connection: A connection that has each coil end connected end-to-end to form a closed loop.

depletion mode: The operation of a MOSFET with a negative gate voltage.

DeviceNet: A network with a controller connected to field devices over a two-wire network in a trunk line configuration with either single drops off the trunk or branched drops through multiport interfaces at the device locations.

diac: A three-layer, two-terminal bidirectional device that is typically used as a triggering device to control the gate current of a triac.

diaphragm: A deflecting mechanism that moves when a force (pressure) is applied.

dielectric: A nonconductor of direct electric current.

dielectric absorption test: A test that checks the absorption characteristics of moist or contaminated insulation.

dielectric material: A medium in which an electric field is maintained with little or no outside energy supply.

dielectric variation: The range at which a material can sustain an electric field with a minimum dissipation of power.

diffused mode: A method of ultrasonic sensor operation in which the emitter and receiver are housed in the same enclosure.

diffuse scan (proximity scan): A method of scanning in which the transmitter and receiver are housed in the same enclosure and a small percentage of the transmitted light beam is reflected back to the receiver from the target.

digital display: An electronic device that displays readings on a meter as numerical values.

digital logic probe: A special DC voltmeter that detects the presence or absence of a signal.

digital multimeter (DMM): A meter that can measure two or more electrical properties and displays the measured properties as numerical values.

diode: An electronic component that allows current to pass through it in only one direction.

direct current (DC): Current that flows in only one direction.

direct current (DC) motor: A motor that uses direct current connected to the field and armature to produce shaft rotation.

directional control valve: A valve that is used to direct the flow of fluid throughout a fluid power system.

direct mode: A method of ultrasonic sensor operation in which the emitter and receiver are placed opposite each other so that the sound waves from the emitter are received directly by the receiver.

direct scan (transmitted beam, thru-beam, opposed scan): A method of scanning in which the transmitter and receiver are placed opposite each other so that the light beam from the transmitter shines directly at the receiver.

disconnect: A device used only periodically to remove electrical circuits from their supply source.

display parameters: Parameters that allow drive or motor operating conditions, such as applied voltage, current draw, and internal drive temperature, to be viewed but not changed.

distribution substation: An outdoor facility located close to the point of electrical service use and is used for changing voltage levels, providing a central place for system switching, monitoring protection, and redistributing power.

done bit: The controlled contact (or output) that changes from open or closed (OFF to ON) after the preset time (or number of counts) has been reached

doping: The addition of impurities to the crystal structure of a semiconductor.

double-break contacts: Contacts that break an electrical circuit in two places.

drop-out voltage: The voltage that exists when voltage is reduced sufficiently to allow the solenoid to open.

drum switch: A manual switch made up of moving contacts mounted on an insulated rotating shaft.

dynamic braking: A method of motor braking in which a motor is reconnected to act as a generator immediately after it is turned off.

E

earmuff: An ear protection device worn over the ears.

earplug: An ear protection device made of moldable rubber, foam, or plastic and inserted into the ear canal.

ear protection: Any device worn to limit the noise entering the ear and includes earplugs and earmuffs.

eddy current: An unwanted current induced in the metal structure of a device due to the rate of change in the induced magnetic field.

editing parameters: Parameters that can be programmed or adjusted to set a drive for a specific application and motor.

effective light beam: The area of light that travels directly from the transmitter to the receiver.

electrical analysis: An analysis that uses electrical monitoring equipment and/or test instruments to evaluate the quality of electrical power delivered to equipment and the performance of electrical equipment.

electrical circuit: An assembly of conductors and electrical devices through which current flows.

electrical grid: A network that delivers electrical power from the power plants where it is generated to customers.

electrical noise: Unwanted signals that are present on a power line.

electrical shock: A shock that results any time a body becomes part of an electrical circuit.

electrical warning signal word: A word used to indicate a high-voltage location and conditions that could result in death or serious personal injury from an electrical shock if proper precautions are not taken.

electric arc: A discharge of electric current across an air gap.

electric braking: A method of braking in which a DC voltage is applied to the stationary windings of a motor after the AC voltage is removed.

electrolyte: A conducting medium in which the current flow occurs by ion migration.

electromagnet: A magnet whose magnetic energy is produced by the flow of electric current.

electromagnetism: The magnetism produced when electric current passes through a conductor.

electromechanical relay (EMR): A switching device that has sets of contacts that are closed by a magnetic effect.

electron: A negatively charged particle that whirls around the nucleus at great speed in shells.

electron current flow: The movement of electrons from negative to positive.

electronic overload: A device that has built-in circuitry to sense changes in current and temperature.

electrostatic discharge (ESD): The movement of electrons from a source to an object across a gap.

emergency work: Work performed to correct an unexpected malfunction on equipment that has received some scheduled maintenance.

enable bit: A controlling contact or input that sends power to the timer (or counter) and starts the timing of the timer (or the counting of the counter).

encoder: A sensor (transducer) that produces discrete electrical pulses during each increment of shaft rotation.

energy: The capacity to do work.

enhancement mode: The operation of a MOSFET with a positive gate voltage.

equipment grounding conductor (EGC): An electrical conductor that provides a low-impedance ground path between electrical equipment and enclosures within the distribution system.

eutectic alloy: A metal that has a fixed temperature at which it changes directly from a solid to a liquid state.

exact replacement method: A method of SSR replacement in which a bad relay is replaced with a relay of the same type and size.

explosion warning signal word: A word used to indicate locations and conditions where exploding parts may cause death or serious personal injury if proper precautions and procedures are not followed.

extended button operator: A pushbutton that has the button extended beyond the guard.

F

face shield: An eye and face protection device that covers the entire face with a plastic shield and is used for protection from flying objects.

facilities maintenance: Maintenance performed on systems and equipment in hotels, schools, office buildings, and hospitals.

fiber optics: A technology that uses a thin, flexible glass or plastic optical fiber to transmit light.

fieldbus: An open, serial, two-way communication network that connects with high-level information devices.

field-effect transistor (FET): A three- or four-terminal device in which output current is controlled by an input voltage.

field windings: Electromagnets used to produce the magnetic field in a generator.

filter: A circuit in a power supply section that smooths the pulsating DC to make it more consistent.

flexible belt drive: A system in which resilient flexible belts are used to drive one or more shafts.

flexible coupling: A coupling within a resilient center, such as rubber or oil, that flexes under temporary torque or misalignment due to thermal expansion.

flow: The travel of fluid in response to a force caused by pressure or gravity.

flow chart: A diagram that shows a logical sequence of steps for a given set of conditions.

flow detection sensor: A sensor that detects the movement (flow) of liquid or gas using a solid-state device.

flow switch: A control switch that detects the movement of a fluid.

flush button operator: A pushbutton with a guard ring surrounding the button that prevents accidental operation.

foot protection: Shoes worn to prevent foot injuries that are typically caused by objects falling less than 4′ and having an average weight of less than 65 lb.

foot switch: A control switch that is operated by a person's foot.

force: Any cause that changes the position, motion, direction, or shape of an object.

fork lever actuator: An actuator operated by either one of two roller arms.

forward-bias voltage: The application of the proper polarity to a diode.

foundation: An underlying base or support.

friction: The resistance to motion that occurs when two surfaces slide against each other.

fuel cell: An energy source that transforms the chemical energy from fuel into electrical energy.

full-load current (FLC): The current required by a motor to produce full-load torque at the motor's rated speed.

full-load torque (FLT): The torque required to produce the rated power at the full speed of the motor.

full-wave bridge rectifier: An electrical circuit containing four diodes that allow both halves of a sine wave to be changed into pulsating DC.

full-wave rectifier: An electrical circuit containing two diodes and a center-tapped transformer used to produce pulsating DC.

fundamental frequency: The frequency of the voltage used to control motor speed.

fuse: An overcurrent protection device with a fusible link that melts and opens the circuit when an overload condition or short circuit occurs.

G

gain: A ratio of the amplitude of an output signal to the amplitude of an input signal.

general-purpose relay: A mechanical switch operated by a magnetic coil.

generator: A machine that converts mechanical energy into electrical energy by means of electromagnetic induction.

ghost voltage: A voltage that appears on a meter not connected to a circuit.

glazing: A slick, polished surface caused by dirt and other debris being rubbed on the surface of a belt.

goggles: An eye protection device with a flexible frame that is secured on the face with an elastic headband.

graph: A diagram that shows a variable in comparison to other variables.

grounded circuit: A circuit in which current leaves its normal path and travels to the frame of the motor.

grounded conductor: A conductor that has been intentionally grounded.

ground fault circuit interrupter (GFCI): A device that protects against electrical shock by detecting an imbalance of current in the normal conductor pathways and opening the circuit.

grounding: The connection of all exposed non-current-carrying metal parts to the earth.

grounding electrode: A conducting object through which a direct connection to earth is established.

grounding electrode conductor (GEC): A conductor that connects grounded parts of a power distribution system (equipment grounding conductors, grounded conductors, and all metal parts) to the NEC®-approved earth grounding system.

ground resistance tester: A device used to measure ground connection resistance of electrical installations such as power plants, industrial plants, high-tension towers, and lightning arrestors.

H

half-shrouded button operator: A pushbutton with a guard ring that extends over the top half of the button.

half-wave phase control circuit: An SCR that has the ability to turn on at different points of the conducting cycle of a half-wave rectifier.

half-wave rectifier: An electrical circuit containing an AC source, a load resistor (RL), and a diode that permits only the positive half cycles of the AC sine wave to pass, which creates pulsating DC.

half-waving: A phenomenon that occurs when a relay fails to turn off because the current and voltage in the circuit reach zero at different times.

Hall effect sensor: A sensor that detects the proximity of a magnetic field.

Hall generator: A thin strip of semiconductor material through which a constant control current is passed.

handheld oscilloscope: A test instrument that displays the shape of a voltage waveform and is typically used for field testing.

harmonic filter: A device used to reduce harmonic frequencies and harmonic distortion in a power distribution system.

hasp: A multiple lockout/tagout device.

head-on actuation: An active method of Hall effect sensor activation in which a magnet is oriented perpendicular to the surface of the sensor and is usually centered over the point of maximum sensitivity.

heater coil: A sensing device used to monitor the heat generated by excessive current and the heat created through ambient temperature rise.

hipot tester: A test instrument that measures insulation resistance by measuring leakage current.

holes: The missing electrons in the structure of a crystal.

horsepower (HP): A unit of power equal to 746 W or 33,000 lb-ft per min (550 lb-ft per sec).

hot trip: The trip point after the overloads have tripped and have been reset (motor operating near or over nameplate rated current).

I

impedance (Z): The combined opposition to the flow of current in circuits that contain resistance and reactance.

improper phase sequence: The changing of the sequence of any two phases (phase reversal) in a 3ϕ motor circuit.

impulse transient voltage: A transient voltage commonly caused by lightning strikes and when loads with coils (motor starters and motors) are turned off.

increment current: The maximum current permitted by the utility in any one step of an incremental start.

inductance (L): The property of a circuit that causes it to oppose a change in current due to energy stored in a magnetic field.

induction motor: A motor that has no physical electrical connection to the rotor.

inductive circuit: A circuit in which current lags voltage.

inductive proximity sensor: A sensor that detects only conductive substances.

inductive reactance (X_L): The opposition to current flow of an inductor in an AC circuit.

industrial maintenance: Maintenance performed on production systems and equipment in industrial settings.

inertia: The property of matter by which a mass persists in its state of rest or motion until acted upon by an external force.

infrared light: Light that is not visible to the human eye.

infrared temperature meter: A noncontact temperature probe that senses the infrared energy emitted by a material.

inherent motor protector: An overload device located directly on or in a motor to provide overload protection.

input circuit: The part of an SSR to which the control component is connected.

instant-on switching relay: An SSR that turns on the load immediately when the control voltage is present.

insulated gate bipolar transistor (IGBT): A three-terminal switching device that combines an FET for control with a bipolar transistor for switching.

insulation spot test: A test that checks motor insulation over the life of the motor.

insulation step voltage test: A test that creates electrical stress on internal insulation cracks to reveal aging or damage not found during other motor insulation tests.

insulator: A material with an atomic structure that allows few free electrons to pass through it.

integrated circuit (IC): A circuit composed of thousands of semiconductor devices, providing a complete circuit function in one small semiconductor package.

International Electrotechnical Commission (IEC): An organization that develops international safety standards for electrical equipment.

inventory control: The organization and management of commonly used parts, vendors and suppliers, and purchasing records in a preventive maintenance system.

inverter: An electronic device that changes DC voltage into AC voltage.

J

jogging: The frequent starting and stopping of a motor for short periods of time.

journal: The part of a shaft, such as an axle or spindle, that moves in a plain bearing.

joystick: An operator that selects one to eight different circuit conditions when the joystick is shifted from the center position into one of the other positions.

jumbo mushroom button operator: A pushbutton that has a large curved operator extending beyond the guard.

junction field-effect transistor (JFET): A simple FET with a PN junction in which output current is controlled by an input voltage.

K

kinetic energy: The energy of motion.

knee pad: A rubber, leather, or plastic pad strapped onto the knees for protection.

L

laser diode: A diode similar to an LED but with an optical cavity that is required for lasing production (emitting coherent light).

leakage current: Current that flows through insulation.

leather protectors: Gloves worn over rubber insulating gloves to prevent penetration of the rubber insulating gloves and provide added protection against electrical shock.

left-hand generator rule: The relationship between the current in a conductor and the magnetic field existing around the conductor.

legend plate: The part of a switch that includes the written description of the switch's operation.

level switch: A switch that detects the height of a liquid or solid (gases cannot be detected by level switches) inside a tank.

lever actuator: An actuator operated by means of a lever that is attached to the shaft of the limit switch.

light-activated SCR (LASCR): An SCR that is activated by light.

light-emitting diode (LED): A semiconductor diode that produces light when current flows through it.

lightning arrester: A device that protects transformers and other electrical equipment from voltage surges caused by lightning.

light-operated photoelectric control: A photoelectric control that energizes the output switch when the target is missing (removed from the beam).

limited approach boundary: The distance from an exposed energized conductor or circuit part at which a person can get an electric shock and is the closest distance an unqualified person can approach.

limit switch: A mechanical input that requires physical contact of the object with the switch actuator.

linear scale: A scale that is divided into equally spaced segments.

line (ladder) diagram: A diagram that shows the logic of an electrical circuit or system using standard symbols.

load: Any device that converts electrical energy to motion, heat, light, or sound.

load current: The amount of current drawn by a load when energized.

locked rotor: A condition when a motor is loaded so heavily that the motor shaft cannot turn.

locked rotor current (LRC): The steady-state current taken from the power line with the rotor locked (stopped) and with the voltage applied.

locked rotor torque (LRT): The torque a motor produces when its rotor is stationary and full power is applied to the motor.

lockout: The process of removing the source of electrical power and installing a lock that prevents the power from being turned on.

logbook: A book or electronic file that documents all work performed during a shift and lists information needed to complete work by maintenance technicians on other shifts.

L-section inductive filter: A filter that reduces surge currents by using a current-limiting inductor and a capacitor.

L-section resistive filter: A filter that reduces or eliminates the amount of DC ripple at the output of a circuit by using a resistor and capacitor as an RC time constant.

lubricating oil analysis: A predictive maintenance technique that detects and analyzes the presence of acids, dirt, fuel, and wear particles in lubricating oil to predict equipment failure.

M

machine control relay: An EMR that includes several sets (usually two to eight) of NO and NC replaceable contacts (typically rated at 10 A to 20 A) that are activated by a coil.

magnet: A substance that produces a magnetic field and attracts iron.

magnetic level switch: A switch that contains a float, a moving magnet, and a magnetically operated reed switch to detect the level of a liquid.

magnetic motor starter: An electrically operated switch (contactor) that includes motor overload protection.

main bonding jumper (MBJ): A connection at the service equipment that connects the equipment grounding conductor, the grounding electrode conductor, and the grounded conductor (neutral conductor).

manual contactor: A control device that uses pushbuttons to energize or de-energize the load connected to it.

manual control circuit: Any circuit that requires a person to initiate an action for the circuit to operate.

manual starter: A contactor with an added overload protective device.

mechanical interlock: The arrangement of contacts in such a way that both sets of contacts cannot be closed at the same time.

mechanical level switch: A level switch that uses a float that moves up and down with the level of the liquid and activates electrical contacts at a set height.

mechanical life: The number of times the mechanical parts of a relay operate before malfunctioning.

megohmmeter: A device that detects insulation deterioration by measuring high resistance values under high test voltage conditions.

metal-oxide semiconductor field-effect transistor (MOSFET): A three-terminal or four-terminal electronic switching device with metal-oxide or polysilicon insulating material that can be used for amplification.

minimum holding current: The minimum amount of current required to keep a sensor operating.

misalignment: The condition where the centerlines of two machine shafts are not aligned within specified tolerances.

MODBUS: A messaging structure with master-slave communication between smart devices and is independent of the physical interconnecting method.

molecular theory of magnetism: The theory that states that all substances are made up of an infinite number of molecular magnets that can be arranged in either an organized or disorganized manner.

momentary power interruption: A decrease to 0 V on one or more power lines lasting from 0.5 cycles up to 3 sec.

motor: A machine that converts electrical energy into mechanical energy by means of electromagnetic induction.

motor drive: An electronic unit designed to control the speed of a motor using solid-state components.

motor drive current rating: The maximum continuous output current a motor drive can deliver for an extended period of time.

motor nameplate current: The amount of current a motor draws when the motor is operating at its nameplate power (HP or kW) rating.

N

NAND gate: A device that provides a low output when both inputs are high.

National Fire Protection Association (NFPA): A national organization that provides guidance in assessing the hazards of the products of combustion.

needle bearing: An antifriction roller-type bearing with long rollers of small diameter.

neutron: A particle contained in the nucleus of an atom that has no electrical charge.

noncontact temperature probe: A device used for taking temperature measurements on energized circuits or on moving parts.

nonlinear scale: A scale that is divided into unequally spaced segments.

nonparallel misalignment: Misalignment where two pulleys or shafts are not parallel.

non-permit confined space: A confined space that does not contain or, with respect to atmospheric hazards, have the potential to contain any hazards capable of causing death or serious physical harm.

nonretentive timer: A timer that does not maintain its current accumulated time value when its control input signal is interrupted or power to the timer is removed.

NOR gate: A device that provides a low output when either or both inputs are high.

N-type material: Material created by doping a region of a crystal with atoms of a material that have more electrons in their outer shells than the crystal.

nucleus: The heavy, dense center of an atom and has a positive electrical charge.

O

off-delay (delay-on-release) timer: A device that does not start its timing function until the power is removed from the timer.

off-line programming: The use of a PC to program a programmable controller that is not in the run mode.

offset error: A slight mismatch between internal components.

offset misalignment: 1. A condition where two shafts are parallel but are not on the same axis. **2.** A condition where two shafts are parallel but the pulleys are not on the same axis.

Ohm's law: The relationship between voltage, current, and resistance in a circuit.

on-delay (delay-on-operate) timer: A device that has a preset time period that must pass after the timer has been energized before any action occurs on the timer contacts.

one-line diagram: A diagram that uses single lines and graphic symbols to indicate the path and components of an electrical circuit.

one-shot (interval) timer: A device in which the contacts change position immediately and remain changed for the set period of time after the timer has received power.

open circuit: An electrical circuit that has an incomplete path that prevents current flow.

open circuit transition switching: A process in which power is momentarily disconnected when switching a circuit from one voltage supply (or level) to another.

operating current (residual or leakage current): The amount of current a sensor draws from the power lines to develop a field that can detect a target.

operational amplifier (op amp): A high-gain, directly coupled amplifier that uses external feedback to control response characteristics.

operator: The device that is pressed, pulled, or rotated by the individual operating the circuit.

optical time domain reflectometer (OTDR): A test instrument that is used to measure fiber optic cable attenuation.

optocoupler: An electrically isolated device that consists of an IRED as the input stage and an NPN phototransistor as the output stage.

OR gate: A device with an output that is high when either or both inputs are high.

oscillatory transient voltage: A transient voltage commonly caused by turning off high inductive loads and by switching off large utility power factor correction capacitors.

oscilloscope: A test instrument that provides a visual display of voltages.

overcurrent protection device (OCPD): A disconnect switch with circuit breakers (CBs) or fuses added to provide overcurrent protection for the switched circuit.

overcycling: The process of turning a motor on and off repeatedly.

overhead power lines: Electrical conductors designed to deliver electrical power and located in an above-ground aerial position.

overhead service: Electrical service in which service-entrance conductors are run through the air from the utility pole to the building.

overload: The application of excessive load to a motor.

overloaded motor: A motor that has a current reading greater than 105% of the nameplate current rating.

overvoltage: An increase of voltage of more than 10% above the normal rated line voltage for a period of time longer than 1 min.

P

panelboard: A wall-mounted distribution cabinet containing a group of overcurrent and short-circuit protective devices for lighting, appliance, or power distribution branch circuits.

parallel connection: A connection that has two or more components connected so there is more than one path for current flow.

part-winding starting: A method of starting a motor by first applying power to part of the motor coil windings for starting and then applying power to the remaining coil windings for normal running.

peak load: The maximum output required of a transformer.

peak switching relay: An SSR that turns on the load when the control voltage is present and the voltage at the load is at its peak.

pendulum actuation: A method of Hall effect sensor activation that is a combination of the head-on and slide-by actuation methods.

periodic maintenance: Tasks completed at specific time intervals to prevent breakdowns and production inefficiency.

permanent magnet: A magnet that can retain its magnetism after the magnetizing force has been removed.

personal protective equipment (PPE): Clothing and/or equipment worn by a technician to reduce the possibility of injury in the work area.

phase control: The control of the time relationship between two events when dealing with voltage and current.

phase sequence indicator: A device used to determine phase sequence and open phases.

phase unbalance: The unbalance that occurs when power lines are out of phase.

phasor measurement unit (PMU): A device that measures electrical waveforms on the electrical grid.

photoconductive cell (photocell): A device that conducts current when energized by light.

photoconductive diode (photodiode): A diode that is switched on and off by light.

photoelectric sensor (photoelectric switch): A solid-state sensor that can detect the presence of an object without touching the object.

phototransistor: A device that combines the effect of a photodiode and the switching capability of a transistor.

phototriac: A triac that is activated by light.

photovoltaic effect: The production of electrical energy due to the absorption of light photons in a semiconductor material.

pick-up voltage: The minimum voltage that causes the armature to start to move.

pictorial drawing: A drawing that shows the length, height, and depth of an object in one view as well as physical details of that object as seen by the eye.

pigtail: An extended, flexible connection or a braided copper conductor.

PIN photodiode: A diode with a large intrinsic region sandwiched between P-type and N-type regions.

pi-section filter: A filter made with two capacitors and an inductor or resistor to smooth out the AC ripple in a rectified waveform.

piston: A cylinder that is moved back and forth in a tight-fitting chamber by the pressure applied in the chamber.

plain bearing: A bearing in which the shaft turns and is lubricated by a sleeve.

plugging: A method of motor braking in which the motor connections are reversed so that the motor develops a countertorque that acts as a braking force.

point-to-point wiring: Wiring in which each component in a circuit is connected (wired) directly to the next component as specified on the wiring and line diagrams.

polarized scan: A method of scanning in which the receiver responds only to the depolarized reflected light from corner cube reflectors or polarized sensitive reflective tape.

pole: The number of completely isolated circuits that a relay can switch.

position: The number of locations within the valve in which the spool can be placed to direct fluid through the valve.

potential energy: The stored energy a body has due to its position, chemical state, or physical condition.

potentiometer: A variable-resistance electric device that divides voltage proportionally between two circuits.

power circuit: The part of an electrical circuit that connects the loads to the main power lines.

power distribution: The process of delivering electrical power from a substation to the customer's service-entrance equipment.

power formula: The relationship between power (P), voltage (E), and current (I) in an electrical circuit.

power harvesting: The process of obtaining power from the surrounding environment.

power source: A device that converts various forms of energy into electricity.

power transmission: The process of delivering electrical power from a power-generating plant to a substation.

predictive maintenance (PDM): The monitoring of wear conditions and equipment characteristics against a predetermined tolerance to predict possible malfunctions or failures.

press fit: A bearing installation where the bore of the inner rotating ring is smaller than the diameter of the shaft and considerable force must be used to press the bearing onto the shaft.

pressure: Force exerted over a surface divided by its area.

pressure sensor: A transducer that outputs a voltage or current with a corresponding change in pressure.

pressure switch: A switch that detects a set amount of force and activates electrical contacts when the set amount of force is reached.

preventive maintenance (PM): Maintenance performed to keep machines, assembly lines, production operations, and plant operations running with little or no downtime.

primary division: A division with a listed value.

primary resistor starting: A reduced-voltage starting method that uses a resistor connected in each motor line (in one line for a 1φ starter) to produce a voltage drop.

primary winding: The coil of a transformer that draws power from the source.

processor section: The section of a programmable controller that organizes all control activity by receiving inputs, performing logical decisions according to the program, and controlling the outputs.

PROFIBUS: A process bus network capable of communicating information between a master controller (or host) and a smart slave process field device, as well as from one host to another.

programmable automation controller (PAC): A combination of a PLC and a PC-based CPU control device that is designed to operate in an industrial environment.

programmable logic controller (PLC): A solid-state control device that is designed to be programmed and reprogrammed to automatically control industrial processes or machine circuits.

programmable logic relay (PLR): A solid-state control device that includes internal relays, timers, counters, and other control functions that can be programmed and reprogrammed to automatically control small residential, commercial, and industrial circuits.

programming diagram: A line diagram that better matches the language of the programmable controller.

programming section: The section of a programmable controller that allows input into the controller through a keyboard.

prohibited approach boundary: The distance from an exposed energized conductor or circuit part inside which any work performed is considered the same as making contact with the energized conductor or circuit part.

project work: Work on long-term projects that require advanced planning and more time than typical maintenance tasks.

protective clothing: Clothing that provides protection from contact with sharp objects, hot equipment, and harmful materials.

protective helmet: A hard hat that is used in the workplace to prevent injury from the impact of falling and flying objects and from electrical shock.

proton: A particle contained in the nucleus of an atom that has a positive electrical charge.

proximity sensor (proximity switch): A solid-state sensor that detects the presence of an object by means of an electronic sensing field.

P-type material: Material with empty spaces (holes) in its crystal structure.

pull-up torque (PUT): The torque required to bring a load up to its rated speed.

pulsating DC: Direct current that varies in amplitude but does not change polarity.

pulse closing technology: A unique means for verifying that the line is clear of faults before power interrupters are reclosed.

pulse width modulation (PWM): A method of controlling the amount of voltage sent to a motor by converting the DC voltage into fixed values of individual DC pulses.

pure DC power: Power obtained from a battery or DC generator.

pushbutton station: An enclosure that protects the pushbutton, contact block, and wiring from dust, dirt, water, and corrosive fluids.

push fit: A bearing installation where the diameter of the outer fixed ring is smaller than the diameter of the bearing housing and the ring can be pushed in by hand.

push-roller actuator: An actuator operated by direct forward movement into the limit switch.

Q

qualified person: A person who is trained and has special knowledge of the construction and operation of electrical equipment or a specific task, and is trained to recognize and avoid electrical hazards that might be present with respect to the equipment or specific task.

R

radial load: A load in which the applied force is perpendicular to the axis of rotation.

ramp stop: A stopping method in which the level of voltage applied to a motor is reduced as the motor decelerates.

random monitoring: The unscheduled monitoring of equipment as required.

RC circuit: A circuit in which resistance (R) and capacitance (C) are used to help filter the power in a circuit.

reactance: The opposition to the flow of alternating current in a circuit due to inductance.

reactive power: Power absorbed and returned to a load due to its inductive and/or capacitive properties.

receptacle tester: A device that is plugged into a standard receptacle to determine if the receptacle is properly wired and energized.

rectifier: An electrical circuit that changes AC into DC.

recycle timer: A device in which the contacts cycle open and closed repeatedly once the timer has received power.

reed relay: A fast-operating, single-pole, single-throw switch with normally open (NO) contacts hermetically sealed in a glass envelope.

regenerative braking: A method of motor braking in which the regenerated power of a DC motor that is coming to a stop is returned to the input power supply.

regulated power supply: A power supply that maintains a constant voltage across an output even when loads vary.

relay: A device that controls one electrical circuit by opening and closing contacts in another circuit.

resistance (*R*): 1. The opposition to the flow of electrons. **2.** Any force that tends to hinder the movement of an object.

resolution: The degree of measurement precision a test instrument is capable of making as it is used.

resonance: The magnification of vibration and its noise by 20% or more.

response time: The number of pulses (objects) per second a controller can detect.

restricted approach boundary: The distance from an exposed energized conductor or circuit part where an increased risk of electric shock exists due to the close proximity of the person to the energized conductor or circuit part.

retentive timer: A timer that maintains its current accumulated time value when its control input signal is interrupted or power to the timer is removed.

retroreflective scan (retro scan): A method of scanning in which the transmitter and receiver are housed in the same enclosure and the transmitted light beam is reflected back to the receiver from a reflector.

reverse-bias voltage: The application of the opposite polarity to a diode.

rigid coupling: A device that joins two precisely aligned shafts within a common frame.

ripple voltage: The amount of varying voltage present in a DC power supply.

roller bearing: An antifriction bearing that has parallel or tapered steel rollers confined between inner and outer rings.

rolling-contact (antifriction) bearing: A bearing that contains rolling elements that provide a low-friction support surface for rotating or sliding surfaces.

rotor: The rotating part of an AC motor.

rubber insulating gloves: Gloves made of latex rubber and are used to provide maximum insulation from electrical shock.

rubber insulating matting: A floor covering that provides technicians protection from electrical shock when working on live electrical circuits.

S

safety glasses: An eye protection device with special impact-resistant glass or plastic lenses, reinforced frames, and side shields.

safety label: A label that indicates areas or tasks that can pose a hazard to personnel and/or equipment.

saturation region: The maximum current that can flow in a transistor circuit.

scaffold: A temporary or movable platform and structure for workers to stand on when working at a height above the floor.

scan: The process of evaluating the I/O status, executing the program, and updating the system.

scanning: The process of using a light source and photosensor together to measure a change in light intensity when a target is present in, or absent from, the transmitted light beam.

scan time: The time it takes a programmable controller to make a sweep of the program.

scheduled maintenance: Work that is planned and scheduled for completion.

scheduled monitoring: The monitoring of equipment at specific time intervals.

schematic diagram: A diagram that shows the electrical connections and functions of a specific circuit arrangement with graphic symbols.

seal-in voltage: The minimum control voltage required to cause the armature to seal against the pole faces of the magnet.

secondary division: A division that divides primary divisions in halves, thirds, fourths, fifths, etc.

secondary winding: The coil of a transformer that delivers the energy at the transformed or changed voltage to the load.

selector switch: A switch with an operator that is rotated (instead of pushed) to activate the electrical contacts.

self-excited shunt field: A shunt field connected to the same power supply as the armature.

sensor-controlled timer: A timer controlled by an external sensor in which the timer supplies the power required to operate the sensor.

separately derived system (SDS): A system that supplies electrical power derived (taken) from transformers, storage batteries, solar photovoltaic systems, or generators.

separately excited shunt field: A shunt field connected to a different power supply than the armature.

series connection: A connection that has two or more components connected so there is only one path for current flow.

series/parallel connection: A combination of series- and parallel-connected components.

series-wound generator: A generator that has its field windings connected in series with the armature and the external circuit (load).

service factor (SF): A number designation that represents the percentage of extra demand that can be placed on a motor for short intervals without damaging the motor.

service lateral: An electrical service in which service-entrance conductors are run underground from the utility system to the service point.

service point: The point of connection between the facilities of the utility and the premises wiring.

shaded-pole motor: A 1ϕ AC motor that uses a shaded stator pole for starting.

shading coil: A single turn of conducting material (normally copper or aluminum) mounted on the face of the magnetic laminate assembly or armature.

sheave: A grooved wheel used to hold a V-belt.

short circuit: A circuit in which current takes a shortcut around the normal path of current flow.

shunt-wound generator: A generator that has its field windings connected in parallel (shunt) with the armature and the external circuit (load).

silicon-controlled rectifier (SCR): A four-layer (PNPN) semiconductor device that uses three electrodes for normal operation.

single phasing: The operation of a motor that is designed to operate on three phases but is only operating on two phases because one phase is lost.

single-voltage motor: A motor that operates at only one voltage level.

slide-by actuation: An active method of Hall effect sensor activation in which a magnet is moved across the face of a sensor at a constant distance (gap).

slip rings: Metallic rings connected to the ends of the armature and are used to connect the induced voltage to the brushes.

smart grid: An electrical grid that uses computer-based remote control and automation to deliver electrical power from where it is generated to customers

snubber circuit: A circuit that suppresses noise and high voltage on the power lines.

soft foot: A condition that occurs when one or more machine feet do not make complete contact with a base plate.

soft stop (S-curve): A stopping method in which the programmed deceleration time is doubled and the stop function is changed from a ramp slope to an S-curve slope.

solenoid: An electric output device that converts electrical energy into a linear mechanical force.

solid-state motor starter: An electronically operated switch (contactor) that uses solid-state components to eliminate mechanical contacts and includes motor overload protection.

solid-state programmable timer: A timer that is programmed within a programmable logic relay (PLR) or other programmable logic device (PLD).

solid-state relay (SSR): A switching device that has no contacts and switches entirely by electronic means.

solid-state timer: A timer with a time delay that is provided by solid-state electronic devices enclosed within the timing device.

specular scan: A method of scanning in which the transmitter and receiver are placed at equal angles from a highly reflective surface.

split-phase motor: A 1ϕ AC motor that includes a running winding (main winding) and a starting winding (auxiliary winding).

static electricity: An electrical charge at rest.

stator: The stationary part of an AC motor.

stepper motor: A motor that divides shaft rotation into discrete distances (steps).

subdivision: A division that divides secondary divisions in halves, thirds, fourths, fifths, etc.

sulfidation: The formation of film on the contact surface.

supply voltage-controlled timer: A timer that requires the control switch to be connected so that it controls power to the timer coil.

surge protection device: A device that limits the intensity of voltage surges that occur on the power lines of a power distribution system.

surge suppressor: An electrical device that provides protection from transient voltages by limiting the level of voltage allowed downstream from the surge suppressor.

sustained power interruption: A decrease to 0 V on all power lines for a period of more than 1 min.

switchboard: A large floor-mounted panel or assembly of panels in which electrical switches, OCPDs, buses, and instruments are mounted.

switchgear: Any high-powered electrical device that switches or interrupts devices or circuits in a building distribution system.

symbol: A graphic element that represents a quantity or unit.

symmetrical recycle timer: A timer that operates with equal on and off time periods.

synchronous clock timer: A timer that opens and closes a circuit depending on the position of the hands of a clock.

T

tagout: The process of placing a danger tag on the source of electrical power, which indicates that the equipment may not be operated until the danger tag is removed.

tap: A connection brought out of a winding at a point between its endpoints to allow the voltage or current ratio to be changed.

temperature switch: A control device that reacts to heat intensity.

temporary magnet: A magnet that retains only trace amounts of magnetism after the magnetizing force has been removed.

temporary power interruption: A decrease to 0 V on one or more power lines lasting for more than 3 sec up to 1 min.

thermal expansion: A dimensional change in a substance due to a change in temperature.

thermal imager: A device that detects heat patterns in the infrared-wavelength spectrum without making direct contact with equipment.

thermal resistance (R_{TH}): The ability of a device to impede the flow of heat.

thermistor: A temperature-sensitive resistor whose resistance changes with a change in temperature.

thermography: The use of temperature-indicating devices to detect temperature changes in operating equipment.

throw: The number of closed contact positions per pole.

tie-down troubleshooting method: A testing method in which one DMM probe is connected to either the L2 (neutral) or L1 (hot) side of a circuit and the other DMM probe is moved along a section of the circuit to be tested.

torque: The force that produces rotation.

totalizer: A counting device that keeps track of the total number of units or events and displays the total counted value.

transformer: An electric device that uses electromagnetism to change voltage from one level to another or to isolate one voltage from another.

transient voltage (voltage spike): A temporary, unwanted voltage in an electrical circuit.

transistor: A three-terminal device that controls current through the device depending on the amount of voltage applied to the base.

transistor-controlled timer: A timer that is controlled by an external transistor from a separately powered electronic circuit.

transistor outline (TO) number: A number determined by the manufacturer that represents the shape and configuration of a transistor.

transistor-transistor logic (TTL) ICs: A broad family of ICs that employ a two-transistor arrangement.

transmission line: A conductor that carries large amounts of electrical power at high voltages over long distances.

triac: A three-electrode, bidirectional AC switch that allows electrons to flow in either direction.

trip class setting: The length of time it takes for an overload relay to trip and remove power from the motor.

troubleshooting: The systematic elimination of the various parts of a system, circuit, or process to locate a malfunctioning part.

troubleshooting report: A record of a specific problem that occurs with a particular piece of equipment.

true power (P_T): The actual power used in an electrical circuit.

turns ratio: The ratio of the number of turns in the primary winding to the number of turns in the secondary winding of a transformer.

U

ultrasonic analysis: An analysis that uses high vibration frequencies to create an image or reading.

ultrasonic sensor: A solid-state sensor that can detect the presence of an object by emitting and receiving high-frequency sound waves.

undervoltage: A drop in voltage of more than 10% (but not to 0 V) below the normal rated line voltage for a period of time longer than 1 min.

unijunction transistor (UJT): A three-electrode device that contains one PN junction consisting of a bar of N-type material with a region of P-type material doped within the N-type material.

uninterruptible power system (UPS): A power supply that provides constant on-line power when the primary power supply is interrupted.

unregulated power supply: A power supply with an output that varies depending on changes of line voltage or load.

unscheduled maintenance: Unplanned service performed by a maintenance technician that includes emergency work and breakdown maintenance.

up counter: A device used to count inputs and provide an output (contacts) after the preset count value is reached.

up/down counter: A device used to count input from two different inputs, one input that adds a count and another input that subtracts a count.

vane actuation: A passive method of Hall effect sensor activation in which an iron vane shunts or redirects the magnetic field in the air gap away from the sensor.

variable torque/variable horsepower (VT/VH) load: A load that requires a varying torque and horsepower at different speeds.

V-belt: A continuous-power transmission belt with a trapezoidal cross section.

vibration analysis: The monitoring of equipment vibration characteristics to analyze the equipment condition.

visual adjustment method: A belt tension method in which the tension is adjusted by observing the slight sag at the slack side of the belt.

visual and auditory inspection: The analysis of the appearance and sound of operating equipment.

voltage (*E*): The amount of electrical pressure in a circuit.

voltage fluctuation: An increase or decrease in the normal line voltage within the range of +5% to −10%.

voltage regulator: An electrical circuit that is used to maintain a relatively constant value of output voltage over a wide range of operating situations.

voltage sag: A voltage drop of more than 10% (but not to 0 V) below the normal rated line voltage that lasts from 0.5 cycles up to 1 min.

voltage surge: A higher than normal voltage that temporarily exists on one or more power lines.

voltage swell: A voltage increase of more than 10% above the normal rated line voltage that lasts from 0.5 cycles up to 1 min.

voltage tester: A device that indicates approximate voltage level and type (AC or DC) by the movement and vibration of a pointer on a scale.

voltage unbalance: Also known as voltage imbalance, is the unbalance that occurs when voltages at the terminals of an electric motor or other 3ϕ load are not equal.

volts-per-hertz (*V/Hz*) ratio: The relationship between voltage and frequency that exists in a motor.

warning signal word: A word used to indicate a potentially hazardous situation which, if not avoided, could result in death or serious injury.

watt (*W*): A unit of measure equal to the power produced by a current of 1 A across a potential difference of 1 V.

wear particle analysis: The study of wear particles present in lubricating oil.

wiring diagram: A diagram that shows the connection of all components in a piece of equipment.

wobble-stick actuator: An actuator operated by means of any movement into the switch, except a direct pull.

work: The application of force over a distance.

work order: A document that details work required to complete specific maintenance tasks.

work priority: The order in which work is done based on its importance.

wraparound bar graph: A bar graph that displays a fraction of the full range on the graph at one time.

wye configuration: A transformer connection that has one end of each transformer coil connected together.

wye (Y) connection: A connection that has one end of each coil connected together and the other end of each coil left open for external connections.

zero switching relay: An SSR that turns on the load when the control voltage is applied and the voltage at the load crosses zero (or within a few volts of zero).

Index